STUDENT'S SOLUTIONS MANUAL

CINDY TRIMBLE & ASSOCIATES

INTERMEDIATE ALGEBRA: A GRAPHING APPROACH

FIFTH EDITION

Elayn Martin-Gay

University of New Orleans

Margaret Greene

Boston   Columbus   Indianapolis   New York   San Francisco   Upper Saddle River
Amsterdam   Cape Town   Dubai   London   Madrid   Milan   Munich   Paris   Montreal   Toronto
Delhi   Mexico City   Sao Paulo   Sydney   Hong Kong   Seoul   Singapore   Taipei   Tokyo

The author and publisher of this book have used their best efforts in preparing this book. These efforts include the development, research, and testing of the theories and programs to determine their effectiveness. The author and publisher make no warranty of any kind, expressed or implied, with regard to these programs or the documentation contained in this book. The author and publisher shall not be liable in any event for incidental or consequential damages in connection with, or arising out of, the furnishing, performance, or use of these programs.

Reproduced by Pearson from electronic files supplied by the author.

ISBN-13: 978-0-321-88242-4
ISBN-10: 0-321-88242-3

1 2 3 4 5 6 EBM 17 16 15 14 13

www.pearsonhighered.com

Contents

Chapter 1   1

Chapter 2   59

Chapter 3   123

Chapter 4   170

Chapter 5   225

Chapter 6   279

Chapter 7   356

Chapter 8   409

Chapter 9   500

Chapter 10   554

Chapter 11   604

Appendices   641

Practice Final Exam   650

Chapter 1

Section 1.2 Practice Exercises

1. Let $b = 3.5$ and $h = 8$.
$$A = \frac{1}{2}bh$$
$$A = \frac{1}{2}(3.5)(8) = 14$$
The area is 14 square centimeters.

2. Let $p = 17$ and $q = 3$.
$2p - q = 2(17) - 3 = 34 - 3 = 31$

3. a. $\{6, 7, 8, 9\}$

   b. $\{41, 42, 43, ...\}$

4. a. True, since 7 is a natural number and therefore an element of the set.

   b. True, since 6 is not an element of the set $\{1, 3, 5, 7\}$.

5. a. True; every integer is a real number.

   b. False; $\sqrt{8}$ is an irrational number.

   c. True; every whole number is a rational number.

   d. False; since the element 2 in the first set is not an element of the second set.

6. a. $|4| = 4$ since 4 is located 4 units from 0 on the number line.

   b. $\left|-\frac{1}{2}\right| = \frac{1}{2}$ since $-\frac{1}{2}$ is $\frac{1}{2}$ unit from 0 on the number line.

   c. $|1| = 1$ since 1 is 1 unit from 0 on the number line.

   d. $-|6.8| = -6.8$
The negative sign outside the absolute value bars means to take the opposite of the absolute value of 6.8.

   e. $-|-4| = -4$
Since $|-4| = 4$, $-|-4| = -4$.

7. a. The opposite of 5.4 is -5.4.

   b. The opposite of $-\frac{3}{5}$ is $\frac{3}{5}$.

   c. The opposite of 18 is -18.

8. a. $3 \cdot x$ or $3x$

   b. $2x - 5$

   c. $3\frac{5}{8} + x$

   d. $x \div 2$ or $\frac{x}{2}$

   e. $x - 14$

   f. $5(x + 10)$

Vocabulary, Readiness & Video Check 1.2

1. Letters that represent numbers are called <u>variables</u>.

2. Finding the <u>value</u> of an expression means evaluating the expression.

3. The <u>absolute value</u> of a number is that number's distance from 0 on the number line.

4. An <u>expression</u> is formed by numbers and variables connected by operations such as addition, subtraction, multiplication, division, raising to powers, and/or taking roots.

5. The <u>natural numbers</u> are $\{1, 2, 3, ...\}$.

6. The <u>whole numbers</u> are $\{0, 1, 2, 3, ...\}$.

7. The <u>integers</u> are $\{..., -3, -2, -1, 0, 1, 2, 3, ...\}$.

8. The number $\sqrt{5}$ is an <u>irrational number</u>.

9. The number $\frac{5}{7}$ is a <u>rational number</u>.

10. The opposite of a is <u>$-a$</u>.

11. This is an application, so the answer needs to be put in context. The decimal actually represents money.

12. Every real number is either a rational number or an irrational number. There are no numbers that are both rational and irrational.

13. The absolute value of a number is that number's <u>distance</u> from zero on a number line. Or, more formally, $|a| = a$ if a is <u>0</u> or a <u>positive</u> number. Also, $|a| = -a$ if a is a <u>negative</u> number.

14. The absolute value of a number is its *distance* from 0 on the number line, regardless of direction—distance must be positive or zero. The opposite of a number is the *number* that lies the same distance from 0 on the number line as the original number, but on the other side of 0—the opposite of a number can be negative.

15. Order is important in subtraction so you must read the phrase carefully to determine the order of subtraction in your algebraic expression.

Exercise Set 1.2

1. $5x = 5(7) = 35$

3. $9.8z = 9.8(3.1) = 30.38$

5. $ab = \left(\dfrac{1}{2}\right)\left(\dfrac{3}{4}\right) = \dfrac{3}{8}$

7. $3x + y = 3(6) + (4) = 18 + 4 = 22$

9. $400t = 400(5) = 2000$
   The B737-400 travels 2000 miles in 5 hours.

11. $lw = (5.1)(4) = 20.4$
   The display needs 20.4 sq ft of floor space.

13. $2948t = 2948(3.6) = 10{,}612.80$
   It costs \$10,612.80 to operate the B737-400 for 3.6 hours.

15. $\{1, 2, 3, 4, 5\}$

17. $\{11, 12, 13, 14, 15, 16\}$

19. $\{0\}$

21. $\{0, 2, 4, 6, 8\}$

23.
   −2 −1 0 1 2 3 4 5 6 7 8 9

25.

27.
   −12 −10 −8 −6 −4 −2

29.
   −2 −1 0

31. $\left\{3, 0, \sqrt{36}\right\}$

33. $\left\{3, \sqrt{36}\right\}$

35. $\left\{\sqrt{7}\right\}$

37. $-11 \in \{x | x \text{ is an integer}\}$

39. $-6 \notin \{2, 4, 6, \ldots\}$

41. $12 \subseteq \{1, 3, 5, \ldots\}$

43. $\dfrac{1}{2} \notin \{x | x \text{ is an irrational number}\}$

45. True; every integer is a real number.

47. True; -1 is an integer.

49. False; 0 is not a natural number.

51. False; $\sqrt{5}$ is an irrational number.

53. True; every natural number is an integer.

55. False; the number $\sqrt{7}$, for example, is a real number, but it is not a rational number.

57. $-|2| = -2$ (the opposite of $|2|$)

59. $|-4| = 4$ since -4 is located 4 units from 0 on the number line.

61. $|0| = 0$ since 0 is located 0 units from 0 on the number line.

63. $-|-3| = -3$ (the opposite of $|-3|$)

65. The opposite of -6.2 is $-(-6.2) = 6.2$.

67. The opposite of $\dfrac{4}{7}$ is $-\dfrac{4}{7}$.

69. The opposite of $-\dfrac{2}{3}$ is $\dfrac{2}{3}$.

71. The opposite of 0 is 0.

73. $2x$

75. $2x + 5$

77. $x - 10$

79. $x + 2$

81. $\dfrac{x}{11}$

83. $12 - 3x$

85. $x + 2.3$ or $x + 2\dfrac{3}{10}$

87. $1\dfrac{1}{3} - x$

89. $\dfrac{5}{4 - x}$

91. $2(x + 3)$

93. The height of the bar representing China is about 137. Therefore, 137 million tourists are predicted for China.

95. The height of the bar representing Spain is about 69. Therefore, 69 million tourists are predicted for Spain.

97. answers may vary

99. answers may vary

101. answers may vary

103. answers may vary

Section 1.3 Practice Exercises

1. a. $-6 + (-2) = -(6 + 2) = -8$

   b. $5 + (-8) = -3$

   c. $-4 + 9 = 5$

   d. $(-3.2) + (-4.9) = -8.1$

   e. $-\dfrac{3}{5} + \dfrac{2}{3} = -\dfrac{3}{5} \cdot \dfrac{3}{3} + \dfrac{2}{3} \cdot \dfrac{5}{5} = -\dfrac{9}{15} + \dfrac{10}{15} = \dfrac{1}{15}$

   f. $-\dfrac{5}{11} + \dfrac{3}{22} = -\dfrac{5}{11} \cdot \dfrac{2}{2} + \dfrac{3}{22} = -\dfrac{10}{22} + \dfrac{3}{22} = -\dfrac{7}{22}$

2. a. $3 - 11 = 3 + (-11) = -8$

   b. $-6 - (-3) = -6 + (3) = -3$

   c. $-7 - 5 = -7 + (-5) = -12$

   d. $4.2 - (-3.5) = 4.2 + 3.5 = 7.7$

   e. $-\dfrac{5}{7} - \dfrac{1}{3} = -\dfrac{5 \cdot 3}{7 \cdot 3} - \dfrac{1 \cdot 7}{3 \cdot 7}$
$$= -\dfrac{15}{21} + \left(-\dfrac{7}{21}\right)$$
$$= -\dfrac{22}{21}$$

   f. $3 - 1.2 = 3 + (-1.2) = 1.8$

   g. $2 - 9 = 2 + (-9) = -7$

3. a. $13 + 5 - 6 = 18 - 6 = 12$

   b. $-6 - 2 + 4 = -8 + 4 = -4$

4. a. Since the signs of the two numbers are different or unlike, the product is negative. $(-5)(3) = -15$

   b. Since the signs of the two numbers are the same, the product is positive.
$$(-7)\left(-\dfrac{1}{14}\right) = \dfrac{7}{14} = \dfrac{1}{2}$$

   c. $5.1(-2) = -10.2$

   d. $14(0) = 0$

   e. $\left(-\dfrac{1}{4}\right)\left(\dfrac{8}{13}\right) = -\dfrac{8}{52} = -\dfrac{2}{13}$

   f. $6(-1)(-2)(3) = -6(-2)(3) = 12(3) = 36$

   g. $5(-2.3) = -11.5$

5. a. Since the signs are different or unlike, the quotient is negative.
$$\dfrac{-16}{8} = -2$$

b. Since the signs are the same, the quotient is positive.

$$\frac{-15}{-3} = 5$$

c. $-\dfrac{2}{3} \div 4 = -\dfrac{2}{3} \cdot \dfrac{1}{4} = -\dfrac{1}{6}$

d. $\dfrac{54}{-9} = -6$

e. $-\dfrac{1}{12} \div \left(-\dfrac{3}{4}\right) = -\dfrac{1}{12} \cdot -\dfrac{4}{3} = \dfrac{1}{9}$

f. $\dfrac{0}{-7} = 0$

6. a. $2^3 = 2 \cdot 2 \cdot 2 = 8$

b. $\left(\dfrac{1}{3}\right)^2 = \left(\dfrac{1}{3}\right)\left(\dfrac{1}{3}\right) = \dfrac{1}{9}$

c. $-12^2 = -(12 \cdot 12) = -144$

d. $(-12)^2 = (-12)(-12) = 144$

e. $-4^3 = -(4 \cdot 4 \cdot 4) = -64$

f. $(-4)^3 = (-4)(-4)(-4) = -64$

7. a. $\sqrt{49} = 7$ since 7 is positive and $7^2 = 49$.

b. $\sqrt{\dfrac{1}{16}} = \dfrac{1}{4}$ since $\left(\dfrac{1}{4}\right)^2 = \dfrac{1}{16}$.

c. $-\sqrt{64} = -8$

d. $\sqrt{-64}$ is not a real number.

e. $\sqrt{100} = 10$ since $10^2 = 100$.

8. a. $\sqrt[3]{64} = 4$ since $4^3 = 64$.

b. $\sqrt[5]{-1} = -1$ since $(-1)^5 = -1$.

c. $\sqrt[4]{10,000} = 10$ since $10^4 = 10,000$.

9. a. $14 - 3 \cdot 4 = 14 - 12 = 2$

b. $3(5-8)^2 = 3(-3)^2 = 3(9) = 27$

c. $\dfrac{|-5|^2 + 4}{\sqrt{4} - 3} = \dfrac{5^2 + 4}{2 - 3} = \dfrac{25 + 4}{-1} = \dfrac{29}{-1} = -29$

10.
$$\begin{aligned}
5 - [(3-5) + 6(2-4)] &= 5 - [-2 + 6(-2)] \\
&= 5 - [-2 + (-12)] \\
&= 5 - [-14] \\
&= 5 + 14 \\
&= 19
\end{aligned}$$

11.
$$\begin{aligned}
\frac{-2\sqrt{12+4} - (-3)^2}{6^2 + |1-9|} &= \frac{-2\sqrt{16} - (-3)^2}{6^2 + |-8|} \\
&= \frac{-2(4) - 9}{36 + 8} \\
&= \frac{-8 - 9}{44} \\
&= -\frac{17}{44}
\end{aligned}$$

12. For each expression, replace x with 16 and y with -5.

a. $2x - 7y = 2(16) - 7(-5) = 32 + 35 = 67$

```
16→X: -5→Y
                  -5
2X-7Y
                  67
■
```

b. $-4y^2 = -4(-5)^2 = -4(25) = -100$

```
16→X: -5→Y
                  -5
-4Y²
               -100
```

c.
$$\begin{aligned}
\frac{\sqrt{x}}{y} - \frac{y}{x} &= \frac{\sqrt{16}}{-5} - \frac{-5}{16} \\
&= -\frac{4}{5} + \frac{5}{16} \\
&= -\frac{4}{5} \cdot \frac{16}{16} + \frac{5}{16} \cdot \frac{5}{5} \\
&= -\frac{64}{80} + \frac{25}{80} \\
&= -\frac{39}{80}
\end{aligned}$$

```
16→X: -5→Y
                  -5
√(X)/Y-Y/X
             -.4875
■
```

13. a. $5x^2 + 1;\ x = 3$

$3^2 = 9$

$5 \cdot 9 = 45$

$45 + 1 = 46$

b. $5x^2 + 1;\ x = 11$

$11^2 = 121$

$5 \cdot 121 = 605$

$605 + 1 = 606$

c. $5x^2 + 1;\ x = 4.86$

```
4.86→X
              4.86
5X²+1
              119.098
■
```

The value of $5x^2 + 1$ when $x = 4.86$ is 119.098.

14. When $x = -5$,

$\dfrac{9}{5}x + 32 = \dfrac{9}{5}(-5) + 32 = -9 + 32 = 23.$

When $x = 10$,

$\dfrac{9}{5}x + 32 = \dfrac{9}{5}(10) + 32 = 18 + 32 = 50.$

When $x = 25$,

$\dfrac{9}{5}x + 32 = \dfrac{9}{5}(25) + 32 = 45 + 32 = 77.$

The completed table is

Degrees Celsius	x	-5	10	25
Degrees Fahrenheit	$\dfrac{9}{5}x + 32$	23	50	77

Vocabulary, Readiness & Video Check 1.3

1. $0 \cdot a = \underline{0}$

2. $\dfrac{0}{4}$ simplifies to $\underline{0}$ while $\dfrac{4}{0}$ is <u>undefined</u>.

3. The <u>reciprocal</u> of the nonzero number b is $\dfrac{1}{b}$.

4. The fraction $-\dfrac{a}{b} = \dfrac{-a}{\underline{b}} = \dfrac{a}{\underline{-b}}.$

5. An <u>exponent</u> is a shorthand notation for repeated multiplication of the same number.

6. In $(-5)^2$, the 2 is the <u>exponent</u> and the -5 is the <u>base</u>.

7. The opposite of squaring a number is taking the <u>square root</u> of a number.

8. Using order of operations, $9 \div 3 \cdot 3 = \underline{9}$.

9. addition

10. The signs of the numbers determine the sign of the product or quotient. Same signs mean a positive product or quotient and different signs mean a negative product or quotient.

11. The parentheses, or lack of them, determine the base of the expression. In Example 7, -7^2, the base is 7 and only 7 is squared. In Example 8, $(-7)^2$, the base is -7 and all of -7 is squared.

12. the positive or principal square root

13. It allows each expression to evaluate to a single number.

14. The 2 is part of a multiplication which must happen before addition in the order of operations.

Exercise Set 1.3

1. $-3 + 8 = 5$

3. $-14 + (-10) = -24$

5. $-4.3 - 6.7 = -11$

7. $13 - 17 = -4$

9. $\dfrac{11}{15} - \left(-\dfrac{3}{5}\right) = \dfrac{11}{15} + \dfrac{9}{15} = \dfrac{20}{15} = \dfrac{4}{3}$

11. $19 - 10 - 11 = 9 - 11 = -2$

13. $-\dfrac{4}{5} - \left(-\dfrac{3}{10}\right) = -\dfrac{4}{5} + \dfrac{3}{10} = -\dfrac{8}{10} + \dfrac{3}{10} = -\dfrac{5}{10} = -\dfrac{1}{2}$

15. $8 - 14 = -6$

17. $-5 \cdot 12 = -60$

19. $-7 \cdot 0 = 0$

21. $\dfrac{0}{-2} = 0$

23. $\dfrac{-9}{3} = -3$

25. $\dfrac{-12}{-4} = 3$

27. $3\left(-\dfrac{1}{18}\right) = -\dfrac{3}{18} = -\dfrac{1}{6}$

29. $(-0.7)(-0.8) = 0.56$

31. $9.1 \div -1.3 = \dfrac{9.1}{1} \cdot \dfrac{1}{-1.3} = -7$

33. Multiplying from left to right gives
$(-4)(-2)(-1) = 8(-1) = -8.$

35. $-7^2 = -(7 \cdot 7) = -49$

37. $(-6)^2 = (-6)(-6) = 36$

39. $(-2)^3 = (-2)(-2)(-2) = 4(-2) = -8$

41. $\left(-\dfrac{1}{3}\right)^3 = \left(-\dfrac{1}{3}\right)\left(-\dfrac{1}{3}\right)\left(-\dfrac{1}{3}\right) = -\dfrac{1}{27}$

43. $\sqrt{49} = 7$ since 7 is positive and $7^2 = 49.$

45. $-\sqrt{\dfrac{4}{9}} = -\dfrac{2}{3}$ since $\left(\dfrac{2}{3}\right)^2 = \dfrac{4}{9}.$

47. $\sqrt[3]{64} = 4$ since $4^3 = 64.$

49. $\sqrt[4]{81} = 3$ since $3^4 = 81.$

51. $\sqrt{-100}$ is not a real number.

53. $3(5-7)^4 = 3(-2)^4 = 3(16) = 48$

55. $-3^2 + 2^3 = -9 + 8 = -1$

57. $\dfrac{3.1 - (-1.4)}{-0.5} = \dfrac{3.1 + 1.4}{-0.5} = \dfrac{4.5}{-0.5} = -9$

59. $(-3)^2 + 2^3 = 9 + 8 = 17$

61. $-8 \div 4 \cdot 2 = -2 \cdot 2 = -4$

63. $-8\left(-\dfrac{3}{4}\right) - 8 = 6 - 8 = -2$

65. $2 - [(7-6) + (9-19)] = 2 - [1 + (-10)]$
$= 2 - (-9)$
$= 11$

67. $\dfrac{(-9+6)(-1^2)}{-2-2} = \dfrac{(-3)(-1)}{-4} = \dfrac{3}{-4} = -\dfrac{3}{4}$

69. $\left(\sqrt[3]{8}\right)(-4) - \left(\sqrt{9}\right)(-5) = (2)(-4) - (3)(-5)$
$= -8 - (-15)$
$= -8 + 15$
$= 7$

71. $25 - [(3-5) + (14-18)]^2 = 25 - [(-2) + (-4)]^2$
$= 25 - (-6)^2$
$= 25 - 36$
$= -11$

73. $\dfrac{\left(3 - \sqrt{9}\right) - (-5 - 1.3)}{-3} = \dfrac{(3-3) - (-6.3)}{-3}$
$= \dfrac{0 + 6.3}{-3}$
$= \dfrac{6.3}{-3}$
$= -2.1$

75. $\dfrac{|3-9| - |-5|}{-3} = \dfrac{6-5}{-3} = \dfrac{1}{-3} = -\dfrac{1}{3}$

77. $\dfrac{3(-2+1)}{5} - \dfrac{-7(2-4)}{1-(-2)} = \dfrac{3(-1)}{5} - \dfrac{-7(-2)}{1+2}$
$= \dfrac{-3}{5} - \dfrac{14}{3}$
$= -\dfrac{9}{15} - \dfrac{70}{15}$
$= -\dfrac{79}{15}$

79. $\dfrac{\frac{1}{3} \cdot 9 - 7}{3 + \frac{1}{2} \cdot 4} = \dfrac{3-7}{3+2} = \dfrac{-4}{5} = -\dfrac{4}{5}$

81.
$$\begin{aligned}
3\{-2+5[1-2(-2+5)]\} &= 3\{-2+5[1-2(3)]\} \\
&= 3\{-2+5[1-6]\} \\
&= 3\{-2+5[-5]\} \\
&= 3\{-2+[-25]\} \\
&= 3\{-27\} \\
&= -81
\end{aligned}$$

83.
$$\begin{aligned}
\frac{-4\sqrt{80+1}+(-4)^2}{3^3+\left|-2(3)\right|} &= \frac{-4\sqrt{81}+(-4)^2}{3^3+\left|-6\right|} \\
&= \frac{-4(9)+16}{27+6} \\
&= \frac{-36+16}{27+6} \\
&= \frac{-20}{33} \\
&= -\frac{20}{33}
\end{aligned}$$

85. Let $x = 9$, $y = -2$.
$$9x - 6y = 9(9) - 6(-2) = 81 + 12 = 93$$

87. Let $y = -2$.
$$-3y^2 = -3(-2)^2 = -3(4) = -12$$

89. Let $x = 9$, $y = -2$.
$$\begin{aligned}
\frac{\sqrt{x}}{y} - \frac{y}{x} &= \frac{\sqrt{9}}{-2} - \frac{-2}{9} \\
&= \frac{3}{-2} + \frac{2}{9} \\
&= -\frac{27}{18} + \frac{4}{18} \\
&= -\frac{23}{18}
\end{aligned}$$

91. Let $x = 9$, $y = -2$.
$$\begin{aligned}
\frac{3+2\left|x-y\right|}{x+2y} &= \frac{3+2\left|9-(-2)\right|}{9+2(-2)} \\
&= \frac{3+2\left|9+2\right|}{9+2(-2)} \\
&= \frac{3+2\left|11\right|}{9+(-4)} \\
&= \frac{3+22}{5} \\
&= \frac{25}{5} \\
&= 5
\end{aligned}$$

93. Let $x = 9$, $y = -2$.
$$\begin{aligned}
\frac{y^3+\sqrt{x-5}}{\left|4x-y\right|} &= \frac{(-2)^3+\sqrt{9-5}}{\left|4\cdot 9-(-2)\right|} \\
&= \frac{-8+\sqrt{4}}{\left|36+2\right|} \\
&= \frac{-8+2}{\left|38\right|} \\
&= \frac{-6}{38} \\
&= -\frac{3}{19}
\end{aligned}$$

95.

```
1.4→X: -6.2→Y
            -6.2
3X-2Y
            16.6
```

When $x = 1.4$ and $y = -6.2$, $3x - 2y = 16.6$.

97.

```
1.4→X: -6.2→Y
            -6.2
abs(X-Y)/(2Y)
      -.6129032258
```

When $x = 1.4$ and $y = -6.2$, $\dfrac{\left|x-y\right|}{2y} \approx -0.61$.

99.

```
1.4→X: -6.2→Y
            -6.2
-3(X²+Y²)
           -121.2
```

When $x = 1.4$ and $y = -6.2$,
$$-3(x^2 + y^2) = -121.2.$$

101. When $x^2 + 2y$ is evaluated for $x = -2.7$ and $y = 0.9$, the result is 9.09.

103. When $A^2 + B^2$ is evaluated for $A = 4$ and $B = -2.5$, the result is 22.25.

105. a.

```
2→X
            2
(1/2)X
            1
1/2X
            1
```

b. They are the same; answers may vary.

107. a. $y = 5: 8 + 2y = 8 + 2(5) = 8 + 10 = 18$
$y = 7: 8 + 2y = 8 + 2(7) = 8 + 14 = 22$
$y = 10: 8 + 2y = 8 + 2(10) = 8 + 20 = 28$
$y = 100: 8 + 2y = 8 + 2(100) = 8 + 200 = 208$
The completed table is:

Length	y	5	7	10	100
Perimeter	$8 + 2y$	18	22	28	208

b. The perimeter increases as length increases; answers may vary.

109. a. $x = 10:$ $\dfrac{100x + 5000}{x} = \dfrac{100(10) + 5000}{10}$
$= \dfrac{1000 + 5000}{10}$
$= \dfrac{6000}{10}$
$= 600$

$x = 100:$ $\dfrac{100x + 5000}{x} = \dfrac{100(100) + 5000}{100}$
$= \dfrac{10,000 + 5000}{100}$
$= \dfrac{15,000}{100}$
$= 150$

$x = 1000:$ $\dfrac{100x + 5000}{x} = \dfrac{100(1000) + 5000}{1000}$
$= \dfrac{100,000 + 5000}{1000}$
$= \dfrac{105,000}{1000}$
$= 105$

The completed table is:

Number of Bookshelves	x	10	100	1000
Cost per Bookshelf	$\frac{100x+5000}{x}$	600	150	105

b. The cost per bookshelf decreases as the number of bookshelves increases; answers may vary.

111. $-\dfrac{1}{7} = \dfrac{-1}{7} = \dfrac{1}{-7}$; b, c

113. $\dfrac{5}{-(x+y)} = \dfrac{-5}{(x+y)} = -\dfrac{5}{(x+y)}$; b, d

115. $\dfrac{-9x}{-2y} = \dfrac{9x}{2y}$; b

117. Let $x_1 = 2$, $x_2 = 4$, $y_1 = -3$, $y_2 = 2$.

$$\frac{y_2 - y_1}{x_2 - x_1} = \frac{2 - (-3)}{4 - 2} = \frac{2 + 3}{4 - 2} = \frac{5}{2}$$

119. $1 - \dfrac{1}{5} - \dfrac{3}{7} = \dfrac{35}{35} - \dfrac{7}{35} - \dfrac{15}{35} = \dfrac{13}{35}$

121. $10{,}203 - 5998 = 4205$ meters

123. $(2 + 7) \cdot (1 + 3) = 9 \cdot 4 = 36$

125. answers may vary

127. $\sqrt{10} \approx 3.1623$

129. $\sqrt{7.9} \approx 2.8107$

131. $\dfrac{-1.682 - 17.895}{(-7.102)(-4.691)} \approx -0.5876$

Section 1.4 Practice Exercises

1. The product of -4 and x is 20.

$$-4x = 20$$

2. Three times the difference of z and 3 equals 9.

$$3 \qquad (z - 3) \qquad = \quad 9$$

3. The sum of x and 5 is the same as 3 less than twice x.

$$x + 5 \qquad = \qquad 2x - 3$$

4. The sum of y and 2 is 4 more than the quotient of z and 8.

$$y + 2 \qquad = \qquad 4 + \frac{z}{8}$$

5. a. $-6 < -5$ since -6 lies to the left of -5 on the number line.

$$-10\ -9\ -8\ -7\ -6\ -5\ -4\ -3\ -2\ -1\ 0\ 1$$

   b. $\dfrac{24}{3} = 8$

   c. $0 > -7$ since 0 lies to the right of -7 on the number line.

$$-9\ -8\ -7\ -6\ -5\ -4\ -3\ -2\ -1\ 0\ 1\ 2$$

   d. $-2.76 < -2.67$ since -2.76 lies to the left of -2.67 on the number line.

$$-2.76 \quad -2.67$$
$$3 \qquad 2 \qquad 1$$

e. $\dfrac{9}{10} > \dfrac{7}{10}$

The denominators are the same, so $\dfrac{9}{10} > \dfrac{7}{10}$ since $9 > 7$.

f. $\dfrac{2}{3} < \dfrac{7}{9}$

By dividing, we see that $\dfrac{2}{3} = 0.666...$ and $\dfrac{7}{9} = 0.777....$ Thus, $\dfrac{2}{3} < \dfrac{7}{9}$ since $0.666... < 0.777....$

6.   a. $x - 3 \le 5$

b. $y \ne -4$

c. $2 < 4 + \dfrac{1}{2}z$

7.   a.   The opposite of -7 is $-(-7) = 7$.

b.   The opposite of 4.7 is -4.7.

c.   The opposite of $-\dfrac{3}{8}$ is $-\left(-\dfrac{3}{8}\right) = \dfrac{3}{8}$.

8.   a.   The reciprocal of $-\dfrac{5}{3}$ is $-\dfrac{3}{5}$ because $-\dfrac{5}{3}\left(-\dfrac{3}{5}\right) = 1$.

b.   The reciprocal of 14 is $\dfrac{1}{14}$.

c.   The reciprocal of -2 is $-\dfrac{1}{2}$.

9. $8 + 13x = 13x + 8$

10. $3 \cdot (11b) = (3 \cdot 11)b = 33b$

11.   a. $4(x + 5y) = 4 \cdot x + 4 \cdot 5y = 4x + 20y$

b. $-(3 - 2z) = -1(3 - 2z)$
$\qquad\qquad = -1 \cdot 3 + (-1)(-2z)$
$\qquad\qquad = -3 + 2z$

c. $0.3x(y - 3) = 0.3x \cdot y - 0.3x \cdot 3$
$\qquad\qquad\quad = 0.3xy - 0.9x$

12.   a. $-(-11.7) = 11.7$ illustrates the double negative property.

b. $7(y + 5) = 7 \cdot y + 7 \cdot 5$ illustrates the distributive property.

c. $2 + (a + b) = (2 + a) + b$ illustrates the associative property of addition.

13. a. In words: Value of a dime   ·   number of dimes

$\downarrow$ $\downarrow$

Translate: 0.10 x , or $0.10x$

b. In words: number of grams of carbohydrates in one cookie   ·   number of cookies

$\downarrow$ $\downarrow$

Translate: 26 y , or $26y$

c. In words: cost of one birthday card   ·   number of cards

$\downarrow$ $\downarrow$

Translate: 1.75 z , or $1.75z$

d. In words: Discount   ·   purchase price

$\downarrow$ $\downarrow$

Translate: 0.15 t , or $0.15t$

14. a. If two numbers have a sum of 16 and one number is x, the other number is the rest of 16.

In words: Sixteen   minus x

$\downarrow$ $\downarrow$ $\downarrow$

Translate: 16 $-$ x

b. In words: One hundred eighty   minus   one angle, x

$\downarrow$ $\downarrow$ $\downarrow$

Translate: 180 $-$ x

c. The next consecutive even integer is always two more than the previous even integer.

In words: first integer   plus   two

$\downarrow$ $\downarrow$ $\downarrow$

Translate: x $+$ 2

d. In words: younger brother's age   plus   nine

$\downarrow$ $\downarrow$ $\downarrow$

Translate: x $+$ 9

15. a. $6ab - ab = 6ab - 1ab = (6-1)ab = 5ab$

b. $4x - 5 + 6x = 4x + 6x - 5$
$$= (4+6)x - 5$$
$$= 10x - 5$$

c. $17p - 9$ cannot be simplified further since $17p$ and -9 are not like terms.

16. a. $5pq - 2pq - 11 - 4pq + 18 = 5pq - 2pq - 4pq - 11 + 18$
$$= (5 - 2 - 4)pq + (-11 + 18)$$
$$= -1pq + (7)$$
$$= -pq + 7$$

b. $3x^2 + 7 - 2(x^2 - 6) = 3x^2 + 7 - 2x^2 + 12$
$$= 3x^2 - 2x^2 + 7 + 12$$
$$= x^2 + 19$$

c. $(3.7x + 2.5) - (-2.1x - 1.3) = 3.7x + 2.5 + 2.1x + 1.3$
$$= 3.7x + 2.1x + 2.5 + 1.3$$
$$= 5.8x + 3.8$$

d. $\frac{1}{5}(15c - 25d) - \frac{1}{2}(8c + 6d + 1) + \frac{3}{4} = 3c - 5d - 4c - 3d - \frac{1}{2} + \frac{3}{4}$
$$= -c - 8d + \frac{1}{4}$$

Vocabulary, Readiness & Video Check 1.4

	Symbol	Meaning
1.	$<$	is less than
2.	$>$	is greater than
3.	$\neq$	is not equal to
4.	$=$	is equal to
5.	$\geq$	is greater than or equal to
6.	$\leq$	is less than or equal to

7. The opposite or <u>additive inverse</u> of nonzero number a is <u>$-a$</u>.

8. The reciprocal or <u>multiplicative inverse</u> of nonzero number a is <u>$\frac{1}{a}$</u>.

9. The <u>commutative</u> property has to do with "order."

10. The <u>associative</u> property has to do with "grouping."

11. $a(b + c) = ab + ac$ illustrates the <u>distributive</u> property.

12. The <u>terms</u> of an expression are the addends of the expression.

13. The x^{-1} key on a calculator is used to find the multiplicative inverse of a number.

14. $=, \neq, <, \leq, >, \geq$

15. *Reciprocal* is the same as <u>multiplicative</u> inverse and *opposite* is the same as <u>additive</u> inverse.

16. order; grouping

17. Understand the difference between the *number* of coins you have and the value the coins represent.

18. by combining like terms; distributive property

Exercise Set 1.4

1. The sum of 10 and x is -12.

$$10 + x \quad = \quad -12$$

or $10 + x = -12$

3. Twice x plus 5 is the same as -14.

$$2x + 5 \quad = \quad -14$$

or $2x + 5 = -14$

5. The quotient of n and 5 is 4 times n.

$$\frac{n}{5} \quad = \quad 4n$$

or $\dfrac{n}{5} = 4n$

7. The difference of z and one-half is the same as the product of z and one-half.

$$z - \frac{1}{2} \quad = \quad \frac{1}{2}z$$

or $z - \dfrac{1}{2} = \dfrac{1}{2}z$

9. The product of 7 and x is less than or equal to -21.

$$7x \quad \leq \quad -21$$

or $7x \leq -21$

11. Twice the difference of x and 6 is greater than the reciprocal of 11.

$$2(x - 6) \quad > \quad \frac{1}{11}$$

or $2(x - 6) > \dfrac{1}{11}$

13. $\underbrace{\text{Twice the difference of } x \text{ and } 6}\ \underset{\downarrow}{\text{is}}\ \underbrace{-27}.$
$$2(x-6) \qquad\qquad =-27$$
or $2(x-6)=-27$

15. $-16 > -17$ since -16 is to the right of -17 on the number line.

17. $7.4 = 7.40$

19. $\dfrac{7}{11} < \dfrac{9}{11}$ since $7 < 9$.

21. $\dfrac{1}{2} < \dfrac{5}{8}$ since $\dfrac{1}{2}$ is to the left of $\dfrac{5}{8}$ on the number line.

23. $-7.9 < -7.09$ since -7.9 is to the left of -7.09 on the number line.

	Number	Opposite	Reciprocal
25.	5	-5	$\dfrac{1}{5}$
27.	-8	8	$-\dfrac{1}{8}$
29.	$-\dfrac{1}{7}$	$\dfrac{1}{7}$	-7
31.	0	0	Undefined
33.	$\dfrac{7}{8}$	$-\dfrac{7}{8}$	$\dfrac{8}{7}$

35. $7x + y = y + 7x$

37. $z \cdot w = w \cdot z$

39. $\dfrac{1}{3} \cdot \dfrac{x}{5} = \dfrac{x}{5} \cdot \dfrac{1}{3}$

41. $5 \cdot (7x) = (5 \cdot 7)x$

43. $(x + 1.2) + y = x + (1.2 + y)$

45. $(14z) \cdot y = 14(z \cdot y)$

47. $3(x + 5) = 3 \cdot x + 3 \cdot 5 = 3x + 15$

49. $-(2a + b) = -1(2a + b)$
$$= -1 \cdot 2a + (-1) \cdot b$$
$$= -2a - b$$

51. $2(6x + 5y + 2z) = 2 \cdot 6x + 2 \cdot 5y + 2 \cdot 2z$
$$= 12x + 10y + 4z$$

53. $-4(x - 2y + 7) = -4 \cdot x + (-4)(-2y) + (-4)(7)$
$$= -4x + 8y - 28$$

55. $0.5x(6y - 3) = 0.5x \cdot 6y - 0.5x \cdot 3 = 3xy - 1.5x$

57. $3x + 6 = 6 + 3x$

59. $\dfrac{2}{3} + \left(-\dfrac{2}{3}\right) = 0$

61. $7 \cdot 1 = 7$

63. $10(2y) = (10 \cdot 2)y$

65. The screen illustrates the associative property of addition.

67. The screen illustrates the commutative property of multiplication.

69. In words: $\boxed{\text{Value of a dime}} \cdot \boxed{\text{Number of dimes}}$
Translate: $0.1 \cdot d$ or $0.1d$

71. If two numbers have a sum of 112 and one number is x, then the other number is the "rest of 112." So, in other words, we have
$\boxed{\text{One hundred twelve}} - \boxed{x}$
Translate: $112 - x$

73. In words: $\boxed{\text{One hundred eighty}} - \boxed{x}$
Translate: $180 - x$

75. In words:
$\boxed{\text{Cost of compact disc}} \cdot \boxed{\text{Number of discs}}$
Translate: $\$6.49 \cdot x$ or $\$6.49x$

77. The next odd integer would be 2 more than the given odd integer.
In words: $\boxed{\text{Odd integer}} + \boxed{\text{Two}}$
Translate: $x + 2$

79. $-9 + 4x + 18 - 10x = 4x - 10x + 18 - 9$
$$= (4 - 10)x + (18 - 9)$$
$$= -6x + 9$$

81. $5k - (3k - 10) = 5k - 3k + 10$
$$= (5 - 3)k + 10$$
$$= 2k + 10$$

83. $(3x + 4) - (6x - 1) = 3x + 4 - 6x + 1$
$$= 3x - 6x + 4 + 1$$
$$= (3 - 6)x + 5$$
$$= -3x + 5$$

85. $3(xy-2)+xy+15-x^2 = 3xy-6+xy+15-x^2$
$$= 4xy+9-x^2$$
$$= -x^2+4xy+9$$

87. $-(n+5)+(5n-3) = -1(n+5)+(5n-3)$
$$= -n-5+5n-3$$
$$= 4n-8$$

89. $4(6n^2-3)-3(8n^2+4) = 24n^2-12-24n^2-12$
$$= 24n^2-24n^2-12-12$$
$$= (24-24)n^2-24$$
$$= 0n^2-24$$
$$= -24$$

91. $3x-2(x-5)+x = 3x-2x+10+x$
$$= 3x-2x+x+10$$
$$= (3-2+1)x+10$$
$$= 2x+10$$

93. $1.5x+2.3-0.7x-5.9 = (1.5-0.7)x+(2.3-5.9)$
$$= 0.8x-3.6$$

95. $\dfrac{3}{4}b-\dfrac{1}{2}+\dfrac{1}{6}b-\dfrac{2}{3} = \left(\dfrac{3}{4}+\dfrac{1}{6}\right)b+\left(-\dfrac{1}{2}-\dfrac{2}{3}\right)$
$$= \dfrac{11}{12}b-\dfrac{7}{6}$$

97. $2(3x+7) = 6x+14$

99. $\dfrac{1}{2}(10x-2)-\dfrac{1}{6}(60x-5y) = 5x-1-10x+\dfrac{5}{6}y$
$$= 5x-10x+\dfrac{5}{6}y-1$$
$$= -5x+\dfrac{5}{6}y-1$$

101. $\dfrac{1}{6}(24a-18b)-\dfrac{1}{7}(7a-21b-2)-\dfrac{1}{5}$
$$= 4a-3b-a+3b+\dfrac{2}{7}-\dfrac{1}{5}$$
$$= 3a+\dfrac{3}{35}$$

103. $3(x+4) = 3 \cdot x + 3 \cdot 4 = 3x+12$

105. $5(7y) = (5 \cdot 7)y$

107. $a(b+c) = ab+ac$

109. zero; answers may vary

111. no; answers may vary

113. answers may vary

115. $8.1z+7.3(z+5.2)-6.85$
$$= 8.1z+7.3z+37.96-6.85$$
$$= 15.4z+31.11$$

Integrated Review

1. Let $z = -4$.
$$z^2 = (-4)^2 = (-4)(-4) = 16$$

2. Let $z = -4$.
$$-z^2 = -(-4)^2 = -(-4)(-4) = -16$$

3. Let $x = -1$, $y = 3$, $z = -4$.
$$\dfrac{4x-z}{2y} = \dfrac{4(-1)-(-4)}{2(3)} = \dfrac{-4+4}{6} = \dfrac{0}{6} = 0$$

4. Let $x = -1$, $y = 3$, $z = -4$.
$$x(y-2z) = -1[3-2(-4)]$$
$$= -1[3+8]$$
$$= -1[11]$$
$$= -11$$

5. $-7-(-2) = -7+2 = -5$

6. $\dfrac{9}{10}-\dfrac{11}{12} = \dfrac{9}{10} \cdot \dfrac{6}{6} - \dfrac{11}{12} \cdot \dfrac{5}{5} = \dfrac{54}{60}-\dfrac{55}{60} = -\dfrac{1}{60}$

7. $\dfrac{-13}{2-2} = \dfrac{-13}{0}$ is undefined.

8. $(1.2)^2-(2.1)^2 = 1.44-4.41 = -2.97$

9. $\sqrt{64}-\sqrt[3]{64} = 8-4 = 4$

10. $-5^2-(-5)^2 = -25-25 = -50$

11. $9+2[(8-10)^2+(-3)^2] = 9+2[(-2)^2+(-3)^2]$
$$= 9+2(4+9)$$
$$= 9+2(13)$$
$$= 9+26$$
$$= 35$$

12.
$$8 - 6\left[\sqrt[3]{8}(-2) + \sqrt{4}(-5)\right] = 8 - 6[2(-2) + 2(-5)]$$
$$= 8 - 6[(-4) + (-10)]$$
$$= 8 - 6(-14)$$
$$= 8 + 84$$
$$= 92$$

13. $-15 - 2x$

14. $3x + 5$

15. 0 is a whole number that is not a natural number.

16. True

17. $-5(9x) = (-5 \cdot 9)x = -45x$

18.
$$(3x - 7) - (4x + 1) = 3x - 7 - 4x - 1$$
$$= 3x - 4x - 7 - 1$$
$$= (3 - 4)x - 7 - 1$$
$$= -x - 8$$

19.
$$8.6a + 2.3b - a + 4.9b = 8.6a - a + 2.3b + 4.9b$$
$$= (8.6 - 1)a + (2.3 + 4.9)b$$
$$= 7.6a + 7.2b$$

20.
$$\frac{2}{3}y - \frac{2}{3} + y - \frac{1}{9}y + \frac{9}{10}$$
$$= \frac{2}{3}y + y - \frac{1}{9}y - \frac{2}{3} + \frac{9}{10}$$
$$= \left(\frac{2}{3} + 1 - \frac{1}{9}\right)y + \left(-\frac{2}{3} + \frac{9}{10}\right)$$
$$= \left(\frac{6}{9} + \frac{9}{9} - \frac{1}{9}\right)y + \left(-\frac{20}{30} + \frac{27}{30}\right)$$
$$= \frac{6 + 9 - 1}{9}y + \frac{-20 + 27}{30}$$
$$= \frac{14}{9}y + \frac{7}{30}$$

Section 1.5 Practice Exercises

1.
$$3x + 7 = 22$$
$$3x + 7 - 7 = 22 - 7$$
$$3x = 15$$
$$\frac{3x}{3} = \frac{15}{3}$$
$$x = 5$$

2.
$$2.5 = 3 - 2.5t$$
$$2.5 - 3 = 3 - 2.5t - 3$$
$$-0.5 = -2.5t$$
$$\frac{-0.5}{-2.5} = \frac{-2.5t}{-2.5}$$
$$0.2 = t$$

3.
$$-8x - 4 + 6x = 5x + 11 - 4x$$
$$-2x - 4 = x + 11$$
$$-2x - 4 - x = x + 11 - x$$
$$-3x - 4 = 11$$
$$-3x - 4 + 4 = 11 + 4$$
$$-3x = 15$$
$$\frac{-3x}{-3} = \frac{15}{-3}$$
$$x = -5$$

4.
$$3(x - 5) = 6x - 3$$
$$3x - 15 = 6x - 3$$
$$3x - 15 - 6x = 6x - 3 - 6x$$
$$-3x - 15 = -3$$
$$-3x - 15 + 15 = -3 + 15$$
$$-3x = 12$$
$$\frac{-3x}{-3} = \frac{12}{-3}$$
$$x = -4$$

5.
$$\frac{y}{2} - \frac{y}{5} = \frac{1}{4}$$
$$20\left(\frac{y}{2} - \frac{y}{5}\right) = 20\left(\frac{1}{4}\right)$$
$$20\left(\frac{y}{2}\right) - 20\left(\frac{y}{5}\right) = 5$$
$$10y - 4y = 5$$
$$6y = 5$$
$$\frac{6y}{6} = \frac{5}{6}$$
$$y = \frac{5}{6}$$

6.
$$x - \frac{x-2}{12} = \frac{x+3}{4} + \frac{1}{4}$$
$$12\left(x - \frac{x-2}{12}\right) = 12\left(\frac{x+3}{4} + \frac{1}{4}\right)$$
$$12 \cdot x - 12\left(\frac{x-2}{12}\right) = 12\left(\frac{x+3}{4}\right) + 12 \cdot \frac{1}{4}$$
$$12x - (x-2) = 3(x+3) + 3$$
$$12x - x + 2 = 3x + 9 + 3$$
$$11x + 2 = 3x + 12$$
$$11x + 2 - 3x = 3x + 12 - 3x$$
$$8x + 2 = 12$$
$$8x + 2 - 2 = 12 - 2$$
$$8x = 10$$
$$\frac{8x}{8} = \frac{10}{8}$$
$$x = \frac{5}{4}$$

7.
$$0.15x - 0.03 = 0.2x + 0.12$$
$$100(0.15x - 0.03) = 100(0.2x + 0.12)$$
$$100(0.15x) - 100(0.03) = 100(0.2x) + 100(0.12)$$
$$15x - 3 = 20x + 12$$
$$15x - 20x = 12 + 3$$
$$-5x = 15$$
$$\frac{-5x}{-5} = \frac{15}{-5}$$
$$x = -3$$

8.
$$4x - 3 = 4(x+5)$$
$$4x - 3 = 4x + 20$$
$$4x - 3 - 4x = 4x + 20 - 4x$$
$$-3 = 20$$
This equation is false no matter what value the variable x might have. Thus, there is no solution. The solution set is $\{\ \}$ or $\varnothing$.

9.
$$5x - 2 = 3 + 5(x-1)$$
$$5x - 2 = 3 + 5x - 5$$
$$5x - 2 = -2 + 5x$$
$$5x - 2 + 2 = -2 + 5x + 2$$
$$5x = 5x$$
$$5x - 5x = 5x - 5x$$
$$0 = 0$$
Since $0 = 0$ is a true statement for every value of x, all real numbers are solutions. The solution set is $\{x | x \text{ is a real number}\}$.

Vocabulary, Readiness & Video Check 1.5

1. Equations with the same solution set are called <u>equivalent</u> equations.

2. A value for the variable in an equation that makes the equation a true statement is called a <u>solution</u> of the equation.

3. By the <u>addition</u> property of equality, $y = -3$ and $y - 7 = -3 - 7$ are equivalent equations.

4. By the <u>multiplication</u> property of equality, $2y = -3$ and $\frac{2y}{2} = \frac{-3}{2}$ are equivalent equations.

5. $\frac{1}{3}x - 5$ <u>expression</u>

6. $2(x - 3) = 7$ <u>equation</u>

7. $\frac{5}{9}x + \frac{1}{3} = \frac{2}{9} - x$ <u>equation</u>

8. $\frac{5}{9}x + \frac{1}{3} - \frac{2}{9} - x$ <u>expression</u>

9. The addition property of equality allows us to add the same number to (or subtract the same number from) <u>both sides</u> of an equation and have an equivalent equation. The multiplication property of equality allows us to multiply (or divide) both sides of an equation by the <u>same</u> nonzero number and have an equivalent equation.

10. distributive property

11. to make the calculations less tedious

12. When solving a linear equation and all variable terms subtract out and:

  a. you have a <u>true</u> statement, then the equation has all real numbers for which the equation is defined as solutions.

  b. you have a <u>false</u> statement, then the equation has no solution.

Exercise Set 1.5

1.
$$-5x = -30$$
$$\frac{-5x}{-5} = \frac{-30}{-5}$$
$$x = 6$$

The solution is 6.

3. $-10 = x + 12$
$-10 - 12 = x + 12 - 12$
$-22 = x$

The solution is -22.

5. $x - 2.8 = 1.9$
$x - 2.8 + 2.8 = 1.9 + 2.8$
$x = 4.7$

The solution is 4.7.

7. $5x - 4 = 26 + 2x$
$5x - 2x = 26 + 4$
$3x = 30$
$\dfrac{3x}{3} = \dfrac{30}{3}$
$x = 10$

The solution is 10.

9. $-4.1 - 7z = 3.6$
$-4.1 - 7z + 4.1 = 3.6 + 4.1$
$-7z = 7.7$
$\dfrac{-7z}{-7} = \dfrac{7.7}{-7}$
$z = -1.1$

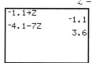

The solution is -1.1.

11. $5y + 12 = 2y - 3$
$5y + 12 - 2y = 2y - 3 - 2y$
$3y + 12 = -3$
$3y + 12 - 12 = -3 - 12$
$3y = -15$
$\dfrac{3y}{3} = \dfrac{-15}{3}$
$y = -5$

The solution is -5.

13. $3x - 4 - 5x = x + 4 + x$
$-4 - 2x = 2x + 4$
$-2x - 2x = 4 + 4$
$-4x = 8$
$x = -2$

The solution is -2.

15. $8x - 5x + 3 = x - 7 + 10$
$3x + 3 = x + 3$
$2x = 0$
$x = 0$

The solution is 0.

17. $5x + 12 = 2(2x + 7)$
$5x + 12 = 4x + 14$
$x + 12 = 14$
$x = 2$

The solution is 2.

19. $3(x - 6) = 5x$
$3x - 18 = 5x$
$-18 = 2x$
$-9 = x$

The solution is -9.

21. $-2(5y - 1) - y = -4(y - 3)$
$-10y + 2 - y = -4y + 12$
$-7y + 2 = 12$
$-7y = 10$
$y = -\dfrac{10}{7}$

The solution is $-\dfrac{10}{7}$.

23.
$$\frac{x}{2}+\frac{x}{3}=\frac{3}{4}$$
$$12\left(\frac{x}{2}+\frac{x}{3}\right)=12\left(\frac{3}{4}\right)$$
$$6x+4x=9$$
$$10x=9$$
$$x=\frac{9}{10}$$

```
9/10→X              .9
X/2+X/3            .75
3/4                .75
■
```

The solution is $\dfrac{9}{10}$.

25.
$$\frac{3t}{4}-\frac{t}{2}=1$$
$$4\left(\frac{3t}{4}-\frac{t}{2}\right)=4(1)$$
$$3t-2t=4$$
$$t=4$$

```
4→T                 4
3T/4-T/2            1
```

The solution is 4.

27.
$$\frac{n-3}{4}+\frac{n+5}{7}=\frac{5}{14}$$
$$28\left(\frac{n-3}{4}\right)+28\left(\frac{n+5}{7}\right)=28\left(\frac{5}{14}\right)$$
$$7(n-3)+4(n+5)=2(5)$$
$$7n-21+4n+20=10$$
$$11n-1=10$$
$$11n=11$$
$$n=1$$

```
1→N                 1
(N-3)/4+(N+5)/7
            .3571428571
5/14        .3571428571
■
```

The solution is 1.

29.
$$0.6x-10=1.4x-14$$
$$-0.8x-10=-14$$
$$-0.8x=-4$$
$$x=5$$

```
5→X                 5
.6X-10             -7
1.4X-14            -7
■
```

The solution is 5.

31.
$$\frac{3x-1}{9}+x=\frac{3x+1}{3}+4$$
$$9\left(\frac{3x-1}{9}+x\right)=9\left(\frac{3x+1}{3}+4\right)$$
$$(3x-1)+9x=3(3x+1)+36$$
$$3x-1+9x=9x+3+36$$
$$12x=9x+40$$
$$3x=40$$
$$x=\frac{40}{3}$$

```
40/3→X
           13.33333333
(3X-1)/9+X
           17.66666667
(3X+1)/3+4
           17.66666667
```

The solution is $\dfrac{40}{3}$.

33.
$$1.5(4-x)=1.3(2-x)$$
$$10[1.5(4-x)]=10[1.3(2-x)]$$
$$15(4-x)=13(2-x)$$
$$60-15x=26-13x$$
$$-2x=-34$$
$$x=17$$

```
17→X                17
1.5(4-X)
                 -19.5
1.3(2-X)
                 -19.5
■
```

The solution is 17.

35. $4(n+3)=2(6+2n)$
$$4n+12=12+4n$$
$$0=0$$
This is true for all x. Therefore, all real numbers are solutions.

37. $3(x+1)+5=3x+2$
$$3x+3+5=3x+2$$
$$3x+8=3x+2$$
$$8=2$$
This is false for any x. Therefore, no solution exists, $\varnothing$.

39. $2(x-8)+x = 3(x-6)+2$
$2x-16+x = 3x-18+2$
$3x-16 = 3x-16$
$0 = 0$
This is true for all x. Therefore, all real numbers are solutions.

41. $4(x+5) = 3(x-4)+x$
$4x+20 = 3x-12+x$
$4x+20 = 4x-12$
$20 = -12$
This is false for any x. Therefore, no solution exists, $\varnothing$.

43. The screen shows that the solution of $6x-5 = 4x-21$ is $x = -8$.

45. The screen shows that the solution of $3(N-3)+2 = -1+N$ is $N = 3$.

47.
$$\frac{3}{8}+\frac{b}{3} = \frac{5}{12}$$
$$24\left(\frac{3}{8}\right)+24\left(\frac{b}{3}\right) = 24\left(\frac{5}{12}\right)$$
$$9+8b = 10$$
$$8b = 1$$
$$b = \frac{1}{8}$$

49. $x-10 = -6x-10$
$x+6x = -10+10$
$7x = 0$
$x = 0$

51. $5(x-2)+2x = 7(x+4)-38$
$5x-10+2x = 7x+28-38$
$7x-10 = 7x-10$
$0 = 0$
This is true for all x. Therefore, all real numbers are solutions.

53. $y+0.2 = 0.6(y+3)$
$y+0.2 = 0.6y+1.8$
$0.4y = 1.6$
$y = 4$

55.
$$\frac{1}{4}(a+2) = \frac{1}{6}(5-a)$$
$$\frac{1}{4}a+\frac{1}{2} = \frac{5}{6}-\frac{1}{6}a$$
$$\frac{1}{4}a+\frac{1}{6}a = \frac{5}{6}-\frac{1}{2}$$
$$\frac{3}{12}a+\frac{2}{12}a = \frac{5}{6}-\frac{3}{6}$$
$$\frac{5}{12}a = \frac{2}{6}$$
$$a = \frac{2}{6}\cdot\frac{12}{5}$$
$$a = \frac{4}{5}$$

57. $2y+5(y-4) = 4y-2(y-10)$
$2y+5y-20 = 4y-2y+20$
$7y-20 = 2y+20$
$7y-2y = 20+20$
$5y = 40$
$y = 8$

59. $6x-2(x-3) = 4(x+1)+4$
$6x-2x+6 = 4x+4+4$
$4x+6 = 4x+8$
$4x-4x = 8-6$
$0 = 2$
This is false for any x. Therefore, the solution set is $\varnothing$.

61.
$$\frac{m-4}{3}-\frac{3m-1}{5} = 1$$
$$15\left(\frac{m-4}{3}\right)-15\left(\frac{3m-1}{5}\right) = 15(1)$$
$$5(m-4)-3(3m-1) = 15$$
$$5m-20-9m+3 = 15$$
$$-4m-17 = 15$$
$$-4m = 32$$
$$m = -8$$

63. $8x-12-3x = 9x-7$
$5x-12 = 9x-7$
$5x-9x = -7+12$
$-4x = 5$
$$x = -\frac{5}{4}$$

65. $-(3x-5)-(2x-6)+1=-5(x-1)-(3x+2)+3$
$-3x+5-2x+6+1=-5x+5-3x-2+3$
$-5x+12=-8x+6$
$3x=-6$
$x=-2$

67. $\dfrac{1}{3}(y+4)+6=\dfrac{1}{4}(3y-1)-2$
$12\left[\dfrac{1}{3}(y+4)+6\right]=12\left[\dfrac{1}{4}(3y-1)-2\right]$
$4(y+4)+12\cdot6=3(3y-1)-12\cdot2$
$4y+16+72=9y-3-24$
$4y+88=9y-27$
$-5y=-115$
$y=23$

69. $2[7-5(1-n)]+8n=-16+3[6(n+1)-3n]$
$2[7-5+5n]+8n=-16+3[6n+6-3n]$
$2(2+5n)+8n=-16+3(3n+6)$
$4+10n+8n=-16+9n+18$
$4+18n=2+9n$
$9n=-2$
$n=-\dfrac{2}{9}$

71. Subtract 19 instead of adding;
$3x+19=13$
$2x=-6$
$\dfrac{2x}{2}=\dfrac{-6}{2}$
$x=-3$

73. $0.4-1.6=-1.2$, not 1.2;
$9x+1.6=4x+0.4$
$5x=-1.2$
$\dfrac{5x}{5}=\dfrac{-1.2}{5}$
$x=-0.24$

75. $2x+3=2x+3$
Since the two sides of the equation are identical, the equation is true for any value of *x*. All real numbers are solutions.

77. $2x+1=2x+3$
Adding 1 o a number and adding 3 to the same number will not result in equal numbers for any value of *x*. There is no solution.

79. a. $4(x+1)+1=4x+4+1=4x+5$

b. $4(x+1)+1=-7$
$4x+4+1=-7$
$4x+5=-7$
$4x=-12$
$x=-3$
The solution is -3.

c. answers may vary

81. answers may vary

83. $3.2x+4=5.4x-7$
$3.2x+4-4=5.4x-7-4$
$3.2x=5.4x-11$
From this we see that $K=-11$.

85. $\dfrac{7}{11}x+9=\dfrac{3}{11}x-14$
$\dfrac{7}{11}x+9-9=\dfrac{3}{11}x-14-9$
$\dfrac{7}{11}x=\dfrac{3}{11}x-23$
From this we see that $K=-23$.

87. answers may vary

89. $x(x-6)+7=x(x+1)$
$x^2-6x+7=x^2+x$
$-6x+7=x$
$7=7x$
$1=x$

91. $3x(x+5)-12=3x^2+10x+3$
$3x^2+15x-12=3x^2+10x+3$
$15x-12=10x+3$
$5x=15$
$x=3$

93. $2.569x=-12.48534$
$\dfrac{2.569x}{2.569}=\dfrac{-12.48534}{2.569}$
$x=-4.86$

```
-4.86→X
            -4.86
2.569X
        -12.48534
```

The solution is -4.86.

95. $2.86z - 8.1258 = -3.75$

$$2.86z = 4.3758$$

$$\frac{2.86z}{2.86} = \frac{4.3758}{2.86}$$

$$z = 1.53$$

```
1.53→Z
             1.53
2.86Z-8.1258
             -3.75
```

The solution is 1.53.

Section 1.6 Practice Exercises

1. a. In words:

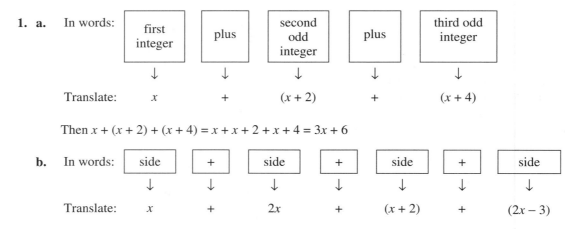

| first integer | plus | second odd integer | plus | third odd integer |
| ↓ | ↓ | ↓ | ↓ | ↓ |

Translate: x + $(x + 2)$ + $(x + 4)$

Then $x + (x + 2) + (x + 4) = x + x + 2 + x + 4 = 3x + 6$

b. In words:

| side | + | side | + | side | + | side |
| ↓ | ↓ | ↓ | ↓ | ↓ | ↓ | ↓ |

Translate: x + $2x$ + $(x + 2)$ + $(2x - 3)$

Then $x + 2x + (x + 2) + (2x - 3) = x + 2x + x + 2 + 2x - 3 = 6x - 1$

2. If x = number of arrivals and departures at Frankfurt airport, then $x + 12.9$ = number at London, and $x + 5.2$ = number at Paris.

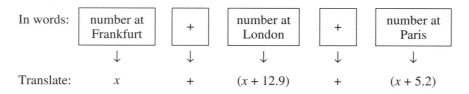

| number at Frankfurt | + | number at London | + | number at Paris |
| ↓ | ↓ | ↓ | ↓ | ↓ |

Translate: x + $(x + 12.9)$ + $(x + 5.2)$

Then $x + (x + 12.9) + (x + 5.2) = x + x + 12.9 + x + 5.2 = 3x + 18.1$

3. Let x = the first number, then $3x - 8$ = the second number, and $5x$ = the third number. The sum of the three numbers is 118.

$$x + (3x - 8) + 5x = 118$$

$$x + 3x + 5x - 8 = 118$$

$$9x - 8 = 118$$

$$9x = 126$$

$$x = 14$$

The numbers are 14, $3x - 8 = 3(14) - 8 = 34$, and $5x = 5(14) = 70$.

4. Let x = the original price. Then $0.4x$ = the discount. The original price, minus the discount, is equal to \$270.

$$x - 0.4x = 270$$
$$0.6x = 270$$
$$x = \frac{270}{0.6} = 450$$

The original price was \$450.

5. Let x = width, then $2x - 16$ = length.
   The perimeter is 160 inches.

$$2(x) + 2(2x - 16) = 160$$
$$2x + 4x - 32 = 160$$
$$6x - 32 = 160$$
$$6x = 192$$
$$x = 32$$

$2x - 16 = 2(32) - 16 = 48$

The width is 32 inches and the length is 48 inches.

6. Let x = first odd integer, then $x + 2$ = second odd integer, and $x + 4$ = third odd integer.
   The sum of the integers is 81.

$$x + (x + 2) + (x + 4) = 81$$
$$3x + 6 = 81$$
$$3x = 75$$
$$x = 25$$

$x + 2 = 27$
$x + 4 = 29$
The integers are 25, 27, and 29.

Vocabulary, Readiness & Video Check 1.6

1. 130% of a number _>_ the number.

2. 70% of a number _<_ the number.

3. 100% of a number _=_ the number.

4. 200% of a number _>_ the number.

	First Integer	All Described Integers
5. Four consecutive integers	31	31, 32, 33, 34
6. Three consecutive odd integers	31	31, 33, 35
7. Three consecutive even integers	18	18, 20, 22
8. Four consecutive even integers	92	92, 94, 96, 98
9. Three consecutive integers	y	$y, y+1, y+2$
10. Three consecutive even integers	z (z is even)	$z, z+2, z+4$
11. Four consecutive integers	p	$p, p+1, p+2, p+3$
12. Three consecutive odd integers	s (s is odd)	$s, s+2, s+4$

13. distributive property

14. The original application asks you to find three numbers. The solution $x = 45$ only gives you the first number. You need to INTERPRET this result.

Exercise Set 1.6

1. The perimeter is the sum of the lengths of the four sides.
   $y + y + y + y = 4y$

3. Let z = first integer, then $z + 1$ = second integer, and $z + 2$ = third integer.
   $z + (z+1) + (z+2) = z + z + z + 1 + 2 = 3z + 3$

5. Find the sum of x nickels worth 5¢ each, and $(x + 3)$ dimes worth 10¢ each, and $2x$ quarters worth 25¢ each.
   $$5x + 10(x+3) + 25(2x) = 5x + 10x + 30 + 50x$$
   $$= 65x + 30$$
   The total amount is $(65x + 30)$ cents.

7. $4x + 3(2x + 1) = 4x + 6x + 3 = 10x + 3$

9. The length of the side denoted by ? is $10 - 2 = 8$. Similarly, the length of the unmarked side is
   $(x - 3) - (x - 10) = x - 3 - x + 10 = 7$.
   Thus the perimeter of the floor plan is given by $(x - 10) + 2 + 7 + 8 + (x - 3) + 10 = 2x + 14$.

11. Let x = the number.
    $$4(x-2) = 2 + 4x + 2x$$
    $$4x - 8 = 2 + 6x$$
    $$-2x = 10$$
    $$x = -5$$
    The number is -5.

13. Let x = the first number, then
$5x$ = the second number, and
$x + 100$ = the third number.
$$x + 5x + (x + 100) = 415$$
$$7x + 100 = 415$$
$$7x = 315$$
$$x = 45$$
$5x = 225$
$x + 100 = 145$
The numbers are 45, 225, and 145.

15. 29% of $2271 = 0.29 \cdot 2271 = 658.59$;
$2271 - 658.59 = 1612.41$.
Approximately 1612.41 million acres are not federally owned.

17. 91.4% of $8476 = 0.914 \cdot 8476 \approx 7747$
Approximately 7747 minor earthquakes occurred in the United States in 2010.

19. 15% of $1500 = 0.15 \cdot 1500 = 225$
$1500 - 225 = 1275$
1275 are willing to do business with any size retailer.

21. Let x be the percent of email users that spend less than 15 minutes on email at work per day.
$$x + 50 + 8 + 10 + 9 = 100$$
$$x + 77 = 100$$
$$x = 23$$
23% of email users spend less than 15 minutes on email each day.

23. Let x be the number of employees who use email for more than 3 hours per day. Then x is 9% of 4633.
$x = 0.09(4633) = 416.97$
417 employees would be expected to use email for more than 3 hours per day.

25. $\quad x + 4x + (x + 6) = 180$
$$6x + 6 = 180$$
$$6x = 174$$
$$x = 29$$
$4x = 4(29) = 116$
$x + 6 = 29 + 6 = 35$
The angles measure 29°, 35°, and 116°.

27. $\quad (4x) + (5x + 1) + (5x + 3) = 102$
$$14x + 4 = 102$$
$$14x = 98$$
$$x = 7$$

$4x = 4(7) = 28$
$5x + 1 = 5(7) + 1 = 36$
$5x + 3 = 5(7) + 3 = 38$
The sides measure 28 meters, 36 meters, and 38 meters.

29. $\quad x + (2.5x - 9) + x + 1.5x = 99$
$$6x - 9 = 99$$
$$6x = 108$$
$$x = 18$$
$1.5x = 1.5(18) = 27$
$2.5x - 9 = 2.5(18) - 9 = 36$
The sides measure 18 inches, 18 inches, 27 inches, and 36 inches.

31. Let x = first integer; then
$x + 1$ = next integer and
$x + 2$ = third integer.
$$x + (x + 1) + (x + 2) = 228$$
$$3x + 3 = 228$$
$$3x = 225$$
$$x = 75$$
$x + 1 = 75 + 1 = 76$
$x + 2 = 75 + 2 = 77$
The integers are 75, 76, and 77.

33. Let x = first even integer, then
$x + 2$ = second even integer, and
$x + 4$ = third even integer.
$$2x + (x + 4) = 268,222$$
$$3x + 4 = 268,222$$
$$3x = 268,218$$
$$x = 89,406$$
$x + 2 = 89,408$
$x + 4 = 89,410$
Fallon's ZIP code is 89406, Fernley's ZIP code is 89408, and Gardnerville Ranchos's ZIP code is 89410.

35. $\quad (2x - 21) + \left(\dfrac{5}{2}x + 2 \right) + (3x + 24) = 290$
$$2x + \frac{5}{2}x + 3x - 21 + 2 + 24 = 290$$
$$\frac{15}{2}x + 5 = 290$$
$$\frac{15}{2}x = 285$$
$$x = 38$$
$2x - 21 = 2(38) - 21 = 55$
$\dfrac{5}{2}x + 2 = \dfrac{5}{2}(38) + 2 = 97$
$3x + 24 = 3(38) + 24 = 138$

Year	Increase in Wi-Fi-Enabled Cell Phones	Predicted Number
2010	$2x - 21$	55 million
2012	$\frac{5}{2}x + 2$	97 million
2014	$3x + 24$	138 million
Total	290 million	

37. Let x be the growth in the number of biomedical engineer jobs (in thousands). Then $3x - 7$ is the growth in the number of physician assistant jobs, and $\frac{1}{2}x + 9$ is the growth in the number of skin care specialist jobs.
$$x + (3x - 7) + \left(\frac{1}{2}x + 9\right) = 56$$
$$x + 3x + \frac{1}{2}x - 7 + 9 = 56$$
$$\frac{9}{2}x + 2 = 56$$
$$\frac{9}{2}x = 54$$
$$x = 12$$
$$3x - 7 = 3(12) - 7 = 29$$
$$\frac{1}{2}x + 9 = \frac{1}{2}(12) + 9 = 15$$
The growth predictions are as follows:
biomedical engineer: 12 thousand
skin care specialist: 15 thousand
physician assistant: 29 thousand

39. Let x be the number of seats in a B737-200 aircraft. Then the number of seats in a B767-300ER is $x + 88$, and the number of seats in a F-100 is $x - 32$.
$$x + (x + 88) + (x - 32) = 413$$
$$3x + 56 = 413$$
$$3x = 357$$
$$x = 119$$
$$x + 88 = 119 + 88 = 207$$
$$x - 32 = 119 - 32 = 87$$
The B737-200 has 119 seats, the B767-300ER has 207 seats, and the F-100 has 87 seats.

41. Let x be the price of the fax machine before tax.
$$x + 0.08x = 464.40$$
$$1.08x = 464.40$$
$$x = 430$$
The fax machine cost $430 before tax.

43. The new salary is 1.023 times the current salary.
$436,000(1.023) = 446,028$
The new salary is $446,028.

45. Let x be the expected population of Swaziland in 2050. Then x is 1,200,000 plus the increase of 50% of 1,200,000.
$$x = 1,200,000 + 0.50(1,200,000)$$
$$= 1,200,000 + 600,000$$
$$= 1,800,000$$
The population of Swaziland is expected to be 1,800,000 in 2050.

47. Let $x =$ measure of the angle; then $180 - x =$ measure of its supplement.
$$x = 3(180 - x) + 20$$
$$x = 540 - 3x + 20$$
$$4x = 560$$
$$x = 140$$
$$180 - x = 180 - 140 = 40$$
The angles measure 140° and 40°.

49. Let $x =$ measure of second angle; then $2x =$ measure of first angle and $3x - 12 =$ measure of third angle.
$$x + 2x + (3x - 12) = 180$$
$$6x - 12 = 180$$
$$6x = 192$$
$$x = 32$$
$$2x = 2(32) = 64$$
$$3x - 12 = 3(32) - 12 = 84$$
The angles measure 64°, 32°, and 84°.

51. Let $x =$ the length of a side of the square. Then $x + 6 =$ the length of a side of the triangle.
$$4x = 3(x + 6)$$
$$4x = 3x + 18$$
$$x = 18$$
The sides of the square are 18 cm and the sides of the triangle are 24 cm.

53. Let $x =$ first even integer, then $x + 2 =$ second even integer, and $x + 4 =$ third even integer.
$$x + (x + 4) = 156$$
$$2x + 4 = 156$$
$$2x = 152$$
$$x = 76$$
$$x + 2 = 78$$
$$x + 4 = 80$$
The integers are 76, 78, and 80.

55. Let x be the number of grandstand seats at Darlington Motor Raceway. Then $2x + 37{,}000$ is the number of grandstand seats at Daytona International Speedway.

$$x + (2x + 37{,}000) = 220{,}000$$
$$3x + 37{,}000 = 220{,}000$$
$$3x = 183{,}000$$
$$x = 61{,}000$$

$2x + 37{,}000 = 2(61{,}000) + 37{,}000 = 159{,}000$

Darlington has 61,000 grandstand seats and Daytona has 159,000 grandstand seats.

57. Let x be the population of the New York metropolitan region. Then Mexico City's population is $x + 0.03$, and Tokyo's population is $2x - 2.19$.

$$x + (x + 0.03) + (2x - 2.19) = 75.56$$
$$4x - 2.16 = 75.56$$
$$4x = 77.72$$
$$x = 19.43$$

$x + 0.03 = 19.46$
$2x - 2.19 = 36.67$

The population of New York is 19.43 million, the population of Mexico City is 19.46 million, and the population of Tokyo is 36.67 million.

59.
$$x + 5x + (6x - 3) = 483$$
$$12x - 3 = 483$$
$$12x = 486$$
$$x = 40.5$$
$5x = 5(40.5) = 202.5$
$6x - 3 = 6(40.5) - 3 = 240$

The sides measure 40.5 feet, 202.5 feet, and 240 feet.

61. Let x be the number of bulb hours for a halogen bulb. Then $25x$ is the number of bulb hours for a fluorescent bulb, and $x - 2500$ is the number of bulb hours for an incandescent bulb.

$$x + 25x + (x - 2500) = 105{,}500$$
$$27x - 2500 = 105{,}500$$
$$27x = 108{,}000$$
$$x = 4000$$

$25x = 25(4000) = 100{,}000$
$x - 2500 = 4000 - 2500 = 1500$

A halogen bulb lasts 4000 bulb hours. A fluorescent bulb lasts 100,000 bulb hours. An incandescent bulb lasts 1500 bulb hours.

63. Let x = number of wins for the Chicago Cubs. Then $x + 1$ = number of wins for the Houston Astros, and $x + 2$ = number of wins for the Milwaukee Brewers.

$$x + (x + 1) + (x + 2) = 228$$
$$3x + 3 = 228$$
$$3x = 225$$
$$x = 75$$

$x + 1 = 76$
$x + 2 = 77$

The number of wins are as follows:
Chicago Cubs: 75 wins
Houston Astros: 76 wins
Milwaukee Brewers: 77 wins

65. Let x = height of Galter Pavilion; then $x + 67$ = height of Guy's Tower and $x + 47$ = height of Queen Mary

$$x + (x + 67) + (x + 47) = 1320$$
$$3x + 114 = 1320$$
$$3x = 1206$$
$$x = 402$$

$x + 67 = 402 + 67 = 469$
$x + 47 = 402 + 47 = 449$

Galter Pavilion: 402 ft
Guy's Tower: 469 ft
Queen Mary: 449 ft

67. yes; answers may vary

69. Let $x°$ be the measure of an angle. Then its complement measures $(90 - x)°$ and its supplement measures $(180 - x)°$.

$$180 - x = 10(90 - x)$$
$$180 - x = 900 - 10x$$
$$180 + 9x = 900$$
$$9x = 720$$
$$x = 80$$

The angle measures $80°$.

71.
$$y = -94.8x + 2049$$
$$0 = -94.8x + 2049$$
$$94.8x = 2049$$
$$x \approx 22$$

The average number of cigarettes smoked will be zero 22 years after 2000, or in 2022.

73. $y = -94.8(20) + 2049 = 153$

The average number of cigarettes smoked by an American adult in 2020 will be 153.

75. Let x be the first integer. Then $x + 1$ and $x + 2$ are the next two consecutive integers.

$$x + (x + 1) + (x + 2) = 3(x + 1)$$
$$3x + 3 = 3x + 3$$
$$3 = 3 \quad \text{True}$$

The sum of any three consecutive integers is equal to three times the second integer.

77. $R = C$
$24x = 100 + 20x$
$4x = 100$
$x = 25$
25 skateboards are needed to break even.

79. The company loses money if it makes and sells fewer products than the break-even number.

Section 1.7 Practice Exercises

1. In words: $2000 plus 6% of sales
Translate: 2000 +   0.06x

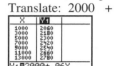

The completed table is:

Sales, x	1000	3000	5000	7000
Gross pay	2060	2180	2300	2420

Find the 2540 entry in the y_1 (gross pay) column. The corresponding x-value (sales) figure is 9000. She must average $9000 in sales per month to earn a gross pay of $2540 per month.

2. Let x be the number of hours. The charge for the first plumber is $60 + 35x$, while the charge for the second plumber is $45x$.

The first plumber charges $235 for 5 hours and the second plumber charges $225, so the second plumber charges $10 less for 5 hours.

3. First, find the discounted price, which is 40% off the original price.

In words: Discounted price   is   60%   of   original price

Translate: Discounted price   =   0.60   · x

Thus $y_1 = 0.60x$ is the discounted price.
The sales tax rate is 6.5%.

In words: Sales tax   is   6.5%   of   discounted price

Translate: Sales tax   =   0.065   · $(0.60x)$

The total price for the discounted items was $87.00. Including tax, she paid $92.66 for the items.

4. Let x be the number of items produced and sold. The cost is $5600 + 4x$ and the revenue is $11x$.

No. of items	600	700	800	900	1000
Cost (x items)	8000	8400	8800	9200	9600
Revenue (x items)	6600	7700	8800	9900	11,000

Since the table entries for 800 items are the same for cost and revenue, the sale and production of 800 items produces the break-even point.

5. $y = -16x^2 + 112x$

The maximum height, 196 feet, occurs 3.5 seconds after the rocket is launched.

a. Since the window height is 160 feet, the rocket first passes the window 2 seconds after it is shot off.

b. The rocket passes the window again at 5 seconds.

c. At 7 seconds, the height is 0, indicating that the rocket has hit the ground.

d. The rocket appears to reach its maximum height between 3 and 4 seconds.

The maximum height, 196 feet, occurs 3.5 seconds after the rocket is launched.

Vocabulary, Readiness & Video Check 1.7

1. Each y_1-value is -2 times the x-value, so $y = -2x$; f.

2. Each y_1-value is 5 times the x-value, so $y = 5x$; a.

3. Each y_1-value is 1 more than 3 times the x-value, so $y = 3x + 1$; b.

4. Each y_1-value is 5 more than -1 times the x-value, so $y = -x + 5$; e.

5. Each y_1-value is 1 more than the square of the x-value, so $y = x^2 + 1$; d.

6. Each y_1-value is the square of the x-value, so $y = x^2$; c.

7. To use the table feature of a calculator, give the table a specific value to <u>start</u> with, tell the table by what <u>increment</u> to <u>increase</u>, and give the <u>expression</u> you want evaluated.

Exercise Set 1.7

1.

Side Length of a Cube	x	7	8	9	10	11
Volume	x^3	343	512	729	1000	1331

  a. The volume of a cube whose side measures 9 centimeters is 729 cubic centimeters.

  b. The volume of a cube whose side measures 11 feet is 1331 cubic feet.

  c. A cube with a volume of 343 cubic inches has sides of length 7 inches.

3.

Hours Worked	x	39	39.5	40	40.5
Gross Pay	$25 + 7x$	298	301.50	305	308.50

  a. 44 hours worked gives a gross pay of $333.

  b. 35 hours worked gives a gross pay of $270.

  c. A gross pay of $284 corresponds to 37 hours worked.

5. **a.** $y_1 = 15.25x$

  b. $y_2 = 17x$

  c.

Hours	1	1.25	1.5	1.75	2	2.25	2.5	2.75	3
Rough Draft	15.25	19.06	22.88	26.69	30.50	34.31	38.13	41.94	45.75
Manuscript	17	21.25	25.50	29.75	34	38.25	42.50	46.75	51

    d. If Emily can type 2 pages in 15 minutes, then she can type $10 = 5 \cdot 2$ pages in $75 = 5 \cdot 15$ minutes or 1.25 hours. From the table, she charges $19.06 for a 10-page rough draft.

    e.

    Emily charges $182.75 for the job.

7.

Radius of Circle	x	2	5	21	94.2
Circumference	$2\pi x$	12.57	31.42	131.95	591.88
Area	πx^2	12.57	78.54	1385.44	27,877.36

    a. If the radius of the circle is 30 yards, then the circumference is 188.50 yards and the area is 2827.43 square yards.

    b. If the radius of the circle is 78.5 millimeters, then the circumference is 493.23 millimeters and the area is 19,359.28 square millimeters.

9. a. Let x be the price of a home sold.
      Salary $= 500 + 0.04x$

Sales (in thousands)	125	200	250	350	400
Gross Pay	5500	8500	10,500	14,500	16,500

    b. If she sells one home per month, her gross salary would be between $5500 and $16,500.

11. a. $y_1 = 12.96 + 1.10x$

    b.

Gallons of Water Used (Thousands)	0	1	2	3	4	5	6
Monthly Cost (Dollars)	12.96	14.06	15.16	16.26	17.36	18.46	19.56

13. a. $y_1 = 500 - 8x$

b.

No. of Minutes	0	10	20	30	40	50	60	70
No. of Computers Not Infected	500	420	340	260	180	100	20	−60

c.   In 30 minutes, there will be 260 uninfected computers.

d.

In 62.5 minutes, all the computers will be infected.

15. a.   Use $y_1 = 14,500 + 1000(x - 1)$ and $y_2 = 20,000 + 575(x - 1)$.

Year	5	10	15	20
First Job	$18,500	$23,500	$28,500	$33,500
Second Job	$22,300	$25,175	$28,050	$30,925

b.   answers may vary

17. a. $y_1 = 15 + 0.08924x$

b.

Kilowatt-hours	0	500	1000	1500	2000
Monthly Cost (Dollars)	15	59.62	104.24	148.86	193.48

19. a. $y_1 = 1000 - 2.5x$

b.

X	Y1
0	1000
60	850
120	700
180	550
240	400
300	250
360	100

Y1■1000-2.5X

Minutes	0	60	120	180	240	300	360
ml Solution Remaining	1000	850	700	550	400	250	100

c.

X	Y1
360	100
370	75
380	50
390	25
400	0
410	-25
420	-50

Y1■1000-2.5X

She should return in 380 minutes, which is 6 hours, 20 minutes. She should return at 4:20 P.M.

21. a. A rate of 15 miles per gallon corresponds to $\dfrac{1}{15}$ gallon per mile.

$$y_1 = 24 - \frac{x}{15}$$

b. A rate of 25 miles per gallon corresponds to $\dfrac{1}{25}$ gallon per mile.

$$y_2 = 24 - \frac{x}{25}$$

c.

X	Y1	Y2
0	24	24
100	17.333	20
200	10.667	16
300	4	12
400	-2.667	8
500	-9.333	4
600	-16	0

Y1■24-X/15

Miles	0	100	200	300	400	500	600
City	24	17.3	10.7	4	−2.7	−9.3	−16
Highway	24	20	16	12	8	4	0

d. After 100 miles of city driving, there are 17.3 gallons left in the tank.

e. After 100 miles of highway driving, there are 20 gallons left in the tank.

f. She can drive 600 miles on the highway before her tank is empty.

g.

X	Y1	Y2
300	4	12
310	3.3333	11.6
320	2.6667	11.2
330	2	10.8
340	1.3333	10.4
350	.66667	10
360	0	9.6

Y1■24-X/15

She can drive 360 miles in the city before her tank is empty.

h. answers may vary

23. a. If the regular price is x dollars, the sale price is $x - 0.35x = 0.65x$ dollars.
$y_1 = 0.65x$

b.

X	Y1
29.95	19.468
35.95	23.368
19.25	12.513
39.95	25.968
17.95	11.668
9.95	6.4675
25.75	16.738

Y1■.65X

Item	Blouse	Skirt	Shorts	Shoes	Purse	Earrings	Backpack
Price tag	$29.95	$35.95	$19.25	$39.95	$17.95	$9.95	$25.75
Sale price	$19.47	$23.37	$12.51	$25.97	$11.67	$6.47	$16.74

c.

19.47+23.37+12.5
1+25.97+11.67+6.
47+16.74
                116.2

Her total bill before taxes is $116.20 if she purchases all the items on sale.

d.

116.2*.07
              8.134
116.2+8.13
              124.33

Her total after taxes is $124.33.

e.

29.95+35.95+19.2
5+39.95+17.95+9.
95+25.75
              178.75
178.75-116.20
              62.55

She saves $62.55 by waiting for the sale.

25. Let $y_1 = 3500 + 55x$, the cost equation, and $y_2 = 75x$, the revenue equation.

X	Y1	Y2
100	9000	7500
125	10375	9375
150	11750	11250
175	13125	13125
200	14500	15000
225	15875	16875
250	17250	18750

Y1■3500+55X

x Calculators	100	125	150	175	200
Cost	9000	10,375	11,750	13,125	14,500
Revenue	7500	9375	11,250	13,125	15,000

The break-even point is when 175 calculators are produced and sold.

27. When 200 calculators are produced and sold, the profit is $15,000 − $14,500 = $500.

29. a.

X	Y1
0	4050
5	4380
10	4710
15	5040
20	5370
25	5700
30	6030

Y1■66X+4050

years after 2010	0	5	10	15	20
projected billions of kilowatt-hours	4050	4380	4710	5040	5370

    b.   2025 is 15 years after 2010. In 2025, the demand for electricity is projected to be 5040 billion kilowatt-hours.

31. a.

Seconds	1	2	3	4	5
Height (Feet)	1569	1606	1611	1584	1525

    b.

The rocket's maximum height is about 1613 feet.

33. a.   Use $y_1 = 25 + 0.08x$ for Anne's cost, and $y_2 = 35 + 0.05x$ for Michelle's cost.

For 300 miles, Anne's cost is \$49.00 and Michelle's cost is \$50.00.

    b.

Anne's rental is cheaper for distances less than $333\frac{1}{3}$ miles.

35. a. $C = 90 + 75x$

    b.

No. of Hours	2	4	6	8
Total Fee (Dollars)	240	390	540	690

37.

$Y_1 = -.14X^2 + 1.4X$

   a. The frog reaches a maximum height of 3.5 feet.

   b. The frog has traveled 5 feet horizontally when it reaches its maximum height.

   c. The frog lands 10 feet from its starting point.

39. Let x be the length of the sides perpendicular to the patio. Then $92 - 2x$ is the length of the side parallel to the patio, and the area is $x(92 - 2x)$.

$Y_1 = X*(92-2X)$

The largest area, 1058 square feet, occurs when $x = 23$. Then
$92 - 2x = 92 - 2(23) = 92 - 46 = 46$ and the dimensions are 23 feet by 46 feet.

41.

$Y_1 = -.14X^2 + 1.09X...$

   a. On May 1, sunrise is at 6.73 or 6:44 A.M.

   b. Sunrise is the latest for $x = 4$, which corresponds to July 1.

   c. The latest sunrise is at 7:14 A.M.

   d.

$X = 6$

September 15 corresponds to $x = 6.5$.
Sunrise on September 15 is at 6:17 A.M.

   e. answers may vary

Section 1.8 Practice Exercises

   1.
$$I = PRT$$
$$\frac{I}{PR} = \frac{PRT}{PR}$$
$$\frac{I}{PR} = T \text{ or } T = \frac{I}{PR}$$

2.
$$7x - 2y = 5$$
$$7x - 2y - 7x = 5 - 7x$$
$$-2y = 5 - 7x$$
$$\frac{-2y}{-2} = \frac{5 - 7x}{-2}$$
$$y = \frac{7}{2}x - \frac{5}{2}$$

3.
$$A = P + Prt$$
$$A - P = P + Prt - P$$
$$A - P = Prt$$
$$\frac{A - P}{Pt} = \frac{Prt}{Pt}$$
$$\frac{A - P}{Pt} = r \text{ or } r = \frac{A - P}{Pt}$$

4. Let $P = 8000$, $r = 6\% = 0.06$, $t = 4$, $n = 2$.
$$A = P\left(1 + \frac{r}{n}\right)^{nt}$$
$$A = 8000\left(1 + \frac{0.06}{2}\right)^{2 \cdot 4}$$
$$A = 8000(1.03)^8$$
$$A \approx 8000(1.266770081)$$
$$A \approx 10,134.16$$
Russ will have \$10,134.16 in his account.

5. Let $d = 192$ and $r = 7.5$.
$$d = rt$$
$$192 = 7.5t$$
$$\frac{192}{7.5} = \frac{7.5t}{7.5}$$
$$25.6t = t$$
They spent 25.6 hours cycling, or 25 hours 36 minutes.

6.
$$C = \frac{5}{9}(F - 32)$$
$$\frac{9}{5}C = F - 32$$
$$\frac{9}{5}C + 32 = F$$

Use $y_1 = \frac{9}{5}x + 32$.

$Y_1 = 9/5X+32$

Celsius	−40	−30	10	30
Fahrenheit	−40	−22	50	86

7. Let x be the width of the pen. Then the length of the pen along the shed is $(88 − 2x)$ feet, and the area is $x(88 − 2x)$.

The largest area is 968 square feet when the width is 22 feet and the length is $88 − 2 \cdot 22 = 44$ feet.

Vocabulary, Readiness & Video Check 1.8

1. $2x + y = 5$
$$y = 5 − 2x$$

2. $7x − y = 3$
$$−y = 3 − 7x$$
$$y = −3 + 7x \text{ or } y = 7x − 3$$

3. $a − 5b = 8$
$$a = 5b + 8$$

4. $7r + s = 10$
$$s = 10 − 7r$$

5. $5j + k − h = 6$
$$5j + k = h + 6$$
$$k = h − 5j + 6$$

6. $w − 4y + z = 0$
$$w + z = 4y$$
$$z = 4y − w$$

7. That the specified variable will equal some expression and that this expression should not contain the specified variable.

8. The only way to check the solution is in the formula used, because if the wrong formula is used, a wrong answer may seem to check correctly.

Exercise Set 1.8

1. $D = rt$
$$\frac{D}{r} = \frac{rt}{r}$$
$$\frac{D}{r} = t$$
$$t = \frac{D}{r}$$

3. $I = PRT$
$$\frac{I}{PT} = \frac{PRT}{PT}$$
$$\frac{I}{PT} = R$$
$$R = \frac{I}{PT}$$

5. $9x − 4y = 16$
$$9x − 4y − 9x = 16 − 9x$$
$$−4y = 16 − 9x$$
$$\frac{−4y}{−4} = \frac{16 − 9x}{−4}$$
$$y = \frac{9x − 16}{4}$$

7. $P = 2L + 2W$
$$P − 2L = 2W$$
$$\frac{P − 2L}{2} = \frac{2W}{2}$$
$$\frac{P − 2L}{2} = W$$
$$W = \frac{P − 2L}{2}$$

9. $J = AC − 3$
$$J + 3 = AC$$
$$\frac{J + 3}{C} = \frac{AC}{C}$$
$$\frac{J + 3}{C} = A$$
$$A = \frac{J + 3}{C}$$

11.
$$W = gh - 3gt^2$$
$$W = g(h - 3t^2)$$
$$\frac{W}{h - 3t^2} = \frac{g(h - 3t^2)}{h - 3t^2}$$
$$\frac{W}{h - 3t^2} = g$$
$$g = \frac{W}{h - 3t^2}$$

13.
$$T = C(2 + AB)$$
$$T = 2C + ABC$$
$$T - 2C = 2C + ABC - 2C$$
$$T - 2C = ABC$$
$$\frac{T - 2C}{AC} = \frac{ABC}{AC}$$
$$\frac{T - 2C}{AC} = B$$
$$B = \frac{T - 2C}{AC}$$

15.
$$C = 2\pi r$$
$$\frac{C}{2\pi} = \frac{2\pi r}{2\pi}$$
$$\frac{C}{2\pi} = r$$
$$r = \frac{C}{2\pi}$$

17.
$$E = I(r + R)$$
$$E = Ir + IR$$
$$E - IR = Ir + IR - IR$$
$$E - IR = Ir$$
$$\frac{E - IR}{I} = \frac{Ir}{I}$$
$$\frac{E - IR}{I} = r$$
$$r = \frac{E - IR}{I}$$

19.
$$s = \frac{n}{2}(a + L)$$
$$2s = 2 \cdot \frac{n}{2}(a + L)$$
$$2s = n(a + L)$$
$$2s = na + nL$$
$$2s - na = na + nL - na$$
$$2s - na = nL$$
$$\frac{2s - na}{n} = \frac{nL}{n}$$
$$\frac{2s - na}{n} = L$$
$$L = \frac{2s - na}{n}$$

21.
$$N = 3st^4 - 5sv$$
$$N - 3st^4 = 3st^4 - 5sv - 3st^4$$
$$N - 3st^4 = -5sv$$
$$\frac{N - 3st^4}{-5s} = \frac{-5sv}{-5s}$$
$$\frac{3st^4 - N}{5s} = v$$
$$v = \frac{3st^4 - N}{5s}$$

23.
$$S = 2LW + 2LH + 2WH$$
$$S - 2LW = 2LW + 2LH + 2WH - 2LW$$
$$S - 2LW = 2LH + 2WH$$
$$S - 2LW = H(2L + 2W)$$
$$\frac{S - 2LW}{2L + 2W} = \frac{H(2L + 2W)}{2L + 2W}$$
$$\frac{S - 2LW}{2L + 2W} = H$$
$$H = \frac{S - 2LW}{2L + 2W}$$

25. $A = P\left(1 + \dfrac{r}{n}\right)^{nt} = 3500\left(1 + \dfrac{0.03}{n}\right)^{10n}$

n	1	2	4
A	\$4703.71	\$4713.99	\$4719.22

n	12	365
A	\$4722.74	\$4724.45

27. $A = P\left(1 + \dfrac{r}{n}\right)^{nt} = 6000\left(1 + \dfrac{0.04}{n}\right)^{5n}$

  a. $n = 2$

  $A = 6000\left(1 + \dfrac{0.04}{2}\right)^{5 \cdot 2} \approx 7313.97$

  $\$7313.97$

  b. $n = 4$

  $A = 6000\left(1 + \dfrac{0.04}{4}\right)^{5 \cdot 4} \approx 7321.14$

  $\$7321.14$

  c. $n = 12$

  $A = 6000\left(1 + \dfrac{0.04}{12}\right)^{5 \cdot 12} \approx 7325.98$

  $\$7325.98$

29. Roundtrip distance $= 90 + 90 = 180$ miles
  $d = r \cdot t$
  $180 = 50t$
  $\dfrac{180}{50} = t$
  $t = 3.6$
  3.6 hours = 3 hours 36 minutes
  The trip will take 3.6 hours or 3 hours 36 minutes.

31. $C = \dfrac{5}{9}(F - 32)$

  $C = \dfrac{5}{9}(104 - 32)$

  $C = \dfrac{5}{9}(72)$

  $C = 40°$
  The day's high temperature was 40°C.

33. $A = s^2 = (64)^2 = 4096 \text{ ft}^2;\ \dfrac{4096}{24} \approx 171$
  There should be 171 packages of tiles bought.

35. $A = \dfrac{1}{2}bh$

  $18 = \dfrac{1}{2}(4)h$

  $18 = 2h$
  $9 = h$
  The height is 9 feet.

37. The area of one pair of walls is $2 \cdot 14 \cdot 8 = 224 \text{ ft}^2$ and the area of the other walls is $2 \cdot 16 \cdot 8 = 256 \text{ ft}^2$ for a total of 480 ft^2. Multiplying by 2, the number of coats, yields 960 ft^2. Dividing this by 500 yields 1.92. Thus, 2 gallons should be purchased.

39. $V = \pi r^2 h$
  $980\pi = \pi(7)^2 h$
  $980\pi = 49\pi h$
  $980 = 49h$
  $20 = h$
  The height is 20 meters.

41. a. $V = \dfrac{4}{3}\pi r^3;\ r = \dfrac{d}{2} = \dfrac{12}{2} = 6$

  $V = \dfrac{4}{3}\pi(6)^3$

  $V = \dfrac{4}{3}\pi(216)$

  $V = 288\pi$
  The volume is 288π cubic mm.

  b. $V = 288\pi \approx 904.78$ cubic mm.

43. a. $V = \pi r^2 h$
  $V = \pi(4.2)^2(2.12)$
  $V \approx 1174.86$
  The volume of the cylinder is 1174.86 cubic meters.

  b. $V = \dfrac{4}{3}\pi r^3$

  $V = \dfrac{4}{3}\pi(4.2)^3$

  $V \approx 310.34$
  The volume of the sphere is 310.34 cubic meters.

  c. $V = 1174.86 + 310.34 = 1485.20$
  The volume of the tank is 1485.20 cubic meters.

45. 19 hours 5 minutes $= 19\dfrac{5}{60}$ hours

  $d = rt$

  $2447.8 = r\left(19\dfrac{5}{60}\right)$

  $128.3 \approx r$
  Her average speed was 128.3 miles per hour.

47. $V = \pi r^2 h$
1 mile = 5280 feet
1.3 miles = 6864 feet
$3800 = \pi r^2 (6864)$
$0.42 \approx r$
The radius of the hole is 0.42 feet.

49. $C = \pi d = \pi(41.125) = 41.125\pi$ ft ≈ 129.1325 ft
The circumference of Earth is
$41.125\pi \approx 129.1325$ feet.

51. Let x be the width of the garden. Then the length of the garden along the patio is $(52 - 2x)$ feet, and the area is $x(52 - 2x)$ square feet.

X	Y1
10	320
11	330
12	336
13	338
14	336
15	330
16	320

Y1▊X(52-2X)

The largest area is 338 square feet when the width is 13 feet and the length is
$52 - 2 \cdot 13 = 26$ feet.

53. $\qquad C = 4h + 9f + 4p$
$C - 4h - 4p = 9f$
$\dfrac{C - 4h - 4p}{9} = f$
$f = \dfrac{C - 4h - 4p}{9}$

55. $C = 4h + 9f + 4p$
$C = 4(7) + 9(14) + 4(6)$
$C = 178$
There are 178 calories in this serving.

57. $\qquad C = 4h + 9f + 4p$
$130 = 4(31) + 9(0) + 4p$
$130 = 124 + 4p$
$6 = 4p$
$\dfrac{6}{4} = p$
$p = 1.5$
There are 1.5 g of protein provided by this serving of raisins.

	Planet	AU from Sun
59.	Venus	$\frac{67.2}{92.9} = 0.723$
61.	Mars	$\frac{141.5}{92.9} = 1.523$
63.	Saturn	$\frac{886.1}{92.9} \approx 9.538$
65.	Neptune	$\frac{279.3}{92.9} \approx 30.065$

67. answers may vary

69. answers may vary;
$$W = gh - 3gt^2$$
$$W = g(h - 3t^2)$$
$$\frac{W}{h - 3t^2} = \frac{g(h - 3t^2)}{h - 3t^2}$$
$$\frac{W}{h - 3t^2} = g \text{ or } g = \frac{W}{h - 3t^2}$$

71. Only one of the 8 sectors is green.
$$P(\text{green}) = \frac{1}{8}$$

73. Only one of the 8 sectors is black.
$$P(\text{black}) = \frac{1}{8}$$

75. Four of the sectors are green or blue.
$$P(\text{green or blue}) = \frac{4}{8} = \frac{1}{2}$$

77. Three of the sectors are red, green, or black.
$$P(\text{red, green, or black}) = \frac{3}{8}$$

79. None of the sectors is white.
$P(\text{white}) = 0$

81. $P(\text{impossible event}) = 0$

Chapter 1 Vocabulary Check

1. An <u>algebraic expression</u> is formed by numbers and variables connected by the operations of addition, subtraction, multiplication, division, raising to powers, and/or taking roots.

2. The <u>opposite</u> of a number a is $-a$.

3. $3(x-6) = 3x - 18$ by the <u>distributive</u> property.

4. The <u>absolute value</u> of a number is the distance between the number and 0 on the number line.

5. An <u>exponent</u> is a shorthand notation for repeated multiplication of the same factor.

6. A letter that represents a number is called a <u>variable</u>.

7. The symbols $<$ and $>$ are called <u>inequality</u> symbols.

8. If a is not 0, then a and $\dfrac{1}{a}$ are called <u>reciprocals</u>.

9. $A + B = B + A$ by the <u>commutative</u> property.

10. $(A + B) + C = A + (B + C)$ by the <u>associative</u> property.

11. The numbers 0, 1, 2, 3, ... are called <u>whole</u> numbers.

12. If a number corresponds to a point on the number line, we know that number is a <u>real</u> number.

13. An equation in one variable that has no solution is called a <u>contradiction</u>.

14. An equation in one variable that has every number (for which the equation is defined) as a solution is called an <u>identity</u>.

15. The equation $d = rt$ is also called a <u>formula</u>.

16. When a variable in an equation is replaced by a number and the resulting equation is true, then that number is called a <u>solution</u> of the equation.

17. The integers 17, 18, 19 are examples of <u>consecutive integers</u>.

18. The statement $5x - 0.2 = 7$ is an example of a <u>linear equation in one variable</u>.

Chapter 1 Review

1. $7x = 7(3) = 21$

2. $st = (1.6)(5) = 8$

3. One minute = 60 seconds.
$70t = 70(60) = 4200$
4200 wingbeats per minute.

4. One hour is $60(60) = 3600$ seconds.
$70t = 70(3600) = 252,000$
252,000 wingbeats per hour.

5. $\{x|x$ is an odd integer between -2 and $4\}$
$= \{-1, 1, 3\}$

6. $\{x|x$ is an even integer between -3 and $7\}$
$= \{-2, 0, 2, 4, 6\}$

7. There are no whole numbers that are negative.
$\varnothing$

8. All natural numbers are rational numbers.
$\varnothing$

9. $\{x|x$ is a whole number greater than $5\}$
$= \{6, 7, 8, ...\}$

10. $\{x|x$ is an integer less than $3\} = \{..., -1, 0, 1, 2\}$

11. Since $D = \{2, 4, 6, 8, 10, ..., 16\}$, $10 \in D$ is true.

12. Since $B = \{5, 9, 11\}$, $B \in 9$ is false.

13. $\sqrt{169} = 13$, which is a rational number. So $\sqrt{169} \notin G$ is true.

14. Since $F = \{ \ \ \}$ and 0 is not an element of the empty set, then $0 \notin F$ is true.

15. Since $E = \{x|x$ is a rational number$\}$ and π is irrational, then $\pi \in E$ is false.

16. Since $H = \{x|x$ is a real number$\}$ and π is a real number, then $\pi \in H$ is true.

17. Since $\sqrt{4} = 2$ and $G = \{x|x$ is an irrational number$\}$, and 2 is a rational number, then $\sqrt{4} \in G$ is false.

18. Since $E = \{x|x$ is a rational number$\}$ and -9 is a rational number, then $-9 \in E$ is true.

19. Since $A = \{6, 10, 12\}$ and $D = \{2, 4, 6, 8, 10, 12, 14, 16\}$, then $A \subseteq D$ is true.

20. Since $C = \{..., -3, -2, -1, 0, 1, 2, 3, ...\}$ and $B = \{5, 9, 11\}$, then $C \nsubseteq B$ is true.

21. Since $C = \{..., -3, -2, -1, 0, 1, 2, 3, ...\}$ and $E = \{x | x$ is a rational number$\}$, and all integers are rational numbers, then $C \not\subseteq E$ is false.

22. Since $F = \{\ \}$ and $H = \{x | x$ is a real number$\}$, and the empty set is a subset of all sets, then $F \subseteq H$ is true.

23. Whole numbers: $\left\{5, \dfrac{8}{2}, \sqrt{9}\right\}$

24. Natural numbers: $\left\{5, \dfrac{8}{2}, \sqrt{9}\right\}$

25. Rational numbers: $\left\{5, -\dfrac{2}{3}, \dfrac{8}{2}, \sqrt{9}, 0.3, 1\dfrac{5}{8}, -1\right\}$

26. Irrational numbers: $\left\{\sqrt{7}, \pi\right\}$

27. Real numbers:
$\left\{5, -\dfrac{2}{3}, \dfrac{8}{2}, \sqrt{9}, 0.3, \sqrt{7}, 1\dfrac{5}{8}, -1, \pi\right\}$

28. Integers: $\left\{5, \dfrac{8}{2}, \sqrt{9}, -1\right\}$

29. The opposite of $-\dfrac{3}{4}$ is $-\left(-\dfrac{3}{4}\right) = \dfrac{3}{4}$.

30. The opposite of 0.6 is -0.6.

31. The opposite of 0 is $-0 = 0$.

32. The opposite of 1 is -1.

33. The reciprocal of $-\dfrac{3}{4}$ is $\dfrac{1}{\left(-\dfrac{3}{4}\right)} = -\dfrac{4}{3}$.

34. The reciprocal of 0.6 is $\dfrac{1}{0.6}$.

35. The reciprocal of 0 is $\dfrac{1}{0}$ which is undefined.

36. The reciprocal of 1 is $\dfrac{1}{1} = 1$.

37. $-7 + 3 = -4$

38. $-10 + (-25) = -35$

39. $5(-0.4) = -2$

40. $(-3.1)(-0.1) = 0.31$

41. $-7 - (-15) = -7 + 15 = 8$

42. $9 - (-4.3) = 9 + 4.3 = 13.3$

43. $(-6)(-4)(0)(-3) = 0$

44. $(-12)(0)(-1)(-5) = 0$

45. $(-24) \div 0$ is undefined.

46. $0 \div (-45) = 0$

47. $(-36) \div (-9) = 4$

48. $60 \div (-12) = -5$

49. $\left(-\dfrac{4}{5}\right) - \left(-\dfrac{2}{3}\right) = -\dfrac{4}{5} + \dfrac{2}{3} = -\dfrac{12}{15} + \dfrac{10}{15} = -\dfrac{2}{15}$

50. $\left(\dfrac{5}{4}\right) - \left(-2\dfrac{3}{4}\right) = \dfrac{5}{4} + \dfrac{11}{4} = \dfrac{16}{4} = 4$

51. $1 - \dfrac{1}{4} - \dfrac{1}{3} = \dfrac{12}{12} - \dfrac{3}{12} - \dfrac{4}{12} = \dfrac{5}{12}$

52. $31,441 - 1589 = 29,852$
The elevation relative to sea level is 29,852 feet below sea level.

53. $-5 + 7 - 3 - (-10) = 2 - 3 + 10 = -1 + 10 = 9$

54. $8 - (-3) + (-4) + 6 = 8 + 3 - 4 + 6$
$= 11 - 4 + 6$
$= 7 + 6$
$= 13$

55. $3(4 - 5)^4 = 3(-1)^4 = 3(1) = 3$

56. $6(7 - 10)^2 = 6(-3)^2 = 6(9) = 54$

57. $\left(-\dfrac{8}{15}\right) \cdot \left(-\dfrac{2}{3}\right)^2 = -\dfrac{8}{15} \cdot \dfrac{4}{9} = -\dfrac{32}{135}$

58. $\left(-\dfrac{3}{4}\right)^2 \cdot \left(-\dfrac{10}{21}\right) = \left(\dfrac{9}{16}\right)\left(-\dfrac{10}{21}\right) = -\dfrac{15}{56}$

59. $-\dfrac{6}{15} \div \dfrac{8}{25} = -\dfrac{6}{15} \cdot \dfrac{25}{8} = -\dfrac{150}{120} = -\dfrac{5}{4}$

60. $\dfrac{4}{9} \div \left(-\dfrac{8}{45} \right) = \dfrac{4}{9} \cdot \left(-\dfrac{45}{8} \right) = -\dfrac{180}{72} = -\dfrac{5}{2}$

61. $-\dfrac{3}{8} + 3(2) \div 6 = -\dfrac{3}{8} + 6 \div 6 = -\dfrac{3}{8} + 1 = -\dfrac{3}{8} + \dfrac{8}{8} = \dfrac{5}{8}$

62. $\begin{aligned}5(-2)-(-3)-\dfrac{1}{6}+\dfrac{2}{3} &= -10+3-\dfrac{1}{6}+\dfrac{2}{3} \\ &= -7-\dfrac{1}{6}+\dfrac{2}{3} \\ &= -\dfrac{42}{6}-\dfrac{1}{6}+\dfrac{4}{6} \\ &= -\dfrac{39}{6} \\ &= -6\dfrac{1}{2}\end{aligned}$

63. $\begin{aligned}\left| 2^3 - 3^2 \right| - \left| 5-7 \right| &= \left| 8-9 \right| - \left| -2 \right| \\ &= \left| -1 \right| - 2 \\ &= 1-2 \\ &= -1\end{aligned}$

64. $\begin{aligned}\left| 5^2 - 2^2 \right| + \left| 9 \div (-3) \right| &= \left| 25-4 \right| + \left| -3 \right| \\ &= \left| 21 \right| + 3 \\ &= 21+3 \\ &= 24\end{aligned}$

65. $(2^3 - 3^2) - (5-7) = (8-9) - (-2) = -1+2 = 1$

66. $\begin{aligned}(5^2 - 2^4) + [9 \div (-3)] &= (25-16) + (-3) \\ &= 9 + (-3) \\ &= 6\end{aligned}$

67. $\begin{aligned}\dfrac{(8-10)^3 - (-4)^2}{2+8(2) \div 4} &= \dfrac{(-2)^3 - 16}{2 + 16 \div 4} \\ &= \dfrac{-8-16}{2+4} \\ &= \dfrac{-24}{6} \\ &= -4\end{aligned}$

68. $\begin{aligned}\dfrac{(2+4)^2 + (-1)^5}{12 \div 2 \cdot 3 - 3} &= \dfrac{(6)^2 + (-1)}{6 \cdot 3 - 3} \\ &= \dfrac{36-1}{18-3} \\ &= \dfrac{35}{15} \\ &= \dfrac{7}{3}\end{aligned}$

69. $\begin{aligned}\dfrac{(4-9)+4-9}{10-12 \div 4 \cdot 8} &= \dfrac{(-5)+4-9}{10-3 \cdot 8} \\ &= \dfrac{-1-9}{10-24} \\ &= \dfrac{-10}{-14} \\ &= \dfrac{5}{7}\end{aligned}$

70. $\dfrac{3-7-(7-3)}{15+30 \div 6 \cdot 2} = \dfrac{-4-(4)}{15+5 \cdot 2} = \dfrac{-8}{15+10} = \dfrac{-8}{25} = -\dfrac{8}{25}$

71. $\dfrac{\sqrt{25}}{4+3 \cdot 7} = \dfrac{5}{4+21} = \dfrac{5}{25} = \dfrac{1}{5}$

72. $\dfrac{\sqrt{64}}{24-8 \cdot 2} = \dfrac{8}{24-16} = \dfrac{8}{8} = 1$

73. Let $x=0$, $y=3$, $z=-2$.
$\begin{aligned}x^2 - y^2 + z^2 &= (0)^2 - (3)^2 + (-2)^2 \\ &= 0-9+4 \\ &= -5\end{aligned}$

74. Let $x=0$, $y=3$, $z=-2$.
$\dfrac{5x+z}{2y} = \dfrac{5(0)+(-2)}{2(3)} = \dfrac{0-2}{6} = \dfrac{-2}{6} = -\dfrac{1}{3}$

75. Let $y=3$, $z=-2$.
$\dfrac{-7y-3z}{-3} = \dfrac{-7(3)-3(-2)}{-3} = \dfrac{-21+6}{-3} = \dfrac{-15}{-3} = 5$

76. Let $x=0$, $y=3$, $z=-2$.
$\begin{aligned}(x-y+z)^2 &= (0-3+(-2))^2 \\ &= (-3-2)^2 \\ &= (-5)^2 \\ &= 25\end{aligned}$

77. When $r = 1$, $2\pi r = 2\pi(1) = 2(3.14) = 6.28$.
When $r = 10$,
$2\pi r = 2\pi(10) = 20(3.14) = 62.8$.
When $r = 100$,
$2\pi r = 2\pi(100) = 200(3.14) = 628$.

r	1	10	100
$2\pi r$	6.28	62.8	628

78. As the radius increases, the circumference increases.

79. $5xy - 7xy + 3 - 2 + xy = 5xy - 7xy + xy + 3 - 2$
$$= (5 - 7 + 1)xy + (3 - 2)$$
$$= (-1)xy + 1$$
$$= -xy + 1$$

80. $4x + 10x - 19x + 10 - 19 = (4 + 10 - 19)x + (10 - 19)$
$$= -5x + (-9)$$
$$= -5x - 9$$

81. $6x^2 + 2 - 4(x^2 + 1) = 6x^2 + 2 - 4x^2 - 4$
$$= 6x^2 - 4x^2 + 2 - 4$$
$$= (6 - 4)x^2 + (2 - 4)$$
$$= 2x^2 + (-2)$$
$$= 2x^2 - 2$$

82. $-7(2x^2 - 1) - x^2 - 1 = -14x^2 + 7 - x^2 - 1$
$$= -14x^2 - x^2 + 7 - 1$$
$$= (-14 - 1)x^2 + (7 - 1)$$
$$= -15x^2 + 6$$

83. $(3.2x - 1.5) - (4.3x - 1.2) = 3.2x - 1.5 - 4.3x + 1.2$
$$= 3.2x - 4.3x - 1.5 + 1.2$$
$$= (3.2 - 4.3)x - 0.3$$
$$= -1.1x - 0.3$$

84. $(7.6x + 4.7) - (1.9x + 3.6) = 7.6x + 4.7 - 1.9x - 3.6$
$$= 7.6x - 1.9x + 4.7 - 3.6$$
$$= (7.6 - 1.9)x + 4.7 - 3.6$$
$$= 5.7x + 1.1$$

85. Twelve is the product of x and negative 4.
$$12 \quad = \quad -4x$$
or $12 = -4x$

86. The sum of n and twice n is negative fifteen.
$$n + 2n \quad = \quad -15$$

87. Four times the sum of y and three is -1.

$$4 \quad \cdot \quad (y+3) \quad = -1$$

or $4(y + 3) = -1$

88. The difference of t and 5, multiplied by six is four.

$$(t-5) \quad \cdot \quad 6 = 4$$

or $6(t - 5) = 4$

89. Seven subtracted from z is six.

$$z - 7 \quad = 6$$

or $z - 7 = 6$

90. Ten less than the product of x and nine is five.

$$9x - 10 \quad = 5$$

or $9x - 10 = 5$

91. The difference of x and 5 is at least 12.

$$x - 5 \quad \geq \quad 12$$

or $x - 5 \geq 12$

92. The opposite of four is less than the product of y and seven.

$$-4 \quad < \quad 7y$$

or $-4 < 7y$

93. Two-thirds is not equal to twice the sum of n and one-fourth.

$$\frac{2}{3} \quad \neq \quad 2 \cdot \quad \left(n + \frac{1}{4}\right)$$

or $\dfrac{2}{3} \neq 2\left(n + \dfrac{1}{4}\right)$

94. The sum of t and six is not more than negative twelve.

$$t + 6 \quad \leq \quad -12$$

or $t + 6 \leq -12$

95. $(M + 5) + P = M + (5 + P)$: Associative Property of Addition

96. $5(3x - 4) = 15x - 20$: Distributive Property

97. $(-4) + 4 = 0$: Additive Inverse Property

98. $(3 + x) + 7 = 7 + (3 + x)$: Commutative Property of Addition

99. Associative and Commutative Properties of Multiplication
To see this: $(XY)Z = X(YZ) = (YZ)X$

45

100. $\left(-\frac{3}{5}\right) \cdot \left(-\frac{5}{3}\right) = 1$: Multiplicative Inverse Property

101. $T \cdot 0 = 0$: Multiplication Property of Zero

102. $(ab)c = a(bc)$: Associative Property of Multiplication

103. $A + 0 = A$: Additive Identity Property

104. $8 \cdot 1 = 8$: Multiplicative Identity Property

105. $5x - 15z = 5(x - 3z)$

106. $(7 + y) + (3 + x) = (3 + x) + (7 + y)$

107. $0 = 2 + (-2)$, for example

108. $1 = 2 \cdot \frac{1}{2}$, for example

109. $[(3.4)(0.7)]5 = (3.4)[(0.7)(5)]$

110. $7 = 7 + 0$

111. $-9 > -12$, since -9 is to the right of -12 on the number line.

112. $0 > -6$, since 0 is to the right of -6 on the number line.

113. $-3 < -1$, since -3 is to the left of -1 on the number line.

114. $7 = |-7|$

115. $-5 < -(-5)$, since $-(-5) = 5$.

116. $-(-2) > -2$, since $-(-2) = 2$.

117. $4(x - 5) = 2x - 14$
$4x - 20 = 2x - 14$
$2x = 6$
$x = 3$

118. $x + 7 = -2(x + 8)$
$x + 7 = -2x - 16$
$3x = -23$
$x = -\frac{23}{3}$

119. $3(2y - 1) = -8(6 + y)$
$6y - 3 = -48 - 8y$
$14y = -45$
$y = -\frac{45}{14}$

120. $-(z + 12) = 5(2z - 1)$
$-z - 12 = 10z - 5$
$-11z = 7$
$z = -\frac{7}{11}$

121. $n - (8 + 4n) = 2(3n - 4)$
$n - 8 - 4n = 6n - 8$
$-3n = 6n$
$-9n = 0$
$n = 0$

122. $4(9v + 2) = 6(1 + 6v) - 10$
$36v + 8 = 6 + 36v - 10$
$36v + 8 = 36v - 4$
$8 = -4$
No solution, or $\varnothing$

123. $\quad 0.3(x - 2) = 1.2$
$10[0.3(x - 2) = 10(1.2)$
$3(x - 2) = 12$
$3x - 6 = 12$
$3x = 18$
$x = 6$

124. $\quad 1.5 = 0.2(c - 0.3)$
$1.5 = 0.2c - 0.06$
$100(1.5) = 100(0.2c - 0.06)$
$150 = 20c - 6$
$156 = 20c$
$7.8 = c$

125. $-4(2 - 3x) = 2(3x - 4) + 6x$
$-8 + 12x = 6x - 8 + 6x$
$-8 + 12x = 12x - 8$
$-8 = -8$
All real numbers

126. $6(m - 1) + 3(2 - m) = 0$
$6m - 6 + 6 - 3m = 0$
$3m = 0$
$m = 0$

127. $6 - 3(2g + 4) - 4g = 5(1 - 2g)$
$6 - 6g - 12 - 4g = 5 - 10g$
$-6 - 10g = 5 - 10g$
$-6 = 5$
No solution, $\varnothing$

128. $20 - 5(p + 1) + 3p = -(2p - 15)$
$20 - 5p - 5 + 3p = -2p + 15$
$15 - 2p = -2p + 15$
$15 = 15$
All real numbers

129. $\dfrac{x}{3} - 4 = x - 2$
$3\left(\dfrac{x}{3} - 4\right) = 3(x - 2)$
$x - 12 = 3x - 6$
$-2x = 6$
$x = -3$

130. $\dfrac{9}{4}y = \dfrac{2}{3}y$
$12\left(\dfrac{9}{4}y\right) = 12\left(\dfrac{2}{3}y\right)$
$27y = 8y$
$19y = 0$
$y = 0$

131. $\dfrac{3n}{8} - 1 = 3 + \dfrac{n}{6}$
$24\left(\dfrac{3n}{8} - 1\right) = 24\left(3 + \dfrac{n}{6}\right)$
$9n - 24 = 72 + 4n$
$5n = 96$
$n = \dfrac{96}{5}$

132. $\dfrac{z}{6} + 1 = \dfrac{z}{2} + 2$
$6\left(\dfrac{z}{6} + 1\right) = 6\left(\dfrac{z}{2} + 2\right)$
$z + 6 = 3z + 12$
$-2z = 6$
$z = -3$

133. $\dfrac{y}{4} - \dfrac{y}{2} = -8$
$4\left(\dfrac{y}{4} - \dfrac{y}{2}\right) = 4(-8)$
$y - 2y = -32$
$-y = -32$
$y = 32$

134. $\dfrac{2x}{3} - \dfrac{8}{3} = x$
$2x - 8 = 3x$
$-8 = x$

135. $\dfrac{b - 2}{3} = \dfrac{b + 2}{5}$
$5(b - 2) = 3(b + 2)$
$5b - 10 = 3b + 6$
$2b = 16$
$b = 8$

136. $\dfrac{2t - 1}{3} = \dfrac{3t + 2}{15}$
$15\left(\dfrac{2t - 1}{3}\right) = 15\left(\dfrac{3t + 2}{15}\right)$
$5(2t - 1) = 3t + 2$
$10t - 5 = 3t + 2$
$7t = 7$
$t = 1$

137. $\dfrac{2(t + 1)}{3} = \dfrac{2(t - 1)}{3}$
$3\left[\dfrac{2(t + 1)}{3}\right] = 3\left[\dfrac{2(t - 1)}{3}\right]$
$2(t + 1) = 2(t - 1)$
$2t + 2 = 2t - 2$
$2 = -2$
No solution, $\varnothing$

138. $\dfrac{3a - 3}{6} = \dfrac{4a + 1}{15} + 2$
$30\left(\dfrac{3a - 3}{6}\right) = 30\left(\dfrac{4a + 1}{15} + 2\right)$
$5(3a - 3) = 2(4a + 1) + 30(2)$
$15a - 15 = 8a + 2 + 60$
$15a - 15 = 8a + 62$
$7a = 77$
$a = 11$

139. Let x = the number.
$$2(x-3) = 3x+1$$
$$2x-6 = 3x+1$$
$$-7 = x$$
The number is -7.

140. Let x = smaller number, then $x + 5$ = larger number.
$$x+x+5 = 285$$
$$2x = 280$$
$$x = 140$$
$$x + 5 = 145$$
The numbers are 140 and 145.

141. Let n = the first integer, then $n + 1$ = the second integer, $n + 2$ = the third integer, and $n + 3$ = the fourth integer.
$$(n+1)+(n+2)+(n+3)-2n = 16$$
$$n+6 = 16$$
$$n = 10$$
Therefore, the integers are 10, 11, 12, and 13.

142. Let x = smaller odd integer, then $x + 2$ = larger odd integer.
$$5x = 3(x+2)+54$$
$$5x = 3x+6+54$$
$$2x = 60$$
$$x = 30$$
Since this is not odd, no such consecutive odd integers exist.

143. Let x = width of the playing field, then $2x - 5$ = length of the playing field.
$$2x+2(2x-5) = 230$$
$$2x+4x-10 = 230$$
$$6x = 240$$
$$x = 40$$
Then $2x - 5 = 2(40) - 5 = 75$. The field is 75 meters long and 40 meters wide.

144. Let m = number of miles of driven.
$$2(19.95)+0.12(m-200) = 46.86$$
$$39.90+0.12m-24 = 46.86$$
$$0.12m+15.90 = 46.86$$
$$0.12m = 30.96$$
$$m = 258$$

145.

X	Y1
1	10.472
1.5	23.562
2	41.888
2.5	65.45
3	94.248
3.5	128.28
4	167.55

Y₁☐10/3πX²

Radius of Cone	x	1	1.5	2	2.5	3
Volume (If Height is 10 Inches)	$\frac{10}{3}\pi x^2$	10.47	23.56	41.89	65.45	94.25

146.

Width of Rectangle	x	2	4.68	9.5	12.68
Perimeter (If Length is 15 Units)	$30 + 2x$	34	39.36	49	55.36

147. a.

Seconds	0	1	2	3	4	5
Height in Feet	500	524	516	476	404	300

b.

The maximum height of the rock is 525 feet.

c.

The rock strikes the ground after 7.0 seconds.

148. Use $y_1 = 8 + 1.10(x - 4)$ for Coast Waterworks and $y_2 = 12 + 1.50(x - 5)$ for Cross Gates Water Company.

Gallons (In Thousands Used)	5	6	7	8
Coast Charge (Dollars)	9.1	10.2	11.3	12.4
Cross Gates Charge (Dollars)	12	13.5	15	16.5

149. $V = lwh$

$$w = \frac{V}{lh}$$

150. $C = 2\pi r$

$$\frac{C}{2\pi} = r$$

151. $5x - 4y = -12$

$\quad 5x + 12 = 4y$

$\quad\quad y = \dfrac{5x + 12}{4}$

152. $5x - 4y = -12$

$\quad 5x = 4y - 12$

$\quad\quad x = \dfrac{4y - 12}{5}$

153. $y - y_1 = m(x - x_1)$

$\quad m = \dfrac{y - y_1}{x - x_1}$

154. $\quad\quad y - y_1 = m(x - x_1)$

$\quad\quad y - y_1 = mx - mx_1$

$\quad y - y_1 + mx_1 = mx$

$\quad \dfrac{y - y_1 + mx_1}{m} = x$

155. $\quad\quad E = I(R + r)$

$\quad\quad E = IR + Ir$

$\quad I - IR = Ir$

$\quad \dfrac{E - IR}{I} = r$

156. $\quad\quad S = vt + gt^2$

$\quad\quad S - vt = gt^2$

$\quad \dfrac{S - vt}{t^2} = g$

157. $T = gr + gvt$

$\quad T = g(r + vt)$

$\quad g = \dfrac{T}{r + vt}$

158. $\quad I = Prt + P$

$\quad\quad I = P(rt + 1)$

$\quad \dfrac{I}{rt + 1} = P$

159. $\quad\quad A = \dfrac{h}{2}(B + b)$

$\quad\quad 2A = hB + hb$

$\quad 2A - hb = hB$

$\quad \dfrac{2A - hb}{h} = B$

160. $\quad V = \dfrac{1}{3}\pi r^2 h$

$\quad\quad 3V = \pi r^2 h$

$\quad\quad \dfrac{3V}{\pi r^2} = h$

161. $\quad R = \dfrac{r_1 + r_2}{2}$

$\quad\quad 2R = r_1 + r_2$

$\quad\quad 2R - r_2 = r_1$

162. $\dfrac{V_1}{T_1} = \dfrac{V_2}{T_2}$

$\quad V_1 T_2 = V_2 T_1$

$\quad T_2 = \dfrac{V_2 T_1}{V_1}$

163. $A = P\left(1 + \dfrac{r}{n}\right)^{nt} = 3000\left(1 + \dfrac{0.03}{n}\right)^{7n}$

  a. $A = 3000\left(1 + \dfrac{0.03}{2}\right)^{14} \approx \$3695.27$

  b. $A = 3000\left(1 + \dfrac{0.03}{52}\right)^{364} \approx \$3700.81$

164. Let x = original width, then

$x + 2$ = original length.

$\quad (x + 4)(x + 2 + 4) = x(x + 2) + 88$

$\quad\quad (x + 4)(x + 6) = x^2 + 2x + 88$

$\quad\quad x^2 + 10x + 24 = x^2 + 2x + 88$

$\quad\quad\quad\quad\quad 8x = 64$

$\quad\quad\quad\quad\quad\quad x = 8$

$x + 2 = 10$

The original width is 8 in. and the original length is 10 in.

165.

X	Y1
-40	-40
-15	5
10	50
60	140
0	32
100	212

Y1■(9X+160)/5

  a.

Celsius	−40	−15	10	60
Fahrenheit	−40	5	50	140

  b. 100 degrees Celsius is 212 degrees Fahrenheit.

    c. 0 degrees Celsius is 32 degrees Fahrenheit.

166. a.

Hours	0	0.25	0.50	0.75	1	1.25	1.5
Miles	0	13.75	27.5	41.25	55	68.75	82.5

    b.

It takes 3.5 hours to travel 192.5 miles at 55 mph.

167. $8 \cdot 21 = 378$

The area of the floor is 378 square feet.

$$\frac{378}{24} = 15.75$$

It takes 16 packages to tile the floor.

168. $V_{\text{box}} = lwh = 8 \cdot 5 \cdot 3 = 120 \text{ in}^3$, while $V_{\text{cyl}} = \pi r^2 h = \pi \cdot 3^2 \cdot 6 = 54\pi \approx 170 \text{ in}^3$

Therefore, the cylinder holds more ice cream.

169. The opposite of $-\dfrac{3}{4}$ is $\dfrac{3}{4}$.

The reciprocal of $-\dfrac{3}{4}$ is $-\dfrac{4}{3}$.

170. If the opposite of the number is -5, then the number is $-(-5) = 5$. The reciprocal of 5 is $\dfrac{1}{5}$.

171. $-2\left(5x + \dfrac{1}{2}\right) + 7.1 = -2 \cdot 5x + (-2) \cdot \dfrac{1}{2} + 7.1$

$$= -10x - 1 + 7.1$$
$$= -10x + 6.1$$

172. $\sqrt{36} \div 2 \cdot 3 = 6 \div 2 \cdot 3 = 3 \cdot 3 = 9$

173. $-\dfrac{7}{11} - \left(-\dfrac{1}{11}\right) = -\dfrac{7}{11} + \dfrac{1}{11} = -\dfrac{6}{11}$

174. $10 - (-1) + (-2) + 6 = 10 + 1 + (-2) + 6$

$$= 11 + (-2) + 6$$
$$= 9 + 6$$
$$= 15$$

175.

$$\left(-\frac{2}{3}\right)^3 \div \frac{10}{9} = -\frac{8}{27} \div \frac{10}{9}$$

$$= -\frac{8}{27} \cdot \frac{9}{10}$$

$$= -\frac{2 \cdot 4 \cdot 9}{3 \cdot 9 \cdot 2 \cdot 5}$$

$$= -\frac{4}{15}$$

176.

$$\frac{(3-5)^2 + (-1)^3}{1 + 2(3 - (-1))^2} = \frac{(-2)^2 + (-1)^3}{1 + 2(3+1)^2}$$

$$= \frac{4 + (-1)}{1 + 2(4)^2}$$

$$= \frac{3}{1 + 2(16)}$$

$$= \frac{3}{1 + 32}$$

$$= \frac{3}{33}$$

$$= \frac{1}{11}$$

177.

$$\frac{x-2}{5} + \frac{x+2}{2} = \frac{x+4}{3}$$

$$30\left(\frac{x-2}{5} + \frac{x+2}{2}\right) = 30\left(\frac{x+4}{3}\right)$$

$$6(x-2) + 15(x+2) = 10(x+4)$$

$$6x - 12 + 15x + 30 = 10x + 40$$

$$21x + 18 = 10x + 40$$

$$11x = 22$$

$$x = 2$$

178.

$$\frac{2z-3}{4} - \frac{4-z}{2} = \frac{z+1}{3}$$

$$12\left(\frac{2z-3}{4} - \frac{4-z}{2}\right) = 12\left(\frac{z+1}{3}\right)$$

$$3(2z-3) - 6(4-z) = 4(z+1)$$

$$6z - 9 - 24 + 6z = 4z + 4$$

$$12z - 33 = 4z + 4$$

$$8z = 37$$

$$z = \frac{37}{8}$$

179.
$$A = \frac{h}{2}(B + b)$$
$$2A = hB + hb$$
$$2A - hb = hB$$
$$\frac{2A - hb}{h} = B$$

180.
$$V = \frac{1}{3}\pi r^2 h$$
$$3V = \pi r^2 h$$
$$\frac{3V}{\pi r^2} = h$$

181. Let x = number of tourists for France, then
$x + 9$ = number of tourists for United States, and
$x + 44$ = number of tourists for China.
$$x + (x + 9) + (x + 44) = 332$$
$$3x + 53 = 332$$
$$3x = 279$$
$$x = 93$$
$x + 9 = 102$
$x + 44 = 137$
China is predicted to have 137 million tourists, whereas the United States is predicted to have 102 million and France, 93 million.

182. Since $h = 10$, use $y_1 = \frac{1}{3}\pi x^2 \cdot 10 = \frac{10}{3}\pi x^2$.

X	Y1
1.5	23.562
2.1	46.181
2.75	79.194
3	94.248
3.5	128.28

Y1■10/3πX²

Height h	10	10	10	10	10
Radius x	1.5	2.1	2.75	3	3.5
Volume	23.56	46.18	79.19	94.25	128.28

183. Since $h = 6$, use $y_1 = \frac{1}{3}\pi x^2 \cdot 6 = 2\pi x^2$.

X	Y1
1	6.2832
1.5	14.137
2	25.133
2.25	31.809
3	56.549

Y1■2πX²

Height h	6	6	6	6	6
Radius x	1	1.5	2	2.25	3
Volume	6.28	14.14	25.13	31.81	56.55

Chapter 1 Test

1. True; -2.3 lies to the right of -2.33 on the number line.

2. False; $-6^2 = -36,$ while $(-6)^2 = 36.$

3. False; $-5 - 8 = -13,$ while $-(5 - 8) = -(-3) = 3.$

4. False; $(-2)(-3)(0) = 0,$ while $\dfrac{(-4)}{0}$ is undefined.

5. True

6. False; for example, $\dfrac{1}{2}$ is a rational number that is not an integer.

7. $5 - 12 \div 3(2) = 5 - 4(2) = 5 - 8 = -3$

8. $5^2 - 3^4 = 25 - 81 = -56$

9. $(4-9)^3 - |-4-6|^2 = (-5)^3 - |-10|^2$
$$\begin{aligned}
&= -125 - 10^2 \\
&= -125 - 100 \\
&= -225
\end{aligned}$$

10. $12 + \{6 - [5 - 2(-5)]\} = 12 + \{6 - [5 + 10]\}$
$$\begin{aligned}
&= 12 + (6 - 15) \\
&= 12 + (-9) \\
&= 12 - 9 \\
&= 3
\end{aligned}$$

11. $\dfrac{6(7-9)^3 + (-2)}{(-2)(-5)(-5)} = \dfrac{6(-2)^3 - 2}{10(-5)}$
$$\begin{aligned}
&= \dfrac{6(-8) - 2}{-50} \\
&= \dfrac{-48 - 2}{-50} \\
&= \dfrac{-50}{-50} \\
&= 1
\end{aligned}$$

12. $\dfrac{\left(4 - \sqrt{16}\right) - (-7 - 20)}{-2(1-4)^2} = \dfrac{(4-4) - (-27)}{-2(-3)^2}$
$$\begin{aligned}
&= \dfrac{0 + 27}{-2(9)} \\
&= \dfrac{27}{-18} \\
&= -\dfrac{3}{2}
\end{aligned}$$

13. Let $q = 4$ and $r = -2$.
$$q^2 - r^2 = (4)^2 - (-2)^2 = 16 - 4 = 12$$

14. Let $q = 4$, $r = -2$, and $t = 1$.
$$\frac{5t - 3q}{3r - 1} = \frac{5(1) - 3(4)}{3(-2) - 1} = \frac{5 - 12}{-6 - 1} = \frac{-7}{-7} = 1$$

15. a.   When $x = 1$, $8.75x = 8.75(1) = 8.75$.
When $x = 3$, $8.75x = 8.75(3) = 26.25$.
When $x = 10$, $8.75x = 8.75(10) = 87.50$.
When $x = 20$, $8.75x = 8.75(20) = 175.00$.

x	1	3	10	20
$8.75x$	8.75	26.25	87.50	175.00

 b.   As the number of adults increases the total cost increases.

16.   Twice the sum of x and five is 30.
$$2 \cdot \qquad (x + 5) \qquad = 30$$
or $2(x + 5) = 30$

17.   The square of the difference of six and y divided by seven is not equal to 10.
$$(6 - y)^2 \qquad \div \quad 7 \quad \neq \quad 10$$
or $\dfrac{(6 - y)^2}{7} \neq 10$

18.   The product of nine and z, divided by the absolute value of -12 is not equal to 10.
$$9z \qquad \div \qquad |-12| \qquad \neq \quad 10$$
or $\dfrac{9z}{|-12|} \neq 10$

19.   Three times the quotient of n and five is the opposite of n.
$$3 \qquad \cdot \qquad \frac{n}{5} \qquad = \qquad -n$$
or $3\left(\dfrac{n}{5}\right) = -n$

20.   Twenty is equal to 6 subtracted from twice x.
$$20 \qquad = \qquad 2x - 6$$
or $20 = 2x - 6$

21. Negative two is equal to x divided by the sum of x and five.

$$-2 \quad = \quad x \quad \div \quad (x+5)$$

or $-2 = \dfrac{x}{x+5}$

22. $6(x-4) = 6x - 24$: Distributive Property

23. $(4 + x) + z = 4 + (x + z)$: Associative Property of Addition

24. $(-7) + 7 = 0$: Additive Inverse Property

25. $(-18)(0) = 0$: Multiplication Property of Zero

26. Let 0.05 be the value of each nickel and 0.1 be the value of each dime. If there are n nickels, and d dimes, then the total amount of money is $0.05n + 0.1d$.

27. The reciprocal of $-\dfrac{7}{11}$ is $-\dfrac{11}{7}$.

The opposite of $-\dfrac{7}{11}$ is $\dfrac{7}{11}$.

28.
$$\frac{1}{3}a - \frac{3}{8} + \frac{1}{6}a - \frac{3}{4} = \frac{1}{3}a + \frac{1}{6}a - \frac{3}{8} - \frac{3}{4}$$
$$= \left(\frac{1}{3} + \frac{1}{6}\right)a - \frac{3}{8} - \frac{3}{4}$$
$$= \left(\frac{2}{6} + \frac{1}{6}\right)a - \frac{3}{8} - \frac{6}{8}$$
$$= \left(\frac{3}{6}\right)a - \frac{9}{8}$$
$$= \frac{1}{2}a - \frac{9}{8}$$

29.
$$4y + 10 - 2(y + 10) = 4y + 10 - 2y - 20$$
$$= 4y - 2y + 10 - 20$$
$$= (4 - 2)y - 10$$
$$= 2y - 10$$

30.
$$(8.3x - 2.9) - (9.6x - 4.8) = 8.3x - 2.9 - 9.6x + 4.8$$
$$= 8.3x - 9.6x - 2.9 + 4.8$$
$$= (8.3 - 9.6)x + 1.9$$
$$= -1.3x + 1.9$$

31.

```
-3.1→X
              -3.1
.2X^3+5X²-6.2X+3
            64.3118
■
```

The value is 64.3118.

32. Let x be the diameter. Use $y_1 = \dfrac{1}{2}x$ for the radius, $y_2 = \pi x$ for the circumference, and $y_3 = \pi(y_1)^2$ for the area.

Diameter	d	2	3.8	10	14.9
Radius	r	1	1.9	5	7.45
Circumference	πd	6.28	11.94	31.42	46.81
Area	πr^2	3.14	11.34	78.54	174.37

33. $8x + 14 = 5x + 44$
$3x = 30$
$x = 10$

34. $3(x+2) = 11 - 2(2-x)$
$3x + 6 = 11 - 4 + 2x$
$3x + 6 = 7 + 2x$
$x + 6 = 7$
$x = 1$

35. $3(y-4) + y = 2(6+2y)$
$3y - 12 + y = 12 + 4y$
$4y - 12 = 12 + 4y$
$-12 = 12$
No solution, $\varnothing$

36. $7n - 6 + n = 2(4n-3)$
$8n - 6 = 8n - 6$
$-6 = -6$
All real numbers

37. $\dfrac{7w}{4} + 5 = \dfrac{3w}{10} + 1$
$20\left(\dfrac{7w}{4} + 5\right) = 20\left(\dfrac{3w}{10} + 1\right)$
$35w + 100 = 6w + 20$
$29w = -80$
$w = -\dfrac{80}{29}$

38. $3x - 4y = 8$
$3x - 8 = 4y$
$y = \dfrac{3x-8}{4}$

39. $S = gt^2 + gvt$
$S = g(t^2 + vt)$
$g = \dfrac{S}{t^2 + vt}$

40.
$$F = \frac{9}{5}C + 32$$
$$F - 32 = \frac{9}{5}C$$
$$C = \frac{5}{9}(F - 32)$$

41. Let x be the number of people employed as registered nurses in 2008. The number of people employed in this field in 2018 is x increased by 22%.
$$x + 0.22x = 3,200,000$$
$$1.22x = 3,200,000$$
$$x \approx 2,623,000$$
In 2008, there were 2,623,000 registered nurses employed.

42.
$$C = 2\pi r$$
$$78.5 = 2(3.14)r$$
$$\frac{78.5}{6.28} = r$$
$$12.5 = r$$
$$A = \pi r^2 = 3.14(12.5)^2 = 490.625$$
$$\frac{490.625}{60} \approx 8.2$$
Approximately 8 dogs can be kept in the pen.

43.
$$A = P\left(1 + \frac{r}{n}\right)^{nt}$$
$$= 2500\left(1 + \frac{0.035}{4}\right)^{4(10)}$$
$$= 2500(1.00875)^{40}$$
$$\approx 3542.27$$
There will be $3542.27 in the account.

44. Let x be the amount of money international travelers spend in New York. Then $x + 4$ is the amount of money international travelers spend in California and $2x - 1$ is the amount of money international travelers spend in Florida.
$$x + (x + 4) + (2x - 1) = 39$$
$$4x + 3 = 39$$
$$4x = 36$$
$$x = 9$$
$$x + 4 = 9 + 4 = 13$$
$$2x - 1 = 2(9) - 1 = 18 - 1 = 17$$
International travelers spend $9 billion in New York, $13 billion in California, and $17 billion in Florida.

45. a. Use $y_1 = 1500 + 0.05x_0$

Sales (Dollars)	8000	9000	10,000	11,000	12,000
Gross Monthly Pay (Dollars)	1900	1950	2000	2050	2100

b. 12(2050) = 24,600
If her sales are $11,000 per month, her gross annual income is $24,600.

c. To achieve a monthly pay of $2200, her sales must be $14,000.

46.

a. The height of the rocket is 116 feet at 2 seconds and 3 seconds.

b. The rocket's maximum height is 120 feet.

c. The rocket reaches it maximum height after 2.5 seconds.

d. The rocket explodes 2.5 + 0.5 = 3.0 seconds after being launched.

Chapter 2

Section 2.1 Practice Exercises

1. The six points are graphed as shown.

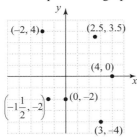

a. $(3, -4)$ lies in quadrant IV.

b. $(0, -2)$ is on the y-axis.

c. $(-2, 4)$ lies in quadrant II.

d. $(4, 0)$ is on the x-axis.

e. $\left(-1\dfrac{1}{2}, -2\right)$ is in quadrant III.

f. $(2.5, 3.5)$ is in quadrant I.

2.

3. Let $x = 1$ and $y = 4$.

$$4x + y = 8$$
$$4(1) + 4 \overset{?}{=} 8$$
$$4 + 4 \overset{?}{=} 8$$
$$8 = 8 \quad \text{True}$$

Let $x = 0$ and $y = 6$.

$$4x + y = 8$$
$$4(0) + 6 \overset{?}{=} 8$$
$$0 + 6 \overset{?}{=} 8$$
$$6 = 8 \quad \text{False}$$

Let $x = 3$ and $y = -4$.

$$4x + y = 8$$
$$4(3) + (-4) \overset{?}{=} 8$$
$$12 - 4 \overset{?}{=} 8$$
$$8 = 8 \quad \text{True}$$

Thus, $(0, 6)$ is not a solution, but both $(1, 4)$ and $(3, -4)$ are solutions.

4. a. Since x is products sold, find 6000 along the x-axis and move vertically up until you reach a point on the line. From this point on the line, move horizontally to the left until you reach the y-axis. Its value on the y-axis is 4200, which means if \$6000 worth of products is sold, the salary for the month is \$4200.

b. Since y is monthly salary, find 4800 along the y-axis and move horizontally to the right until you reach a point on the line. Move vertically downward until you reach the x-axis. The corresponding x-value is 9000. This means that \$9000 worth of products sold gives a salary of \$4800 for the month. For the salary to be greater than \$4800, products sold must be greater than \$9000.

5. $y = -3x - 2$

This is a linear equation. (In standard form, it is $3x + y = -2$.) Since the equation is solved for y, we choose three x-values.

Let $x = 0$.
$$y = -3x - 2$$
$$y = -3 \cdot 0 - 2$$
$$y = -2$$

Let $x = -1$.
$$y = -3x - 2$$
$$y = -3(-1) - 2$$
$$y = 1$$

Let $x = -2$.
$y = -3x - 2$
$y = -3(-2) - 2$
$y = 4$

The three ordered pairs $(0, -2)$, $(-1, 1)$, and $(-2, 4)$ are listed in the table.

x	y
0	-2
-1	1
-2	4

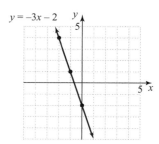

$y = -3x - 2$

6. $y = -\dfrac{1}{2}x$

To avoid fractions, we choose x-values that are multiples of 2. To find the y-intercept, we let $x = 0$.

If $x = 0$, then $y = -\dfrac{1}{2}(0)$, or 0.

If $x = 2$, then $y = -\dfrac{1}{2}(2)$, or -1.

If $x = -2$, then $y = -\dfrac{1}{2}(-2)$, or 1.

x	y
0	0
2	-1
-2	1

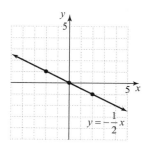

$y = -\dfrac{1}{2}x$

7. First solve the equation for y.
$$7x - 2y = 21$$
$$7x - 7x - 2y = -7x + 21$$
$$-2y = -7x + 21$$
$$y = \frac{-7x + 21}{-2}$$

8. $y = 2x^2$

This equation is not linear because of the x^2 term. Its graph is not a line.

If $x = -3$, then $y = 2(-3)^2$, or 18.

If $x = -2$, then $y = 2(-2)^2$, or 8.

If $x = -1$, then $y = 2(-1)^2$, or 2.

If $x = 0$, then $y = 2(0)^2$, or 0.

If $x = 1$, then $y = 2(1)^2$, or 2.

If $x = 2$, then $y = 2(2)^2$, or 8.

If $x = 3$, then $y = 2(3)^2$, or 18.

x	y
-3	18
-2	8
-1	2
0	0
1	2
2	8
3	18

9.

For each integer x-value from -3 to 3, the y-values in the Practice 9 table are 1 unit greater than those in the Example 9 table.

10. $y = -|x|$

This equation is not linear because it cannot be written in the form $Ax + By = C$. Its graph is not a line.

If $x = -3$, then $y = -|-3|$, or -3.

If $x = -2$, then $y = -|-2|$, or -2.

If $x = -1$, then $y = -|-1|$, or -1.

If $x = 0$, then $y = -|0|$, or 0.

If $x = 1$, then $y = -|1|$, or -1.

If $x = 2$, then $y = -|2|$, or -2.

If $x = 3$, then $y = -|3|$, or -3.

x	y
-3	-3
-2	-2
-1	-1
0	0
1	-1
2	-2
3	-3

11.

The graph of $y = |x| + 5$ is the same as the graph of $y = |x|$, except raised 5 units.

Vocabulary, Readiness & Video Check 2.1

1. The intersection of the x-axis and y-axis is a point called the <u>origin</u>.

2. The rectangular coordinate system has <u>4</u> quadrants and <u>2</u> axes.

3. The graph of a single ordered pair of numbers is how many points? <u>1</u>

4. The graph of $Ax + By = C$, where A and B are not both 0 is a <u>line</u>.

5. The graph of $y = |x|$ looks <u>V-shaped</u>.

6. The graph of $y = x^2$ is a <u>parabola</u>.

7. origin; left (if negative) or right (if positive); up (if positive) or down (if negative)

8. The notation $[-10, 10, 1]$ by $[-10, 10, 1]$ refers to [<u>Xmin</u>, <u>Xmax</u>, <u>Xscl</u>] by [<u>Ymin</u>, <u>Ymax</u>, <u>Yscl</u>]. This window setting is called the <u>standard</u> window.

9. An ordered pair is a solution of an equation in <u>two</u> variables if, when the variables are replaced with their ordered pair values, a <u>true</u> statement results.

10. The graph of an equation is a picture of its ordered pair solutions; a third point is found as a check to make sure the points line up and a mistake hasn't been made.

11. When graphing a nonlinear equation, first recognize it as a nonlinear equation and know that the graph is <u>not</u> a line. If you don't know the <u>shape</u> of the graph, plot enough points until you see a pattern.

   61

Exercise Set 2.1

1. (3, 2) is in quadrant I

3. (−5, 3) is in quadrant II.

5. $\left(5\frac{1}{2}, -4\right)$ is in quadrant IV.

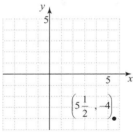

7. (0, 3.5) is on the *y*-axis.

9. (−2, −4) is in quadrant III.

11. Point *C* is (−5, −2).

13. Point *E* is (−1, 0).

15. Point *B* is (3, 0).

17. Each tick mark on the *x*-axis represents 1 unit, while each tick mark on the *y*-axis represents 2 units. Thus, the point shown is (2, 8) in quadrant I.

19. Each tick mark on both the *x*-axis and *y*-axis represents 1 unit. Thus, the point shown is (1, −4) in quadrant IV.

21. Possible answer: [−10, 10, 1] by [0, 20, 1]

23. Possible answer: [−100, 100, 10] by [−100, 100, 10]

25. The screen shows window setting C.

27. The screen shows window setting D.

29.

X	Y1
-3	-14
-2	-11
-1	-8
0	-5
1	-2
2	1
3	4

Y1❚3X−5

(0, 5) is not a solution; (−1, −8) is a solution.

31.

X	Y1
-3	-4.8
-2	-3.6
-1	-2.4
0	-1.2
1	0
2	1.2
3	2.4

Y1❚(−6+6X)/5

Both (1, 0) and $\left(2, \frac{6}{5}\right)$ are solutions.

33.

X	Y1
-3	18
-2	8
-1	2
0	0
1	2
2	8
3	18

Y1❚2X²

Both (1, 2) and (3, 18) are solutions.

35.

Equation	Linear or nonlinear	Shape (Line, Parabola, Cubic, V-shaped)
$y - x = 8$	linear	line
$y = 6x$	linear	line
$y = x^2 + 3$	nonlinear	parabola
$y = 6x - 5$	linear	line
$y = -|x| + 2$	nonlinear	V-shaped
$y = 3x^2$	nonlinear	parabola
$y = -4x + 2$	linear	line
$y = -|x|$	nonlinear	V-shaped
$y = x^3$	nonlinear	cubic

37. $$2x + y = 10$$
$$2x + y - 2x = -2x + 10$$
$$y = -2x + 10$$

39. $$-7x - 3y = 4$$
$$-7x - 3y + 7x = 7x + 4$$
$$-3y = 7x + 4$$
$$y = \frac{7x + 4}{-3}$$

41.

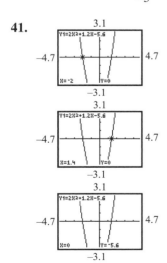

The intercepts are $(-2, 0)$, $(1.4, 0)$, and $(0, -5.6)$.

43. An integer window only displays coordinates with integer x-values, thus $(1.2, -4.2)$ would not be displayed; d.

45. $y = x^2 - 4$ is a parabola; C.

47. $y = x^3 + x^2 - 4$ is a cubic; A.

49. $y = x - 5$ is a line; D.

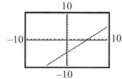

51. $y = -x^3 + x^2 + 2x - 3$ is a cubic; C.

53. The graph shows a parabola that is the same as $y = x^2$ but lowered by 3 units; A.

55. The graph is V-shaped and is the same as $y = |x|$ but raised by 5 units; C.

57.

59.

61. $x + y = 3$

x	y
0	3
3	0
1	2

63. $y = 4x$

x	y
−1	−4
0	0
1	4

65. $y = 4x - 2$

x	y
0	−2
$\frac{1}{2}$	0
1	2

67. $y = |x| + 3$

x	y
−3	6
−2	5
−1	4
0	3
1	4
2	5
3	6

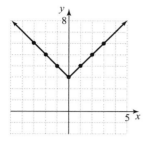

69. $2x - y = 5$

x	y
$2\frac{1}{2}$	0
0	-5
1	-3

71. $y = 2x^2$

x	y
-3	18
-2	8
-1	2
0	0
1	2
2	8
3	18

73. $y = x^2 - 3$

x	y
-3	6
-2	1
-1	-2
0	-3
1	-2
2	1
3	6

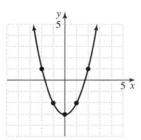

75. $y = -2x$

x	y
-1	2
0	0
1	-2

77. $y = -2x + 3$

x	y
-1	5
0	3
1	1

79. $y = |x + 2|$

x	y
-4	2
-3	1
-2	0
-1	1
0	2
1	3

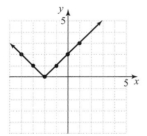

81. $y = x^3$

x	y
-3	-27
-2	-8
-1	-1
0	0
1	1
2	8

83. $y = -|x|$

x	y
-3	-3
-2	-2
-2	-1
0	0
1	-1
2	-2
3	-3

85.

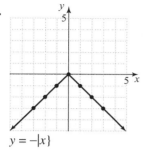

$y = -|x|$

87. a. The shape of the graph from 55°N to 15°N only resembles a <u>parabola</u>.

   b. The shape of the graph from 30°N to 15°N resembles a <u>line</u>.

89. a. The shape of the graph from 1650 to 1800 resembles a <u>line</u>.

   b. The shape of the graph from 1900 to 2000 resembles a <u>parabola</u>.

91. a. $y = 2x + 6$

x	y
0	6
1	8
2	10

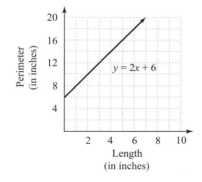

   b. When x is 4, y is 14. Thus, when the length is 4 inches, the perimeter is 14 inches.

93. $3(x-2)+5x = 6x-16$
$3x-6+5x = 6x-16$
$8x-6 = 6x-16$
$2x = -10$
$x = -5$
The solution is -5.

95. $3x + \dfrac{2}{5} = \dfrac{1}{10}$
$30x + 4 = 1$
$30x = -3$
$x = -\dfrac{1}{10}$

The solution is $-\dfrac{1}{10}$.

97. (2, 3); QI

99. (−2, 7); QII

101. (−1, −4); QIII

103. (0, −100); y-axis

105. (−10, −30); QIII

107. (−87, 0); x-axis

109. The first coordinate, −1, indicates that the point is 1 unit to the left of the y-axis. The second coordinate, 5.3, indicates that the point is 5.3 units above the x-axis. The answer is b.

111. Look for the graph where the only nonzero y-values are 40 and 60. The answer is b.

113. Look for the graph where all the y-values are between 10 and 30. The answer is c.

115. The first segment in the graph with y-coordinate greater than 5.0 begins in 1997. Thus, 1997 is the first year that the minimum hourly wage rose above $5.00.

117. answers may vary

119. When $x = 0$, $y = 7$. Thus, the purchase price was $7000.

121. 7000 − 6500 = $500

123. Depreciation is the same from year to year.

125. "The y-value is 5 more than three times the x-value" is written as $y = 3x + 5$.

$y = 3x + 5$

127. "The y-value is 2 more than the square of the x-value" is written as $y = x^2 + 2$.

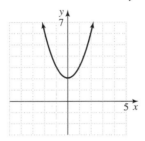

Section 2.2 Practice Exercises

1. a. The domain is the set of all first coordinates, $\{4, 5\}$. The range is the set of all second coordinates, $\{1, -3, -2, 6\}$.

b. Ordered pairs are not listed here but are given in graph form. The relation is $\{(3, -4), (3, -3), (3, -2), (3, -1), (3, 0),$ $(3, 1), (3, 2), (3, 3), (3, 4)\}$. The domain is $\{3\}$. The range is $\{-4, -3, -2, -1, 0, 1, 2, 3, 4\}$.

c. The domain is the set of inputs, $\{$Administrative Secretary, Game Developer, Engineer, Restaurant Manager, Marketing$\}$. The range is comprised of the numbers in the set of outputs that correspond to elements in the set of inputs, $\{27, 73, 50, 35\}$.

2. a. Although the ordered pairs (3, 1) and (9, 1) have the same y-value, each x-value is assigned to only one y-value, so this set of ordered pairs is a function.

b. The x-value -2 is assigned to two y-values, -3 and 4, in this graph, so this relation does not define a function.

c. This relation is a function because although two different people may have the same birth date, each person has only one birth date. This means that each element in the first set is assigned to only one element in the second set.

3. The relation $y = -3x + 5$ is a function if each x-value corresponds to just one y-value. For each x-value substituted into the equation $y = -3x + 5$, the multiplication and addition performed on each gives a single result, so only one y-value will be associated with each x-value. Thus, $y = -3x + 5$ is a function.

4. The relation $y = -x^2$ is a function if each x-value corresponds to just one y-value. For each x-value substituted into the equation $y = -x^2$, squaring each gives a single result, so only one y-value will be associated with each x-value. Thus, $y = -x^2$ is a function.

5. a. This is the graph of a function since no vertical line will intersect this graph more than once.

b. This is the graph of a function since no vertical line will intersect this graph more than once.

c. This is not the graph of a function. Note that vertical lines can be drawn that intersect the graph in two points.

d. This is the graph of a function since no vertical line will intersect this graph more than once.

e. This is not the graph of a function. A vertical line can be drawn that intersects this line at every point.

f. This is the graph of a function since no vertical line will intersect this graph more than once.

6. a. By the vertical line test, the graph is the graph of a function. The x-values are graphed from -1 to 2, so the domain is $\{x | -1 \le x \le 2\}$. The y-values are graphed from -2 to 9, so the range is $\{y | -2 \le y \le 9\}$.

b. By the vertical line test, the graph is not the graph of a function. The *x*-values are graphed from −1 to 1, so the domain is $\{x|-1 \le x \le 1\}$. The *y*-values are graphed from −4 to 4, so the range is $\{y|-4 \le y \le 4\}$.

c. By the vertical line test, the graph is the graph of a function. The arrows indicate that the graph continues forever. All *x*-values are graphed, so the domain is all real numbers. The *y*-values for 4 and numbers less than 4 are graphed, so the range is $\{y|y \le 4\}$.

d. By the vertical line test, the graph is the graph of a function. The arrows indicate that the graph continues forever. All *x*-values and all *y*-values are graphed, so the domain is all real numbers and the range is all real numbers.

7. a. Substitute 1 for *x* in *f*(*x*).
$$f(x) = 3x - 2$$
$$f(1) = 3(1) - 2 = 3 - 2 = 1$$

b. Substitute 1 for *x* in *g*(*x*).
$$g(x) = 5x^2 + 2x - 1$$
$$g(1) = 5(1)^2 + 2(1) - 1$$
$$= 5 + 2 - 1$$
$$= 6$$

c. Substitute 0 for *x* in *f*(*x*).
$$f(x) = 3x - 2$$
$$f(0) = 3(0) - 2 = 0 - 2 = -2$$

d. Substitute −2 for *x* in *g*(*x*).
$$g(x) = 5x^2 + 2x - 1$$
$$g(-2) = 5(-2)^2 + 2(-2) - 1$$
$$= 5(4) - 4 - 1$$
$$= 20 - 4 - 1 = 15$$

8. a. Define $y_1 = 2.6x^2 - 4$ and graph using a decimal window. Move the cursor to find *x* = 2.

Since 2 is paired with 6.4, *f*(2) = 6.4.

b. Move the cursor to *x* = −4 and see that *y* = 37.6, so *f*(−4) = 37.6.

c. Move the cursor to *x* = 0 and see that *y* = −4, so *f*(0) = −4.

d. Move the cursor to *x* = 0.7 and see that *y* = −2.726, so *f*(0.7) = −2.726.

9. a. To find *f*(1), find the *y*-value when *x* = 1. We see from the graph that when *x* = 1, *y* or *f*(*x*) = −3. Thus, *f*(1) = −3.

b. *f*(0) = −2 from the ordered pair (0, −2).

c. *g*(−2) = 3 from the ordered pair (−2, 3).

d. *g*(0) = 1 from the ordered pair (0, 1).

e. To find *x*-values such that *f*(*x*) = 1, we are looking for any ordered pairs on the graph of *f* whose *f*(*x*) or *y*-value is 1. They are (−1, 1) and (3, 1). Thus, *f*(−1) = 1 and *f*(3) = 1. The *x*-values are −1 and 3.

f. Find ordered pairs on the graph of *g* whose *g*(*x*) or *y*-value is −2. There is one such ordered pair, (−3, −2). Thus, *g*(−3) = −2. The only *x*-value is −3.

10. Use $R(w) = w + 0.15w$.

```
15→W:W+.15W
          17.25
22.75→W:W+.15W
          26.1625
■
```

```
38.50→W:W+.15W
          44.275
53→W:W+.15W
          60.95
■
```

Wholesale Cost	w	$15.00	$22.75	$38.50	$53.00
Retail Price	$R(w)$	$17.25	$26.16	$44.28	$60.95

11. Find the semester Fall 2010 and move upward until you reach the graph representing online enrollment. From the point on the graph, move horizontally to the left until the vertical axis is reached. In fall 2010, approximately 6 million students, or 6,000,000 students were enrolled in online programs.

12. Since x represents the number of years since 2000, use $x = 11$ and find $f(11)$.
$f(x) = 0.57x + 0.2$
$f(11) = 0.57(11) + 0.2 = 6.47$
We predict that the online enrollment in fall 2011 will be 6.47 million or 6,470,000 students.

Vocabulary, Readiness & Video Check 2.2

1. A relation is a set of ordered pairs.

2. The range of a relation is the set of all second components of the ordered pairs.

3. The domain of a relation is the set of all first components of the ordered pairs.

4. A function is a relation in which each first component in the ordered pairs corresponds to *exactly* one second component.

5. By the vertical line test, all linear equations are functions except those whose graphs are vertical lines.

6. If $f(-2) = 1.7$, the corresponding ordered pair is (−2, 1.7).

7. An equation is one way of writing down a correspondence that produces ordered pair solutions; a relation between two sets of coordinates, x's and y's.

8. Yes. The function definition restricts each first component to correspond to exactly one second component, but it makes no such restriction on each second component.

9. If a vertical line intersects a graph two (or more) times, then there's an x-value corresponding to two (or more) different y-values and is thus not a function.

10. No, equations of the form $x = c$, whose graphs are vertical lines, are not functions.

11. Using function notation, the replacement value for x and the resulting $f(x)$ or y-value corresponds to an ordered pair (x, y) solution to the function.

Exercise Set 2.2

1. The domain is the set of all first coordinates and the range is the set of all second coordinates.
Domain = {−1, 0, −2, 5}
Range = {7, 6, 2}
Function since each x-value corresponds to exactly one y-value.

3. The domain is the set of all first coordinates and the range is the set of all second coordinates.
Domain = $\{-2, 6, -7\}$
Range = $\{4, -3, -8\}$
The relation is not a function since -2 is paired with both 4 and -3.

5. The domain is the set of all first coordinates and the range is the set of all second coordinates.
Domain = $\{1\}$
Range = $\{1, 2, 3, 4\}$
The relation is not a function since 1 is paired with both 1 and 2 for example.

7. The domain is the set of all first coordinates and the range is the set of all second coordinates.

Domain = $\left\{\dfrac{3}{2}, 0\right\}$

Range = $\left\{\dfrac{1}{2}, -7, \dfrac{4}{5}\right\}$

The relation is not a function since $\dfrac{3}{2}$ is paired with both $\dfrac{1}{2}$ and -7.

9. The domain is the set of all first coordinates and the range is the set of all second coordinates.
Domain = $\{-3, 0, 3\}$
Range = $\{-3, 0, 3\}$
Function since each x-value corresponds to exactly one y-value.

11. Points on graph: $(-1, 2), (1, 1), (2, 1), (3, 1)$
Domain = $\{-1, 1, 2, 3\}$
Range = $\{2, 1\}$
Function since each x-value corresponds to exactly one y-value.

13. Domain = {Iowa, Alaska, Delaware, Illinois, Connecticut, New York}
Range = $\{5, 1, 19, 29\}$
Function since each input corresponds to exactly one output.

15. Domain = $\{32°, 104°, 212°, 50°\}$
Range = $\{0°, 40°, 10°, 100°\}$
Function since each input corresponds to exactly one output.

17. Domain = $\{0\}$
Range = $\{2, -1, 5, 100\}$
Not a function since the input 0 corresponds to more than one output.

19. This relation is a function because although two different students may have the same final grade average, each student has only one final grade average. This means that each element in the first set is assigned to only one element in the second set.

21. This relation is not a function because more than one person in Cincinnati has blue eyes. This means that one element in the first set is assigned to more than one element in the second set.

23. Yes, this is the graph of a function since no vertical line will intersect this graph more than once.

25. No, this is not the graph of a function. Note that vertical lines can be drawn that intersect the graph in two points. The y-axis is such a vertical line.

27. Yes, this is the graph of a function since no vertical line will intersect this graph more than once.

29. Yes, this is the graph of a function since no vertical line will intersect this graph more than once.

31. No, this is not the graph of a function. The vertical line graphed intersects itself at more than one point.

33. The x-values are graphed from for all positive values and 0; domain = $\{x | x \geq 0\}$
The arrows indicate the y-values continue forever; range = all real numbers
The relation is not a function since it fails the vertical line test (try $x = 1$).

35. The x-values are graphed from -1 to 1; domain = $\{x | -1 \leq x \leq 1\}$
The arrows indicate the y-values continue forever; range = all real numbers
The relation is not a function since it fails the vertical line test (try $x = 0$).

37. The arrows indicate the x-values continue forever; domain = all real numbers
The y-values do not include -3 to 3; range = $\{y | y \leq -3 \text{ and } y \geq 3\}$
The relation is not a function since it fails the vertical line test (try $x = 2$).

39. The *x*-values are graphed from 2 to 7;
domain = $\{x | 2 \leq x \leq 7\}$
The *y*-values are graphed from 1 to 6;
range = $\{y | 1 \leq y \leq 6\}$
The relation is not a function since it fails the
vertical line test (try $x = 4$).

41. The only *x*-value is 2; domain = $\{-2\}$
The *y*-values continue forever;
range = all real numbers
The relation is not a function since it fails the
vertical line test (try $x = -2$).

43. The *x*-values continue forever;
domain = all real numbers
The *y*-values are 3 and less; range = $\{y | y \leq 3\}$
Function since it passes the vertical line test.

45. The *x*-values continue forever;
domain = all real numbers
The *y*-values continue forever;
range = all real numbers
Function since it passes the vertical line test.

47. $y = x + 1$
For each *x*-value substituted into the equation
$y = x + 1$, the addition performed gives a single
result, so only one *y*-value will be associated
with each *x*-value. Thus, $y = x + 1$ is a function.

49. $x = 2y^2$
The *x*-value 8 is associated with two *y*-values, -2
and 2. Thus, $x = 2y^2$ is not a function.

51. $y - x = 7$
For each *x*-value substituted into the equation
$y - x = 7$, the process of solving for *y* gives a
single result, so only one *y*-value will be
associated with each *x*-value. Thus, $y - x = 7$ is a
function.

53. $y = \dfrac{1}{x}$

For each *x*-value substituted into the equation
$y = \dfrac{1}{x}$, the division performed gives a single
result, so only one *y*-value will be associated
with each *x*-value. Thus, $y = \dfrac{1}{x}$ is a function.

55. $y = 5x - 12$
For each *x*-value substituted into the equation
$y = 5x - 12$, the multiplication and addition
performed on each gives a single result, so only
one *y*-value will be associated with each *x*-value.
Thus, $y = 5x - 12$ is a function.

57. $x = y^2$
The *x*-value 4 is associated with two *y*-values, -2
and 2. Thus, $x = y^2$ is not a function.

59. $f(x) = 3x + 3$
$f(4) = 3(4) + 3 = 12 + 3 = 15$

61. $h(x) = 5x^2 - 7$
$$\begin{aligned}
h(-3) &= 5(-3)^2 - 7 \\
&= 5(9) - 7 \\
&= 45 - 7 \\
&= 38
\end{aligned}$$

63. $g(x) = 4x^2 - 6x + 3$
$$\begin{aligned}
g(2) &= 4(2)^2 - 6(2) + 3 \\
&= 4(4) - 12 + 3 \\
&= 16 - 12 + 3 \\
&= 7
\end{aligned}$$

65. $g(x) = 4x^2 - 6x + 3$
$$\begin{aligned}
g(0) &= 4(0)^2 - 6(0) + 3 \\
&= 4(0) - 0 + 3 \\
&= 0 - 0 + 3 \\
&= 3
\end{aligned}$$

67. $f(x) = \dfrac{1}{2}x$

   a. $f(0) = \dfrac{1}{2}(0) = 0$

   b. $f(2) = \dfrac{1}{2}(2) = 1$

   c. $f(-2) = \dfrac{1}{2}(-2) = -1$

69. $g(x) = 2x^2 + 4$

    a. $g(-11) = 2(-11)^2 + 4$
$$= 2(121) + 4$$
$$= 242 + 4$$
$$= 246$$

    b. $g(-1) = 2(-1)^2 + 4$
$$= 2(1) + 4$$
$$= 2 + 4$$
$$= 6$$

    c. $g\left(\dfrac{1}{2}\right) = 2\left(\dfrac{1}{2}\right)^2 + 4$
$$= 2\left(\dfrac{1}{4}\right) + 4$$
$$= \dfrac{1}{2} + \dfrac{8}{2}$$
$$= \dfrac{9}{2}$$

71. $f(x) = -5$

    a. $f(2) = -5$

    b. $f(0) = -5$

    c. $f(606) = -5$

73. $f(x) = 1.3x^2 - 2.6x + 5.1$

    a. $f(2) = 1.3(2)^2 - 2.6(2) + 5.1$
$$= 1.3(4) - 5.2 + 5.1$$
$$= 5.2 - 5.2 + 5.1$$
$$= 5.1$$

    b. $f(-2) = 1.3(-2)^2 - 2.6(-2) + 5.1$
$$= 1.3(4) + 5.2 + 5.1$$
$$= 5.2 + 5.2 + 5.1$$
$$= 15.5$$

    c. $f(3.1) = 1.3(3.1)^2 - 2.6(3.1) + 5.1$
$$= 1.3(9.61) - 8.06 + 5.1$$
$$= 12.493 - 8.06 + 5.1$$
$$= 9.533$$

75. a. $f(-5) = 23$

    b. $f(10) = 23$

c. $f(15) = 33$

77. $f(2) = 7$

79. If $f(1) = -10$, then $y = -10$ when $x = 1$. The ordered pair is $(1, -10)$.

81. If $g(4) = 56$, then $y = 56$ when $x = 4$. The ordered pair is $(4, 56)$.

83. The ordered pair $(-1, -2)$ is on the graph of f. Thus, $f(-1) = -2$.

85. The ordered pair $(2, 0)$ is on the graph of g. Thus, $g(2) = 0$.

87. There are two ordered pairs on the graph of f with a y-value of -5, $(-4, -5)$ and $(0, -5)$. The x-values are -4 and 0.

89. To the right of the y-axis, there is one ordered pair on the graph of g with a y-value of -5, $(-4, -5)$. The x-value is -4.

91. $f(x) = 2.7x + 3.6$
$f(10) = 2.7(10) + 3.6 = 27 + 3.6 = 30.6$
In 2010, about 30.6% of students took at least one online course.

93. $g(x) = 0.06x^2 + 1.9x + 5.4$
$$g(16) = 0.06(16)^2 + 1.9(16) + 5.4$$
$$= 15.36 + 30.4 + 5.4$$
$$= 51.16$$
In 2016, we predict that 51.16% of students will take at least one online course.

95. answers may vary

97.

Time in Hours	t	1	2	3	4	5
Total Cost	$C(t)$	55	90	125	160	195

99.

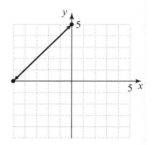

    a. $h(0.5) = 36$

    b. $h(1) = 64$

    c. $h(1.5) = 84$

    d. $h(2) = 96$

    e. $h(2.5) = 100$

    f. $h(5) = 0$

101. Since 2012 is 12 years after 2000, find $f(12)$.
$f(x) = 0.42x + 10.5$
$$f(12) = 0.42(12) + 10.5$$
$$= 15.54$$
We predict that diamond production will be $15.54 billion in 2012.

103. $f(x) = x + 7$

105. $A(r) = \pi r^2$
$$A(5) = \pi(5)^2 = 25\pi \text{ square centimeters}$$

107. $V(x) = x^3$
$$V(14) = (14)^3 = 2744 \text{ cubic inches}$$

109. $H(f) = 2.59f + 47.24$
$$H(46) = 2.59(46) + 47.24$$
$$= 166.38 \text{ centimeters}$$

111. $D(x) = \dfrac{136}{25}x$
$$D(30) = \dfrac{136}{25}(30) = 163.2 \text{ milligrams}$$

113. $C(x) = -0.74x + 67.1$

    a. $C(4) = -0.74(4) + 67.1 = 64.14$
        The per capita consumption of beef was
        64.14 pounds in 2004.

    b. $C(14) = -0.74(14) + 67.1 = 56.74$
        In 2014, the per capita consumption is
        predicted to be 56.74 pounds.

115. $x - y = -5$

x	0	−5	1
y	5	0	6

117. $7x + 4y = 8$

x	0	$\dfrac{8}{7}$	$\dfrac{12}{7}$
y	2	0	−1

119. $y = 6x$

x	0	0	−1
y	0	0	−6

121. Yes, it is possible to find the perimeter. The sum of the lengths of the two sides parallel to the side measuring 45 meters is also 45 meters. The sum of the lengths of the two sides parallel to the side measuring 40 meters is also 40 meters. The perimeter is $45 + 45 + 40 + 40 = 170$ meters.

123. $f(7) = 50$ means that $y = 50$ when $x = 7$. We conclude that the given statement is false, since it does not follow from $f(7) = 50$.

125. Since $f(7) = 50$ when $f(x) = 10x - 20$, the statement is true.

127. $f(x) = x^2 - 12$

    a. $f(12) = (12)^2 - 12 = 144 - 12 = 132$

    b. $f(a) = a^2 - 12$

129. $f(x) = 2x + 7$

    a. $f(2) = 2(2) + 7 = 4 + 7 = 11$

    b. $f(a) = 2a + 7$

    c. $f(-x) = 2(-x) + 7 = -2x + 7$

    d. $f(x + h) = 2(x + h) + 7 = 2x + 2h + 7$

131. 1

133. answers may vary

135. answers may vary

Section 2.3 Practice Exercises

    1. $f(x) = 4x$, $g(x) = 4x - 3$

x	$f(x)$	$g(x)$
0	0	−3
−1	−4	−7
1	4	1

2.

The graph of $g(x) = -3x + 2$ can be obtained from the graph of $f(x) = -3x$ by <u>shifting the graph of $f(x)$ up 2 units</u>.

3.

The graph of $y_2 = -x - 5$ can be obtained from the graph of $y_1 = -x$ by <u>shifting the graph of y_1 down 5 units</u>.

4. $4x - 5y = -20$

Let $x = 0$.
$$4x - 5y = -20$$
$$4 \cdot 0 - 5y = -20$$
$$-5y = -20$$
$$y = 4$$

Let $y = 0$.
$$4x - 5y = -20$$
$$4x - 5 \cdot 0 = -20$$
$$4x = -20$$
$$x = -5$$

Let $x = -2$.
$$4x - 5y = -20$$
$$4(-2) - 5y = -20$$
$$-8 - 5y = -20$$
$$-5y = -12$$
$$y = \frac{12}{5} = 2\frac{2}{5}$$

The ordered pairs are in the table.

x	y
0	4
−5	0
−2	$2\frac{2}{5}$

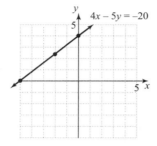

Solve for *y* to check.
$$4x - 5y = -20$$
$$-5y = -4x - 20$$
$$y = \frac{-4x - 20}{-5}$$

5. a. The *y*-intercept of $f(x) = \frac{3}{4}x - \frac{2}{5}$ is

$\left(0, -\frac{2}{5}\right)$.

b. The *y*-intercept of $y = 2.6x + 4.1$ is $(0, 4.1)$.

6. $y = -3x$

If $x = 0$, then $y = -3(0) = 0$.

If $x = 1$, then $y = -3(1) = -3$.

If $x = -1$, then $y = -3(-1) = 3$.

The ordered pairs are in the table.

x	y
0	0
1	−3
−1	3

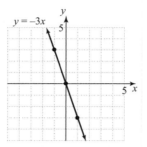

7. $x = -4$

The equation can be written as $x + 0y = -4$. For any *y*-value chosen, notice that *x* is −4.

x	y
−4	−3
−4	0
−4	3

8. $y = 4$

The equation can be written as $0x + y = 4$. For any *x*-value chosen, notice that *y* is 4.

x	y
−3	4
0	4
3	4

9. Define $y_1 = 79 + 0.10x$ and graph in a [0, 500, 50] by [0, 150, 25] window. The graph shows the cost for 50 miles.

No. of miles (x)	50	175	230	450
Cost C(x)	84	96.50	102	124

Vocabulary, Readiness & Video Check 2.3

1. A <u>linear</u> function can be written in the form $f(x) = mx + b$.

2. In the form $f(x) = mx + b$, the *y*-intercept is <u>(0, *b*)</u>.

3. The graph of $x = c$ is a <u>vertical</u> line with *x*-intercept <u>(*c*, 0)</u>.

4. The graph of $y = c$ is a <u>horizontal</u> line with *y*-intercept <u>(0, *c*)</u>.

5. To find an *x*-intercept, let <u>*y*</u> = 0 or <u>*f*(*x*)</u> = 0 and solve for <u>*x*</u>.

6. To find a *y*-intercept, let <u>*x*</u> = 0 and solve for <u>*y*</u>.

7. $f(x) = mx + b$, or slope-intercept form

8. A third point is found as a check that the points lie along the same line.

9. For a horizontal line, the coefficient of *x* is 0 and the coefficient of *y* is 1; for a vertical line, the coefficient of *y* is 0 and the coefficient of *x* is 1.

10. Graph the function. Then enter the *x*-values from the table to find the corresponding values of the function.

Exercise Set 2.3

1. $f(x) = -2x$

x	0	-1	1
y	0	2	-2

Plot the points to obtain the graph.

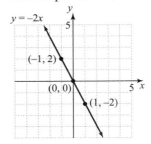

3. $f(x) = -2x + 3$

x	0	1	-1
y	3	1	5

Plot the points to obtain the graph.

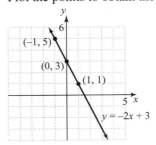

5. $f(x) = \dfrac{1}{2}x$

x	0	2	-2
y	0	1	-1

Plot the points to obtain the graph.

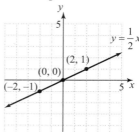

7. $f(x) = \dfrac{1}{2}x - 4$

x	0	2	4
y	-4	-3	-2

Plot the points to obtain the graph.

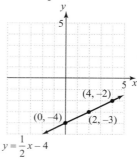

9. The graph of $f(x) = 5x - 6$ is the graph of $f(x) = 5x$ shifted down 6 units. The correct graph is C.

11. The graph of $f(x) = 5x + 7$ is the graph of $f(x) = 5x$ shifted up 7 units. The correct graph is D.

13. $x - y = 3$

Let $x = 0$.   Let $y = 0$.   Let $x = 2$.
$0 - y = 3$ $x - 0 = 3$ $2 - y = 3$
$\quad y = -3$ $x = 3$ $y = -1$

x	0	3	2
y	-3	0	-1

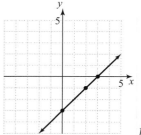

$f(x) = x - 3$

15. $x = 5y$

Let $x = 0$.   Let $x = 5$.   Let $x = -5$.
$0 = 5y$ $5 = 5y$ $-5 = 5y$
$y = 0$ $y = 1$ $y = -1$

x	0	5	-5
y	0	1	-1

$f(x) = \dfrac{x}{5}$

17. $-x + 2y = 6$

Let $x = 0$.   Let $y = 0$.   Let $x = 2$.
$-0 + 2y = 6$ $-x + 2(0) = 6$ $-2 + 2y = 6$
$\quad y = 3$ $x = -6$ $y = 4$

x	0	-6	2
y	3	0	4

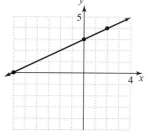

$f(x) = \dfrac{1}{2}x + 3$

19. $2x - 4y = 8$

Let $x = 0$.   Let $y = 0$.
$2(0) - 4y = 8$ $2x - 4(0) = 8$
        $y = -2$ $x = 4$

Let $x = 2$.
$2(2) - 4y = 8$
        $y = -1$

x	0	4	2
y	−2	0	−1

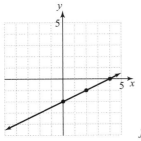

$$f(x) = \frac{1}{2}x - 2$$

21. $x = -1$
Vertical line with x-intercept at −1

23. $y = 0$
Horizontal line with y-intercept at 0

25. $y + 7 = 0$
    $y = -7$
Horizontal line with y-intercept at −7

27. The graph of $y = 2$ is a horizontal line with y-intercept $(0, 2)$. The correct graph is C.

29. The graph of $x - 2 = 0$ or $x = 2$ is a vertical line with x-intercept $(2, 0)$. The correct graph is A.

31. $x + 2y = 8$

33. $3x + 5y = 7$

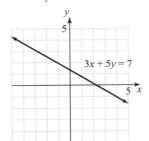

35. $x + 8y = 8$

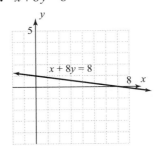

37. $5 = 6x - y$

39. $-x + 10y = 11$

41. $y = \dfrac{3}{2}$

43. $2x + 3y = 6$

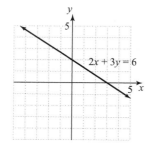

45. $x + 3 = 0$, or $x = -3$

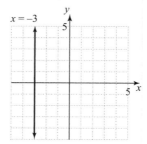

47. $f(x) = \dfrac{3}{4}x + 2$

49. $f(x) = x$

51. $f(x) = \dfrac{1}{2}x$

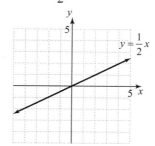

53. $f(x) = 4x - \dfrac{1}{3}$

55. $x = -3$

57. The graph shows the *y*-intercept of (0, 7), so $f(0) = 7$. Statements b and d are true.

59.

No. of Overtime Hours	x	2	3.25	7.5	9.75
Salary in Dollars	$S(x)$	677.25	706.78	807.19	860.34

61. $\dfrac{-6-3}{2-8} = \dfrac{-9}{-6} = \dfrac{3}{2}$

63. $\dfrac{-8-(-2)}{-3-(-2)} = \dfrac{-8+2}{-3+2} = \dfrac{-6}{-1} = 6$

65. $\dfrac{0-6}{5-0} = \dfrac{-6}{5} = -\dfrac{6}{5}$

67. no; answers may vary

69. yes; answers may vary

71. $2x + 3y = 1500$

   a. $2(0) + 3y = 1500$
                $3y = 1500$
                 $y = 500$

      (0, 500); If no tables are produced, 500 chairs can be produced.

   b. $2x + 3(0) = 1500$
              $2x = 1500$
               $x = 750$

      (750, 0); If no chairs are produced, 750 tables can be produced.

   c. $2(50) + 3y = 1500$
        $100 + 3y = 1500$
            $3y = 1400$
             $y = 466.7$

      466 chairs

73. $C(x) = 0.2x + 24$

   a. $C(200) = 0.2(200) + 24$
                 $= 40 + 24$
                 $= 64$

      $64

   b.

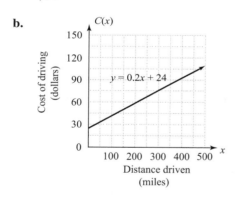

   c. The line moves upward from left to right.

75. $f = 64x + 2083$

   a. Since 2016 is 16 years after 2000, find $f(16)$.
      $f(16) = 64(16) + 2083 = 1024 + 2083 = 3107$
      In 2016, the yearly cost of attending a two-year college will be approximately \$3107.

   b. Let $f(x) = 3200$.
      $3200 = 64x + 2083$
      $1117 = 64x$
      $\dfrac{1117}{64} = x$
      $17.45 \approx x$
      Round up to 18. The yearly cost of attending a two-year college will first exceed \$3200 in 2018, 18 years after 2000.

   c. answers may vary

77. answers may vary

79. The vertical line $x = 0$ has y-intercepts.

81. a. $y = -4x + 2$ is a line parallel to $y = -4x$ but with y-intercept (0, 2).

   b. $y = -4x - 5$ is a line parallel to $y = -4x$ but with y-intercept $(0, -5)$.

83. The graph shows the graph of $y = |x|$ shifted up 1 unit. Its equation is $y = |x| + 1$. The correct answer is b.

85. The graph shows the graph of $y = |x|$ shifted down 1 unit. Its equation is $y = |x| - 1$. The correct answer is a.

Section 2.4 Practice Exercises

  1. Let $(x_1, y_1) = (4, 0)$ and $(x_2, y_2) = (-2, 3)$.
$$m = \frac{y_2 - y_1}{x_2 - x_1}$$
$$= \frac{3 - 0}{-2 - 4}$$
$$= \frac{3}{-6}$$
$$= -\frac{1}{2}$$

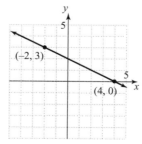

2. Let $(x_1, y_1) = (-5, -4)$ and $(x_2, y_2) = (5, 2)$.

$$m = \frac{y_2 - y_1}{x_2 - x_1}$$

$$= \frac{2 - (-4)}{5 - (-5)}$$

$$= \frac{6}{10}$$

$$= \frac{3}{5}$$

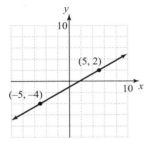

3. $2x - 3y = 9$

Write the equation in slope-intercept form by solving for y.

$$2x - 3y = 9$$
$$-3y = -2x + 9$$
$$\frac{-3y}{-3} = \frac{-2x}{-3} + \frac{9}{-3}$$
$$y = \frac{2}{3}x - 3$$

The coefficient of x, $\frac{2}{3}$, is the slope, and the y-intercept is $(0, -3)$.

4. a. To find the number of people in 2010, find y when x is $2010 - 1980 = 30$.

$$y = 110{,}520x + 2{,}127{,}400$$
$$= 110{,}520(30) + 2{,}127{,}400$$
$$= 5{,}433{,}000$$

There were an estimated 5,433,000 people age 85 or older in the U.S. in 2010.

b. The number of people age 85 or older in the U.S. increases at a rate of 110,520 per year.

c. At year $x = 0$, or 1980, there were 2,127,400 people age 85 or older in the U.S.

5. $x = 4$

The graph of $x = 4$ is a vertical line. We choose two points on the line, $(4, 0)$ and $(4, 3)$. Let $(x_1, y_1) = (4, 0)$ and $(x_2, y_2) = (4, 3)$.

$$m = \frac{y_2 - y_1}{x_2 - x_1}$$

$$= \frac{3 - 0}{4 - 4}$$

$$= \frac{3}{0}$$

Since $\frac{3}{0}$ is undefined, the slope of the vertical line $x = 4$ is undefined.

6. $y = -3$

The graph of $y = -3$ is a horizontal line. We choose two points on the line, $(0, -3)$ and $(4, -3)$. Let $(x_1, y_1) = (0, -3)$ and $(x_2, y_2) = (4, -3)$.

$$m = \frac{y_2 - y_1}{x_2 - x_1}$$

$$= \frac{-3 - (-3)}{4 - 0}$$

$$= \frac{0}{4}$$

$$= 0$$

The slope of the horizontal line $y = -3$ is 0.

7. a. Find the slope of each line.

$$x - 2y = 3$$
$$-2y = -x + 3$$
$$\frac{-2y}{-2} = \frac{-x}{-2} + \frac{3}{-2}$$
$$y = \frac{1}{2}x - \frac{3}{2}$$

The slope is $\frac{1}{2}$.

$$2x + y = 3$$
$$y = -2x + 3$$

The slope is -2. The product of the slopes is $-1\left[\frac{1}{2}(-2) = -1\right]$. The lines are perpendicular.

b. Find the slope of each line.

$$4x - 3y = 2$$
$$-3y = -4x + 2$$
$$\frac{-3y}{-3} = \frac{-4x}{-3} + \frac{2}{-3}$$
$$y = \frac{4}{3}x - \frac{2}{3}$$

The slope is $\frac{4}{3}$. The y-intercept is

$\left(0, -\dfrac{2}{3}\right).$

$$-8x + 6y = -6$$
$$6y = 8x - 6$$
$$\frac{6y}{6} = \frac{8x}{6} - \frac{6}{6}$$
$$y = \frac{4}{3}x - 1$$

The slope is $\dfrac{4}{3}$. The y-intercept is $(0, -1)$.

The slopes of both lines are $\dfrac{4}{3}$. The y-intercepts are different, so the lines are not the same. Therefore, the lines are parallel.

Vocabulary, Readiness & Video Check 2.4

1. The measure of the steepness or tilt of a line is called <u>slope</u>.

2. The slope of a line through two points is measured by the ratio of <u>vertical</u> change to <u>horizontal</u> change.

3. If a linear equation is in the form $y = mx + b$, or $f(x) = mx + b$, the slope of the line is <u>m</u> and the y-intercept is <u>$(0, b)$</u>.

4. The form $y = mx + b$ or $f(x) = mx + b$ is the <u>slope-intercept</u> form.

5. The slope of a <u>horizontal</u> line is 0.

6. The slope of a <u>vertical</u> line is undefined.

7. Two perpendicular lines have slopes whose product is <u>-1</u>.

8. Two non-vertical lines are parallel if they have <u>the same</u> slope and different <u>y-intercepts</u>.

9. A positive slope means the slope <u>increases</u> from left to right. A negative slope means the line <u>decreases</u> from left to right.

10. Solve the equation for y; the slope is the coefficient of x.

11. Slope-intercept form allows us to see the slope and y-intercept so we can discuss them in the context of the application. Slope is a rate of change and tells us the rate of increase of physician assistants (2.9 thousand per year). The y-intercept tells us the number of physician assistants at "zero" year (75,000 in 2008).

12. Slope is 0 indicates $m = 0$ and a horizontal line; undefined slope indicates m is undefined and a vertical line; no slope refers to an undefined slope.

13. The slopes of vertical lines are undefined and we can't mathematically work with undefined slopes.

Exercise Set 2.4

1. $m = \dfrac{11 - 2}{8 - 3} = \dfrac{9}{5}$

3. $m = \dfrac{3 - 8}{4 - (-2)} = \dfrac{-5}{6} = -\dfrac{5}{6}$

5. $m = \dfrac{-4 - (-6)}{4 - (-2)} = \dfrac{-4 + 6}{4 + 2} = \dfrac{2}{6} = \dfrac{1}{3}$

7. $m = \dfrac{5 - 5}{3 - (-2)} = \dfrac{0}{5} = 0$

9. $m = \dfrac{-5 - 1}{-1 - (-1)} = \dfrac{-6}{0}$
   undefined slope

11. $m = \dfrac{4 - 2}{-3 - (-1)} = \dfrac{2}{-2} = -1$

13. A line with slope $m = \dfrac{7}{6}$ slants upward.

15. Since $m = 0$, the line is horizontal.

17. $m = \dfrac{-13 - (-5)}{15 - 3} = \dfrac{-8}{12} = -\dfrac{2}{3}$

19. Use $(0, 5)$ and $(1, 6.5)$.
    $$m = \frac{6.5 - 5}{1 - 0} = \frac{1.5}{1} = 1.5$$

21. Use $(0, -8)$ and $(10, -6)$.
    $$m = \frac{-6 - (-8)}{10 - 0} = \frac{2}{10} = \frac{1}{5} = 0.2$$

23. The slope of l_1 is negative, and the slope of l_2 is positive. Since a positive number is greater than any negative number, l_2 has the greater slope.

25. The slope of l_1 is negative, and the slope of l_2 is 0. Since 0 is greater than any negative number, l_2 has the greater slope.

27. Both lines have positive slope. Since l_2 is steeper, it has the greater slope.

29. $f(x) = 5x - 2$

$m = 5$, $b = -2$ so y-intercept is $(0, -2)$.

31. $2x + y = 7$

$\quad\quad y = -2x + 7$

$m = -2$, $b = 7$ so y-intercept is $(0, 7)$.

33. $2x - 3y = 10$

$\quad\quad -3y = -2x + 10$

$\quad\quad\quad y = \dfrac{2}{3}x - \dfrac{10}{3}$

$m = \dfrac{2}{3}$, $b = -\dfrac{10}{3}$ so y-intercept is $\left(0, -\dfrac{10}{3}\right)$.

35. $f(x) = \dfrac{1}{2}x$

$m = \dfrac{1}{2}$, $b = 0$ so y-intercept is $(0, 0)$.

37. $f(x) = 2x + 3$

The slope is 2, and the y-intercept is $(0, 3)$. The correct graph is A.

39. $f(x) = -2x + 3$

The slope is -2, and the y-intercept is $(0, 3)$. The correct graph is B.

41. $x = 1$ is a vertical line.

m is undefined.

43. $y = -3$ is a horizontal line.

$m = 0$

45. $x + 2 = 0$

$\quad\quad x = -2$

This is a vertical line. m is undefined.

47. $f(x) = -x + 5$ or $y = -1x + 5$

$m = -1$, $b = 5$ so y-intercept is $(0, 5)$.

49. $-6x + 5y = 30$

$\quad\quad 5y = 6x + 30$

$\quad\quad\; y = \dfrac{6}{5}x + 6$

$m = \dfrac{6}{5}$, $b = 6$ so y-intercept is $(0, 6)$.

51. $3x + 9 = y$

$\quad\quad y = 3x + 9$

$m = 3$, $b = 9$ so y-intercept is $(0, 9)$.

53. $y = 4$

$m = 0$, $b = 4$ so y-intercept is $(0, 4)$.

55. $f(x) = 7x$

$m = 7$, $b = 0$ so y-intercept is $(0, 0)$.

57. $6 + y = 0$

$\quad\quad y = -6$

$m = 0$, $b = -6$ so y-intercept is $(0, -6)$.

59. $2 - x = 3$

$\quad\quad -x = -1$

m is undefined. There is no y-intercept.

61. $y = 12x + 6$ $y = 12x - 2$

$m = 12$ $m = 12$

Parallel, since they have the same slope.

63. $y = -9x + 3$ $y = \dfrac{3}{2}x - 7$

$m = -9$ $m = \dfrac{3}{2}$

Neither, since their slopes are not equal nor does their product equal -1.

65. $f(x) = -3x + 6$ $g(x) = \dfrac{1}{3}x + 5$

$m = -3$ $m = \dfrac{1}{3}$

Perpendicular, since the product of their slopes is -1.

67. $-4x + 2y = 5$ $2x - y = 7$

$\quad\quad y = 2x + \dfrac{5}{2}$ $y = 2x - 7$

$\quad\quad\quad\quad\quad\quad\quad\quad\quad\; m = 2$

$m = 2$

Parallel, since they have the same slope.

69. $-2x+3y=1$　　　　　　$3x+2y=12$

$$y=\frac{2}{3}x+\frac{1}{3}\qquad\qquad y=-\frac{3}{2}x+6$$

$$m=\frac{2}{3}\qquad\qquad\qquad m=-\frac{3}{2}$$

Perpendicular, since the product of their slopes is -1.

71. Two points on the line: (0, 0), (2, 3)

$$m=\frac{3-0}{2-0}=\frac{3}{2}$$

73. Two points on the line: (4, 0), (0, 2)

$$m=\frac{2-0}{0-4}=\frac{2}{-4}=-\frac{1}{2}$$

75. $m=\dfrac{8}{12}=\dfrac{2}{3}$

77. $m=\dfrac{-1600\text{ ft}}{2.5\text{ mi}}$

$$=\frac{-1600\text{ ft}}{2.5(5280\text{ ft})}$$

$$=\frac{-1600}{13,200}\approx -0.12$$

79. a. Use $y = 0.16x + 71.6$ with
$x = 1980 - 1950 = 30$.
$y = 0.16(30) + 71.6 = 76.4$
The life expectancy of an American female born in 1980 is 76.4 years.

b. From $y = 0.16x + 71.6$, $m = 0.16$.
The life expectancy of a female born in the United States is approximately 0.16 year more than a female born one year before.

c. From $y = 0.16x + 71.6$, the y-intercept is (0, 71.6). At year = 0, or 1950, the life expectancy of an American female was 71.6 years.

81. a. $29x-10y=-750$
$$-10y=-29x-750$$
$$y=2.9x+75$$
$m = 2.9$, y-intercept (0, 75)

b. The slope means that the number of people employed as physician assistants increases 2.9 thousand for every 1 year.

c. The y-intercept means that 75 thousand physician assistants were employed in 2008.

83. a. Use $y = -0.04x + 1.62$ with
$x = 2007 - 2002 = 5$.
$y = -0.04(5) + 1.62 = 1.42$
In 2007, there were an estimated 1.42 billion movie admissions.

b. Let $y = 1$ and solve for x.
$$y=-0.04x+1.62$$
$$1=-0.04x+1.62$$
$$-0.62=-0.04x$$
$$15.5=x$$
$2002 + 15 = 2017$
During 2017, the number of movie admissions is predicted to be below 1 billion.

c. answers may vary

85. $y-2=5(x+6)$
$$y-2=5x+30$$
$$y=5x+32$$

87. $y-(-1)=2(x-0)$
$$y+1=2x$$
$$y=2x-1$$

89. The denominator in the first fraction should be $7 - (-2)$.
$$m=\frac{-14-6}{7-(-2)}=\frac{-20}{9}=-\frac{20}{9}$$

91. The numerator in the first fraction should be $-10 - (-5)$. The denominator in the first fraction should be $-8 - (-11)$.
$$m=\frac{-10-(-5)}{-8-(-11)}=\frac{-5}{3}=-\frac{5}{3}$$

93. $f(x)=-\dfrac{7}{2}x-6$ or $y=-\dfrac{7}{2}x-6$

$$m=-\frac{7}{2}$$

The slope of a parallel line is $-\dfrac{7}{2}$.

95. $f(x)=-\dfrac{7}{2}x-6$ or $y=-\dfrac{7}{2}x-6$

$$m=-\frac{7}{2}$$

The slope of a perpendicular line is $\dfrac{2}{7}$.

97. $5x - 2y = 6 \Rightarrow y = \dfrac{5}{2}x - 3$

$m = \dfrac{5}{2}$

The slope of a parallel line is $\dfrac{5}{2}$.

99. $\dfrac{-3}{-4} = \dfrac{3}{4}$

$\dfrac{-8}{6} = -\dfrac{4}{3}$

Since $\dfrac{3}{4}\left(-\dfrac{4}{3}\right) = 1$, the lines are perpendicular.

101. Since $0.5 = \dfrac{1}{2}$, the lines are parallel.

103. (6, 20)

105. $B(6, 20)$, $C(10, 13)$

$m = \dfrac{13 - 20}{10 - 6} = \dfrac{-7}{4} = -\dfrac{7}{4} = -1.75$

The rate of change is -1.75 yd per sec.

107. $m_1 = \dfrac{-2 - 4}{2 - (-1)} = \dfrac{-6}{3} = -2$

$m_2 = \dfrac{2 - 6}{-4 - (-8)} = \dfrac{-4}{4} = -1$

$m_3 = \dfrac{-4 - 0}{0 - (-6)} = \dfrac{-4}{6} = -\dfrac{2}{3}$

109. answers may vary

111. answers may vary

Section 2.5 Practice Exercises

1. $y = \dfrac{3}{4}x + 2$

The slope is $\dfrac{3}{4}$, and the y-intercept is (0, 2). Plot (0, 2). Then plot a second point by starting at (0, 2), rising 3 units up, and running 4 units to the right. The second point is (4, 5).

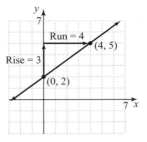

2. $x + 2y = 6$

Solve for y.

$x + 2y = 6$

$2y = -x + 6$

$y = -\dfrac{1}{2}x + 3$

The slope is $-\dfrac{1}{2}$, and the y-intercept is (0, 3).

Plot (0, 3). Then plot a second point by starting at (0, 3), moving 1 unit down, and moving 2 units to the right. The second point is (2, 2).

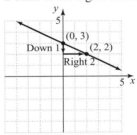

3. We are given the slope, $-\dfrac{3}{4}$, and the y-intercept, (0, 4).

Let $m = -\dfrac{3}{4}$ and $b = 4$.

$y = mx + b$

$y = -\dfrac{3}{4}x + 4$

4. Use the point-slope form with $m = -4$ and $(x_1, y_1) = (-2, 5)$.

$y - y_1 = m(x - x_1)$

$y - 5 = -4[x - (-2)]$

$y - 5 = -4(x + 2)$

$y - 5 = -4x - 8$

$y = -4x - 3$

5. First find the slope.

$$m = \frac{0-2}{2-(-1)} = \frac{-2}{3} = -\frac{2}{3}$$

Use the slope and one of the points in the point-slope form. We use (2, 0).

$$y - y_1 = m(x - x_1)$$

$$y - 0 = -\frac{2}{3}(x-2)$$

$$y = -\frac{2}{3}x + \frac{4}{3}$$

$$f(x) = -\frac{2}{3}x + \frac{4}{3}$$

6. The points on the graph have coordinates (−2, 3) and (1, 1). Find the slope.

$$m = \frac{1-3}{1-(-2)} = \frac{-2}{3} = -\frac{2}{3}$$

Use the slope and one of the points in the point-slope form. We use (1, 1).

$$y - y_1 = m(x - x_1)$$

$$y - 1 = -\frac{2}{3}(x-1)$$

$$3(y-1) = -2(x-1)$$

$$3y - 3 = -2x + 2$$

$$2x + 3y = 5$$

7. Let x = the number of years after 2000 and y = the number of houses sold in the year corresponding to x. We have two ordered pairs, (8, 7513) and (12, 9198). Find the slope.

$$m = \frac{9198 - 7513}{12 - 8}$$

$$= \frac{1685}{4}$$

$$= 421.25$$

Use the slope and one of the points in the point-slope form. We use (8, 7513).

$$y - y_1 = m(x - x_1)$$

$$y - 7513 = 421.25(x - 8)$$

$$y - 7513 = 421.25x - 3370$$

$$y = 421.25x + 4143$$

The year 2020 corresponds to $x = 20$.

$$y = 421.25(20) + 4143$$

$$= 8425 + 4143$$

$$= 12,568$$

We predict that there will 12,568 house sales in 2020.

8. A horizontal line has an equation of the form $y = b$. Since the line contains the point (6, −2), the equation is $y = -2$.

9. Since the line has undefined slope, the line must be vertical. A vertical line has an equation of the form $x = c$. Since the line contains the point (6, −2), the equation is $x = 6$.

10. Solve the given equation for y.

$$3x + 4y = 1$$

$$4y = -3x + 1$$

$$y = -\frac{3}{4}x + \frac{1}{4}$$

The slope of this line is $-\frac{3}{4}$, so the slope of any line parallel to it is also $-\frac{3}{4}$. Use this slope and the point (8, −3) in the point-slope form.

$$y - y_1 = m(x - x_1)$$

$$y - (-3) = -\frac{3}{4}(x - 8)$$

$$4(y + 3) = -3(x - 8)$$

$$4y + 12 = -3x + 24$$

$$3x + 4y = 12$$

11. Solve the given equation for y.

$$3x + 4y = 1$$

$$4y = -3x + 1$$

$$y = -\frac{3}{4}x + \frac{1}{4}$$

The slope of this line is $-\frac{3}{4}$, so the slope of any line perpendicular to it is the negative reciprocal of $-\frac{3}{4}$, or $\frac{4}{3}$. Use this slope and the point (8, −3) in the point-slope form.

$$y - y_1 = m(x - x_1)$$

$$y - (-3) = \frac{4}{3}(x - 8)$$

$$3(y + 3) = 4(x - 8)$$

$$3y + 9 = 4x - 32$$

$$3y = 4x - 41$$

$$y = \frac{4}{3}x - \frac{41}{3}$$

$$f(x) = \frac{4}{3}x - \frac{41}{3}$$

Vocabulary, Readiness & Video Check 2.5

1. $m = -4$, $b = 12$ so y-intercept is (0, 12).

2. $m = \frac{2}{3}$, $b = -\frac{7}{2}$ so y-intercept is $\left(0, -\frac{7}{2}\right)$.

3. $m = 5$, $b = 0$ so y-intercept is $(0, 0)$.

4. $m = -1$, $b = 0$ so y-intercept is $(0, 0)$.

5. $m = \dfrac{1}{2}$, $b = 6$ so y-intercept is $(0, 6)$.

6. $m = -\dfrac{2}{3}$, $b = 5$ so y-intercept is $(0, 5)$.

7. The lines both have slope 12 and they have different y-intercepts, $(0, 6)$ and $(0, -2)$, so they are parallel.

8. The lines both have slope -5 and they have different y-intercepts, $(0, 8)$ and $(0, -8)$, so they are parallel.

9. The line have slopes -9 and $\dfrac{3}{2}$. The slopes are not equal and their product is not -1, so the lines are neither parallel nor perpendicular.

10. The line have slopes 2 and $\dfrac{1}{2}$. The slopes are not equal and their product is not -1, so the lines are neither parallel nor perpendicular.

11. To graph a line using its slope and y-intercept, first write the equation in <u>slope-intercept</u> form. Graph the one point you now know, the <u>y-intercept</u>. Use the <u>slope</u> to find a second point.

12. The y-intercept is of the form $(0, b)$, so the b value is the y-coordinate of the y-intercept.

13. if one of the two points given is the y-intercept

14. Example 5: $y = 3$; Example 6: $x = -1$

15. $f(x) = \dfrac{1}{3}x - \dfrac{17}{3}$

Exercise Set 2.5

1. $y = 5x - 2$
possible points: $(0, -2)$, $(1, 3)$

3. $4x + y = 7$
$\qquad y = -4x + 7$
possible points: $(0, 7)$, $(1, 3)$

5. $-3x + 2y = 3$
$\qquad 2y = 3x + 3$
$\qquad y = \dfrac{3}{2}x + \dfrac{3}{2}$
possible points: $\left(0, \dfrac{3}{2}\right)$, $\left(2, \dfrac{9}{2}\right)$

7. $m = -1$, $b = 1$
$y = mx + b$
$y = -1x + 1$
$y = -x + 1$

9. $m = 2$, $b = \dfrac{3}{4}$
$y = mx + b$
$y = 2x + \dfrac{3}{4}$

11. $m = \dfrac{2}{7}, \ b = 0$

$y = mx + b$

$y = \dfrac{2}{7}x + 0$

$y = \dfrac{2}{7}x$

13. $y - y_1 = m(x - x_1)$

$y - 2 = 3(x - 1)$

$y - 2 = 3x - 3$

$y = 3x - 1$

15. $y - y_1 = m(x - x_1)$

$y - (-3) = -2(x - 1)$

$y + 3 = -2x + 2$

$y = -2x - 1$

17. $y - y_1 = m(x - x_1)$

$y - 2 = \dfrac{1}{2}[x - (-6)]$

$y - 2 = \dfrac{1}{2}(x + 6)$

$y - 2 = \dfrac{1}{2}x + 3$

$y = \dfrac{1}{2}x + 5$

19. $y - y_1 = m(x - x_1)$

$y - 0 = -\dfrac{9}{10}[x - (-3)]$

$y = -\dfrac{9}{10}(x + 3)$

$y = -\dfrac{9}{10}x - \dfrac{27}{10}$

21. $m = \dfrac{6 - 0}{4 - 2} = \dfrac{6}{2} = 3$

$y - 0 = 3(x - 2)$

$y = 3x - 6$

$f(x) = 3x - 6$

23. $m = \dfrac{13 - 5}{-6 - (-2)} = \dfrac{8}{-4} = -2$

$y - 5 = -2[x - (-2)]$

$y - 5 = -2(x + 2)$

$y - 5 = -2x - 4$

$y = -2x + 1$

$f(x) = -2x + 1$

25. $m = \dfrac{-3 - (-4)}{-4 - (-2)} = \dfrac{1}{-2} = -\dfrac{1}{2}$

$y - (-4) = -\dfrac{1}{2}[x - (-2)]$

$y + 4 = -\dfrac{1}{2}(x + 2)$

$2y + 8 = -(x + 2)$

$2y + 8 = -x - 2$

$2y = -x - 10$

$y = -\dfrac{1}{2}x - 5$

$f(x) = -\dfrac{1}{2}x - 5$

27. $m = \dfrac{-9 - (-8)}{-6 - (-3)} = \dfrac{-1}{-3} = \dfrac{1}{3}$

$y - (-8) = \dfrac{1}{3}[x - (-3)]$

$y + 8 = \dfrac{1}{3}(x + 3)$

$3y + 24 = x + 3$

$3y = x - 21$

$y = \dfrac{1}{3}x - 7$

$f(x) = \dfrac{1}{3}x - 7$

29. $m = \dfrac{\frac{7}{10} - \frac{4}{10}}{-\frac{1}{5} - \frac{3}{5}} = \dfrac{\frac{3}{10}}{-\frac{4}{5}} = \dfrac{3}{10}\left(-\dfrac{5}{4}\right) = -\dfrac{3}{8}$

$y - \dfrac{4}{10} = -\dfrac{3}{8}\left(x - \dfrac{3}{5}\right)$

$y - \dfrac{4}{10} = -\dfrac{3}{8}x + \dfrac{9}{40}$

$y = -\dfrac{3}{8}x + \dfrac{5}{8}$

$f(x) = -\dfrac{3}{8}x + \dfrac{5}{8}$

31. $(0, 3), (1, 1)$

$m = \dfrac{1 - 3}{1 - 0} = \dfrac{-2}{1} = -2$

$b = 3$

$y = -2x + 3$

$2x + y = 3$

33. $(-2, 1), (4, 5)$

$$m = \frac{5-1}{4-(-2)} = \frac{4}{6} = \frac{2}{3}$$

$$y - 1 = \frac{2}{3}(x + 2)$$

$$3y - 3 = 2(x + 2)$$

$$3y - 3 = 2x + 4$$

$$2x - 3y = -7$$

35. $f(0) = -2$

37. $f(2) = 2$

39. $f(x) = -6$

$$f(-2) = -6$$

$$x = -2$$

41. $y = mx + b$

$$-4 = 0(-2) + b$$

$$-4 = b$$

$$y = -4$$

43. Every vertical line is in the form $x = c$. Since the line passes through the point $(4, 7)$, its equation is $x = 4$.

45. Every horizontal line is in the form $y = c$. Since the line passes through the point $(0, 5)$, its equation is $y = 5$.

47. $y = 4x - 2$ so $m = 4$

$$y - 8 = 4(x - 3)$$

$$y - 8 = 4x - 12$$

$$y = 4x - 4$$

$$f(x) = 4x - 4$$

49. $3y = x - 6$ or $y = \frac{1}{3}x - 2$ so

$$m = \frac{1}{3} \text{ and } m_\perp = -3$$

$$y - (-5) = -3(x - 2)$$

$$y + 5 = -3x + 6$$

$$y = -3x + 1$$

$$f(x) = -3x + 1$$

51. $3x + 2y = 5$

$$2y = -3x + 5$$

$$y = -\frac{3}{2}x + \frac{5}{2} \text{ so } m = -\frac{3}{2}$$

$$y - (-3) = -\frac{3}{2}[x - (-2)]$$

$$2(y + 3) = -3(x + 2)$$

$$2y + 6 = -3(x + 2)$$

$$2y + 6 = -3x - 6$$

$$y = -\frac{3}{2}x - 6$$

$$f(x) = -\frac{3}{2}x - 6$$

53.

$$y - 3 = 2[x - (-2)]$$

$$y - 3 = 2(x + 2)$$

$$y - 3 = 2x + 4$$

$$2x - y = -7$$

55. $m = \dfrac{2 - 6}{5 - 1} = \dfrac{-4}{4} = -1$

$$y - 6 = -1(x - 1)$$

$$y - 6 = -x + 1$$

$$y = -x + 7$$

$$f(x) = -x + 7$$

57.

$$y = -\frac{1}{2}x + 11$$

$$2y = -x + 22$$

$$x + 2y = 22$$

59. $m = \dfrac{-6 - (-4)}{0 - (-7)} = \dfrac{-2}{7} = -\dfrac{2}{7}$

$$y = -\frac{2}{7}x - 6$$

$$7y = -2x - 42$$

$$2x + 7y = -42$$

61.

$$y - 0 = -\frac{4}{3}[x - (-5)]$$

$$3y = -4(x + 5)$$

$$3y = -4x - 20$$

$$4x + 3y = -20$$

63. Every vertical line is in the form $x = c$. Since the line passes through the point $(-2, -10)$, its equation is $x = -2$.

65. $2x + 4y = 8$

$$4y = -2x + 8$$

$$y = -\frac{1}{2}x + 2 \text{ so } m = -\frac{1}{2}$$

$$y - (-2) = -\frac{1}{2}(x - 6)$$

$$2(y + 2) = -(x - 6)$$

$$2y + 4 = -x + 6$$

$$x + 2y = 2$$

67. Lines with slopes of 0 are horizontal. Every horizontal line is in the form $y = c$. Since the line passes through $(-9, 12)$, its equation is $y = 12$.

69. $8x - y = 9$

$$y = 8x - 9 \text{ so } m = 8$$

$$y - 1 = 8(x - 6)$$

$$y - 1 = 8x - 48$$

$$8x - y = 47$$

71. A line perpendicular to $y = 9$ will have the form $x = c$. Since the line passes through the point $(5, -6)$, its equation is $x = 5$.

73. $m = \dfrac{-5 - (-8)}{-6 - 2} = \dfrac{3}{-8} = -\dfrac{3}{8}$

$$y - (-8) = -\frac{3}{8}(x - 2)$$

$$8(y + 8) = -3(x - 2)$$

$$8y + 64 = -3x + 6$$

$$y = -\frac{3}{8}x - \frac{29}{4}$$

$$f(x) = -\frac{3}{8}x - \frac{29}{4}$$

75. a. Use the points $(1, 32)$ and $(3, 96)$.

$$m = \frac{96 - 32}{3 - 1} = \frac{64}{2} = 32$$

$$y - y_1 = m(x - x_1)$$

$$y - 32 = 32(x - 1)$$

$$y - 32 = 32x - 32$$

$$y = 32x$$

b. Let $x = 4$. $y = 32x = 32(4) = 128$
The rock is traveling 128 feet per second at 4 seconds.

77. a. Use the points $(6, 2000)$ and $(8, 1500)$.

$$m = \frac{1500 - 2000}{8 - 6} = \frac{-500}{2} = -250$$

$$y - y_1 = m(x - x_1)$$

$$y - 2000 = -250(x - 6)$$

$$y - 2000 = -250x + 1500$$

$$y = -250x + 3500$$

b. Let $x = 7.50$.
$y = -250x + 3500$
$y = -250(7.50) + 3500 = 1625$
The daily sales of Frisbees at $7.50 each is predicted to be 1625 Frisbees.

79. a. Use the points $(0, 2619)$ and $(10, 3200)$.

$$m = \frac{3200 - 2619}{10 - 0} = \frac{581}{10} = 58.1$$

$$y - y_1 = m(x - x_1)$$

$$y - 2619 = 58.1(x - 0)$$

$$y = 58.1x + 2619$$

b. Let $x = 2012 - 2008 = 4$.
$y = 58.1(4) + 2619 = 2851.4$
The estimated number of people who will be employed as registered nurses in 2012 is 2851.4 thousand.

81. a. Use the points $(0, 297{,}000)$ and $(5, 272{,}900)$.

$$m = \frac{272{,}900 - 297{,}000}{5 - 0} = \frac{-24{,}100}{5} = -4820$$

$$y - y_1 = m(x - x_1)$$

$$y - 297{,}000 = -4820(x - 0)$$

$$y = -4820x + 297{,}000$$

b. Let $x = 2013 - 2005 = 8$.
$y = -4820(8) + 297{,}000 = 258{,}440$
The average price of a new home in 2013 is predicted to be $258,440.

83. $2x - 7 = 21$

$$2x = 28$$

$$x = 14$$

85. $5(x - 2) = 3(x - 1)$

$$5x - 10 = 3x - 3$$

$$2x = 7$$

$$x = \frac{7}{2}$$

87.
$$\frac{x}{2} + \frac{1}{4} = \frac{1}{8}$$
$$8\left(\frac{x}{2} + \frac{1}{4}\right) = 8\left(\frac{1}{8}\right)$$
$$4x + 2 = 1$$
$$4x = -1$$
$$x = -\frac{1}{4}$$

89. Since any vertical line intersects any horizontal line in a right angle, the statement is true.

91. $m = \dfrac{1-(-1)}{-5-3} = \dfrac{2}{-8} = -\dfrac{1}{4}$ so $m_\perp = 4$

$M((3,-1),(5,1)) = \left(\dfrac{3-5}{2}, \dfrac{-1+1}{2}\right) = (1,0)$

$$y - 0 = 4[x - (-1)]$$
$$y = 4(x+1)$$
$$y = 4x + 4$$
$$-4x + y = 4$$

93. $m = \dfrac{-4-6}{-22-(-2)} = \dfrac{-10}{-20} = \dfrac{1}{2}$ so $m_\perp = -2$

$M((-2,6),(-22,-4)) = \left(\dfrac{-2-22}{2}, \dfrac{6-4}{2}\right)$
$$= (-12, 1)$$

$$y - 1 = -2[x - (-12)]$$
$$y - 1 = -2(x + 12)$$
$$y - 1 = -2x - 24$$
$$2x + y = -23$$

95. $m = \dfrac{7-3}{-4-2} = \dfrac{4}{-6} = -\dfrac{2}{3}$ so $m_\perp = \dfrac{3}{2}$

$M((2,3),(-4,7)) = \left(\dfrac{2-4}{2}, \dfrac{3+7}{2}\right)$
$$= (-1, 5)$$

$$y - 5 = \frac{3}{2}[x - (-1)]$$
$$2(y-5) = 3(x+1)$$
$$2y - 10 = 3x + 3$$
$$3x - 2y = -13$$

97. answers may vary

Integrated Review

1. $y = -2x$

x	-1	0	1
y	2	0	-2

2. $3x - 2y = 6$

x	0	2	4
y	-3	0	3

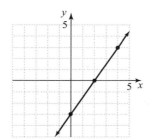

3. $x = -3$

The graph of $x = -3$ is a vertical line.

4. $y = 1.5$

5. $m = \dfrac{-5-(-5)}{3-(-2)} = \dfrac{0}{5} = 0$

6. $m = \dfrac{5-2}{0-5} = \dfrac{3}{-5} = -\dfrac{3}{5}$

7. $y = 3x - 5$
$m = 3;\ (0, -5)$

8. $5x - 2y = 7$
$\quad -2y = -5x + 7$
$\qquad y = \dfrac{5}{2}x - \dfrac{7}{2}$
$\quad m = \dfrac{5}{2};\ \left(0, -\dfrac{7}{2}\right)$

9. $y = 8x - 6 \quad y = 8x + 6$
$m = 8 \qquad m = 8$
Parallel, since their slopes are equal.

10. $y = \dfrac{2}{3}x + 1 \qquad\qquad 2y + 3x = 1$
$\quad m = \dfrac{2}{3} \qquad\qquad\qquad 2y = -3x + 1$
$\qquad\qquad\qquad\qquad\qquad y = -\dfrac{3}{2}x + \dfrac{1}{2}$
$\qquad\qquad\qquad\qquad\qquad m = -\dfrac{3}{2}$

Perpendicular, since the product of their slopes is -1.

11. $m = \dfrac{2-6}{5-1} = \dfrac{-4}{4} = -1$
$y - 6 = -1(x - 1)$
$y - 6 = -x + 1$
$\quad\ y = -x + 7$

12. Every vertical line is in the form $x = c$. Since the line passes through the point $(-2, -10)$, its equation is $x = -2$.

13. Every horizontal line is in the form $y = c$. Since the line passes through the point $(1, 0)$, its equation is $y = 0$.

14. $m = \dfrac{-5-(-9)}{-6-2} = \dfrac{4}{-8} = -\dfrac{1}{2}$
$y - (-9) = -\dfrac{1}{2}(x - 2)$
$2(y + 9) = -1(x - 2)$
$2y + 18 = -x + 2$
$\quad 2y = -x - 16$
$\quad\ y = -\dfrac{1}{2}x - 8$
$\ f(x) = -\dfrac{1}{2}x - 8$

15. $y - 4 = -5[x - (-2)]$
$y - 4 = -5(x + 2)$
$y - 4 = -5x - 10$
$\quad\ y = -5x - 6$
$\ f(x) = -5x - 6$

16. $y = -4x + \dfrac{1}{3}$
$f(x) = -4x + \dfrac{1}{3}$

17. $y = \dfrac{1}{2}x - 1$
$f(x) = \dfrac{1}{2}x - 1$

18. $y - 0 = 3\left(x - \dfrac{1}{2}\right)$
$\quad\ y = 3x - \dfrac{3}{2}$

19. $3x - y = 5$
$\quad y = 3x - 5$
$\quad m = 3$
$y - (-5) = 3[x - (-1)]$
$\quad y + 5 = 3(x + 1)$
$\quad y + 5 = 3x + 3$
$\qquad y = 3x - 2$

Copyright © 2014 Pearson Education, Inc.

20. $4x - 5y = 10$

$-5y = -4x + 10$

$y = \frac{4}{5}x - 2;\ \ m = \frac{4}{5}\ \ \text{so}\ \ m_{\perp} = -\frac{5}{4}$

Therefore, $y = -\frac{5}{4}x + 4$.

21. $4x + y = \frac{2}{3}$

$y = -4x + \frac{2}{3};\ \ m = -4\ \ \text{so}\ \ m_{\perp} = \frac{1}{4}$

$y - (-3) = \frac{1}{4}(x - 2)$

$4(y + 3) = x - 2$

$4y + 12 = x - 2$

$4y = x - 14$

$y = \frac{1}{4}x - \frac{7}{2}$

22. $5x + 2y = 2$

$2y = -5x + 2$

$y = -\frac{5}{2}x + 1$

$m = -\frac{5}{2}$

$y - 0 = -\frac{5}{2}[x - (-1)]$

$y = -\frac{5}{2}(x + 1)$

$2y = -5(x + 1)$

$2y = -5x - 5$

$y = -\frac{5}{2}x - \frac{5}{2}$

23. A line having undefined slope is vertical. Therefore, the equation is $x = -1$.

24. $y - 3 = 0[x - (-1)]$

$y - 3 = 0$

$y = 3$

Section 2.6 Practice Exercises

1. $2020 - 1995 = 25$

Trace to $x = 25$ or evaluate the function for $x = 25$.

The number of passengers in 2020 is expected to be about 12.06 billion.

2.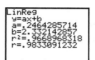

We predict tablet PC sales of 81.4 million in 2020. The slope indicates that the average increase in the sales of tablet PCs was 6.5 million per year.

3. b.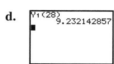

If a and b are rounded to two decimal places, the regression equation is $y = 0.25x + 2.33$. Since the slope (0.25) is positive, the line is increasing.

  c. The number of sick or injured workers is increasing at a rate of 0.25 per 100 workers per year.

  d.

Y1(28) is about 9.23 or approximately 9.23 per 100 workers in 2018.

4. a.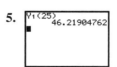

The regression equation is $y = 195.91x + 38{,}438.48$.

  b.

A home with 1662 square feet of living area is expected to sell for \$364,033.02.

5.

It is expected that 46.22% of workers will be 55 or older in 2025.

Vocabulary, Readiness & Video Check 2.6

1. Regression analysis is the process of fitting a line or a curve to a set of data points.

2. Enter the data into the calculator using lists.

3. Use the <u>Stat Plot</u> feature to indicate the <u>scatter plot</u> as the type of graph.

4. When finding an appropriate window, make sure the <u>domain</u> and <u>range</u> are both included in your choice.

5. Use a <u>linear regression equation</u> to find the line of best fit.

6. In the linear regression equation, $x = 0$ represents the year 1970, and $2013 - 1970 = 43$.

Exercise Set 2.6

1.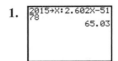

   Approximately \$65.03 billion will be spent in 2015.

3.

   The price was approximately \$50 in 2000.

5. **a.**

   The regression equation is
   $y = 102.054x - 758.049$.

   b. The number of smart phones shipped worldwide is increasing by 102.054 million per year.

   c.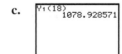

   The number of smart phones shipped worldwide is expected to be 1078.929 million or 1.078929 billion in 2018.

7. **a.**

   The regression equation is
   $y = -0.725x + 52.566$

b.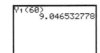

The percent of male smokers is predicted to be 9.0% in 2020.

c. The percent of male smokers is decreasing by 0.725% per year.

9. **a.**

   The regression equation is
   $y = 47.143x + 269.929$.

   b.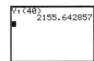

   There are predicted to be 2156 thousand or 2,156,000 male prisoners in 2020.

   c. The number of male prisoners is increasing by 47 thousand or 47,000 per year.

11. **a.**

   b.

   The regression equation is
   $y = 0.212x + 12.04$.

   c. The slope indicates that the percentage of workers over 65 is increasing at the rate of 0.212% per year.

   d.

   About 18.4% of the workforce will be over 65 in 2020.

13.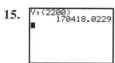

The regression equation is
$y = 114.453x - 81{,}378.675$.

15.

A house with 2200 square feet is predicted to sell for $170,418.

17.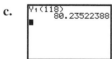

The regression equation is $y = 0.08x + 700$.

19. a.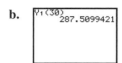

The regression equation is
$y = 0.268x + 48.587$.

b. The life expectancy at birth of males is increasing at the rate of 0.268 year annually.

c.

The life expectancy of a male born in 2018 is predicted to be 80.2 years.

21. a.

The regression equation is
$y = 8.212x + 41.160$.

b.

It is predicted that there will be 287.51 million PPO members in 2020.

c.

There will be over 240 million PPO members in 2014.

d. The number of PPO members is increasing at a rate of 8.212 million per year.

23. a.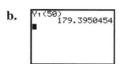

The regression equation is
$y = 3.786x - 9.899$.

b.

The average top ticket price is predicted to be $179.40 per ticket in 2020.

c. The average top ticket price is increasing at the rate of $3.79 per year.

25. a.

b.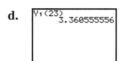

The regression equation is
$y = 0.124x + 0.505$.

c. The cost is increasing at the rate of $0.124 million or $124,000 per year.

d.

A 30-second commercial during the Super Bowl is predicted to cost $3.357 million or $3,357,000 in 2013.

27.
$$7x + 2 = 9x - 14$$
$$7x - 7x + 2 = 9x - 7x - 14$$
$$2 = 2x - 14$$
$$2 + 14 = 2x - 14 + 14$$
$$16 = 2x$$
$$\frac{16}{2} = \frac{2x}{2}$$
$$8 = x$$

29.
$$\frac{y}{7} + \frac{y}{3} = \frac{1}{21}$$
$$21\left(\frac{y}{7} + \frac{y}{3}\right) = 21\left(\frac{1}{21}\right)$$
$$3y + 7y = 1$$
$$10y = 1$$
$$y = \frac{1}{10}$$

31.

The "full service" regression equation is
$y = 7.157x + 62.484$.

The "fast food" regression equation is
$y = 5.971x + 69.452$.

The two types of dining occurred the same number of times in 1996.

Section 2.7 Practice Exercises

1. $f(x) = \begin{cases} -4x - 2 & \text{if } x \le 0 \\ x + 1 & \text{if } x > 0 \end{cases}$

Since $4 > 0$, $f(4) = 4 + 1 = 5$.
Since $-2 \le 0$, $f(-2) = -4(-2) - 2 = 8 - 2 = 6$.
Since $0 \le 0$, $f(0) = -4(0) - 2 = 0 - 2 = -2$.

2. $f(x) = \begin{cases} -4x - 2 & \text{if } x \le 0 \\ x + 1 & \text{if } x > 0 \end{cases}$

For $x \le 0$:

x	$f(x)$
-2	6
-1	2
0	-2

For $x > 0$:

x	$f(x)$
1	2
2	3
3	4

Graph a closed circle at $(0, -2)$. Graph an open circle at $(0, 1)$, which is found by substituting 0 for x in $f(x) = x + 1$.

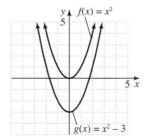

3. $f(x) = x^2$ and $g(x) = x^2 - 3$

The graph of $g(x) = x^2 - 3$ is the graph of $f(x) = x^2$ moved downward 3 units.

4. $f(x) = \sqrt{x}$ and $g(x) = \sqrt{x} + 1$

The graph of $g(x) = \sqrt{x} + 1$ is the graph of $f(x) = \sqrt{x}$ moved upward 1 unit.

5. $f(x) = |x|$ and $g(x) = |x-3|$

x	$f(x)$	$g(x)$
-2	2	5
-1	1	4
0	0	3
1	1	2
2	2	1
3	3	0
4	4	1
5	5	2

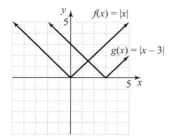

6. $f(x) = |x|$ and $g(x) = |x-2| + 3$

The graph of $g(x)$ is the same as the graph of $f(x)$ shifted 2 units to the right and 3 units up.

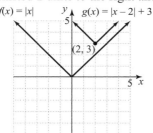

7. $h(x) = -(x+2)^2 - 1$

The graph of $h(x) = -(x+2)^2 - 1$ is the same as the graph of $f(x) = x^2$ reflected about the *x*-axis, then moved 2 units to the left and 1 unit downward.

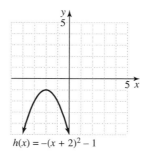

Vocabulary, Readiness & Video Check 2.7

1. The graph that corresponds to $y = \sqrt{x}$ is C.

2. The graph that corresponds to $y = x^2$ is B.

3. The graph that corresponds to $y = x$ is D.

4. The graph that corresponds to $y = |x|$ is A.

5. Although $f(x) = x + 3$ isn't defined for $x = -1$, we need to clearly indicate the point where this piece of the graph ends. Therefore, we find this point and graph it as an open circle.

6. Once you know the shapes, use the shifting rules to tell you where to move the vertex or starting point for each graph, then you can easily draw in the appropriate basic shape.

7. The graph of $f(x) = -\sqrt{x+6}$ has the same shape as the graph of $f(x) = \sqrt{x+6}$ but it is reflected about the <u>x-axis</u>.

Exercise Set 2.7

1. $f(x) = \begin{cases} 2x & \text{if } x < 0 \\ x+1 & \text{if } x \geq 0 \end{cases}$

For $x < 0$:   For $x \geq 0$:

x	$f(x)$
-3	-6
-2	-4
-1	-2

x	$f(x)$
0	1
1	2
2	3

Graph a closed circle at (0, 1). Graph an open circle at (0, 0), which is found by substituting 0 for *x* in $f(x) = 2x$.

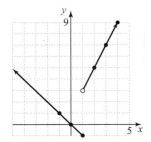

3. $f(x) = \begin{cases} 4x+5 & \text{if } x \le 0 \\ \frac{1}{4}x+2 & \text{if } x > 0 \end{cases}$

For $x \le 0$:   For $x > 0$:

x	f(x)
−2	−3
−1	1
0	5

x	f(x)
1	$2\frac{1}{4}$
2	$2\frac{1}{2}$
4	3

Graph a closed circle at (0, 5). Graph an open circle at (0, 2), which is found by substituting 0 for x in $f(x) = \frac{1}{4}x+2$.

5. $g(x) = \begin{cases} -x & \text{if } x \le 1 \\ 2x+1 & \text{if } x > 1 \end{cases}$

For $x \le 1$:   For $x > 1$:

x	g(x)
−1	1
0	0
1	−1

x	g(x)
2	5
3	7
4	9

Graph a closed circle at (1, −1). Graph an open circle at (1, 3), which is found by substituting 1 for x in $g(x) = 2x+1$.

7. $f(x) = \begin{cases} 5 & \text{if } x < -2 \\ 3 & \text{if } x \ge -2 \end{cases}$

For $x < -2$:   For $x \ge -2$:

x	f(x)
−5	5
−4	5
−3	5

x	f(x)
−2	3
−1	3
0	3

Graph a closed circle at (−2, 3). Graph an open circle at (−2, 5), which is found by substituting −2 for x in $f(x) = 5$.

9. $f(x) = \begin{cases} -2x & \text{if } x \le 0 \\ 2x+1 & \text{if } x > 0 \end{cases}$

For $x \le 0$:   For $x > 0$:

x	f(x)
−1	2
0	0

x	f(x)
1	3
2	5

Graph a closed circle at (0, 0). Graph an open circle at (0, 1), which is found by substituting 0 for x in $f(x) = 2x+1$.

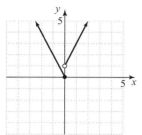

The function is defined for all real numbers, so the domain is all real numbers. The function takes on all y-values greater than or equal to 0, so the range is $\{y|y \geq 0\}$.

11. $h(x) = \begin{cases} 5x-5 & \text{if } x < 2 \\ -x+3 & \text{if } x \geq 2 \end{cases}$

For $x < 2$:

x	$h(x)$
0	-5
1	0

For $x \geq 2$:

x	$h(x)$
2	1
3	0

Graph a closed circle at $(2, 1)$. Graph an open circle at $(2, 5)$, which is found by substituting 2 for x in $h(x) = 5x - 5$.

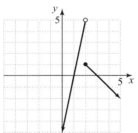

The function is defined for all real numbers, so the domain is all real numbers. The function takes on all y-values less than 5, so the range is $\{y|y < 5\}$.

13. $f(x) = \begin{cases} x+3 & \text{if } x < -1 \\ -2x+4 & \text{if } x \geq -1 \end{cases}$

For $x < -1$:

x	$f(x)$
-4	-1
-3	0
-2	1

For $x \geq -1$:

x	$f(x)$
-1	6
0	4
1	2

Graph a closed circle at $(-1, 6)$. Graph an open circle at $(-1, 2)$, which is found by substituting -1 for x in $f(x) = x + 3$.

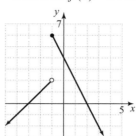

The function is defined for all real numbers, so the domain is all real numbers. The function takes on all y-values less than or equal to 6, so the range is $\{y|y \leq 6\}$.

15. $g(x) = \begin{cases} -2 & \text{if } x \leq 0 \\ -4 & \text{if } x \geq 1 \end{cases}$

For $x \leq 0$:

x	$g(x)$
-2	-2
-1	-2
0	-2

For $x \geq 1$:

x	$g(x)$
1	-4
2	-4
3	-4

Graph closed circles at $(0, -2)$ and $(1, -4)$.

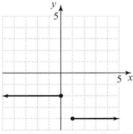

The function is defined for $x \leq 0$ or $x \geq 1$, so the domain is $\{x|x \leq 0 \text{ or } x \geq 1\}$. The function takes on two y-values, -2 and -4, so the range is $\{-2, -4\}$.

17. $f(x) = |x| + 3$ is graph A.

19. $f(x) = \sqrt{x} - 2$ is graph B.

21. $f(x) = |x - 4|$ is graph A.

23. $f(x) = \sqrt{x+2}$ is graph C.

25. $y = (x-4)^2$

The graph of $y = (x-4)^2$ is the same as the graph of $y = x^2$ shifted right 4 units.

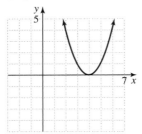

27. $f(x) = x^2 + 4$

The graph of $f(x) = x^2 + 4$ is the same as the graph of $y = x^2$ shifted up 4 units.

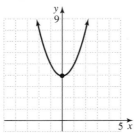

29. $f(x) = \sqrt{x-2} + 3$

The graph of $f(x) = \sqrt{x-2} + 3$ is the same as the graph of $y = \sqrt{x}$ shifted right 2 units and up 3 units.

31. $f(x) = |x-1| + 5$

The graph of $f(x) = |x-1| + 5$ is the same as the graph of $y = |x|$ shifted right 1 unit and up 5 units.

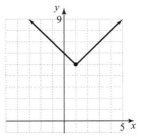

33. $f(x) = \sqrt{x+1} + 1$

The graph of $f(x) = \sqrt{x+1} + 1$ is the same as the graph of $y = \sqrt{x}$ shifted left 1 unit and up 1 unit.

35. $f(x) = |x+3| - 1$

The graph of $f(x) = |x+3| - 1$ is the same as the graph of $y = |x|$ shifted left 3 units and down 1 unit.

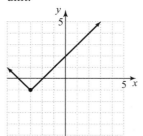

37. $g(x) = (x-1)^2 - 1$

The graph of $g(x) = (x-1)^2 - 1$ is the same as

the graph of $y = x^2$ shifted right 1 unit and

down 1 unit.

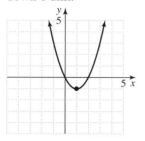

39. $f(x) = (x+3)^2 - 2$

The graph of $f(x) = (x+3)^2 - 2$ is the same as

the graph of $y = x^2$ shifted left 3 units and down

2 units.

41. $f(x) = -(x-1)^2$

The graph of $f(x) = -(x-1)^2$ is the same as the

graph of $y = x^2$ reflected about the *x*-axis and

then shifted right 1 unit.

43. $h(x) = -\sqrt{x} + 3$

The graph of $h(x) = -\sqrt{x} + 3$ is the same as the

graph of $y = \sqrt{x}$ reflected about the *x*-axis and

then shifted up 3 units.

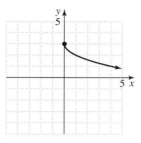

45. $h(x) = -|x+2| + 3$

The graph of $h(x) = -|x+2| + 3$ is the same as

the graph of $y = |x|$ reflected about the

x-axis and then shifted left 2 units and up 3 units.

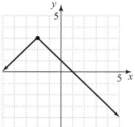

47. $f(x) = (x-3) + 2$

Since the function can be simplified to

$f(x) = x - 1$, we see that its graph is a line with

slope $m = 1$ and *y*-intercept $(0, -1)$.

49. The graph of $y = -1$ is a horizontal line with *y*-intercept $(0, -1)$. The correct graph is A.

51. The graph of $x = 3$ is a vertical line with *x*-intercept $(3, 0)$. The correct graph is D.

53. answers may vary

55. $f(x) = \begin{cases} -\frac{1}{2}x & \text{if } x \leq 0 \\ x+1 & \text{if } 0 < x \leq 2 \\ 2x-1 & \text{if } x > 2 \end{cases}$

Some points for $x \leq 0$: $(-4, 2)$, $(-2, 1)$, $(0, 0)$
Closed dot at $(0, 0)$
Some points for $0 < x \leq 2$: $(1, 2)$, $(2, 3)$

Open dot at (0, 1), closed dot at (2, 3)
Some points for $x > 2$: (3, 5), (4, 7)
There would be an open dot at (2, 3) except that
it gets filled by the middle piece of the graph.

57. $f(x) = \sqrt{x-2} + 3$

The function is defined when $x - 2 \geq 0$, or
$x \geq 2$, so the domain is $\{x | x \geq 2\}$. The function
takes on all y-values greater than or equal to 3,
so the range is $\{y | y \geq 3\}$.

59. $h(x) = -|x+2| + 3$

The function is defined for all real numbers, so
the domain is all real numbers. The function
takes on all y-values less than or equal to 3, so
the range is $\{y | y \leq 3\}$.

61. $f(x) = 5\sqrt{x-20} + 1$

The function is defined when $x - 20 \geq 0$, or
$x \geq 20$, so the domain is $\{x | x \geq 20\}$.

63. $h(x) = 5|x-20| + 1$

The function is defined for all real numbers, so
the domain is all real numbers.

65. $g(x) = 9 - \sqrt{x+103}$

The function is defined when $x + 103 \geq 0$, or
$x \geq -103$, so the domain is $\{x | x \geq 103\}$.

67. $f(x) = \begin{cases} |x| & \text{if } x \leq 0 \\ x^2 & \text{if } x > 0 \end{cases}$

For $x \leq 0$:   For $x > 0$:

x	$f(x)$
−2	2
−1	1
0	0

x	$f(x)$
1	1
2	4
3	9

Graph a closed circle at (0, 0). The graph of
$f(x) = x^2$ for $x > 0$ also approaches the point
(0, 0).

The function is defined for all real numbers, so
the domain is all real numbers. The function
takes on all y-values greater than or equal to 0,
so the range is $\{y | y \geq 0\}$.

69. $g(x) = \begin{cases} |x-2| & \text{if } x < 0 \\ -x^2 & \text{if } x \geq 0 \end{cases}$

For $x < 0$:   For $x \geq 0$:

x	$g(x)$
−3	5
−2	4
−1	3

x	$g(x)$
0	0
1	−1
2	−4

Graph an open circle at (0, 2). Graph a closed
circle at (0, 0).

The function is defined for all real numbers, so
the domain is all real numbers. The function
takes on all y-values such $y > 2$ or $y \leq 0$, so the
range is $\{y | y > 2 \text{ or } y \leq 0\}$.

Chapter 2 Vocabulary Check

1. A <u>relation</u> is a set of ordered pairs.

2. The graph of every linear equation in two
   variables is a <u>line</u>.

3. The equation $y - 8 = -5(x + 1)$ is written in
   <u>point-slope</u> form.

4. Standard form of linear equation in two variables is $Ax + By = C$.

5. The range of a relation is the set of all second components of the ordered pairs of the relation.

6. Parallel lines have the same slope and different y-intercepts.

7. Slope-intercept form of a linear equation in two variables is $y = mx + b$.

8. A function is a relation in which each first component in the ordered pairs corresponds to exactly one second component.

9. In the equation $y = 4x - 2$, the coefficient of x is the slope of its corresponding graph.

10. Two lines are perpendicular if the product of their slopes is -1.

11. To find the x-intercept of a linear equation, let $y = 0$ and solve for the other variable.

12. The domain of a relation is the set of all first components of the ordered pairs of the relation.

13. A linear function is a function that can be written in the form $f(x) = mx + b$.

14. To find the y-intercept of a linear equation, let $x = 0$ and solve for the other variable.

Chapter 2 Review

1. $A(2, -1)$, quadrant IV
$B(-2, 1)$, quadrant II
$C(0, 3)$, y-axis
$D(-3, -5)$, quadrant III

2. $A(-3, 4)$, quadrant II
$B(4, -3)$, quadrant IV
$C(-2, 0)$, x-axis
$D(-4, 1)$, quadrant II

3. $7x - 8y = 56$
$(0, 56)$; No
$7(0) - 8(56) \overset{?}{=} 56$
$\qquad -448 = 56$, False
$(8, 0)$; Yes
$7(8) - 8(0) \overset{?}{=} 56$
$\qquad 56 = 56$, True

4. $-2x + 5y = 10$
$(-5, 0)$; Yes
$-2(-5) + 5(0) \overset{?}{=} 10$
$\qquad\qquad 10 = 10$, True
$(1, 1)$, No
$-2(1) + 5(1) \overset{?}{=} 10$
$\qquad\qquad 3 = 10$, False

5. $x = 13$
$(13, 5)$; Yes
$13 = 13$, True
$(13, 13)$; Yes
$13 = 13$, True

6. $y = 2$
$(7, 2)$; Yes
$2 = 2$, True
$(2, 7)$; No
$7 = 2$, False

7. $y = 3x$; Linear

x	-1	0	1
y	-3	0	3

8. $y = 5x$; Linear

x	-1	0	1
y	-5	0	5

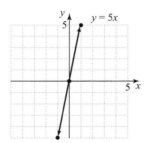

9. $3x - y = 4$; Linear
Find three ordered pair solutions, or find x- and y-intercepts, or find m and b.

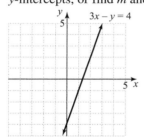

10. $x - 3y = 2$; Linear
Find three ordered pair solutions, or find x- and y-intercepts, or find m and b.

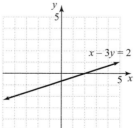

11. $y = |x| + 4$; Nonlinear

x	-3	-2	-1	0	1	2	3
y	7	6	5	4	5	6	7

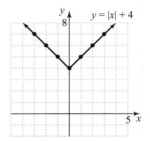

12. $y = x^2 + 4$; Nonlinear

x	-3	-2	-1	0	1	2	3
y	13	8	5	4	5	8	13

13. $y = -\dfrac{1}{2}x + 2$; Linear

Find three ordered pair solutions, or find x- and y-intercepts, or find m and b.

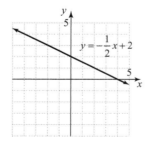

14. $y = -x + 5$; Linear
Find three ordered pair solutions, or find x- and y-intercepts, or find m and b.

15. The graph shown is the graph of $y = x^2 + 2$; d.

16. The graph shown is the graph of $y = x^2 - 4$; a.

17. The graph shown is the graph of $y = |x| + 2$; c.

18. The graph shown is the graph of $y = -|x| + 2$; b.

19. The domain is the set of all first coordinates (or inputs) and the range is the set of all second coordinates (or outputs).

Domain: $\left\{ -\dfrac{1}{2}, 6, 0, 25 \right\}$

Range: $\left\{ \dfrac{3}{4}, -12, 25 \right\}$

Function since each x-value corresponds to exactly one y-value.

20. The domain is the set of all first coordinates (or inputs) and the range is the set of all second coordinates (or outputs).

Domain: $\left\{ \dfrac{3}{4}, -12, 25 \right\}$

Range: $\left\{ -\dfrac{1}{2}, 6, 0, 25 \right\}$

Not a function since $\dfrac{3}{4}$ (or 0.75) is paired with

both $-\dfrac{1}{2}$ and 6.

21. The domain is the set of all first coordinates (or inputs) and the range is the set of all second coordinates (or outputs).
Domain: {2, 4, 6, 8}
Range: {2, 4, 5, 6}
Not a function since 2 is paired with both 2 and 4.

22. The domain is the set of all first coordinates (or inputs) and the range is the set of all second coordinates (or outputs).
Domain:
{Triangle, Square, Rectangle, Parallelogram}
Range: {3, 4}
Function since each input is paired with exactly one output.

23. Domain: all real numbers
Range: $\{y | y \leq -1 \text{ or } y \geq 1\}$
Not a function since it fails the vertical line test.

24. Domain: $\{-3\}$
Range: all real numbers
Not a function since it fails the vertical line test.

25. Domain: all real numbers
Range: $\{4\}$
Function since it passes the vertical line test.

26. Domain: $\{x | -1 \leq x \leq 1\}$
Range: $\{y | -1 \leq y \leq 1\}$
Not a function since it fails the vertical line test.

27. $f(x) = x - 5$
$f(2) = (2) - 5 = -3$

28. $g(x) = -3x$
$g(0) = -3(0) = 0$

29. $g(x) = -3x$
$g(-6) = -3(-6) = 18$

30. $h(x) = 2x^2 - 6x + 1$
$h(-1) = 2(-1)^2 - 6(-1) + 1$
$ = 2(1) + 6 + 1$
$ = 9$

31. $h(x) = 2x^2 - 6x + 1$
$h(1) = 2(1)^2 - 6(1) + 1 = 2 - 6 + 1 = -3$

32. $f(x) = x - 5$
$f(5) = (5) - 5 = 0$

33. $J(x) = 2.54x$
$J(150) = 2.54(150) = 381$ pounds

34. $J(x) = 2.54x$
$J(2000) = 2.54(2000) = 5080$ pounds

35. The point $(-1, 0)$ is on the graph, so $f(-1) = 0$.

36. The point (1, –2) is on the graph, so $f(1) = -2$.

37. $f(x) = 1$
$f(-2) = f(4) = 1$
$x = -2, 4$

38. $f(x) = -1$
$f(0) = f(2) = -1$
$x = 0, 2$

39. $f(x) = \dfrac{1}{5}x$ or $y = \dfrac{1}{5}x$

$m = \dfrac{1}{5}, \ b = 0$

40. $f(x) = -\dfrac{1}{3}x$ or $y = -\dfrac{1}{3}x$

$m = -\dfrac{1}{3}, \ b = 0$

41. $g(x) = -2x + 3$ or $y = -2x + 3$
$m = -2, b = 3$

42. $g(x) = 4x - 1$ or $y = 4x - 1$
$m = 4, \ b = -1$

43. $f(x) = 3x + 1$
The *y*-intercept should be (0, 1). The correct graph is C.

44. $f(x) = 3x - 2$
The *y*-intercept should be (0, –2). The correct graph is A.

45. $f(x) = 3x + 2$
The *y*-intercept should be (0, 2). The correct graph is B.

46. $f(x) = 3x - 5$
The *y*-intercept should be (0, –5). The correct graph is D.

47. $4x + 5y = 20$

Let $x = 0$ Let $y = 0$

$4(0) + 5y = 20$ $4x + 5(0) = 20$
$\quad\quad\quad y = 4$ $\quad\quad x = 5$
$(0, 4)$ $(5, 0)$

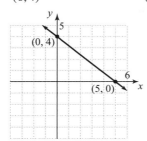

48. $3x - 2y = -9$

Let $x = 0$

$3(0) - 2y = -9$

$y = \dfrac{9}{2}$

$\left(0, \dfrac{9}{2}\right)$

Let $y = 0$

$3x - 2(0) = -9$

$x = -3$

$(-3, 0)$

49. $4x - y = 3$

Let $x = 0$

$4(0) - y = 3$

$y = -3$

$(0, -3)$

Let $y = 0$

$4x - (0) = 3$

$x = \dfrac{3}{4}$

$\left(\dfrac{3}{4}, 0\right)$

50. $2x + 6y = 9$

Let $x = 0$

$2(0) + 6y = 9$

$y = \dfrac{3}{2}$

$\left(0, \dfrac{3}{2}\right)$

Let $y = 0$

$2x + 6(0) = 9$

$x = \dfrac{9}{2}$

$\left(\dfrac{9}{2}, 0\right)$

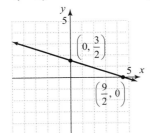

51. $y = 5$

Horizontal line with y-intercept 5.

52. $x = -2$

Vertical line with x-intercept -2.

53. $x - 2 = 0$

$x = 2$

54. $y + 3 = 0$

$y = -3$

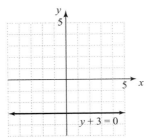

55. $m = \dfrac{-4 - 8}{6 - 2} = \dfrac{-12}{4} = -3$

56. $m = \dfrac{13 - 9}{5 - (-3)} = \dfrac{4}{8} = \dfrac{1}{2}$

57. $m = \dfrac{6-(-4)}{-3-(-7)} = \dfrac{10}{4} = \dfrac{5}{2}$

58. $m = \dfrac{7-(-2)}{-5-7} = \dfrac{9}{-12} = -\dfrac{3}{4}$

59. $f(x) = -3x + \dfrac{1}{2}$ or $y = -3x + \dfrac{1}{2}$

$m = -3$, $b = \dfrac{1}{2}$, y-intercept $\left(0, \dfrac{1}{2}\right)$

60. $g(x) = 2x + 4$ or $y = 2x + 4$
$m = 2$, $b = 4$, y-intercept $(0, 4)$

61. $6x - 15y = 20$

$-15y = -6x + 20$

$y = \dfrac{2}{5}x - \dfrac{4}{3}$

$m = \dfrac{2}{5}$, $b = -\dfrac{4}{3}$, y-intercept $\left(0, -\dfrac{4}{3}\right)$

62. $4x + 14y = 21$

$14y = -4x + 21$

$y = -\dfrac{2}{7}x + \dfrac{3}{2}$

$m = -\dfrac{2}{7}$, $b = \dfrac{3}{2}$, y-intercept $\left(0, \dfrac{3}{2}\right)$

63. $y - 3 = 0$

$y = 3$; Slope $= 0$

64. $x = -5$; Vertical line
Slope is undefined.

65. The slope of l_1 is negative, and the slope of l_2 is positive. Since a positive number is greater than any negative number, l_2 has the greater slope.

66. The slope of l_1 is 0, and the slope of l_2 is positive. Since a positive number is greater than 0, l_2 has the greater slope.

67. The slope of l_1 and the slope of l_2 are both positive. Since l_2 is steeper, it has the greater slope.

68. The slope of l_1 is 0, and the slope of l_2 is negative. Since a negative number is less than 0, l_1 has the greater slope.

69. $C = 0.3x + 42$

a. Let $x = 150$.
$C(150) = 0.3(150) + 42 = 87$
It would cost $87 to rent the minivan.

b. The slope is 0.3 which means the cost of renting increases by $0.30 for each additional mile driven.

c. The y-intercept is $(0, 42)$. The cost of renting for 0 miles driven is $42.

70. $C = 0.4x + 19$

a. Let $x = 325$.
$C = 0.4(325) + 19 = 149$
It will cost $149 to rent the car.

b. The slope is 0.4; the cost increases by $0.40 for each additional mile driven.

c. The y-intercept is $(0, 19)$; the cost for 0 miles driven is $19.

71. $f(x) = -2x + 6 \qquad g(x) = 2x - 1$
$m = -2 \qquad\qquad m = 2$
Neither; The slopes are not the same and their product is not -1.

72. $y = \dfrac{3}{4}x + 1 \qquad y = -\dfrac{4}{3}x + 1$

$m = \dfrac{3}{4} \qquad\qquad m = -\dfrac{4}{3}$

The lines are perpendicular since the product of their slopes is -1.

73. $-x + 3y = 2 \qquad\quad 6x - 18y = 3$

$y = \dfrac{1}{3}x + \dfrac{2}{3} \qquad y = \dfrac{1}{3}x - \dfrac{1}{6}$

$m = \dfrac{1}{3} \qquad\qquad m = \dfrac{1}{3}$

Parallel, since their slopes are equal.

74. $x - 2y = 6 \qquad\quad 4x + y = 8$

$x - 6 = 2y \qquad\qquad y = -4x + 8$

$\dfrac{1}{2}x - 3 = y$

$m = \dfrac{1}{2} \qquad\qquad m = -4$

Neither, since their slopes are not equal, nor does their product equal -1.

75.

76.

77. $y = -x + 1$
$m = -1,\ b = 1$

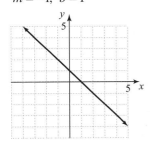

78. $y = 4x - 3$
$m = 4,\ b = -3$

79. $3x - y = 6$
$\quad\quad y = 3x - 6$
$m = 3,\ b = -6$

80. $y = -5x$
$m = -5,\ b = 0$

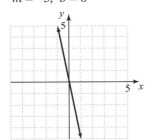

81. Every horizontal line is in the form $y = c$. Since the line passes through the point $(3, -1)$, its equation is $y = -1$.

82. Every vertical line has the form $x = c$. Since the line passes through the point $(-2, -4)$, its equation is $x = -2$.

83. A line parallel to $x = 6$ has the form $x = c$. Since the line passes through $(-4, -3)$, its equation is $x = -4$.

84. Lines with slope 0 are horizontal, and have the form $y = c$. Since it passes through $(2, 5)$, its equation is $y = 5$.

85. $y - y_1 = m(x - x_1)$
$\quad\ y - 5 = 3[x - (-3)]$
$\quad\ y - 5 = 3(x + 3)$
$\quad\ y - 5 = 3x + 9$
$\ 3x - y = -14$

86. $\quad y - y_1 = m(x - x_1)$
$y - (-2) = 2(x - 5)$
$\quad\ y + 2 = 2x - 10$
$\quad 2x - y = 12$

87. $m = \dfrac{-2-(-1)}{-4-(-6)} = \dfrac{-1}{2} = -\dfrac{1}{2}$

$y - y_1 = m(x - x_1)$

$y - (-1) = -\dfrac{1}{2}[x - (-6)]$

$2(y+1) = -(x+6)$

$2y + 2 = -x - 6$

$x + 2y = -8$

88. $m = \dfrac{-8-3}{-4-(-5)} = \dfrac{-11}{1} = -11$

$y - y_1 = m(x - x_1)$

$y - 3 = -11[x - (-5)]$

$y - 3 = -11(x+5)$

$y - 3 = -11x - 55$

$11x + y = -52$

89. $x = 4$ has undefined slope.
A line perpendicular to $x = 4$ has slope = 0 and is therefore horizontal.
$y = 3$

90. $y = 8$ has slope = 0.
A line parallel to $y = 8$ has slope = 0.
$y = -5$

91. $y = mx + b$

$y = -\dfrac{2}{3}x + 4$

$f(x) = -\dfrac{2}{3}x + 4$

92. $y = mx + b$

$y = -x - 2$

$f(x) = -x - 2$

93. $6x + 3y = 5$

$3y = -6x + 5$

$y = -2x + \dfrac{5}{3}$ so $m = -2$

$y - y_1 = m(x - x_1)$

$y - (-6) = -2(x - 2)$

$y + 6 = -2x + 4$

$y = -2x - 2$

$f(x) = -2x - 2$

94. $3x + 2y = 8$

$2y = -3x + 8$

$y = -\dfrac{3}{2}x + 4$ so $m = -\dfrac{3}{2}$

$y - y_1 = m(x - x_1)$

$y - (-2) = -\dfrac{3}{2}[x - (-4)]$

$2(y+2) = -3(x+4)$

$2y + 4 = -3x - 12$

$2y = -3x - 16$

$y = -\dfrac{3}{2}x - 8$

$f(x) = -\dfrac{3}{2}x - 8$

95. $4x + 3y = 5$

$3y = -4x + 5$

$y = -\dfrac{4}{3}x + \dfrac{5}{3}$

so $m = -\dfrac{4}{3}$ and $m_\perp = \dfrac{3}{4}$

$y - y_1 = m(x - x_1)$

$y - (-1) = \dfrac{3}{4}[x - (-6)]$

$4(y+1) = 3(x+6)$

$4y + 4 = 3x + 18$

$4y = 3x + 14$

$y = \dfrac{3}{4}x + \dfrac{7}{2}$

$f(x) = \dfrac{3}{4}x + \dfrac{7}{2}$

96. $2x - 3y = 6$

$-3y = -2x + 6$

$y = \dfrac{2}{3}x - 2$

so $m = \dfrac{2}{3}$ and $m_\perp = -\dfrac{3}{2}$

$y - y_1 = m(x - x_1)$

$y - 5 = -\dfrac{3}{2}[x - (-4)]$

$2(y-5) = -3(x+4)$

$2y - 10 = -3x - 12$

$2y = -3x - 2$

$y = -\dfrac{3}{2}x - 1$

$f(x) = -\dfrac{3}{2}x - 1$

97. a. Use the points (7, 210,000) and
(12, 270,000).
$$m = \frac{270,000 - 210,000}{12 - 7} = \frac{60,000}{5} = 12,000$$
$$y - y_1 = m(x - x_1)$$
$$y - 210,000 = 12,000(x - 7)$$
$$y - 210,000 = 12,000x - 84,000$$
$$y = 12,000x + 126,000$$

b. Let $x = 18$ since 2018 is 18 years after 2000.
$$y = 12,000x + 126,000$$
$$y = 12,000(18) + 126,000$$
$$= 216,000 + 126,000$$
$$= 342,000$$
In 2018, the value of the building is estimated to be $342,000.

98. Use the points (2, 17,500) and (4, 14,300).

a. $$m = \frac{14,300 - 17,500}{4 - 2} = \frac{-3200}{2} = -1600$$
$$y - y_1 = m(x - x_1)$$
$$y - 17,500 = -1600(x - 2)$$
$$y - 17,500 = -1600x + 3200$$
$$y = -1600x + 20,700$$

b. Let $x = 6$, since 2012 is 6 years after 2006.
$$y = -1600x + 20,700$$
$$y = -1600(6) + 20,700$$
$$= -9600 + 20,700$$
$$= 11,100$$
In 2012 the value of the automobile was $11,100.

99. a.

b. The number of U.S. student Internet users is increasing by about 0.309 million per year.

c.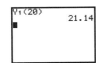

The number of U.S. college students using the Internet is expected to be 21.14 million or 21,140,000 in 2020.

100. a.

The regression equation is
$y = 5.831x + 147.451$.

b. The number of college students worldwide using the Internet is increasing by about 5.831 million per year.

c.

The number of college students worldwide using the Internet is expected to be 264.08 million or 264,080,000 in 2020.

101. $g(x) = \begin{cases} -\dfrac{1}{5}x & \text{if } x \le -1 \\ -4x + 2 & \text{if } x > -1 \end{cases}$

For $x \le -1$:　　　For $x > -1$:

x	$g(x)$
-5	1
-3	$\frac{3}{5}$
-1	$\frac{1}{5}$

x	$g(x)$
0	2
1	-2
2	-6

Graph a closed circle at $\left(-1, \dfrac{1}{5}\right)$. Graph an open circle at (–1, 6), which is found by substituting –1 for x in $g(x) = -4x + 2$.

102. $f(x) = \begin{cases} -3x & \text{if } x < 0 \\ x-3 & \text{if } x \geq 0 \end{cases}$

For $x < 0$:   For $x \geq 0$:

x	$f(x)$
-3	9
-2	6
-1	3

x	$f(x)$
0	-3
1	-2
2	-1

Graph a closed circle at $(0, -3)$. Graph an open circle at $(0, 0)$, which is found by substituting 0 for x in $f(x) = -3x$.

103. $f(x) = \sqrt{x-4}$

The graph of $f(x) = \sqrt{x-4}$ is the same as the graph of $y = \sqrt{x}$ shifted right 4 units.

104. $y = \sqrt{x} - 4$

The graph of $f(x) = \sqrt{x} - 4$ is the same as the graph of $y = \sqrt{x}$ shifted down 4 units.

105. $h(x) = -(x+3)^2 - 1$

The graph of $h(x) = -(x+3)^2 - 1$ is the same as the graph of $y = x^2$ reflected about the x-axis and then shifted left 3 units and down 1 unit.

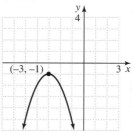

106. $g(x) = |x-2| - 2$

The graph of $g(x) = |x-2| - 2$ is the same as the graph of $y = |x|$ shifted right 2 units and down 2 units.

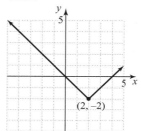

107. $x = -4y$ or $y = -\dfrac{1}{4}x$

The slope is $-\dfrac{1}{4}$, and the y-intercept is $(0, 0)$.

108. $3x - 2y = -9$

Let $x = 0$.
$$3x - 2y = -9$$
$$3(0) - 2y = -9$$
$$-2y = -9$$
$$y = \frac{9}{2}$$

Let $y = 0$.
$$3x - 2y = -9$$
$$3x - 2(0) = -9$$
$$3x = -9$$
$$x = -3$$

The intercepts are $\left(0, \frac{9}{2}\right)$ and $(-3, 0)$.

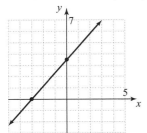

109. Vertical; through $\left(-7, -\frac{1}{2}\right)$

A vertical line has an equation of the form $x = a$, where a is the x-coordinate of any point on the line. The equation is $x = -7$.

110. Slope 0; through $\left(-4, \frac{9}{2}\right)$

A line with slope 0 is horizontal, and a horizontal line has an equation of the form $y = b$, where b is the y-coordinate of any point on the line. The equation is $y = \frac{9}{2}$.

111. Slope $\frac{3}{4}$; through $(-8, -4)$

$$y - y_1 = m(x - x_1)$$
$$y - (-4) = \frac{3}{4}(x - (-8))$$
$$y + 4 = \frac{3}{4}(x + 8)$$
$$4(y + 4) = 3(x + 8)$$
$$4y + 16 = 3x + 24$$
$$4y = 3x + 8$$
$$y = \frac{3}{4}x + 2$$

112. Through $(-3, 8)$ and $(-2, 3)$
Find the slope.
$$m = \frac{3 - 8}{-2 - (-3)} = \frac{-5}{1} = -5$$

Use the slope and one of the points in the point-slope form. We use $(-2, 3)$.
$$y - y_1 = m(x - x_1)$$
$$y - 3 = -5(x - (-2))$$
$$y - 3 = -5(x + 2)$$
$$y - 3 = -5x - 10$$
$$y = -5x - 7$$

113. Through $(-6, 1)$; parallel to $y = -\frac{3}{2}x + 11$

The slope of a line parallel to $y = -\frac{3}{2}x + 11$ will have the same slope, $-\frac{3}{2}$.
$$y - y_1 = m(x - x_1)$$
$$y - 1 = -\frac{3}{2}(x - (-6))$$
$$y - 1 = -\frac{3}{2}(x + 6)$$
$$2(y - 1) = -3(x + 6)$$
$$2y - 2 = -3x - 18$$
$$2y = -3x - 16$$
$$y = -\frac{3}{2}x - 8$$

114. Through $(-5, 7)$; perpendicular to $5x - 4y = 10$
Find the slope of $5x - 4y = 10$.
$$5x - 4y = 10$$
$$-4y = -5x + 10$$
$$y = \frac{5}{4}x - \frac{5}{2}$$

The slope is $\frac{5}{4}$. The slope of any line perpendicular to this line is the negative reciprocal of $\frac{5}{4}$, or $-\frac{4}{5}$.
$$y - y_1 = m(x - x_1)$$
$$y - 7 = -\frac{4}{5}(x - (-5))$$
$$y - 7 = -\frac{4}{5}(x + 5)$$
$$5(y - 7) = -4(x + 5)$$
$$5y - 35 = -4x - 20$$
$$5y = -4x + 15$$
$$y = -\frac{4}{5}x + 3$$

115. $f(x) = \begin{cases} x - 2 & \text{if } x \le 0 \\ -\frac{x}{3} & \text{if } x \ge 3 \end{cases}$

For $x \le 0$:   For $x \ge 3$:

x	$f(x)$
-2	-4
-1	-3
0	-2

x	$f(x)$
3	-1
4	$-\frac{4}{3}$
6	-2

Graph closed circles at (0, –2) and (3, –1).

116. $g(x) = \begin{cases} 4x - 3 & \text{if } x \leq 1 \\ 2x & \text{if } x > 1 \end{cases}$

For $x \leq 1$:

x	$g(x)$
–1	–7
0	–3
1	1

For $x > 1$:

x	$g(x)$
2	4
3	6
4	8

Graph a closed circle at (1, 1). Graph an open circle at (1, 2), which is found by substituting 1 for x in $g(x) = 2x$.

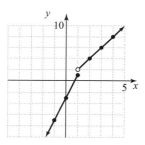

117. $f(x) = \sqrt{x - 2}$

The graph of $f(x) = \sqrt{x - 2}$ is the same as the graph of $y = \sqrt{x}$ shifted right 2 units.

118. $f(x) = |x + 1| - 3$

The graph of $f(x) = |x + 1| - 3$ is the same as the graph of $y = |x|$ shifted left 1 unit and down 3 units.

Chapter 2 Test

1.

A is in quadrant IV.
B is on the x-axis, no quadrant.
C is in quadrant II.

2. $2x - 3y = -6$

$-3y = -2x - 6$

$y = \dfrac{2}{3}x + 2$

$m = \dfrac{2}{3}$, $b = 2$, y-intercept (0, 2)

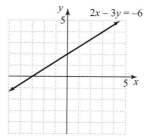

3. $4x + 6y = 7$

$6y = -4x + 7$

$y = -\dfrac{2}{3}x + \dfrac{7}{6}$

$m = -\dfrac{2}{3}$, $b = \dfrac{7}{6}$

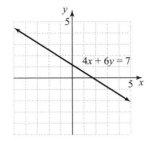

4. $f(x) = \dfrac{2}{3}x$ or $y = \dfrac{2}{3}x$

5. $y = -3$

Horizontal line with y-intercept at -3.

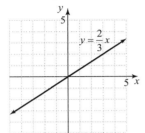

6. $m = \dfrac{10-(-8)}{-7-5} = \dfrac{18}{-12} = -\dfrac{3}{2}$

7. $3x + 12y = 8$

$$12y = -3x + 8$$
$$y = -\dfrac{1}{4}x + \dfrac{2}{3}$$

$m = -\dfrac{1}{4}$, $b = \dfrac{2}{3}$, so y-intercept is $\left(0, \dfrac{2}{3}\right)$.

8. Horizontal; through $(2, -8)$
A horizontal line has an equation of the form $y = b$, where b is the y-coordinate of any point on the line. The equation is $y = -8$.

9. Vertical; through $(-4, -3)$
A vertical line has an equation of the form $x = a$, where a is the x-coordinate of any point on the line. The equation is $x = -4$.

10. Perpendicular to $x = 5$; through $(3, -2)$
The line $x = 5$ is vertical, so any line perpendicular to it is horizontal. A horizontal line has an equation of the form $y = b$, where b is the y-coordinate of any point on the line. The equation is $y = -2$.

11.
$$y - y_1 = m(x - x_1)$$
$$y - (-1) = -3(x - 4)$$
$$y + 1 = -3x + 12$$
$$3x + y = 11$$

12.
$$y - y_1 = m(x - x_1)$$
$$y - (-2) = 5(x - 0)$$
$$y + 2 = 5x$$
$$5x - y = 2$$

13. $m = \dfrac{-3-(-2)}{6-4} = \dfrac{-1}{2} = -\dfrac{1}{2}$
$$y - y_1 = m(x - x_1)$$
$$y - (-2) = -\dfrac{1}{2}(x - 4)$$
$$2(y + 2) = -(x - 4)$$
$$2y + 4 = -x + 4$$
$$2y = -x$$
$$y = -\dfrac{1}{2}x$$
$$f(x) = -\dfrac{1}{2}x$$

14. $3x - y = 4$
$$y = 3x - 4$$
$$m = 3 \text{ so } m_\perp = -\dfrac{1}{3}$$
$$y - y_1 = m(x - x_1)$$
$$y - 2 = -\dfrac{1}{3}[x - (-1)]$$
$$3(y - 2) = -(x + 1)$$
$$3y - 6 = -x - 1$$
$$3y = -x + 5$$
$$y = -\dfrac{1}{3}x + \dfrac{5}{3}$$
$$f(x) = -\dfrac{1}{3}x + \dfrac{5}{3}$$

15. $2y + x = 3$

$2y = -x + 3$

$y = -\dfrac{1}{2}x + 3$ so $m = -\dfrac{1}{2}$

$y - y_1 = m(x - x_1)$

$y - (-2) = -\dfrac{1}{2}(x - 3)$

$2(y + 2) = -(x - 3)$

$2y + 4 = -x + 3$

$2y = -x - 1$

$y = -\dfrac{1}{2}x - \dfrac{1}{2}$

$f(x) = -\dfrac{1}{2}x - \dfrac{1}{2}$

16. $2x - 5y = 8$

$-5y = -2x + 8$

$y = \dfrac{2}{5}x - \dfrac{8}{5}$ so $m_1 = \dfrac{2}{5}$

$m_2 = \dfrac{-1-4}{-1-1} = \dfrac{-5}{-2} = \dfrac{5}{2}$

Therefore, lines L_1 and L_2 are neither parallel nor perpendicular since their slopes are not equal and the product of their slopes is not -1.

17. The graph shown is a parabola. It is the graph of $y = x^2 + 2x + 3$; B.

18. The graph shown is V-shaped. It is the graph of $y = 2|x - 1| + 3$; A.

19. The graph shown is linear. It is the graph of $y = 2x + 3$; D.

20. The graph shown is cubic. It is the graph of $y = 2(x - 1)^3 + 3$; C.

21. Domain: all real numbers
Range: {5}
Function since it passes the vertical line test.

22. Domain: {−2}
Range: all real numbers
Not a function since it fails the vertical line test.

23. Domain: all real numbers
Range: $\{y | y \geq 0\}$
Function since it passes the vertical line test.

24. Domain: all real numbers
Range: all real numbers
Function since it passes the vertical line test.

25. $y = 0.096x + 72.81$

a. $y = 0.096x + 72.81$
Let $x = 90$.
$y = 0.096(90) + 72.81$
$y = 81.45 \approx 81$
81 games would be won in the 2009 season.

b. Let $x = 67$.
$y = 0.096(67) + 72.81$
$y = 79.242 \approx 79$
79 games would be won in the 2009 season.

c. Let $y = 95$ and solve for x.
$95 = 0.096x + 72.81$
$22.19 = 0.096x$
$231 \approx x$
A payroll of \$231 million would be necessary.

d. The slope is 0.096. Every million dollars spent on payroll increases winnings by 0.096 game.

26. $f(x) = \begin{cases} -\dfrac{1}{2}x & \text{if } x \leq 0 \\ 2x - 3 & \text{if } x > 0 \end{cases}$

For $x \leq 0$:   For $x > 0$:

x	$f(x)$
−4	2
−2	1
0	0

x	$f(x)$
1	−1
2	1
3	3

Graph a closed circle at (0, 0). Graph an open circle at (0, −3), which is found by substituting 0 for x in $f(x) = 2x - 3$.

The domain is all real numbers.
The range is $\{y | y > -3\}$.

27. $f(x) = (x-4)^2$

The graph of $f(x) = (x-4)^2$ is the same as the

graph of $y = x^2$ shifted right 4 units.

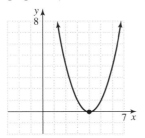

28. $g(x) = |x| + 2$

The graph of $g(x) = |x| + 2$ is the same as the
graph of $y = |x|$ shifted up 2 units.

29. $g(x) = -|x+2| - 1$

The graph of $g(x) = -|x+2| - 1$ is the same as

the graph of $y = |x|$ reflected about the

x-axis and then shifted left 2 units and down
1 unit.

The domain is all real numbers.
The range is $\{y | y \leq -1\}$.

30. $h(x) = \sqrt{x} - 1$

The graph of $h(x) = \sqrt{x} - 1$ is the same as the

graph of $y = \sqrt{x}$ shifted down 1 unit.

31. a.

b.

The regression equation is
$y = 295.07x + 6800.4$.

c.

The population of Florida is predicted to be
21,553.9 thousand, or 21,553,900, in 2020.

Chapter 2 Cumulative Review

1. $3x - y = 3(15) - (4) = 45 - 4 = 41$

2. a. $-4 + (-3) = -7$

b. $\dfrac{1}{2} - \left(-\dfrac{1}{3}\right) = \dfrac{3}{6} + \dfrac{2}{6} = \dfrac{5}{6}$

c. $7 - 20 = -13$

3. a. True, 3 is a real number

b. False, $\dfrac{1}{5}$ is not an irrational number.

c. False, every rational number is not an

integer, for example, $\dfrac{2}{3}$.

d. False, since 1 is not in the second set.

4. a. The opposite of -7 is 7.

b. The opposite of 0 is 0.

c. The opposite of $\dfrac{1}{4}$ is $-\dfrac{1}{4}$.

5. a. $2 - 8 = -6$

b. $-8 - (-1) = -8 + 1 = -7$

c. $-11 - 5 = -16$

d. $10.7 - (-9.8) = 10.7 + 9.8 = 20.5$

e. $\dfrac{2}{3} - \dfrac{1}{2} = \dfrac{4}{6} - \dfrac{3}{6} = \dfrac{1}{6}$

f. $1 - 0.06 = 0.94$

g. $4 - 7 = -3$

6. a. $\dfrac{-42}{-6} = 7$

b. $\dfrac{0}{14} = 0$

c. $-1(-5)(-2) = 5(-2) = -10$

7. a. $3^2 = 3 \cdot 3 = 9$

b. $\left(\dfrac{1}{2}\right)^4 = \dfrac{1}{2^4} = \dfrac{1}{16}$

c. $-5^2 = -(5 \cdot 5) = -25$

d. $(-5)^2 = (-5)(-5) = 25$

e. $-5^3 = -(5 \cdot 5 \cdot 5) = -125$

f. $(-5)^3 = (-5)(-5)(-5) = -125$

8. a. Distributive Property

b. Commutative Property of Addition

9. a. $-1 > -2$ since -1 is to the right of -2 on the number line.

b. $\dfrac{12}{4} = 3$

c. $-5 < 0$ since -5 is to the left of 0 on the number line.

d. $-3.5 \le -3.05$ since -3.5 is to the left of -3.05 on the number line.

e. $\dfrac{5}{8} > \dfrac{3}{8}$

The denominators are the same, so $\dfrac{5}{8} > \dfrac{3}{8}$ since $5 > 3$.

f. $\dfrac{2}{3} < \dfrac{3}{4}$

By dividing, we see that $\dfrac{3}{4} = 0.75$ and $\dfrac{2}{3} = 0.666....$ Thus, $\dfrac{2}{3} < \dfrac{3}{4}$ since $0.666... < 0.75$.

10. $2x^2$

a. $2(7)^2 = 2(49) = 98$

b. $2(-7)^2 = 2(49) = 98$

11. a. The reciprocal of 11 is $\dfrac{1}{11}$.

b. The reciprocal of -9 is $-\dfrac{1}{9}$.

c. The reciprocal of $\dfrac{7}{4}$ is $\dfrac{4}{7}$.

12. $-2 + 3[5 - (7 - 10)] = -2 + 3[5 - (-3)]$
$= -2 + 3(8)$
$= -2 + 24$
$= 22$

13. $0.6 = 2 - 3.5c$
$-1.4 = -3.5c$
$\dfrac{-1.4}{-3.5} = \dfrac{-3.5c}{-3.5}$
$0.4 = c$

14. $2(x-3) = -40$
$$2x - 6 = -40$$
$$2x = -34$$
$$x = -17$$

15. $3x + 5 = 3(x + 2)$
$$3x + 5 = 3x + 6$$
$$5 = 6 \text{ False}$$
The solution is $\varnothing$.

16. $5(x - 7) = 4x - 35 + x$
$$5x - 35 = 5x - 35$$
$$-35 = -35 \text{ True for any number}$$
The solution is all real numbers.

17. a.   If x is the first integer, then the next two consecutive integers are $x + 1$ and $x + 2$. The sum is $x + (x + 1) + (x + 2) = 3x + 3$.

     b.   The perimeter is found by adding the lengths of the sides.
$$x + 5x + (6x - 3) = 12x - 3$$

18.   25% of $16 = 0.25(16) = 4$

19.   Let x = the lowest of the scores. Then the other two scores are $x + 2$ and $x + 4$.
$$x + (x + 2) + (x + 4) = 264$$
$$3x + 6 = 264$$
$$3x = 258$$
$$x = 86$$
$$x + 2 = 86 + 2 = 88$$
$$x + 4 = 86 + 4 = 90$$
The scores are 86, 88, and 90.

20.   Let x = first odd integer, then $x + 2$ = next odd integer and $x + 4$ = third odd integer.
$$x + (x + 2) + (x + 4) = 213$$
$$3x + 6 = 213$$
$$3x = 207$$
$$x = 69$$
$$x + 2 = 69 + 2 = 71$$
$$x + 4 = 69 + 4 = 73$$
The integers are 69, 71, and 73.

21. $V = lwh$
$$\frac{V}{lw} = \frac{lwh}{lw}$$
$$\frac{V}{lw} = h$$

22. $7x + 3y = 21$
$$3y = -7x + 21$$
$$y = -\frac{7}{3}x + 7$$

23. a. $(2, -1)$ is in quadrant IV.

     b. $(0, 5)$ is not in any quadrant, it is on the y-axis.

     c. $(-3, 5)$ is in quadrant II.

     d. $(-2, 0)$ is not in any quadrant, it is on the x-axis.

     e. $\left(-\dfrac{1}{2}, -4\right)$ is in quadrant III.

     f. $(1.5, 1.5)$ is in quadrant I.

24. a. $(0, -2)$ is on the y-axis.

     b. $(-1, -2.5)$ is in quadrant III.

     c. $\left(\dfrac{1}{2}, 0\right)$ is on the x-axis.

     d. $(4, -0.5)$ is in quadrant IV.

25. $3x - y = 12$
$$3(0) - (-12) \stackrel{?}{=} 12$$
$$12 = 12 \text{ True}$$
$(0, -12)$ is a solution.
$$3(1) - 9 \stackrel{?}{=} 12$$
$$-6 = 12 \text{ False}$$
$(1, 9)$ is not a solution.
$$3(2) - (-6) \stackrel{?}{=} 12$$
$$6 + 6 \stackrel{?}{=} 12$$
$$12 = 12 \text{ True}$$
$(2, -6)$ is a solution.

26. $7x + 2y = 10$
$$2y = -7x + 10$$
$$y = -\frac{7}{2}x + 5$$
$$m = -\frac{7}{2}, \ y\text{-intercept} = (0, 5)$$

27.   Yes, $y = 2x + 1$ is a function (graph the function and use the vertical line test).

28.   No, it is not a function (by the vertical line test).

29. a. $f(x) = \frac{1}{2}x + \frac{3}{7}$

$y = mx + b$

$b = \frac{3}{7}$

$y\text{-intercept} = \left(0, \frac{3}{7}\right)$

b. $y = -2.5x - 3.2$

$y = mx + b$

$b = -3.2$

$y\text{-intercept} = (0, -3.2)$

30. $m = \frac{y_2 - y_1}{x_2 - x_1} = \frac{9 - 6}{0 - (-1)} = \frac{3}{1} = 3$

31. $f(x) = \frac{2}{3}x + 4$

$y = mx + b$

The slope of the line is m, the coefficient of x,

$\frac{2}{3}$.

32. Vertical; through $\left(-2, -\frac{3}{4}\right)$

A vertical line has an equation of the form $x = a$, where a is the x-coordinate of any point on the line. The equation is $x = -2$.

33. y-intercept = $(0, -3)$ means that $b = -3$. Using the equation $y = mx + b$, we have $y = \frac{1}{4}x - 3$.

34. Horizontal; through $\left(-2, -\frac{3}{4}\right)$

A horizontal line has an equation of the form $y = b$, where b is the y-coordinate of any point on the line. The equation is $y = -\frac{3}{4}$.

Chapter 3

Section 3.1 Practice Exercises

1. $2(x - 1) + 6 = 8$

Algebraic Solution:

$2(x-1)+6 = 8$

$2x-2+6 = 8$

$2x+4 = 8$

$2x = 4$

$x = 2$

Check: $2(x-1)+6 = 8$

$2(2-1)+6 \stackrel{?}{=} 8$

$2(1)+6 \stackrel{?}{=} 8$

$8 = 8$ True

The solution is 2.

Graphical Solution:

Graph $y_1 = 2(x-1)+6$ and $y_2 = 8$. Use a

$[-10, 10, 1]$ by $[-10, 10, 1]$ window.

The x-coordinate of the point of intersection is 2. Thus, the solution is 2.

2. $3(x - 10) + 5(x + 2) - 25 = 9 - x$

Graph $y_1 = 3(x-10)+5(x+2)-25$ and

$y_2 = 9 - x$ in an integer window.

The point of intersection $(6, 3)$ indicates that the solution is 6.

Check:

3. Define $y_1 = 3.51x + 5.728$ and $y_2 = x - 3.41$ and

graph in a standard window.

The solution is approximately -3.641.

4. Two years is $2 \cdot 365 \cdot 24 = 17{,}520$ hours.

Use a $[0, 20{,}000, 1000]$ by $[0, 2000, 100]$ window.

The \$575 refrigerator costs \$1801.40 at the end of 2 years.

The \$825 refrigerator costs \$1525.80 at the end of 2 years.

The \$575 refrigerator costs

\$1801.40 − \$1525.80 = \$275.60 more to run for 2 years.

5.

In 2018, the average annual expenditure for cell phone usage is predicted to be \$1315.29.

In 2018, the average annual expenditure for residential phone usage is predicted to be \$127.93.

6. $5(x - 1) - 3(x + 7) = 2x - 26$

Algebraic Solution:

$5(x-1)-3(x+7) = 2x-26$

$5x-5-3x-21 = 2x-26$

$2x-26 = 2x-26$

Since both sides are the same, all real numbers are solutions.

Graphical Solution:

Graph $y_1 = 5(x-1)-3(x+7)$ and $y_2 = 2x-26$.

By tracing along y_1 and y_2, we see that the lines are identical. The equation is an identity; all real numbers are solutions and the solution set is $\{x \mid x \text{ is a real number}\}$.

7. $2x - 7 = 7(x + 1) - 5(x - 2)$
Algebraic Solution:
$2x - 7 = 7(x + 1) - 5(x - 2)$
$2x - 7 = 7x + 7 - 5x + 10$
$2x - 7 = 2x + 17$
$\quad\quad -7 = 17$
This equation is false no matter what value the variable x has.
Graphical Solution:
Graph $y_1 = 2x - 7$ and $y_2 = 7(x + 1) - 5(x - 2)$.

The graphs appear to be parallel lines. Check the slope of each graph.
$y_1 = 2x - 7$
$y_2 = 7(x + 1) - 5(x - 2) = 2x + 17$
The lines both have slope 2, but are distinct, so they are parallel.
The equation is a contradiction and the solution set is $\{\ \}$ or $\varnothing$.

8. Write the equation so that one side equals 0 by subtracting $2x$ from and adding 1 to both sides.
$$2.5(x - 1) + 3.4 = 2x - 1$$
$2.5(x - 1) + 3.4 - 2x + 1 = 0$
Define $y_1 = 2.5(x - 1) + 3.4 - 2x + 1$ and graph in a standard window. Then use the root or zero feature.

The x-intercept is $(-3.8, 0)$, which means that the solution is -3.8.

Vocabulary, Readiness & Video Check 3.1

1. To solve an equation by the <u>intersection-of-graphs</u> method, graph the left-hand side of the equation in y_1, the right-hand side of the equation in y_2, and find the intersection point of the two graphs.

2. To solve an equation by the <u>x-intercept</u> method, first write the equation as an equivalent equation with one side 0.

3. for the intersection-of-graphs method, the <u>x</u>-coordinate of the point of intersection is the solution of the equation.

4. For the intersection-of-graphs method, the <u>y</u>-coordinate of the point of intersection is the value of each side of the equation when the solution is substituted into the equation.

5. Two other terms for the solution of an equation are <u>root</u> and <u>zero</u>.

6. $4x = 24$
$\dfrac{4x}{4} = \dfrac{24}{4}$
$x = 6$

7. $6x = -12$
$\dfrac{6x}{6} = \dfrac{-12}{6}$
$x = -2$

8. $\quad\quad 2x + 10 = 20$
$2x + 10 - 10 = 20 - 10$
$\quad\quad\quad\ 2x = 10$
$\quad\quad\quad\ \dfrac{2x}{2} = \dfrac{10}{2}$
$\quad\quad\quad\quad x = 5$

9. $\quad\quad 5x + 25 = 30$
$5x + 25 - 25 = 30 - 25$
$\quad\quad\quad\ 5x = 5$
$\quad\quad\quad\ \dfrac{5x}{5} = \dfrac{5}{5}$
$\quad\quad\quad\quad x = 1$

10. $-3x = 0$

$$\frac{-3x}{-3} = \frac{0}{-3}$$

$$x = 0$$

11. $-2x = -14$

$$\frac{-2x}{-2} = \frac{-14}{-2}$$

$$x = 7$$

12. $2x + 3 = 2x - 1$

$$2x - 2x + 3 = 2x - 2x - 1$$

$$3 = -1$$

This statement is false, so the equation has no solution; $\varnothing$.

13. $2(x - 5) = 2x - 10$

$$2x - 10 = 2x - 10$$

Both sides of the equation are the same, so the equation is an identity; all real numbers are solutions.

14. The x-coordinate of a point of intersection is a solution of the equation.

15. The solution of a linear equation is the x-value of the point of intersection of the graph of $y =$ the left side of the equation and the graph of $y =$ the right side of the equation.

16. Graph the cost function in y_1 and the revenue function in y_2. The point of intersection of the graphs is the break-even point.

17. When using the intersection of graphs method, if the same line is graphed twice, the equation is an <u>identity</u>. If the two lines are parallel, the equation is a <u>contradiction</u>.

18. Since the first step of the x-intercept method is to write the equation so that one side is 0, and the graph of $y = 0$ is the x-axis, finding the x-intercepts of the equation with one side equal to 0 is the same as finding the intersection of the graphs of each side of the original equation.

Exercise Set 3.1

1. $5x + 2 = 3x + 6$

Algebraic Solution:

$$5x + 2 = 3x + 6$$

$$2x + 2 = 6$$

$$2x = 4$$

$$x = 2$$

The solution is 2.

Graphical Solution:

Graph $y_1 = 5x + 2$ and $y_2 = 3x + 6$ in a $[-20, 20, 5]$ by $[-20, 20, 5]$ viewing window and find the intersection point.

The point of intersection (2, 12) indicates that the solution is 2.

3. $9 - x = 2x + 12$

Algebraic Solution:

$$9 - x = 2x + 12$$

$$9 = 3x + 12$$

$$-3 = 3x$$

$$-1 = x$$

The solution is -1.

Graphical Solution:

Graph $y_1 = 9 - x$ and $y_2 = 2x + 12$ in a $[-20, 20, 5]$ by $[-20, 20, 5]$ viewing window and find the intersection point.

The point of intersection $(-1, 10)$ indicates that the solution is -1.

5. $8 - (2x - 1) = 13$

Algebraic Solution:

$$8 - (2x - 1) = 13$$

$$8 - 2x + 1 = 13$$

$$-2x + 9 = 13$$

$$-2x = 4$$

$$x = -2$$

The solution is -2.

Graphical Solution:

Graph $y_1 = 8 - (2x - 1)$ and $y_2 = 13$ in a $[-20, 20, 5]$ by $[-20, 20, 5]$ viewing window and find the intersection point.

The point of intersection $(-2, 13)$ indicates that the solution is -2.

7.
$$y_1 = y_2$$
$$2(x-5)+3 = 3x-13$$
The point of intersection $(6, 5)$ indicates that the solution is 6.

9.
$$y_1 = y_2$$
$$-(2x+3)+2 = 5x-1-7x$$
The two equations have the same graph, so the solution is the set of all real numbers.

11. $7(x-6) = 5(x+2)+2x$

Algebraic Solution:
$$7(x-6) = 5(x+2)+2x$$
$$7x-42 = 5x+10+2x$$
$$7x-42 = 7x+10$$
$$-42 = 10 \quad \text{False}$$
This equation is a false statement no matter what value the variable x might have. The equation is a contradiction and has no solution. The solution set is $\varnothing$ or { }.
Graphical Solution:
Graph $y_1 = 7(x-6)$ and $y_2 = 5(x+2)+2x$ in a $[-20, 20, 5]$ by $[-20, 20, 5]$ viewing window.

The parallel lines indicate that the equation is a contradiction and has no solution. The solution set is $\varnothing$ or { }.

13. $3x-(6x+2) = -(3x+2)$

Algebraic Solution:
$$3x-(6x+2) = -(3x+2)$$
$$3x-6x-2 = -3x-2$$
$$-3x-2 = -3x-2$$
Since both sides are the same, this equation is a true statement no matter what value the variable x might have. The equation is an identity and the solution set is all real numbers.
Graphical Solution:
Graph $y_1 = 3x-(6x+2)$ and $y_2 = -(3x+2)$ in a $[-20, 20, 5]$ by $[-20, 20, 5]$ viewing window.

The graphs of y_1 and y_2 are the same. When we trace along y_1, we have the same ordered

pairs as when we trace along points of y_2. We know that the lines are identical because two points uniquely determine a line. The equation is an identity and the solution is the set of all real numbers.

15. $5(x+1)-3(x-7) = 2(x+4)-3$

Algebraic Solution:
$$5(x+1)-3(x-7) = 2(x+4)-3$$
$$5x+5-3x+21 = 2x+8-3$$
$$2x+26 = 2x+5$$
$$26 = 5 \quad \text{False}$$
This equation is a false statement no matter what value the variable x might have. The equation is a contradiction and has no solution. The solution set is $\varnothing$ or { }.
Graphical Solution:
Graph $y_1 = 5(x+1)-3(x-7)$ and $y_2 = 2(x+4)-3$ in a $[-20, 20, 5]$ by $[-20, 20, 5]$ viewing window.

The parallel lines indicate that the equation is a contradiction and has no solution. The solution set is $\varnothing$ or { }.

17. $3(x+2)-6(x-5) = 36-3x$

Algebraic Solution:
$$3(x+2)-6(x-5) = 36-3x$$
$$3x+6-6x+30 = 36-3x$$
$$-3x+36 = -3x+36$$
Since both sides are the same, this equation is a true statement no matter what value the variable x might have. The equation is an identity and the solution set is all real numbers.
Graphical Solution:
Graph $y_1 = 3(x+2)-6(x-5)$ and $y_2 = 36-3x$ in an integer viewing window centered at the origin.

The graphs of y_1 and y_2 are the same. When we trace along y_1, we have the same ordered pairs as when we trace along points of y_2. We know that the lines are identical because two points uniquely determine a line. The equation is an identity and the solution is the set of all real numbers.

19. $(x+2.1)-(0.5x+3)=12$

Algebraic Solution:
$$(x+2.1)-(0.5x+3)=12$$
$$x+2.1-0.5x-3=12$$
$$0.5x-0.9=12$$
$$0.5x=12.9$$
$$x=25.8$$
The solution is 25.8.
Graphical Solution:
Graph $y_1=(x+2.1)-(0.5x+3)$ and $y_2=12$ in an integer viewing window centered at the origin and find the intersection point.

The point of intersection (25.8, 12) indicates that the solution is 25.8.

21. $5(a-12)+2(a+15)=a-9$

Algebraic Solution:
$$5(a-12)+2(a+15)=a-9$$
$$5a-60+2a+30=a-9$$
$$7a-30=a-9$$
$$6a-30=-9$$
$$6a=21$$
$$a=\frac{21}{6}=\frac{7}{2}=3.5$$

The solution is $\frac{7}{2}=3.5$.

Graphical Solution:
Graph $y_1=5(x-12)+2(x+15)$ and $y_2=x-9$ in an integer viewing window centered at the origin and find the intersection point.

The point of intersection (3.5, −5.5) indicates that the solution is 3.5.

23. $8(p-4)-5(2p+3)=3.5(2p-5)$

Algebraic Solution:
$$8(p-4)-5(2p+3)=3.5(2p-5)$$
$$8p-32-10p-15=7p-17.5$$
$$-2p-47=7p-17.5$$
$$-9p-47=-17.5$$
$$-9p=29.5$$
$$p=\frac{29.5}{-9}\approx-3.28$$

The solution is approximately −3.28.
Graphical Solution:
Graph $y_1=8(x-4)-5(2x+3)$ and $y_2=3.5(2x-5)$ in a $[-10,10,1]$ by $[-100,100,10]$ viewing window and find the intersection point.

Rounded to the nearest hundredth, the point of intersection $(-3.28,-40.44)$ indicates that the solution is approximately −3.28.

25. $5(x-2)+2x=7(x+4)$

Algebraic Solution:
$$5(x-2)+2x=7(x+4)$$
$$5x-10+2x=7x+28$$
$$7x-10=7x+28$$
$$-10=28 \quad \text{False}$$
This equation is a false statement no matter what value the variable x might have. The equation is a contradiction and has no solution. The solution set is $\varnothing$ or { }.
Graphical Solution:
Graph $y_1=5(x-2)+2x$ and $y_2=7(x+4)$ in a $[-20,20,5]$ by $[-20,20,5]$ viewing window.

The parallel lines indicate that the equation is a contradiction and has no solution. The solution set is $\varnothing$ or { }.

27. $y + 0.2 = 0.6(y + 3)$

Algebraic Solution:
$$y + 0.2 = 0.6(y + 3)$$
$$y + 0.2 = 0.6y + 1.8$$
$$0.4y = 1.6$$
$$y = 4$$

The solution is 4.

Graphical Solution:
Graph $y_1 = x + 0.2$ and $y_2 = 0.6(x + 3)$ in a standard viewing window and find the intersection point.

The point of intersection $(4, 4.2)$ indicates that the solution is 4.

29. $2y + 5(y - 4) = 4y - 2(y - 10)$

Algebraic Solution:
$$2y + 5(y - 4) = 4y - 2(y - 10)$$
$$2y + 5y - 20 = 4y - 2y + 20$$
$$7y - 20 = 2y + 20$$
$$5y - 20 = 20$$
$$5y = 40$$
$$y = 8$$

The solution is 8.

Graphical Solution:
Graph $y_1 = 2x + 5(x - 4)$ and
$y_2 = 4x - 2(x - 10)$ in a $[-20, 20, 5]$ by
$[-100, 100, 10]$ viewing window and find the intersection point.

The point of intersection $(8, 36)$ indicates that the solution is 8.

31. $2(x - 8) + x = 3(x - 6) + 2$

Algebraic Solution:
$$2(x - 8) + x = 3(x - 6) + 2$$
$$2x - 16 + x = 3x - 18 + 2$$
$$3x - 16 = 3x - 16$$

Since both sides are the same, this equation is a true statement no matter what value the variable x might have. The equation is an identity and the solution set is all real numbers.

Graphical Solution:
Graph $y_1 = 2(x - 8) + x$ and $y_2 = 3(x - 6) + 2$ in a $[-20, 20, 5]$ by $[-20, 20, 5]$ viewing window.

The graphs of y_1 and y_2 are the same. When we trace along y_1, we have the same ordered pairs as when we trace along points of y_2. We know that the lines are identical because two points uniquely determine a line. The equation is an identity and the solution is the set of all real numbers.

33. $\dfrac{5x - 1}{6} - 3x = \dfrac{1}{3} + \dfrac{4x + 3}{9}$

Algebraic Solution:
$$\frac{5x - 1}{6} - 3x = \frac{1}{3} + \frac{4x + 3}{9}$$
$$18\left(\frac{5x - 1}{6} - 3x\right) = 18\left(\frac{1}{3} + \frac{4x + 3}{9}\right)$$
$$3(5x - 1) - 18(3x) = 6 + 2(4x + 3)$$
$$15x - 3 - 54x = 6 + 8x + 6$$
$$-39x - 3 = 8x + 12$$
$$-47x - 3 = 12$$
$$-47x = 15$$
$$x = \frac{15}{-47} = -\frac{15}{47} \approx -0.32$$

The solution is $-\dfrac{15}{47} \approx -0.32$.

Graphical Solution:
Graph $y_1 = \dfrac{5x - 1}{6} - 3x$ and $y_2 = \dfrac{1}{3} + \dfrac{4x + 3}{9}$ in a standard viewing window and find the intersection point.

Rounded to the nearest hundredth, the point of intersection $(-0.32, 0.52)$ indicates that the solution is approximately -0.32.

35. $-2(b-4)-(3b-1)=5b+3$

Algebraic Solution:
$$-2(b-4)-(3b-1)=5b+3$$
$$-2b+8-3b+1=5b+3$$
$$-5b+9=5b+3$$
$$-10b+9=3$$
$$-10b=-6$$
$$b=\frac{-6}{-10}=\frac{3}{5}=0.6$$

The solution is $\frac{3}{5}=0.6$.

Graphical Solution:
Graph $y_1=-2(x-4)-(3x-1)$ and $y_2=5x+3$ in a standard viewing window and find the intersection point.

The point of intersection $(0.6, 6)$ indicates that the solution is 0.6.

37. $1.5(4-x)=1.3(2-x)$

Algebraic Solution:
$$1.5(4-x)=1.3(2-x)$$
$$6-1.5x=2.6-1.3x$$
$$-0.2x+6=2.6$$
$$-0.2x=-3.4$$
$$x=17$$

The solution is 17.

Graphical Solution:
Graph $y_1=1.5(4-x)$ and $y_2=1.3(2-x)$ in a $[-50,50,5]$ by $[-50,50,5]$ viewing window and find the intersection point.

The point of intersection $(17,-19.5)$ indicates that the solution is 17.

39. $\frac{1}{4}(a+2)=\frac{1}{6}(5-a)$

Algebraic Solution:
$$\frac{1}{4}(a+2)=\frac{1}{6}(5-a)$$
$$24\cdot\frac{1}{4}(a+2)=24\cdot\frac{1}{6}(5-a)$$
$$6(a+2)=4(5-a)$$
$$6a+12=20-4a$$
$$10a+12=20$$
$$10a=8$$
$$a=\frac{8}{10}=\frac{4}{5}=0.8$$

The solution is $\frac{4}{5}=0.8$.

Graphical Solution:
Graph $y_1=\frac{1}{4}(x+2)$ and $y_2=\frac{1}{6}(5-x)$ in a standard viewing window and find the intersection point.

The point of intersection $(0.8, 0.7)$ indicates that the solution is 0.8.

41. a. $C_1(2)=30+20(2)=\$70$
$C_2(2)=25(2)=\$50$

If the job takes 2 hours, then the second consultant will have a lower cost.

b. $C_1(8)=30+20(8)=\$190$
$C_2(8)=25(8)=\$200$

If the job takes 8 hours, then the first consultant will have a lower cost.

c. Graph $y_1=30+20x$ and $y_2=25x$ and find the point of intersection.

The intersection point $(6, 150)$ indicates that the cost of hiring each consultant will be the same if the job takes 6 hours.

43. a. $R_1(50) = 25 + 0.30(50) = \$40$

$R_2(50) = 28 + 0.25(50) = \$40.50$

If 50 miles are driven, then the first agency will have a lower cost.

b. $R_1(100) = 25 + 0.30(100) = \$55$

$R_2(100) = 28 + 0.25(100) = \$53$

If 100 miles are driven, then the second agency will have a lower cost.

c.   Graph $y_1 = 25 + 0.30x$ and $y_2 = 28 + 0.25x$ and find the point of intersection.

The intersection point (60, 43) indicates that the cost of using each agency will be the same if 60 miles are driven.

45. Let x be Tara's monthly sales. Graph $y_1 = 1000 + 0.04x$ and $y_2 = 0.08x$.

The intersection point (25,000, 2000) indicates that her gross pay will be the same for sales of $25,000.

47. Let x be the number of calculators produced and sold. Graph $y_1 = 125,000 + 75x$ and $y_2 = 150x$.

The intersection point of approximately (1667, 250,000) indicates that when the company produces 1667 calculators, the cost and the revenue will be approximately $250,000.

49. $x < 0$

$\{-3, -2, -1\}$

51. $x + 5 \le 6$

$x \le 1$

$\{-3, -2, -1, 0, 1\}$

53. answers may vary

55. From the intersection point shown on the screen, the ordered pair is (12, 22) .

57. When $x > 12$, the graph of y_1 is above the graph of y_2. Thus, if x is greater than 12, y_1 is greater than y_2.

59. Graph $y_1 = 1.75x - 2.5$ and $y_2 = 0$ in a standard viewing window and find the intersection point.

Rounding to the nearest hundredth, the point of intersection (1.43, 0) indicates that the solution is approximately 1.43.

Check.

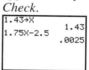

Since $0.0025 \approx 0$, the approximate solution 1.43 checks.

61. Graph $y_1 = 2.5x + 3$ and $y_2 = 7.8x - 5$ in a standard viewing window and find the intersection point.

Rounding to the nearest hundredth, the point of intersection (1.51, 6.77) indicates that the solution is approximately 1.51.

Check.

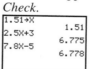

Since $6.775 \approx 6.778$, the approximate solution 1.51 checks.

63. Graph $y_1 = 3x + \sqrt{5}$ and $y_2 = 7x - \sqrt{2}$ in a standard viewing window and find the intersection point.

Rounding to the nearest hundredth, the point of intersection $(0.91, 4.97)$ indicates that the solution is approximately 0.91.

Check.

Since $4.966067977 \approx 4.955786438$, the approximate solution 0.91 checks.

65. Graph $y_1 = 2\pi x - 5.6$ and $y_2 = 7(x - \pi)$ in a $[-50, 50, 10]$ by $[-300, 300, 50]$ viewing window and find the intersection point.

Rounding to the nearest hundredth, the point of intersection $(22.87, 138.08)$ indicates that the solution is approximately 22.87.

Check.

```
22.87→X
            22.87
2πX-5.6
         138.096448
7(X-π)
         138.0988514
```

Since $138.096448 \approx 138.0988514$, the approximate solution 22.87 checks.

67. If the intersection-of-graphs method leads to parallel lines, then the equation has no solution.

Section 3.2 Practice Exercises

1. a. $\{x | x < 3.5\}$　$(-\infty, 3.5)$

　　　　←————)————→
　　　　　　　　3.5

b. $\{x | x \geq -3\}$　$[-3, \infty)$

　　←————[————→
　　　　　-3

c. $\{x | -1 \leq x < 4\}$　$[-1, 4)$

　←——[————)——→
　　　-1　　　4

2. $x + 5 > 9$

$x + 5 - 5 > 9 - 5$

$x > 4$

$(4, \infty)$

　←————(————→
　　　　　4

3. $8x + 21 \leq 2x - 3$

$8x + 21 - 2x \leq 2x - 3 - 2x$

$6x + 21 \leq -3$

$6x + 21 - 21 \leq -3 - 21$

$6x \leq -24$

$\dfrac{6x}{6} \leq \dfrac{-24}{6}$

$x \leq -4$

$(-\infty, -4]$

　←————]————→
　　　　-4

4. a. $\dfrac{2}{5}x \geq \dfrac{4}{15}$

$\dfrac{5}{2} \cdot \dfrac{2}{5}x \geq \dfrac{5}{2} \cdot \dfrac{4}{15}$

$x \geq \dfrac{2}{3}$

$\left[\dfrac{2}{3}, \infty\right)$

　←————[————→
　　　　　$\frac{2}{3}$

b. $-2.4x < 9.6$

$\dfrac{-2.4x}{-2.4} > \dfrac{9.6}{-2.4}$

$x > -4$

$(-4, \infty)$

　←————(————→
　　　　-4

5. $-(4x + 6) \leq 2(5x + 9) + 2x$

$-4x - 6 \leq 10x + 18 + 2x$

$-4x - 6 \leq 12x + 18$

$-4x - 6 + 4x \leq 12x + 18 + 4x$

$-6 \leq 16x + 18$

$-6 - 18 \leq 16x + 18 - 18$

$-24 \leq 16x$

$\dfrac{-24}{16} \leq \dfrac{16x}{16}$

$-\dfrac{3}{2} \leq x$

$\left[-\dfrac{3}{2}, \infty\right)$

Check:

6. $\dfrac{3}{5}(x-3) \geq x-7$

$$5\left[\dfrac{3}{5}(x-3)\right] \geq 5(x-7)$$
$$3(x-3) \geq 5(x-7)$$
$$3x-9 \geq 5x-35$$
$$3x-9-5x \geq 5x-35-5x$$
$$-2x-9 \geq -35$$
$$-2x-9+9 \geq -35+9$$
$$-2x \geq -26$$
$$\dfrac{-2x}{-2} \leq \dfrac{-26}{-2}$$
$$x \leq 13$$

$(-\infty, 13]$

Check:

7. **a.** $4(x-2) < 4x+5$
$$4x-8 < 4x+5$$
$$4x-8-4x < 4x+5-4x$$
$$-8 < 5$$

This is a true statement for all values of x.

$(-\infty, \infty)$

    b.   Solving $4(x-2) > 4x+5$ in a similar fashion leads to the statement $-8 > 5$. This statement is false for all values of x. Thus, the solution set is $\varnothing$.

8.   In words:

900	+	commission (15% of sales)	$\geq$	2400
↓	↓	↓	↓	↓

Translate: 900   +   0.15x $\geq$ 2400

Sales must be greater than or equal to $10,000 per month.

9. $-9.4t + 431 < 275$
$$-9.4t < -156$$
$$\frac{-9.4t}{-9.4} > \frac{-156}{-9.4}$$
$$t > \text{approximately } 16.6$$

The annual consumption of cigarettes will be less than 275 billion more than 16.6 years after 2000, or in approximately $17 + 2000 = 2017$ and after.

Vocabulary, Readiness & Video Check 3.2

1. d. $(-\infty, -5)$

2. c. $[-11, \infty)$

3. b. $\left(-2.5, \dfrac{7}{4}\right]$

4. a. $\left[-\dfrac{10}{3}, 0.2\right)$

5. The set $\{x|x \geq -0.4\}$ written in interval notation is $[-0.4, \infty)$.

6. The set $\{x|x < -0.4\}$ written in interval notation is $(-\infty, -0.4)$.

7. The set $\{x|x \leq -0.4\}$ written in interval notation is $(-\infty, -0.4]$.

8. The set $\{x|x > -0.4\}$ written in interval notation is $(-0.4, \infty)$.

9. The graph of Example 1 is shaded from $-\infty$ to, but not including, -3, as indicated by a parenthesis. To write interval notation, write down what is shaded for the inequality from left to right. A parenthesis is always used with $-\infty$, so from the graph, the interval notation is $(-\infty, -3)$.

10. We can add the same number to (or subtract the same number from) both sides of a linear inequality in one variable and have an equivalent inequality; addition property of equality.

11. If you multiply or divide both sides of an inequality by the <u>same</u> nonzero negative number, you must <u>reverse</u> the direction of the inequality symbol.

12. maximum, or less

Exercise Set 3.2

1. $\{x|x < -3\}$
$(-\infty, -3)$

3. $\{x|x \geq 0.3\}$
$[0.3, \infty)$

5. $\{x|-2 < x < 5\}$
$(-2, 5)$

7. $\{x|5 \geq x > -1\}$
$(-1, 5]$

9. $x - 7 \geq -9$
$$x \geq -2$$
$[-2, \infty)$

11. $8x - 7 \leq 7x - 5$
$$x - 7 \leq -5$$
$$x \leq 2$$
$(-\infty, 2]$

13. $\dfrac{3}{4}x \geq 6$
$$\dfrac{4}{3} \cdot \dfrac{3}{4}x \geq \dfrac{4}{3} \cdot 6$$
$$x \geq 8$$
$[8, \infty)$

15. $5x < -23.5$
$\quad x < -4.7$
$\quad (-\infty, -4.7)$

$\qquad$ -4.7

17. The graph of y_1 is below the graph of y_2 when $x < 4$, so the solution set of $y < y_2$ is $(-\infty, 4)$.

19. The graph of y_1 is above the graph of y_2 for $x > -3$, so the solution set of $y_1 \geq y_2$ is $[-3, \infty)$.

21. The graph of y_1 is never above the graph of y_2, so the solution set is $\varnothing$.

23. $-2x + 7 \geq 9$
$\quad -2x \geq 2$
$\quad\quad x \leq -1$
$\quad (-\infty, -1]$

25. $15 + 2x \geq 4x - 7$
$\quad 15 \geq 2x - 7$
$\quad 22 \geq 2x$
$\quad 11 \geq x \text{ or } x \leq 11$
$\quad (-\infty, 11]$

27. $4(2x + 1) > 4$
$\quad 8x + 4 > 4$
$\quad\quad 8x > 0$
$\quad\quad\ x > 0$
$\quad (0, \infty)$

29. $3(x - 5) < 2(2x - 1)$
$\quad 3x - 15 < 4x - 2$
$\quad\ -15 < x - 2$
$\quad\ -13 < x$
$\quad\quad x > -13$
$\quad (-13, \infty)$

31. $\dfrac{5x + 1}{7} - \dfrac{2x - 6}{4} \geq -4$

$28\left(\dfrac{5x + 1}{7} - \dfrac{2x - 6}{4}\right) \geq 28(-4)$

$4(5x + 1) - 7(2x - 6) \geq -112$

$20x + 4 - 14x + 42 \geq -112$

$6x + 46 \geq -112$

$6x \geq -158$

$x \geq -\dfrac{79}{3}$

$\left[-\dfrac{79}{3}, \infty\right)$

33. $-3(2x - 1) < -4[2 + 3(x + 2)]$
$\quad -6x + 3 < -4(2 + 3x + 6)$
$\quad -6x + 3 < -4(8 + 3x)$
$\quad -6x + 3 < -32 - 12x$
$\quad\ 6x + 3 < -32$
$\quad\ 6x < -35$
$\quad\quad x < -\dfrac{35}{6}$

$\left(-\infty, -\dfrac{35}{6}\right)$

35. $x + 9 < 3$
$\quad x + 9 - 9 < 3 - 9$
$\quad\quad x < -6$
$\quad (-\infty, -6)$

37. $-x < -4$
$\quad \dfrac{-x}{-1} > \dfrac{-4}{-1}$
$\quad\ x > 4$
$\quad (4, \infty)$

39. $-7x \leq 3.5$
$\quad \dfrac{-7x}{-7} \geq \dfrac{3.5}{-7}$
$\quad\ x \geq -0.5$
$\quad [-0.5, \infty)$

41. $\dfrac{1}{2} + \dfrac{2}{3} \geq \dfrac{x}{6}$

$6\left(\dfrac{1}{2} + \dfrac{2}{3}\right) \geq 6\left(\dfrac{x}{6}\right)$

$3 + 4 \geq x$

$7 \geq x$

$x \leq 7$

$(-\infty, 7]$

43. $-5x + 4 \le -4(x - 1)$

$\quad -5x + 4 \le -4x + 4$

$\quad\quad\quad\quad -x \le 0$

$\quad\quad\quad\quad\quad x \ge 0$

$[0, \infty)$

45. $\dfrac{3}{4}(x - 7) \ge x + 2$

$\quad 4\left[\dfrac{3}{4}(x - 7)\right] \ge 4(x + 2)$

$\quad\quad 3(x - 7) \ge 4(x + 2)$

$\quad\quad 3x - 21 \ge 4x + 8$

$\quad\quad\quad -x - 21 \ge 8$

$\quad\quad\quad\quad\quad -x \ge 29$

$\quad\quad\quad\quad\quad\quad x \le -29$

$(-\infty, -29]$

47. $0.8x + 0.6x \ge 4.2$

$\quad\quad\quad 1.4x \ge 4.2$

$\quad\quad\quad\quad x \ge 3$

$[3, \infty)$

49. $4(x - 6) + 2x - 4 \ge 3(x - 7) + 10x$

$\quad 4x - 24 + 2x - 4 \ge 3x - 21 + 10x$

$\quad\quad\quad 6x - 28 \ge 13x - 21$

$\quad\quad\quad\quad\quad -28 \ge 7x - 21$

$\quad\quad\quad\quad\quad\quad -7 \ge 7x$

$\quad\quad\quad\quad\quad\quad -1 \ge x$

$\quad\quad\quad\quad\quad\quad\quad x \le -1$

$(-\infty, -1]$

51. $14 - (5x - 6) \ge -6(x + 1) - 5$

$\quad 14 - 5x + 6 \ge -6x - 6 - 5$

$\quad\quad -5x + 20 \ge -6x - 11$

$\quad\quad\quad x + 20 \ge -11$

$\quad\quad\quad\quad\quad x \ge -31$

$[-31, \infty)$

53. $4(x - 1) \ge 4x - 8$

$\quad 4x - 4 \ge 4x - 8$

$\quad\quad\quad -4 \ge -8$ (True for all x)

All real numbers

$(-\infty, \infty)$

55. $3x + 1 < 3(x - 2)$

$\quad 3x + 1 < 3x - 6$

$\quad\quad\quad 1 < -6$ False

$\varnothing$

57. $0.4(4x - 3) < 1.2(x + 2)$

$\quad 10[0.4(4x - 3)] < 10[1.2(x + 2)]$

$\quad\quad 4(4x - 3) < 12(x + 2)$

$\quad\quad 16x - 12 < 12x + 24$

$\quad\quad\quad 4x - 12 < 24$

$\quad\quad\quad\quad 4x < 36$

$\quad\quad\quad\quad\quad x < 9$

$(-\infty, 9)$

59. $\dfrac{2}{5}x - \dfrac{1}{4} \le \dfrac{3}{10}x - \dfrac{4}{5}$

$\quad 20\left[\dfrac{2}{5}x - \dfrac{1}{4}\right] \le 20\left[\dfrac{3}{10}x - \dfrac{4}{5}\right]$

$\quad 4 \cdot 2x - 5 \le 2 \cdot 3x - 4 \cdot 4$

$\quad\quad 8x - 5 \le 6x - 16$

$\quad\quad 2x - 5 \le -16$

$\quad\quad\quad 2x \le -11$

$\quad\quad\quad\quad x \le -\dfrac{11}{2}$

$\left(-\infty, -\dfrac{11}{2}\right]$

61. $\dfrac{1}{2}(3x - 4) \le \dfrac{3}{4}(x - 6) + 1$

$\quad 4\left[\dfrac{1}{2}(3x - 4)\right] \le 4\left[\dfrac{3}{4}(x - 6) + 1\right]$

$\quad\quad 2(3x - 4) \le 3(x - 6) + 4$

$\quad\quad 6x - 8 \le 3x - 18 + 4$

$\quad\quad 6x - 8 \le 3x - 14$

$\quad\quad 3x - 8 \le -14$

$\quad\quad\quad 3x \le -6$

$\quad\quad\quad\quad x \le -2$

$(-\infty, -2]$

63. $\dfrac{-x + 2}{2} - \dfrac{1 - 5x}{8} < -1$

$\quad 8\left(\dfrac{-x + 2}{2} - \dfrac{1 - 5x}{8}\right) < 8(-1)$

$\quad 4(-x + 2) - (1 - 5x) < -8$

$\quad\quad -4x + 8 - 1 + 5x < -8$

$\quad\quad\quad\quad x + 7 < -8$

$\quad\quad\quad\quad\quad x < -15$

$(-\infty, -15)$

65.
$$\frac{x+5}{5} - \frac{3+x}{8} \geq -\frac{3}{10}$$
$$40\left(\frac{x+5}{5} - \frac{3+x}{8}\right) \geq 40\left(-\frac{3}{10}\right)$$
$$8(x+5) - 5(3+x) \geq -12$$
$$8x + 40 - 15 - 5x \geq -12$$
$$3x + 25 \geq -12$$
$$3x \geq -37$$
$$x \geq -\frac{37}{3}$$
$$\left[-\frac{37}{3}, \infty\right)$$

67.
$$\frac{x+3}{12} + \frac{x-5}{15} < \frac{2}{3}$$
$$60\left(\frac{x+3}{12} + \frac{x-5}{15}\right) < 60\left(\frac{2}{3}\right)$$
$$5(x+3) + 4(x-5) < 20(2)$$
$$5x + 15 + 4x - 20 < 40$$
$$9x - 5 < 40$$
$$9x < 45$$
$$x < 5$$
$$(-\infty, 5)$$

69. a. Let x be the final exam score.
$$\frac{72 + 67 + 82 + 79 + x + x}{6} \geq 77$$
$$6\left(\frac{72 + 67 + 82 + 79 + 2x}{6}\right) \geq 6(77)$$
$$72 + 67 + 82 + 79 + 2x \geq 462$$
$$300 + 2x \geq 462$$
$$300 + 2x - 300 \geq 462 - 300$$
$$2x \geq 162$$
$$\frac{2x}{2} \geq \frac{162}{2}$$
$$x \geq 81$$
The solution is $\{x|x \geq 81\}$.

b. A final exam grade of 81 or higher will result in an average of 77 or higher.

71. a. Let x be the number of whole boxes on the elevator.
$$147 + 66x \leq 1500$$
$$147 + 66x - 147 \leq 1500 - 147$$
$$66x \leq 1353$$
$$x \leq 20.5$$
The solution is $\{x|x \leq 20.5\}$ or since x is a whole number $\{x|x \leq 20\}$.

b. She can move at most 20 whole boxes at one time.

73. a. Let x be the weight of luggage and cargo. The six passengers have a total weight of $6(160)$ pounds.
$$x + 6(160) \leq 2000$$
$$x + 960 \leq 2000$$
$$x \leq 1040$$
The solution is $\{x|x \leq 1040\}$.

b. The luggage and cargo must weigh no more than 1040 pounds.

75. Let x be the number of daily miles driven.
$$36 < 24 + 0.15x$$
$$12 < 0.15x$$
$$\frac{12}{0.15} < \frac{0.15x}{0.15}$$
$$80 < x$$
$$\{x|x > 80\}$$
If you drive more than 80 miles a day, plan A is more economical.

77. $F \geq \dfrac{9}{5}C + 32$
$$F \geq \frac{9}{5}(500) + 32$$
$$F \geq 932$$
$$\{F|F \geq 932\}$$
Glass is a liquid at temperatures of 932°F or higher.

79. a.
$$1284.7t + 48,133 > 70,000$$
$$1284.7t > 21,867$$
$$t > \text{approximately } 17.02$$
t is more than 17.02, so $t \geq 18$.
$2000 + 18 = 2018$
Salaries will be greater than $70,000 in 2018.

b. answers may vary

81. Consumption of whole milk is decreasing over time; answers may vary.

83. 2015 is 15 years after 2000, so 2015 corresponds to $t = 15$.
$$w = -2.11t + 69.2$$
$$w = -2.11(15) + 69.2$$
$$w = -31.65 + 69.2$$
$$w = 37.55$$
The average consumption of whole milk is expected to be 37.55 pounds per person per year in 2015.

85. $-2.11t + 69.2 < 45$
$-2.11t < -24.2$
$t >$ approximately 11.5

11.5 years after 2000 corresponds to some time during the year 2012. During 2012, the consumption of whole milk will be less than 45 pounds per person per year.

87. answers may vary

89. a. $s > w$
$-0.42t + 29.9 > -2.11t + 69.2$
$1.69t > 39.3$
$t >$ approximately 23.3
$\{t \mid t > 23.3\}$

   b. $2000 + 24 = 2024$
2024 is the first year when the consumption of skim milk is predicted to be greater than the consumption of whole milk.

91. $x < 5$ and $x > 1$
The integers are 2, 3, and 4.

93. $x \geq -2$ and $x \geq 2$
The integers are 2, 3, 4, ...

95. $2x - 6 = 4$
$2x - 6 + 6 = 4 + 6$
$2x = 10$
$\dfrac{2x}{2} = \dfrac{10}{2}$
$x = 5$

97. $-x + 7 = 5x - 6$
$-x - 5x = -6 - 7$
$-6x = -13$
$\dfrac{-6x}{-6} = \dfrac{-13}{-6}$
$x = \dfrac{13}{6}$

99. $\{x \mid x \geq 2\}$, $[2, \infty)$

101.

$(-\infty, 0)$

103. $\{x \mid -2 < x \leq 1.5\}$

105. To solve $-3x \leq 14$, both sides must be divided by -3, so the inequality symbol will be reversed.

107. To solve $-x \geq -23$, both sides must be divided by -1, so the inequality symbol will be reversed.

109. $2x - 3 = 5$
$2x = 8$
$x = 4$
The solution set is $\{4\}$.

111. $2x - 3 < 5$
$2x < 8$
$x < 4$
The solution set is $(-\infty, 4)$.

113. answers may vary

115. answers may vary

Integrated Review

1. $-4x = 20$
$\dfrac{-4x}{-4} = \dfrac{20}{-4}$
$x = -5$

2. $-4x < 20$
$\dfrac{-4x}{-4} > \dfrac{20}{-4}$
$x > -5$
$(-5, \infty)$

3. $\dfrac{3x}{4} \geq 2$
$4\left(\dfrac{3x}{4}\right) \geq 4(2)$
$3x \geq 8$
$x \geq \dfrac{8}{3}$
$\left[\dfrac{8}{3}, \infty\right)$

4. $5x + 3 \geq 2 + 4x$
$x + 3 \geq 2$
$x \geq -1$
$[-1, \infty)$

5. $6(y - 4) = 3(y - 8)$
$6y - 24 = 3y - 24$
$3y = 0$
$y = 0$

6. $-4x \le \dfrac{2}{5}$

$-20x \le 2$

$x \ge -\dfrac{1}{10}$

$\left[-\dfrac{1}{10}, \infty \right)$

7. $-3x \ge \dfrac{1}{2}$

$2(-3x) \ge 2\left(\dfrac{1}{2} \right)$

$-6x \ge 1$

$x \le -\dfrac{1}{6}$

$\left(-\infty, -\dfrac{1}{6} \right]$

8. $5(y+4) = 4(y+5)$

$5y + 20 = 4y + 20$

$y = 0$

9. $7x < 7(x-2)$

$7x < 7x - 14$

$0 < -14$ (False)

No Solution; $\varnothing$

10. $\dfrac{-5x+11}{2} \le 7$

$2\left(\dfrac{-5x+11}{2} \right) \le 2(7)$

$-5x + 11 \le 14$

$-5x \le 3$

$x \ge -\dfrac{3}{5}$

$\left[-\dfrac{3}{5}, \infty \right)$

11. $-5x + 1.5 = -19.5$

$-5x + 1.5 - 1.5 = -19.5 - 1.5$

$-5x = -21$

$\dfrac{-5x}{-5} = \dfrac{-21}{-5}$

$x = 4.2$

12. $-5x + 4 = -26$

$-5x = -30$

$x = 6$

13. $5 + 2x - x = -x + 3 - 14$

$5 + x = -x - 11$

$5 + 2x = -11$

$2x = -16$

$x = -8$

14. $12x + 14 < 11x - 2$

$x + 14 < -2$

$x < -16$

$(-\infty, -16)$

15. $\dfrac{x}{5} - \dfrac{x}{4} = \dfrac{x-2}{2}$

$20\left(\dfrac{x}{5} - \dfrac{x}{4} \right) = 20\left(\dfrac{x-2}{2} \right)$

$4x - 5x = 10(x-2)$

$-x = 10x - 20$

$-11x = -20$

$x = \dfrac{20}{11}$

16. $12x - 12 = 8(x-1)$

$12x - 12 = 8x - 8$

$4x - 12 = -8$

$4x = 4$

$x = 1$

17. $2(x-3) > 70$

$2x - 6 > 70$

$2x > 76$

$x > 38$

$(38, \infty)$

18. $-3x - 4.7 = 11.8$

$-3x - 4.7 + 4.7 = 11.8 + 4.7$

$-3x = 16.5$

$\dfrac{-3x}{-3} = \dfrac{16.5}{-3}$

$x = -5.5$

19. $-2(b-4) - (3b-1) = 5b + 3$

$-2b + 8 - 3b + 1 = 5b + 3$

$-5b + 9 = 5b + 3$

$-10b = -6$

$b = \dfrac{-6}{-10} = \dfrac{3}{5}$

20. $8(x+3) < 7(x+5)+x$
$8x+24 < 7x+35+x$
$8x+24 < 8x+35$
$\quad\quad 24 < 35 \quad$ (True for all x)
All real numbers; $(-\infty, \infty)$

21. $\dfrac{3t+1}{8} = \dfrac{5+2t}{7}+2$
$56\left(\dfrac{3t+1}{8}\right) = 56\left(\dfrac{5+2t}{7}\right)+56(2)$
$7(3t+1) = 8(5+2t)+112$
$21t+7 = 40+16t+112$
$21t+7 = 16t+152$
$5t = 145$
$t = 29$

22. $4(x-6)-x = 8(x-3)-5x$
$4x-24-x = 8x-24-5x$
$3x-24 = 3x-24$
$\quad -24 = -24 \quad$ (True for all x)
The solution is all real numbers.

23. $\dfrac{x}{6}+\dfrac{3x-2}{2} < \dfrac{2}{3}$
$6\left(\dfrac{x}{6}+\dfrac{3x-2}{2}\right) < 6\left(\dfrac{2}{3}\right)$
$x+3(3x-2) < 4$
$x+9x-6 < 4$
$10x-6 < 4$
$10x < 10$
$x < 1$
$(-\infty, 1)$

24. $\dfrac{y}{3}+\dfrac{y}{5} = \dfrac{y+3}{10}$
$30\left(\dfrac{y}{3}\right)+30\left(\dfrac{y}{5}\right) = 30\left(\dfrac{y+3}{10}\right)$
$10y+6y = 3(y+3)$
$16y = 3y+9$
$13y = 9$
$y = \dfrac{9}{13}$

25. $5(x-6)+2x > 3(2x-1)-4$
$5x-30+2x > 6x-3-4$
$7x-30 > 6x-7$
$x > 23$
$(23, \infty)$

26. $14(x-1)-7x \le 2(3x-6)+4$
$14x-14-7x \le 6x-12+4$
$7x-14 \le 6x-8$
$x \le 6$
$(-\infty, 6]$

27. $\dfrac{1}{4}(3x+2)-x \ge \dfrac{3}{8}(x-5)+2$
$8\left[\dfrac{1}{4}(3x+2)-x\right] \ge 8\left[\dfrac{3}{8}(x-5)+2\right]$
$2(3x+2)-8x \ge 3(x-5)+16$
$6x+4-8x \ge 3x-15+16$
$-2x+4 \ge 3x+1$
$3 \ge 5x$
$\dfrac{3}{5} \ge x \quad$ or $\quad x \le \dfrac{3}{5}$
$\left(-\infty, \dfrac{3}{5}\right]$

28. $\dfrac{1}{3}(x-10)-4x > \dfrac{5}{6}(2x+1)-1$
$6\left[\dfrac{1}{3}(x-10)-4x\right] > 6\left[\dfrac{5}{6}(2x+1)-1\right]$
$2(x-10)-24x > 5(2x+1)-6$
$2x-20-24x > 10x+5-6$
$-22x-20 > 10x-1$
$-19 > 32x$
$-\dfrac{19}{32} > x \quad$ or $\quad x < -\dfrac{19}{32}$
$\left(-\infty, -\dfrac{19}{32}\right)$

Section 3.3 Practice Exercises

1. $A = \{1, 3, 5, 7, 9\}$ and $B = \{1, 2, 3, 4\}$
The numbers 1 and 3 are in sets A and B.
The intersection is $\{1, 3\}$. $A \cap B = \{1, 3\}$.

2. $x+3 < 8 \quad$ and $\quad 2x-1 < 3$
$\quad\quad x < 5 \quad$ and $\quad\quad 2x < 4$
$\quad\quad x < 5 \quad$ and $\quad\quad\quad x < 2$
$\{x \mid x < 5\}, (-\infty, 5)$

$\{x \mid x < 2\}, (-\infty, 2)$

$\{x \mid x < 5 \text{ and } x < 2\} = \{x \mid x < 2\}$

The solution set is $(-\infty, 2)$.

3. $4x \le 0$ and $3x + 2 > 8$

     $x \le 0$ and $3x > 6$

     $x \le 0$ and $x > 2$

$\{x | x \le 0\}$, $(-\infty, 0]$

$\{x | x > 2\}$, $(2, \infty)$

$\{x | 4x \le 0 \text{ and } 3x + 2 > 8\} = \{\quad\}$ or $\varnothing$

4. $3 < 5 - x < 9$

    $3 - 5 < 5 - x - 5 < 9 - 5$

         $-2 < -x < 4$

       $\dfrac{-2}{-1} > \dfrac{-x}{-1} > \dfrac{4}{-1}$

         $2 > x > -4$

or $-4 < x < 2$

The solution set is $(-4, 2)$.

5. $-4 \le \dfrac{x}{2} - 1 \le 3$

    $2(-4) \le 2\left(\dfrac{x}{2} - 1\right) \le 2(3)$

       $-8 \le x - 2 \le 6$

  $-8 + 2 \le x - 2 + 2 \le 6 + 2$

        $-6 \le x \le 8$

The solution set is $[-6, 8]$.

Check:

6. $A = \{1, 3, 5, 7, 9\}$ and $B = \{2, 3, 4, 5, 6\}$.

The numbers that are in either set or both sets are $\{1, 2, 3, 4, 5, 6, 7, 9\}$. This set is the union, $A \cup B$.

7. $8x + 5 \le 8$ or $x - 1 \ge 2$

    $8x \le 3$ or $x \ge 3$

     $x \le \dfrac{3}{8}$ or $x \ge 3$

$\left\{x \Big| x \le \dfrac{3}{8}\right\}$, $\left(-\infty, \dfrac{3}{8}\right]$

$\{x | x \ge 3\}$, $[3, \infty)$

$\left\{x \Big| x \le \dfrac{3}{8} \text{ or } x \ge 3\right\} = \left(-\infty, \dfrac{3}{8}\right] \cup [3, \infty)$

The solution set is $\left(-\infty, \dfrac{3}{8}\right] \cup [3, \infty)$.

8. $-3x - 2 > -8$ or $5x > 0$

     $-3x > -6$ or $x > 0$

       $x < 2$ or $x > 0$

$\{x | x < 2\}$, $(-\infty, 2)$

$\{x | x > 0\}$, $(0, \infty)$

$\{x | x < 2 \text{ or } x > 0\}$, $(-\infty, \infty)$

The solution set is $(-\infty, \infty)$.

Vocabulary, Readiness & Video Check 3.3

1. Two inequalities joined by the words "and" or "or" are called <u>compound</u> inequalities.

2. The word <u>and</u> means intersection.

3. The word <u>or</u> means union.

4. The symbol $\cap$ means intersection.

5. The symbol $\cup$ represents union.

6. The symbol $\varnothing$ is the empty set.

7. For an element to be in the intersection of sets A and B, the element must be in set A <u>and</u> in set B.

8. Graph the two intervals, each on its own number line, so you can see their intersection. Graph this intersection on the third number line—this intersection is the solution set.

9. For an element to be in the union of sets A and B, the element must be in set A <u>or</u> in set B.

10. Graph the two intervals, each on its own number line, so you can see their union. Graph this union on the third number line—this union is the solution set.

Exercise Set 3.3

1. $C \cup D = \{2, 3, 4, 5, 6, 7\}$

3. $A \cap D = \{4, 6\}$

5. $A \cup B = \{..., -2, -1, 0, 1, ...\}$

7. $B \cap D = \{5, 7\}$

9. $B \cup C = \{x | x$ is an odd integer or $x = 2$ or $x = 4\}$

11. $A \cap C = \{2, 4\}$

13. $x < 1$ and $x > -3$
$-3 < x < 1$
$(-3, 1)$

15. $x \le -3$ and $x \ge -2$
$\varnothing$

17. $x < -1$ and $x < 1$
$x < -1$
$(-\infty, -1)$

19. $x + 1 \ge 7$ and $3x - 1 \ge 5$
$\quad x \ge 6$ and $\quad 3x \ge 6$
$\qquad\qquad\qquad x \ge 2$

$x \ge 6$
$[6, \infty)$

21. $4x + 2 \le -10$ and $2x \le 0$
$\quad 4x \le -12$ and $\quad x \le 0$
$\quad\ x \le -3$

$x \le -3$
$(-\infty, -3]$

23. $-2x < -8$ and $x - 5 < 5$
$\quad x > 4$ and $\quad x < 10$
$(4, 10)$

25. $5 < x - 6 < 11$
$11 < x < 17$
$(11, 17)$

27. $-2 \le 3x - 5 \le 7$
$3 \le 3x \le 12$
$1 \le x \le 4$
$[1, 4]$

29. $1 \le \dfrac{2}{3}x + 3 \le 4$
$-2 \le \dfrac{2}{3}x \le 1$
$-3 \le x \le \dfrac{3}{2}$
$\left[-3, \dfrac{3}{2}\right]$

31. $-5 \le \dfrac{-3x + 1}{4} \le 2$
$4(-5) \le 4\left(\dfrac{-3x + 1}{4}\right) \le 4(2)$
$-20 \le -3x + 1 \le 8$
$-21 \le -3x \le 7$
$7 \ge x \ge -\dfrac{7}{3}$
$-\dfrac{7}{3} \le x \le 7$
$\left[-\dfrac{7}{3}, 7\right]$

33. $x < 4$ or $x < 5$
$(-\infty, 5)$

35. $x \le -4$ or $x \ge 1$
$(-\infty, -4] \cup [1, \infty)$

37. $x > 0$ or $x < 3$
$(-\infty, \infty)$

39. $-2x \le -4$ or $5x - 20 \ge 5$
$\quad x \ge 2$ or $\quad 5x \ge 25$
$\qquad\qquad\qquad x \ge 5$

$x \ge 2$
$[2, \infty)$

41. $x + 4 < 0$ or $6x > -12$
$\quad x < -4$ or $\quad x > -2$
$(-\infty, -4) \cup (-2, \infty)$

43. $3(x-1) < 12$ or $x + 7 > 10$
$x - 1 < 4$ or $x > 3$
$x < 5$
$(-\infty, \infty)$

45. $x < \dfrac{2}{3}$ and $x > -\dfrac{1}{2}$

$-\dfrac{1}{2} < x < \dfrac{2}{3}$

$\left(-\dfrac{1}{2}, \dfrac{2}{3}\right)$

47. $x < \dfrac{2}{3}$ or $x > -\dfrac{1}{2}$

$(-\infty, \infty)$

49. $0 \le 2x - 3 \le 9$
$3 \le 2x \le 12$
$\dfrac{3}{2} \le x \le 6$

$\left[\dfrac{3}{2}, 6\right]$

51. $\dfrac{1}{2} < x - \dfrac{3}{4} < 2$

$4\left(\dfrac{1}{2}\right) < 4\left(x - \dfrac{3}{4}\right) < 4(2)$

$2 < 4x - 3 < 8$
$5 < 4x < 11$
$\dfrac{5}{4} < x < \dfrac{11}{4}$

$\left(\dfrac{5}{4}, \dfrac{11}{4}\right)$

53. $x + 3 \ge 3$ and $x + 3 \le 2$
$x \ge 0$ and $x \le -1$
No solution exists.
$\varnothing$

55. $3x \ge 5$ or $-\dfrac{5}{8}x - 6 > 1$

$x \ge \dfrac{5}{3}$ or $-\dfrac{5}{8}x > 7$

$x < -\dfrac{56}{5}$

$\left(-\infty, -\dfrac{56}{5}\right) \cup \left[\dfrac{5}{3}, \infty\right)$

57. $0 < \dfrac{5 - 2x}{3} < 5$

$0 < 5 - 2x < 15$

$\dfrac{-5}{-2} > \dfrac{-2x}{-2} > \dfrac{10}{-2}$

$\dfrac{5}{2} > x > -5$

$-5 < x < \dfrac{5}{2}$

$\left(-5, \dfrac{5}{2}\right)$

59. $-6 < 3(x - 2) \le 8$
$-6 < 3x - 6 \le 8$
$0 < 3x \le 14$
$0 < x < \dfrac{14}{3}$

$\left(0, \dfrac{14}{3}\right]$

61. $-x + 5 > 6$ and $1 + 2x \le -5$
$-x > 1$ and $2x \le -6$
$x < -1$ and $x \le -3$
$x \le -3$
$(-\infty, -3]$

63. $3x + 2 \le 5$ or $7x > 29$

$3x \le 3$ or $x > \dfrac{29}{7}$

$x \le 1$ or $x > \dfrac{29}{7}$

$(-\infty, 1] \cup \left(\dfrac{29}{7}, \infty\right)$

65. $5 - x > 7$ and $2x + 3 \ge 13$
$-x > 2$ and $2x \ge 10$
$x < -2$ and $x \ge 5$
No solution exists.
$\varnothing$

67. $-\dfrac{1}{2} \le \dfrac{4x-1}{6} < \dfrac{5}{6}$

$6\left(-\dfrac{1}{2}\right) \le 6\left(\dfrac{4x-1}{6}\right) < 6\left(\dfrac{5}{6}\right)$

$-3 \le 4x - 1 < 5$

$-2 \le 4x < 6$

$-\dfrac{1}{2} \le x < \dfrac{3}{2}$

$\left[-\dfrac{1}{2}, \dfrac{3}{2}\right)$

69. $\dfrac{1}{15} < \dfrac{8-3x}{15} < \dfrac{4}{5}$

$15\left(\dfrac{1}{15}\right) < 15\left(\dfrac{8-3x}{15}\right) < 15\left(\dfrac{4}{5}\right)$

$1 < 8 - 3x < 12$

$-7 < -3x < 4$

$-\dfrac{4}{3} < x < \dfrac{7}{3}$

$\left(-\dfrac{4}{3}, \dfrac{7}{3}\right)$

71. $0.3 < 0.2x - 0.9 < 1.5$

$1.2 < 0.2x < 2.4$

$6 < x < 12$

$(6, 12)$

73. a. The graph of y_2 is between the graphs of y_1 and y_2 when $-5 < x < 2.5$, so the solution of $y_1 < y_2 < y_3$ is $(-5, 2.5)$.

b. The graph of y_2 is below the graph of y_1 when $x > 2.5$. The graph of y_2 is above the graph of y_3 when $x < -5$. Thus the solution of $y_2 < y_1$ or $y_2 > y_3$ is $(-\infty, -5) \cup (2.5, \infty)$.

75. The graph of y_2 is between the graphs of y_1 and y_3 when $2 < x < 9$, so the solution of $y_1 \le y_2 \le y_3$ is $[2, 9]$. The graph of y_2 is below the graph of y_1 when $x < 2$. The graph of y_2 is above the graph of y_3 when $x > 9$. Thus the solution of $y_2 \le y_1$ or $y_2 \ge y_3$ is $(-\infty, 2] \cup [9, \infty)$.

77. $|-7| - |19| = 7 - 19 = -12$

79. $-(-6) - |-10| = 6 - 10 = -4$

81. $|x| = 7$

$x = -7, 7$

83. $|x| = 0$

$x = 0$

85. Both lines are above the level representing 1500 for the years 2004 and 2005.

87. answers may vary

89. $-29 \le C \le 35$

$-29 \le \dfrac{5}{9}(F - 32) \le 35$

$-52.5 \le F - 32 \le 63$

$-20.2 \le F \le 95$

$-20.2° \le F \le 95°$

91. $70 \le \dfrac{68 + 65 + 75 + 78 + 2x}{6} \le 79$

$420 \le 286 + 2x \le 474$

$134 \le 2x \le 188$

$67 \le x \le 94$

If Christian scores between 67 and 94 inclusive on his final exam, he will receive a C in the course.

93. $2x - 3 < 3x + 1 < 4x - 5$

$\begin{array}{lll} 2x - 3 < 3x + 1 & \text{and} & 3x + 1 < 4x - 5 \\ -x < 4 & \text{and} & -x < -6 \\ x > -4 & \text{and} & x > 6 \end{array}$

$x > 6$

$(6, \infty)$

95. $-3(x - 2) \le 3 - 2x \le 10 - 3x$

$\begin{array}{lll} -3x + 6 \le 3 - 2x & \text{and} & 3 - 2x \le 10 - 3x \\ -x \le -3 & \text{and} & x \le 7 \\ x \ge 3 & & \end{array}$

$3 \le x \le 7$

$[3, 7]$

97. $5x - 8 < 2(2 + x) < -2(1 + 2x)$

$\begin{array}{lll} 5x - 8 < 4 + 2x & \text{and} & 4 + 2x < -2 - 4x \\ 3x < 12 & \text{and} & 6x < -6 \\ x < 4 & \text{and} & x < -1 \end{array}$

$x < -1$

$(-\infty, -1)$

Section 3.4 Practice Exercises

1. $|q| = 3$
   $q = 3$ or $q = -3$
   The solution set is $\{-3, 3\}$.

2. $|2x - 3| = 5$
   $2x - 3 = 5$ or $2x - 3 = -5$
      $2x = 8$ or $2x = -2$
       $x = 4$ or $x = -1$
   The solution set is $\{-1, 4\}$.
   Check:

3. $\left|\dfrac{x}{5} + 1\right| = 15$

   $\dfrac{x}{5} + 1 = 15$ or $\dfrac{x}{5} + 1 = -15$

      $\dfrac{x}{5} = 14$ or $\dfrac{x}{5} = -16$

       $x = 70$ or $x = -80$

   The solutions are -80 and 70.

4. $|3x| + 8 = 14$
       $|3x| = 6$
   $3x = 6$ or $3x = -6$
     $x = 2$ or $x = -2$
   The solutions are -2 and 2.

5. $|z| = 0$
   The solution is 0.

6. $3|z| + 9 = 7$
   Algebraic Solution:
   $3|z| + 9 = 7$
      $3|z| = -2$
       $|z| = -\dfrac{2}{3}$

   The absolute value of a number is never
   negative, so there is no solution. The solution set
   is $\{\ \}$ or $\varnothing$.

Graphical Solution:

Since the graphs do not intersect, there is no
solution. The solution set is $\{\ \}$ or $\varnothing$.

7. $\left|\dfrac{5x + 3}{4}\right| = -8$

   The absolute value of a number is never
   negative, so there is no solution. The solution set
   is $\{\ \}$ or $\varnothing$.

8. $|2x + 4| = |3x - 1|$
   $2x + 4 = 3x - 1$ or $2x + 4 = -(3x - 1)$
   $-x + 4 = -1$ $2x + 4 = -3x + 1$
      $-x = -5$ $5x + 4 = 1$
        $x = 5$ $5x = -3$
                          $x = -\dfrac{3}{5}$

   The solutions are $-\dfrac{3}{5}$ and 5.

   Check:

9. $|x - 2| = |8 - x|$
   $x - 2 = 8 - x$ or $x - 2 = -(8 - x)$
   $2x - 2 = 8$ $x - 2 = -8 + x$
      $2x = 10$ $-2 = -8$ False
        $x = 5$
   The solution is 5.

Vocabulary, Readiness & Video Check 3.4

1. $|x - 2| = 5$
   C. $x - 2 = 5$ or $x - 2 = -5$

2. $|x - 2| = 0$
   A. $x - 2 = 0$

3. $|x - 2| = |x + 3|$
   B. $x - 2 = x + 3$ or $x - 2 = -(x + 3)$

4. $|x + 3| = 5$
   E. $x + 3 = 5$ or $x + 3 = -5$

5. $|x + 3| = -5$
   D. $\varnothing$

6. If a is negative, $|X| = a$ has no solution. (Also, if a is 0, we solve $X = 0$.)

Exercise Set 3.4

1. $|x| = 7$
   $x = 7$ or $x = -7$

3. $|3x| = 12.6$
   $3x = 12.6$ or $3x = -12.6$
   $x = 4.2$ or $x = -4.2$

5. $|2x - 5| = 9$
   $2x - 5 = 9$ or $2x - 5 = -9$
    $2x = 14$ or $2x = -4$
     $x = 7$ or $x = -2$

7. $\left|\dfrac{x}{2} - 3\right| = 1$

$$\dfrac{x}{2} - 3 = 1 \quad \text{or} \quad \dfrac{x}{2} - 3 = -1$$

$$2\left(\dfrac{x}{2} - 3\right) = 2(1) \quad \text{or} \quad 2\left(\dfrac{x}{2} - 3\right) = 2(-1)$$

$$x - 6 = 2 \quad \text{or} \quad x - 6 = -2$$
$$x = 8 \quad \text{or} \quad x = 4$$

9. $|z| + 4 = 9$
    $|z| = 5$
  $z = -5$ or $z = -5$

11. $|3x| + 5 = 14$
    $|3x| = 9$
  $3x = 9$ or $3x = -9$
   $x = 3$ or $x = -3$

13. $|2x| = 0$
   $2x = 0$
    $x = 0$

15. $|4n + 1| + 10 = 4$
    $|4n + 1| = -6$ which is impossible.

The solution set is $\varnothing$.

17. $|5x - 1| = 0$
   $5x - 1 = 0$
     $5x = 1$
      $x = \dfrac{1}{5}$

19. $|5x - 7| = |3x + 11|$
  $5x - 7 = 3x + 11$ or $5x - 7 = -(3x + 11)$
    $2x = 18$ or $5x - 7 = -3x - 11$
     $x = 9$ or $8x = -4$
                       $x = -\dfrac{1}{2}$

21. $|z + 8| = |z - 3|$
  $z + 8 = z - 3$ or $z + 8 = -(z - 3)$
    $8 = -3$ or $z + 8 = -z + 3$
                     $2z = -5$
                      $z = -\dfrac{5}{2}$

The only solution is $-\dfrac{5}{2}$.

23. The graphs intersect when $x = -1$ and $x = 4$, so the solution set is $\{-1, 4\}$.

25. The graphs intersect when $x = 2.5$, so the solution set is $\{2.5\}$.

27. $|x| = 4$
   $x = 4$ or $x = -4$

29. $|y| = 0$; $y = 0$

31. $|z| = -2$ is impossible. The solution set is $\varnothing$.

33. $|7 - 3x| = 7$
  $7 - 3x = 7$ or $7 - 3x = -7$
   $-3x = 0$ or $-3x = -14$
     $x = 0$ or $x = \dfrac{14}{3}$

35. $|6x| - 1 = 11$
    $|6x| = 12$
  $6x = 12$ or $6x = -12$
   $x = 2$ or $x = -2$

37. $|4p| = -8$ is impossible. The solution set is $\varnothing$.

39. $|x-3|+3=7$

$\qquad |x-3|=4$

$\qquad x-3=4 \quad$ or $\quad x-3=-4$

$\qquad\qquad x=7 \quad$ or $\qquad x=-1$

41. $\left|\dfrac{z}{4}+5\right|=-7$ is impossible. The solution set is $\varnothing$.

43. $|9v-3|=-8$ is impossible. The solution set is $\varnothing$.

45. $|8n+1|=0$

$\qquad 8n+1=0$

$\qquad\quad 8n=-1$

$\qquad\quad\ n=-\dfrac{1}{8}$

47. $|1-6c|-7=-3$

$\qquad |1-6c|=4$

$\qquad 1-6c=4 \quad$ or $\quad 1-6c=-4$

$\qquad\quad 6c=3 \quad$ or $\qquad 6c=-5$

$\qquad\quad\ c=\dfrac{1}{2} \quad$ or $\qquad c=-\dfrac{5}{6}$

49. $|5x+1|=11$

$\qquad 5x+1=11 \quad$ or $\quad 5x+1=-11$

$\qquad\quad 5x=10 \quad$ or $\qquad 5x=-12$

$\qquad\quad\ x=2 \quad$ or $\qquad x=-\dfrac{12}{5}$

51. $|4x-2|=|-10|$

$\qquad |4x-2|=10$

$\qquad 4x-2=10 \quad$ or $\quad 4x-2=-10$

$\qquad\quad 4x=12 \quad$ or $\qquad 4x=-8$

$\qquad\quad\ x=3 \quad$ or $\qquad x=-2$

53. $|5x+1|=|4x-7|$

$\qquad 5x+1=4x-7 \quad$ or $\quad 5x+1=-(4x-7)$

$\qquad\qquad x=-8 \qquad$ or $\quad 5x+1=-4x+7$

$\qquad\qquad\qquad\qquad\qquad\quad 9x=6$

$\qquad\qquad\qquad\qquad\qquad\quad\ x=\dfrac{2}{3}$

55. $|6+2x|=-|-7|$

$\qquad |6+2x|=-7$ which is impossible. The solution set is $\varnothing$.

57. $|2x-6|=|10-2x|$

$\qquad 2x-6=10-2x \quad$ or $\quad 2x-6=-(10-2x)$

$\qquad\qquad 4x=16 \qquad$ or $\quad 2x-6=-10+2x$

$\qquad\qquad\ x=4 \qquad$ or $\qquad -6=-10$

$\qquad -6=-10$ is impossible. The only solution is 4.

59. $\left|\dfrac{2x-5}{3}\right|=7$

$\qquad \dfrac{2x-5}{3}=7 \quad$ or $\quad \dfrac{2x-5}{3}=-7$

$\qquad 2x-5=21 \quad$ or $\quad 2x-5=-21$

$\qquad\quad 2x=26 \quad$ or $\qquad 2x=-16$

$\qquad\quad\ x=13 \quad$ or $\qquad x=-8$

61. $2+|5n|=17$

$\qquad |5n|=15$

$\qquad 5n=15 \quad$ or $\quad 5n=-15$

$\qquad\ n=3 \quad$ or $\qquad n=-3$

63. $\left|\dfrac{2x-1}{3}\right|=|-5|$

$\qquad \left|\dfrac{2x-1}{3}\right|=5$

$\qquad \dfrac{2x-1}{3}=5 \quad$ or $\quad \dfrac{2x-1}{3}=-5$

$\qquad 2x-1=15 \quad$ or $\quad 2x-1=-15$

$\qquad\quad 2x=16 \quad$ or $\qquad 2x=-14$

$\qquad\quad\ x=8 \quad$ or $\qquad x=-7$

65. $|2y-3|=|9-4y|$

$\qquad 2y-3=9-4y \quad$ or $\quad 2y-3=-(9-4y)$

$\qquad\quad 6y=12 \qquad$ or $\quad 2y-3=-9+4y$

$\qquad\quad\ y=2 \qquad$ or $\qquad -2y=-6$

$\qquad\qquad\qquad\qquad\qquad\quad y=3$

67. $\left|\dfrac{3n+2}{8}\right|=|-1|$

$\qquad \left|\dfrac{3n+2}{8}\right|=1$

$\qquad \dfrac{3n+2}{8}=1 \quad$ or $\quad \dfrac{3n+2}{8}=-1$

$\qquad 3n+2=8 \quad$ or $\quad 3n+2=-8$

$\qquad\quad 3n=6 \quad$ or $\qquad 3n=-10$

$\qquad\quad\ n=2 \quad$ or $\qquad n=-\dfrac{10}{3}$

69. $|x + 4| = |7 - x|$

$x + 4 = 7 - x$ or $x + 4 = -(7 - x)$

$2x = 3$ or $x + 4 = -7 + x$

$x = \dfrac{3}{2}$ or $4 = -7$

$4 = -7$ is impossible. The only solution is $\dfrac{3}{2}$.

71. $\left|\dfrac{8c - 7}{3}\right| = -|-5|$

$\left|\dfrac{8c - 7}{3}\right| = -5$ which is impossible.

The solution set is $\varnothing$.

73. 31% of cheese production came from cheddar cheese.

75. 9% of 360° is 0.09(360°) = 32.4°

77. answers may vary

79. answers may vary

81. Since absolute value is never negative, the solution set is $\varnothing$.

83. All numbers whose distance from 0 is 5 units is written as $|x| = 5$.

85. answers may vary

87. $|x - 1| = 5$

89. answers may vary

91. $|x| = 6$

93. $|x - 2| = |3x - 4|$

Section 3.5 Practice Exercises

1. $|x| < 5$

The solution set of this inequality contains all numbers whose distance from 0 is less than 5. The solution set is (−5, 5).

2. $|b + 1| < 3$

Algebraic Solution:

$|b + 1| < 3$

$-3 < b + 1 < 3$

$-3 - 1 < b + 1 - 1 < 3 - 1$

$-4 < b < 2$

(−4, 2)

Graphical Solution:

Graph $y_1 = |x + 1|$ and $y_2 = 3$ in a standard window.

y_1 is below y_2 for x-values between −4 and 2.

Thus the solution set of $|x + 1| < 3$ is (−4, 2).

3. $|3x - 2| + 5 \le 9$

$|3x - 2| \le 9 - 5$

$|3x - 2| \le 4$

$-4 \le 3x - 2 \le 4$

$-4 + 2 \le 3x - 2 + 2 \le 4 + 2$

$-2 \le 3x \le 6$

$-\dfrac{2}{3} \le x \le 2$

$\left[-\dfrac{2}{3}, 2\right]$

4. $\left|3x + \dfrac{5}{8}\right| < -4$

The absolute value of a number is always nonnegative and can never be less than −4. The solution set is { } or $\varnothing$.

The absolute value graph is always above the horizontal line $y = -4$, so the solution set is { } or $\varnothing$.

5. $\left|\dfrac{3(x-2)}{5}\right| \le 0$

$\dfrac{3(x-2)}{5} = 0$

$5\left[\dfrac{3(x-2)}{5}\right] = 5(0)$

$3(x-2) = 0$

$3x - 6 = 0$

$3x = 6$

$x = 2$

The solution set is {2}.

6. $|y+4| \ge 6$
Algebraic Solution:
$|y+4| \ge 6$

$y+4 \le -6 \qquad$ or $\qquad y+4 \ge 6$

$y+4-4 \le -6-4 \quad$ or $\quad y+4-4 \ge 6-4$

$y \le -10 \qquad$ or $\qquad y \ge 2$

$(-\infty, -10] \cup [2, \infty)$

$-10 \qquad 2$

Graph $y_1 = |x+4|$ and $y_2 = 6$.

y_1 intersects or is above y_2 for x-values less than or equal to -10 and also x-values greater than or equal to 2, or $(-\infty, -10] \cup [2, \infty)$.

7. $|4x+3|+5 > 3$

$|4x+3|+5-5 > 3-5$

$|4x+3| > -2$

The absolute value of any number is always nonnegative and thus is always greater than -2.
$(-\infty, \infty)$

Graph $y_1 = |4x+3|+5$ and $y_2 = 3$.

$y_1 > y_2$ or $|4x+3|+5 > 3$ for all real numbers.

8. $\left|\dfrac{x}{2}-3\right| - 3 > 0$

Graph $y_1 = \left|\dfrac{x}{2}-3\right| - 3$ and find x-values where

the graph of y_1 is above the x-axis.

The graph of $y = \left|\dfrac{x}{2}-3\right| - 3$ is above the x-axis

for x-values less than 0 or greater than 12.
The solution set is $(-\infty, 0) \cup (12, \infty)$.

Vocabulary, Readiness & Video Check 3.5

1. D

2. E

3. C

4. B

5. A

6. The left side of the inequality is an absolute value, which must be nonnegative—it must be 0 or positive. Therefore, there is no value of x that can make the value of this absolute value be less than the negative value on the right side of the inequality.

7. The solution set involves "or" and "or" means "union."

Exercise Set 3.5

1. $|x| \le 4$

$-4 \le x \le 4$

$[-4, 4]$

$\xleftarrow{\hspace{1cm}} \underset{-4}{\rule[0.3ex]{0pt}{0pt}} \rule[0.3ex]{1.5cm}{1pt} \underset{4}{\rule[0.3ex]{0pt}{0pt}} \xrightarrow{\hspace{1cm}}$

3. $|x-3| < 2$

$-2 < x-3 < 2$

$1 < x < 5$

$(1, 5)$

$\xleftarrow{\hspace{1cm}} \underset{1}{\rule[0.3ex]{0pt}{0pt}} \rule[0.3ex]{1.5cm}{1pt} \underset{5}{\rule[0.3ex]{0pt}{0pt}} \xrightarrow{\hspace{1cm}}$

5. $|x + 3| < 2$
$-2 < x + 3 < 2$
$-5 < x < -1$
$(-5, -1)$

7. $|2x + 7| \leq 3$
$-13 \leq 2x + 7 \leq 13$
$-20 \leq 2x \leq 6$
$-10 \leq x \leq 3$
$[-10, 3]$

9. $|x| + 7 \leq 12$
$|x| \leq 5$

$-5 \leq x \leq 5$
$[-5, 5]$

11. $|3x - 1| < -5$
No real solutions; $\varnothing$

13. $|x - 6| - 7 \leq -1$
$|x - 6| \leq 6$

$-6 \leq x - 6 \leq 6$
$0 \leq x \leq 12$
$[0, 12]$

15. $|x| > 3$
$x < -3$ or $x > 3$
$(-\infty, -3) \cup (3, \infty)$

17. $|x + 10| \geq 14$
$x + 10 \leq -14$ or $x + 10 \geq 14$
$x \leq -24$ or $x \geq 4$
$(-\infty, -24] \cup [4, \infty)$

19. $|x| + 2 > 6$
$|x| > 4$

$x < -4$ or $x > 4$
$(-\infty, -4) \cup (4, \infty)$

21. $|5x| > -4$
All real numbers
$(-\infty, \infty)$

23. $|6x - 8| + 3 > 7$
$\qquad |6x - 8| > 4$

$6x - 8 < -4$ or $6x - 8 > 4$
$\quad 6x < 4$ or $\quad 6x > 12$
$\quad x < \dfrac{2}{3}$ or $\quad x > 2$

$\left(-\infty, \dfrac{2}{3}\right) \cup (2, \infty)$

25. $|x| \leq 0$
$|x| = 0$
$\ x = 0$

27. $|8x + 3| > 0$ only excludes $|8x + 3| = 0$
$8x + 3 = 0$
$\quad 8x = -3$
$\quad x = -\dfrac{3}{8}$

All real numbers except $-\dfrac{3}{8}$.

$\left(-\infty, -\dfrac{3}{8}\right) \cup \left(-\dfrac{3}{8}, \infty\right)$

29. $|x| \leq 2$
$-2 \leq x \leq 2$
$[-2, 2]$

31. $|y| > 1$
$y < -1$ or $y > 1$
$(-\infty, -1) \cup (1, \infty)$

33. $|x - 3| < 8$
$-8 < x - 3 < 8$
$-5 < x < 11$
$(-5, 11)$

35. $|0.6x - 3| > 0.6$

$\quad 0.6x - 3 < -0.6 \quad$ or $\quad 0.6x - 3 > 0.6$

$\qquad 0.6x < 2.4 \quad$ or $\qquad 0.6x > 3.6$

$\qquad\quad x < 4 \quad$ or $\qquad\quad x > 6$

$(-\infty, 4) \cup (6, \infty)$

37. $5 + |x| \le 2$

$\qquad |x| \le -3$

No real solution

$\varnothing$

39. $|x| > -4$

All real numbers

$(-\infty, \infty)$

41. $|2x - 7| \le 11$

$\quad -11 \le 2x - 7 \le 11$

$\quad -4 \le 2x \le 18$

$\quad -2 \le x \le 9$

$[-2, 9]$

43. $|x + 5| + 2 \ge 8$

$\qquad |x + 5| \ge 6$

$\quad x + 5 \le -6 \quad$ or $\quad x + 5 \ge 6$

$\qquad x \le -11 \quad$ or $\qquad x \ge 1$

$(-\infty, -11] \cup [1, \infty)$

45. $|x| > 0$ only excludes $|x| = 0$, or $x = 0$.

All real numbers except $x = 0$

$(-\infty, 0) \cup (0, \infty)$

47. $9 + |x| > 7$

$\qquad |x| > -2$

All real numbers

$(-\infty, \infty)$

49. $6 + |4x - 1| \le 9$

$\qquad |4x - 1| \le 3$

$\quad -3 \le 4x - 1 \le 3$

$\quad -2 \le 4x \le 4$

$\quad -\dfrac{1}{2} \le x \le 1$

$\left[-\dfrac{1}{2}, 1\right]$

51. $\left|\dfrac{2}{3}x + 1\right| > 1$

$\quad \dfrac{2}{3}x + 1 < -1 \quad$ or $\quad \dfrac{2}{3}x + 1 > 1$

$\qquad \dfrac{2}{3}x < -2 \quad$ or $\qquad \dfrac{2}{3}x > 0$

$\qquad\quad x < -3 \quad$ or $\qquad\quad x > 0$

$(-\infty, -3) \cup (0, \infty)$

53. $|5x + 3| < -6$

No real solution

$\varnothing$

55. $\left|\dfrac{8x - 3}{4}\right| \le 0$

$\quad \dfrac{8x - 3}{4} = 0$

$\quad 8x - 3 = 0$

$\quad 8x = 3$

$\quad x = \dfrac{3}{8}$

$\left\{\dfrac{3}{8}\right\}$

57. $|1 + 3x| + 4 < 5$

$\qquad |1 + 3x| < 1$

$\quad -1 < 1 + 3x < 1$

$\quad -2 < 3x < 0$

$\quad -\dfrac{2}{3} < x < 0$

$\left(-\dfrac{2}{3}, 0\right)$

59. $\left|\dfrac{x+6}{3}\right| > 2$

$\dfrac{x+6}{3} < -2$ or $\dfrac{x+6}{3} > 2$

$x+6 < -6$ or $x+6 > 6$

$x < -12$ or $x > 0$

$(-\infty, -12) \cup (0, \infty)$

61. $-15 + |2x - 7| \le -6$

$\qquad |2x - 7| \le 9$

$-9 \le 2x - 7 \le 9$

$-2 \le 2x \le 16$

$-1 \le x \le 8$

$[-1, 8]$

63. $\left|2x + \dfrac{3}{4}\right| - 7 \le -2$

$\left|2x + \dfrac{3}{4}\right| \le 5$

$-5 \le 2x + \dfrac{3}{4} \le 5$

$-20 \le 8x + 3 \le 20$

$-23 \le 8x \le 17$

$-\dfrac{23}{8} \le x \le \dfrac{17}{8}$

$\left[-\dfrac{23}{8}, \dfrac{17}{8}\right]$

65. $|2x - 3| < 7$

$-7 < 2x - 3 < 7$

$-4 < 2x < 10$

$-2 < x < 5$

$(-2, 5)$

67. $|2x - 3| = 7$

$2x - 3 = 7$ or $2x - 3 = -7$

$2x = 10$ or $2x = -4$

$x = 5$ or $x = -2$

69. $|x - 5| \ge 12$

$x - 5 \le -12$ or $x - 5 \ge 12$

$x \le -7$ or $x \ge 17$

$(-\infty, -7] \cup [17, \infty)$

71. $|9 + 4x| = 0$

$9 + 4x = 0$

$4x = -9$

$x = -\dfrac{9}{4}$

73. $|2x + 1| + 4 < 7$

$|2x + 1| < 3$

$-3 < 2x + 1 < 3$

$-4 < 2x < 2$

$-2 < x < 1$

$(-2, 1)$

75. $|3x - 5| + 4 = 5$

$|3x - 5| = 1$

$3x - 5 = 1$ or $3x - 5 = -1$

$3x = 6$ or $3x = 4$

$x = 2$ or $x = \dfrac{4}{3}$

77. $|x + 11| = -1$ is impossible. The solution set is $\varnothing$.

79. $\left|\dfrac{2x - 1}{3}\right| = 6$

$\dfrac{2x - 1}{3} = 6$ or $\dfrac{2x - 1}{3} = -6$

$2x - 1 = 18$ or $2x - 1 = -18$

$2x = 19$ or $2x = -17$

$x = \dfrac{19}{2}$ or $x = -\dfrac{17}{2}$

81. $\left|\dfrac{3x - 5}{6}\right| > 5$

$\dfrac{3x - 5}{6} < -5$ or $\dfrac{3x - 5}{6} > 5$

$3x - 5 < -30$ or $3x - 5 > 30$

$3x < -25$ or $3x > 35$

$x < -\dfrac{25}{3}$ or $x > \dfrac{35}{3}$

$\left(-\infty, -\dfrac{25}{3}\right) \cup \left(\dfrac{35}{3}, \infty\right)$

83. a. The graphs of $y_1 = |x-3| - 2$ and $y_2 = 6$
intersect at the points $(-5, 6)$ and $(11, 6)$.
Thus the solution set of $|x - 3| - 2 = 6$ is
$\{-5, 11\}$.

b. The graph of y_1 is below the graph of y_2
for x-values between -5 and 11. Thus the
solution set of $|x - 3| - 2 < 6$ is $(-5, 11)$.

c. The graph of y_1 is above the graph of y_2
for x-values less than -5 or greater than 11.
Thus the solution set of $|x - 3| - 2 \geq 6$ is
$(-\infty, -5] \cup [11, \infty)$.

85. a. The graphs of $y_1 = |x+2| - 10$ and $y_2 = -4$
intersect at the points $(-8, -4)$ and $(4, -4)$.
Thus the solution set of $|x - 2| - 10 = -4$ is
$\{-8, 4\}$.

b. The graph of y_1 is below the graph of y_2
for x-values between -8 and 4. Thus the
solution set of $|x - 2| - 10 \leq -4$ is $[-8, 4]$.

c. The graph of y_1 is above the graph of y_2
for x-values less than -8 or greater than 4.
Thus the solution set of $|x - 2| < 10 > -4$ is
$(-\infty, -8) \cup (4, \infty)$.

87. $P(\text{rolling a } 2) = \dfrac{1}{6}$

89. $P(\text{rolling a } 7) = 0$

91. $P(\text{rolling a 1 or 3}) = \dfrac{1}{3}$

93.
$$3x - 4y = 12$$
$$3(2) - 4y = 12$$
$$6 - 4y = 12$$
$$-4y = 6$$
$$y = -\frac{3}{2} = -1.5$$

95.
$$3x - 4y = 12$$
$$3x - 4(-3) = 12$$
$$3x + 12 = 12$$
$$3x = 0$$
$$x = 0$$

97. $|x| < 7$

99. $|x| \leq 5$

101. answers may vary

103. $|3.5 - x| < 0.05$
$$-0.05 < 3.5 - x < 0.05$$
$$-3.55 < -x < -3.45$$
$$3.55 > x > 3.45$$
$$3.45 < x < 3.55$$

Section 3.6 Practice Exercises

1. $3x + y < 8$
The boundary line is $3x + y = 8$. Graph a dashed
boundary line because the inequality symbol is
$<$. The point $(0, 0)$ is not on the boundary line, so
we use it as a test point. Replace x with 0 and y
with 0 in the original inequality.
$$3x + y < 8$$
$$3(0) + 0 < 8$$
$$0 < 8 \quad \text{True}$$
Since $(0, 0)$ satisfies the inequality, shade the
half-plane that contains $(0, 0)$. Every point in the
shaded half-plane satisfies the original
inequality.

2. $x \geq 3y$
First graph the boundary line $x = 3y$. Graph a
solid boundary line because the inequality
symbol is $\geq$. We choose $(0, 1)$ as a test point.
$$x \geq 3y$$
$$0 \geq 3(1)$$
$$0 \geq 3 \quad \text{False}$$
Since this point does not satisfy the inequality,
shade the half-plane on the opposite side of the
boundary line from $(0, 1)$. The graph of $x \geq 3y$ is
the boundary line together with the shaded
region.

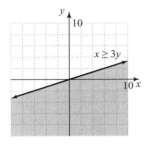

3. The intersection of $x \le 3$ and $y \le x - 2$

Graph each inequality. The intersection of the two graphs is all points common to both regions, as shown by the darker shading in the graph.

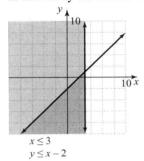

$x \le 3$
$y \le x - 2$

4. The union of $2x - 3y \le -2$ or $y \ge 1$

Graph each inequality. The union of the two inequalities is both shaded regions, including the boundary lines, as shown in the graph.

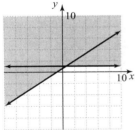

Vocabulary, Readiness & Video Check 3.6

1. You find the boundary line equation by replacing the inequality symbol with =. The points on this line are solutions (line is solid) if the inequality is $\ge$ or $\le$; the points on this line are not solutions (line is dashed) if the inequality is $>$ or $<$.

2. Graph each inequality separately on the same rectangular coordinate system. The intersection is where their solution regions overlap—that is, where you've shaded twice.

Exercise Set 3.6

1. $x < 2$

Graph $x = 2$ as a dashed line. Shade to the left of the line.

3. $x - y \ge 7$

Graph $x - y = 7$ as a solid line.
Test: $(0, 0)$
$0 - 0 \ge 7$
$\quad 0 \ge 7$ False
Shade the half-plane that does not contain $(0, 0)$.

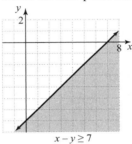

$x - y \ge 7$

5. $3x + y > 6$

Graph $3x + y = 6$ as a dashed line.
Test: $(0, 0)$
$3(0) + 0 > 6$
$\quad 0 > 6$ False
Shade the half-plane that does not contain $(0, 0)$.

7. $y \le -2x$

Graph $y = -2x$ as a solid line.
Test: $(1, 1)$
$1 \le -2(1)$
$1 \le -2$ False
Shade the half-plane that does not contain $(1, 1)$.

9. $2x + 4y \ge 8$

Graph $2x + 4y = 8$ as a solid line.
Test: $(0, 0)$
$2(0) + 4(0) \ge 8$
$0 \ge 8$ False
Shade the half-plane that does not contain $(0, 0)$.

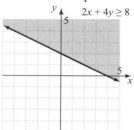

11. $5x + 3y > -15$

Graph $5x + 3y = -15$ as a dashed line.
Test: $(0, 0)$
$5(0) + 3(0) > -15$
$0 > -15$ True
Shade the half-plane that contains $(0, 0)$.

13. $x \ge 3$ and $y \le -2$

The intersection is shown by the darker shading in the graph.

15. $x \le -2$ or $y \ge 4$

The union is both shaded regions, including the boundary line.

17. $x - y < 3$ and $x > 4$

The intersection is shown by the darker shading in the graph.

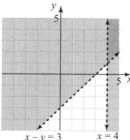

19. $x + y \le 3$ or $x - y \ge 5$

The union is both shaded regions, including the boundary lines.

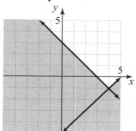

21. $y \geq -2$

Graph $y = -2$ as a solid line.

Test: (0, 0)

$0 \geq -2$ True

Shade the half-plane that contains (0, 0).

23. $x - 6y < 12$

$-6y < 12 - x$

$y > \dfrac{1}{6}x - 2$

Graph the boundary line as a dashed line.

Test: (0, 0)

$0 > \dfrac{1}{6}(0) - 2$ True

Shade the half-plane that contains (0, 0).

25. $x > 5$

Graph $x = 5$ as a dashed line.

Test: (0, 0)

$0 > 5$ False

Shade the half-plane that does not contain (0, 0).

27. $-2x + y \leq 4$

$y \leq 2x + 4$

Graph the boundary line as a solid line.

Test: (0, 0)

$0 \leq 2(0) + 4$ True

Shade the half-plane that contains (0, 0).

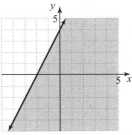

29. $x - 3y < 0$

$-3y < -x$

$y > \dfrac{x}{3}$

Graph the boundary line as a dashed line.

Test: (0, 1)

$1 > \dfrac{0}{3}$ True

Shade the half-plane that contains (0, 1).

31. $3x - 2y \leq 12$

$-2y \leq -3x + 12$

$y \geq \dfrac{3}{2}x - 6$

Graph the boundary line as a solid line.

Test: (0, 0)

$3(0) - 2(0) \leq 12$ True

Shade the half-plane that contains (0, 0).

33. $x - y > 2$ or $y < 5$
$y < x - 2$ or $y < 5$

Graph each inequality. The union of the two inequalities is both shaded regions, as shown by the shading in the graph below.

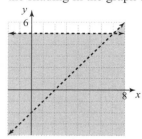

35. $x + y \le 1$ and $y \le -1$
$y \le -x + 1$ and $y \le -1$

Graph each inequality. The intersection of the two inequalities is all points common to both regions, as shown by the shading in the graph below.

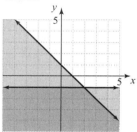

37. $2x + y > 4$ or $x \ge 1$
$y > -2x + 4$ or $x \ge 1$

Graph each inequality. The union of the two inequalities is both shaded regions, as shown by the shading in the graph below.

39. $x \ge -2$ and $x \le 1$
$-2 \le x \le 1$

Graph each inequality. The intersection of the two inequalities is all points common to both regions, as shown by the shading in the graph below.

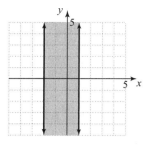

41. $x + y \le 0$ or $3x - 6y \ge 12$
$y \le -x$ or $-6y \ge -3x + 12$
$y \le -x$ or $y \le \dfrac{1}{2}x - 2$

Graph each inequality. The union of the two inequalities is both shaded regions, as shown by the shading in the graph below.

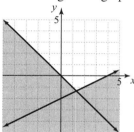

43. $2x - y > 3$ and $x > 0$
$y < 2x - 3$ and $x > 0$

Graph each inequality. The intersection of the two inequalities is all points common to both regions, as shown by the shading in the graph below.

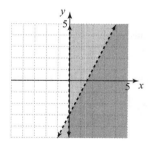

45. $y \le 2x + 3$

The boundary line should be solid, and the half-plane below the boundary line should be shaded. The correct graph is D.

47. $y > 2x + 3$

The boundary line should be dashed, and the half-plane above the boundary line should be shaded. The correct graph is A.

49. The boundary line, $x = 2$, is solid, and the half-plane with x-values greater than 2 is shaded. The inequality is $x \geq 2$.

51. The boundary line, $y = -3$, is solid, and the half-plane with y-values less than -3 is shaded. The inequality is $y \leq -3$.

53. The boundary line, $y = 4$, is dashed, and the half-plane with y-values greater than 4 is shaded. The inequality is $y > 4$.

55. The boundary line, $x = 1$, is dashed, and the half-plane with x-values less than 1 is shaded. The inequality is $x < 1$.

57. $2^3 = 2 \cdot 2 \cdot 2 = 8$

59. $-5^2 = -(5 \cdot 5) = -25$

61. $(-2)^4 = (-2)(-2)(-2)(-2) = 16$

63. $\left(\dfrac{3}{5}\right)^3 = \dfrac{3^3}{5^3} = \dfrac{27}{125}$

65. Domain: [1, 5]
Range: [1, 3]
Since it fails the vertical line test it is not a function.

67. A dashed boundary line should be used when the inequality contains a < or >.

69. $\begin{cases} x \geq 0 \\ y \geq 0 \\ 2x + 4y \leq 40 \end{cases}$

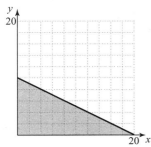

Chapter 3 Review

1. $4(x - 6) + 3 = 27$
Algebraic Solution:
$4(x - 6) + 3 = 27$
$4x - 24 + 3 = 27$
$4x - 21 = 27$
$4x = 48$
$x = 12$
The solution is 12.
Graphical Solution:
Graph $y_1 = 4(x - 6) + 3$ and $y_2 = 27$ in a $[-50, 50, 5]$ by $[-50, 50, 5]$ viewing window and find the intersection point.

The point of intersection (12, 27) indicates that the solution is 12.

2. $15(x + 2) - 6 = 18$
Algebraic Solution:
$15(x + 2) - 6 = 18$
$15x + 30 - 6 = 18$
$15x + 24 = 18$
$15x = -6$
$x = \dfrac{-6}{15} = -\dfrac{2}{5} = -0.4$

The solution is $-\dfrac{2}{5} = -0.4$.

Graphical Solution:
Graph $y_1 = 15(x + 2) - 6$ and $y_2 = 18$ in a $[-20, 20, 5]$ by $[-20, 20, 5]$ viewing window and find the intersection point.

The point of intersection $(-0.4, 18)$ indicates that the solution is $-0.4 = -\dfrac{2}{5}$.

3. $5x + 15 = 3(x + 2) + 2(x - 3)$

Algebraic Solution:
$5x + 15 = 3x + 6 + 2x - 6$
$5x + 15 = 5x$
$\quad 15 = 0 \quad$ False

This equation is a false statement no matter what value the variable x might have. The equation is a contradiction and has no solution. The solution set is $\varnothing$ or { }.

Graphical Solution:
Graph $y_1 = 5x + 15$ and $y_2 = 3(x + 2) + 2(x - 3)$ in a $[-20, 20, 5]$ by $[-20, 20, 5]$ viewing window.

The parallel lines indicate that the equation is a contradiction and has no solution. The solution set is $\varnothing$ or { }.

4. $2x - 5 + 3(x - 4) = 5(x + 2) - 27$

Algebraic Solution:
$2x - 5 + 3(x - 4) = 5(x + 2) - 27$
$2x - 3 + 3x - 12 = 5x + 10 - 27$
$\quad\quad 5x - 15 = 5x - 15$

Since both sides are the same, this equation is a true statement no matter what value the variable x might have. The equation is an identity and the solution is the set of all real numbers.

Graphical Solution:
Graph $y_1 = 2x - 5 + 3(x - 4)$ and $y_2 = 5(x + 2) - 27$ in a $[-20, 20, 5]$ by $[-20, 20, 5]$ viewing window.

The graphs of y_1 and y_2 are the same. When we trace along y_1, we have the same ordered pairs as when we trace along points of y_2. We know that the lines are identical because two points uniquely determine a line. The equation is an identity and the solution is the set of all real numbers.

5. $14 - 2(x + 3) = 3(x - 9) + 18$

Algebraic Solution:
$14 - 2(x + 3) = 3(x - 9) + 18$
$14 - 2x - 6 = 3x - 27 + 18$
$\quad -2x + 8 = 3x - 9$
$\quad\quad -5x + 8 = -9$
$\quad\quad\quad -5x = -17$
$$x = \frac{-17}{-5} = \frac{17}{5} = 3.4$$

The solution is $\dfrac{17}{5} = 3.4$.

Graphical Solution:
Graph $y_1 = 14 - 2(x + 3)$ and $y_2 = 3(x - 9) + 18$ in a $[-20, 20, 5]$ by $[-20, 20, 5]$ viewing window and find the intersection point.

The point of intersection $(3.4, 1.2)$ indicates that the solution is $3.4 = \dfrac{17}{5}$.

6. $16 + 2(5 - x) = 19 - 3(x + 2)$

Algebraic Solution:
$16 + 2(5 - x) = 19 - 3(x + 2)$
$16 + 10 - 2x = 19 - 3x - 6$
$\quad -2x + 26 = -3x + 13$
$\quad\quad x + 26 = 13$
$\quad\quad\quad x = -13$

The solution is -13.

Graphical Solution:
Graph $y_1 = 16 + 2(5 - x)$ and $y_2 = 19 - 3(x + 2)$ in a $[-100, 100, 10]$ by $[-100, 100, 10]$ viewing window and find the intersection point.

The point of intersection $(-13, 52)$ indicates that the solution is -13.

7. $0.4(x-6) = \pi x + \sqrt{3}$

Graph $y_1 = 0.4(x-6)$ and $y_2 = \pi x + \sqrt{3}$ in a standard viewing window and find the intersection point.

Rounding to the nearest hundredth, the point of intersection $(-1.51, -3.00)$ indicates that the solution is approximately -1.51.

8. $1.7x + \sqrt{7} = -0.4x - \sqrt{6}$

Graph $y_1 = 1.7x + \sqrt{7}$ and $y_2 = -0.4x - \sqrt{6}$ in a standard viewing window and find the intersection point.

Rounding to the nearest hundredth, the point of intersection $(-2.43, -1.48)$ indicates that the solution is approximately -2.43.

9. $3(x-5) > -(x+3)$
$3x - 15 > -x - 3$
$4x > 12$
$x > 3$
$(3, \infty)$

10. $-2(x+7) \geq 3(x+2)$
$-2x - 14 \geq 3x + 6$
$-5x \geq 20$
$x \leq -4$
$(-\infty, -4]$

11. $4x - (5 + 2x) < 3x - 1$
$4x - 5 - 2x < 3x - 1$
$2x - 5 < 3x - 1$
$-x < 4$
$x > -4$
$(-4, \infty)$

12. $3(x-8) < 7x + 2(5 - x)$
$3x - 24 < 7x + 10 - 2x$
$3x - 24 < 5x + 10$
$-2x < 34$
$x > -17$
$(-17, \infty)$

13. $24 \geq 6x - 2(3x - 5) + 2x$
$24 \geq 6x - 6x + 10 + 2x$
$24 \geq 10 + 2x$
$14 \geq 2x$
$7 \geq x$
$(-\infty, 7]$

14. $\dfrac{x}{3} + \dfrac{1}{2} > \dfrac{2}{3}$
$6\left(\dfrac{x}{3} + \dfrac{1}{2}\right) > 6\left(\dfrac{2}{3}\right)$
$2x + 3 > 4$
$2x > 1$
$x > \dfrac{1}{2}$
$\left(\dfrac{1}{2}, \infty\right)$

15. $x + \dfrac{3}{4} < -\dfrac{x}{2} + \dfrac{9}{4}$
$4\left(x + \dfrac{3}{4}\right) < 4\left(-\dfrac{x}{2} + \dfrac{9}{4}\right)$
$4x + 3 < -2x + 9$
$6x < 6$
$x < 1$
$(-\infty, 1)$

16. $\dfrac{x-5}{2} \leq \dfrac{3}{8}(2x + 6)$
$8\left(\dfrac{x-5}{2}\right) \leq 8\left[\dfrac{3}{8}(2x + 6)\right]$
$4(x-5) \leq 3(2x + 6)$
$4x - 20 \leq 6x + 18$
$-2x \leq 38$
$x \geq -19$
$[-19, \infty)$

17. Let n = number of pounds of laundry.
$15 < 0.5(10) + 0.4(n - 10)$
$15 < 5 + 0.4n - 4$
$15 < 1 + 0.4n$
$14 < 0.4n$
$35 < n$
It is more economical to use the housekeeper for more than 35 pounds of laundry per week.

18. Let x = the score from the last judge.

$$\frac{9.5+9.7+9.9+9.7+9.7+9.6+9.5+x}{8} \geq 9.65$$

$$67.6 + x \geq 77.2$$

$$x \geq 9.6$$

The last judge must give Nana at least a 9.6 for her to win the silver medal.

19. $1 \leq 4x - 7 \leq 3$

$8 \leq 4x \leq 10$

$2 \leq x \leq \dfrac{5}{2}$

$\left[2, \dfrac{5}{2}\right]$

20. $-2 \leq 8 + 5x < -1$

$-10 \leq 5x \leq -9$

$-2 \leq x \leq -\dfrac{9}{5}$

$\left[-2, \dfrac{9}{5}\right)$

21. $-3 < 4(2x - 1) < 12$

$-3 < 8x - 4 < 12$

$1 < 8x < 16$

$\dfrac{1}{8} < x < 2$

$\left(\dfrac{1}{8}, 2\right)$

22. $-6 < x - (3 - 4x) < -3$

$-6 < x - 3 + 4x < -3$

$-6 < 5x - 3 < -3$

$-3 < 5x < 0$

$-\dfrac{3}{5} < x < 0$

$\left(-\dfrac{3}{5}, 0\right)$

23.

$$\frac{1}{6} < \frac{4x-3}{3} \leq \frac{4}{5}$$

$$30\left(\frac{1}{6}\right) < 30\left(\frac{4x-3}{3}\right) \leq 30\left(\frac{4}{5}\right)$$

$$5 < 10(4x - 3) \leq 24$$

$$5 < 40x - 30 \leq 24$$

$$35 < 40x < 54$$

$$\frac{7}{8} < x \leq \frac{27}{20}$$

$$\left(\frac{7}{8}, \frac{27}{20}\right]$$

24. $x \leq 2$ and $x > -5$

$-5 < x \leq 2$

$(-5, 2]$

25. $3x - 5 > 6$ or $-x < -5$

$3x > 11$ or $x > 5$

$x > \dfrac{11}{3}$ or $x > 5$

$x > \dfrac{11}{3}$

$\left(\dfrac{11}{3}, \infty\right)$

26. $500 \leq F \leq 1000$

$500 \leq \dfrac{9}{5}C + 32 \leq 1000$

$468 \leq \dfrac{9}{5}C \leq 968$

$260 \leq C \leq 538$

Rounded to the nearest degree, firing temperatures range from 260°C to 538°C.

27. Let x = the amount saved each summer.

$4000 \leq 2x + 500 \leq 8000$

$3500 \leq 2x \leq 7500$

$1750 \leq x \leq 3750$

She must save between \$1750 and \$3750 each summer.

28. $|x - 7| = 9$

$x - 7 = 9$ or $x - 7 = -9$

$x = 16$ or $x = -2$

29. $|8 - x| = 3$

$8 - x = 3$ or $8 - x = -3$

$-x = -5$ or $-x = -11$

$x = 5$ or $x = 11$

30. $|2x + 9| = 9$
$$2x + 9 = 9 \quad \text{or} \quad 2x + 9 = -9$$
$$2x = 0 \quad \text{or} \quad 2x = -18$$
$$x = 0 \quad \text{or} \quad x = -9$$

31. $|-3x + 4| = 7$
$$-3x + 4 = 7 \quad \text{or} \quad -3x + 4 = -7$$
$$-3x = 3 \quad \text{or} \quad -3x = -11$$
$$x = -1 \quad \text{or} \quad x = \frac{11}{3}$$

32. $|3x - 2| + 6 = 10$
$$|3x - 2| = 4$$
$$3x - 2 = 4 \quad \text{or} \quad 3x - 2 = -4$$
$$3x = 6 \quad \text{or} \quad 3x = -2$$
$$x = 2 \quad \text{or} \quad x = -\frac{2}{3}$$

33. $5 + |6x + 1| = 5$
$$|6x + 1| = 0$$
$$6x + 1 = 0$$
$$6x = -1$$
$$x = -\frac{1}{6}$$

34. $-5 = |4x - 3|$
The solution set is $\varnothing$.

35. $|5 - 6x| + 8 = 3$
$$|5 - 6x| = -5$$
The solution set is $\varnothing$.

36. $-8 = |x - 3| - 10$
$$2 = |x - 3|$$
$$x - 3 = 2 \quad \text{or} \quad x - 3 = -2$$
$$x = 5 \quad \text{or} \quad x = 1$$

37. $\left| \dfrac{3x - 7}{4} \right| = 2$
$$\frac{3x - 7}{4} = 2 \quad \text{or} \quad \frac{3x - 7}{4} = -2$$
$$3x - 7 = 8 \quad \text{or} \quad 3x - 7 = -8$$
$$3x = 15 \quad \text{or} \quad 3x = -1$$
$$x = 5 \quad \text{or} \quad x = -\frac{1}{3}$$

38. $|6x + 1| = |15 + 4x|$
$$6x + 1 = 15 + 4x \quad \text{or} \quad 6x + 1 = -(15 + 4x)$$
$$2x = 14 \quad \text{or} \quad 6x + 1 = -15 - 4x$$
$$x = 7 \quad \text{or} \quad 10x = -16$$
$$x = -\frac{8}{5}$$

39. $|5x - 1| < 9$
$$-9 < 5x - 1 < 9$$
$$-8 < 5x < 10$$
$$-\frac{8}{5} < x < 2$$
$$\left(-\frac{8}{5},\ 2 \right)$$

40. $|6 + 4x| \geq 10$
$$6 + 4x \leq -10 \quad \text{or} \quad 6 + 4x \geq 10$$
$$4x \leq -16 \quad \text{or} \quad 4x \geq 4$$
$$x \leq -4 \quad \text{or} \quad x \geq 1$$
$$(-\infty, -4] \cup [1, \infty)$$

41. $|3x| - 8 > 1$
$$|3x| > 9$$
$$3x < -9 \quad \text{or} \quad 3x > 9$$
$$x < -3 \quad \text{or} \quad x > 3$$
$$(-\infty, -3) \cup (3, \infty)$$

42. $9 + |5x| < 24$
$$|5x| < 15$$
$$-15 < 5x < 15$$
$$-3 < x < 3$$
$$(-3, 3)$$

43. $|6x - 5| \leq -1$
The solution set is $\varnothing$.

44. $\left|3x + \dfrac{2}{5}\right| \ge 4$

$$3x + \dfrac{2}{5} \le -4 \quad \text{or} \quad 3x + \dfrac{2}{5} \ge 4$$

$$5\left(3x + \dfrac{2}{5}\right) \le 5(-4) \quad \text{or} \quad 5\left(3x + \dfrac{2}{5}\right) \ge 5(4)$$

$$15x + 2 \le -20 \quad \text{or} \quad 15x + 2 \ge 20$$

$$15x \le -22 \quad \text{or} \quad 15x \ge 18$$

$$x \le -\dfrac{22}{15} \quad \text{or} \quad x \ge \dfrac{6}{5}$$

$$\left(-\infty, -\dfrac{22}{15}\right] \cup \left[\dfrac{6}{5}, \infty\right)$$

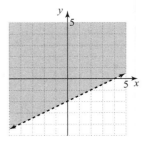

45. $\left|\dfrac{x}{3} + 6\right| - 8 > -5$

$$\left|\dfrac{x}{3} + 6\right| > 3$$

$$\dfrac{x}{3} + 6 < -3 \quad \text{or} \quad \dfrac{x}{3} + 6 > 3$$

$$\dfrac{x}{3} < -9 \quad \text{or} \quad \dfrac{x}{3} > -3$$

$$x < -27 \quad \text{or} \quad x > -9$$

$$(-\infty, -27) \cup (-9, \infty)$$

46. $\left|\dfrac{4(x-1)}{7}\right| + 10 < 2$

$$\left|\dfrac{4(x-1)}{7}\right| < -8$$

The solution set is $\varnothing$.

47. $\dfrac{1}{2}x - y < 2$

$$y > \dfrac{1}{2}x - 2$$

Graph the boundary line as dashed.
Test: $(0, 0)$

$$\dfrac{1}{2}(0) - 0 < 2 \quad \text{True}$$

Shade the half-plane that contains $(0, 0)$.

48. $3x + y > 4$

$$y > -3x + 4$$

Graph the boundary line as dashed.
Test: $(0, 0)$
$3(0) + 0 > 4$ False
Shade the half-plane that does not include $(0, 0)$.

49. $3y \ge x$

$$y \ge \dfrac{x}{3}$$

Graph the boundary line as solid.
Test: $(0, 1)$
$3(1) \ge 0$ True
Shade the half-plane that contains $(0, 1)$.

50. $5x - 2y \le 9$

$$-2y \le -5x + 9$$

$$y \ge \dfrac{5}{2}x - \dfrac{9}{2}$$

Graph the boundary line as solid.
Test: $(0, 0)$
$5(0) - 2(0) \le 9$ True
Shade the half-plane that contains $(0, 0)$.

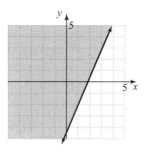

51. $x > -2$

Graph the boundary line as dashed. Shade the half-plane to the right of $x = -2$.

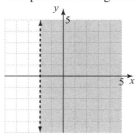

52. $y < 1$

Graph the boundary line as dashed. Shade the half-plane below $y = 1$.

53. $2x < 3y + 8$ and $y \geq -2$

Graph each inequality. The intersection of the two inequalities is all points common to both regions, as shown by the shading in the graph below.

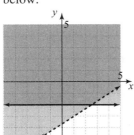

54. $y > 2x + 3$ or $x \leq -3$

Graph each inequality. The union of the two inequalities is both shaded regions, as shown by the graph below.

55.
$$\frac{x-2}{5} + \frac{x+2}{2} = \frac{x+4}{3}$$
$$30\left(\frac{x-2}{5} + \frac{x+2}{2}\right) = 30\left(\frac{x+4}{3}\right)$$
$$6(x-2) + 15(x+2) = 10(x+4)$$
$$6x - 12 + 15x + 30 = 10x + 40$$
$$21x + 18 = 10x + 40$$
$$11x = 22$$
$$x = 2$$

56.
$$\frac{2z-3}{4} - \frac{4-z}{2} = \frac{z+1}{3}$$
$$12\left(\frac{2z-3}{4} - \frac{4-z}{2}\right) = 12\left(\frac{z+1}{3}\right)$$
$$3(2z-3) - 6(4-z) = 4(z+1)$$
$$6z - 9 - 24 + 6z = 4z + 4$$
$$12z - 33 = 4z + 4$$
$$8z = 37$$
$$z = \frac{37}{8}$$

57.
$$\frac{3(x-2)}{5} > \frac{-5(x-2)}{3}$$
$$15\left[\frac{3(x-2)}{5}\right] > 15\left[\frac{-5(x-2)}{3}\right]$$
$$9(x-2) > -25(x-2)$$
$$9x - 18 > -25x + 50$$
$$34x > 68$$
$$x > 2$$
$$(2, \infty)$$

58. $0 \le \dfrac{2(3x+4)}{5} \le 3$

$5(0) \le 5\left[\dfrac{2(3x+4)}{5}\right] \le 5(3)$

$0 \le 2(3x+4) \le 15$

$0 \le 6x+8 \le 15$

$-8 \le 6x \le 7$

$-\dfrac{4}{3} \le x \le \dfrac{7}{6}$

$\left[-\dfrac{4}{3}, \dfrac{7}{6}\right]$

59. $x \le 2$ or $x > -5$

$(-\infty, \infty)$

60. $-2x \le 6$ and $-2x+3 < -7$

$x \ge -3$ and $-2x < -10$

$x \ge -3$ and $x > 5$

$x > 5$

$(5, \infty)$

61. $|7x| - 26 = -5$

$|7x| = 21$

$7x = 21$ or $7x = -21$

$x = 3$ or $x = -3$

62. $\left|\dfrac{9-2x}{5}\right| = -3$

The solution set is $\varnothing$.

63. $|x - 3| = |7 + 2x|$

$x - 3 = 7 + 2x$ or $x - 3 = -(7 + 2x)$

$-10 = x$ or $x - 3 = -7 - 2x$

$3x = -4$

$x = -\dfrac{4}{3}$

64. $|6x - 5| \ge -1$

Since $|6x - 5|$ is nonnegative for all numbers x, the solution set is $(-\infty, \infty)$.

65. $\left|\dfrac{4x-3}{5}\right| < 1$

$-1 < \dfrac{4x-3}{5} < 1$

$-5 < 4x - 3 < 5$

$-2 < 4x < 8$

$-\dfrac{1}{2} < x < 2$

$\left(-\dfrac{1}{2}, 2\right)$

66. $48 + x \ge 5(2x+4) - 2x$

$48 + x \ge 10x + 20 - 2x$

$48 + x \ge 8x + 20$

$28 \ge 7x$

$4 \ge x$

$(-\infty, 4]$

Chapter 3 Test

1. $15x + 26 = -2(x+1) - 1$

Algebraic Solution:

$15x + 26 = -2(x+1) - 1$

$15x + 26 = -2x - 2 - 1$

$15x + 26 = -2x - 3$

$17x + 26 = -3$

$17x = -29$

$x = -\dfrac{29}{17} \approx -1.71$

The solution is $-\dfrac{29}{17} \approx -1.71$.

Graphical Solution:

Graph $y_1 = 15x + 26$ and $y_2 = -2(x+1) - 1$ in a $[-10, 10, 1]$ by $[-10, 10, 1]$ viewing window and find the intersection point.

Rounding to the nearest hundredth, the point of intersection $(-1.71, 0.41)$ indicates that the solution is approximately -1.71.

2. $-3x - \sqrt{5} = \pi(x - 1)$

Graph $y_1 = -3x - \sqrt{}(5)$ and $y_2 = \pi(x-1)$ in a standard viewing window and find the intersection point.

Copyright © 2014 Pearson Education, Inc.

Rounding to the nearest hundredth, the point of intersection $(0.15, -2.68)$ indicates that the solution is approximately 0.15.

3. $|6x - 5| - 3 = -2$

$|6x - 5| = 1$

$6x - 5 = 1$ or $6x - 5 = -1$

$6x = 6$ or $6x = 4$

$x = 1$ or $x = \dfrac{2}{3}$

4. $|8 - 2t| = -6$

No solution, $\varnothing$

5. $3(2x - 7) - 4x > -(x + 6)$

$6x - 21 - 4x > -x - 6$

$2x - 21 > -x - 6$

$3x > 15$

$x > 5$

$(5, \infty)$

6. $8 - \dfrac{x}{2} \geq 7$

$2\left(8 - \dfrac{x}{2}\right) \geq 2(7)$

$16 - x \geq 14$

$-x \geq -2$

$x \leq 2$

The solution set is $(-\infty, 2]$.

7. $-3 < 2(x - 3) \leq 4$

$-3 < 2x - 6 \leq 4$

$3 < 2x \leq 10$

$\dfrac{3}{2} < x \leq 5$

$\left(\dfrac{3}{2}, 5\right]$

8. $|3x + 1| > 5$

$3x + 1 < -5$ or $3x + 1 > 5$

$3x < -6$ or $3x > 4$

$x < -2$ or $x > \dfrac{4}{3}$

$(-\infty, -2) \cup \left(\dfrac{4}{3}, \infty\right)$

9. $x \leq -2$ and $x \leq -5$

$x \leq -5$

The solution set is $(-\infty, -5]$.

10. $x \leq -2$ or $x \leq -5$

$x \leq -2$

The solution set is $(-\infty, -2]$.

11. $-x > 1$ and $3x + 3 \geq x - 3$

$x < -1$ and $2x \geq -6$

$x \geq -3$

$-3 \leq x < -1$

The solution set is $[-3, -1)$.

12. $6x + 1 > 5x + 4$ or $1 - x > -4$

$x > 3$ or $5 > x$

$(-\infty, \infty)$

13. $|x - 5| - 4 < -2$

$|x - 5| < 2$

$-2 < x - 5 < 2$

$3 < x < 7$

$(3, 7)$

14. $\left|\dfrac{5x - 7}{2}\right| = 4$

$\dfrac{5x - 7}{2} = 4$ or $\dfrac{5x - 7}{2} = -4$

$5x - 7 = 8$ or $5x - 7 = -8$

$5x = 15$ or $5x = -1$

$x = 3$ or $x = -\dfrac{1}{5}$

The solution set is $\left\{-\dfrac{1}{5}, 3\right\}$.

15. $\left|17x - \dfrac{1}{5}\right| > -2$ is true for all real values of x.

The solution set is $(-\infty, \infty)$.

16. $|x - 5| = |x + 2|$

$x - 5 = x + 2$ or $x - 5 = -(x + 2)$

$-5 = 2$ False   or $x - 5 = -x - 2$

$2x = 3$

$x = \dfrac{3}{2}$

Since $-5 = 2$ is not possible, the only solution is $\dfrac{3}{2}$.

17. $x \le -4$

Graph a solid boundary line and shade the half-plane to the left of $x = -4$.

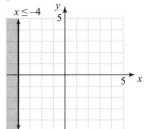

18. $2x - y > 5$

$\quad y < 2x - 5$

Graph the boundary line as a dashed line.
Test: $(0, 0)$
$2(0) - 0 > 5$ False
Shade the half-plane that does not contain the point $(0, 0)$.

19. $2x + 4y < 6$ and $y \le -4$

$\quad 4y < -2x + 6$ and $y \le -4$

$\quad y < -\dfrac{1}{2}x + \dfrac{3}{2}$ and $y \le -4$

Graph each inequality. The intersection of the two inequalities is all points common to both regions, as shown by the shading in the graph below.

20. The graph of y_1 is below the graph of y_2 when $x < -3$. Thus, the solution of the inequality $y_1 < y_2$ is $(-\infty, -3)$.

21. The graph of y_1 is above the graph of y_2 when $x > -3$. Thus, the solution of the inequality $y_1 > y_2$ is $(-3, \infty)$.

22. Solve $R > C$.
$\quad 7.4x > 3910 + 2.8x$
$\quad 4.6x > 3910$
$\qquad x > 850$
Therefore, more than 850 sunglasses must be produced and sold in order for them to yield a profit.

Chapter 3 Cumulative Review

1. a. $\{101, 102, 103, \dots\}$

b. $\{2, 3, 4, 5\}$

2. a. $\{-2, -1, 0, 1, 2, 3, 4\}$

b. $\{4\}$

3. a. $|3| = 3$

b. $|-5| = 5$

c. $-|2| = -2$

d. $-|-8| = -8$

e. $|0| = 0$

4. a. The opposite of $\dfrac{2}{3}$ is $-\dfrac{2}{3}$.

b. The opposite of -9 is 9.

c. The opposite of 1.5 is -1.5.

5. a. $-3 + (-11) = -14$

b. $3 + (-7) = -4$

c. $-10 + 15 = 5$

d. $-8.3 + (-1.9) = -10.2$

e. $-\dfrac{1}{4} + \dfrac{1}{2} = -\dfrac{1}{4} + \dfrac{2}{4} = \dfrac{1}{4}$

f. $-\dfrac{2}{3} + \dfrac{3}{7} = -\dfrac{14}{21} + \dfrac{9}{21} = -\dfrac{5}{21}$

6. a. $-2 - (-10) = -2 + 10 = 8$

b. $1.7 - 8.9 = -7.2$

c. $-\dfrac{1}{2} - \dfrac{1}{4} = -\dfrac{2}{4} - \dfrac{1}{4} = -\dfrac{3}{4}$

7. a. $\sqrt{9} = 3$ since $3^2 = 9$.

b. $\sqrt{25} = 5$ since $5^2 = 25$.

c. $\sqrt{\dfrac{1}{4}} = \dfrac{1}{2}$ since $\left(\dfrac{1}{2}\right)^2 = \dfrac{1}{4}$.

d. $-\sqrt{36} = -6$ since $6^2 = 36$.

e. $\sqrt{-36}$ is not a real number.

8. a. $-3(-2) = 6$

b. $-\dfrac{3}{4}\left(-\dfrac{4}{7}\right) = \dfrac{3}{7}$

c. $\dfrac{0}{-2} = 0$

d. $\dfrac{-20}{-2} = 10$

9. a. $z - y = (-3) - (-1) = -2$

b. $z^2 = (-3)^2 = 9$

c. $\dfrac{2x + y}{z} = \dfrac{2(2) + (-1)}{-3} = \dfrac{3}{-3} = -1$

10. a. $\sqrt[4]{1} = 1$ since $1^4 = 1$.

b. $\sqrt[3]{8} = 2$ since $2^3 = 8$.

c. $\sqrt[4]{81} = 3$ since $3^4 = 81$.

11. a. $x + 5 = 20$

b. $2(3 + y) = 4$

c. $x - 8 = 2x$

d. $\dfrac{z}{9} = 9 + z$

12. a. $-3 > -5$ since -3 is to the right of -5 on the number line.

b. $\dfrac{-12}{-4} = 3$

c. $0 > -2$ since 0 is to the right of -2 on the number line.

13. $7x + 5 = 5 + 7x$

14. $5 \cdot (7x) = (5 \cdot 7)x = 35x$

15. $2x + 5 = 9$
$2x = 4$
$x = 2$

16. $11.2 = 1.2 - 5x$
$10 = -5x$
$-2 = x$

17. $6x - 4 = 2 + 6(x - 1)$
$6x - 4 = 2 + 6x - 6$
$6x - 4 = 6x - 4$
$-4 = -4,$ which is always true.
All real numbers

18. $2x + 1.5 = -0.2 + 1.6x$
$0.4x = -1.7$
$x = -4.25$

19. a. Let $x =$ the first integer. Then
$x + 1 =$ the second integer and
$x + 2 =$ the third integer.
$x + (x + 1) + (x + 2) = 3x + 3$

b. $x + (5x) + (6x - 3) = 12x - 3$

20. a. Let $x =$ the first integer. Then
$x + 1 =$ the second integer and
$x + 2 =$ the third integer.
$x + (x + 1) + (x + 2) = 3x + 3$

b. $4(3x + 1) = 12x + 4$

21. Let $x =$ first number, then
$2x + 3 =$ second number and
$4x =$ third number.
$x + (2x + 3) + 4x = 164$
$7x + 3 = 164$
$7x = 161$
$x = 23$
$2x + 3 = 2(23) + 3 = 49$
$4x = 4(23) = 92$
The three numbers are 23, 49 and 92.

22. Let x = first number, then
$3x + 2$ = second number.
$$(3x+2) - x = 24$$
$$2x + 2 = 24$$
$$2x = 22$$
$$x = 11$$
$3x + 2 = 3(11) + 2 = 35$
The two numbers are 11 and 35.

23. $3y - 2x = 7$
$$3y = 2x + 7$$
$$y = \frac{2x+7}{3}, \text{ or } y = \frac{2x}{3} + \frac{7}{3}$$

24. $7x - 4y = 10$
$$7x = 4y + 10$$
$$x = \frac{4y+10}{7}, \text{ or } x = \frac{4y}{7} + \frac{10}{7}$$

25.
$$A = \frac{1}{2}(B+b)h$$
$$2A = (B+b)h$$
$$2A = Bh + bh$$
$$2A - Bh = bh$$
$$\frac{2A - Bh}{h} = b$$

26.
$$P = 2l + 2w$$
$$P - 2w = 2l$$
$$\frac{P - 2w}{2} = l$$

27. a. $\{x | x \geq 2\}$

$[2, \infty)$

b. $\{x | x < -1\}$

$(-\infty, -1)$

c. $\{x | 0.5 < x \leq 3\}$

$(0.5, 3]$

28. a. $\{x | x \leq -3\}$

$(-\infty, -3]$

b. $\{x | -2 \leq x < 0.1\}$

$[-2, 0.1)$

29. $-(x-3) + 2 \leq 3(2x - 5) + x$
$$-x + 3 + 2 \leq 6x - 15 + x$$
$$-x + 5 \leq 7x - 15$$
$$20 \leq 8x$$
$$\frac{5}{2} \leq x$$

$\left[\dfrac{5}{2}, \infty\right)$

30. $2(7x-1) - 5x > -(-7x) + 4$
$$14x - 2 - 5x > 7x + 4$$
$$9x - 2 > 7x + 4$$
$$2x > 6$$
$$x > 3$$

$(3, \infty)$

31. a. $2(x+3) > 2x + 1$
$$2x + 6 > 2x + 1$$
$$6 > 1; \text{ True for all real numbers } x.$$

$(-\infty, \infty)$

b. $2(x+3) < 2x + 1$
$$2x + 6 < 2x + 1$$
$$6 < 1; \text{ False}$$

$\varnothing$

32. $4(x+1) - 3 < 4x + 1$
$$4x + 4 - 3 < 4x + 1$$
$$4x + 1 < 4x + 1$$
$$1 < 1 \quad \text{Never true}$$

$\varnothing$

33. $A = \{2, 4, 6, 8\}$, $B = \{3, 4, 5, 6\}$; the numbers 4 and 6 are in both sets so the intersection of A and B is $\{4, 6\}$.

34. The elements in either set or both sets are $-2, -1,$ $0, 1, 2, 3, 4,$ and 5, so the union is $\{-2, -1, 0, 1, 2, 3, 4, 5\}$.

35. $x - 7 < 2 \quad$ and $\quad 2x + 1 < 9$
$\qquad x < 9 \quad$ and $\qquad 2x < 8$
$\qquad\qquad\qquad\qquad\qquad\qquad x < 4$

$x < 4$
$(-\infty, 4)$

36. $x + 3 \le 1 \quad$ or $\quad 3x - 1 < 8$

$\quad\quad x \le -2 \quad$ or $\quad\quad 3x < 9$

$\quad\quad\quad\quad\quad\quad\quad\quad\quad\quad x < 3$

$x < 3$

$(-\infty, 3)$

37. $A = \{2, 4, 6, 8\}$ and $B = \{3, 4, 5, 6\}$, so the union of A and B is $\{2, 3, 4, 5, 6, 8\}$.

38. $\varnothing$; there are no elements in common.

39. $-2x - 5 < -3 \quad$ or $\quad 6x < 0$

$\quad\quad -2x < 2 \quad$ or $\quad\quad x < 0$

$\quad\quad\quad x > -1$

All real numbers

$(-\infty, \infty)$

40. $-2x - 5 < -3 \quad$ and $\quad 6x < 0$

$\quad\quad -2x < 2 \quad$ and $\quad\quad x < 0$

$\quad\quad\quad x > -1$

$-1 < x < 0$

$(-1, 0)$

41. $|p| = 2$

$p = 2 \quad$ or $\quad p = -2$

42. $|x| = 5$

$x = 5 \quad$ or $\quad x = -5$

43. $\left| \dfrac{x}{2} - 1 \right| = 11$

$\dfrac{x}{2} - 1 = 11 \quad$ or $\quad \dfrac{x}{2} - 1 = -11$

$\quad \dfrac{x}{2} = 12 \quad$ or $\quad\quad \dfrac{x}{2} = -10$

$\quad\quad x = 24 \quad$ or $\quad\quad\quad x = -20$

44. $\left| \dfrac{y}{3} + 2 \right| = 10$

$\dfrac{y}{3} + 2 = 10 \quad$ or $\quad \dfrac{y}{3} + 2 = -10$

$\quad \dfrac{y}{3} = 8 \quad$ or $\quad\quad \dfrac{y}{3} = -12$

$\quad\quad y = 24 \quad$ or $\quad\quad\quad y = -36$

45. $|x - 3| = |5 - x|$

$x - 3 = 5 - x \quad$ or $\quad x - 3 = -(5 - x)$

$\quad 2x = 8 \quad$ or $\quad x - 3 = -5 + x$

$\quad\quad x = 4 \quad$ or $\quad\quad -3 = -5$

Since $-3 = -5$ is not possible, the only solution is 4.

46. $|x + 3| = |7 - x|$

$x + 3 = 7 - x \quad$ or $\quad x + 3 = -(7 - x)$

$\quad 2x = 4 \quad$ or $\quad x - 3 = -7 + x$

$\quad\quad x = 2 \quad$ or $\quad\quad -3 = -7$

Since $-3 = -7$ is not possible, the only solution is 2.

47. $|x| \le 3$

$-3 \le x \le 3$

$[-3, 3]$

48. $|x| > 1$

$x < -1 \quad$ or $\quad x > 1$

$(-\infty, -1) \cup (1, \infty)$

49. $|2x + 9| + 5 > 3$

$\quad |2x + 9| > -2$

Since $|2x + 9|$ is nonnegative for all numbers x, the solution set is $(-\infty, \infty)$.

50. $|3x + 1| + 9 < 1$

$\quad |3x + 1| < -8$

The solution set is $\varnothing$.

Chapter 4

Section 4.1 Practice Exercises

1. a. $\begin{cases} -x - 4y = 1 \\ 2x + y = 5 \end{cases}$

Replace x with 3 and y with -1 in each equation.
$$-x - 4y = 1$$
$$-3 - 4(-1) \overset{?}{=} 1$$
$$-3 + 4 \overset{?}{=} 1$$
$$1 = 1 \quad \text{True}$$

$$2x + y = 5$$
$$2(3) + (-1) \overset{?}{=} 5$$
$$6 - 1 \overset{?}{=} 5$$
$$5 = 5 \quad \text{True}$$

Since $(3, -1)$ makes both equations true, it is a solution.

b. $\begin{cases} 4x + y = -4 \\ -x + 3y = 8 \end{cases}$

Replace x with -2 and y with 4 in each equation.
$$4x + y = -4$$
$$4(-2) + 4 \overset{?}{=} -4$$
$$-8 + 4 \overset{?}{=} -4$$
$$-4 = -4 \quad \text{True}$$

$$-x + 3y = 8$$
$$-(-2) + 3(4) \overset{?}{=} 8$$
$$2 + 12 \overset{?}{=} 8$$
$$14 = 8 \quad \text{False}$$

Since $(-2, 4)$ does not make both equations true, it is not a solution.

2. a. $\begin{cases} 3x - 2y = 4 \\ -9x + 6y = -12 \end{cases}$

Solve each equation for y.
$$\begin{cases} y = \dfrac{4 - 3x}{-2} \\ y = \dfrac{-12 + 9x}{6} \end{cases}$$

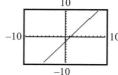

The graphs appear to be the same line. Notice that if both sides of the first equation are multiplied by -3, the result is the second equation. Any solution of one equation satisfies the other equation as well. There are infinitely many solutions of the form $\{(x, y)\,|\,3x - 2y = 4\}$ or $\{(x, y)\,|-9x + 6y = -12\}$.

b. $\begin{cases} y = 5x \\ 2x + y = 7 \end{cases}$

Solve the second equation for y.
$$\begin{cases} y = 5x \\ y = 7 - 2x \end{cases}$$

$(1, 5)$ is the solution of the system.

c. $\begin{cases} y = \dfrac{3}{4}x + 1 \\ 3x - 4y = 12 \end{cases}$

Solve the second equation for y.
$$\begin{cases} y = \dfrac{3}{4}x + 1 \\ y = \dfrac{12 - 3x}{4} \end{cases}$$

The lines appear to be parallel. The first equation is in point-slope. Write the second equation in point-slope form.
$$3x - 4y = 12$$
$$-4y = -3x + 12$$
$$y = \dfrac{3}{4}x - 3$$

The graphs have the same slope, $\dfrac{3}{4}$, but different y-intercepts, so the lines are parallel. The system has no solution.

3. $\begin{cases} y - 0.25x = 1.6 \\ y + 1.03x = -5.1 \end{cases}$

Solve both equations for y.

$\begin{cases} y = 1.6 + 0.25x \\ y = -5.1 - 1.03x \end{cases}$

The approximate point of intersection is $(-5.23, 0.29)$.

4. $\begin{cases} y = 4x + 7 \\ 2x + y = 4 \end{cases}$

In the first equation, we are told that y is equal to $4x + 7$, so we substitute $4x + 7$ for y in the second equation and solve for x.

$$2x + y = 4$$
$$2x + (4x + 7) = 4$$
$$6x + 7 = 4$$
$$6x = -3$$
$$x = \frac{-3}{6} = -\frac{1}{2}$$

To find the y-coordinate, we replace x with $-\frac{1}{2}$ in the first equation.

$$y = 4x + 7$$
$$y = 4\left(-\frac{1}{2}\right) + 7$$
$$y = -2 + 7 = 5$$

The solution is $\left(-\frac{1}{2}, 5\right)$.

5. $\begin{cases} -\dfrac{x}{3} + \dfrac{y}{4} = \dfrac{1}{2} \\ \dfrac{x}{4} - \dfrac{y}{2} = -\dfrac{1}{4} \end{cases}$

Multiply each equation by its LCD to clear fractions.

$$\begin{cases} 12\left(-\dfrac{x}{3} + \dfrac{y}{4}\right) = 12\left(\dfrac{1}{2}\right) \\ 4\left(\dfrac{x}{4} - \dfrac{y}{2}\right) = 4\left(-\dfrac{1}{4}\right) \end{cases}$$

$$\begin{cases} -4x + 3y = 6 \\ x - 2y = -1 \end{cases}$$

Solve the second equation for x.

$$x - 2y = -1$$
$$x = 2y - 1$$

Replace x with $2y - 1$ in the first equation.

$$-4x + 3y = 6$$
$$-4(2y - 1) + 3y = 6$$
$$-8y + 4 + 3y = 6$$
$$-5y + 4 = 6$$
$$-5y = 2$$
$$y = -\frac{2}{5}$$

To find the x-coordinate, replace y with $-\frac{2}{5}$ in $x = 2y - 1$.

$$x = 2\left(-\frac{2}{5}\right) - 1 = -\frac{4}{5} - 1 = -\frac{4}{5} - \frac{5}{5} = -\frac{9}{5}$$

The solution is $\left(-\dfrac{9}{5}, -\dfrac{2}{5}\right)$.

6. $\begin{cases} 3x - y = 5 \\ 5x + y = 11 \end{cases}$

We add the equations.

$$\begin{array}{r} 3x - y = 5 \\ 5x + y = 11 \\ \hline 8x \quad\ = 16 \\ x = 2 \end{array}$$

Replace x with 2 in the second equation to find y.

$$5x + y = 11$$
$$5(2) + y = 11$$
$$10 + y = 11$$
$$y = 1$$

The solution is $(2, 1)$.

7. $\begin{cases} 3x - 2y = -6 \\ 4x + 5y = -8 \end{cases}$

We can eliminate y if we multiply both sides of the first equation by 5 and both sides of the second equation by 2.

$$\begin{cases} 5(3x - 2y) = 5(-6) \\ 2(4x + 5y) = 2(-8) \end{cases}$$

$$\begin{array}{r} 15x - 10y = -30 \\ 8x + 10y = -16 \\ \hline 23x \qquad\ = -46 \\ x = -2 \end{array}$$

To find y, replace x with -2 in either equation.

$$4x + 5y = -8$$
$$4(-2) + 5y = -8$$
$$-8 + 5y = -8$$
$$5y = 0$$
$$y = 0$$

The solution is $(-2, 0)$.

8. $\begin{cases} 8x + y = 6 \\ 2x + \dfrac{y}{4} = -2 \end{cases}$

If we multiply the second equation by −4, the coefficients of x will be opposites.

$\begin{cases} 8x + y = 6 \\ -4\left(2x + \dfrac{y}{4}\right) = -4(-2) \end{cases}$

$\begin{cases} 8x + y = 6 \\ -8x - y = 8 \end{cases}$

$\overline{\qquad 0 = 14 \quad \text{False}}$

The system has no solution. The solution set is $\{\ \}$ or $\varnothing$.

9. $\begin{cases} -3x + 2y = -1 \\ 9x - 6y = 3 \end{cases}$

To eliminate x, we multiply both sides of the first equation by 3.

$\begin{cases} 3(-3x + 2y) = 3(-1) \\ 9x - 6y = 3 \end{cases}$

$\begin{cases} -9x + 6y = -3 \\ 9x - 6y = 3 \end{cases}$

$\overline{\qquad 0 = 0 \quad \text{True}}$

There are infinitely many solutions. The solution set is $\{(x, y)|-3x + 2y = -1\}$ or $\{(x, y)|9x - 6y = 3\}$.

10. The break-even point is the solution of the system.

$\begin{cases} y = 17x \\ y = 6x + 8030 \end{cases}$

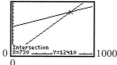

15,000

0 · · · · · · · · · · · · · 1000
Intersection R=730 Y=12410
0

The company breaks even when 730 units are produced and sold.

Vocabulary, Readiness & Video Check 4.1

1. A system with no solution has lines that are parallel. The correct graph is B.

2. A system with an infinite number of solutions has lines that are the same. The correct graph is C.

3. A system with solution (1, −2) has lines that intersect at (1, −2). The correct graph is A.

4. A system with solution (−3, 0) has lines that intersect at (−3, 0). The correct graph is D.

5. The ordered pair must be a solution of *both* equations of the system in order to be a solution of the system.

6. If the system contains two equations in two variables and has a single solution, this solution is the intersection of two lines. An intersection point with noninteger coordinates would be very difficult to graph/guess correctly.

7. Solve one equation for a variable. Next be sure to substitute this expression for the variable into the *other* equation.

8. Step 2 says to multiply one or both equations by a nonzero number to get opposite coefficients on a variable. However, once we eliminate fractions, the variable x already has opposite coefficients, so we proceed to Step 3.

Exercise Set 4.1

1. $\begin{cases} x - y = 3 \\ 2x - 4y = 8 \end{cases}$

$x - y = 3$
$2 - (-1) \overset{?}{=} 3$
$\qquad\quad 3 = 3 \quad \text{True}$
$2x - 4y = 8$
$2(2) - 4(-1) \overset{?}{=} 8$
$\qquad\quad 4 + 4 \overset{?}{=} 8$
$\qquad\qquad\quad 8 = 8 \quad \text{True}$
Yes, (2, −1) is a solution.

3. $\begin{cases} 2x - 3y = -9 \\ 4x + 2y = -2 \end{cases}$

$2x - 3y = -9$
$2(3) - 3(5) \overset{?}{=} -9$
$\quad\ 6 - 15 \overset{?}{=} -9$
$\qquad\quad -9 = -9 \quad \text{True}$
$4x + 2y = -2$
$4(3) + 2(5) \overset{?}{=} -2$
$\quad\ 12 + 10 \overset{?}{=} -2$
$\qquad\quad 22 = -2 \quad \text{False}$
No, (3, 5) is not a solution.

5. $\begin{cases} y = -5x \\ x = -2 \end{cases}$

$y = -5x$

$10 \overset{?}{=} -5(-2)$ $x = -2$

$10 = 10$ True $-2 = -2$ True

Yes, $(-2, 10)$ is a solution of the system.

7. $\begin{cases} 3x + 7y = -19 \\ -6x = 5y + 8 \end{cases}$

$3x + 7y = -19$

$3\left(\dfrac{2}{3}\right) + 7(-3) \overset{?}{=} -19$

$2 + (-21) \overset{?}{=} -19$

$-19 = -19$ True

$-6x = 5y + 8$

$-6\left(\dfrac{2}{3}\right) \overset{?}{=} 5(-3) + 8$

$-4 \overset{?}{=} -15 + 8$

$-4 = -7$ False

No, $\left(\dfrac{2}{3}, -3\right)$ is not a solution.

9. The solution appears to be $(1.5, 1)$.

The solution is $(1.5, 1)$.

11. The solution appears to be $(-2, 0.75)$.

The solution is $(-2, 0.75)$.

13. $\begin{cases} x + y = 1 \\ x - 2y = 4 \end{cases}$

Solve both equations for y.

$\begin{cases} y = 1 - x \\ y = \dfrac{4 - x}{-2} \end{cases}$

The solution is $(2, -1)$.

15. $\begin{cases} 2y - 4x = 0 \\ x + 2y = 5 \end{cases}$

Solve both equations for y.

$\begin{cases} y = \dfrac{4x}{2} \\ y = \dfrac{5 - x}{2} \end{cases}$

The solution is $(1, 2)$.

17. $\begin{cases} 3x - y = 4 \\ 6x - 2y = 4 \end{cases}$

Solve both equations for y.

$\begin{cases} y = 3x - 4 \\ y = \dfrac{4 - 6x}{-2} \end{cases}$

The solution set is $\varnothing$.

19. $\begin{cases} x + y = 10 \\ y = 4x \end{cases}$

Replace y with $4x$ in E1.

$x + (4x) = 10$

$5x = 10$

$x = 2$

Replace x with 2 in E2.

$y = 4(2)$

$y = 8$

The solution is $(2, 8)$.

21. $\begin{cases} 4x - y = 9 \\ 2x + 3y = -27 \end{cases}$

Solve E1 for y.

$4x - y = 9$

$y = 4x - 9$

Replace y with $4x - 9$ in E2.

$2x + 3(4x - 9) = -27$

$2x + 12x - 27 = -27$

$14x = 0$

$x = 0$

Replace x with 0 in E1.

$4(0) - y = 9$
$y = -9$
The solution is $(0, -9)$.

23. $\begin{cases} \dfrac{1}{2}x + \dfrac{3}{4}y = -\dfrac{1}{4} \\ \dfrac{3}{4}x - \dfrac{1}{4}y = 1 \end{cases}$

Clear fractions by multiplying each equation by 4.
$\begin{cases} 2x + 3y = -1 \\ 3x - y = 4 \end{cases}$

Now solve E2 for y.
$3x - y = 4$
$y = 3x - 4$

Replace y with $3x - 4$ in E1.
$2x + 3(3x - 4) = -1$
$2x + 9x - 12 = -1$
$11x = 11$
$x = 1$

Replace x with 1 in equation $y = 3x - 4$.
$y = 3(1) - 4$
$y = -1$
The solution is $(1, -1)$.

25. $\begin{cases} \dfrac{x}{3} + y = \dfrac{4}{3} \\ -x + 2y = 11 \end{cases}$

Clear fractions by multiplying the first equation by 3.
$\begin{cases} x + 3y = 4 \\ -x + 2y = 11 \end{cases}$

Solve E2 for x.
$2y - 11 = x$
$x = 2y - 11$

Replace x with $2y - 11$ in E1.
$(2y - 11) + 3y = 4$
$5y = 15$
$y = 3$

Replace y with 3 in equation $x = 2y - 11$.
$x = 2(3) - 11$
$x = -5$
The solution is $(-5, 3)$.

27. $\begin{cases} -x + 2y = 0 \\ x + 2y = 5 \end{cases}$

$\begin{array}{r} -x + 2y = 0 \\ x + 2y = 5 \\ \hline \end{array}$
E1 + E2: $\quad 4y = 5$
$y = \dfrac{5}{4}$

Replace y with $\dfrac{5}{4}$ in E1.

$-x + 2\left(\dfrac{5}{4}\right) = 0$

$-x + \dfrac{5}{2} = 0$

$\dfrac{5}{2} = x$

The solution is $\left(\dfrac{5}{2}, \dfrac{5}{4}\right)$.

29. $\begin{cases} 5x + 2y = 1 \\ x - 3y = 7 \end{cases}$

Multiply E2 by -5.
$\begin{cases} 5x + 2y = 1 \\ -5x + 15y = -35 \end{cases}$

E1 + E2: $\begin{array}{r} 5x + 2y = 1 \\ -5x + 15y = -35 \\ \hline 17y = -34 \\ y = -2 \end{array}$

Replace y with -2 in E2.
$x - 3(-2) = 7$
$x + 6 = 7$
$x = 1$
The solution is $(1, -2)$.

31. $\begin{cases} \dfrac{3}{4}x + \dfrac{5}{2}y = 11 \\ \dfrac{1}{16}x - \dfrac{3}{4}y = -1 \end{cases}$

Clear fractions by multiplying E1 by 4 and E2 by 16.
$\begin{cases} 3x + 10y = 44 \\ x - 12y = -16 \end{cases}$

Multiply E2 by -3.
$\begin{cases} 3x + 10y = 44 \\ -3x + 36y = 48 \end{cases}$

E1 + E2: $\begin{array}{r} 3x + 10y = 44 \\ -3x + 36y = 48 \\ \hline 46y = 92 \\ y = 2 \end{array}$

Replace y with 2 in the equation $x - 12y = -16$.
$$x - 12(2) = -16$$
$$x - 24 = -16$$
$$x = 8$$
The solution is (8, 2).

33. $\begin{cases} 3x - 5y = 11 \\ 2x - 6y = 2 \end{cases}$

Multiply E1 by 2 and E2 by –3.
$$\begin{cases} 6x - 10y = 22 \\ -6x + 18y = -6 \end{cases}$$

E1 + E2: $6x - 10y = 22$
$$\underline{\quad\quad -6x + 18y = -6}$$
$$8y = 16$$
$$y = 2$$

Replace y with 2 in E2.
$$2x - 6(2) = 2$$
$$2x - 12 = 2$$
$$2x = 14$$
$$x = 7$$
The solution is (7, 2).

35. $\begin{cases} x - 2y = 4 \\ 2x - 4y = 4 \end{cases}$

Multiply E1 by –2.
$$\begin{cases} -2x + 4y = -8 \\ 2x - 4y = 4 \end{cases}$$

E1 + E2: $-2x + 4y = -8$
$$\underline{\quad\quad 2x - 4y = 4}$$
$$0 = -4 \quad \text{False}$$

Inconsistent system; the solution set is $\varnothing$.

37. $\begin{cases} 3x + y = 1 \\ 2y = 2 - 6x \end{cases}$

$\begin{cases} 3x + y = 1 \\ 6x + 2y = 2 \end{cases}$

Multiply E1 by –2.
$$\begin{cases} -6x - 2y = -2 \\ 6x + 2y = 2 \end{cases}$$

E1 + E2: $-6x - 2y = -2$
$$\underline{\quad\quad 6x + 2y = 2}$$
$$0 = 0 \quad \text{True}$$

Dependent system; the solution set is $\{(x, y) | 3x + y = 1\}$.

39. $\begin{cases} 2x + 5y = 8 \\ 6x + y = 10 \end{cases}$

Multiply E1 by –3.
$$\begin{cases} -6x - 15y = -24 \\ 6x + y = 10 \end{cases}$$

E1 + E2: $-6x - 15y = -24$
$$\underline{\quad\quad 6x + y = 10}$$
$$-14y = -14$$
$$y = 1$$

Replace y with 1 in E2.
$$6x + 1 = 10$$
$$6x = 9$$
$$x = \frac{9}{6} = \frac{3}{2}$$

The solution is $\left(\frac{3}{2}, 1\right)$.

41. $\begin{cases} x + y = 1 \\ x - 2y = 4 \end{cases}$

Multiply E1 by –1.
$$\begin{cases} -x - y = -1 \\ x - 2y = 4 \end{cases}$$

E1 + E2: $-x - y = -1$
$$\underline{\quad\quad x - 2y = 4}$$
$$-3y = 3$$
$$y = -1$$

Replace y with –1 in E1.
$$x + (-1) = 1$$
$$x - 1 = 1$$
$$x = 2$$
The solution is (2, –1).

43. $\begin{cases} \dfrac{1}{3}x + y = \dfrac{4}{3} \\ -\dfrac{1}{4}x - \dfrac{1}{2}y = -\dfrac{1}{4} \end{cases}$

Clear fractions by multiplying E1 by 3 and E2 by 4.
$$\begin{cases} x + 3y = 4 \\ -x - 2y = -1 \end{cases}$$

E1 + E2: $x + 3y = 4$
$$\underline{\quad\quad -x - 2y = -1}$$
$$y = 3$$

Replace y with 3 in the equation $x + 3y = 4$.
$$x + 3(3) = 4$$
$$x + 9 = 4$$
$$x = -5$$
The solution is (–5, 3).

45. $\begin{cases} 2x+6y=8 \\ 3x+9y=12 \end{cases}$

Multiply E1 by –3 and E2 by 2.

$\begin{cases} -6x-18y=-24 \\ 6x+18y=24 \end{cases}$

E1 + E2: $\quad -6x-18y=-24$

$\underline{\quad\quad 6x+18y=24 \quad}$

$\quad\quad\quad\quad 0=0 \quad$ True

Dependent system; the solution is
$\{(x, y)|3x + 9y = 12\}$.

47. $\begin{cases} 4x+2y=5 \\ 2x+y=-1 \end{cases}$

Multiply E2 by –2.

$\begin{cases} 4x+2y=5 \\ -4x-2y=2 \end{cases}$

E1 + E2: $\quad 4x+2y=5$

$\underline{\quad -4x-2y=2 \quad}$

$\quad\quad\quad\quad 0=7 \quad$ False

Inconsistent system; the solution set is $\varnothing$.

49. $\begin{cases} 10y-2x=1 \\ \quad 5y=4-6x \end{cases}$

$\begin{cases} 10y-2x=1 \\ 5y+6x=4 \end{cases}$

Multiply E2 by –2.

$\begin{cases} 10y-2x=1 \\ -10y-12x=-8 \end{cases}$

E1 + E2: $\quad 10y-2x=1$

$\underline{\quad -10y-12x=-8 \quad}$

$\quad\quad\quad\quad -14x=-7$

$\quad\quad\quad\quad\quad x=\dfrac{1}{2}$

Replace x with $\dfrac{1}{2}$ in the equation $5y = 4 - 6x$.

$5y=4-6\left(\dfrac{1}{2}\right)$

$5y=4-3$

$5y=1$

$y=\dfrac{1}{5}$

The solution is $\left(\dfrac{1}{2},\dfrac{1}{5}\right)$.

51. $\begin{cases} 5x-2y=27 \\ -3x+5y=18 \end{cases}$

Multiply E1 by 3 and E2 by 5.

$\begin{cases} 15x-6y=81 \\ -15x+25y=90 \end{cases}$

E1 + E2: $\quad 15x-6y=81$

$\underline{\quad -15x+25y=90 \quad}$

$\quad\quad\quad\quad 19y=171$

$\quad\quad\quad\quad\quad y=9$

Replace y with 9 in E1.

$5x-2(9)=27$

$5x-18=27$

$5x=45$

$x=9$

The solution is $(9, 9)$.

53. $\begin{cases} \quad\quad x=3y+2 \\ 5x-15y=10 \end{cases}$

Replace x with $3y+2$ in E2.

$5(3y+2)-15y=10$

$15y+10-15y=10$

$\quad\quad\quad\quad 10=10 \quad$ True

The system is dependent. The solution is
$\{(x, y)|x = 3y + 2\}$.

55. $\begin{cases} 2x-y=-1 \\ \quad y=-2x \end{cases}$

Replace y with $-2x$ in E1.

$2x-(-2x)=-1$

$4x=-1$

$x=-\dfrac{1}{4}$

Replace x with $-\dfrac{1}{4}$ in E2.

$y=-2\left(-\dfrac{1}{4}\right)$

$y=\dfrac{1}{2}$

The solution is $\left(-\dfrac{1}{4},\dfrac{1}{2}\right)$.

57. $\begin{cases} 2x=6 \\ y=5-x \end{cases}$

E1 yields $x = 3$.
Replace x with 3 in E2.

$y=5-3$

$y=2$

The solution is $(3, 2)$.

59. $\begin{cases} \dfrac{x+5}{2} = \dfrac{6-4y}{3} \\ \dfrac{3x}{5} = \dfrac{21-7y}{10} \end{cases}$

Multiply E1 by 6 and E2 by 10.

$\begin{cases} 3x+15 = 12-8y \\ 6x = 21-7y \end{cases}$

$\begin{cases} 3x+8y = -3 \\ 6x+7y = 21 \end{cases}$

Multiply E1 by –2.

$\begin{cases} -6x-16y = 6 \\ 6x+7y = 21 \end{cases}$

E1 + E2:
$\begin{aligned} -6x-16y &= 6 \\ \underline{6x+7y} &= \underline{21} \\ -9y &= 27 \\ y &= -3 \end{aligned}$

Replace y with –3 in the equation $3x + 8y = -3$.

$3x+8(-3) = -3$
$3x-24 = -3$
$3x = 21$
$x = 7$

The solution is (7, –3).

61. $\begin{cases} 4x-7y = 7 \\ 12x-21y = 24 \end{cases}$

Multiply E1 by –3.

$\begin{cases} -12x+21y = -21 \\ 12x-21y = 24 \end{cases}$

$ 0 = 3 \quad \text{False}$

Inconsistent system; the solution set is $\varnothing$.

63. $\begin{cases} \dfrac{2}{3}x - \dfrac{3}{4}y = -1 \\ -\dfrac{1}{6}x + \dfrac{3}{8}y = 1 \end{cases}$

Multiply E1 by 12 and E2 by 24.

$\begin{cases} 8x-9y = -12 \\ -4x+9y = 24 \end{cases}$

E1 + E2:
$\begin{aligned} 8x-9y &= -12 \\ \underline{-4x+9y} &= \underline{24} \\ 4x &= 12 \\ x &= 3 \end{aligned}$

Replace x with 3 in the equation $-4x + 9y = 24$.

$-4(3)+9y = 24$
$-12+9y = 24$
$9y = 36$
$y = 4$

The solution is (3, 4).

65. $\begin{cases} 0.7x-0.2y = -1.6 \\ 0.2x-y = -1.4 \end{cases}$

Multiply both equations by 10.

$\begin{cases} 7x-2y = -16 \\ 2x-10y = -14 \end{cases}$

Multiply E1 by –5.

$\begin{cases} -35x+10y = 80 \\ 2x-10y = -14 \end{cases}$

E1 + E2:
$\begin{aligned} -35x+10y &= 80 \\ \underline{2x-10y} &= \underline{-14} \\ -33x &= 66 \\ x &= -2 \end{aligned}$

Replace x with –2 in the equation $7x - 2y = -16$.

$7(-2)-2y = -16$
$-14-2y = -16$
$-2y = -2$
$y = 1$

The solution is (–2, 1).

67. $\begin{cases} 4x-1.5y = 10.2 \\ 2x+7.8y = -25.68 \end{cases}$

Multiply E2 by –2.

$\begin{cases} 4x-1.5y = 10.2 \\ -4x-15.6y = 51.36 \end{cases}$

E1 + E2:
$\begin{aligned} 4x-1.5y &= 10.2 \\ \underline{-4x-15.6y} &= \underline{51.36} \\ -17.1y &= 61.56 \\ y &= -3.6 \end{aligned}$

Replace y with –3.6 in E1.

$4x-1.5(-3.6) = 10.2$
$4x+5.4 = 10.2$
$4x = 4.8$
$x = 1.2$

The solution is (1.2, –3.6).

69. $\begin{aligned} 3x-4y+2z &= 5 \\ 3(1)-4(2)+2(5) &\stackrel{?}{=} 5 \\ 3-8+10 &\stackrel{?}{=} 5 \\ 5 &= 5 \end{aligned}$

True

71. $\begin{aligned} -x-5y+3z &= 15 \\ -(0)-5(-1)+3(5) &\stackrel{?}{=} 15 \\ 0+5+15 &\stackrel{?}{=} 15 \\ 20 &= 15 \end{aligned}$

False

73. $\begin{aligned} 3x + 2y - 5z &= 10 \\ -3x + 4y + z &= 15 \\ \hline 6y - 4z &= 25 \end{aligned}$

75. $\begin{aligned} 10x + 5y + 6z &= 14 \\ -9x + 5y - 6z &= -12 \\ \hline x + 10y \quad\;\; &= 2 \end{aligned}$

77. $\begin{cases} y = 2x - 5 \\ y = 2x + 1 \end{cases}$

The lines have the same slope and different y-intercepts, so they are parallel. There is no solution.

79. $\begin{cases} x + y = 3 \\ 5x + 5y = 15 \end{cases}$

The result of multiplying the first equation by 5 is $5x + 5y = 15$, which is the second equation. The system is dependent, so it has an infinite number of solutions.

81. no; answers may vary

83. The lines intersect at the point (5, 21), which corresponds to 5000 DVDs at a price of $21 per DVD.

85. The graph for supply is higher than that for demand when x is greater than 6. Therefore, supply is greater than demand.

87. $\begin{cases} y = 2.5x \\ y = 0.9x + 3000 \end{cases}$

Replace y with $2.5x$ in the second equation.
$y = 0.9x + 3000$
$2.5x = 0.9x + 3000$
$1.6x = 3000$
$x = 1875$
Replace x with 1875 in the first equation.
$y = 2.5x = 2.5(1875) = 4687.5$
The break-even point is (1875, 4687.5).

89. At $x = 2000$, the graph for revenue is higher than that for cost, so the company makes money.

91. The revenue graph is above the cost graph to the right of the intersection point. Thus, for x-values greater than 1875, the company will make a profit.

93. answers may vary; one possibility:
$\begin{cases} -2x + y = 1 \\ x - 2y = -8 \end{cases}$

95. a. Since $1.47 > 0.13$, consumption of bottled water is growing faster than that of carbonated diet soft drink.

b. $\begin{cases} y = 1.47x + 9.26 \\ y = 0.13x + 13.55 \end{cases}$

Replace y with $1.47x + 9.26$ in the second equation.
$1.47x + 9.26 = 0.13x + 13.55$
$1.47x = 0.13x + 4.29$
$1.34x = 4.29$
$x \approx 3.20 \approx 3$
Replace x with 3.20 in the first equation.
$y = 1.47(3.20) + 9.26 \approx 13.96 \approx 14$
The solution of the system is approximately (3, 14).

c. $1995 + 3 = 1998$
In the year 1998, per capita bottled water consumption was about the same as per capita carbonated diet soft drink consumption.

97. $\begin{cases} \dfrac{1}{x} + y = 12 \\ \dfrac{3}{x} - y = 4 \end{cases}$

Replacing $\dfrac{1}{x}$ with a, we have
$\begin{cases} a + y = 12 \\ 3a - y = 4 \end{cases}$
Adding the two new equations we get
$4a = 16$
$a = 4$
Replace a with 4 in the equation $a + y = 12$.
$4 + y = 12$
$y = 8$

Since $a = 4$, $x = \dfrac{1}{4}$.

The solution is $\left(\dfrac{1}{4}, 8 \right)$.

99. $\begin{cases} \dfrac{1}{x} + \dfrac{1}{y} = 5 \\ \dfrac{1}{x} - \dfrac{1}{y} = 1 \end{cases}$

Replace $\dfrac{1}{x}$ with a and $\dfrac{1}{y}$ with b.

$$\begin{cases} a+b=5 \\ a-b=1 \end{cases}$$

Adding the two new equations we get

$2a = 6$

$a = 3$

Replace a with 3 in the equation $a + b = 5$.

$3 + b = 5$

$b = 2$

Since $a = 3$, $x = \dfrac{1}{3}$. Similarly, $y = \dfrac{1}{2}$.

The solution is $\left(\dfrac{1}{3}, \dfrac{1}{2}\right)$.

101. $\begin{cases} \dfrac{2}{x} - \dfrac{4}{y} = 5 \\ \dfrac{1}{x} - \dfrac{2}{y} = \dfrac{3}{2} \end{cases}$

Replace $\dfrac{1}{x}$ with a and $\dfrac{1}{y}$ with b.

$\begin{cases} 2a - 4b = 5 \\ a - 2b = \dfrac{3}{2} \end{cases}$

Multiply E2 by 2.

$\begin{cases} 2a - 4b = 5 \\ 2a - 4b = 3 \end{cases}$

This system is inconsistent. The solution set is $\varnothing$.

103. $\begin{cases} \dfrac{3}{x} - \dfrac{2}{y} = -18 \\ \dfrac{2}{x} + \dfrac{3}{y} = 1 \end{cases}$

Replace $\dfrac{1}{x}$ with a and $\dfrac{1}{y}$ with b.

$\begin{cases} 3a - 2b = -18 \\ 2a + 3b = 1 \end{cases}$

Multiply E1 by 3 and E2 by 2 to obtain

$\begin{cases} 9a - 6b = -54 \\ 4a + 6b = 2 \end{cases}$

Add these equations to get $13a = -52$

$a = -4$.

Replace a with -4 in the equation $2a + 3b = 1$.

$2(-4) + 3b = 1$

$3b = 9$

$b = 3$

Since $a = -4$, $x = -\dfrac{1}{4}$. Similarly, $y = \dfrac{1}{3}$. The

solution is $\left(-\dfrac{1}{4}, \dfrac{1}{3}\right)$.

Section 4.2 Practice Exercises

1. $\begin{cases} 3x + 2y - z = 0 & (1) \\ x - y + 5z = 2 & (2) \\ 2x + 3y + 3z = 7 & (3) \end{cases}$

Multiply equation (2) by 2 and add to equation (1) to eliminate y.

$\begin{cases} 3x + 2y - z = 0 \\ 2(x - y + 5z) = 2(2) \end{cases}$

$\begin{cases} 3x - 2y - z = 0 \\ 2x - 2y + 10z = 4 \end{cases}$

$5x + 9z = 4 \quad (4)$

Multiply equation (2) by 3 and add to equation (3) to eliminate y again.

$\begin{cases} 3(x - y + 5z) = 3(2) \\ 2x + 3y + 3z = 7 \end{cases}$

$\begin{cases} 3x - 3y + 15z = 6 \\ 2x + 3y + 3z = 7 \end{cases}$

$5x + 18z = 13 \quad (5)$

Multiply equation (4) by -1 and add to equation (5) to eliminate x.

$\begin{cases} -1(5x + 9z) = -1(4) \\ 5x + 18z = 13 \end{cases}$

$\begin{cases} -5x - 9z = -4 \\ 5x + 18z = 13 \end{cases}$

$9z = 9$

$z = 1$

Replace z with 1 in equation (4) or (5).

$5x + 9z = 4$

$5x + 9(1) = 4$

$5x = -5$

$x = -1$

Replace x with -1 and z with 1 in equation (1), (2), or (3).

$x - y + 5z = 2$

$-1 - y + 5(1) = 2$

$-y + 4 = 2$

$-y = -2$

$y = 2$

The solution is $(-1, 2, 1)$. To check, let $x = -1$, $y = 2$, and $z = 1$ in all three original equations of the system.

2. $\begin{cases} 6x - 3y + 12z = 4 & (1) \\ -6x + 4y - 2z = 7 & (2) \\ -2x + y - 4z = 3 & (3) \end{cases}$

Multiply equation (3) by 3 and add to equation (1) to eliminate x.

$\begin{cases} 6x - 3y + 12z = 4 \\ 3(-2x + y - 4z) = 3(3) \end{cases}$

$\begin{cases} 6x - 3y + 12z = 4 \\ -6x + 3y - 12z = 9 \end{cases}$

$$0 = 13 \quad \text{False}$$

Since the statement is false, this system is inconsistent and has no solution. The solution set is { } or $\varnothing$.

3. $\begin{cases} 3x + 4y = 0 & (1) \\ 9x - 4z = 6 & (2) \\ -2y + 7z = 1 & (3) \end{cases}$

Equation (2) has no term containing the variable y. Eliminate y using equations (1) and (3). Multiply equation (3) by 2 and add to equation (1).

$\begin{cases} 3x + 4y = 0 \\ 2(-2y + 7z) = 2(1) \end{cases}$

$\begin{cases} 3x + 4y = 0 \\ -4y + 14z = 2 \end{cases}$

$$3x + 14z = 2 \quad (4)$$

Multiply equation (4) by -3 and add to equation (2) to eliminate x.

$\begin{cases} 9x - 4z = 6 \\ -3(3x + 14z) = -3(2) \end{cases}$

$\begin{cases} 9x - 4z = 6 \\ -9x - 52z = -6 \end{cases}$

$$-56z = 0$$
$$z = 0$$

Replace z with 0 in equation (2) and solve for x.
$$9x - 4z = 6$$
$$9x - 4(0) = 6$$
$$9x = 6$$
$$x = \frac{6}{9} = \frac{2}{3}$$

Replace z with 0 in equation (3) and solve for y.
$$-2y + 7z = 1$$
$$-2y + 7(0) = 1$$
$$-2y = 1$$
$$y = -\frac{1}{2}$$

The solution is $\left(\frac{2}{3}, -\frac{1}{2}, 0 \right)$.

4. $\begin{cases} 2x + y - 3z = 6 & (1) \\ x + \dfrac{1}{2}y - \dfrac{3}{2}z = 3 & (2) \\ -4x - 2y + 6z = -12 & (3) \end{cases}$

Multiply both sides of equation (2) by 2 to eliminate fractions, and multiply both sides of equation (3) by $-\dfrac{1}{2}$ since all coefficients in equation (3) are divisible by 2 and the coefficient of x is negative. The resulting system is
$\begin{cases} 2x + y - 3z = 6 \\ 2x + y - 3z = 6 \\ 2x + y - 3z = 6 \end{cases}$

Since the three equations are identical, there are infinitely many solutions of the system. The equations are dependent. The solution set can be written as $\{(x, y, z) | 2x + y - 3z = 6\}$.

5. $\begin{cases} x + 2y + 4z = 16 & (1) \\ x + 2z = -4 & (2) \\ y - 3z = 30 & (3) \end{cases}$

Solve equation (2) for x and equation (3) for y.
$$x + 2z = -4 \qquad\qquad y - 3z = 30$$
$$x = -2z - 4 \qquad\qquad y = 3z + 30$$

Substitute $-2z - 4$ for x and $3z + 30$ for y in equation (1) and solve for z.
$$x + 2y + 4z = 16$$
$$(-2z - 4) + 2(3z + 30) + 4z = 16$$
$$-2z - 4 + 6z + 60 + 4z = 16$$
$$8z + 56 = 16$$
$$8z = -40$$
$$z = -5$$

Use $x = -2z - 4$ to find x:
$x = -2(-5) - 4 = 10 - 4 = 6$.
Use $y = 3z + 30$ to find y:
$y = 3(-5) + 30 = -15 + 30 = 15$.
The solution is $(6, 15, -5)$.

Vocabulary, Readiness & Video Check 4.2

1. a. $\quad x + y + z = 3$
$$-1 + 3 + 1 \overset{?}{=} 3$$
$$3 = 3 \quad \text{True}$$

b. $\quad -x + y + z = 5$
$$-(-1) + 3 + 1 \overset{?}{=} 5$$
$$1 + 3 + 1 \overset{?}{=} 5$$
$$5 = 5 \quad \text{True}$$

c. $-x + y + 2z = 0$

$-(-1) + 3 + 2(1) \overset{?}{=} 0$

$1 + 3 + 2 \overset{?}{=} 0$

$6 = 0$ False

d. $x + 2y - 3z = 2$

$-1 + 2(3) - 3(1) \overset{?}{=} 2$

$-1 + 6 - 3 \overset{?}{=} 2$

$2 = 2$ True

$(-1, 3, 1)$ is a solution to the equations a, b, and d.

2. a. $x + y + z = -1$

$2 + 1 + (-4) \overset{?}{=} -1$

$-1 = -1$ True

b. $x - y - z = -3$

$2 - 1 - (-4) \overset{?}{=} -3$

$2 - 1 + 4 \overset{?}{=} -3$

$5 = -3$ False

c. $2x - y + z = -1$

$2(2) - 1 + (-4) \overset{?}{=} -1$

$4 - 1 - 4 \overset{?}{=} -1$

$-1 = -1$ True

d. $-x - 3y - z = -1$

$-2 - 3(1) - (-4) \overset{?}{=} -1$

$-2 - 3 + 4 \overset{?}{=} -1$

$-1 = -1$ True

$(2, 1, -4)$ is a solution to the equations a, c, and d.

3. yes; answers may vary

4. no; answers may vary

5. Once we have one equation in two variables, we need to get another equation in the *same* two variables, giving us a system of two equations in two variables. We solve this new system to find the values of two variables. We then substitute these values into an original equation to find the value of the third.

Exercise Set 4.2

1. $\begin{cases} x - y + z = -4 & (1) \\ 3x + 2y - z = 5 & (2) \\ -2x + 3y - z = 15 & (3) \end{cases}$

Add the equations (1) and (2) to eliminate z.

$x \ - y + z = -4$

$\underline{3x + 2y - z = 5}$

$4x \ + y \quad = 1$ (4)

Add equations (1) and (3) to eliminate z.

$x \ - y + z = -4$

$\underline{-2x + 3y - z = 15}$

$-x + 2y \quad = 11$ (5)

Multiply equation (5) by 4 and add it to equation (4).

$4x + y = 1$

$\underline{-4x + 8y = 44}$

$9y = 45$

$y = 5$

Replace y with 5 in equation (4).

$4x + y = 1$

$4x + 5 = 1$

$4x = -4$

$x = -1$

Replace x with -1 and y with 5 in equation (1).

$x - y + z = -4$

$-1 - 5 + z = -4$

$-6 + z = -4$

$z = 2$

The solution of the system is $(-1, 5, 2)$.

3. $\begin{cases} x + y \quad = 3 & (1) \\ 2y \quad = 10 & (2) \\ 3x + 2y - 3z = 1 & (3) \end{cases}$

Solve equation (2) for y.

$2y = 10$

$y = 5$

Replace y with 5 in equation (1).

$x + y = 3$

$x + 5 = 3$

$x = -2$

Replace x with -2 and y with 5 in equation (3).

$3x + 2y - 3z = 1$

$3(-2) + 2(5) - 3z = 1$

$-6 + 10 - 3z = 1$

$4 - 3z = 1$

$-3z = -3$

$z = 1$

The solution of the system is $(-2, 5, 1)$.

5. $\begin{cases} 2x+2y+z=1 & (1) \\ -x+y+2z=3 & (2) \\ x+2y+4z=0 & (3) \end{cases}$

Add equations (2) and (3) to eliminate x.

$\begin{aligned} -x+y+2z &= 3 \\ x+2y+4z &= 0 \\ \hline 3y+6z &= 3 \\ y+2z &= 1 \quad (4) \end{aligned}$

Multiply equation (2) by 2 and add it to equation (1) to eliminate x.

$\begin{aligned} 2x+2y+z &= 1 \\ -2x+2y+4z &= 6 \\ \hline 4y+5z &= 7 \quad (5) \end{aligned}$

Multiply equation (4) by -4 and add it to equation (5).

$\begin{aligned} -4y-8z &= -4 \\ 4y+5z &= 7 \\ \hline -3z &= 3 \\ z &= -1 \end{aligned}$

Replace z with -1 in equation (4).

$\begin{aligned} y+2z &= 1 \\ y+2(-1) &= 1 \\ y-2 &= 1 \\ y &= 3 \end{aligned}$

Replace y with 3 and z with -1 in equation (3).

$\begin{aligned} x+2y+4z &= 0 \\ x+2(3)+4(-1) &= 0 \\ x+6-4 &= 0 \\ x+2 &= 0 \\ x &= -2 \end{aligned}$

The solution of the system is $(-2, 3, -1)$.

7. $\begin{cases} x-2y+z=-5 & (1) \\ -3x+6y-3z=15 & (2) \\ 2x-4y+2z=-10 & (3) \end{cases}$

Multiply equation (1) by -3.
$-3x + 6y - 3z = 15$
This is equation (2).
Multiply equation (1) by 2.
$2x - 4y + 2z = -10$
This is equation (3).
The system is dependent and the solution set is
$\{(x, y, z)|x - 2y + z = -5\}$.

9. $\begin{cases} 4x-y+2z=5 & (1) \\ 2y+z=4 & (2) \\ 4x+y+3z=10 & (3) \end{cases}$

Multiply equation (1) by -1 and add it to equation (3) to eliminate x.

$\begin{aligned} -4x+y-2z &= -5 \\ 4x+y+3z &= 10 \\ \hline 2y+z &= 5 \quad (4) \end{aligned}$

Multiply equation (2) by -1 and add it to equation (4).

$\begin{aligned} -2y-z &= -4 \\ 2y+z &= 5 \\ \hline 0 &= 1 \end{aligned}$

The statement $0 = 1$ is false, so the system has no solution.

11. $\begin{cases} x+5z=0 & (1) \\ 5x+y=0 & (2) \\ y-3z=0 & (3) \end{cases}$

Solve equation (1) for x.
$\begin{aligned} x+5z &= 0 \\ x &= -5z \end{aligned}$
Solve equation (3) for y.
$\begin{aligned} y-3z &= 0 \\ y &= 3z \end{aligned}$

Replace x with $-5z$ and y with $3z$ in equation (2).

$\begin{aligned} 5x+y &= 0 \\ 5(-5z)+3z &= 0 \\ -25z+3z &= 0 \\ -22z &= 0 \\ z &= 0 \end{aligned}$

Replace z with 0 in equation (1).

$\begin{aligned} x+5z &= 0 \\ x+5(0) &= 0 \\ x+0 &= 0 \\ x &= 0 \end{aligned}$

Replace x with 0 in equation (2).

$\begin{aligned} 5x+y &= 0 \\ 5(0)+y &= 0 \\ 0+y &= 0 \\ y &= 0 \end{aligned}$

The solution of the system is $(0, 0, 0)$.

13. $\begin{cases} 6x-5z=17 & (1) \\ 5x-y+3z=-1 & (2) \\ 2x+y=-41 & (3) \end{cases}$

Add equations (2) and (3) to eliminate y.

$\begin{aligned} 5x-y+3z &= -1 \\ 2x+y &= -41 \\ \hline 7x+3z &= -42 \quad (4) \end{aligned}$

Multiply equation (1) by 3 and equation (4) by 5, then add the resulting equations.

$\begin{aligned} 18x-15z &= 51 \\ 35x+15z &= -210 \\ \hline 53x &= -159 \\ x &= -3 \end{aligned}$

Replace x with -3 in equation (1).
$$6x - 5z = 17$$
$$6(-3) - 5z = 17$$
$$-18 - 5z = 17$$
$$-5z = 35$$
$$z = -7$$
Replace x with -3 in equation (3).
$$2x + y = -41$$
$$2(-3) + y = -41$$
$$-6 + y = -41$$
$$y = -35$$
The solution of the system is $(-3, -35, -7)$.

15. $\begin{cases} x + y + z = 8 & (1) \\ 2x - y - z = 10 & (2) \\ x - 2y - 3z = 22 & (3) \end{cases}$

Add equations (1) and (2) to eliminate y and z.
$$\begin{array}{r} x + y + z = 8 \\ 2x - y - z = 10 \\ \hline 3x \qquad\quad = 18 \\ x = 6 \end{array}$$
Multiply equation (1) by 2 and add it to equation (3) to eliminate y.
$$\begin{array}{r} 2x + 2y + 2z = 16 \\ x - 2y - 3z = 22 \\ \hline 3x \qquad\quad - z = 38 \quad (4) \end{array}$$
Replace x with 6 in equation (4).
$$3x - z = 38$$
$$3(6) - z = 38$$
$$18 - z = 38$$
$$z = -20$$
Replace x with 6 and z with -20 in equation (1).
$$x + y + z = 8$$
$$6 + y + (-20) = 8$$
$$y - 14 = 8$$
$$y = 22$$
The solution of the system is $(6, 22, -20)$.

17. $\begin{cases} x + 2y - z = 5 & (1) \\ 6x + y + z = 7 & (2) \\ 2x + 4y - 2z = 5 & (3) \end{cases}$

Add equations (1) and (2) to eliminate z.
$$\begin{array}{r} x + 2y - z = 5 \\ 6x + y + z = 7 \\ \hline 7x + 3y \quad\;\; = 12 \quad (4) \end{array}$$
Multiply equation (2) by 2 and add to equation (3) to eliminate z.
$$\begin{array}{r} 12x + 2y + 2z = 14 \\ 2x + 4y - 2z = 5 \\ \hline 14x + 6y \qquad = 19 \quad (5) \end{array}$$

Multiply equation (4) by -2 and add to equation (5).
$$\begin{array}{r} -14x - 6y = -24 \\ 14x + 6y = 19 \\ \hline 0 = -5 \end{array}$$
The statement $0 = -5$ is false, so the system has no solution.

19. $\begin{cases} 2x - 3y + z = 2 & (1) \\ x - 5y + 5z = 3 & (2) \\ 3x + y - 3z = 5 & (3) \end{cases}$

Multiply equation (1) by -5 and add it to equation (2) to eliminate z.
$$\begin{array}{r} -10x + 15y - 5z = -10 \\ x - 5y + 5z = 3 \\ \hline -9x + 10y \qquad = -7 \quad (4) \end{array}$$
Multiply equation (1) by 3 and add it to equation (3) to eliminate z.
$$\begin{array}{r} 6x - 9y + 3z = 6 \\ 3x + y - 3z = 5 \\ \hline 9x - 8y \qquad = 11 \quad (5) \end{array}$$
Add equations (4) and (5) to eliminate x.
$$\begin{array}{r} -9x + 10y = -7 \\ 9x - 8y = 11 \\ \hline 2y = 4 \\ y = 2 \end{array}$$
Replace y with 2 in equation (5).
$$9x - 8y = 11$$
$$9x - 8(2) = 11$$
$$9x - 16 = 11$$
$$9x = 27$$
$$x = 3$$
Replace x with 3 and y with 2 in equation (1).
$$2x - 3y + z = 2$$
$$2(3) - 3(2) + z = 2$$
$$6 - 6 + z = 2$$
$$z = 2$$
The solution of the system is $(3, 2, 2)$.

21. $\begin{cases} -2x - 4y + 6z = -8 & (1) \\ x + 2y - 3z = 4 & (2) \\ 4x + 8y - 12z = 16 & (3) \end{cases}$

Multiply equation (2) by -2.
$$-2x - 4y + 6z = -8$$
This is equation (1).
Multiply equation (2) by 4.
$$4x + 8y - 12z = 16$$
This is equation (3).
The system is dependent and the solution set is $\{(x, y, z) | x + 2y - 3z = 4\}$.

23. $\begin{cases} 2x+2y-3z=1 & (1) \\ \quad\ y+2z=-14 & (2) \\ 3x-2y\quad\ =-1 & (3) \end{cases}$

Multiply equation (1) by -3 and equation (3) by 2, then add the results to eliminate x.

$$\begin{array}{r} -6x-6y+9z=-3 \\ 6x-4y\quad\ =-2 \\ \hline -10y+9z=-5 \quad (4) \end{array}$$

Multiply equation (2) by 10 and add the result to equation (4) to eliminate y.

$$\begin{array}{r} 10y+20z=-140 \\ -10y+9z=-5 \\ \hline 29z=-145 \\ z=-5 \end{array}$$

Replace z with -5 in equation (2).

$$y+2z=-14$$
$$y+2(-5)=-14$$
$$y-10=-14$$
$$y=-4$$

Replace y with -4 in equation (3).

$$3x-2y=-1$$
$$3x-2(-4)=-1$$
$$3x+8=-1$$
$$3x=-9$$
$$x=-3$$

The solution of the system is $(-3, -4, -5)$.

25. $\begin{cases} \quad\ x+2y\ -z=5 & (1) \\ -3x-2y-3z=11 & (2) \\ \quad 4x+4y+5z=-18 & (3) \end{cases}$

Add equations (1) and (2) to eliminate y.

$$\begin{array}{r} x+2y\ -z=5 \\ -3x-2y-3z=11 \\ \hline -2x\quad\ -4z=16 \\ x+2z=-8 \quad (4) \end{array}$$

Multiply equation (2) by 2 and add to equation (3) to eliminate y.

$$\begin{array}{r} -6x-4y-6z=22 \\ 4x+4y+5z=-18 \\ \hline -2x\quad\ -z=4 \quad (5) \end{array}$$

Multiply equation (5) by 2 and add to equation (4).

$$\begin{array}{r} x+2z=-8 \\ -4x-2z=8 \\ \hline -3x\quad\ =0 \\ x=0 \end{array}$$

Replace x with 0 in equation (4).

$$x+2z=-8$$
$$0+2z=-8$$
$$2z=-8$$
$$z=-4$$

Replace x with 0 and z with -4 in equation (1).

$$x+2y-z=5$$
$$0+2y-(-4)=5$$
$$2y+4=5$$
$$2y=1$$
$$y=\frac{1}{2}$$

The solution of the system is $\left(0, \dfrac{1}{2}, -4\right)$.

27. $\begin{cases} \dfrac{3}{4}x-\dfrac{1}{3}y+\dfrac{1}{2}z=9 & (1) \\ \dfrac{1}{6}x+\dfrac{1}{3}y-\dfrac{1}{2}z=2 & (2) \\ \dfrac{1}{2}x-y+\dfrac{1}{2}z=2 & (3) \end{cases}$

Multiply equation (1) by 12, equation (2) by 6, and equation (3) by 2 to clear fractions.

$$\begin{cases} 9x-4y+6z=108 & (4) \\ \ x+2y-3z=12 & (5) \\ \ x-2y+z=4 & (6) \end{cases}$$

Multiply equation (5) by 2 and add it to equation (4) to eliminate y and z.

$$\begin{array}{r} 9x-4y+6z=108 \\ 2x+4y-6z=24 \\ \hline 11x\quad\ =132 \\ x=12 \end{array}$$

Add equations (5) and (6) to eliminate y.

$$\begin{array}{r} x+2y-3z=12 \\ x-2y\ +z=4 \\ \hline 2x\quad\ -2z=16 \quad (7) \end{array}$$

Replace x with 12 in equation (7).

$$2x-2z=16$$
$$2(12)-2z=16$$
$$24-2z=16$$
$$-2z=-8$$
$$z=4$$

Replace x with 12 and z with 4 in equation (6).

$$x-2y+z=4$$
$$12-2y+4=4$$
$$-2y+16=4$$
$$-2y=-12$$
$$y=6$$

The solution of the system is $(12, 6, 4)$.

29. Let x = the first number, then
$2x$ = the second number.
$$x + 2x = 45$$
$$3x = 45$$
$$x = 15$$
$$2x = 2(15) = 30$$
The numbers are 15 and 30.

31. $2(x-1) - 3x = x - 12$
$$2x - 2 - 3x = x - 12$$
$$-x - 2 = x - 12$$
$$-2x = -10$$
$$x = 5$$

33. $-y - 5(y+5) = 3y - 10$
$$-y - 5y - 25 = 3y - 10$$
$$-6y - 25 = 3y - 10$$
$$-9y = 15$$
$$y = -\frac{15}{9} = -\frac{5}{3}$$

35. answers may vary

37. answers may vary

39. $\begin{cases} x + y + z = 1 & (1) \\ 2x - y + z = 0 & (2) \\ -x + 2y + 2z = -1 & (3) \end{cases}$

Add E1 and E3.
$3y + 3z = 0$ or $y + z = 0$ (4)
Add -2 times E1 to E2.
$$\begin{aligned} -2x - 2y - 2z &= -2 \\ \underline{2x - y + z} &= \underline{0} \\ -3y - z &= -2 \quad (5) \end{aligned}$$
Add E4 and E5.
$$-2y = -2$$
$$y = 1$$
Replace y with 1 in E4.
$$1 + z = 0$$
$$z = -1$$
Replace y with 1 and z with -1 in E1.
$$x + 1 + (-1) = 1$$
$$x = 1$$
The solution is $(1, 1, -1)$, and
$$\begin{aligned} \frac{x}{8} + \frac{y}{4} + \frac{z}{3} &= \frac{1}{8} + \frac{1}{4} - \frac{1}{3} \\ &= \frac{3}{24} + \frac{6}{24} - \frac{8}{24} \\ &= \frac{1}{24}. \end{aligned}$$

41. $\begin{cases} x + y \quad\;\; - w = 0 & (1) \\ y + 2z + w = 3 & (2) \\ x \quad\;\; - z \quad\;\; = 1 & (3) \\ 2x - y \quad\;\; - w = -1 & (4) \end{cases}$

Add E1 and E2.
$x + 2y + 2z = 3$ (5)
Add E2 and E4.
$2x + 2z = 2$ or $x + z = 1$ (6)
Add E3 and E6.
$$\begin{aligned} x - z &= 1 \\ \underline{x + z} &= \underline{1} \\ 2x &= 2 \end{aligned}$$
Replace x with 1 in E3.
$$1 - z = 1$$
$$z = 0$$
Replace x with 1 and z with 0 in E5.
$$1 + 2y + 2(0) = 3$$
$$1 + 2y = 3$$
$$2y = 2$$
$$y = 1$$
Replace y with 1, and z with 0 in E2.
$$1 + 2(0) + w = 3$$
$$1 + w = 3$$
$$w = 2$$
The solution is $(1, 1, 0, 2)$.

43. $\begin{cases} x + y + z + w = 5 & (1) \\ 2x + y + z + w = 6 & (2) \\ x + y + z \quad\;\; = 2 & (3) \\ x + y \quad\;\;\;\; = 0 & (4) \end{cases}$

Add -1 times E4 to E3.
$$\begin{aligned} -x - y &= 0 \\ \underline{x + y + z} &= \underline{2} \\ z &= 2 \end{aligned}$$
Replace z with 2 in E1 and E2.
$\begin{cases} x + y + w = 3 & (5) \\ 2x + y + w = 4 & (6) \end{cases}$
Add -1 times E5 to E6.
$$\begin{aligned} -x - y - w &= -3 \\ \underline{2x + y + w} &= \underline{4} \\ x &= 1 \end{aligned}$$
Replace x with 1 in E4.
$$1 + y = 0$$
$$y = -1$$
Replace x with 1, y with -1, and z with 2 in E1.
$$1 + (-1) + 2 + w = 5$$
$$2 + w = 5$$
$$w = 3$$
The solution is $(1, -1, 2, 3)$.

45. answers may vary

Section 4.3 Practice Exercises

1. a. $\begin{cases} y = -0.54x + 110.6 \\ y = -0.36x + 74.1 \end{cases}$

Use substitution.
$$-0.54x + 110.6 = -0.36x + 74.1$$
$$-0.18x = -36.5$$
$$x \approx 203$$

The predicted year is 203 years from 2005 or in 2208.

b. yes; answers may vary

2. Let x = first number
y = second number
"A first number is five more than a second number" is translated as $x = y + 5$. "Twice the first number is 2 less than 3 times the second number" is translated as $2x = 3y - 2$.
We solve the following system.
$$\begin{cases} x = y + 5 \\ 2x = 3y - 2 \end{cases}$$

Since the first equation is solved for x, we use substitution. Substitute $y + 5$ for x in the second equation.
$$2(y + 5) = 3y - 2$$
$$2y + 10 = 3y - 2$$
$$12 = y$$

Replace y with 12 in the equation $x = y + 5$ and solve for x.
$$x = 12 + 5 = 17$$
The numbers are 17 and 12.

3. Let x = speed of the V150
y = speed of the Atlantique
We summarize the information in a chart. Both trains have traveled two hours.

	Rate	•	Time	=	Distance
V150	x		2		$2x$
Atlantique	y		2		$2y$

The trains are 2150 kilometers apart, so the sum of the distances is 2150: $2x + 2y = 2150$.

The V150 is 75 k/h faster than the Atlantique: $x = y + 75$.

We solve the following system.
$$\begin{cases} 2x + 2y = 2150 \\ x = y + 75 \end{cases}$$

Since the second equation is solved for x, we use substitution. Substitute $y + 75$ for x in the first equation.
$$2(y + 75) + 2y = 2150$$
$$2y + 150 + 2y = 2150$$
$$4y + 150 = 2150$$
$$4y = 2000$$
$$y = 500$$

To find x, we replace y with 500 in the second equation.
$$x = 500 + 75 = 575$$
The speed of the V150 is 575 k/h, and the speed of the Atlantique is 500 k/h.

4. Let x = amount of 99% acid
y = amount of water (0%)
Both x and y are measured in liters. We use a table to organize the given data.

	Amount	Acid Strength	Amount of Pure Acid
99% acid	x	99%	$0.99x$
Water	y	0%	$0y$

The amount of 99% acid and water combined must equal 1 liter, so $x + y = 1$.
The amount of pure acid in the mixture must equal the sum of the amounts of pure acid in the 99% acid and in the water, so $0.99x + 0y = 0.05(1)$, which simplifies to $0.99x = 0.05$.
We solve the following system.
$$\begin{cases} x + y = 1 \\ 0.99x = 0.05 \end{cases}$$

Since the second equation does not contain y, we solve it for x.
$$0.99x = 0.05$$
$$x = \frac{0.05}{0.99} \approx 0.05$$

To find y, we replace x with 0.05 in the first equation.
$$x + y = 1$$
$$0.05 + y = 1$$
$$y = 0.95$$
The teacher should use 0.05 liter of the 99% HCL solution and 0.95 liter of water.

5. Let x = the number of packages.
The firm charges the customer $4.50 for each package, so the revenue equation is $R(x) = 4.5x$.
Each package costs $2.50 to produce and the

equipment costs $3000, so the cost equation is $C(x) = 2.5x + 3000$.

Since the break-even point is when $R(x) = C(x)$, we solve the equation $4.5x = 2.5x + 3000$.

$$4.5x = 2.5x + 3000$$
$$2x = 3000$$
$$x = 1500$$

The company must sell 1500 packages to break even.

6. Let x = measure of smallest angle
y = measure of largest angle
z = measure of third angle
The sum of the measures is 180°:
$x + y + z = 180$.

The measure of the largest angle is 40° more than the measure of the smallest angle:
$y = x + 40$.

The measure of the remaining angle is 20° more than the measure of the smallest angle:
$z = x + 20$.
We solve the following system.
$$\begin{cases} x + y + z = 180 \\ y = x + 40 \\ z = x + 20 \end{cases}$$
We substitute $x + 40$ for y and $x + 20$ for z in the first equation.
$$x + (x + 40) + (x + 20) = 180$$
$$3x + 60 = 180$$
$$3x = 120$$
$$x = 40$$
Then $y = x + 40 = 40 + 40 = 80$ and
$z = x + 20 = 40 + 20 = 60$.
The angle measures are 40°, 60°, and 80°.

Vocabulary, Readiness & Video Check 4.3

1. Up to now we've been choosing one variable/unknown and translating to one equation. To solve by a system of equations, we'll choose two variables to represent two unknowns and translate to two equations.

2. The break-even point occurs when revenue equals cost—money has not been lost or made; set the revenue function equal to the cost function and solve for the variable.

3. The ordered triple still needs to be interpreted in the context of the application. Each value actually represents the angle measure of a triangle, in degrees.

Exercise Set 4.3

1. Let x = the first number, y = the second number.
$$\begin{cases} x = y + 2 \\ 2x = 3y - 4 \end{cases}$$
Substitute $x = y + 2$ in the second equation.
$$2(y + 2) = 3y - 4$$
$$2y + 4 = 3y - 4$$
$$y = 8$$
Replace y with 8 in the first equation.
$x = 8 + 2 = 10$
The numbers are 10 and 8.

3. **a.** Let e = length of the Enterprise class,
n = length of the Nimitz class.
$$\begin{cases} e + n = 2193 \\ e - n = 9 \end{cases}$$
Add the equations.
$$2e = 2202$$
$$e = 1101$$
Replace e with 1101 in the first equation.
$$1101 + n = 2193$$
$$n = 1092$$
The Enterprise class is 1101 feet and the Nimitz class is 1092 feet.

b. There are 3 feet in each yard, so there are 300 feet in 100 yards.
$$\frac{1101}{300} = 3.67$$
The length of the Enterprise class carrier is 3.67 football fields.

5. With the wind, the plane was moving at
$\frac{2520}{4.5} = 560$ mph. Against the wind, the plane was moving at $\frac{2160}{4.5} = 480$ mph.
Let p = speed of the plane in still air,
w = speed of the wind.
$$\begin{cases} p + w = 560 \\ p - w = 480 \end{cases}$$
Add the equations.
$$2p = 1040$$
$$p = 520$$
Replace p with 520 in the first equation.
$$520 + w = 560$$
$$w = 40$$
The speed of the plane is 520 mph and the speed of the wind is 40 mph.

7. Let x = amount of 4% butterfat milk, and y = amount of 1% butterfat milk.

qt	strength	amount of butterfat
x	4%	$0.04x$
y	1%	$0.01y$
60	2%	$0.02(60) = 1.2$

$$\begin{cases} x+y=60 \\ 0.04x+0.01y=1.2 \end{cases}$$

Multiply the second equation by -100 and add the result to the first equation.

$$\begin{array}{r} x+y=60 \\ -4x-y=-120 \\ \hline -3x=-60 \\ x=20 \end{array}$$

Replace x with 20 in the first equation.

$$20+y=60$$
$$y=40$$

Thus, mix 20 quarts of 4% butterfat milk with 40 quarts of 1% butterfat milk.

9. Let x be the number of students studying in the United Kingdom, and y be the number studying in Italy.

$$\begin{cases} x+y=58,704 \\ x=y+3980 \end{cases}$$

Replace x with $y + 3980$ in the first equation.

$$x+y=58,704$$
$$y+3980+y=58,704$$
$$2y+3980=58,704$$
$$2y=54,724$$
$$y=27,362$$

Replace y with 27,362 in the second equation.

$x = y + 3980 = 27,362 + 3980 = 31,342$
31,342 students studied in the United Kingdom, and 27,362 studied in Italy.

11. Let l be the number of large frames and s be the number of small frames.

$$\begin{cases} l+s=22 \\ 15l+8s=239 \end{cases}$$

Solve the first equation for l.

$$l=22-s$$

Replace l with $22 - s$ in the second equation.

$$15(22-s)+8s=239$$
$$330-15s+8s=239$$
$$-7s=-91$$
$$s=13$$

Replace s with 13 in the first equation.

$$l+13=22$$
$$l=9$$

She bought 9 large frames and 13 small frames.

13. Let m = the first number, n = the second number.

$$\begin{cases} m=n-2 \\ 2m=3n+4 \end{cases}$$

Substitute $m = n - 2$ in the second equation.

$$2(n-2)=3n+4$$
$$2n-4=3n+4$$
$$-8=n$$

Replace n with -8 in the first equation.

$$m=-8-2=-10$$

The numbers are -10 and -8.

15. a.

$$\begin{cases} y=-4.5x+24 \\ y=2x+7 \end{cases}$$

Replace y with $-4.5x + 24$ in the second equation.

$$y=2x+7$$
$$-4.5x+24=2x+7$$
$$-6.5x+24=7$$
$$-6.5x=-17$$
$$x\approx 3$$

$2007 + 3 = 2010$

The predicted year is 2010 where the percent of adults under 30 and the percent of adults over 30 will blog at the same rate.

b. answers may vary

17. Let p be the price of a pen and w be the price of a writing tablet.

$$\begin{cases} 7w+4p=6.40 \\ 2w+19p=5.40 \end{cases}$$

Multiply the first equation by -2 and the second equation by 7 and add the resulting equations.

$$\begin{array}{r} -14w-8p=-12.8 \\ 14w+133p=37.8 \\ \hline 125p=25 \\ p=0.2 \end{array}$$

Replace p with 0.2 in the first equation.

$$7w+4p=6.40$$
$$7w+4(0.2)=6.40$$
$$7w+0.8=6.4$$
$$7w=5.6$$
$$w=0.8$$

The price of a writing tablet is $0.80 and the price of a pen is $0.20.

19. Let p be the speed of the plane and w be the speed of the wind.
$$\begin{cases} 3p + 3w = 2160 \\ 4p - 4w = 2160 \end{cases}$$

Multiply the first equation by $\dfrac{1}{3}$ and the second equation by $\dfrac{1}{4}$, and add the results.
$$\begin{array}{r} p + w = 720 \\ p - w = 540 \\ \hline 2p = 1260 \\ p = 630 \end{array}$$
Replace p with 630 in the first equation.
$$3(630) + 3w = 2160$$
$$1890 + 3w = 2160$$
$$3w = 270$$
$$w = 90$$
The plane's speed is 630 mph and the wind's speed is 90 mph.

21. a. answers may vary

b. $\begin{cases} y = 0.06x + 9.7 \\ y = 0.21x + 9.3 \end{cases}$

Replace y with $0.21x + 9.3$ in the first equation.
$$y = 0.06x + 9.7$$
$$0.21x + 9.3 = 0.06x + 9.7$$
$$0.15x + 9.3 = 9.7$$
$$0.15x = 0.4$$
$$x \approx 2.7$$
The pounds of each cheese consumed were the same 3 years after 2000, or in 2003.

23. Let x be the length of each of the equal sides and y be the length of the third side.
$$\begin{cases} 2x + y = 93 \\ y = x + 9 \end{cases}$$
Replace y with $x + 9$ in the first equation.
$$2x + y = 93$$
$$2x + x + 9 = 93$$
$$3x = 84$$
$$x = 28$$
Replace x with 28 in the second equation.
$$y = x + 9 = 28 + 9 = 37$$
The lengths of the sides are 28 cm, 28 cm, and 37 cm.

25. Let m be the number of miles.
Hertz $= 25 + 0.10m$
Budget $= 20 + 0.25m$
Using Budget $= 2 \cdot$ Hertz gives
$$20 + 0.25m = 2(25 + 0.10m)$$
$$20 + 0.25m = 50 + 0.20m$$
$$0.25m = 30 + 0.20m$$
$$0.05m = 30$$
$$m = \frac{30}{0.05} = 600$$
The Budget charge is twice the Hertz charge for a daily mileage of 600 miles.

27. $\begin{cases} x = y - 30 \\ x + y = 180 \end{cases}$

Replace x with $y - 30$ in the second equation.
$$x + y = 180$$
$$y - 30 + y = 180$$
$$2y = 210$$
$$y = 105$$
Replace y with 105 in the first equation.
$$x = y - 30 = 105 - 30 = 75$$
The values are $x = 75$ and $y = 105$.

29. The break-even point is where $C(x) = R(x)$.
$$30x + 10,000 = 46x$$
$$10,000 = 16x$$
$$625 = x$$
625 units must be sold to break even.

31. The break-even point is where $C(x) = R(x)$.
$$1.2x + 1500 = 1.7x$$
$$1500 = 0.5x$$
$$3000 = x$$
3000 units must be sold to break even.

33. The break-even point is where $C(x) = R(x)$.
$$75x + 160,000 = 200x$$
$$160,000 = 125x$$
$$1280 = x$$
1280 units must be sold to break even.

35. a. Let x be the number of desks. The revenue from each desk is \$450, so $R(x) = 450x$.

b. The cost is \$6000 plus \$200 for each desk, so $C(x) = 200x + 6000$.

c. $\begin{aligned} R(x) &= C(x) \\ 450x &= 200x + 6000 \\ 250x &= 6000 \\ x &= 24 \end{aligned}$

The break-even point is 24 desks.

37. Let x = number of units of Mix A,
y = number of units of Mix B,
z = number of units of Mix C.
$$\begin{cases} 4x+6y+4z=30 & (1) \\ 6x+y+z=16 & (2) \\ 3x+2y+12z=24 & (3) \end{cases}$$
Multiply equation (2) by -6 and add to equation (1).
$-32x-2z=-66$ or
$16x+z=33$ (4)
Multiply equation (2) by -2 and add to equation (3).
$-9x+10z=-8$ (5)
Multiply equation (4) by -10 and add to equation (5).
$-169x=-338$
$\quad x=2$
Replace x with 2 in equation (4).
$16(2)+z=33$
$\quad 32+z=33$
$\qquad z=1$
Replace x with 2 and z with 1 in equation (2).
$6(2)+y+1=16$
$\quad 12+y+1=16$
$\qquad y+13=16$
$\qquad\quad y=3$
Combine 2 units of Mix A, 3 units of Mix B, and 1 unit of Mix C.

39. Let x = length of shortest side,
y = length of equal sides,
z = length of longest side.
$$\begin{cases} x+2y+z=29 \\ z=2x \\ y=x+2 \end{cases}$$
Replace y with $x+2$ and z with $2x$ in the first equation.
$x+2(x+2)+2x=29$
$\quad x+2x+4+2x=29$
$\qquad\qquad\quad 5x=25$
$\qquad\qquad\quad\ x=5$
Replace x with 5 in the third equation.
$y=5+2=7$
Replace x with 5 in the second equation.
$z=2(5)=10$
The four sides measure 5 inches, 7 inches, 7 inches, and 10 inches.

41. Let x = the first number, y = the second number, and z = the third number.
$$\begin{cases} x+y+z=40 \\ x=y+5 \\ x=2z \end{cases}$$
$$\begin{cases} x+y+z=40 \\ y=x-5 \\ z=\dfrac{1}{2}x \end{cases}$$
Substitute $y=x-5$ and $z=\dfrac{1}{2}x$ in the first equation.
$x+x-5+\dfrac{1}{2}x=40$
$\qquad \dfrac{5}{2}x-5=40$
$\qquad\quad \dfrac{5}{2}x=45$
$\qquad\qquad x=\dfrac{2}{5}(45)=18$
$y=x-5=18-5=13$
$z=\dfrac{1}{2}x=\dfrac{1}{2}(18)=9$
The three numbers are 18, 13, and 9.

43. Let x = number of free throws,
y = number of two-point field goals, and
z = number of three-point field goals.
$$\begin{cases} x+2y+3z=2161 \\ x=4z+14 \\ y=x-28 \end{cases}$$
Substitute $4z+14$ for x in the third equation.
$y=x-28$
$y=(4z+14)-28$
$y=4z-14$
Substitute $4z+14$ for x and $4z-14$ for y in the first equation.
$$x+2y+3z=2161$$
$$(4z+14)+2(4z-14)+3z=2161$$
$$4z+14+8z-28+3z=2161$$
$$15z-14=2161$$
$$15z=2175$$
$$z=145$$
$x=4z+14=4(145)+14=594$
$y=4z-14=4(145)-14=566$
He scored 594 free throws, 566 2-point field goals, and 145 3-point field goals.

45. $\begin{cases} x+y+z=180 \\ 2x+5+y=180 \\ 2x-5+z=180 \end{cases}$

$z=185-2x$

$y=175-2x$

Replace y with $175-2x$ and z with $185-2x$ in the first equation.

$$x+(175-2x)+(185-2x)=180$$
$$360-3x=180$$
$$-3x=-180$$
$$x=60$$

$z=185-2(60)=185-120=65$

$y=175-2(60)=175-120=55$

The values are $x=60$, $y=55$, and $z=65$.

47. $\begin{array}{r} 6x-2y+2z=4 \\ -x+2y+3z=6 \\ \hline 5x \qquad +5z=10 \end{array}$

49. $\begin{array}{r} -3x-6y+3z=0 \\ 3x+y-z=2 \\ \hline -5y+2z=2 \end{array}$

51. Let x = number filed in 2007 and y = number filed in 2010.

$y=2x-30{,}000$

$y=x+70{,}000$

Use substitution.

$$2x-30{,}000=x+750{,}000$$
$$x-30{,}000=750{,}000$$
$$x=780{,}000$$

$y=x+750{,}000$

$=780{,}000+750{,}000$

$=1{,}530{,}000$

There were 780,000 personal bankruptcy petitions filed in 2007 and 1,530,000 filed in 2010.

53. $y=ax^2+bx+c$

For $(1, 6)$, use $x=1$ and $y=6$.

$6=a+b+c$ (1)

For $(-1, -2)$, use $x=-1$ and $y=-2$.

$-2=a-b+c$ (2)

For $(0, -1)$, use $x=0$ and $y=-1$.

$-1=a\cdot 0+b\cdot 0+c$

$-1=c$ (3)

The system is

$\begin{cases} 6=a+b+c & (1) \\ -2=a-b+c & (2) \\ -1=c & (3) \end{cases}$

From equation (3), we see that $c=-1$. Multiply

equation (2) by -1 and add to equation (1).

$8=2b$

$4=b$

Replace b with 4 and c with -1 in equation (1).

$6=a+4-1$

$6=a+3$

$3=a$

The solution is $a=3$, $b=4$, and $c=-1$.

55. $y=ax^2+bx+c$

For $(3, 927)$, $x=3$ and $y=927$.

$927=a(3)^2+b(3)+c$

$927=9a+3b+c$ (1)

For $(11, 1179)$, $x=11$ and $y=1179$.

$1179=a(11)^2+b(11)+c$

$1179=121a+11b+c$ (2)

For $(19, 1495)$, $x=19$ and $y=1495$.

$1495=a(19)^2+b(19)+c$

$1495=361a+19b+c$ (3)

The system is

$\begin{cases} 927=9a+3b+c & (1) \\ 1179=121a+11b+c & (2) \\ 1495=361a+19b+c & (3) \end{cases}$

Multiply equation (1) by -1 and add to equation (2).

$252=112a+8b$ (4)

Multiply equation (2) by -1 and add to equation (3).

$316=240a+8b$ (5)

Multiply equation (4) by -1 and add to equation (5).

$\begin{array}{r} -252=-112a-8b \\ 316=240a+8b \\ \hline 64=128a \\ \dfrac{1}{2}=a \end{array}$

Replace a with $\dfrac{1}{2}=0.5$ in equation (4).

$252=112(0.5)+8b$

$252=56+8b$

$196=8b$

$b=\dfrac{49}{2}=24.5$

Replace a with 0.5 and b with 24.5 in equation (1).

$927=9(0.5)+3(24.5)+c$

$927=4.5+73.5+c$

$927=78+c$

$849=c$

The solution is $a = 0.5$, $b = 24.5$, and $c = 849$.
For 2015, $x = 25$.

$$y = 0.5(25)^2 + 24.5(25) + 849 = 1774$$

According to the model, 1774 thousand students will take the ACT in 2015.

57. $f(x) = -8.6x + 275$
$f(x) = 204.9x - 1217$
Use substitution.
$$-8.6x + 275 = 204.9x - 1217$$
$$275 = 213.5x - 1217$$
$$1492 = 213.5x$$
$$7 \approx x$$
$f(7) = -8.6(7) + 275$
$f(7) \approx 215$
The solution is $(7, 215)$.

Integrated Review

1. A system with solution $(1, 2)$ has lines that intersect at $(1, 2)$. The correct graph is C.

2. A system with solution $(-2, 3)$ has lines that intersect at $(-2, 3)$. The correct graph is D.

3. A system with no solution has lines that are parallel. The correct graph is A.

4. A system with an infinite number of solutions has lines that are the same. The correct graph is B.

5. $\begin{cases} x + y = 4 & (1) \\ \quad\ y = 3x & (2) \end{cases}$

Substitute $y = 3x$ in E1.
$$x + (3x) = 4$$
$$4x = 4$$
$$x = 1$$
Replace x with 1 in E2.
$y = 3x = 3(1) = 3$
The solution is $(1, 3)$.

6. $\begin{cases} x - y = -4 & (1) \\ \quad\ y = 4x & (2) \end{cases}$

Substitute $y = 4x$ in E1.
$$x - (4x) = -4$$
$$-3x = -4$$
$$x = \frac{4}{3}$$
Replace x with $\frac{4}{3}$ in E2.

$$y = 4x = 4\left(\frac{4}{3}\right) = \frac{16}{3}$$

The solution is $\left(\frac{4}{3}, \frac{16}{3}\right)$.

7. $\begin{cases} x + y = 1 & (1) \\ x - 2y = 4 & (2) \end{cases}$

Multiply E1 by -1 and add to E2.
$$-x - y = -1$$
$$\underline{x - 2y = 4}$$
$$-3y = 3$$
$$y = -1$$
Replace y with -1 in E1.
$$x + (-1) = 1$$
$$x - 1 = 1$$
$$x = 2$$
The solution is $(2, -1)$.

8. $\begin{cases} 2x - y = 8 & (1) \\ x + 3y = 11 & (2) \end{cases}$

Multiply E1 by 3 and add to E2.
$$6x - 3y = 24$$
$$\underline{x + 3y = 11}$$
$$7x \qquad = 35$$
$$x = 5$$
Replace x with 5 in E1.
$$2(5) - y = 8$$
$$10 - y = 8$$
$$y = 2$$
The solution is $(5, 2)$.

9. $\begin{cases} 2x + 5y = 8 & (1) \\ 6x + \ y = 10 & (2) \end{cases}$

Multiply E2 by -5 and add to E1.
$$2x + 5y = 8$$
$$\underline{-30x - 5y = -50}$$
$$-28x \qquad = -42$$
$$x = \frac{3}{2}$$
Replace x with $\frac{3}{2}$ in E2.

$$6\left(\frac{3}{2}\right) + y = 10$$
$$9 + y = 8$$
$$y = 1$$

The solution is $\left(\frac{3}{2}, 1\right)$.

10. $\begin{cases} \dfrac{1}{8}x - \dfrac{1}{2}y = -\dfrac{5}{8} & (1) \\ -3x - 8y = 0 & (2) \end{cases}$

Multiply E1 by -16 and add to E2.

$-2x + 8y = 10$

$\underline{-3x - 8y = 0}$

$-5x \quad\quad = 10$

$\quad\quad x = -2$

Replace x with -2 in E2.

$-3(-2) - 8y = 0$

$6 - 8y = 0$

$-8y = -6$

$y = \dfrac{-6}{-8} = \dfrac{3}{4}$

The solution is $\left(-2, \dfrac{3}{4}\right)$.

11. $\begin{cases} 4x - 7y = 7 & (1) \\ 12x - 21y = 24 & (2) \end{cases}$

Multiply E1 by -3 and add to E2.

$-12x + 21y = -21$

$\underline{12x - 21y = 24}$

$0 = 3 \quad$ False

The system is inconsistent. The solution set is $\varnothing$.

12. $\begin{cases} 2x - 5y = 3 & (1) \\ -4x + 10y = -6 & (2) \end{cases}$

Multiply E1 by 2 and add to E2.

$4x - 10y = 6$

$\underline{-4x + 10y = -6}$

$0 = 0 \quad$ True

The system is dependent. The solution set is $\{(x, y) | 2x - 5y = 3\}$.

13. $\begin{cases} y = \dfrac{1}{3}x \\ 5x - 3y = 4 \end{cases}$

Substitute $\dfrac{1}{3}x$ for y in E2.

$5x - 3\left(\dfrac{1}{3}x\right) = 4$

$5x - x = 4$

$4x = 4$

$x = 1$

Replace x with 1 in E1.

$y = \dfrac{1}{3}(1) = \dfrac{1}{3}$

The solution is $\left(1, \dfrac{1}{3}\right)$.

14. $\begin{cases} y = \dfrac{1}{4}x \\ 2x - 4y = 3 \end{cases}$

Substitute $\dfrac{1}{4}x$ for y in E2.

$2x - 4\left(\dfrac{1}{4}x\right) = 3$

$2x - x = 3$

$x = 3$

Replace x with 3 in E1.

$y = \dfrac{1}{4}(3) = \dfrac{3}{4}$

The solution is $\left(3, \dfrac{3}{4}\right)$.

15. $\begin{cases} x + y \quad\quad = 2 & (1) \\ \quad -3y + z = -7 & (2) \\ 2x + y - z = -1 & (3) \end{cases}$

Add E2 and E3.

$2x - 2y = -8 \quad$ or $\quad x - y = -4 \quad (4)$

Add E1 and E4.

$2x = -2$

$x = -1$

Replace x with -1 in E1.

$-1 + y = 2$

$y = 3$

Replace y with 3 in E2.

$-3(3) + z = -7$

$-9 + z = -7$

$z = 2$

The solution is $(-1, 3, 2)$.

16. $\begin{cases} \quad\quad y + 2z = -3 & (1) \\ x - 2y \quad\quad = 7 & (2) \\ 2x - y + z = 5 & (3) \end{cases}$

Multiply E2 by -2 and add to E3.

$-2x + 4y \quad\quad = -14$

$\underline{2x - y + z = 5}$

$3y + z = -9 \quad (4)$

Multiply E4 by -2 and add to E1.

$-6y - 2z = 18$

$\underline{y + 2z = -3}$

$-5y \quad\quad = 15$

$y = -3$

Replace y with -3 in E4.

$3(-3) + z = -9$

$z = 0$

Replace y with -3 in E2.

$$x - 2(-3) = 7$$
$$x + 6 = 7$$
$$x = 1$$

The solution is $(1, -3, 0)$.

17. $\begin{cases} 2x + 4y - 6z = 3 \quad (1) \\ -x + \ y - \ z = 6 \quad (2) \\ x + 2y - 3z = 1 \quad (3) \end{cases}$

Multiply E3 by -2 and add to E1.

$$-2x - 4y + 6z = -2$$
$$\underline{\ 2x + 4y - 6z = 3\ }$$
$$0 = 1 \ \text{False}$$

The system is inconsistent. The solution set is $\varnothing$.

18. $\begin{cases} x - y + 3z = 2 \quad (1) \\ -2x + 2y - 6z = -4 \quad (2) \\ 3x - 3y + 9z = 6 \quad (3) \end{cases}$

Multiply E1 by 2 and add to E2.

$$2x - 2y + 6z = 4$$
$$\underline{-2x + 2y - 6z = -4}$$
$$0 = 0 \ \text{True}$$

The system is dependent. The solution set is
$\{(x, y) | x - y + 3z = 2\}$.

19. $\begin{cases} x + y - 4z = 5 \quad (1) \\ x - y + 2z = -2 \quad (2) \\ 3x + 2y + 4z = 18 \quad (3) \end{cases}$

Add E1 and E2.

$$2x - 2z = 3 \quad (4)$$

Multiply E2 by 2 and add to E3.

$$2x - 2y + 4z = -4$$
$$\underline{3x + 2y + 4z = 18}$$
$$5x \qquad + 8z = 14 \quad (5)$$

Multiply E4 by 4 and add to E5.

$$8x - 8z = 12$$
$$\underline{5x + 8z = 14}$$
$$13x \qquad = 26$$
$$x = 2$$

Replace x with 2 in E4.

$$2(2) - 2z = 3$$
$$-2z = -1$$
$$z = \frac{1}{2}$$

Replace x with 2 and z with $\frac{1}{2}$ in E1.

$$2 + y - 4\left(\frac{1}{2}\right) = 5$$
$$2 + y - 2 = 5$$
$$y = 5$$

The solution is $\left(2, 5, \frac{1}{2}\right)$.

20. $\begin{cases} 2x - y + 3z = 2 \quad (1) \\ x + y - 6z = 0 \quad (2) \\ 3x + 4y - 3z = 6 \quad (3) \end{cases}$

Add E1 and E3.

$$5x + 3y = 8 \quad (4)$$

Multiply E1 by 2 and add to E2.

$$4x - 2y + 6z = 4$$
$$\underline{x + \ y - 6z = 0}$$
$$5x - y \qquad = 4 \quad (5)$$

Multiply E5 by 3 and add to E4.

$$15x - 3y = 12$$
$$\underline{5x + 3y = 8}$$
$$20x \qquad = 20$$
$$x = 1$$

Replace x with 1 in E5.

$$5(1) - y = 4$$
$$-y = -1$$
$$y = 1$$

Replace both x and y with 1 in E1.

$$2(1) - 1 + 3z = 2$$
$$1 + 3z = 2$$
$$3z = 1$$
$$z = \frac{1}{3}$$

The solution is $\left(1, 1, \frac{1}{3}\right)$.

21. Let $x = $ the first number and
$y = $ the second number.
$\begin{cases} x = y - 8 \quad (1) \\ 2x = y + 11 \quad (2) \end{cases}$

Substitute $x = y - 8$ in E2.

$$2(y - 8) = y + 11$$
$$2y - 16 = y + 11$$
$$y = 27$$

Replace y with 27 in E1.

$$x = 27 - 8 = 19$$

The numbers are 19 and 27.

22. Let x = measure of the two smallest angles, y = measure of the third angle, and z = measure of the fourth angle.
$$\begin{cases} 2x + y + z = 360 \\ y = x + 30 \\ z = x + 50 \end{cases}$$
Substitute $y = x + 30$ and $z = x + 50$ in the first equation.
$$2x + (x + 30) + (x + 50) = 360$$
$$4x + 80 = 360$$
$$4x = 280$$
$$x = 70$$
so $y = 70 + 30 = 100$ and $z = 70 + 50 = 120$

The two smallest angles are 70°, the third angle is 100°, and the fourth angle is 120°.

Section 4.4 Practice Exercises

1. $\begin{cases} x + 4y = -2 \\ 3x - y = 7 \end{cases}$

The corresponding matrix is $\begin{bmatrix} 1 & 4 & | & -2 \\ 3 & -1 & | & 7 \end{bmatrix}$. The element in the first row, first column is already 1. Multiply row 1 by -3 and add to row 2 to get a 0 below the 1.
$$\begin{bmatrix} 1 & 4 & | & -2 \\ -3(1)+3 & -3(4)+(-1) & | & -3(-2)+7 \end{bmatrix}$$
$$\begin{bmatrix} 1 & 4 & | & -2 \\ 0 & -13 & | & 13 \end{bmatrix}$$
We change -13 to a 1 by dividing row 2 by -13.
$$\begin{bmatrix} 1 & 4 & | & -2 \\ 0 & \frac{-13}{-13} & | & \frac{13}{-13} \end{bmatrix}$$
$$\begin{bmatrix} 1 & 4 & | & -2 \\ 0 & 1 & | & -1 \end{bmatrix}$$

The last matrix corresponds to $\begin{cases} x + 4y = -2 \\ y = -1 \end{cases}$

To find x, we let $y = -1$ in the first equation.
$$x + 4y = -2$$
$$x + 4(-1) = -2$$
$$x - 4 = -2$$
$$x = 2$$
The solution is $(2, -1)$.

2. $\begin{cases} x - 3y = 3 \\ -2x + 6y = 4 \end{cases}$

The corresponding matrix is $\begin{bmatrix} 1 & -3 & | & 3 \\ -2 & 6 & | & 4 \end{bmatrix}$. The element in the first row, first column is already

1. Multiply row 1 by 2 and add to row 2 to get a 0 below the 1.
$$\begin{bmatrix} 1 & -3 & | & 3 \\ 2(1)+(-2) & 2(-3)+6 & | & 2(3)+4 \end{bmatrix}$$
$$\begin{bmatrix} 1 & -3 & | & 3 \\ 0 & 0 & | & 10 \end{bmatrix}$$

The corresponding system is $\begin{cases} x - 3y = 3 \\ 0 = 10 \end{cases}$

The equation $0 = 10$ is false. Hence, the system is inconsistent and has no solution. The solution set is $\varnothing$.

3. $\begin{cases} x + 3y - z = 0 \\ 2x + y + 3z = 5 \\ -x - 2y + 4z = 7 \end{cases}$

The corresponding matrix is $\begin{bmatrix} 1 & 3 & -1 & | & 0 \\ 2 & 1 & 3 & | & 5 \\ -1 & -2 & 4 & | & 7 \end{bmatrix}$.

The element in the first row, first column is already 1. Multiply row 1 by -2 and add to row 2 to get a 0 below the 1 in row 2. Add row 1 to row 3 to get a 0 below the 1 in row 3.
$$\begin{bmatrix} 1 & 3 & -1 & | & 0 \\ -2(1)+2 & -2(3)+1 & -2(-1)+3 & | & -2(0)+5 \\ 1+(-1) & 3+(-2) & -1+4 & | & 0+7 \end{bmatrix}$$
$$\begin{bmatrix} 1 & 3 & -1 & | & 0 \\ 0 & -5 & 5 & | & 5 \\ 0 & 1 & 3 & | & 7 \end{bmatrix}$$

Now we want a 1 where the -5 is now. Interchange rows 2 and 3.
$$\begin{bmatrix} 1 & 3 & -1 & | & 0 \\ 0 & 1 & 3 & | & 7 \\ 0 & -5 & 5 & | & 5 \end{bmatrix}$$

Now we want a 0 below the 1. Multiply row 2 by 5 and add to row 3.
$$\begin{bmatrix} 1 & 3 & -1 & | & 0 \\ 0 & 1 & 3 & | & 7 \\ 5(0)+0 & 5(1)+(-5) & 5(3)+5 & | & 5(7)+5 \end{bmatrix}$$
$$\begin{bmatrix} 1 & 3 & -1 & | & 0 \\ 0 & 1 & 3 & | & 7 \\ 0 & 0 & 20 & | & 40 \end{bmatrix}$$

Finally, divide row 3 by 20.
$$\begin{bmatrix} 1 & 3 & -1 & | & 0 \\ 0 & 1 & 3 & | & 7 \\ 0 & 0 & \frac{20}{20} & | & \frac{40}{20} \end{bmatrix}$$

$$\begin{bmatrix} 1 & 3 & -1 & | & 0 \\ 0 & 1 & 3 & | & 7 \\ 0 & 0 & 1 & | & 2 \end{bmatrix}$$

This matrix corresponds to the system
$$\begin{cases} x + 3y - z = 0 \\ y + 3z = 7 \\ z = 2 \end{cases}$$

The z-coordinate is 2. Replace z with 2 in the second equation and solve for y.
$$y + 3z = 7$$
$$y + 3(2) = 7$$
$$y + 6 = 7$$
$$y = 1$$

To find x, we let $z = 2$ and $y = 1$ in the first equation.
$$x + 3y - z = 0$$
$$x + 3(1) - 2 = 0$$
$$x + 1 = 0$$
$$x = -1$$
The solution is $(-1, 1, 2)$.

4. The augmented matrix from Practice 1 is
$$\begin{bmatrix} 1 & 4 & | & -2 \\ 3 & -1 & | & 7 \end{bmatrix}.$$

 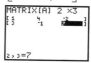

This corresponds to the system
$$\begin{cases} 1x + 0y = 2 \\ 0x + 1y = -1 \end{cases} \quad \text{or} \quad \begin{cases} x = 2 \\ y = -1 \end{cases}.$$

The solution of the system is $(2, -1)$.

5. The augmented matrix from Practice 3 is
$$\begin{bmatrix} 1 & 3 & -1 & | & 0 \\ 2 & 1 & 3 & | & 5 \\ -1 & -2 & 4 & | & 7 \end{bmatrix}.$$

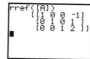

This corresponds to the system

$$\begin{cases} 1x + 0y + 0z = -1 \\ 0x + 1y + 0z = 1 \\ 0x + 0y + 1z = 2 \end{cases} \quad \text{or} \quad \begin{cases} x = -1 \\ y = 1 \\ z = 2 \end{cases}.$$

The solution of the system is $(-1, 1, 2)$.

Vocabulary, Readiness & Video Check 4.4

1. A <u>matrix</u> is a rectangular array of numbers.

2. Each of the numbers in a matrix is called an <u>element</u>.

3. The numbers aligned horizontally in a matrix are in the same <u>row</u>.

4. The numbers aligned vertically in a matrix are in the same <u>column</u>.

5. Any two columns may be interchanged. <u>false</u>

6. Any two rows may be interchanged. <u>true</u>

7. The elements in a row may be added to their corresponding elements in another row. <u>true</u>

8. The elements of a column may be multiplied by any nonzero number. <u>false</u>

9. Two rows may be interchanged, the elements of any row may be multiplied/divided by the same nonzero number, the elements of any row may be multiplied/divided by the same nonzero number and added to their corresponding elements in any other row; rows were not interchanged in Example 1.

10. Consider the possible confusion or errors that might occur if you're careless and it's unclear which row or column numbers belong in, especially when working with larger matrices and/or with a lot of calculations.

11. The matrix in reduced row echelon form has 0s above and below the main diagonal, so the last column contains the solutions of the system of equations.

Exercise Set 4.4

1. $$\begin{cases} x + y = 1 \\ x - 2y = 4 \end{cases}$$
$$\begin{bmatrix} 1 & 1 & | & 1 \\ 1 & -2 & | & 4 \end{bmatrix}$$
Multiply R1 by -1 and add to R2.

$$\begin{bmatrix} 1 & 1 & | & 1 \\ 0 & -3 & | & 3 \end{bmatrix}$$

Divide R2 by −3.

$$\begin{bmatrix} 1 & 1 & | & 1 \\ 0 & 1 & | & -1 \end{bmatrix}$$

This corresponds to $\begin{cases} x + y = 1 \\ y = -1 \end{cases}$.

$$x + (-1) = 1$$
$$x - 1 = 1$$
$$x = 2$$

The solution is (2, −1).

3. $\begin{cases} x + 3y = 2 \\ x + 2y = 0 \end{cases}$

$$\begin{bmatrix} 1 & 3 & | & 2 \\ 1 & 2 & | & 0 \end{bmatrix}$$

Multiply R1 by −1 and add to R2.

$$\begin{bmatrix} 1 & 3 & | & 2 \\ 0 & -1 & | & -2 \end{bmatrix}$$

Multiply R2 by −1.

$$\begin{bmatrix} 1 & 3 & | & 2 \\ 0 & 1 & | & 2 \end{bmatrix}$$

This corresponds to $\begin{cases} x + 3y = 2 \\ y = 2 \end{cases}$.

$$x + 3(2) = 2$$
$$x + 6 = 2$$
$$x = -4$$

The solution is (−4, 2).

5. $\begin{cases} x - 2y = 4 \\ 2x - 4y = 4 \end{cases}$

$$\begin{bmatrix} 1 & -2 & | & 4 \\ 2 & -4 & | & 4 \end{bmatrix}$$

Multiply R1 by −2 and add to R2.

$$\begin{bmatrix} 1 & -2 & | & 4 \\ 0 & 0 & | & -4 \end{bmatrix}$$

This corresponds to $\begin{cases} x - 2y = 4 \\ 0 = -4 \end{cases}$.

This is an inconsistent system. The solution is ∅.

7. $\begin{cases} 3x - 3y = 9 \\ 2x - 2y = 6 \end{cases}$

$$\begin{bmatrix} 3 & -3 & | & 9 \\ 2 & -2 & | & 6 \end{bmatrix}$$

Divide R1 by 3.

$$\begin{bmatrix} 1 & -1 & | & 3 \\ 2 & -2 & | & 6 \end{bmatrix}$$

Multiply R1 by −2 and add to R2.

$$\begin{bmatrix} 1 & -1 & | & 3 \\ 0 & 0 & | & 0 \end{bmatrix}$$

This corresponds to $\begin{cases} x - y = 3 \\ 0 = 0 \end{cases}$.

This is a dependent system. The solution is $\{(x, y) | x - y = 3\}$.

9. $\begin{cases} x + y = 3 \\ 2y = 10 \\ 3x + 2y - 4z = 12 \end{cases}$

$$\begin{bmatrix} 1 & 1 & 0 & | & 3 \\ 0 & 2 & 0 & | & 10 \\ 3 & 2 & -4 & | & 12 \end{bmatrix}$$

Multiply R1 by −3 and add to R3.

$$\begin{bmatrix} 1 & 1 & 0 & | & 3 \\ 0 & 2 & 0 & | & 10 \\ 0 & -1 & -4 & | & 3 \end{bmatrix}$$

Divide R2 by 2.

$$\begin{bmatrix} 1 & 1 & 0 & | & 3 \\ 0 & 1 & 0 & | & 5 \\ 0 & -1 & -4 & | & 3 \end{bmatrix}$$

Add R2 to R3.

$$\begin{bmatrix} 1 & 1 & 0 & | & 3 \\ 0 & 1 & 0 & | & 5 \\ 0 & 0 & -4 & | & 8 \end{bmatrix}$$

Divide R3 by −4.

$$\begin{bmatrix} 1 & 1 & 0 & | & 3 \\ 0 & 1 & 0 & | & 5 \\ 0 & 0 & 1 & | & -2 \end{bmatrix}$$

This corresponds to $\begin{cases} x + y = 3 \\ y = 5 \\ z = -2 \end{cases}$.

$$x + 5 = 3$$
$$x = -2$$

The solution is (−2, 5, −2).

11. $\begin{cases} 2y - z = -7 \\ x + 4y + z = -4 \\ 5x - y + 2z = 13 \end{cases}$

$$\begin{bmatrix} 0 & 2 & -1 & | & -7 \\ 1 & 4 & 1 & | & -4 \\ 5 & -1 & 2 & | & 13 \end{bmatrix}$$

Interchange R1 and R2.

$$\begin{bmatrix} 1 & 4 & 1 & | & -4 \\ 0 & 2 & -1 & | & -7 \\ 5 & -1 & 2 & | & 13 \end{bmatrix}$$

Multiply R1 by −5 and add to R3.

$$\begin{bmatrix} 1 & 4 & 1 & | & -4 \\ 0 & 2 & -1 & | & -7 \\ 0 & -21 & -3 & | & 33 \end{bmatrix}$$

Divide R2 by 2.

$$\begin{bmatrix} 1 & 4 & 1 & | & -4 \\ 0 & 1 & -\frac{1}{2} & | & -\frac{7}{2} \\ 0 & -21 & -3 & | & 33 \end{bmatrix}$$

Multiply R2 by 21 and add to R3.

$$\begin{bmatrix} 1 & 4 & 1 & | & -4 \\ 0 & 1 & -\frac{1}{2} & | & -\frac{7}{2} \\ 0 & 0 & -\frac{27}{2} & | & -\frac{81}{2} \end{bmatrix}$$

Multiply R2 by $-\dfrac{2}{27}$.

$$\begin{bmatrix} 1 & 4 & 1 & | & -4 \\ 0 & 1 & -\frac{1}{2} & | & -\frac{7}{2} \\ 0 & 0 & 1 & | & 3 \end{bmatrix}$$

This corresponds to $\begin{cases} x + 4y + z = -4 \\ \quad y - \dfrac{1}{2}z = -\dfrac{7}{2} \\ \quad\quad\quad z = 3 \end{cases}$.

$$y - \frac{1}{2}(3) = -\frac{7}{2}$$
$$y - \frac{3}{2} = -\frac{7}{2}$$
$$y = -2$$
$$x + 4(-2) + 3 = -4$$
$$x - 8 + 3 = -4$$
$$x = 1$$

The solution is (1, –2, 3).

13. $\begin{cases} x - 4 = 0 \\ x + y = 1 \end{cases}$ or $\begin{cases} x = 4 \\ x + y = 1 \end{cases}$

$$\begin{bmatrix} 1 & 0 & | & 4 \\ 1 & 1 & | & 1 \end{bmatrix}$$

Multiply R1 by –1 and add to R2.

$$\begin{bmatrix} 1 & 0 & | & 4 \\ 0 & 1 & | & -3 \end{bmatrix}$$

This corresponds to $\begin{cases} x = 4 \\ y = -3 \end{cases}$

The solution is (4, –3).

15. $\begin{cases} x + y + z = 2 \\ 2x \quad\quad - z = 5 \\ \quad\quad 3y + z = 2 \end{cases}$

$$\begin{bmatrix} 1 & 1 & 1 & | & 2 \\ 2 & 0 & -1 & | & 5 \\ 0 & 3 & 1 & | & 2 \end{bmatrix}$$

Multiply R1 by –2 and add to R2.

$$\begin{bmatrix} 1 & 1 & 1 & | & 2 \\ 0 & -2 & -3 & | & 1 \\ 0 & 3 & 1 & | & 2 \end{bmatrix}$$

Divide R2 by –2.

$$\begin{bmatrix} 1 & 1 & 1 & | & 2 \\ 0 & 1 & \frac{3}{2} & | & -\frac{1}{2} \\ 0 & 3 & 1 & | & 2 \end{bmatrix}$$

Multiply R2 by –3 and add to R3.

$$\begin{bmatrix} 1 & 1 & 1 & | & 2 \\ 0 & 1 & \frac{3}{2} & | & -\frac{1}{2} \\ 0 & 0 & -\frac{7}{2} & | & \frac{7}{2} \end{bmatrix}$$

Multiply R3 by $-\dfrac{2}{7}$.

$$\begin{bmatrix} 1 & 1 & 1 & | & 2 \\ 0 & 1 & \frac{3}{2} & | & -\frac{1}{2} \\ 0 & 0 & 1 & | & -1 \end{bmatrix}$$

This corresponds to $\begin{cases} x + y + z = 2 \\ \quad y + \dfrac{3}{2}z = -\dfrac{1}{2} \\ \quad\quad\quad z = -1 \end{cases}$.

$$y + \frac{3}{2}(-1) = -\frac{1}{2}$$
$$y - \frac{3}{2} = -\frac{1}{2}$$
$$y = 1$$
$$x + 1 + (-1) = 2$$
$$x = 2$$

The solution is (2, 1, –1).

17. $\begin{cases} 5x - 2y = 27 \\ -3x + 5y = 18 \end{cases}$

$$\begin{bmatrix} 5 & -2 & | & 27 \\ -3 & 5 & | & 18 \end{bmatrix}$$

Divide R1 by 5.

$$\begin{bmatrix} 1 & -\frac{2}{5} & | & \frac{27}{5} \\ -3 & 5 & | & 18 \end{bmatrix}$$

Multiply R1 by 3 and add to R2.

$$\begin{bmatrix} 1 & -\frac{2}{5} & | & \frac{27}{5} \\ 0 & \frac{19}{5} & | & \frac{171}{5} \end{bmatrix}$$

Multiply R2 by $\frac{5}{19}$.

$$\begin{bmatrix} 1 & -\frac{2}{5} & | & \frac{27}{5} \\ 0 & 1 & | & 9 \end{bmatrix}$$

This corresponds to $\begin{cases} x - \dfrac{2}{5}y = \dfrac{27}{5} \\ \qquad\quad y = 9 \end{cases}$.

$$x - \frac{2}{5}(9) = \frac{27}{5}$$
$$x - \frac{18}{5} = \frac{27}{5}$$
$$x = 9$$

The solution is (9, 9).

19. $\begin{cases} 4x - 7y = 7 \\ 12x - 21y = 24 \end{cases}$

$$\begin{bmatrix} 4 & -7 & | & 7 \\ 12 & -21 & | & 24 \end{bmatrix}$$

Divide R1 by 4.

$$\begin{bmatrix} 1 & -\frac{7}{4} & | & \frac{7}{4} \\ 12 & -21 & | & 24 \end{bmatrix}$$

Multiply R1 by –12 and add to R2.

$$\begin{bmatrix} 1 & -\frac{7}{4} & | & \frac{7}{4} \\ 0 & 0 & | & 3 \end{bmatrix}$$

This corresponds to $\begin{cases} x - \dfrac{7}{4}y = \dfrac{7}{4} \\ \qquad\quad 0 = 3 \end{cases}$.

This is an inconsistent system. The solution set is $\varnothing$.

21. $\begin{cases} 4x - y + 2z = 5 \\ \quad\;\; 2y + z = 4 \\ 4x + y + 3z = 10 \end{cases}$

$$\begin{bmatrix} 4 & -1 & 2 & | & 5 \\ 0 & 2 & 1 & | & 4 \\ 4 & 1 & 3 & | & 10 \end{bmatrix}$$

Divide R1 by 4.

$$\begin{bmatrix} 1 & -\frac{1}{4} & \frac{1}{2} & | & \frac{5}{4} \\ 0 & 2 & 1 & | & 4 \\ 4 & 1 & 3 & | & 10 \end{bmatrix}$$

Multiply R1 by –4 and add to R3.

$$\begin{bmatrix} 1 & -\frac{1}{4} & \frac{1}{2} & | & \frac{5}{4} \\ 0 & 2 & 1 & | & 4 \\ 0 & 2 & 1 & | & 5 \end{bmatrix}$$

Divide R2 by 2.

$$\begin{bmatrix} 1 & -\frac{1}{4} & \frac{1}{2} & | & \frac{5}{4} \\ 0 & 1 & \frac{1}{2} & | & 2 \\ 0 & 2 & 1 & | & 5 \end{bmatrix}$$

Multiply R2 by –2 and add to R3.

$$\begin{bmatrix} 1 & -\frac{1}{4} & \frac{1}{2} & | & \frac{5}{4} \\ 0 & 1 & \frac{1}{2} & | & 2 \\ 0 & 0 & 0 & | & 1 \end{bmatrix}$$

This corresponds to $\begin{cases} x - \dfrac{1}{4}y + \dfrac{1}{2}z = \dfrac{5}{4} \\ \qquad\quad y + \dfrac{1}{2}z = 2 \\ \qquad\qquad\qquad 0 = 1 \end{cases}$.

This is an inconsistent system. The solution set is $\varnothing$.

23. $\begin{cases} 4x + y + z = 3 \\ -x + y - 2z = -11 \\ x + 2y + 2z = -1 \end{cases}$

$$\begin{bmatrix} 4 & 1 & 1 & | & 3 \\ -1 & 1 & -2 & | & -11 \\ 1 & 2 & 2 & | & -1 \end{bmatrix}$$

Interchange R1 and R3.

$$\begin{bmatrix} 1 & 2 & 2 & | & -1 \\ -1 & 1 & -2 & | & -11 \\ 4 & 1 & 1 & | & 3 \end{bmatrix}$$

Add R1 to R2. Multiply R1 by –4 and add to R3.

$$\begin{bmatrix} 1 & 2 & 2 & | & -1 \\ 0 & 3 & 0 & | & -12 \\ 0 & -7 & -7 & | & 7 \end{bmatrix}$$

Divide R2 by 3.

$$\begin{bmatrix} 1 & 2 & 2 & | & -1 \\ 0 & 1 & 0 & | & -4 \\ 0 & -7 & -7 & | & 7 \end{bmatrix}$$

Multiply R2 by 7 and add to R3.

$$\begin{bmatrix} 1 & 2 & 2 & | & -1 \\ 0 & 1 & 0 & | & -4 \\ 0 & 0 & -7 & | & -21 \end{bmatrix}$$

Divide R3 by –7.

$$\begin{bmatrix} 1 & 2 & 2 & | & -1 \\ 0 & 1 & 0 & | & -4 \\ 0 & 0 & 1 & | & 3 \end{bmatrix}$$

This corresponds to $\begin{cases} x + 2y + 2z = -1 \\ \quad\quad y = -4. \\ \quad\quad\quad z = 3 \end{cases}$

$$x + 2(-4) + 2(3) = -1$$
$$x - 8 + 6 = -1$$
$$x = 1$$

The solution is $(1, -4, 3)$.

25. No vertical line intersects the graph more than once. It is the graph of a function.

27. The y-axis is a vertical line that intersects the graph more than once. It is not the graph of a function.

29. $(-1)(-5) - (6)(3) = 5 - 18 = -13$

31. $(4)(-10) - (2)(-2) = -40 + 4 = -36$

33. $(-3)(-3) - (-1)(-9) = 9 - 9 = 0$

35. The matrix should have four columns, so (a) is not the correct matrix. The matrix should have a 0 in the first column, second row since the coefficient of x in the second equation is 0, so (b) is not the correct matrix. The correct matrix is (c).

37. a. $\begin{cases} y - 0.11x = 14.05 \\ 5.13x - y = 0.65 \end{cases}$ or $\begin{cases} -0.11x + y = 14.05 \\ 5.13x - y = 0.65 \end{cases}$

$$\begin{bmatrix} -0.11 & 1 & | & 14.05 \\ 5.13 & -1 & | & 0.65 \end{bmatrix}$$

Multiply row 1 by $-\dfrac{1}{0.11}$.

$$\begin{bmatrix} 1 & -9.09 & | & -127.73 \\ -5.13 & -1 & | & 0.65 \end{bmatrix}$$

Multiply row 1 by -5.13 and add to row 2.

$$\begin{bmatrix} 1 & -9.09 & | & -127.73 \\ 0 & 45.64 & | & 655.89 \end{bmatrix}$$

Multiply row 2 by $\dfrac{1}{45.64}$.

$$\begin{bmatrix} 1 & -9.09 & | & -127.73 \\ 0 & 1 & | & 14.37 \end{bmatrix}$$

This corresponds to $\begin{cases} x - 9.09y = -127.73 \\ \quad\quad y = 14.37 \end{cases}$

$$x - 9.09(14.37) = -127.73$$
$$x \approx 2.9$$
$$2000 + 2.9 = 2002.9$$

At the end of 2002, the electricity was the same.

b. no; answers may vary

c. no; it has a positive slope

d. answers may vary

39. answers may vary

Section 4.5 Practice Exercises

1. $\begin{cases} 4x \geq y \\ x + 3y \geq 6 \end{cases}$

Graph both inequalities on the same set of axes. The solution is the intersection of the solution regions.

For $4x \geq y$, the boundary line is the graph of $4x = y$. The boundary line is solid since the inequality means $4x > y$ or $4x = y$. The test point $(1, 0)$ satisfies the inequality, so we shade the half-plane that includes $(1, 0)$.

For $x + 3y \geq 6$, sketch the solid boundary line $x + 3y = 6$. The test point $(0, 0)$ does not satisfy the inequality, so shade the half-plane that does not include $(0, 0)$.

The solution of the system is the darker shaded region. This solution includes parts of both boundary lines.

2. $\begin{cases} x - y < 1 \\ \quad y < 4 \\ 3x + y > -3 \end{cases}$

Graph all three inequalities on the same set of axes. All boundary lines are dashed since the inequality symbols are $<$ and $>$. The solution set of the system is the shaded region. The boundary lines are not a part of the solution.

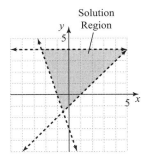

Solution
Region

3. $\begin{cases} -2x+5y \le 10 \\ x \le 4 \\ x \ge 0 \\ y \ge 0 \end{cases}$

Graph the inequalities on the same set of axes. The intersection of the inequalities is the solution region. It is the only shaded region in this graph and includes the portion of all four boundary lines that border the shaded region.

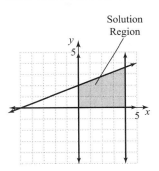

Solution
Region

Vocabulary, Readiness & Video Check 4.5

1. Two or more linear inequalities form a <u>system</u> of linear inequalities.

2. An ordered pair that satisfies each inequality in a system is a <u>solution</u> of the system.

3. The point where two boundary lines intersect is a <u>corner</u> point.

4. The solution region of a system of inequalities consists of the <u>intersection</u> of the solution regions of the inequalities in the system.

5. No; we can choose any test point except a point on the second inequality's own boundary line.

Exercise Set 4.5

1. $\begin{cases} y \ge x+1 \\ y \ge 3-x \end{cases}$

Graph both inequalities on the same set of axes. The solution is the intersection of the solution regions. The solution of the system is the darker shaded region. This solution includes parts of both boundary lines.

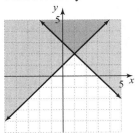

3. $\begin{cases} y < 3x-4 \\ y \le x+2 \end{cases}$

Graph both inequalities on the same set of axes. The solution is the intersection of the solution regions. The solution of the system is the darker shaded region. This solution includes the part of the solid boundary line that borders the region but not the dashed boundary line or the point where the boundary lines intersect.

5. $\begin{cases} y < -2x-2 \\ y > x+4 \end{cases}$

Graph both inequalities on the same set of axes. The solution is the intersection of the solution regions. The solution of the system is the darker shaded region. The boundary lines are not a part of the solution.

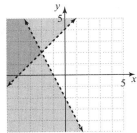

7. $\begin{cases} y \geq -x + 2 \\ y \leq 2x + 5 \end{cases}$

Graph both inequalities on the same set of axes. The solution is the intersection of the solution regions. The solution of the system is the darker shaded region. This solution includes parts of both boundary lines.

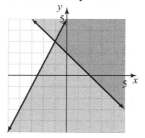

9. $\begin{cases} x \geq 3y \\ x + 3y \leq 6 \end{cases}$

Graph both inequalities on the same set of axes. The solution is the intersection of the solution regions. The solution of the system is the darker shaded region. This solution includes parts of both boundary lines.

11. $\begin{cases} x \leq 2 \\ y \geq -3 \end{cases}$

Graph both inequalities on the same set of axes. The solution is the intersection of the solution regions. The solution of the system is the darker shaded region. This solution includes parts of both boundary lines.

13. $\begin{cases} y \geq 1 \\ x < -3 \end{cases}$

Graph both inequalities on the same set of axes. The solution is the intersection of the solution regions. The solution of the system is the darker shaded region. This solution includes the part of the solid boundary line that borders the region but not the dashed boundary line or the point where the boundary lines intersect.

15. $\begin{cases} y + 2x \geq 0 \\ 5x - 3y \leq 12 \\ y \leq 2 \end{cases}$

Graph all three inequalities on the same set of axes. The solution set of the system is the shaded region. The parts of the boundary lines that border the shaded region are part of the solution.

17. $\begin{cases} 3x - 4y \geq -6 \\ 2x + y \leq 7 \\ y \geq -3 \end{cases}$

Graph all three inequalities on the same set of axes. The solution set of the system is the shaded region. The parts of the boundary lines that border the shaded region are part of the solution.

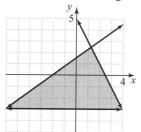

19. $\begin{cases} 2x+y \le 5 \\ \quad x \le 3 \\ \quad x \ge 0 \\ \quad y \ge 0 \end{cases}$

Graph the inequalities on the same set of axes. The intersection of the inequalities is the solution region. It is the only shaded region in this graph and includes the portion of all four boundary lines that border the shaded region.

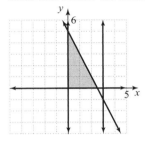

21. $\begin{cases} y < 5 \\ x > 3 \end{cases}$

Both boundary lines should be dashed. The region including (5, 0) should be shaded since (5, 0) satisfies both inequalities. The correct graph is C.

23. $\begin{cases} y \le 5 \\ x < 3 \end{cases}$

The boundary line $y = 5$ should be solid. The correct graph is D.

25. $(-3)^2 = (-3)(-3) = 9$

27. $\left(\dfrac{2}{3}\right)^2 = \dfrac{2}{3} \cdot \dfrac{2}{3} = \dfrac{2 \cdot 2}{3 \cdot 3} = \dfrac{4}{9}$

29. $(-2)^2 - (-3) + 2(-1) = 4 - (-3) + 2(-1)$
$$= 4 - (-3) + (-2)$$
$$= 4 + 3 - 2$$
$$= 7 - 2$$
$$= 5$$

31. $8^2 + (-13) - 4(-2) = 64 + (-13) - 4(-2)$
$$= 64 + (-13) - (-8)$$
$$= 64 - 13 + 8$$
$$= 51 + 8$$
$$= 59$$

33. $\begin{cases} y \le 3 \\ y \ge 3 \end{cases}$

The only y-values that satisfy both inequalities are those that equal 3. The solution of the system is the line $y = 3$.

35. answers may vary.

Chapter 4 Vocabulary Check

1. Two or more linear equations in two variables form a <u>system of equations</u>.

2. A <u>solution</u> of a system of two equations in two variables is an ordered pair that makes both equations true.

3. A <u>consistent</u> system of equations has at least one solution.

4. A solution of a system of three equations in three variables is an ordered <u>triple</u> that makes all three equations true.

5. An <u>inconsistent</u> system of equations has no solution.

6. A <u>matrix</u> is a rectangular array of numbers.

7. Each of the numbers in a matrix is called an <u>element</u>.

8. The numbers aligned horizontally in a matrix are in the same <u>row</u>.

9. The numbers aligned vertically in a matrix are in the same <u>column</u>.

Chapter 4 Review

1. $\begin{cases} 3x + 10y = 1 \quad (1) \\ \quad x + 2y = -1 \quad (2) \end{cases}$

(1)

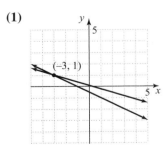

(2) From E2: $x = -2y - 1$

Replace x with $-2y - 1$ in E1.

$$3(-2y - 1) + 10y = 1$$
$$-6y - 3 + 10y = 1$$
$$4y = 4$$
$$y = 1$$

Replace y with 1 in the equation $x = -2y - 1$.

$$x = -2(1) - 1 = -3$$

The solution is $(-3, 1)$.

(3) Multiply E2 by -3 and add to E1.

$$3x + 10y = 1$$
$$\underline{-3x - 6y = 3}$$
$$4y = 4$$
$$y = 1$$

Replace y with 1 in E2.

$$x + 2(1) = -1$$
$$x + 2 = -1$$
$$x = -3$$

The solution is $(-3, 1)$.

2. $\begin{cases} y = \dfrac{1}{2}x + \dfrac{2}{3} & (1) \\ 4x + 6y = 4 & (2) \end{cases}$

(1)

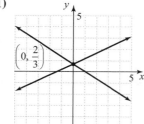

(2) Replace y with $\dfrac{1}{2}x + \dfrac{2}{3}$ in E2.

$$4x + 6\left(\dfrac{1}{2}x + \dfrac{2}{3}\right) = 4$$
$$4x + 3x + 4 = 4$$
$$x = 0$$

Replace x with 0 in E1.

$$y = \dfrac{1}{2}(0) + \dfrac{2}{3} = \dfrac{2}{3}$$

The solution is $\left(0, \dfrac{2}{3}\right)$.

(3) Rewrite the system: $\begin{cases} -\dfrac{1}{2}x + y = \dfrac{2}{3} \\ 4x + 6y = 4 \end{cases}$.

Multiply the first equation by -6.

$$\begin{cases} 3x - 6y = -4 \\ 4x + 6y = 4 \end{cases}$$

Add these equations.

$$7x = 0$$
$$x = 0$$

Replace x with 0 in second equation.

$$4(0) + 6y = 4$$
$$6y = 4$$
$$y = \dfrac{4}{6} = \dfrac{2}{3}$$

The solution is $\left(0, \dfrac{2}{3}\right)$.

3. $\begin{cases} 2x - 4y = 22 & (1) \\ 5x - 10y = 15 & (2) \end{cases}$

(1)

(2) Solve E1 for x.

$$2x - 4y = 22$$
$$2x = 4y + 22$$
$$x = 2y + 11$$

Replace x with $2y + 11$ in E2.

$$5(2y + 11) - 10y = 15$$
$$10y + 55 - 10y = 15$$
$$55 = 15 \quad \text{False}$$

This is an inconsistent system. The solution is $\varnothing$.

(3) Multiply E1 by 5 and E2 by -2.

$$\begin{cases} 10x - 20y = 110 \\ -10x + 20y = -30 \end{cases}$$

Add these equations.

$$10x - 20y = 110$$
$$\underline{-10x + 20y = -30}$$
$$0 = 80 \quad \text{False}$$

This is an inconsistent system. The solution is $\varnothing$.

4. $\begin{cases} 3x - 6y = 12 & (1) \\ 2y = x - 4 & (2) \end{cases}$

(1)

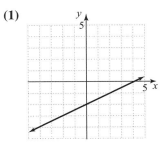

(2) Solve E2 for x.

$x = 2y + 4$

Replace x with $2y + 4$ in E1.

$3(2y + 4) - 6y = 12$
$6y + 12 - 6y = 12$
$12 = 12$ True

This is a dependent system. The solution is $\{(x, y)| 3x - 6y = 12\}$.

(3) $\begin{cases} 3x - 6y = 12 & (1) \\ -x + 2y = -4 & (2) \end{cases}$

Multiply E2 by 3.

$\begin{cases} 3x - 6y = 12 \\ -3x + 6y = -12 \end{cases}$

Add these equations.

$\begin{array}{r} 3x - 6y = 12 \\ -3x + 6y = -12 \\ \hline 0 = 0 \quad \text{True} \end{array}$

This is a dependent system. The solution is $\{(x, y)| 3x - 6y = 12\}$.

5. $\begin{cases} \dfrac{1}{2}x - \dfrac{3}{4}y = -\dfrac{1}{2} & (1) \\ \dfrac{1}{8}x + \dfrac{3}{4}y = \dfrac{19}{8} & (2) \end{cases}$

(1)

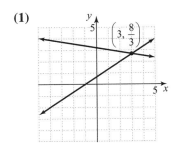

(2) Clear fractions by multiplying E1 by 4 and E2 by 8.

(1) $\begin{cases} 2x - 3y = -2 & (1) \\ x + 6y = 19 & (2) \end{cases}$

Solve the new E2 for x.

$x = -6y + 19$

Replace x with $-6y + 19$ in new E1.

$2(-6y + 19) - 3y = -2$
$-12y + 38 - 3y = -2$
$-15y = -40$
$y = \dfrac{-40}{-15} = \dfrac{8}{3}$

Replace y with $\dfrac{8}{3}$ in the equation

$x = -6y + 19$.

$x = -6\left(\dfrac{8}{3}\right) + 19$

$x = -16 + 19$

$x = 3$

The solution is $\left(3, \dfrac{8}{3}\right)$.

(3) Add the equations.

$\begin{array}{r} \dfrac{1}{2}x - \dfrac{3}{4}y = -\dfrac{1}{2} \\ \dfrac{1}{8}x + \dfrac{3}{4}y = \dfrac{19}{8} \\ \hline \dfrac{5}{8}x \qquad\quad = \dfrac{15}{8} \end{array}$

$5x = 15$
$x = 3$

Replace x with 3 in E1.

$\dfrac{1}{2}(3) - \dfrac{3}{4}y = -\dfrac{1}{2}$

$-\dfrac{3}{4}y = -2$

$-3y = -8$

$y = \dfrac{8}{3}$

The solution is $\left(3, \dfrac{8}{3}\right)$.

6. $\begin{cases} y = 32x & (1) \\ y = 15x + 25{,}500 & (2) \end{cases}$

Multiply E1 by -1 and add to E2.

$\begin{array}{r} -y = -32 \\ y = 15x + 25{,}500 \\ \hline 0 = -17x + 25{,}500 \end{array}$

$17x = 25{,}500$
$x = 1500$

Replace x with 1500 in E1.
$$y = 32(1500) = 48,000$$
The number of backpacks that the company must sell is 1500.

7. $\begin{cases} x + z = 4 \quad (1) \\ 2x - y = 4 \quad (2) \\ x + y - z = 0 \quad (3) \end{cases}$

Adding E2 and E3 gives $3x - z = 4$ (4)
Adding E1 and E4 gives $4x = 8$ or $x = 2$
Replace x with 2 in E1.
$$2 + z = 4$$
$$z = 2$$
Replace x with 2 and z with 2 in E3.
$$2 + y - 2 = 0$$
$$y = 0$$
The solution is (2, 0, 2).

8. $\begin{cases} 2x + 5y = 4 \quad (1) \\ x - 5y + z = -1 \quad (2) \\ 4x - z = 11 \quad (3) \end{cases}$

Add E2 and E3.
$$5x - 5y = 10 \quad (4)$$
Add E1 and E4.
$$7x = 14$$
$$x = 2$$
Replace x with 2 in E1.
$$2(2) + 5y = 4$$
$$4 + 5y = 4$$
$$5y = 0$$
$$y = 0$$
Replace x with 2 in E3.
$$4(2) - z = 11$$
$$8 - z = 11$$
$$z = -3$$
The solution is (2, 0, –3).

9. $\begin{cases} 4y + 2z = 5 \quad (1) \\ 2x + 8y = 5 \quad (2) \\ 6x + 4z = 1 \quad (3) \end{cases}$

Multiply E1 by –2 and add to E2.
$$\begin{array}{r} -8y - 4z = -10 \\ 2x + 8y = 5 \\ \hline 2x - 4z = -5 \quad (4) \end{array}$$
Add E3 and E4.
$$8x = -4$$
$$x = -\frac{1}{2}$$
Replace x with $-\frac{1}{2}$ in E2.

$$2\left(-\frac{1}{2}\right) + 8y = 5$$
$$-1 + 8y = 5$$
$$8y = 6$$
$$y = \frac{3}{4}$$
Replace x with $-\frac{1}{2}$ in E3.

$$6\left(-\frac{1}{2}\right) + 4z = 1$$
$$-3 + 4z = 1$$
$$4z = 4$$
$$z = 1$$
The solution is $\left(-\frac{1}{2}, \frac{3}{4}, 1\right)$.

10. $\begin{cases} 5x + 7y = 9 \quad (1) \\ 14y - z = 28 \quad (2) \\ 4x + 2z = -4 \quad (3) \end{cases}$

Dividing E3 by 2 gives $2x + z = -2$.
Add this equation to E2.
$$\begin{array}{r} 2x + z = -2 \\ 14y - z = 28 \\ \hline 2x + 14y = 26 \text{ or } x + 7y = 13 \quad (4) \end{array}$$
Multiply E4 by –1 and add to E1.
$$\begin{array}{r} -x - 7y = -13 \\ 5x + 7y = 9 \\ \hline 4x = -4 \\ x = -1 \end{array}$$
Replace x with –1 in E4.
$$-1 + 7y = 13$$
$$7y = 14$$
$$y = 2$$
Replace x with –1 in E3.
$$4(-1) + 2z = -4$$
$$-4 + 2z = -4$$
$$2z = 0$$
$$z = 0$$
The solution is (–1, 2, 0).

11. $\begin{cases} 3x - 2y + 2z = 5 \quad (1) \\ -x + 6y + z = 4 \quad (2) \\ 3x + 14y + 7z = 20 \quad (3) \end{cases}$

Multiply E2 by 3 and add to E1.
$$\begin{array}{r} 3x - 2y + 2z = 5 \\ -3x + 18y + 3z = 12 \\ \hline 16y + 5z = 17 \quad (4) \end{array}$$
Multiply E3 by –1 and add to E1.

$$3x - 2y + 2z = 5$$
$$\underline{-3x - 14y - 7z = -20}$$
$$-16y - 5z = -15 \quad (5)$$

Add E4 and E5.
$$16y + 5z = 17$$
$$\underline{-16y - 5z = -15}$$
$$0 = 2 \quad \text{False}$$

The system is inconsistent. The solution is $\varnothing$.

12. $\begin{cases} x + 2y + 3z = 11 & (1) \\ \quad\quad y + 2z = 3 & (2) \\ 2x \quad\quad + 2z = 10 & (3) \end{cases}$

Multiply E2 by -2 and add to E1.
$$x + 2y + 3z = 11$$
$$\underline{-2y - 4z = -6}$$
$$x \quad\quad - z = 5 \quad (4)$$

Multiply E4 by 2 and add to E3.
$$2x + 2z = 10$$
$$\underline{2x - 2z = 10}$$
$$4x \quad\quad = 20$$
$$x = 5$$

Replace x with 5 in E3.
$$2(5) + 2z = 10$$
$$10 + 2z = 10$$
$$2z = 0$$
$$z = 0$$

Replace z with 0 in E2.
$$y + 2(0) = 3$$
$$y + 0 = 3$$
$$y = 3$$

The solution is $(5, 3, 0)$.

13. $\begin{cases} 7x - 3y + 2z = 0 & (1) \\ 4x - 4y - z = 2 & (2) \\ 5x + 2y + 3z = 1 & (3) \end{cases}$

Multiply E2 by 2 and add to E1.
$$7x - 3y + 2z = 0$$
$$\underline{8x - 8y - 2z = 4}$$
$$15x - 11y \quad = 4 \quad (4)$$

Multiply E2 by 3 and add to E3.
$$12x - 12y - 3z = 6$$
$$\underline{5x + 2y + 3z = 1}$$
$$17x - 10y \quad = 7 \quad (5)$$

Solve the new system.
$$\begin{cases} 15x - 11y = 4 & (4) \\ 17x - 10y = 7 & (5) \end{cases}$$

Multiply E4 by -10, multiply E5 by 11, and add.

$$-150x + 110y = -40$$
$$\underline{187x - 110y = 77}$$
$$37x \quad\quad = 37$$
$$x = 1$$

Replace x with 1 in E4.
$$15(1) - 11y = 4$$
$$15 - 11y = 4$$
$$-11y = -11$$
$$y = 1$$

Replace x with 1 and y with 1 in E1.
$$7(1) - 3(1) + 2z = 0$$
$$4 + 2z = 0$$
$$2z = -4$$
$$z = -2$$

The solution is $(1, 1, -2)$.

14. $\begin{cases} x - 3y - 5z = -5 & (1) \\ 4x - 2y + 3z = 13 & (2) \\ 5x + 3y + 4z = 22 & (3) \end{cases}$

Multiply E1 by -4 and add to E2.
$$-4x + 12y + 20z = 20$$
$$\underline{4x - 2y + 3z = 13}$$
$$10y + 23z = 33 \quad (4)$$

Multiply E1 by -5 and add to E3.
$$-5x + 15y + 25z = 25$$
$$\underline{5x + 3y + 4z = 22}$$
$$18y + 29z = 47 \quad (5)$$

Solve the new system.
$$\begin{cases} 10y + 23z = 33 & (4) \\ 18y + 29z = 47 & (5) \end{cases}$$

Multiply E4 by 9, multiply E5 by -5 and add.
$$90y + 207z = 297$$
$$\underline{-90y - 145z = -235}$$
$$62z = 62$$
$$z = 1$$

Replace z with 1 in E4.
$$10y + 23(1) = 33$$
$$10y = 10$$
$$y = 1$$

Replace y with 1 and z with 1 in E1.
$$x - 3(1) - 5(1) = -5$$
$$x - 8 = -5$$
$$x = 3$$

The solution is $(3, 1, 1)$.

15. Let x = the first number, y = the second number, and z = the third number.

$$\begin{cases} x+y+z=98 & (1) \\ x+y=z+2 & (2) \\ y=4x & (3) \end{cases}$$

Replace y with $4x$ in E1 and E2.

$x+4x+z=98$

$\quad 5x+z=98 \quad (4)$

$x+4x \quad\;\; = z+2$

$\quad 5x-z=2 \quad (5)$

Add E4 and E5.

$\quad 5x+z=98$

$\underline{\quad 5x-z=2\quad}$

$10x \quad\;\; =100$

$\quad\; x=10$

Replace x with 10 in E3.

$y=4(10)=40$

Replace x with 10 and y with 40 in E2.

$10+40=z+2$

$\quad 50=z+2$

$\quad 48=z$

The numbers are 10, 40, and 48.

16. Let x = the first number and y = the second number.

$$\begin{cases} x=3y & (1) \\ 2(x+y)=168 & (2) \end{cases}$$

Replace x with $3y$ in E2.

$2(3y+y)=168$

$\quad 8y=168$

$\quad\; y=21$

Replace y with 21 in E1.

$x=3(21)=63$

The numbers are 63 and 21.

17. Let x = speed of first car and y = speed of the second car.

$$\begin{cases} 4x+4y=492 & (1) \\ y=x+7 & (2) \end{cases}$$

Replace y with $x+7$ in E1.

$4x+4(x+7)=492$

$\quad 8x+28=492$

$\quad 8x=464$

$\quad\; x=58$

Replace x with 58 in E2.

$y=58+7=65$

The cars are going 58 and 65 miles per hour.

18. Let w = the width of the foundation and l = the length of the foundation.

$$\begin{cases} l=3w & (1) \\ 2w+2l=296 & (2) \end{cases}$$

Replace l with $3w$ in E2.

$2w+2(3w)=296$

$\quad 2w+6w=296$

$\quad\; 8w=296$

$\quad\;\; w=37$

Replace w with 37 in E1.

$l=3(37)=111$

The foundation is 37 feet wide and 111 feet long.

19. Let x = liters of 10% solution and y = liters of 60% solution.

$$\begin{cases} x+y=50 & (1) \\ 0.10x+0.60y=0.40(50) & (2) \end{cases}$$

Solve E1 for y.

$y=50-x$

Replace y with $50-x$ in E2.

$0.10x+0.60(50-x)=0.40(50)$

$10[0.10x+0.60(50-x)]=10[0.40(50)]$

$\quad x+6(50-x)=4(50)$

$\quad x+300-6x=200$

$\quad -5x=-100$

$\quad\; x=20$

Replace x with 20 in the equation $y=50-x$.

$y=50-20=30$

He should use 20 liters of 10% solution and 30 liters of 60% solution.

20. Let c = pounds of chocolate used, n = pounds of nuts used, and r = pounds of raisins used.

$$\begin{cases} r=2n & (1) \\ c+n+r=45 & (2) \\ 3.00c+2.70n+2.25r=2.80(45) & (3) \end{cases}$$

Replace r with $2n$ in E2.

$c+n+2n=45$

$\quad c+3n=45$

$\quad\; c=-3n+45$

Replace r with $2n$ and c with $-3n+45$ in E3.

$3.00(-3n+45)+2.70n+2.25(2n)=126$

$\quad -9n+135+2.7n+4.5n=126$

$\quad -1.8n+135=126$

$\quad -1.8n=-9$

$\quad\; n=5$

Replace n with 5 in E1.

$r=2(5)=10$

Replace n with 5 and r with 10 in E2.

$$c + 5 + 10 = 45$$
$$c + 15 = 45$$
$$c = 30$$

She should use 30 pounds of creme-filled chocolates, 5 pounds of chocolate-covered nuts, and 10 pounds of chocolate-covered raisins.

21. Let x = the number of pennies,
y = the number of nickels, and
z = the number dimes.

$$\begin{cases} x + y + z = 53 & (1) \\ 0.01x + 0.05y + 0.10z = 2.77 & (2) \\ y = z + 4 & (3) \end{cases}$$

Clear the decimals from E2 by multiplying by 100.

$$x + 5y + 10z = 277 \quad (4)$$

Replace y with $z + 4$ in E1.

$$x + z + 4 + z = 53$$
$$x + 2z = 49 \quad (5)$$

Replace y with $z + 4$ in E4.

$$x + 5(z + 4) + 10z = 277$$
$$x + 15z = 257 \quad (6)$$

Solve the new system.

$$\begin{cases} x + 2z = 49 & (5) \\ x + 15z = 257 & (6) \end{cases}$$

Multiply E5 by -1 and add to E6.

$$\begin{aligned} -x - 2z &= -49 \\ \underline{x + 15z} &= \underline{257} \\ 13z &= 208 \\ z &= 16 \end{aligned}$$

Replace z with 16 in E3.

$$x + 2(16) = 49$$
$$x + 32 = 49$$
$$x = 17$$

Replace z with 16 in E3.

$$y = 16 + 4 = 20$$

He has 17 pennies, 20 nickels, and 16 dimes in his jar.

22. Let l = rate of interest on the larger investment and s = the rate of interest on the smaller investment, both expressed as decimals.

$$\begin{cases} 10,000l + 4000s = 1250 & (1) \\ l = s + 0.02 & (2) \end{cases}$$

Replace l with $s + 0.02$ in E1.

$$10,000(s + 0.02) + 4000s = 1250$$
$$10,000s + 200 + 4000s = 1250$$
$$14,000s = 1050$$
$$s = \frac{1050}{14,000} = 0.075$$

and $l = 0.075 + 0.02 = 0.095$.
The interest rate on the larger investment is 9.5% and the rate on the smaller investment is 7.5%.

23. Let x = length of the equal side and y = length of the third side.

$$\begin{cases} 2x + y = 73 & (1) \\ y = x + 7 & (2) \end{cases}$$

Replace y with $x + 7$ in E1.

$$2x + x + 7 = 73$$
$$3x = 66$$
$$x = 22$$

Replace x with 22 in E2.

$$y = 22 + 7 = 29$$

Two sides of the triangle have length 22 cm and the third side has length 29 cm.

24. Let f = the first number, s = the second number, and t = the third number.

$$\begin{cases} f + s + t = 295 & (1) \\ f = s + 5 & (2) \\ f = 2t & (3) \end{cases}$$

Solve E2 for s and E3 for t.

$$s = f - 5$$
$$t = \frac{f}{2}$$

Replace s with $f - 5$ and t with $\frac{f}{2}$ in E1.

$$f + f - 5 + \frac{f}{2} = 295$$
$$\frac{5}{2}f = 300$$
$$f = 120$$

Replace f with 300 in the equation $s = f - 5$.

$$s = 120 - 5 = 115$$

Replace f with 120 the equation $\frac{f}{2}$.

$$t = \frac{120}{2} = 60$$

The first number is 120, the second number is 115, and the third number is 60.

25. $$\begin{cases} 3x + 10y = 1 \\ x + 2y = -1 \end{cases}$$

$$\begin{bmatrix} 3 & 10 & \vdots & 1 \\ 1 & 2 & \vdots & -1 \end{bmatrix}$$

Interchange R1 and R2.

$$\begin{bmatrix} 1 & 2 & \vdots & -1 \\ 3 & 10 & \vdots & 1 \end{bmatrix}$$

Multiply R1 by -3 and add to R2.

     209

$$\begin{bmatrix} 1 & 2 & | & -1 \\ 0 & 4 & | & 4 \end{bmatrix}$$

Divide R2 by 4.

$$\begin{bmatrix} 1 & 2 & | & -1 \\ 0 & 1 & | & 1 \end{bmatrix}$$

This corresponds to $\begin{cases} x + 2y = -1 \\ \quad\quad y = 1 \end{cases}$.

$x + 2(1) = -1$

$x = -3$

The solution is $(-3, 1)$.

26. $\begin{cases} 3x - 6y = 12 \\ 2y = x - 4 \end{cases}$, or $\begin{cases} 3x - 6y = 12 \\ -x + 2y = -4 \end{cases}$

$$\begin{bmatrix} 3 & -6 & | & 12 \\ -1 & 2 & | & -4 \end{bmatrix}$$

Divide R1 by 3.

$$\begin{bmatrix} 1 & -2 & | & 4 \\ -1 & 2 & | & -4 \end{bmatrix}$$

Add R1 to R2.

$$\begin{bmatrix} 1 & -2 & | & 4 \\ 0 & 0 & | & 0 \end{bmatrix}$$

This corresponds to $\begin{cases} x - 2y = 4 \\ \quad\quad 0 = 0 \end{cases}$.

This is a dependent system. The solution is $\{(x, y)|3x - 6y = 12\}$.

27. $\begin{cases} 3x - 2y = -8 \\ 6x + 5y = 11 \end{cases}$

$$\begin{bmatrix} 3 & -2 & | & -8 \\ 6 & 5 & | & 11 \end{bmatrix}$$

Divide R1 by 3.

$$\begin{bmatrix} 1 & -\frac{2}{3} & | & -\frac{8}{3} \\ 6 & 5 & | & 11 \end{bmatrix}$$

Multiply R1 by -6 and add to R2.

$$\begin{bmatrix} 1 & -\frac{2}{3} & | & -\frac{8}{3} \\ 0 & 9 & | & 27 \end{bmatrix}$$

Divide R2 by 9.

$$\begin{bmatrix} 1 & -\frac{2}{3} & | & -\frac{8}{3} \\ 0 & 1 & | & 3 \end{bmatrix}$$

This corresponds to $\begin{cases} x - \frac{2}{3}y = -\frac{8}{3} \\ \quad\quad y = 3 \end{cases}$.

$x - \frac{2}{3}(3) = -\frac{8}{3}$

$x - 2 = -\frac{8}{3}$

$x = -\frac{2}{3}$

The solution is $\left(-\frac{2}{3}, 3\right)$.

28. $\begin{cases} 6x - 6y = -5 \\ 10x - 2y = 1 \end{cases}$

$$\begin{bmatrix} 6 & -6 & | & -5 \\ 10 & -2 & | & 1 \end{bmatrix}$$

Divide R1 by 6.

$$\begin{bmatrix} 1 & -1 & | & -\frac{5}{6} \\ 10 & -2 & | & 1 \end{bmatrix}$$

Multiply R1 by -10 and add to R2.

$$\begin{bmatrix} 1 & -1 & | & -\frac{5}{6} \\ 0 & 8 & | & \frac{28}{3} \end{bmatrix}$$

Divide R2 by 8.

$$\begin{bmatrix} 1 & -1 & | & -\frac{5}{6} \\ 0 & 1 & | & \frac{7}{6} \end{bmatrix}$$

Add R2 to R1.

$$\begin{bmatrix} 1 & 0 & | & \frac{1}{3} \\ 0 & 1 & | & \frac{7}{6} \end{bmatrix}$$

This corresponds to $\begin{cases} x = \frac{1}{3} \\ y = \frac{7}{6} \end{cases}$. The solution is $\left(\frac{1}{3}, \frac{7}{6}\right)$.

29. $\begin{cases} 3x - 6y = 0 \\ 2x + 4y = 5 \end{cases}$

$$\begin{bmatrix} 3 & -6 & | & 0 \\ 2 & 4 & | & 5 \end{bmatrix}$$

Divide R1 by 3.

$$\begin{bmatrix} 1 & -2 & | & 0 \\ 2 & 4 & | & 5 \end{bmatrix}$$

Multiply R1 by -2 and add to R2.

$$\begin{bmatrix} 1 & -2 & | & 0 \\ 0 & 8 & | & 5 \end{bmatrix}$$

Divide R2 by 8.

$$\begin{bmatrix} 1 & -2 & | & 0 \\ 0 & 1 & | & \frac{5}{8} \end{bmatrix}$$

This corresponds to $\begin{cases} x - 2y = 0 \\ y = \dfrac{5}{8} \end{cases}$.

$$x - 2\left(\frac{5}{8}\right) = 0$$

$$x - \frac{5}{4} = 0$$

$$x = \frac{5}{4}$$

The solution is $\left(\dfrac{5}{4}, \dfrac{5}{8}\right)$.

30. $\begin{cases} 5x - 3y = 10 \\ -2x + y = -1 \end{cases}$

$$\begin{bmatrix} 5 & -3 & | & 10 \\ -2 & 1 & | & -1 \end{bmatrix}$$

Divide R1 by 5.

$$\begin{bmatrix} 1 & -\frac{3}{5} & | & 2 \\ -2 & 1 & | & -1 \end{bmatrix}$$

Multiply R1 by 2 and add to R2.

$$\begin{bmatrix} 1 & -\frac{3}{5} & | & 2 \\ 0 & -\frac{1}{5} & | & 3 \end{bmatrix}$$

Multiply R2 by –5.

$$\begin{bmatrix} 1 & -\frac{3}{5} & | & 2 \\ 0 & 1 & | & -15 \end{bmatrix}$$

This corresponds to $\begin{cases} x - \dfrac{3}{5}y = 2 \\ \phantom{x - \dfrac{3}{5}}y = -15 \end{cases}$.

$$x - \frac{3}{5}(-15) = 2$$

$$x + 9 = 2$$

$$x = -7$$

The solution is $(-7, -15)$.

31. $\begin{cases} 0.2x - 0.3y = -0.7 \\ 0.5x + 0.3y = 1.4 \end{cases}$

$$\begin{bmatrix} 0.2 & -0.3 & | & -0.7 \\ 0.5 & 0.3 & | & 1.4 \end{bmatrix}$$

Multiply both rows by 10 to clear decimals.

$$\begin{bmatrix} 2 & -3 & | & -7 \\ 5 & 3 & | & 14 \end{bmatrix}$$

Divide R1 by 2.

$$\begin{bmatrix} 1 & -\frac{3}{2} & | & -\frac{7}{2} \\ 5 & 3 & | & 14 \end{bmatrix}$$

Multiply R1 by –5 and add to R2.

$$\begin{bmatrix} 1 & -\frac{3}{2} & | & -\frac{7}{2} \\ 0 & \frac{21}{2} & | & \frac{63}{2} \end{bmatrix}$$

Multiply R2 by $\dfrac{2}{21}$.

$$\begin{bmatrix} 1 & -\frac{3}{2} & | & -\frac{7}{2} \\ 0 & 1 & | & 3 \end{bmatrix}$$

This corresponds to $\begin{cases} x - \dfrac{3}{2}y = -\dfrac{7}{2} \\ \phantom{x - \dfrac{3}{2}}y = 3 \end{cases}$.

$$x - \frac{3}{2}(3) = -\frac{7}{2}$$

$$x - \frac{9}{2} = -\frac{7}{2}$$

$$x = 1$$

The solution is $(1, 3)$.

32. $\begin{cases} 3x + 2y = 8 \\ 3x - y = 5 \end{cases}$

$$\begin{bmatrix} 3 & 2 & | & 8 \\ 3 & -1 & | & 5 \end{bmatrix}$$

Divide R1 by 3.

$$\begin{bmatrix} 1 & \frac{2}{3} & | & \frac{8}{3} \\ 3 & -1 & | & 5 \end{bmatrix}$$

Multiply R1 by –3 and add to R2.

$$\begin{bmatrix} 1 & \frac{2}{3} & | & \frac{8}{3} \\ 0 & -3 & | & -3 \end{bmatrix}$$

Divide R2 by –3.

$$\begin{bmatrix} 1 & \frac{2}{3} & | & \frac{8}{3} \\ 0 & 1 & | & 1 \end{bmatrix}$$

This corresponds to $\begin{cases} x + \dfrac{2}{3}y = \dfrac{8}{3} \\ \phantom{x + \dfrac{2}{3}}y = 1 \end{cases}$.

$$x + \frac{2}{3}(1) = \frac{8}{3}$$

$$x = 2$$

The solution is $(2, 1)$.

33. $\begin{cases} x \quad\;\; + z = 4 \\ 2x - y \quad\;\; = 0 \\ x + y - z = 0 \end{cases}$

$\begin{bmatrix} 1 & 0 & 1 & | & 4 \\ 2 & -1 & 0 & | & 0 \\ 1 & 1 & -1 & | & 0 \end{bmatrix}$

Multiply R1 by –2 and add to R2. Multiply R1 by –1 and add to R3.

$\begin{bmatrix} 1 & 0 & 1 & | & 4 \\ 0 & -1 & -2 & | & -8 \\ 0 & 1 & -2 & | & -4 \end{bmatrix}$

Multiply R2 by –1.

$\begin{bmatrix} 1 & 0 & 1 & | & 4 \\ 0 & 1 & 2 & | & 8 \\ 0 & 1 & -2 & | & -4 \end{bmatrix}$

Multiply R2 by –1 and add to R3.

$\begin{bmatrix} 1 & 0 & 1 & | & 4 \\ 0 & 1 & 2 & | & 8 \\ 0 & 0 & -4 & | & -12 \end{bmatrix}$

Divide R3 by –4.

$\begin{bmatrix} 1 & 0 & 1 & | & 4 \\ 0 & 1 & 2 & | & 8 \\ 0 & 0 & 1 & | & 3 \end{bmatrix}$

This corresponds to $\begin{cases} x + z = 4 \\ y + 2z = 8 \\ \quad\quad z = 3 \end{cases}$.

$y + 2(3) = 8$
$y + 6 = 8$
$\quad\;\; y = 2$
$x + 3 = 4$
$\quad\; x = 1$
The solution is (1, 2, 3).

34. $\begin{cases} 2x + 5y \quad\;\; = 4 \\ x - 5y + z = -1 \\ 4x \quad\quad - z = 11 \end{cases}$

$\begin{bmatrix} 2 & 5 & 0 & | & 4 \\ 1 & -5 & 1 & | & -1 \\ 4 & 0 & -1 & | & 11 \end{bmatrix}$

Interchange R1 and R2.

$\begin{bmatrix} 1 & -5 & 1 & | & -1 \\ 2 & 5 & 0 & | & 4 \\ 4 & 0 & -1 & | & 11 \end{bmatrix}$

Multiply R1 by –2 and add to R2. Multiply R1 by –4 and add to R3.

$\begin{bmatrix} 1 & -5 & 1 & | & -1 \\ 0 & 15 & -2 & | & 6 \\ 0 & 20 & -5 & | & 15 \end{bmatrix}$

Divide R2 by 15.

$\begin{bmatrix} 1 & -5 & 1 & | & -1 \\ 0 & 1 & -\frac{2}{15} & | & \frac{2}{5} \\ 0 & 20 & -5 & | & 15 \end{bmatrix}$

Multiply R2 by –20 and add to R3.

$\begin{bmatrix} 1 & -5 & 1 & | & -1 \\ 0 & 1 & -\frac{2}{15} & | & \frac{2}{5} \\ 0 & 0 & -\frac{7}{3} & | & 7 \end{bmatrix}$

Multiply R3 by $-\dfrac{3}{7}$.

$\begin{bmatrix} 1 & -5 & 1 & | & -1 \\ 0 & 1 & -\frac{2}{15} & | & \frac{2}{5} \\ 0 & 0 & 1 & | & -3 \end{bmatrix}$

This corresponds to $\begin{cases} x - 5y + z = -1 \\ y - \dfrac{2}{15}z = \dfrac{2}{5} \\ \quad\quad\quad z = -3 \end{cases}$.

$y - \dfrac{2}{15}(-3) = \dfrac{2}{5}$
$\quad\;\; y + \dfrac{2}{5} = \dfrac{2}{5}$
$\quad\quad\;\; y = 0$
$x - 5(0) + (-3) = -1$
$\quad\quad\; x - 3 = -1$
$\quad\quad\quad\; x = 2$
The solution is (2, 0, –3).

35. $\begin{cases} 3x - y \quad\;\; = 11 \\ x \quad\;\; + 2z = 13 \\ \quad\; y - z = -7 \end{cases}$

$\begin{bmatrix} 3 & -1 & 0 & | & 11 \\ 1 & 0 & 2 & | & 13 \\ 0 & 1 & -1 & | & -7 \end{bmatrix}$

Interchange R1 and R2.

$\begin{bmatrix} 1 & 0 & 2 & | & 13 \\ 3 & -1 & 0 & | & 11 \\ 0 & 1 & -1 & | & -7 \end{bmatrix}$

Interchange R2 and R3.

$\begin{bmatrix} 1 & 0 & 2 & | & 13 \\ 0 & 1 & -1 & | & -7 \\ 3 & -1 & 0 & | & 11 \end{bmatrix}$

Multiply R1 by –3 and add to R3.

$\begin{bmatrix} 1 & 0 & 2 & | & 13 \\ 0 & 1 & -1 & | & -7 \\ 0 & -1 & -6 & | & -28 \end{bmatrix}$

Add R2 to R3.

$$\begin{bmatrix} 1 & 0 & 2 & | & 13 \\ 0 & 1 & -1 & | & -7 \\ 0 & 0 & -7 & | & -35 \end{bmatrix}$$

Divide R3 by –7.

$$\begin{bmatrix} 1 & 0 & 2 & | & 13 \\ 0 & 1 & -1 & | & -7 \\ 0 & 0 & 1 & | & 5 \end{bmatrix}$$

This corresponds to $\begin{cases} x + 2z = 13 \\ y - z = -7 \\ z = 5 \end{cases}$.

$y - 5 = -7$
$\quad y = -2$
$x + 2(5) = 13$
$\quad\quad x = 3$

The solution is (3, –2, 5).

36. $\begin{cases} 5x + 7y + 3z = 9 \\ \quad\quad 14y - z = 28 \\ 4x \quad\quad + 2z = -4 \end{cases}$

$$\begin{bmatrix} 5 & 7 & 3 & | & 9 \\ 0 & 14 & -1 & | & 28 \\ 4 & 0 & 2 & | & -4 \end{bmatrix}$$

Divide R1 by 5.

$$\begin{bmatrix} 1 & \frac{7}{5} & \frac{3}{5} & | & \frac{9}{5} \\ 0 & 14 & -1 & | & 28 \\ 4 & 0 & 2 & | & -4 \end{bmatrix}$$

Multiply R1 by –4 and add to R3.

$$\begin{bmatrix} 1 & \frac{7}{5} & \frac{3}{5} & | & \frac{9}{5} \\ 0 & 14 & -1 & | & 28 \\ 0 & -\frac{28}{5} & -\frac{2}{5} & | & -\frac{56}{5} \end{bmatrix}$$

Divide R2 by 14.

$$\begin{bmatrix} 1 & \frac{7}{5} & \frac{3}{5} & | & \frac{9}{5} \\ 0 & 1 & -\frac{1}{14} & | & 2 \\ 0 & -\frac{28}{5} & -\frac{2}{5} & | & -\frac{56}{5} \end{bmatrix}$$

Multiply R2 by $\frac{28}{5}$ and add to R3.

$$\begin{bmatrix} 1 & \frac{7}{5} & \frac{3}{5} & | & \frac{9}{5} \\ 0 & 1 & -\frac{1}{14} & | & 2 \\ 0 & 0 & -\frac{4}{5} & | & 0 \end{bmatrix}$$

Multiply R3 by $-\frac{5}{4}$.

$$\begin{bmatrix} 1 & \frac{7}{5} & \frac{3}{5} & | & \frac{9}{5} \\ 0 & 1 & -\frac{1}{14} & | & 2 \\ 0 & 0 & 1 & | & 0 \end{bmatrix}$$

This corresponds to $\begin{cases} x + \dfrac{7}{5}y + \dfrac{3}{5}z = \dfrac{9}{5} \\ \quad y - \dfrac{1}{14}z = 2 \\ \quad\quad z = 0 \end{cases}$.

$y - \dfrac{1}{14}(0) = 2$
$\quad\quad y = 2$
$x + \dfrac{7}{5}(2) + \dfrac{3}{5}(0) = \dfrac{9}{5}$
$\quad\quad x + \dfrac{14}{5} = \dfrac{9}{5}$
$\quad\quad\quad x = -1$

The solution is (–1, 2, 0).

37. $\begin{cases} 7x - 3y + 2z = 0 \\ 4x - 4y - z = 2 \\ 5x + 2y + 3z = 1 \end{cases}$

$$\begin{bmatrix} 7 & -3 & 2 & | & 0 \\ 4 & -4 & -1 & | & 2 \\ 5 & 2 & 3 & | & 1 \end{bmatrix}$$

Interchange R1 and R2.

$$\begin{bmatrix} 4 & -4 & -1 & | & 2 \\ 7 & -3 & 2 & | & 0 \\ 5 & 2 & 3 & | & 1 \end{bmatrix}$$

Divide R1 by 4.

$$\begin{bmatrix} 1 & -1 & -\frac{1}{4} & | & \frac{1}{2} \\ 7 & -3 & 2 & | & 0 \\ 5 & 2 & 3 & | & 1 \end{bmatrix}$$

Multiply R1 by –7 and add to R2. Multiply R1 by –5 and add to R3.

$$\begin{bmatrix} 1 & -1 & -\frac{1}{4} & | & \frac{1}{2} \\ 0 & 4 & \frac{15}{4} & | & -\frac{7}{2} \\ 0 & 7 & \frac{17}{4} & | & -\frac{3}{2} \end{bmatrix}$$

Divide R2 by 4.

$$\begin{bmatrix} 1 & -1 & -\frac{1}{4} & | & \frac{1}{2} \\ 0 & 1 & \frac{15}{16} & | & -\frac{7}{8} \\ 0 & 7 & \frac{17}{4} & | & -\frac{3}{2} \end{bmatrix}$$

Multiply R2 by –7 and add to R3.

$$\begin{bmatrix} 1 & -1 & -\frac{1}{4} & \Big| & \frac{1}{2} \\ 0 & 1 & \frac{15}{16} & \Big| & -\frac{7}{8} \\ 0 & 0 & -\frac{37}{16} & \Big| & -\frac{37}{8} \end{bmatrix}$$

Multiply R3 by $-\dfrac{16}{37}$.

$$\begin{bmatrix} 1 & -1 & -\frac{1}{4} & \Big| & \frac{1}{2} \\ 0 & 1 & \frac{15}{16} & \Big| & -\frac{7}{8} \\ 0 & 0 & 1 & \Big| & -2 \end{bmatrix}$$

This corresponds to $\begin{cases} x - y - \dfrac{1}{4}z = \dfrac{1}{2} \\ y + \dfrac{15}{16}z = -\dfrac{7}{8} \\ z = -2 \end{cases}$.

$$y + \frac{15}{16}(-2) = -\frac{7}{8}$$
$$y - \frac{15}{8} = -\frac{7}{8}$$
$$y = 1$$

$$x - 1 - \frac{1}{4}(-2) = \frac{1}{2}$$
$$x - 1 + \frac{1}{2} = \frac{1}{2}$$
$$x = 1$$

The solution is (1, 1, –2).

38. $\begin{cases} x + 2y + 3z = 14 \\ y + 2z = 3 \\ 2x \quad\quad - 2z = 10 \end{cases}$

$$\begin{bmatrix} 1 & 2 & 3 & | & 14 \\ 0 & 1 & 2 & | & 3 \\ 2 & 0 & -2 & | & 10 \end{bmatrix}$$

Multiply R1 by –2 and add to R3.

$$\begin{bmatrix} 1 & 2 & 3 & | & 14 \\ 0 & 1 & 2 & | & 3 \\ 0 & -4 & -8 & | & -18 \end{bmatrix}$$

Multiply R2 by 4 and add to R3.

$$\begin{bmatrix} 1 & 2 & 3 & | & 14 \\ 0 & 1 & 2 & | & 3 \\ 0 & 0 & 0 & | & -6 \end{bmatrix}$$

This corresponds to $\begin{cases} x + 2y + 3z = 14 \\ y + 2z = 3 \\ 0 = -6 \end{cases}$.

This system is inconsistent. The solution is $\varnothing$.

39. $\begin{cases} y \geq 2x - 3 \\ y \leq -2x + 1 \end{cases}$

Graph both inequalities on the same set of axes. The solution is the intersection of the solution regions. The solution of the system is the darker shaded region. This solution includes parts of both boundary lines.

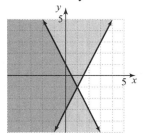

40. $\begin{cases} y \leq -3x - 3 \\ y \leq 2x + 7 \end{cases}$

Graph both inequalities on the same set of axes. The solution is the intersection of the solution regions. The solution of the system is the darker shaded region. This solution includes parts of both boundary lines.

41. $\begin{cases} x + 2y > 0 \\ x - y \leq 6 \end{cases}$

Graph both inequalities on the same set of axes. The solution is the intersection of the solution regions. The solution of the system is the darker shaded region. This solution includes the part of the solid boundary line that borders the region but not the dashed boundary line or the point where the boundary lines intersect.

42. $\begin{cases} x - 2y \ge 7 \\ x + y \le -5 \end{cases}$

Graph both inequalities on the same set of axes. The solution is the intersection of the solution regions. The solution of the system is the darker shaded region. This solution includes parts of both boundary lines.

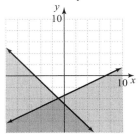

43. $\begin{cases} 3x - 2y \le 4 \\ 2x + y \ge 5 \\ y \le 4 \end{cases}$

Graph all three inequalities on the same set of axes. The solution set of the system is the shaded region. The parts of the boundary lines that border the shaded region are part of the solution.

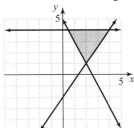

44. $\begin{cases} 4x - y \le 0 \\ 3x - 2y \ge -5 \\ y \ge -4 \end{cases}$

Graph all three inequalities on the same set of axes. The solution set of the system is the shaded region. The parts of the boundary lines that border the shaded region are part of the solution.

45. $\begin{cases} x + 2y \le 5 \\ x \le 2 \\ x \ge 0 \\ y \ge 0 \end{cases}$

Graph the inequalities on the same set of axes. The intersection of the inequalities is the solution region. It is the only shaded region in this graph and includes the portion of all four boundary lines that border the shaded region.

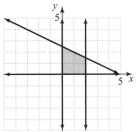

46. $\begin{cases} x + 3y \le 7 \\ y \le 5 \\ x \ge 0 \\ y \ge 0 \end{cases}$

Graph the inequalities on the same set of axes. The intersection of the inequalities is the solution region. It is the only shaded region in this graph and includes the portion of all four boundary lines that border the shaded region.

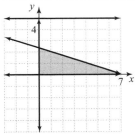

47. $\begin{cases} y = x - 5 \\ y = -2x + 2 \end{cases}$

We substitute $x - 5$ for y in the second equation.

$$x - 5 = -2x + 2$$
$$3x = 7$$
$$x = \frac{7}{3}$$

Replace x with $\frac{7}{3}$ in the first equation.

$$y = \frac{7}{3} - 5 = \frac{7}{3} - \frac{15}{3} = -\frac{8}{3}$$

The solution is $\left(\frac{7}{3}, -\frac{8}{3} \right)$.

48. $\begin{cases} \dfrac{2}{5}x + \dfrac{3}{4}y = 1 \\ x + 3y = -2 \end{cases}$

Multiply both sides of the first equation by 20 to eliminate fractions.

$\begin{cases} 20\left(\dfrac{2}{5}x + \dfrac{3}{4}y\right) = 20(1) \\ x + 3y = -2 \end{cases}$

$\begin{cases} 8x + 15y = 20 \\ x + 3y = -2 \end{cases}$

Multiply both sides of the second equation by -5 and add to the first equation to eliminate y.

$\begin{array}{r} 8x + 15y = 20 \\ -5x - 15y = 10 \\ \hline 3x = 30 \\ x = 10 \end{array}$

To find y, replace x with 10 in the second equation.

$10 + 3y = -2$
$3y = -12$
$y = -4$

The solution is $(10, -4)$.

49. $\begin{cases} 5x - 2y = 10 \\ x = \dfrac{2}{5}y + 2 \end{cases}$

Multiply both sides of the second equation by 5.

$\begin{cases} 5x - 2y = 10 \\ 5x = 5\left(\dfrac{2}{5}y + 2\right) \end{cases}$

$\begin{cases} 5x - 2y = 10 \\ 5x = 2y + 10 \end{cases}$

Subtract $2y$ from both sides of the second equation.

$\begin{cases} 5x - 2y = 10 \\ 5x - 2y = 10 \end{cases}$

The equations are the same. The system has an infinite number of solutions. The solution set can be written as $\{(x, y) \mid 5x - 2y = 10\}$.

50. $\begin{cases} x - 4y = 4 \\ \dfrac{1}{8}x - \dfrac{1}{2}y = 3 \end{cases}$

Multiply the second by -8 and add to the first equation to eliminate x.

$\begin{cases} x - 4y = 4 \\ -x + 4y = -24 \end{cases}$

The equation $0 = -20$ is false. The system has no solution. The solution set is $\{\ \}$ or $\varnothing$.

51. $\begin{cases} x - 3y + 2z = 0 \\ 9y - z = 22 \\ 5x + 3z = 10 \end{cases}$

The corresponding matrix is $\begin{bmatrix} 1 & -3 & 2 & | & 0 \\ 0 & 9 & -1 & | & 22 \\ 5 & 0 & 3 & | & 10 \end{bmatrix}$

Multiply row 1 by -5 and add to row 3.

$\begin{bmatrix} 1 & -3 & 2 & | & 0 \\ 0 & 9 & -1 & | & 22 \\ 0 & 15 & -7 & | & 10 \end{bmatrix}$

Divide row 2 by 9.

$\begin{bmatrix} 1 & -3 & 2 & | & 0 \\ 0 & 1 & -\frac{1}{9} & | & \frac{22}{9} \\ 0 & 15 & -7 & | & 10 \end{bmatrix}$

Multiply row 2 by -15 and add to row 3.

$\begin{bmatrix} 1 & -3 & 2 & | & 0 \\ 0 & 1 & -\frac{1}{9} & | & \frac{22}{9} \\ 0 & 0 & -\frac{48}{9} & | & -\frac{240}{9} \end{bmatrix}$

Multiply row 3 by $-\frac{9}{48}$.

$\begin{bmatrix} 1 & -3 & 2 & | & 0 \\ 0 & 1 & -\frac{1}{9} & | & \frac{22}{9} \\ 0 & 0 & 1 & | & 5 \end{bmatrix}$

This matrix represents the system

$\begin{cases} x - 3y + 2z = 0 \\ y - \dfrac{1}{9}z = \dfrac{22}{9} \\ z = 5 \end{cases}$.

Replace z with 5 in the second equation to find y.

$y - \dfrac{1}{9}(5) = \dfrac{22}{9}$

$y = \dfrac{22}{9} + \dfrac{5}{9} = \dfrac{27}{9} = 3$

Replace y with 3 and z with 5 in the first equation to find x.

$x - 3(3) + 2(5) = 0$
$x - 9 + 10 = 0$
$x + 1 = 0$
$x = -1$

The solution is $(-1, 3, 5)$.

52. Let $x =$ the first number
$y =$ the second number
We solve the system

$\begin{cases} x = 3y - 5 \\ x + y = 127 \end{cases}$

We substitute $3y - 5$ for x in the second equation.

Copyright © 2014 Pearson Education, Inc.

$(3y - 5) + y = 127$
$4y - 5 = 127$
$4y = 132$
$y = 33$

We replace y with 33 in the first equation to find x.

$x = 3(33) - 5 = 99 - 5 = 94$

The numbers are 94 and 33.

53. Let x = length of the shortest side
y = length of the second side
z = length of the third side
We solve the system
$$\begin{cases} x + y + z = 126 \\ y = 2x \\ z = x + 14 \end{cases}$$

We substitute $2x$ for y and $x + 14$ for z in the first equation.
$x + 2x + (x + 14) = 126$
$4x + 14 = 126$
$4x = 112$
$x = 28$

Now we find y and z.
$y = 2x = 2(28) = 56$
$z = x + 14 = 28 + 14 = 42$

The lengths are 28 units, 42 units, and 56 units.

54. $\begin{cases} y \le 3x - \dfrac{1}{2} \\ 3x + 4y \ge 6 \end{cases}$

Graph both inequalities on the same set of axes. The solution is the intersection of the solution regions. The solution of the system is the darker shaded region. This solution includes parts of both boundary lines.

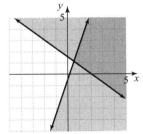

55. We solve the system
$$\begin{cases} y = 12.6x + 716.1 \\ y = -23.8x + 806.3 \end{cases}$$

Substitute $12.6x + 716.1$ for y in the second equation.

$12.6x + 716.1 = -23.8x + 806.3$
$36.4x + 716.1 = 806.3$
$36.4x = 90.2$
$x \approx 2.5$

The number of morning and evening editions of newspapers was the same in $1997 + 3 = 2000$.

Chapter 4 Test

1. $\begin{cases} 2x - y = -1 \quad (1) \\ 5x + 4y = 17 \quad (2) \end{cases}$

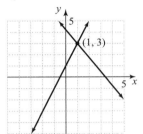

By elimination:
Multiply E1 by 4 and add it to E2.
$8x - 4y = -4$
$\underline{5x + 4y = 17}$
$13x \qquad = 13$
$\qquad\quad x = 1$

Replace x with 1 in E2.
$5(1) + 4y = 17$
$4y = 12$
$y = 3$

The solution is $(1, 3)$.

2. $\begin{cases} 7x - 14y = 5 \quad (1) \\ x = 2y \quad\quad\ (2) \end{cases}$

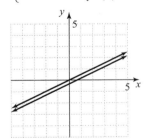

By substitution:
Replace x with $2y$ in E1.
$7(2y) - 14y = 5$
$14y - 14y = 5$
$0 = 5$ False

The system is inconsistent. The solution set is $\varnothing$.

3. $\begin{cases} 4x - 7y = 29 \\ 2x + 5y = -11 \end{cases}$

Multiply E2 by -2 and add to E1.

$-4x - 10y = 22$
$\underline{4x - 7y = 29}$
$-17y = 51$
$y = -3$

Replace y with -3 in E1.
$4x - 7(-3) = 29$
$4x + 21 = 29$
$4x = 8$
$x = 2$

The solution is $(2, -3)$.

4. $\begin{cases} 15x + 6y = 15 \\ 10x + 4y = 10 \end{cases}$

Divide E1 by 3 and E2 by 2.
$\begin{cases} 5x + 2y = 5 \\ 5x + 2y = 5 \end{cases}$

The system is dependent. The solution is $\{(x, y) | 10x + 4y = 10\}$.

5. $\begin{cases} 2x - 3y \quad\;\; = 4 \quad (1) \\ \quad\;\; 3y + 2z = 2 \quad (2) \\ x \quad\quad\;\; - z = -5 \quad (3) \end{cases}$

Add E1 and E2.
$2x + 2z = 6$ or $x + z = 3$ (4)

Add E3 and E4.
$x + z = 3$
$\underline{x - z = -5}$
$2x = -2$
$x = -1$

Replace x with -1 in E3.
$-1 - z = -5$
$-z = -4$ so $z = 4$

Replace x with -1 in E1.
$2(-1) - 3y = 4$
$-2 - 3y = 4$
$-3y = 6$
$y = -2$

The solution is $(-1, -2, 4)$.

6. $\begin{cases} 3x - 2y - z = -1 \quad (1) \\ 2x - 2y \quad\;\; = 4 \quad (2) \\ 2x \quad\;\; - 2z = -12 \quad (3) \end{cases}$

Multiply E2 by -1 and add to E1.
$3x - 2y - z = -1$
$\underline{-2x + 2y \quad\quad = -4}$
$x \quad\quad\;\; - z = -5$ (4)

Multiply E4 by -2 and add to E3.
$2x - 2z = -12$
$\underline{-2x + 2z = 10}$
$0 = -2$ False

The system is inconsistent. The solution set is $\varnothing$.

7. $\begin{cases} \dfrac{x}{2} + \dfrac{y}{4} = -\dfrac{3}{4} \\ x + \dfrac{3}{4}y = -4 \end{cases}$

Clear fractions by multiplying both equations by 4.
$\begin{cases} 2x + \;\; y = -3 \quad (1) \\ 4x + 3y = -16 \quad (2) \end{cases}$

Multiply E1 by -2 and add to E2.
$-4x - 2y = 6$
$\underline{4x + 3y = -16}$
$y = -10$

Replace y with -10 in E1.
$2x + (-10) = -3$
$2x = 7$ so $x = \dfrac{7}{2}$

The solution is $\left(\dfrac{7}{2}, -10\right)$.

8. $\begin{cases} x - \;\; y = -2 \\ 3x - 3y = -6 \end{cases}$

$\begin{bmatrix} 1 & -1 & | & -2 \\ 3 & -3 & | & -6 \end{bmatrix}$

Multiply R1 by -3 and add to R2.

$\begin{bmatrix} 1 & -1 & | & -2 \\ 0 & 0 & | & 0 \end{bmatrix}$

This corresponds to $\begin{cases} x - y = -2 \\ \quad\;\; 0 = 0 \end{cases}$.

This is a dependent system. The solution is $\{(x, y) | x - y = -2\}$.

9. The augmented matrix is $\begin{bmatrix} 1 & 2 & | & -1 \\ 2 & 5 & | & -5 \end{bmatrix}$.

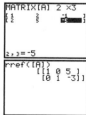

This corresponds to the system $\begin{cases} 1x + 0y = 5 \\ 0x + 1y = -3 \end{cases}$ or

$$\begin{cases} x = 5 \\ y = -3 \end{cases}.$$

The solution of the system is (5, −3).

10. $\begin{cases} x - y - z = 0 \\ 3x - y - 5z = -2 \\ 2x + 3y = -5 \end{cases}$

$$\begin{bmatrix} 1 & -1 & -1 & | & 0 \\ 3 & -1 & -5 & | & -2 \\ 2 & 3 & 0 & | & -5 \end{bmatrix}$$

Multiply R1 by −3 and add to R2. Multiply R1 by −2 and add to R3.

$$\begin{bmatrix} 1 & -1 & -1 & | & 0 \\ 0 & 2 & -2 & | & -2 \\ 0 & 5 & 2 & | & -5 \end{bmatrix}$$

Divide R2 by 2.

$$\begin{bmatrix} 1 & -1 & -1 & | & 0 \\ 0 & 1 & -1 & | & -1 \\ 0 & 5 & 2 & | & -5 \end{bmatrix}$$

Multiply R2 by −5 and add to R3.

$$\begin{bmatrix} 1 & -1 & -1 & | & 0 \\ 0 & 1 & -1 & | & -1 \\ 0 & 0 & 7 & | & 0 \end{bmatrix}$$

Divide R3 by 7.

$$\begin{bmatrix} 1 & -1 & -1 & | & 0 \\ 0 & 1 & -1 & | & -1 \\ 0 & 0 & 1 & | & 0 \end{bmatrix}$$

This corresponds to $\begin{cases} x - y - z = 0 \\ y - z = -1 \\ z = 0 \end{cases}.$

$$y - 0 = -1$$
$$y = -1$$
$$x - (-1) - 0 = 0$$
$$x + 1 = 0$$
$$x = -1$$

The solution is (−1, −1, 0).

11. Let x = double occupancy rooms and y = single occupancy rooms.

$$\begin{cases} x + y = 80 & (1) \\ 90x + 80y = 6930 & (2) \end{cases}$$

Multiply E1 by −80 and add to E2.

$$-80x - 80y = -6400$$
$$\underline{90x + 80y = 6930}$$
$$10x \quad\quad = 530$$
$$x = 53$$

Replace x with 53 in E1.

$$53 + y = 80$$
$$y = 27$$

53 double-occupancy and 27 single-occupancy rooms are occupied.

12. Let x = gallons of 10% solution and y = gallons of 20% solution.

$$\begin{cases} x + y = 20 & (1) \\ 0.10x + 0.20y = 0.175(20) & (2) \end{cases}$$

Multiply E1 by −0.10 add to E2.

$$-0.10x - 0.10y = -2.0$$
$$\underline{0.10x + 0.20y = 3.5}$$
$$0.10y = 1.5$$
$$y = 15$$

Replace y with 15 in E1.

$$x + 15 = 20$$
$$x = 5$$

They should use 5 gallons of 10% fructose solution and 15 gallons of the 20% solution.

13. $R(x) = 4x$ and $C(x) = 1.5x + 2000$

Break even occurs when $R(x) = C(x)$.

$$4x = 1.5x + 2000$$
$$2.5x = 2000$$
$$x = 800$$

The company must sell 800 packages to break even.

14. Let x = measure of the smallest angle. Then the largest angle has a measure of $5x - 3$, and the remaining angle has a measure of $2x - 1$. The sum of the three angles must add to 180°:

$$a + b + c = 180$$
$$x + (5x - 3) + (2x - 1) = 180$$
$$x + 5x - 3 + 2x - 1 = 180$$
$$8x - 4 = 180$$
$$8x = 184$$
$$x = 23$$

$$5x - 3 = 5(23) - 3 = 115 - 3 = 112$$
$$2x - 1 = 2(23) - 1 = 46 - 1 = 45$$

The angle measures are 23°, 45°, and 112°.

15. $\begin{cases} 2y - x \geq 1 \\ x + y \geq -4 \\ y \leq 2 \end{cases}$

Graph all three inequalities on the same set of axes. The solution set of the system is the shaded region. The parts of the boundary lines that border the shaded region are part of the solution.

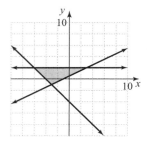

Chapter 4 Cumulative Review

1. **a.** Since 3 is a natural number, the statement is true.

   b. Since 7 is not one of the three numbers listed in the set, the statement is true.

2. **a.** Since 7 is not an element of the second set, the first set is not a subset of the second set. The statement is false.

   b. Since all three numbers in the first set are also elements of the second set, the first set is a subset of the second set. The statement is true.

3. **a.** $11 + 2 - 7 = 13 - 7 = 6$

   b. $-5 - 4 + 2 = -9 + 2 = -7$

4. **a.** $-7 - (-2) = -7 + 2 = -5$

   b. $14 - 38 = -24$

5. **a.** The opposite of 4 is -4.

   b. The opposite of $\dfrac{3}{7}$ is $-\dfrac{3}{7}$.

   c. The opposite of -11.2 is 11.2.

6. **a.** The reciprocal of 5 is $\dfrac{1}{5}$.

   b. The reciprocal of $-\dfrac{2}{3}$ is $-\dfrac{3}{2}$.

7. **a.** $3(2x + y) = 6x + 3y$

   b. $-(3x - 1) = -3x + 1$

   c. $0.7a(b - 2) = 0.7ab - 1.4a$

8. **a.** $7(3x - 2y + 4) = 21x - 14y + 28$

   b. $-(-2s - 3t) = 2s + 3t$

9. **a.** $3x - 5x + 4 = (3 - 5)x + 4 = -2x + 4$

   b. $7yz + yz = (7 + 1)yz = 8yz$

   c. $4z + 6.1 = 4z + 6.1$

10. **a.** $5y^2 - 1 + 2(y^2 + 2) = 5y^2 - 1 + 2y^2 + 4$
    $$= 7y^2 + 3$$

    b. $(7.8x - 1.2) - (5.6x - 2.4)$
    $$= 7.8x - 1.2 - 5.6x + 2.4$$
    $$= 2.2x + 1.2$$

11. $-4x - 1 + 5x = 9x + 3 - 7x$
    $$x - 1 = 2x + 3$$
    $$-x = 4$$
    $$x = -4$$

12. $8y - 14 = 6y - 14$
    $$2y = 0$$
    $$y = 0$$

13. $0.3x + 0.1 = 0.27x - 0.02$
    $$0.03x = -0.12$$
    $$x = -4$$

14. $2(m - 6) - m = 4(m - 3) - 3m$
    $$2m - 12 - m = 4m - 12 - 3m$$
    $$m - 12 = m - 12$$
    $$0 = 0 \quad \text{Always True}$$
    The solution is all real numbers.

15. Let x = length of the third side, then
    $2x + 12$ = length of the two equal sides.
    $$x + (2x + 12) + (2x + 12) = 149$$
    $$5x + 24 = 149$$
    $$5x = 125$$
    $$x = 25$$
    $$2(25) + 12 = 50 + 12 = 62$$
    The sides are 25 cm, 62 cm, and 62 cm.

16. Let x = measure of the equal angles,
$x + 10$ = measure of the third angle, and
$\dfrac{1}{2}x$ = measure of the fourth angle.

$$x + x + (x + 10) + \frac{1}{2}x = 360$$
$$\frac{7}{2}x + 10 = 360$$
$$\frac{7}{2}x = 350$$
$$7x = 700$$
$$x = 100$$
$$x + 10 = 100 + 10 = 110$$
$$\frac{1}{2}x = \frac{1}{2}(100) = 50$$

The measure of the angles are 100°, 100°, 110°, and 50°.

17. $3x + 4 \geq 2x - 6$
$\quad\quad x \geq -10$

18. $5(2x - 1) > -5$
$\quad 10x - 5 > -5$
$\quad\quad 10x > 0$
$\quad\quad\quad x > 0$
$(0, \infty)$

19. $2 < 4 - x < 7$
$\quad -2 < -x < 3$
$\quad\quad 2 > x > -3$
$\quad -3 < x < 2$
$(-3, 2)$

20. $\quad -1 < \dfrac{-2x - 1}{3} < 1$
$$3(-1) < 3\left[\frac{-2x - 1}{3}\right] < 3(1)$$
$$-3 < -2x - 1 < 3$$
$$-2 < -2x < 4$$
$$1 > x > -2$$
$$-2 < x < 1$$
$(-2, 1)$

21. $|2x| + 5 = 7$
$\quad |2x| = 2$
$2x = 2$ or $2x = -2$
$x = 1$ or $\quad x = -1$
The solution set is $\{-1, 1\}$.

22. $|x - 5| = 4$
$x - 5 = 4$ or $x - 5 = -4$
$\quad x = 9$ or $\quad\quad x = 1$
The solution set is $\{1, 9\}$.

23. $|m - 6| < 2$
$\quad -2 < m - 6 < 2$
$\quad\quad 4 < m < 8$
$(4, 8)$

24. $|2x + 1| > 5$
$\quad 2x + 1 < -5$ or $2x + 1 > 5$
$\quad\quad 2x < -6$ or $\quad\quad 2x > 4$
$\quad\quad x < -3$ or $\quad\quad\quad x > 2$
$(-\infty, -3) \cup (2, \infty)$

25.

a. $(2, -1)$ is in Quadrant IV.

b. $(0, 5)$ is on the y-axis.

c. $(-3, 5)$ is in Quadrant II.

d. $(-2, 0)$ is on the x-axis.

e. $\left(-\dfrac{1}{2}, -4\right)$ is in Quadrant III.

f. $(1.5, 1.5)$ is in Quadrant I.

26. a. $(-1, -5)$ is in Quadrant III.

b. $(4, -2)$ is in Quadrant IV.

c. $(0, 2)$ is on the y-axis.

27. No; for the input $x = 4$, there are two outputs, $y = \pm 2$.

28. $-2x + \dfrac{1}{2}y = -2$, or $y = 4x - 4$

29. $f(x) = 7x^2 - 3x + 1$, $g(x) = 3x - 2$, $h(x) = x^2$

   a. $f(1) = 7(1)^2 - 3(1) + 1 = 7 - 3 + 1 = 5$

   b. $g(1) = 3(1) - 2 = 3 - 2 = 1$

   c. $h(-2) = (-2)^2 = 4$

30. $f(x) = 3x^2$

   a. $f(5) = 3(5)^2 = 3(25) = 75$

   b. $f(-2) = 3(-2)^2 = 3(4) = 12$

31. $g(x) = 2x + 1$ and $f(x) = 2x$

The graph of g is the graph of f shifted 1 unit up.

32. $m = \dfrac{9 - 6}{0 - (-2)} = \dfrac{3}{2}$

33. $3x - 4y = 4$
$-4y = -3x + 4$
$y = \dfrac{3}{4}x - 1$
$m = \dfrac{3}{4}$, y-intercept $= (0, -1)$

34. $y = 2$
$m = 0$, y-intercept $= (0, 2)$

35. a. $3x + 7y = 21$
$7y = -3x + 21$
$y = -\dfrac{3}{7}x + 3$
$m = -\dfrac{3}{7}$
$6x + 14y = 7$
$14y = -6x + 7$
$y = -\dfrac{3}{7}x + \dfrac{1}{2}$
$m = -\dfrac{3}{7}$
Parallel, since the slopes are equal.

   b. $-x + 3y = 2$
$3y = x + 2$
$y = \dfrac{1}{3}x + \dfrac{2}{3}$
$m = \dfrac{1}{3}$
$2x + 6y = 5$
$6y = -2x + 5$
$y = -\dfrac{1}{3}x + \dfrac{5}{6}$
$m = -\dfrac{1}{3}$
Neither, since the slopes are not equal and their product is not -1.

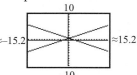

   c. $2x - 3y = 12$
$-3y = -2x + 12$
$y = \dfrac{2}{3}x - 4$
$m = \dfrac{2}{3}$
$6x + 4y = 16$
$4y = -6x + 16$
$y = -\dfrac{3}{2}x + 4$

$$m = -\frac{3}{2}$$

Perpendicular, since the product of the slopes is -1.

36. $y - (-9) = \frac{1}{5}(x - 0)$

$$y + 9 = \frac{1}{5}x$$

$$y = \frac{1}{5}x - 9$$

37. $m = \frac{-5-0}{-4-4} = \frac{-5}{-8} = \frac{5}{8}$

$$y - 0 = \frac{5}{8}(x - 4)$$

$$y = \frac{5}{8}x - \frac{5}{2}$$

$$f(x) = \frac{5}{8}x - \frac{5}{2}$$

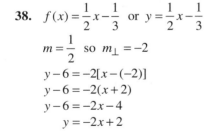

38. $f(x) = \frac{1}{2}x - \frac{1}{3}$ or $y = \frac{1}{2}x - \frac{1}{3}$

$$m = \frac{1}{2} \text{ so } m_\perp = -2$$

$$y - 6 = -2[x - (-2)]$$
$$y - 6 = -2(x + 2)$$
$$y - 6 = -2x - 4$$
$$y = -2x + 2$$

39. $3x \geq y$, or $y \leq 3x$

Graph the boundary line $y = 3x$ with a solid line because the inequality symbol is $\leq$.
Test: (0, 1)

$$3x \geq y$$
$$3(0) \geq 1$$
$$0 \geq 1 \text{ False}$$

Shade the half-plane that does not contain (0, 1).

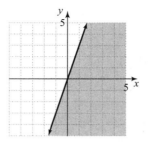

40. $x \geq 1$

Graph the boundary line $x = 1$ with a solid line because the inequality symbol is $\geq$.
Shade the half-plane that does not contain (0, 0).

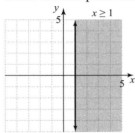

41. a. $\begin{cases} -x + y = 2 \\ 2x - y = -3 \end{cases}$

$$-(-1) + 1 = 2$$
$$1 + 1 = 2$$
$$2 = 2 \text{ True}$$
$$2(-1) - (1) = -3$$
$$-2 - 1 = -3$$
$$-3 = -3 \text{ True}$$

Yes, $(-1, 1)$ is a solution.

b. $\begin{cases} 5x + 3y = -1 \\ x - y = 1 \end{cases}$

$$5(-2) + 3(3) = -1$$
$$-10 + 9 = -1$$
$$-1 = -1 \text{ True}$$
$$-2 - 3 = -1$$
$$-5 = -1 \text{ False}$$

No, $(-2, 3)$ is not a solution.

42. $\begin{cases} 5x + y = -2 \quad (1) \\ 4x - 2y = -10 \quad (2) \end{cases}$

Multiply E1 by 2 and add to E2.

$$10x + 2y = -4$$
$$\underline{4x - 2y = -10}$$
$$14x = -14$$
$$x = -1$$

Replace x with -1 in E1.

$5(-1) + y = -2$

$-5 + y = -2$

$y = 3$

The solution is $(-1, 3)$.

43. $\begin{cases} 3x - y + z = -15 & (1) \\ x + 2y - z = 1 & (2) \\ 2x + 3y - 2z = 0 & (3) \end{cases}$

Add E1 and E2.

$4x + y = -14$ (4)

Multiply E1 by 2 and add to E3.

$6x - 2y + 2z = -30$

$\underline{2x + 3y - 2z = 0}$

$8x + y \qquad = -30$ (5)

Solve the new system:

$\begin{cases} 4x + y = -14 & (4) \\ 8x + y = -30 & (5) \end{cases}$

Multiply E4 by -1 and add to E5.

$-4x - y = 14$

$\underline{8x + y = -30}$

$4x \qquad = -16$

$x = -4$

Replace x with -4 in E4.

$4(-4) + y = -14$

$-16 + y = -14$

$y = 2$

Replace x with -4 and y with 2 in E1.

$3(-4) - (2) + z = -15$

$-12 - 2 + z = -15$

$-14 + z = -15$

$z = -1$

The solution is $(-4, 2, -1)$.

44. $\begin{cases} x - 2y + z = 0 & (1) \\ 3x - y - 2z = -15 & (2) \\ 2x - 3y + 3z = 7 & (3) \end{cases}$

Multiply E1 by 2 and add to E2.

$2x - 4y + 2z = 0$

$\underline{3x - y - 2z = -15}$

$5x - 5y \qquad = -15$ or $x - y = -3$ (4)

Multiply E1 by -3 and add to E3.

$-3x + 6y - 3z = 0$

$\underline{2x - 3y + 3z = 7}$

$-x + 3y \qquad = 7$ (5)

Add E4 and E5.

$2y = 4$

$y = 2$

Replace y with 2 in E4.

$x - 2 = -3$

$x = -1$

Replace x with -1 and y with 2 in E1.

$-1 - 2(2) + z = 0$

$-5 + z = 0$

$z = 5$

The solution is $(-1, 2, 5)$.

45. $\begin{cases} x + 3y = 5 \\ 2x - y = -4 \end{cases}$

$\begin{bmatrix} 1 & 3 & | & 5 \\ 2 & -1 & | & -4 \end{bmatrix}$

Multiply R1 by -2 and add to R2.

$\begin{bmatrix} 1 & 3 & | & 5 \\ 0 & -7 & | & -14 \end{bmatrix}$

Divide R2 by -7.

$\begin{bmatrix} 1 & 3 & | & 5 \\ 0 & 1 & | & 2 \end{bmatrix}$

This corresponds to $\begin{cases} x + 3y = 5 \\ y = 2 \end{cases}$.

$x + 3(2) = 5$

$x + 6 = 5$

$x = -1$

The solution is $(-1, 2)$.

46. The augmented matrix is $\begin{bmatrix} -6 & 8 & | & 0 \\ 9 & -12 & | & 2 \end{bmatrix}$.

This corresponds to the system $\begin{cases} 1x - \dfrac{4}{3}y = 0 \\ 0x + 0y = 1 \end{cases}$.

From the second equation, the system is inconsistent. The solution set is $\varnothing$.

Chapter 5

Section 5.1 Practice Exercises

1. a. $3^4 \cdot 3^2 = 3^{4+2} = 3^6$

b. $x^5 \cdot x^2 = x^{5+2} = x^7$

c. $y \cdot y^3 \cdot y^5 = (y^1 \cdot y^3) \cdot y^5 = y^4 \cdot y^5 = y^9$

2. a. $(5z^3)(7z) = 5(7)z^3 z^1 = 35z^4$

b. $(-4.1t^5 q^3)(5tq^5) = -4.1(5)t^5 t^1 q^3 q^5$
$$= -20.5t^6 q^8$$

3. a. $5^0 = 1$

b. $-5^0 = -(5^0) = -(1) = -1$

c. $(3x - 8)^0 = 1$

d. $3x^0 = 3(1) = 3$

4. a. $\dfrac{z^8}{z^3} = z^{8-3} = z^5$

b. $\dfrac{3^9}{3^3} = 3^{9-3} = 3^6$

c. $\dfrac{45x^7}{5x^3} = 9x^{7-3} = 9x^4$

d. $\dfrac{24a^{14}b^6}{18a^7 b^6} = \dfrac{4}{3}a^{14-7}b^{6-6}$
$$= \dfrac{4}{3}a^7 b^0$$
$$= \dfrac{4}{3}a^7 \text{ or } \dfrac{4a^7}{3}$$

5. a. $6^{-2} = \dfrac{1}{6^2} = \dfrac{1}{36}$

b. $(-2)^{-6} = \dfrac{1}{(-2)^6} = \dfrac{1}{64}$

c. $3x^{-5} = 3 \cdot \dfrac{1}{x^5} = \dfrac{3}{x^5}$

d. $(5y)^{-1} = \dfrac{1}{(5y)^1} = \dfrac{1}{5y}$

e. $\dfrac{k^4}{k^{11}} = k^{4-11} = k^{-7} = \dfrac{1}{k^7}$

f. $\dfrac{5^3}{5^5} = 5^{3-5} = 5^{-2} = \dfrac{1}{5^2} = \dfrac{1}{25}$

g. $5^{-1} + 2^{-2} = \dfrac{1}{5^1} + \dfrac{1}{2^2} = \dfrac{1}{5} + \dfrac{1}{4} = \dfrac{4}{20} + \dfrac{5}{20} = \dfrac{9}{20}$

h. $\dfrac{1}{z^{-8}} = \dfrac{1}{\frac{1}{z^8}} = 1 \div \dfrac{1}{z^8} = 1 \cdot \dfrac{z^8}{1} = z^8$

6. a. $\dfrac{z^{-8}}{z^3} = z^{-8-3} = z^{-11} = \dfrac{1}{z^{11}}$

b. $\dfrac{7t^3}{t^{-5}} = 7 \cdot t^{3-(-5)} = 7t^8$

c. $\dfrac{3^{-2}}{3^{-4}} = 3^{-2-(-4)} = 3^{-2+4} = 3^2 = 9$

d. $\dfrac{5a^{-5}b^3}{15a^2 b^{-4}} = \dfrac{a^{-5-2}b^{3-(-4)}}{3} = \dfrac{a^{-7}b^7}{3} = \dfrac{b^7}{3a^7}$

e. $\dfrac{(2x^{-5})(x^6)}{x^5} = \dfrac{2x^{-5+6}}{x^5}$
$$= \dfrac{2x^1}{x^5}$$
$$= 2x^{1-5}$$
$$= 2x^{-4}$$
$$= \dfrac{2}{x^4}$$

7. a. $x^{3a} \cdot x^4 = x^{3a+4}$

b. $\dfrac{x^{3t-2}}{x^{t-3}} = x^{(3t-2)-(t-3)} = x^{3t-2-t+3} = x^{2t+1}$

8. a. Move the decimal point until the number is between 1 and 10. The decimal point is moved 4 places and the original number is 10 or greater, so the count is positive 4.

$$65,000 = 6.5 \times 10^4$$

b. Move the decimal point until the number is between 1 and 10. The decimal point is moved 5 places and the original number is less than 1, so the count is −5.

$$0.000038 = 3.8 \times 10^{-5}$$

9. a. Since the exponent is positive, move the decimal point 5 places to the right.

$$6.2 \times 10^5 = 620,000$$

b. Since the exponent is negative, more the decimal point 2 places to the left.

$$3.109 \times 10^{-2} = 0.03109$$

Vocabulary, Readiness & Video Check 5.1

1. $9x^5$; base x

2. yz^5; base z

3. -3^5; base 3

4. $(-3)^5$; base −3

5. $(y^7)^5$; base y^7

6. $9 \cdot 2^5$; base 2

7. These properties allow us to reorder and regroup factors to put those with the same bases together so that we may apply the product rule.

8. Since there are no parentheses, the negative is not part of the base and is therefore not raised to the zero power. The evaluation is −1, not 1.

9. Subtract the exponents on like bases when applying the quotient rule.

10. A negative exponent will not make an expression evaluate to a negative number. A negative exponent moves an expression or factor from numerator to denominator or from denominator to numerator.

11. When you move the decimal point to the left, the sign of the exponent is positive; when you move the decimal point to the right, the sign of the exponent is negative.

Exercise Set 5.1

1. $5x^{-1}y^{-2} = \dfrac{5}{xy^2}$

3. $a^2 b^{-1} c^{-5} = \dfrac{a^2}{bc^5}$

5. $\dfrac{y^{-2}}{x^{-4}} = \dfrac{x^4}{y^2}$

7. $4^2 \cdot 4^3 = 4^{2+3} = 4^5$

9. $x^5 \cdot x^3 = x^{5+3} = x^8$

11. $m \cdot m^7 \cdot m^6 = m^{1+7+6} = m^{14}$

13. $(4xy)(-5x) = -20x^{1+1}y = -20x^2 y$

15. $(-4x^3 p^2)(4y^3 x^3) = -16x^{3+3}y^3 p^2$
$$= -16x^6 y^3 p^2$$

17. $-8^0 = -(8^0) = -1$

19. $(4x+5)^0 = 1$

21. $-x^0 = -(x^0) = -(1) = -1$

23. $4x^0 + 5 = 4(1) + 5 = 4 + 5 = 9$

25. $\dfrac{a^5}{a^2} = a^{5-2} = a^3$

27. $-\dfrac{26z^{11}}{2z^7} = -13z^{11-7} = -13z^4$

29. $\dfrac{x^9 y^6}{x^8 y^6} = x^{9-8} y^{6-6} = x^1 y^0 = x$

31. $\dfrac{12x^4 y^7}{9xy^5} = \dfrac{4x^{4-1} y^{7-5}}{3} = \dfrac{4}{3} x^3 y^2$

33. $\dfrac{-36a^5b^7c^{10}}{6ab^3c^4} = -6a^{5-1}b^{7-3}c^{10-4} = -6a^4b^4c^6$

35. $4^{-2} = \dfrac{1}{4^2} = \dfrac{1}{16}$

37. $(-3)^{-3} = \dfrac{1}{(-3)^3} = \dfrac{1}{-27} = -\dfrac{1}{27}$

39. $\dfrac{x^7}{x^{15}} = x^{7-15} = x^{-8} = \dfrac{1}{x^8}$

41. $5a^{-4} = \dfrac{5}{a^4}$

43. $\dfrac{x^{-7}}{y^{-2}} = \dfrac{y^2}{x^7}$

45. $\dfrac{x^{-2}}{x^5} = x^{-2-5} = x^{-7} = \dfrac{1}{x^7}$

47. $\dfrac{8r^4}{2r^{-4}} = 4r^{4-(-4)} = 4r^8$

49. $\dfrac{x^{-9}x^4}{x^{-5}} = \dfrac{x^{-9+4}}{x^{-5}} = \dfrac{x^{-5}}{x^{-5}} = x^{-5-(-5)} = x^0 = 1$

51. $\dfrac{2a^{-6}b^2}{18ab^{-5}} = \dfrac{a^{-6-1}b^{2-(-5)}}{9} = \dfrac{a^{-7}b^7}{9} = \dfrac{b^7}{9a^7}$

53. $\dfrac{(24x^8)(x)}{20x^{-7}} = \dfrac{6x^{8+1}}{5x^{-7}} = \dfrac{6x^{9-(-7)}}{5} = \dfrac{6x^{16}}{5}$

55. $-7x^3 \cdot 20x^9 = -7 \cdot 20x^{3+9} = -140x^{12}$

57. $x^7 \cdot x^8 \cdot x = x^{7+8+1} = x^{16}$

59. $2x^3 \cdot 5x^7 = 2 \cdot 5x^{3+7} = 10x^{10}$

61. $(5x)^0 + 5x^0 = 1 + 5 \cdot 1 = 1 + 5 = 6$

63. $\dfrac{z^{12}}{z^{15}} = z^{12-15} = z^{-3} = \dfrac{1}{z^3}$

65. $3^0 - 3t^0 = 1 - 3 \cdot 1 = 1 - 3 = -2$

67. $\dfrac{y^{-3}}{y^{-7}} = y^{-3-(-7)} = y^4$

69. $4^{-1} + 3^{-2} = \dfrac{1}{4^1} + \dfrac{1}{3^2}$
$= \dfrac{1}{4} + \dfrac{1}{9}$
$= \dfrac{9}{36} + \dfrac{4}{36}$
$= \dfrac{13}{36}$

71. $3x^{-1} = \dfrac{3}{x^1} = \dfrac{3}{x}$

73. $\dfrac{r^4}{r^{-4}} = r^{4-(-4)} = r^8$

75. $\dfrac{x^{-7}y^{-2}}{x^2y^2} = x^{-7-2}y^{-2-2} = x^{-9}y^{-4} = \dfrac{1}{x^9y^4}$

77. $(-4x^2y)(3x^4)(-2xy^5)$
$= -4(3)(-2)x^2 \cdot x^4 \cdot x \cdot y \cdot y^5$
$= 24x^{2+4+1} \cdot y^{1+5}$
$= 24x^7y^6$

79. $2^{-4} \cdot x = \dfrac{x}{2^4} = \dfrac{x}{16}$

81. $\dfrac{5^{17}}{5^{13}} = 5^{17-13} = 5^4 = 625$

83. $\dfrac{8^{-7}}{8^{-6}} = 8^{-7-(-6)} = 8^{-7+6} = 8^{-1} = \dfrac{1}{8}$

85. $\dfrac{9^{-5}a^4}{9^{-3}a^{-1}} = 9^{-5-(-3)}a^{4-(-1)}$
$= 9^{-5+3}a^{4+1}$
$= 9^{-2}a^5$
$= \dfrac{a^5}{9^2}$
$= \dfrac{a^5}{81}$

     227

87. $\dfrac{14x^{-2}yz^{-4}}{2xyz} = \dfrac{14}{2}\cdot x^{-2-1}y^{1-1}z^{-4-1}$

$= 7x^{-3}y^0z^{-5}$

$= \dfrac{7}{x^3z^5}$

89. $x^5\cdot x^{7a} = x^{5+7a}$ or x^{7a+5}

91. $\dfrac{x^{3t-1}}{x^t} = x^{3t-1-t} = x^{2t-1}$

93. $y^{2p}\cdot y^{9p} = y^{2p+9p} = y^{11p}$

95. $\dfrac{z^{6x}}{z^7} = z^{6x-7}$

97. $\dfrac{x^{3t}\cdot x^{4t-1}}{x^t} = \dfrac{x^{3t+(4t-1)}}{x^t} = x^{7t-1-t} = x^{6t-1}$

99. $31,250,000 = 3.125\times10^7$

101. $0.016 = 1.6\times10^{-2}$

103. $67,413 = 6.7413\times10^4$

105. $0.0125 = 1.25\times10^{-2}$

107. $0.000053 = 5.3\times10^{-5}$

109. $778,300,000 = 7.783\times10^8$

111. $16,228,000,000,000 = 1.6228\times10^{13}$

113. $124,000,000,000 = 1.24\times10^{11}$

115. $0.001 = 1.0\times10^{-3}$

117. $3.6\times10^{-9} = 0.0000000036$

119. $9.3\times10^7 = 93,000,000$

121. $1.278\times10^6 = 1,278,000$

123. $7.35\times10^{12} = 7,350,000,000,000$

125. $4.03\times10^{-7} = 0.000000403$

127. $3.0\times10^8 = 300,000,000$

129. $9.5\times10^{-3} = 0.0095$

131. $2.55\times10^{10} = 25,500,000,000$

133. $(5\cdot2)^2 = (10)^2 = 100$

135. $\left(\dfrac{3}{4}\right)^3 = \left(\dfrac{3}{4}\right)\left(\dfrac{3}{4}\right)\left(\dfrac{3}{4}\right) = \dfrac{3\cdot3\cdot3}{4\cdot4\cdot4} = \dfrac{27}{64}$

137. $(2^3)^2 = 8^2 = 64$

139. answers may vary

141. answers may vary

143. a. $x^a\cdot x^a = x^{a+a} = x^{2a}$

b. $x^a + x^a = 2x^a$

c. $\dfrac{x^a}{x^b} = x^{a-b}$

d. $x^a\cdot x^b = x^{a+b}$

e. $x^a + x^b = x^a + x^b$

145. 7^{13}

147. 7^{-11}

Section 5.2 Practice Exercises

1. a. $(z^3)^5 = z^{3\cdot5} = z^{15}$

b. $(5^2)^2 = 5^{2\cdot2} = 5^4 = 625$

c. $(3^{-1})^3 = 3^{-1\cdot3} = 3^{-3} = \dfrac{1}{3^3} = \dfrac{1}{27}$

d. $(x^{-4})^{-6} = x^{-4(-6)} = x^{24}$

2. a. $(2x^3)^5 = 2^5 \cdot (x^3)^5 = 2^5 \cdot x^{3 \cdot 5} = 32x^{15}$

b. $\left(\dfrac{3}{5}\right)^2 = \dfrac{3^2}{5^2} = \dfrac{9}{25}$

c. $\left(\dfrac{2a^5}{b^7}\right)^4 = \dfrac{(2a^5)^4}{(b^7)^4} = \dfrac{2^4 \cdot (a^5)^4}{(b^7)^4} = \dfrac{16a^{20}}{b^{28}}$

d. $\left(\dfrac{3^{-2}}{x}\right)^{-1} = \dfrac{(3^{-2})^{-1}}{x^{-1}} = \dfrac{3^2}{x^{-1}} = 9x$

e. $(a^{-2}b^{-5}c^4)^{-2} = (a^{-2})^{-2} \cdot (b^{-5})^{-2} \cdot (c^4)^{-2}$
$$= a^4 b^{10} c^{-8}$$
$$= \dfrac{a^4 b^{10}}{c^8}$$

3. a. $(3ab^{-5})^{-3} = 3^{-3}a^{-3}(b^{-5})^{-3}$
$$= 3^{-3}a^{-3}b^{15}$$
$$= \dfrac{b^{15}}{3^3 a^3}$$
$$= \dfrac{b^{15}}{27a^3}$$

b. $\left(\dfrac{y^{-7}}{y^{-4}}\right)^{-5} = \dfrac{(y^{-7})^{-5}}{(y^{-4})^{-5}}$
$$= \dfrac{y^{35}}{y^{20}}$$
$$= y^{35-20}$$
$$= y^{15}$$

c. $\left(\dfrac{3}{8}\right)^{-2} = \dfrac{3^{-2}}{8^{-2}} = \dfrac{8^2}{3^2} = \dfrac{64}{9}$

d. $\dfrac{9^{-2}a^{-4}b^3}{a^2 b^{-5}} = 9^{-2}\left(\dfrac{a^{-4}}{a^2}\right)\left(\dfrac{b^3}{b^{-5}}\right)$
$$= 9^{-2}a^{-4-2}b^{3-(-5)}$$
$$= 9^{-2}a^{-6}b^8$$
$$= \dfrac{b^8}{9^2 a^6}$$
$$= \dfrac{b^8}{81a^6}$$

4. a. $\left(\dfrac{5a^4 b}{a^{-8}c}\right)^{-3} = \left(\dfrac{5a^{12}b}{c}\right)^{-3}$
$$= \dfrac{5^{-3}a^{-36}b^{-3}}{c^{-3}}$$
$$= \dfrac{c^3}{5^3 a^{36}b^3}$$
$$= \dfrac{c^3}{125a^{36}b^3}$$

b. $\left(\dfrac{2x^4}{5y^{-2}}\right)^3\left(\dfrac{x^{-4}}{10y^{-2}}\right)^{-1} = \dfrac{8x^{12}}{125y^{-6}} \cdot \dfrac{x^4}{10^{-1}y^2}$
$$= \dfrac{8 \cdot 10 \cdot x^{12}x^4 y^6}{125 y^2}$$
$$= \dfrac{16x^{16}y^4}{25}$$

5. a. $x^{-2a}(3x^a)^3 = x^{-2a} \cdot 3^3 \cdot x^{a \cdot 3}$
$$= 27x^{-2a+3a}$$
$$= 27x^a$$

b. $\dfrac{(y^{3b})^3}{y^{4b-3}} = \dfrac{y^{9b}}{y^{4b-3}}$
$$= y^{9b-(4b-3)}$$
$$= y^{9b-4b+3}$$
$$= y^{5b+3}$$

6. a. $(3.4 \times 10^4)(5 \times 10^{-7})$
$$= 3.4 \times 5 \times 10^4 \times 10^{-7}$$
$$= 17.0 \times 10^{-3}$$
$$= (1.7 \times 10^1) \times 10^{-3}$$
$$= 1.7 \times 10^{-2}$$

b. $\dfrac{1.6 \times 10^8}{4 \times 10^{-2}} = \left(\dfrac{1.6}{4}\right)\left(\dfrac{10^8}{10^{-2}}\right)$
$$= 0.4 \times 10^{8-(-2)}$$
$$= (4 \times 10^{-1}) \times 10^{10}$$
$$= 4 \times 10^9$$

7. $\dfrac{2400 \times 0.0000014}{800}$

$= \dfrac{(2.4 \times 10^3)(1.4 \times 10^{-6})}{8 \times 10^2}$

$= \dfrac{2.4(1.4)}{8} \cdot \dfrac{10^3 \cdot 10^{-6}}{10^2}$

$= 0.42 \times 10^{-5}$

$= (4.2 \times 10^{-1}) \times 10^{-5}$

$= 4.2 \times 10^{-6}$

Vocabulary, Readiness & Video Check 5.2

1. $(x^4)^5 = x^{4(5)} = x^{20}$

2. $(5^6)^2 = 5^{6(2)} = 5^{12}$

3. $x^4 \cdot x^5 = x^{4+5} = x^9$

4. $x^7 \cdot x^8 = x^{7+8} = x^{15}$

5. $(y^6)^7 = y^{6(7)} = y^{42}$

6. $(x^3)^4 = x^{3(4)} = x^{12}$

7. $(z^4)^9 = z^{4(9)} = z^{36}$

8. $(z^3)^7 = z^{3(7)} = z^{21}$

9. $(z^{-6})^{-3} = z^{-6(-3)} = z^{18}$

10. $(y^{-4})^{-2} = y^{-4(-2)} = y^8$

11. The power rule involves a power of a base raised to a power and exponents are multiplied; the product rule involves a product of like bases and exponents are added.

12. power of a quotient, power of a product, power rule, negative exponent, quotient rule

13. We are asked to write the answer in scientific notation. The first product isn't because the number multiplied by the power of 10 is not between 1 and 10.

Exercise Set 5.2

1. $(3^{-1})^2 = 3^{-2} = \dfrac{1}{3^2} = \dfrac{1}{9}$

3. $(x^4)^{-9} = x^{-36} = \dfrac{1}{x^{36}}$

5. $(3x^2 y^3)^2 = 3^2 (x^2)^2 (y^3)^2 = 9x^4 y^6$

7. $\left(\dfrac{2x^5}{y^{-3}}\right)^4 = (2x^5 y^3)^4$

$= 2^4 (x^5)^4 (y^3)^4$

$= 16x^{20} y^{12}$

9. $(2a^2 bc^{-3})^{-6} = 2^{-6} (a^2)^{-6} b^{-6} (c^{-3})^{-6}$

$= 2^{-6} a^{-12} b^{-6} c^{18}$

$= \dfrac{c^{18}}{2^6 a^{12} b^6}$

$= \dfrac{c^{18}}{64 a^{12} b^6}$

11. $\left(\dfrac{x^7 y^{-3}}{z^{-4}}\right)^{-5} = \dfrac{x^{7(-5)} y^{-3(-5)}}{z^{-4(-5)}}$

$= \dfrac{x^{-35} y^{15}}{z^{20}}$

$= \dfrac{y^{15}}{x^{35} z^{20}}$

13. $(-2^{-2} y^{-1})^{-3} = [-1 \cdot (2)^{-2} y^{-1}]^{-3}$

$= (-1)^{-3} (2^{-2})^{-3} (y^{-1})^{-3}$

$= (-1) \cdot 2^6 y^3$

$= -64 y^3$

15. $\left(\dfrac{a^{-4}}{a^{-5}}\right)^{-2} = (a^{-4-(-5)})^{-2}$

$= (a^1)^{-2}$

$= a^{-2}$

$= \dfrac{1}{a^2}$

17. $\left(\dfrac{6p^6}{p^{12}}\right)^2 = (6p^{6-12})^2$

$$= (6p^{-6})^2$$
$$= 6^2 p^{-12}$$
$$= \dfrac{36}{p^{12}}$$

19. $(-8y^3xa^{-2})^{-3} = [-1 \cdot 8y^3xa^{-2}]^{-3}$

$$= (-1)^{-3} \cdot 8^{-3} y^{-9} x^{-3} a^6$$
$$= -\dfrac{a^6}{8^3 x^3 y^9}$$
$$= -\dfrac{a^6}{512 x^3 y^9}$$

21. $\left(\dfrac{3}{4}\right)^{-3} = \dfrac{3^{-3}}{4^{-3}} = \dfrac{4^3}{3^3} = \dfrac{64}{27}$

23. $\left(\dfrac{2a^{-2}b^5}{4a^2b^7}\right)^{-2} = \left(\dfrac{1}{2}a^{-4}b^{-2}\right)^{-2}$

$$= \left(\dfrac{1}{2}\right)^{-2} a^8 b^4$$
$$= \dfrac{1^{-2} a^8 b^4}{2^{-2}}$$
$$= \dfrac{2^2 a^8 b^4}{1^2}$$
$$= 4a^8 b^4$$

25. $\left(\dfrac{x^{-2}y^{-2}}{a^{-3}}\right)^{-7} = \dfrac{(x^{-2})^{-7}(y^{-2})^{-7}}{(a^{-3})^{-7}} = \dfrac{x^{14}y^{14}}{a^{21}}$

27. $(y^{-5})^2 = y^{-5 \cdot 2} = y^{-10} = \dfrac{1}{y^{10}}$

29. $(5^{-1})^3 = 5^{-3} = \dfrac{1}{5^3} = \dfrac{1}{125}$

31. $(x^7)^{-9} = x^{7 \cdot (-9)} = x^{-63} = \dfrac{1}{x^{63}}$

33. $\left(\dfrac{x^4}{y^{-3}}\right)^{-5} = \dfrac{(x^4)^{-5}}{(y^{-3})^{-5}} = \dfrac{x^{-20}}{y^{15}} = \dfrac{1}{x^{20}y^{15}}$

35. $(4x^2)^2 = 4^2(x^2)^2 = 16x^4$

37. $\left(\dfrac{4^{-4}}{y^3x}\right)^{-2} = \dfrac{(4^{-4})^{-2}}{(y^3)^{-2}x^{-2}} = \dfrac{4^8}{y^{-6}x^{-2}} = 4^8 x^2 y^6$

39. $\left(\dfrac{2x^{-3}}{y^{-1}}\right)^{-3} = \dfrac{2^{-3}(x^{-3})^{-3}}{(y^{-1})^{-3}}$

$$= \dfrac{2^{-3}x^9}{y^3}$$
$$= \dfrac{x^9}{2^3 y^3}$$
$$= \dfrac{x^9}{8y^3}$$

41. $\dfrac{4^{-1}x^2yz}{x^{-2}yz^3} = 4^{-1}x^{2-(-2)}y^{1-1}z^{1-3}$

$$= \dfrac{1}{4}x^4 y^0 z^{-2}$$
$$= \dfrac{x^4}{4z^2}$$

43. $\left(\dfrac{3x^5}{6x^4}\right)^4 = \left(\dfrac{1}{2}x\right)^4$

$$= \left(\dfrac{1}{2}\right)^4 x^4$$
$$= \dfrac{1^4}{2^4}x^4$$
$$= \dfrac{1}{16}x^4$$
$$= \dfrac{x^4}{16}$$

45. $\dfrac{(y^3)^{-4}}{y^3} = \dfrac{y^{-12}}{y^3} = y^{-12-3} = y^{-15} = \dfrac{1}{y^{15}}$

47. $\dfrac{3^{-2}a^{-5}b^6}{4^{-2}a^{-7}b^{-3}} = \left(\dfrac{3^{-2}}{4^{-2}}\right)a^{-5-(-7)}b^{6-(-3)}$

$$= \dfrac{4^2}{3^2}a^2 b^9$$
$$= \dfrac{16a^2 b^9}{9}$$

49. $(4x^6 y^5)^{-2}(6x^4 y^3)$

$= 4^{-2}(x^6)^{-2}(y^5)^{-2} \cdot 6x^4 y^3$

$= \dfrac{1}{4^2} x^{-12} y^{-10} \cdot 6x^4 y^3$

$= \dfrac{6}{16} x^{-12+4} y^{-10+3}$

$= \dfrac{3}{8} x^{-8} y^{-7}$

$= \dfrac{3}{8x^8 y^7}$

51. $x^6 (x^6 bc)^{-6} = x^6 \cdot (x^6)^{-6}(b)^{-6}(c)^{-6}$

$= x^6 x^{-36} b^{-6} c^{-6}$

$= x^{6-36} b^{-6} c^{-6}$

$= x^{-30} b^{-6} c^{-6}$

$= \dfrac{1}{x^{30} b^6 c^6}$

53. $\dfrac{2^{-3} x^2 y^{-5}}{5^{-2} x^7 y^{-1}} = \dfrac{5^2}{2^3} x^{2-7} y^{-5-(-1)}$

$= \dfrac{25}{8} x^{-5} y^{-4}$

$= \dfrac{25}{8x^5 y^4}$

55. $\left(\dfrac{2x^2}{y^4}\right)^3 \left(\dfrac{2x^5}{y}\right)^{-2} = \dfrac{2^3 x^6}{y^{12}} \cdot \dfrac{2^{-2} x^{-10}}{y^{-2}}$

$= \dfrac{2^{3+(-2)} x^{6+(-10)}}{y^{12+(-2)}}$

$= \dfrac{2^1 x^{-4}}{y^{10}}$

$= \dfrac{2}{x^4 y^{10}}$

57. $(x^{3a+6})^3 = x^{(3a+6)\cdot 3} = x^{9a+18}$

59. $\dfrac{x^{4a}(x^{4a})^3}{x^{4a-2}} = \dfrac{x^{4a} \cdot x^{12a}}{x^{4a-2}}$

$= \dfrac{x^{4a+12a}}{x^{4a-2}}$

$= x^{16a-(4a-2)}$

$= x^{12a+2}$

61. $(b^{5x-2})^{2x} = b^{(5x-2)\cdot 2x} = b^{10x^2 - 4x}$

63. $\dfrac{(y^{2a})^8}{y^{a-3}} = \dfrac{y^{16a}}{y^{a-3}} = y^{16a-(a-3)} = y^{15a+3}$

65. $\left(\dfrac{2x^{3t}}{x^{2t-1}}\right)^4 = \dfrac{2^4 x^{3t(4)}}{x^{(2t-1)\cdot 4}}$

$= \dfrac{16x^{12t}}{x^{8t-4}}$

$= 16x^{12t-(8t-4)}$

$= 16x^{4t+4}$

67. $\dfrac{25x^{2a+1} y^{a-1}}{5x^{3a+1} y^{2a-3}} = \left(\dfrac{25}{5}\right)\left(\dfrac{x^{2a+1}}{x^{3a+1}}\right)\left(\dfrac{y^{a-1}}{y^{2a-3}}\right)$

$= 5x^{2a+1-(3a+1)} y^{a-1-(2a-3)}$

$= 5x^{2a+1-3a-1} y^{a-1-2a+3}$

$= 5x^{-a} y^{-a+2}$

69. $(5\times10^{11})(2.9\times10^{-3}) = 5\times 2.9\times10^{11+(-3)}$

$= 14.5\times10^8$

$= 1.45\times10^1 \times10^8$

$= 1.45\times10^9$

71. $(2\times10^5)^3 = 2^3 \times10^{5(3)} = 8\times10^{15}$

73. $\dfrac{3.6\times10^{-4}}{9\times10^2} = \dfrac{3.6}{9}\times10^{-4-2}$

$= 0.4\times10^{-6}$

$= 4\times10^{-1}\times10^{-6}$

$= 4\times10^{-7}$

75. $\dfrac{0.0069}{0.023} = \dfrac{6.9\times10^{-3}}{2.3\times10^{-2}}$

$= \dfrac{6.9}{2.3}\times10^{-3-(-2)}$

$= 3\times10^{-1}$

77. $\dfrac{18,200\times100}{91,000} = \dfrac{1.82\times10^4 \times1\times10^2}{9.1\times10^4}$

$= \dfrac{1.82\times10^6}{9.1\times10^4}$

$= 0.2\times10^{6-4}$

$= 2\times10^{-1}\times10^2$

$= 2\times10^{-1+2}$

$= 2\times10^1$

79. $\dfrac{6000 \times 0.006}{0.009 \times 400} = \dfrac{6 \times 10^3 \times 6 \times 10^{-3}}{9 \times 10^{-3} \times 4 \times 10^2}$

$= \dfrac{36 \times 10^0}{36 \times 10^{-1}}$

$= 1 \times 10^{0-(-1)}$

$= 1 \times 10^1$

81. $\dfrac{0.00064 \times 2000}{16,000} = \dfrac{6.4 \times 10^{-4} \times 2 \times 10^3}{1.6 \times 10^4}$

$= \dfrac{12.8 \times 10^{-1}}{1.6 \times 10^4}$

$= 8 \times 10^{-1-4}$

$= 8 \times 10^{-5}$

83. $\dfrac{66,000 \times 0.001}{0.002 \times 0.003} = \dfrac{6.6 \times 10^4 \times 1 \times 10^{-3}}{2 \times 10^{-3} \times 3 \times 10^{-3}}$

$= \dfrac{6.6 \times 10^1}{6 \times 10^{-6}}$

$= 1.1 \times 10^{1-(-6)}$

$= 1.1 \times 10^7$

85. $\dfrac{9.24 \times 10^{15}}{(2.2 \times 10^{-2})(1.2 \times 10^{-5})}$

$= \dfrac{9.24}{(2.2)(1.2)} \cdot 10^{15-(-2)-(-5)}$

$= 3.5 \times 10^{22}$

87. $(6.452 \times 10^{-4})(4 \times 10^{-2}) = 25.808 \times 10^{-6}$

$= (2.5808 \times 10^1) \times 10^{-6}$

$= 2.5808 \times 10^{-5}$

The area of the square is 2.5808×10^{-5} square meter.

89. $-5y + 4y - 18 - y = -2y - 18$

91. $-3x - (4x - 2) = -3x - 4x + 2 = -7x + 2$

93. $3(z-4) - 2(3z+1) = 3z - 12 - 6z - 2$

$= -3z - 14$

95. $\left(\dfrac{2x^{-2}}{y}\right)^3 = \dfrac{2^3 x^{-2(3)}}{y^3} = \dfrac{8x^{-6}}{y^3} = \dfrac{8}{x^6 y^3}$

The volume is $\dfrac{8}{x^6 y^3}$ cubic meters.

97. $D = \dfrac{500,000}{250}$

$= \dfrac{5 \times 10^5}{2.5 \times 10^2}$

$= 2 \times 10^{5-2}$

$= 2 \times 10^3$

Its density is 2×10^3 pounds per cubic feet.

99. $a^{-1} = a^1$

$\dfrac{1}{a} = a$

$1 = a^2$

$(1)^2 = 1$ and $(-1)^2 = 1$

Yes; $a = 1$ and $a = -1$.

101. no; answers may vary

103. $\dfrac{3.15 \times 10^8}{3.536 \times 10^6} = \dfrac{3.15}{3.536} \times \dfrac{10^8}{10^6} \approx 0.89 \times 10^2 = 89$

The population density of the U.S. in 2012 was 89 people per square mile.

105. $\dfrac{2.21 \times 10^{11}}{3.11 \times 10^8} = \dfrac{2.21}{3.11} \times \dfrac{10^{11}}{10^8} \approx 0.711 \times 10^3 = 711$

The average value of imports in the U.S. in March 2011 was $711 per person.

Section 5.3 Practice Exercises

1. a. $4x^5$: The exponent on x is 5, so the degree of the term is 5.

b. $-4^3 y^3$: The exponent on y is 3, so the degree of the term is 3.

c. The degree of z, or z^1, is 1.

d. $65a^3 b^7 c$: The degree is the sum of the exponents on the variables, or $3 + 7 + 1 = 11$.

e. The degree of 36, which can be written as $36x^0$, is 0.

	Polynomial	Degree	Classification
2. a.	$3x^4 + 2x^2 - 3$	4	trinomial
b.	$9abc^3$	5	monomial
c.	$8x^5 + 5x^3$	5	binomial

3.

Term	Degree
$2x^3 y$	4
$-3x^3 y^2$	5
$-9y^5$	5
9.6	0

The largest degree of any term is 5, so the degree of this polynomial is 5.

4. a. Substitute -1 for x and simplify.

$$P(x) = -5x^2 + 2x - 8$$
$$P(-1) = -5(-1)^2 + 2(-1) - 8$$
$$= -5 - 2 - 8$$
$$= -15$$

b. Substitute 3 for x and simplify.

$$P(x) = -5x^2 + 2x - 8$$
$$P(3) = -5(3)^2 + 2(3) - 8$$
$$= -45 + 6 - 8$$
$$= -47$$

5. a. $P(t) = -16t^2 + 290$

Let $t = 0$: $P(0) = -16(0)^2 + 290 = 290$

Let $t = 2$: $P(2) = -16(2)^2 + 290 = 226$

The height of the object at $t = 0$ second is 290 feet and the height at $t = 2$ seconds is 226 feet.

b.

Since $P(4) = 34$ and $P(5) = -110$, and 34 is closer to 0 than -110, the object hits the ground after 4 seconds, to the nearest second.

c. Graph y_1 in a $[-1, 10, 1]$ by $[-100, 500, 100]$ window.

To the nearest tenth, the object hits the ground after 4.3 seconds.

6. a. $8x^4 - 5x^4 - 5x = (8-5)x^4 - 5x$
$$= 3x^4 - 5x$$

b. $4ab - 5b + 3ab + 2b$
$$= 4ab + 3ab - 5b + 2b$$
$$= (4+3)ab + (-5+2)b$$
$$= 7ab - 3b$$

7. $(5x^3 - 3x^2 - 9x - 8) + (x^3 + 9x^2 + 2x)$
$$= 5x^3 + x^3 - 3x^2 + 9x^2 - 9x + 2x - 8$$
$$= 6x^3 + 6x^2 - 7x - 8$$

8. a. $(3a^4 b - 5ab^2 + 7) + (9ab^2 - 12)$
$$= 3a^4 b - 5ab^2 + 7 + 9ab^2 - 12$$
$$= 3a^4 b - 5ab^2 + 9ab^2 + 7 - 12$$
$$= 3a^4 b + 4ab^2 - 5$$

b. $(2x^5 - 3y + x - 6) + (4y - 2x - 3)$
$$= 2x^5 - 3y + x - 6 + 4y - 2x - 3$$
$$= 2x^5 - 3y + 4y + x - 2x - 6 - 3$$
$$= 2x^5 + y - x - 9$$

9. $(13a^4 - 7a^3 - 9) - (-2a^4 + 8a^3 - 12)$
$$= 13a^4 - 7a^3 - 9 + 2a^4 - 8a^3 + 12$$
$$= 13a^4 + 2a^4 - 7a^3 - 8a^3 - 9 + 12$$
$$= 15a^4 - 15a^3 + 3$$

10. $(11x^2 y^2 - 7xy^2) - (5x^2 y^2 - 3xy^2 + 5y^3)$
$$= 11x^2 y^2 - 7xy^2 - 5x^2 y^2 + 3xy^2 - 5y^3$$
$$= 6x^2 y^2 - 4xy^2 - 5y^3$$

11. The degree of $f(x)$ is 3, which means that its graph has the shape of **C**.

12. a. The y-intercept is $(0, -12)$. Since the degree of the function $f(x) = x^2 - x - 12$ is 2, the graph is U-shaped. Since the coefficient of x^2 is 1, the graph opens upward.

b. The y-intercept is $(0, 6)$. Since the degree of the function $g(x) = -3x^2 + 5x + 6$ is 2, the graph is U-shaped. Since the coefficient of x^2 is -3, the graph opens downward.

c. The y-intercept is $(0, 4)$. Since the degree of $h(x) = x^3 + 4$ is 3, the graph is S-shaped.

d. The y-intercept is $(0, -1)$. Since the degree of $f(x) = -x^3 + 2x^2 - 5x - 1$ is 3, the graph is S-shaped.

Vocabulary, Readiness & Video Check 5.3

1. The numerical factor of a term is the <u>coefficient</u>.

2. A <u>polynomial</u> is a finite sum of terms in which all variables are raised to nonnegative integer powers and no variables appear in any denominator.

3. A <u>binomial</u> is a polynomial with 2 terms.

4. A <u>monomial</u> is a polynomial with 1 term.

5. A <u>trinomial</u> is a polynomial with 3 terms.

6. The degree of a term is the sum of the exponents on the <u>variables</u> in the term.

7. The <u>degree</u> of a polynomial is the largest degree of all its terms.

8. <u>Like</u> terms contain the same variables raised to the same powers.

9. We are finding the degree of the polynomial, which is the greatest degree of any of its terms, so we find the degree of each term first.

10. evaluating a polynomial at a given replacement value

11. by combining like terms

12. combine like terms

13. Change the operation from subtraction to addition. Then add the opposite of the polynomial that is being subtracted—that is, change the signs of all terms of the polynomial being subtracted and add.

14. C; degree two tells you it's a parabola and its negative coefficient tells you the parabola opens downward.

Exercise Set 5.3

1. 4 has degree 0.

3. $5x^2$ has degree 2.

5. $-3xy^2$ has degree $1 + 2 = 3$.

7. $-8^7 y^3$ has degree 3 (note: the degree on the *variable* is 3).

9. $3.78ab^3c^5$ has degree $1 + 3 + 5 = 9$.

11. $6x + 0.3$ has degree 1 and is a binomial.

13. $3x^2 - 2x + 5$ has degree 2 and is a trinomial.

15. $-3^4 xy^2$ has degree $1 + 2 = 3$ and is a monomial.

17.

Term	Degree
$x^2 y$	3
$-4xy^2$	3
$5x$	1
y^4	4

19. $P(x) = x^2 + x + 1$
$P(7) = 7^2 + 7 + 1 = 49 + 7 + 1 = 57$

21. $Q(x) = 5x^2 - 1$
$Q(-10) = 5(-10)^2 - 1$
$= 5(100) - 1$
$= 500 - 1$
$= 499$

23. $Q(x) = 5x^2 - 1$

$$Q\left(\frac{1}{4}\right) = 5\left(\frac{1}{4}\right)^2 - 1$$

$$= 5\left(\frac{1}{16}\right) - 1$$

$$= \frac{5}{16} - \frac{16}{16}$$

$$= -\frac{11}{16}$$

25. $P(t) = -16t^2 + 1125$

$P(2) = -16(2)^2 + 1125 = 1061$

After 2 seconds, the height is 1061 feet.

27. $P(t) = -16t^2 + 1125$

$P(6) = -16(6)^2 + 1125 = 549$

After 6 seconds, the height is 549 feet.

29. $5y + y - 6y^2 - y^2 = (5+1)y + (-6-1)y^2$

$$= 6y - 7y^2 \text{ or } -7y^2 + 6y$$

31. $4x^2y + 2x - 3x^2y - \dfrac{1}{2} - 7x$

$$= 4x^2y - 3x^2y + 2x - 7x - \frac{1}{2}$$

$$= x^2y - 5x - \frac{1}{2}$$

33. $7x^2 - 2xy + 5y^2 - x^2 + xy + 11y^2$

$$= 7x^2 - x^2 - 2xy + xy + 5y^2 + 11y^2$$

$$= (7-1)x^2 + (-2+1)xy + (5+11)y^2$$

$$= 6x^2 - xy + 16y^2$$

35. $(9y^2 + y - 8) + (9y^2 - y - 9)$

$$= 9y^2 + y - 8 + 9y^2 - y - 9$$

$$= 9y^2 + 9y^2 + y - y - 8 - 9$$

$$= 18y^2 - 17$$

37. $(x^2 + xy - y^2) + (2x^2 - 4xy + 7y^2)$

$$= x^2 + xy - y^2 + 2x^2 - 4xy + 7y^2$$

$$= x^2 + 2x^2 + xy - 4xy - y^2 + 7y^2$$

$$= 3x^2 - 3xy + 6y^2$$

39.
$$\begin{array}{r} x^2 - 6x + 3 \\ +\quad (2x + 5) \\ \hline x^2 - 4x + 8 \end{array}$$

41. $(7x^3y - 4xy + 8) + (5x^3y + 4xy + 8x)$

$$= 7x^3y - 4xy + 8 + 5x^3y + 4xy + 8x$$

$$= 7x^3y + 5x^3y - 4xy + 4xy + 8x + 8$$

$$= 12x^3y + 8x + 8$$

43. $(0.6x^3 + 1.2x^2 - 4.5x + 9.1) + (3.9x^3 - x^2 + 0.7x)$

$$= 0.6x^3 + 1.2x^2 - 4.5x + 9.1 + 3.9x^3 - x^2 + 0.7x$$

$$= 0.6x^3 + 3.9x^3 + 1.2x^2 - x^2 - 4.5x + 0.7x + 9.1$$

$$= 4.5x^3 + 0.2x^2 - 3.8x + 9.1$$

45. $(9y^2 - 7y + 5) - (8y^2 - 7y + 2)$

$$= 9y^2 - 7y + 5 - 8y^2 + 7y - 2$$

$$= 9y^2 - 8y^2 - 7y + 7y + 5 - 2$$

$$= y^2 + 3$$

47. $(4x^2 + 2x) - (6x^2 - 3x) = 4x^2 + 2x - 6x^2 + 3x$

$$= 4x^2 - 6x^2 + 2x + 3x$$

$$= -2x^2 + 5x$$

49.
$$\begin{array}{r} 6y^2 - 6y + 4 \\ -(-y^2 + 6y + 7) \end{array} \rightarrow \begin{array}{r} 6y^2 - 6y + 4 \\ +\ y^2 - 6y - 7 \\ \hline 7y^2 - 12y - 3 \end{array}$$

51. $(9x^3 - 2x^2 + 4x - 7) - (2x^3 - 6x^2 - 4x + 3)$

$$= 9x^3 - 2x^2 + 4x - 7 - 2x^3 + 6x^2 + 4x - 3$$

$$= 9x^3 - 2x^3 - 2x^2 + 6x^2 + 4x + 4x - 7 - 3$$

$$= 7x^3 + 4x^2 + 8x - 10$$

53. $\left(-19y^2 + 7yx + \dfrac{1}{7}\right) - \left(y^2 + 4yx + \dfrac{1}{7}\right)$

$$= -19y^2 + 7yx + \frac{1}{7} - y^2 - 4yx - \frac{1}{7}$$

$$= -19y^2 - y^2 + 7yx - 4yx + \frac{1}{7} - \frac{1}{7}$$

$$= -20y^2 + 3yx$$

55. $(-3x + 8) + (-3x^2 + 3x - 5)$

$$= -3x + 8 - 3x^2 + 3x - 5$$

$$= -3x^2 - 3x + 3x + 8 - 5$$

$$= -3x^2 + 3$$

57. $(5y^4 - 7y^2 + x^2 - 3) + (-3y^4 + 2y^2 + 4)$
$= 5y^4 - 7y^2 + x^2 - 3 - 3y^4 + 2y^2 + 4$
$= 5y^4 - 3y^4 - 7y^2 + 2y^2 + x^2 - 3 + 4$
$= 2y^4 - 5y^2 + x^2 + 1$

59. $(4x^2 - 6x + 2) - (-x^2 + 3x + 5)$
$= 4x^2 - 6x + 2 + x^2 - 3x - 5$
$= 4x^2 + x^2 - 6x - 3x + 2 - 5$
$= 5x^2 - 9x - 3$

61. $(5x^2 + x + 9) - (2x^2 - 9) = 5x^2 + x + 9 - 2x^2 + 9$
$= 5x^2 - 2x^2 + x + 9 + 9$
$= 3x^2 + x + 18$

63. $(5x - 11) + (-x - 2) = 5x - 11 - x - 2$
$= 5x - x - 11 - 2$
$= 4x - 13$

65. $(3x^3 - b + 2a - 6) + (-4x^3 + b + 6a - 6)$
$= 3x^3 - b + 2a - 6 - 4x^3 + b + 6a - 6$
$= 3x^3 - 4x^3 - b + b + 2a + 6a - 6 - 6$
$= -x^3 + 8a - 12$

67. $(14ab - 10a^2b + 6b^2) - (18a^2 - 20a^2b - 6b^2)$
$= 14ab - 10a^2b + 6b^2 - 18a^2 + 20a^2b + 6b^2$
$= 14ab - 10a^2b + 20a^2b - 18a^2 + 6b^2 + 6b^2$
$= 14ab + 10a^2b - 18a^2 + 12b^2$

69. $\quad 3x^2 + 15x + 8$
$\underline{+ (2x^2 + 7x + 8)}$
$\quad 5x^2 + 22x + 16$

71. $(7x^2 - 5) + (-3x^2 - 2) - (4x^2 - 7)$
$= 7x^2 - 5 - 3x^2 - 2 - 4x^2 + 7$
$= 7x^2 - 3x^2 - 4x^2 - 5 - 2 + 7$
$= 0$

73. $(-3 + 4x^2 + 7xy^2) + (2x^3 - x^2 + xy^2)$
$= -3 + 4x^2 + 7xy^2 + 2x^3 - x^2 + xy^2$
$= 7xy^2 + xy^2 + 2x^3 + 4x^2 - x^2 - 3$
$= 8xy^2 + 2x^3 + 3x^2 - 3$

75. $\quad 3x^2 - 4x + 8$
$\underline{- \quad (5x - 7)}$
Is equivalent to:
$\quad 3x^2 - 4x + 8$
$\underline{+ \quad (-5x + 7)}$
$\quad 3x^2 - 9x + 15$

77. $(7x^2 + 4x + 9) + (8x^2 + 7x - 8) - (3x + 7)$
$= 7x^2 + 4x + 9 + 8x^2 + 7x - 8 - 3x - 7$
$= 7x^2 + 8x^2 + 4x + 7x - 3x + 9 - 8 - 7$
$= 15x^2 + 8x - 6$

79. $\left(\dfrac{2}{3}x^2 - \dfrac{1}{6}x + \dfrac{5}{6}\right) - \left(\dfrac{1}{3}x^2 + \dfrac{5}{6}x - \dfrac{1}{6}\right)$
$= \dfrac{2}{3}x^2 - \dfrac{1}{6}x + \dfrac{5}{6} - \dfrac{1}{3}x^2 - \dfrac{5}{6}x + \dfrac{1}{6}$
$= \dfrac{2}{3}x^2 - \dfrac{1}{3}x^2 - \dfrac{1}{6}x - \dfrac{5}{6}x + \dfrac{5}{6} + \dfrac{1}{6}$
$= \dfrac{1}{3}x^2 - x + 1$

81. If $L = 5$, $W = 4$, and $H = 9$, then
$2HL + 2LW + 2HW$
$= 2(9)(5) + 2(5)(4) + 2(9)(4)$
$= 90 + 40 + 72$
$= 202$
The surface area is 202 square inches.

83. $P(t) = -16t^2 + 300t$

  a. $P(1) = 284$ feet

  b. $P(2) = 536$ feet

  c. $P(3) = 756$ feet

  d. $P(4) = 944$ feet

  e.
  answers may vary

  f.

  Since $P(18) = 216$ and $P(19) = -76$, the projectile hits the ground after 19 seconds, to the nearest second.

85. $P(x) = 45x - 100,000$

$P(4000) = 45(4000) - 100,000$

$\qquad = 180,000 - 100,000$

$\qquad = 80,000$

The profit from selling 4000 computer briefcases is $80,000.

87. $R(x) = 11x$

$R(1500) = 11(1500) = 16,500$

The revenue from selling 1500 boxes is $16,500.

89. A: The degree of $f(x)$ is 2, so the graph has the shape of A or C. The coefficient of x^2 is a positive number, so the graph has the shape of A.

91. D: The degree of $g(x)$ is 3, so the graph has the shape of B or D. The coefficient of x^3 is a negative number, so the graph has the shape of D.

93. $5(3x - 2) = 15x - 10$

95. $-2(x^2 - 5x + 6) = -2x^2 + 10x - 12$

97. The opposite of $8x - 6$ is $-(8x - 6)$ or $-8x + 6$; a and c.

99. $(12x - 1.7) - (15x + 6.2)$

$= 12x - 1.7 - 15x - 6.2$

$= -3x - 7.9$

101. answers may vary

103. answers may vary

105. $(4x^{2a} - 3x^a + 0.5) - (x^{2a} - 5x^a - 0.2)$

$= 4x^{2a} - 3x^a + 0.5 - x^{2a} + 5x^a + 0.2$

$= 3x^{2a} + 2x^a + 0.7$

107. $(8x^{2y} - 7x^y + 3) + (-4x^{2y} + 9x^y - 14)$

$= 8x^{2y} - 7x^y + 3 - 4x^{2y} + 9x^y - 14$

$= 4x^{2y} + 2x^y - 11$

109. $P = 2l + 2w$

$= 2(3x^2 - x + 2y) + 2(x + 5y)$

$= 6x^2 - 2x + 4y + 2x + 10y$

$= 6x^2 + 14y$

The perimeter is $P = (6x^2 + 14y)$ units.

111. $P(x) + Q(x) = (3x + 3) + (4x^2 - 6x + 3)$

$\qquad = 3x + 3 + 4x^2 - 6x + 3$

$\qquad = 4x^2 - 3x + 6$

113. $Q(x) - R(x) = (4x^2 - 6x + 3) - (5x^2 - 7)$

$\qquad = 4x^2 - 6x + 3 - 5x^2 + 7$

$\qquad = -x^2 - 6x + 10$

115. $2[Q(x)] - R(x)$

$= 2(4x^2 - 6x + 3) - (5x^2 - 7)$

$= 8x^2 - 12x + 6 - 5x^2 + 7$

$= 3x^2 - 12x + 13$

117. $3[R(x)] + 4[P(x)] = 3(5x^2 - 7) + 4(3x + 3)$

$\qquad = 15x^2 - 21 + 12x + 12$

$\qquad = 15x^2 + 12x - 9$

119. $P(x) = 2x - 3$

   a. $P(a) = 2a - 3$

   b. $P(-x) = 2(-x) - 3 = -2x - 3$

   c. $P(x + h) = 2(x + h) - 3 = 2x + 2h - 3$

121. $P(x) = 4x$

   a. $P(a) = 4a$

   b. $P(-x) = 4(-x) = -4x$

   c. $P(x + h) = 4(x + h) = 4x + 4h$

123. $P(x) = 4x - 1$

   a. $P(a) = 4a - 1$

   b. $P(-x) = 4(-x) - 1 = -4x - 1$

   c. $P(x + h) = 4(x + h) - 1 = 4x + 4h - 1$

125. $f(x) = 0.19x^2 + 5.67x + 43.7$

   a. 2005 is 35 years after 1970, so $x = 35$.

$\qquad f(35) = 0.19(35)^2 + 5.67(35) + 43.7 = 474.9$

The sales in 2005 were approximately 474.9 billion dollars.

b. 2010 is 40 years after 1970, so $x = 40$.

$f(40) = 0.19(40)^2 + 5.67(40) + 43.7 = 574.5$

The sales in 2010 were approximately 574.5 billion dollars.

c. 2015 is 45 years after 1970, so $x = 45$.

$f(45) = 0.19(45)^2 + 5.67(45) + 43.7 = 683.6$

The sales in 2015 were approximately 683.6 billion dollars.

d. answers may vary

Section 5.4 Practice Exercises

1. a. $(3x^4)(2x^2) = 3(2)(x^4)(x^2) = 6x^6$

b. $(-5m^4np^3)(-8mnp^5)$
$= -5(-8)(m^4m)(n \cdot n)(p^3p^5)$
$= 40m^5n^2p^8$

2. a. $3x(7x-1) = 3x(7x) + 3x(-1)$
$\qquad\qquad = 21x^2 - 3x$

b. $-5a^2(3a^2 - 6a + 5)$
$= -5a^2(3a^2) + (-5a^2)(-6a)$
$\qquad + (-5a^2)(5)$
$= -15a^4 + 30a^3 - 25a^2$

c. $-mn^3(5m^2n^2 + 2mn - 5m)$
$= -mn^3(5m^2n^2) + (-mn^3)(2mn)$
$\qquad + (-mn^3)(-5m)$
$= -5m^3n^5 - 2m^2n^4 + 5m^2n^3$

3. a. $(x+5)(2x+3) = x(2x+3) + 5(2x+3)$
$\qquad\qquad\qquad = 2x^2 + 3x + 10x + 15$
$\qquad\qquad\qquad = 2x^2 + 13x + 15$

b. $(3x-1)(x^2 - 6x + 2)$
$= 3x(x^2 - 6x + 2) + (-1)(x^2 - 6x + 2)$
$= 3x^3 - 18x^2 + 6x - x^2 + 6x - 2$
$= 3x^3 - 19x^2 + 12x - 2$

4.
$$\begin{array}{r} x^2 - 4x - 5 \\ 3x^2 + 2 \\ \hline 2x^2 - 8x - 10 \\ 3x^4 - 12x^3 - 15x^2 \\ \hline 3x^4 - 12x^3 - 13x^2 - 8x - 10 \end{array}$$

5. $(x-5)(x+3)$
$= x \cdot x + 3 \cdot x + (-5)x + (-5)(3)$
$= x^2 + 3x - 5x - 15$
$= x^2 - 2x - 15$

6. a. $(3x-5)(2x-7)$
$= 3x(2x) + 3x(-7) + (-5)(2x)$
$\qquad + (-5)(-7)$
$= 6x^2 - 21x - 10x + 35$
$= 6x^2 - 31x + 35$

b. $(2x^2 - 3y)(4x^2 + y)$
$= 8x^4 + 2x^2y - 12x^2y - 3y^2$
$= 8x^4 - 10x^2y - 3y^2$

7. a. $(x+6)^2 = x^2 + 2 \cdot x \cdot 6 + 6^2$
$\qquad\qquad = x^2 + 12x + 36$

b. $(x-2)^2 = x^2 - 2 \cdot x \cdot 2 + 2^2$
$\qquad\qquad = x^2 - 4x + 4$

c. $(3x+5y)^2$
$= (3x)^2 + 2(3x)(5y) + (5y)^2$
$= 9x^2 + 30xy + 25y^2$

d. $(3x^2 - 8b)^2$
$= (3x^2)^2 - 2(3x^2)(8b) + (8b)^2$
$= 9x^4 - 48x^2b + 64b^2$

8. a. $(x-7)(x+7) = x^2 - 7^2 = x^2 - 49$

b. $(2a+5)(2a-5) = (2a)^2 - 5^2$
$\qquad\qquad\qquad\qquad = 4a^2 - 25$

c. $\left(5x^2 + \dfrac{1}{4}\right)\left(5x^2 - \dfrac{1}{4}\right) = (5x^2)^2 - \left(\dfrac{1}{4}\right)^2$
$\qquad\qquad\qquad\qquad\qquad = 25x^4 - \dfrac{1}{16}$

d. $(a^3 - 4b^2)(a^3 + 4b^2) = (a^3)^2 - (4b^2)^2$
$$= a^6 - 16b^4$$

9. $[2 + (3x - y)]^2$
$$= 2^2 + 2(2)(3x - y) + (3x - y)^2$$
$$= 4 + 4(3x - y) + (3x)^2 - 2(3x) \cdot y + y^2$$
$$= 4 + 12x - 4y + 9x^2 - 6xy + y^2$$

10. $[(3x - y) - 5][(3x - y) + 5]$
$$= (3x - y)^2 - 5^2$$
$$= (3x)^2 - 2(3x)(y) + y^2 - 25$$
$$= 9x^2 - 6xy + y^2 - 25$$

11. $(x + 4)(x - 4)(x^2 - 16)$
$$= (x^2 - 16)(x^2 - 16)$$
$$= (x^2 - 16)^2$$
$$= x^4 - 32x^2 + 256$$

12. $f(x) = x^2 - 3x + 5$
$$f(h + 1) = (h + 1)^2 - 3(h + 1) + 5$$
$$= h^2 + 2h + 1 - 3h - 3 + 5$$
$$= h^2 - h + 3$$

Vocabulary, Readiness & Video Check 5.4

1. $(6x^3)\left(\dfrac{1}{2}x^3\right) = \underline{3x^6}$; choice b

2. $(x + 7)^2 = \underline{x^2 + 14x + 49}$; choice c

3. $(x + 7)(x - 7) = \underline{x^2 - 49}$; choice b

4. The product of $(3x - 1)(4x^2 - 2x + 1)$ is a polynomial of degree 3; choice a.

5. If $f(x) = x^2 + 1$ then $f(a + 1) = \underline{(a + 1)^2 + 1}$; choice d.

6. $[x + (2y + 1)]^2 = \underline{[x + (2y + 1)][x + (2y + 1)]}$; choice b

7. distributive property, product rule

8. multiplying two binomials

9. FOIL order or distributive property

10. Multiplying using the FOIL order gives you four terms and the two middle terms will be opposites and subtract out.

11. Multiply two at a time and multiply as usual, using patterns if you recognize them. Keep multiplying until you are through.

12. No, a variable can also be replaced by a variable expression, so a function can also be evaluated for a variable expression.

Exercise Set 5.4

1. $(-4x^3)(3x^2) = -4(3)(x^3)(x^2)$
$$= -12x^{3+2}$$
$$= -12x^5$$

3. $(8.6a^4b^5c)(10ab^3c^2) = 8.6(10)(a^4a)(b^5b^3)(cc^2)$
$$= 86a^5b^8c^3$$

5. $3x(4x + 7) = 3x(4x) + 3x(7) = 12x^2 + 21x$

7. $-6xy(4x + y) = -6xy(4x) + (-6xy)(y)$
$$= -24x^2y - 6xy^2$$

9. $-4ab(xa^2 + ya^2 - 3)$
$$= -4ab(xa^2) + (-4ab)(ya^2) + (-4ab)(-3)$$
$$= -4a^3bx - 4a^3by + 12ab$$

11. $(x - 3)(2x + 4) = x(2x + 4) - 3(2x + 4)$
$$= 2x^2 + 4x - 6x - 12$$
$$= 2x^2 - 2x - 12$$

13. $(2x + 3)(x^3 - x + 2)$
$$= 2x(x^3 - x + 2) + 3(x^3 - x + 2)$$
$$= 2x^4 - 2x^2 + 4x + 3x^3 - 3x + 6$$
$$= 2x^4 + 3x^3 - 2x^2 + x + 6$$

15.

$$\begin{array}{r} 3x - 2 \\ \times\ 5x + 1 \\ \hline 3x - 2 \\ 15x^2 - 10x \\ \hline 15x^2 - 7x - 2 \end{array}$$

17.
$$
\begin{array}{r}
3m^2 + 2m - 1 \\
\times \qquad 5m + 2 \\
\hline
6m^2 + 4m - 2 \\
15m^3 + 10m^2 - 5m \qquad \\
\hline
15m^3 + 16m^2 - m - 2
\end{array}
$$

19. $-6a^2b^2(5a^2b^2 - 6a - 6b) = -6a^2b^2(5a^2b^2) + (-6a^2b^2)(-6a) + (-6a^2b^2)(-6b)$
$$= -30a^4b^4 + 36a^3b^2 + 36a^2b^3$$

21. $(x-3)(x+4) = x^2 + 4x - 3x - 12 = x^2 + x - 12$

23. $(5x - 8y)(2x - y) = 10x^2 - 5xy - 16xy + 8y^2$
$$= 10x^2 - 21xy + 8y^2$$

25. $\left(4x + \dfrac{1}{3}\right)\left(4x - \dfrac{1}{2}\right) = 16x^2 - 2x + \dfrac{4}{3}x - \dfrac{1}{6}$
$$= 16x^2 - \dfrac{2}{3}x - \dfrac{1}{6}$$

27. $(5x^2 - 2y^2)(x^2 - 3y^2) = 5x^4 - 15x^2y^2 - 2x^2y^2 + 6y^4$
$$= 5x^4 - 17x^2y^2 + 6y^4$$

29. $(x+4)^2 = x^2 + 2 \cdot x \cdot 4 + 4^2 = x^2 + 8x + 16$

31. $(6y - 1)(6y + 1) = (6y)^2 - 1^2 = 36y^2 - 1$

33. $(3x - y)^2 = (3x)^2 - 2 \cdot 3x \cdot y + y^2$
$$= 9x^2 - 6xy + y^2$$

35. $(7ab + 3c)(7ab - 3c) = (7ab)^2 - (3c)^2$
$$= 49a^2b^2 - 9c^2$$

37. $\left(3x + \dfrac{1}{2}\right)\left(3x - \dfrac{1}{2}\right) = (3x)^2 - \left(\dfrac{1}{2}\right)^2 = 9x^2 - \dfrac{1}{4}$

39. $[3 + (4b + 1)]^2 = 3^2 + 2 \cdot 3 \cdot (4b + 1) + (4b + 1)^2$
$$= 9 + 6(4b + 1) + (4b)^2 + 2 \cdot 4b \cdot 1 + 1^2$$
$$= 9 + 24b + 6 + 16b^2 + 8b + 1$$
$$= 16b^2 + 32b + 16$$

41. $[(2s - 3) - 1][(2s - 3) + 1] = (2s - 3)^2 - 1^2$
$$= (2s)^2 - 2 \cdot 2s \cdot 3 + 3^2 - 1$$
$$= 4s^2 - 12s + 9 - 1$$
$$= 4s^2 - 12s + 8$$

     241

43. $[(xy+4)-6]^2 = (xy+4)^2 - 2(xy+4)(6) + 6^2$
$$= x^2y^2 + 8xy + 16 - 12xy - 48 + 36$$
$$= x^2y^2 - 4xy + 4$$

45. $(x+y)(x-y)(x^2-y^2) = (x^2-y^2)(x^2-y^2)$
$$= (x^2)^2 - 2(x^2)(y^2) + (y^2)^2$$
$$= x^4 - 2x^2y^2 + y^4$$

47. $(x-2)^4 = (x-2)(x-2)(x-2)(x-2)$
$$= [(x-2)^2][(x-2)^2]$$
$$= [x^2 - 2\cdot x \cdot 2 + 2^2][x^2 - 2\cdot x \cdot 2 + 2^2]$$
$$= (x^2 - 4x + 4)(x^2 - 4x + 4)$$
$$= x^2(x^2-4x+4) - 4x(x^2-4x+4) + 4(x^2-4x+4)$$
$$= x^4 - 4x^3 + 4x^2 - 4x^3 + 16x^2 - 16x + 4x^2 - 16x + 16$$
$$= x^4 - 8x^3 + 24x^2 - 32x + 16$$

49. $(x-5)(x+5)(x^2+25) = (x^2-25)(x^2+25)$
$$= (x^2)^2 - 25^2$$
$$= x^4 - 625$$

51. $-8a^2b(3b^2 - 5b + 20) = -24a^2b^3 + 40a^2b^2 - 160a^2b$

53. $(6x+1)^2 = (6x)^2 + 2\cdot 6x \cdot 1 + 1^2$
$$= 36x^2 + 12x + 1$$

55. $(5x^3 + 2y)(5x^3 - 2y) = (5x^3)^2 - (2y)^2$
$$= 25x^6 - 4y^2$$

57. $(2x^3 + 5)(5x^2 + 4x + 1) = 2x^3(5x^2 + 4x + 1) + 5(5x^2 + 4x + 1)$
$$= 10x^5 + 8x^4 + 2x^3 + 25x^2 + 20x + 5$$

59. $(3x^2 + 2x - 1)^2 = (3x^2 + 2x - 1)(3x^2 + 2x - 1)$
$$= 3x^2(3x^2 + 2x - 1) + 2x(3x^2 + 2x - 1) - 1(3x^2 + 2x - 1)$$
$$= 9x^4 + 6x^3 - 3x^2 + 6x^3 + 4x^2 - 2x - 3x^2 - 2x + 1$$
$$= 9x^4 + 12x^3 - 2x^2 - 4x + 1$$

61. $(3x-1)(x+3) = 3x(x+3) - 1(x+3)$
$$= 3x^2 + 9x - x - 3$$
$$= 3x^2 + 8x - 3$$

63. $(3x^4 + 1)(3x^2 + 5) = 9x^6 + 15x^4 + 3x^2 + 5$

65. $(3x+1)^2 = (3x)^2 + 2\cdot 3x \cdot 1 + 1^2 = 9x^2 + 6x + 1$

67. $(3b - 6y)(3b + 6y) = (3b)^2 - (6y)^2 = 9b^2 - 36y^2$

69. $(7x - 3)(7x + 3) = (7x)^2 - 3^2 = 49x^2 - 9$

71.
$$
\begin{array}{r}
3x^2 + 4x - 4 \\
\times \qquad 3x + 6 \\
\hline
18x^2 + 24x - 24 \\
9x^3 + 12x^2 - 12x \quad\;\; \\
\hline
9x^3 + 30x^2 + 12x - 24
\end{array}
$$

73. $(4x^2 - 2x + 5)(3x + 1)$
$= 4x^2(3x + 1) - 2x(3x + 1) + 5(3x + 1)$
$= 12x^3 + 4x^2 - 6x^2 - 2x + 15x + 5$
$= 12x^3 - 2x^2 + 13x + 5$

75. $[(xy + 4) - 6]^2 = [xy + 4 - 6]^2$
$\qquad\qquad\qquad = (xy - 2)^2$
$\qquad\qquad\qquad = (xy)^2 - 2 \cdot xy \cdot 2 + 2^2$
$\qquad\qquad\qquad = x^2y^2 - 4xy + 4$

77. $(11a^2 + 1)(2a + 1) = 22a^3 + 11a^2 + 2a + 1$

79. $\left(\dfrac{2}{3}n - 2\right)\left(\dfrac{1}{2}n - 9\right) = \dfrac{1}{3}n^2 - 6n - n + 18$
$\qquad\qquad\qquad\qquad = \dfrac{1}{3}n^2 - 7n + 18$

81. $(3x + 1)(3x - 1)(2y + 5x)$
$= [(3x)^2 - 1^2](2y + 5x)$
$= (9x^2 - 1)(2y + 5x)$
$= 18x^2y + 45x^3 - 2y - 5x$
or $45x^3 + 18x^2y - 5x - 2y$

83. $f(x) = x^2 - 3x$
$f(a) = a^2 - 3a$

85. $f(x) = x^2 - 3x$
$f(a + h) = (a + h)^2 - 3(a + h)$
$\qquad\qquad = a^2 + 2ah + h^2 - 3a - 3h$

87. $f(x) = x^2 - 3x$
$f(b - 2) = (b - 2)^2 - 3(b - 2)$
$\qquad\qquad = b^2 - 4b + 4 - 3b + 6$
$\qquad\qquad = b^2 - 7b + 10$

89. $y = -2x + 7$
$m = -2$

91. $3x - 5y = 14$
$\quad -5y = -3x + 14$
$\qquad y = \dfrac{3}{5}x - \dfrac{14}{5}$
$\quad m = \dfrac{3}{5}$

93. Since any vertical line crosses the graph at most once, it is a function.

95. $7y(3z - 2) + 1 = 21yz - 14y + 1$

97. answers may vary

99. $F(x) = x^2 + 3x + 2$

    a. $F(a + h) = (a + h)^2 + 3(a + h) + 2$
$\qquad\qquad\qquad = a^2 + 2ah + h^2 + 3a + 3h + 2$

    b. $F(a) = a^2 + 3a + 2$

    c. $F(a + h) - F(a)$
$\qquad = a^2 + 2ah + h^2 + 3a + 3h + 2 - (a^2 + 3a + 2)$
$\qquad = 2ah + h^2 + 3h$

101. $5x^2y^n(6y^{n+1} - 2)$
$= 5x^2y^n(6y^{n+1}) + 5x^2y^n(-2)$
$= 30x^2y^{2n+1} - 10x^2y^n$

103. $(x^a + 5)(x^{2a} - 3)$
$= x^a \cdot x^{2a} + x^a(-3) + 5(x^{2a}) + 5(-3)$
$= x^{3a} - 3x^a + 5x^{2a} - 15$

105. Area $= \pi r^2$
$\qquad\quad = \pi(5x - 2)^2$
$\qquad\quad = \pi(25x^2 - 20x + 4)$ square km
or $(25\pi x^2 - 20\pi x + 4\pi)$ square km

107. Area $= (3x - 2)^2 - x^2$
$\qquad\quad = (9x^2 - 12x + 4) - x^2$
$\qquad\quad = (8x^2 - 12x + 4)$ square inches

109. One operation is addition, the other is multiplication.

  a. $(3x+5)+(3x+7)=6x+12$

  b. $(3x+5)(3x+7)$
  $$=9x^2+21x+15x+35$$
  $$=9x^2+36x+35$$

111. $P(x)\cdot R(x)=(5x)(x+5)$
$$=5x\cdot x+5x\cdot 5$$
$$=5x^2+25x$$

113. $[Q(x)]^2=(x^2-2)^2$
$$=(x^2)^2-2(x^2)(2)+2^2$$
$$=x^4-4x^2+4$$

115. $R(x)\cdot Q(x)=(x+5)(x^2-2)$
$$=x^3-2x+5x^2-10$$
$$=x^3+5x^2-2x-10$$

Section 5.5 Practice Exercises

1. $32x^4y^2=2\cdot2\cdot2\cdot2\cdot2\cdot x^4\cdot y\cdot y$
$48x^3y=2\cdot2\cdot2\cdot2\cdot3\cdot x^3\cdot y$
$24y^2=2\cdot2\cdot2\cdot3\cdot y\cdot y$
$\text{GCF}=2\cdot2\cdot2\cdot y=8y$

2. a. The GCF of $6x^2$, 9, and $15x$ is 3.
$$6x^2+9+15x=3(2x^2)+3(3)+3(5x)$$
$$=3(2x^2+3+5x)$$

  b. There is no common factor of the terms $3x$ and $-8y^3$ other than 1 (or -1).

  c. The GCF of $8a^4$ and $-2a^3$ is $2a^3$.
  $$8a^4-2a^3=2a^3\cdot4a-2a^3\cdot1$$
  $$=2a^3(4a-1)$$

3. Factor out the GCF of the two terms, $8x^3y^2$.
$$64x^5y^2-8x^3y^2=8x^3y^2\cdot8x^2-8x^3y^2\cdot1$$
$$=8x^3y^2(8x^2-1)$$

4. Factor out the GCF of the three terms, $-xy^2$.
$$-9x^4y^2+5x^2y^2+7xy^2$$
$$=-xy^2\cdot9x^3-xy^2(-5x)-xy^2(-7)$$
$$=-xy^2(9x^3-5x-7)$$

5. The GCF is $(x+4)$.
$3(x+4)+5b(x+4)=(x+4)(3+5b)$

6. $8b(a^3+2y)-(a^3+2y)$
$$=8b(a^3+2y)-1(a^3+2y)$$
$$=(a^3+2y)(8b-1)$$

7. $xy-5x-10+2y=(xy-5x)+(-10+2y)$
$$=x(y-5)+2(-5+y)$$
$$=x(y-5)+2(y-5)$$
$$=(y-5)(x+2)$$

8. $a^3+2a^2+5a+10=(a^3+2a^2)+(5a+10)$
$$=a^2(a+2)+5(a+2)$$
$$=(a+2)(a^2+5)$$

9. $x^2y^2+3y^2-5x^2-15$
$$=(x^2y^2+3y^2)+(-5x^2-15)$$
$$=y^2(x^2+3)-5(x^2+3)$$
$$=(x^2+3)(y^2-5)$$

10. $pq+3p-q-3=(pq+3p)+(-q-3)$
$$=p(q+3)-1(q+3)$$
$$=(q+3)(p-1)$$

Vocabulary, Readiness & Video Check 5.5

1. The reverse of multiplying is <u>factoring</u>.

2. The greatest common factor (GCF) of x^7, x^3, x^5 is <u>x^3</u>.

3. In general, the GCF of a list of common variables raised to powers is the <u>least</u> exponent in the list.

4. Factoring means writing as a <u>product</u>.

5. True or false: A factored form of $2xy^3+10xy$ is $2xy\cdot y^2+2xy\cdot5$. <u>false</u>

6. True or false: A factored form of $x^3 - 6x^2 + x$ is $x(x^2 - 6x)$. <u>false</u>

7. True or false: A factored form of $5x - 5y + x^3 - x^2y$ is $5(x - y) + x^2(x - y)$. <u>false</u>

8. True or false: A factored form of $5x - 5y + x^3 - x^2y$ is $(x - y)(5 + x^2)$. <u>true</u>

9. The GCF is that common variable raised to the smallest exponent in the list.

10. Multiplication (what times the GCF gives you each term of the polynomial?) or division (what is each term of the polynomial divided by the GCF?).

11. Look for a GCF other than 1 or −1; if you have a four-term polynomial.

Exercise Set 5.5

1. a^8, a^5, and a^3; GCF $= a^3$

3. $x^2y^3z^3$, y^2z^3, and xy^2z^2; GCF $= y^2z^2$

5. $6x^3y, 9x^2y^2$, and $12x^2y$; GCF $= 3x^2y$

7. $10x^3yz^3, 20x^2z^5, 45xz^3$; GCF $= 5xz^3$

9. $18x - 12 = 6 \cdot 3x - 6 \cdot 2 = 6(3x - 2)$

11. $4y^2 - 16xy^3 = 4y^2 \cdot 1 - 4y^2 \cdot 4xy$
$= 4y^2(1 - 4xy)$

13. $6x^5 - 8x^4 + 2x^3 = 2x^3(3x^2) - 2x^3(4x) + 2x^3(1)$
$= 2x^3(3x^2 - 4x + 1)$

15. $8a^3b^3 - 4a^2b^2 + 4ab + 16ab^2$
$= 4ab(2a^2b^2) - 4ab(ab) + 4ab(1) + 4ab(4b)$
$= 4ab(2a^2b^2 - ab + 1 + 4b)$

17. $6(x + 3) + 5a(x + 3) = (x + 3)(6 + 5a)$

19. $2x(z + 7) + (z + 7) = (z + 7)(2x + 1)$

21. $3x(6x^2 + 5) - 2(6x^2 + 5) = (6x^2 + 5)(3x - 2)$

23. $20a^3 + 5a^2 + 1$
There is no common factor other than 1.

25. $39x^3y^3 - 26x^2y^3 = 13x^2y^3 \cdot 3x - 13x^2y^3 \cdot 2$
$= 13x^2y^3(3x - 2)$

27. $ab + 3a + 2b + 6 = a(b + 3) + 2(b + 3)$
$= (a + 2)(b + 3)$

29. $ac + 4a - 2c - 8 = a(c + 4) - 2(c + 4)$
$= (a - 2)(c + 4)$

31. $2xy - 3x - 4y + 6 = x(2y - 3) - 2(2y - 3)$
$= (x - 2)(2y - 3)$

33. $12xy - 8x - 3y + 2 = 4x(3y - 2) - (3y - 2)$
$= (4x - 1)(3y - 2)$

35. $6x^3 + 9 = 3 \cdot 2x^3 + 3 \cdot 3 = 3(2x^3 + 3)$

37. $x^3 + 3x^2 = x^2 \cdot x + x^2 \cdot 3 = x^2(x + 3)$

39. $8a^3 - 4a = 4a \cdot 2a^2 - 4a \cdot 1 = 4a(2a^2 - 1)$

41. $8m^3 + 4m^2 + 1$
There is no common factor other than 1.

43. $-20x^2y + 16xy^3 = 4xy \cdot (-5x) + 4xy \cdot 4y^2$
$= 4xy(-5x + 4y^2)$
or
$-20x^2y + 16xy^3 = -4xy \cdot 5x + (-4xy)(-4y^2)$
$= -4xy(5x - 4y^2)$

45. $10a^2b^3 + 5ab^2 - 15ab^3$
$= 5ab^2 \cdot 2ab + 5ab^2 \cdot 1 - 5ab^2 \cdot 3b$
$= 5ab^2(2ab + 1 - 3b)$

47. $9abc^2 + 6a^2bc - 6ab + 3bc$
$= 3b \cdot 3ac^2 + 3b \cdot 2a^2c + 3b \cdot (-2a) + 3b \cdot c$
$= 3b(3ac^2 + 2a^2c - 2a + c)$

49. $4x(y - 2) - 3(y - 2) = (y - 2)(4x - 3)$

51. $6xy + 10x + 9y + 15 = (6xy + 10x) + (9y + 15)$
$= 2x(3y + 5) + 3(3y + 5)$
$= (3y + 5)(2x + 3)$

53. $xy + 3y - 5x - 15 = (xy + 3y) + (-5x - 15)$
$= y(x + 3) - 5(x + 3)$
$= (x + 3)(y - 5)$

55. $6ab - 2a - 9b + 3 = (6ab - 2a) + (-9b + 3)$
$= 2a(3b - 1) + (-3)(3b - 1)$
$= (3b - 1)(2a - 3)$

57. $12xy + 18x + 2y + 3 = (12xy + 18x) + (2y + 3)$
$= 6x(2y + 3) + 1(2y + 3)$
$= (2y + 3)(6x + 1)$

59. $2m(n - 8) - (n - 8) = 2m(n - 8) - 1(n - 8)$
$= (n - 8)(2m - 1)$

61. $15x^3 y^2 - 18x^2 y^2 = 3x^2 y^2 \cdot 5x - 3x^2 y^2 \cdot 6$
$= 3x^2 y^2 (5x - 6)$

63. $2x^2 + 3xy + 4x + 6y = (2x^2 + 3xy) + (4x + 6y)$
$= x(2x + 3y) + 2(2x + 3y)$
$= (2x + 3y)(x + 2)$

65. $5x^2 + 5xy - 3x - 3y = 5x(x + y) - 3(x + y)$
$= (x + y)(5x - 3)$

67. $x^3 + 3x^2 + 4x + 12 = (x^3 + 3x^2) + (4x + 12)$
$= x^2(x + 3) + 4(x + 3)$
$= (x + 3)(x^2 + 4)$

69. $x^3 - x^2 - 2x + 2 = (x^3 - x^2) + (-2x + 2)$
$= x^2(x - 1) - 2(x - 1)$
$= (x - 1)(x^2 - 2)$

71. $(5x^2)(11x^5) = 5(11)x^2 x^5 = 55x^7$

73. $(5x^2)^3 = 5^3(x^2)^3 = 125x^6$

75. $(x + 2)(x - 5) = x^2 - 5x + 2x - 10$
$= x^2 - 3x - 10$

77. $(x + 3)(x + 2) = x^2 + 3x + 2x + 6$
$= x^2 + 5x + 6$

79. $(y - 3)(y - 1) = y^2 - 1y - 3y + 3$
$= y^2 - 4y + 3$

81. d

   a. $2(5x^2 - x + 1) = 10x^2 - 2x + 2$

   b. $2(5x^2 - x) = 10x^2 - 2x$

   c. $2(5x^2 - x - 2) = 10x^2 - 2x - 4$

   d. $2(5x^2 - x - 1) = 10x^2 - 2x - 2$

83. $2\pi r^2 + 2\pi rh = 2\pi r(r + h)$

85. $A = 5600 + 5600rt$
$A = 5600(1) + 5600(rt)$
$A = 5600(1 + rt)$

87. answers may vary

89. **a.** $(2 - x)(3 - y) = 6 - 2y - 3x + xy$
$= xy - 3x - 2y + 6$

   b. $(-2 + x)(-3 + y) = 6 - 2y - 3x + xy$
$= xy - 3x - 2y + 6$

   c. $(y - 3)(x - 2) = yx - 2y - 3x + 6$
$= xy - 3x - 2y + 6$

   d. $(-x + 2)(-y + 3) = xy - 3x - 2y + 6$

   None; all are factored forms of
$xy - 3x - 2y + 6$.

91. answers may vary

93. $IR_1 + IR_2 = E$
$I(R_1 + R_2) = E$

95. $x^{3n} - 2x^{2n} + 5x^n$
$= x^n \cdot x^{2n} - x^n \cdot 2x^n + x^n \cdot 5$
$= x^n(x^{2n} - 2x^n + 5)$

97. $6x^{8a} - 2x^{5a} - 4x^{3a} = 2x^{3a}(3x^{5a} - x^{2a} - 2)$

99. $h(t) = -16t^2 + 64t$

   a. $h(t) = -16t(t - 4)$

   b. $h(1) = -16(1)^2 + 64(1) = 48$
$h(1) = -16(1)(1 - 4) = 48$ feet

c. answers may vary

101. The greatest common factor is 3.

$$f(x) = 201x^3 - 2517x^2 + 6975x + 83,634$$

$$f(x) = 3(67x^3 - 839x^2 + 2325x + 27,878)$$

Section 5.6 Practice Exercises

1. Find two integers whose product is 6 and whose sum is 5. Since our integers must have a positive product and a positive sum, look for positive factors.

Positive Factors of 6	Sum of Factors
1, 6	$1 + 6 = 7$
3, 2	$3 + 2 = 5$ (correct)

$$x^2 + 5x + 6 = (x+3)(x+2)$$

2. Find two integers whose product is 24 and whose sum is -11. Since our integers must have a positive product and a negative sum, look for negative factors.

Negative Factors of 24	Sum of Factors
$-1, -24$	$-1 + (-24) = -25$
$-2, -12$	$-2 + (-12) = -14$
$-3, -8$	$-3 + (-8) = -11$ (correct)
$-4, -6$	$-4 + (-6) = -10$

$$x^2 - 11x + 24 = (x-3)(x-8)$$

3. $3x^3 - 9x^2 - 30x = 3x(x^2 - 3x - 10)$

Find two integers whose product is -10 and whose sum is -3. The numbers are -5 and 2.

$$3x^3 - 9x^2 - 30x = 3x(x^2 - 3x - 10)$$
$$= 3x(x-5)(x+2)$$

4. $2b^2 - 18b - 22 = 2(b^2 - 9b - 11)$

Find two integers whose product is -11 and whose sum is -9.

Factors	Sum
$-1, 11$	10
$1, -11$	-10

Neither of the pairs has a sum of -9, so no further factoring is possible.

$$2b^2 - 18b - 22 = 2(b^2 - 9b - 11)$$

5. Factors of $2x^2$: $2x \cdot x$

Factors of 6: $1 \cdot 6$ and $2 \cdot 3$

$(2x + 6)(x + 1) \Rightarrow 2x + 6x = 8x$ (incorrect middle term)

$(2x + 1)(x + 6) \Rightarrow 12x + x = 13x$ (correct middle term)

$(2x + 2)(x + 3) \Rightarrow 6x + 2x = 8x$ (incorrect middle term)

$(2x + 3)(x + 2) \Rightarrow 4x + 3x = 7x$ (incorrect middle term)

$$2x^2 + 13x + 6 = (2x+1)(x+6)$$

6. Factors of $4x^2$: $4x \cdot x$ and $2x \cdot 2x$

Factors of -6: $-6 \cdot 1$, $6 \cdot -1$, $2 \cdot -3$, $-2 \cdot 3$

$(4x - 6)(x + 1) \Rightarrow 4x - 6x = -2x$ (incorrect)

$(4x + 6)(x - 1) \Rightarrow -4x + 6x = 2x$ (incorrect)

$(4x + 2)(x - 3) \Rightarrow -12x + 2x = -10x$ (incorrect)

$(4x - 3)(x + 2) \Rightarrow 8x - 3x = 5x$ (correct)

$$4x^2 + 5x - 6 = (4x-3)(x+2)$$

7. $18b^4 - 57b^3 + 30b^2 = 3b^2(6b^2 - 19b + 10)$

Factors of $6b^2$: $2b \cdot 3b$, $6b \cdot b$

Negative factors of 10: $-1 \cdot -10$, $-5 \cdot -2$

$(2b - 1)(3b - 10) \Rightarrow -20b - 3b = -23b$ (incorrect)

$(2b - 10)(3b - 1) \Rightarrow -2b - 30b = -32b$ (incorrect)

$(2b - 5)(3b - 2) \Rightarrow -4b - 15b = -19b$ (correct)

$$18b^4 - 57b^3 + 30b^2 = 3b^2(6b^2 - 19b + 10)$$
$$= 3b^2(2b-5)(3b-2)$$

8. No greatest common factor can be factored out.

Factors of $25x^2$: $25x \cdot x$, $5x \cdot 5x$

Factors of $4y^2$: $4y \cdot y$, $2y \cdot 2y$

Try possible combinations.

$$25x^2 + 20xy + 4y^2 = (5x + 2y)(5x + 2y)$$
$$= (5x + 2y)^2$$

9. $20x^2 + 23x + 6$

$a = 20$, $b = 23$, $c = 6$

Find two numbers whose product is $a \cdot c = 20 \cdot 6 = 120$, and whose sum is b, 23. The two numbers are 8 and 15.

$$20x^2 + 23x + 6 = 20x^2 + 8x + 15x + 6$$
$$= 4x(5x+2) + 3(5x+2)$$
$$= (5x+2)(4x+3)$$

10. $15x^2 + 4x - 3$

    $a = 15, b = 4, c = -3$

    Find two numbers whose product is
    $a \cdot c = 15(-3) = -45$ and whose sum is b, 4. The
    two numbers are 9 and -5.

    $$15x^2 + 4x - 3 = 15x^2 + 9x - 5x - 3$$
    $$= 3x(5x+3) - 1(5x+3)$$
    $$= (5x+3)(3x-1)$$

11. Let $y = x + 1$.

    $$3(x+1)^2 - 7(x+1) - 20 = 3y^2 - 7y - 20$$
    $$= (3y+5)(y-4)$$

    Replace y with $x + 1$.
    $$(3y+5)(y-4) = [3(x+1)+5][(x+1)-4]$$
    $$= (3x+3+5)(x+1-4)$$
    $$= (3x+8)(x-3)$$

    Thus, $3(x+1)^2 - 7(x+1) - 20 = (3x+8)(x-3)$.

12. Let $y = x^2$.

    $$6x^4 - 11x^2 - 10 = 6y^2 - 11y - 10$$
    $$= (3y+2)(2y-5)$$

    Replace y with x^2.
    $$(3y+2)(2y-5) = (3x^2+2)(2x^2-5)$$

Vocabulary, Readiness & Video Check 5.6

1. $10 = 2 \cdot 5$
    $7 = 2 + 5$
    2 and 5

2. $12 = 2 \cdot 2 \cdot 3 = 2 \cdot 6$
    $8 = 2 + 6$
    2 and 6

3. $24 = 2 \cdot 2 \cdot 2 \cdot 3 = 8 \cdot 3$
    $11 = 8 + 3$
    8 and 3

4. $30 = 2 \cdot 3 \cdot 5 = 10 \cdot 3$
    $13 = 10 + 3$
    10 and 3

5. Check by multiplying. If you get a middle term
    of $2x$, not $-2x$, switch the signs of your factors of
    -24.

6. 8 is positive, and since the factors need to sum to
    -6, both factors must be negative.

7. Write down the factors of the first and last terms.
    Try various combinations of these factors and
    look at the sum of the outer and inner products to
    see if you get the middle term of the trinomial. If
    not, try another combination of factors.

8. This gives us a four-term polynomial, which
    may be factored by grouping, if it can be
    factored.

9. Our factoring involves a substitution from the
    original expression, so we need to substitute
    back in order to get a factorization in terms of
    the original variable/expression, and simplify if
    necessary.

Exercise Set 5.6

1. $x^2 + 9x + 18 = (x+6)(x+3)$

3. $x^2 - 12x + 32 = (x-4)(x-8)$

5. $x^2 + 10x - 24 = (x+12)(x-2)$

7. $x^2 - 2x - 24 = (x-6)(x+4)$

9. Note that the GCF is 3.
    $$3x^2 - 18x + 24 = 3(x^2 - 6x + 8)$$
    $$= 3(x-2)(x-4)$$

11. Note that the GCF is $4z$.
    $$4x^2z + 28xz + 40z = 4z(x^2 + 7x + 10)$$
    $$= 4z(x+2)(x+5)$$

13. Since the zeros of the polynomial are -5 and 3,
    factors of the polynomial are $(x + 5)$ and $(x - 3)$.

15. Since the zeros of the polynomial are 2 and 6,
    factors of the polynomial are $(x - 2)$ and $(x - 6)$.

17. $5x^2 + 16x + 3 = (5x+1)(x+3)$

19. $2x^2 - 11x + 12 = (2x-3)(x-4)$

21. $2x^2 + 25x - 20$ is prime.

23. $4x^2 - 12x + 9 = (2x-3)(2x-3)$
$$= (2x-3)^2$$

25. Note that the GCF is 2.
$$12x^2 + 10x - 50 = 2(6x^2 + 5x - 25)$$
$$= 2(3x-5)(2x+5)$$

27. Note that the GCF is y^2.
$$3y^4 - y^3 - 10y^2 = y^2(3y^2 - y - 10)$$
$$= y^2(3y+5)(y-2)$$

29. $2x^2 - 5xy - 3y^2 = (2x+y)(x-3y)$

31. Note that the GCF is 2, so that
$$28y^2 + 22y + 4 = 2(14y^2 + 11y + 2).$$
$ac = 28$; the two numbers are 4 and 7.
$$14y^2 + 11y + 2 = 14y^2 + 7y + 4y + 2$$
$$= 7y(2y+1) + 2(2y+1)$$
$$= (7y+2)(2y+1)$$
So, $28y^2 + 22y + 4 = 2(7y+2)(2y+1)$.

33. $2x^2 + 15x - 27$; $ac = -54$ so the two numbers are 18 and -3.
$$2x^2 + 15x - 27 = 2x^2 + 18x - 3x - 27$$
$$= 2x(x+9) - 3(x+9)$$
$$= (2x-3)(x+9)$$

35. Let $y = x^2$. Then we have
$$x^4 + x^2 - 6 = y^2 + y - 6 = (y+3)(y-2).$$
This yields $(x^2+3)(x^2-2)$.

37. Let $y = 5x + 1$. Then we have
$$(5x+1)^2 + 8(5x+1) + 7 = y^2 + 8y + 7$$
$$= (y+1)(y+7).$$
This yields
$$[(5x+1)+1][(5x+1)+7] = (5x+2)(5x+8).$$

39. Let $y = x^3$. Then we have
$$x^6 - 7x^3 + 12 = y^2 - 7y + 12$$
$$= (y-4)(y-3).$$
This yields $(x^3-4)(x^3-3)$.

41. Let $y = a + 5$. Then we have
$$(a+5)^2 - 5(a+5) - 24 = y^2 - 5y - 24$$
$$= (y-8)(y+3).$$
This yields
$$[(a+5)-8][(a+5)+3] = (a-3)(a+8).$$

43. $x^2 - 24x - 81 = (x-27)(x+3)$

45. $x^2 - 15x - 54 = (x-18)(x+3)$

47. $3x^2 - 6x + 3 = 3(x^2 - 2x + 1)$
$$= 3(x-1)(x-1)$$
$$= 3(x-1)^2$$

49. $3x^2 - 5x - 2 = (3x+1)(x-2)$

51. $8x^2 - 26x + 15 = (4x-3)(2x-5)$

53. $18x^4 + 21x^3 + 6x^2 = 3x^2(6x^2 + 7x + 2)$
$$= 3x^2(3x+2)(2x+1)$$

55. $x^2 + 8xz + 7z^2 = (x+z)(x+7z)$

57. $x^2 - x - 12$; $ac = -12$ so the two numbers are -4 and 3.
$$x^2 - x - 12 = x^2 - 4x + 3x - 12$$
$$= x(x-4) + 3(x-4)$$
$$= (x+3)(x-4)$$

59. $3a^2 + 12ab + 12b^2 = 3(a^2 + 4ab + 4b^2)$
$$= 3(a+2b)(a+2b)$$
$$= 3(a+2b)^2$$

61. $x^2 + 4x + 5$ is prime.

63. Let $y = x + 4$. Then
$$2(x+4)^2 + 3(x+4) - 5$$
$$= 2y^2 + 3y - 5$$
$$= (2y+5)(y-1)$$
$$= [2(x+4)+5][(x+4)-1]$$
$$= (2x+8+5)(x+3)$$
$$= (2x+13)(x+3)$$

65. $6x^2 - 49x + 30 = (3x-2)(2x-15)$

67. Let $y = x^2$. Then
$$x^4 - 5x^2 - 6 = y^2 - 5y - 6$$
$$= (y-6)(y+1)$$
$$= (x^2 - 6)(x^2 + 1)$$

69. $6x^3 - x^2 - x = x(6x^2 - x - 1)$
$$= x(3x+1)(2x-1)$$

71. $12a^2 - 29ab + 15b^2 = (4a - 3b)(3a - 5b)$

73. $9x^2 + 30x + 25 = (3x+5)(3x+5)$
$$= (3x+5)^2$$

75. $3x^2 y - 11xy + 8y = y(3x^2 - 11x + 8)$
$$= y(3x - 8)(x - 1)$$

77. $2x^2 + 2x - 12 = 2(x^2 + x - 6)$
$$= 2(x+3)(x-2)$$

79. Let $y = x - 4$. Then
$$(x-4)^2 + 3(x-4) - 18$$
$$= y^2 + 3y - 18$$
$$= (y+6)(y-3)$$
$$= [(x-4)+6][(x-4)-3]$$
$$= (x+2)(x-7)$$

81. Let $y = x^3$. Then
$$2x^6 + 3x^3 - 9 = 2y^2 + 3y - 9$$
$$= (2y-3)(y+3)$$
$$= (2x^3 - 3)(x^3 + 3)$$

83. $72xy^4 - 24xy^2 z + 2xz^2 = 2x(36y^4 - 12y^2 z + z^2)$
$$= 2x(6y^2 - z)(6y^2 - z)$$
$$= 2x(6y^2 - z)^2$$

85. $2x^3 y + 2x^2 y - 12xy = 2xy(x^2 + x - 6)$
$$= 2xy(x+3)(x-2)$$

87. $x^2 + 6xy + 5y^2 = (x+5y)(x+y)$

89. $(x-3)(x+3) = x^2 - 3^3 = x^2 - 9$

91. $(2x+1)^2 = (2x)^2 + 2(2x)(1) + 1^2$
$$= 4x^2 + 4x + 1$$

93.
$$\begin{array}{r} x^2 + 2x + 4 \\ x - 2 \\ \hline -2x^2 - 4x - 8 \\ x^3 + 2x^2 + 4x \\ \hline x^3 \qquad\qquad -8 \end{array}$$

95. $x^2 + bx + 6$
$6 = 2 \cdot 3$ or $6 = (-2)(-3)$
$6 = 1 \cdot 6$ or $6 = (-1)(-6)$
$(x+2)(x+3) = x^2 + 5x + 6$
$(x-2)(x-3) = x^2 - 5x + 6$
$(x+1)(x+6) = x^2 + 7x + 6$
$(x-1)(x-6) = x^2 - 7x + 6$
$b = \pm 5$ and $b = \pm 7$

97. $V(x) = x^3 + 2x^2 - 8x$
$$= x(x^2 + 2x - 8)$$
$$= x(x+4)(x-2)$$

99. $h(t) = -16t^2 + 80t + 576$

a. $h(0) = -16(0)^2 + 80(0) + 576 = 576$ ft
$h(2) = -16(2)^2 + 80(2) + 576$
$$= -16(4) + 160 + 576$$
$$= -64 + 160 + 576$$
$$= 672 \text{ ft}$$
$h(4) = -16(4)^2 + 80(4) + 576$
$$= -16(16) + 320 + 576$$
$$= -256 + 320 + 576$$
$$= 640 \text{ ft}$$
$h(6) = -16(6)^2 + 80(6) + 576$
$$= -16(36) + 480 + 576$$
$$= -576 + 480 + 576$$
$$= 480 \text{ ft}$$

b. answers may vary

c. $h(t) = -16t^2 + 80t + 576$
$$= -16(t^2 - 5t - 36)$$
$$= -16(t-9)(t+4)$$

101. $x^{2n} + 10x^n + 16 = (x^n + 2)(x^n + 8)$

103. $x^{2n} - 3x^n - 18 = (x^n - 6)(x^n + 3)$

105. $2x^{2n} + 11x^n + 5 = (2x^n + 1)(x^n + 5)$

107. $4x^{2n} - 12x^n + 9 = (2x^n - 3)(2x^n - 3)$
$$= (2x^n - 3)^2$$

Section 5.7 Practice Exercises

1. $b^2 + 16b + 64 = b^2 + 2(b)(8) + 8^2 = (b+8)^2$

2. $45x^2 b - 30xb + 5b = 5b(9x^2 - 6x + 1)$
$$= 5b[(3x)^2 - 2(3x)(1) + 1^2]$$
$$= 5b(3x - 1)^2$$

3. a. $x^2 - 16 = x^2 - 4^2 = (x+4)(x-4)$

   b. $25b^2 - 49 = (5b)^2 - 7^2 = (5b - 7)(5b + 7)$

   c. $45 - 20x^2 = 5(9 - 4x^2)$
$$= 5[3^2 - (2x)^2]$$
$$= 5(3 - 2x)(3 + 2x)$$

   d. $y^2 - \dfrac{1}{81} = y^2 - \left(\dfrac{1}{9}\right)^2 = \left(y - \dfrac{1}{9}\right)\left(y + \dfrac{1}{9}\right)$

4. a. $x^4 - 10{,}000 = (x^2)^2 - 100^2$
$$= (x^2 + 100)(x^2 - 100)$$
$$= (x^2 + 100)(x + 10)(x - 10)$$

   b. $(x+2)^2 - 49 = (x+2)^2 - 7^2$
$$= [(x+2) + 7][(x+2) - 7]$$
$$= (x + 2 + 7)(x + 2 - 7)$$
$$= (x + 9)(x - 5)$$

5. $m^2 + 6m + 9 - n^2 = (m^2 + 6m + 9) - n^2$
$$= (m + 3)^2 - n^2$$
$$= [(m+3) + n][(m+3) - n]$$
$$= (m + 3 + n)(m + 3 - n)$$

6. $x^3 + 64 = x^3 + 4^3$
$$= (x + 4)(x^2 - x \cdot 4 + 4^2)$$
$$= (x + 4)(x^2 - 4x + 16)$$

7. $a^3 + 8b^3 = a^3 + (2b)^3$
$$= (a + 2b)[a^2 - a(2b) + (2b)^2]$$
$$= (a + 2b)(a^2 - 2ab + 4b^2)$$

8. $27 - y^3 = 3^3 - y^3$
$$= (3 - y)(3^2 + 3 \cdot y + y^2)$$
$$= (3 - y)(9 + 3y + y^2)$$

9. $b^3 x^2 - 8x^2 = x^2(b^3 - 8)$
$$= x^2(b^3 - 2^3)$$
$$= x^2(b - 2)(b^2 + b \cdot 2 + 2^2)$$
$$= x^2(b - 2)(b^2 + 2b + 4)$$

Vocabulary, Readiness & Video Check 5.7

1. $81y^2 = (9y)^2$

2. $4z^2 = (2z)^2$

3. $64x^6 = (8x^3)^2$

4. $49y^6 = (7y^3)^2$

5. $8x^3 = (2x)^3$

6. $27y^3 = (3y)^3$

7. $64y^6 = (4x^2)^3$

8. $x^3 y^6 = (xy^2)^3$

9. See if the first term is a square, say a^2, and the last term is a square, say b^2. Check to see if the middle term is $2 \cdot a \cdot b$ or $-2 \cdot a \cdot b$.

10. In order to help you see that the binomial is a difference of squares and also to identify the terms to use in the special factoring formula.

11. First rewrite the original binomial so that each term is some quantity cubed. Your answers will then vary, depending on your interpretation.

Exercise Set 5.7

1. $x^2 + 6x + 9 = (x)^2 + 2 \cdot x \cdot 3 + (3)^2 = (x+3)^2$

3. $4x^2 - 12x + 9 = (2x)^2 - 2 \cdot 2x \cdot 3 + 3^2$
$$= (2x - 3)^2$$

5. $25x^2 + 10x + 1 = (5x)^2 + 2 \cdot 5x \cdot 1 + 1^2 = (5x + 1)^2$

7. $3x^2 - 24x + 48 = 3(x^2 - 8x + 16)$
$$= 3(x^2 - 2 \cdot x \cdot 4 + 4^2)$$
$$= 3(x - 4)^2$$

9. $9y^2x^2 + 12yx^2 + 4x^2$
$$= x^2(9y^2 + 12y + 4)$$
$$= x^2[(3y)^2 + 2 \cdot 3y \cdot 2 + (2)^2]$$
$$= x^2(3y + 2)^2$$

11. $16x^2 - 56xy + 49y^2 = (4x)^2 - 2 \cdot 4x \cdot 7y + (7y)^2$
$$= (4x - 7y)^2$$

13. $x^2 - 25 = (x)^2 - (5)^2 = (x + 5)(x - 5)$

15. $\dfrac{1}{9} - 4z^2 = \left(\dfrac{1}{3}\right)^2 - (2z)^2 = \left(\dfrac{1}{3} + 2z\right)\left(\dfrac{1}{3} - 2z\right)$

17. $(y + 2)^2 - 49 = (y + 2)^2 - (7)^2$
$$= [(y + 2) + 7][(y + 2) - 7]$$
$$= (y + 9)(y - 5)$$

19. $64x^2 - 100 = 4(16x^2 - 25)$
$$= 4[(4x)^2 - 5^2]$$
$$= 4(4x + 5)(4x - 5)$$

21. $(x + 2y)^2 - 9 = (x + 2y)^2 - (3)^2$
$$= [(x + 2y) + 3][(x + 2y) - 3]$$
$$= (x + 2y + 3)(x + 2y - 3)$$

23. $x^2 + 6x + 9 - y^2 = (x + 3)^2 - y^2$
$$= (x + 3 + y)(x + 3 - y)$$

25. $x^2 + 16x + 64 - x^4 = (x^2 + 16x + 64) - x^4$
$$= (x^2 + 2 \cdot x \cdot 8 + 8^2) - x^4$$
$$= (x + 8)^2 - (x^2)^2$$
$$= [(x + 8) + x^2][(x + 8) - x^2]$$
$$= (x + 8 + x^2)(x + 8 - x^2)$$

27. $x^3 + 27 = x^3 + (3)^3 = (x + 3)(x^2 - 3x + 9)$

29. $z^3 - 1 = z^3 - 1^3 = (z - 1)(z^2 + z + 1)$

31. $m^3 + n^3 = (m + n)(m^2 - mn + n^2)$

33. $27y^2 - x^3y^2 = y^2(27 - x^3)$
$$= y^2(3^3 - x^3)$$
$$= y^2(3 - x)(3^2 + 3 \cdot x + x^2)$$
$$= y^2(3 - x)(9 + 3x + x^2)$$

35. $8ab^3 + 27a^4 = a(8b^3 + 27a^3)$
$$= a[(2b)^3 + (3a)^3]$$
$$= a(2b + 3a)[(2b)^2 - 2b \cdot 3a + (3a)^2]$$
$$= a(2b + 3a)(4b^2 - 6ab + 9a^2)$$

37. $250y^3 - 16x^3$
$$= 2(125y^3 - 8x^3)$$
$$= 2[(5y)^3 - (2x)^3]$$
$$= 2(5y - 2x)[(5y)^2 + 5y \cdot 2x + (2x)^2]$$
$$= 2(5y - 2x)(25y^2 + 10xy + 4x^2)$$

39. $x^2 - 12x + 36 = x^2 - 2 \cdot x \cdot 6 + 6^2 = (x - 6)^2$

41. $18x^2y - 2y = 2y(9x^2 - 1)$
$$= 2y[(3x)^2 - 1^2]$$
$$= 2y(3x + 1)(3x - 1)$$

43. $9x^2 - 49 = (3x)^2 - 7^2 = (3x + 7)(3x - 7)$

45. $x^4 - 1 = (x^2)^2 - 1^2$
$$= (x^2 + 1)(x^2 - 1)$$
$$= (x^2 + 1)(x + 1)(x - 1)$$

47. $x^6 - y^3 = (x^2)^3 - y^3$
$$= (x^2 - y)[(x^2)^2 + x^2y + y^2]$$
$$= (x^2 - y)(x^4 + x^2y + y^2)$$

49. $8x^3 + 27y^3$
$$= (2x)^3 + (3y)^3$$
$$= (2x + 3y)[(2x)^2 - 2x \cdot 3y + (3y)^2]$$
$$= (2x + 3y)(4x^2 - 6xy + 9y^2)$$

51. $4x^2 + 4x + 1 - z^2 = (4x^2 + 4x + 1) - z^2$
$$= [(2x)^2 + 2 \cdot 2x \cdot 1 + 1^2] - z^2$$
$$= (2x + 1)^2 - z^2$$
$$= (2x + 1 + z)(2x + 1 - z)$$

53. $3x^6 y^2 + 81y^2 = 3y^2(x^6 + 27)$
$$= 3y^2[(x^2)^3 + 3^3]$$
$$= 3y^2(x^2 + 3)[(x^2)^2 - x^2 \cdot 3 + 3^2]$$
$$= 3y^2(x^2 + 3)(x^4 - 3x^2 + 9)$$

55. $n^3 - \dfrac{1}{27} = n^3 - \left(\dfrac{1}{3}\right)^3$
$$= \left(n - \dfrac{1}{3}\right)\left[n^2 + n \cdot \dfrac{1}{3} + \left(\dfrac{1}{3}\right)^2\right]$$
$$= \left(n - \dfrac{1}{3}\right)\left(n^2 + \dfrac{1}{3}n + \dfrac{1}{9}\right)$$

57. $-16y^2 + 64 = -16(y^2 - 4)$
$$= -16(y^2 - 2^2)$$
$$= -16(y + 2)(y - 2)$$

59. $x^2 - 10x + 25 - y^2 = (x^2 - 10x + 25) - y^2$
$$= (x^2 - 2 \cdot x \cdot 5 + 5^2) - y^2$$
$$= (x - 5)^2 - y^2$$
$$= (x - 5 + y)(x - 5 - y)$$

61. $a^3 b^3 + 125 = (ab)^3 + 5^3$
$$= (ab + 5)[(ab)^2 - ab \cdot 5 + 5^2]$$
$$= (ab + 5)(a^2 b^2 - 5ab + 25)$$

63. $\dfrac{x^2}{25} - \dfrac{y^2}{9} = \left(\dfrac{x}{5}\right)^2 - \left(\dfrac{y}{3}\right)^2 = \left(\dfrac{x}{5} + \dfrac{y}{3}\right)\left(\dfrac{x}{5} - \dfrac{y}{3}\right)$

65. $(x + y)^3 + 125$
$$= (x + y)^3 + 5^3$$
$$= [(x + y) + 5][(x + y)^2 - (x + y) \cdot 5 + 5^2]$$
$$= (x + y + 5)(x^2 + 2xy + y^2 - 5x - 5y + 25)$$

67. $x - 5 = 0$
$$x = 5$$

69. $3x + 1 = 0$
$$3x = -1$$
$$x = -\dfrac{1}{3}$$

71. $-2x = 0$
$$x = 0$$

73. $-5x + 25 = 0$
$$-5x = -25$$
$$x = 5$$

75. No; $x^2 - 4$ can be factored further.
$$5x(x^2 - 4) = 5x(x^2 - 2^2) = 5x(x + 2)(x - 2)$$

77. Yes; $7y(a^2 + a + 1)$ is factored completely.

79. Area $= \pi R^2 - \pi r^2$
$$= \pi(R^2 - r^2)$$
$$= \pi(R + r)(R - r) \text{ sq units}$$

81. Volume $= x^3 - y^2 x$
$$= x(x^2 - y^2)$$
$$= x(x + y)(x - y) \text{ cubic units}$$

83. $\dfrac{1}{2} \cdot b = \dfrac{1}{2} \cdot 6 = 3$ so $c = 3^2 = 9$

85. $\dfrac{1}{2} \cdot b = \dfrac{1}{2}(-14) = -7$ so $c = (-7)^2 = 49$

87. $\dfrac{1}{2} \cdot c = \dfrac{c}{2}$ so $\left(\dfrac{c}{2}\right)^2 = 16$
$$\dfrac{c^2}{4} = 16$$
$$c^2 = 64$$
$$c = \pm 8$$

89. $x^6 - 1$

   a. $(x^3)^2 - 1^2$
   $$= (x^3 + 1)(x^3 - 1)$$
   $$= (x + 1)(x^2 - x + 1)(x - 1)(x^2 + x + 1)$$

   b. $(x^2)^3 - 1^3 = (x^2 - 1)(x^4 + x^2 + 1)$
   $$= (x + 1)(x - 1)(x^4 + x^2 + 1)$$

   c. No; answers may vary

91. $x^{2n} - 25 = (x^n)^2 - 5^2 = (x^n + 5)(x^n - 5)$

93. $36x^{2n} - 49 = (6x^n)^2 - 7^2 = (6x^n + 7)(6x^n - 7)$

95. $x^{4n} - 16 = (x^{2n})^2 - 4^2$

$\qquad = (x^{2n} + 4)(x^{2n} - 4)$

$\qquad = (x^{2n} + 4)[(x^n)^2 - 2^2]$

$\qquad = (x^{2n} + 4)(x^n + 2)(x^n - 2)$

Integrated Review Practice Exercises

1. a. $12x^2 y - 3xy = 3xy(4x) + 3xy(-1)$

$\qquad\qquad\qquad = 3xy(4x - 1)$

  b. $49x^2 - 4 = (7x)^2 - 2^2 = (7x + 2)(7x - 2)$

  c. $5x^2 + 2x - 3 = (5x - 3)(x + 1)$

  d. $3x^2 + 6 + x^3 + 2x = 3(x^2 + 2) + x(x^2 + 2)$

$\qquad\qquad\qquad\qquad = (x^2 + 2)(3 + x)$

  e. $4x^2 + 20x + 25 = (2x)^2 + 2 \cdot 2x \cdot 5 + 5^2$

$\qquad\qquad\qquad\qquad = (2x + 5)^2$

  f. $b^2 + 100$ cannot be factored.

2. a. $64x^3 + y^3 = (4x)^3 + y^3$

$\qquad\qquad\quad = (4x + y)[(4x)^2 - 4x \cdot y + y^2]$

$\qquad\qquad\quad = (4x + y)(16x^2 - 4xy + y^2)$

  b. $7x^2 y^2 - 63y^4 = 7y^2(x^2 - 9y^2)$

$\qquad\qquad\qquad = 7y^2[x^2 - (3y)^2]$

$\qquad\qquad\qquad = 7y^2(x - 3y)(x + 3y)$

  c. $3x^2 + 12x + 12 - 3b^2$

$\qquad = 3(x^2 + 4x + 4 - b^2)$

$\qquad = 3[(x + 2)^2 - b^2]$

$\qquad = 3(x + 2 + b)(x + 2 - b)$

  d. $x^5 y^4 + 27x^2 y$

$\qquad = x^2 y(x^3 y^3 + 27)$

$\qquad = x^2 y[(xy)^3 + 3^3]$

$\qquad = x^2 y(xy + 3)(x^2 y^2 - 3xy + 9)$

  e. $(x + 7)^2 - 81y^2 = (x + 7)^2 - (9y)^2$

$\qquad\qquad\qquad\quad = (x + 7 + 9y)(x + 7 - 9y)$

Integrated Review Exercise Set

1. $(-y^2 + 6y - 1) + (3y^2 - 4y - 10)$

$\qquad = -y^2 + 6y - 1 + 3y^2 - 4y - 10$

$\qquad = 2y^2 + 2y - 11$

2. $(5z^4 - 6z^2 + z + 1) - (7z^4 - 2z + 1)$

$\qquad = 5z^4 - 6z^2 + z + 1 - 7z^4 + 2z - 1$

$\qquad = -2z^4 - 6z^2 + 3z$

3. $(x^2 - 6x + 2) - (x - 5) = x^2 - 6x + 2 - x + 5$

$\qquad\qquad\qquad\qquad\qquad = x^2 - 7x + 7$

4. $(2x^2 + 6x - 5) + (5x^2 - 10x) = 7x^2 - 4x - 5$

5. $(5x - 3)^2 = (5x)^2 - 2(5x)(3) + 3^2$

$\qquad\qquad\quad = 25x^2 - 30x + 9$

6. $(5x^2 - 14x - 3) - (5x + 1) = 5x^2 - 14x - 3 - 5x - 1$

$\qquad\qquad\qquad\qquad\qquad\quad = 5x^2 - 19x - 4$

7. $\dfrac{2x^4}{x} - \dfrac{3x^2}{x} + \dfrac{5x}{x} - \dfrac{2}{2} = 2x^3 - 3x + 5 - 1$

$\qquad\qquad\qquad\qquad\qquad = 2x^3 - 3x + 4$

8. $\begin{array}{r} x^2 - 3x - 2 \\ \times \qquad 4x - 1 \\ \hline -x^2 + 3x + 2 \\ 4x^3 - 12x^2 - 8x \qquad \\ \hline 4x^3 - 13x^2 - 5x + 2 \end{array}$

9. $x^2 - 8x + 16 - y^2 = (x - 4)^2 - y^2$

$\qquad\qquad\qquad\qquad = (x - 4 + y)(x - 4 - y)$

10. $12x^2 - 22x - 20 = 2(6x^2 - 11x - 10)$

$\qquad\qquad\qquad\qquad = 2(3x + 2)(2x - 5)$

11. $x^4 - x = x(x^3 - 1) = x(x - 1)(x^2 + x + 1)$

12. Let $y = 2x + 1$. Then

$$(2x+1)^2 - 3(2x+1) + 2$$
$$= y^2 - 3y + 2$$
$$= (y-2)(y-1)$$
$$= [(2x+1)-2][(2x+1)-1]$$
$$= (2x-1)(2x)$$
$$= 2x(2x-1)$$

13. $14x^2y - 2xy = 2xy(7x-1)$

14. $24ab^2 - 6ab = 6ab(4b-1)$

15. $4x^2 - 16 = 4(x^2 - 4) = 4(x+2)(x-2)$

16. $9x^2 - 81 = 9(x^2 - 9) = 9(x+3)(x-3)$

17. $3x^2 - 8x - 11 = (3x-11)(x+1)$

18. $5x^2 - 2x - 3 = (5x+3)(x-1)$

19. $4x^2 + 8x - 12 = 4(x^2 + 2x - 3)$
$$= 4(x+3)(x-1)$$

20. $6x^2 - 6x - 12 = 6(x^2 - x - 2)$
$$= 6(x-2)(x+1)$$

21. $4x^2 + 36x + 81 = (2x)^2 + 2 \cdot 2x \cdot 9 + 9^2$
$$= (2x+9)^2$$

22. $25x^2 + 40x + 16 = (5x)^2 + 2 \cdot 5x \cdot 4 + 4^2$
$$= (5x+4)^2$$

23. $8x^3 + 125y^3 = (2x)^3 + (5y)^3$
$$= (2x+5y)(4x^2 - 10xy + 25y^2)$$

24. $27x^3 - 64y^3 = (3x)^3 - (4y)^3$
$$= (3x-4y)(9x^2 + 12xy + 16y^2)$$

25. $64x^2y^3 - 8x^2 = 8x^2(8y^3 - 1)$
$$= 8x^2[(2y)^3 - 1^3]$$
$$= 8x^2(2y-1)(4y^2 + 2y + 1)$$

26. $27x^5y^4 - 216x^2y$
$$= 27x^2y(x^3y^3 - 8)$$
$$= 27x^2y[(xy)^3 - 2^3]$$
$$= 27x^2y(xy-2)(x^2y^2 + 2xy + 4)$$

27. $(x+5)^3 + y^3$
$$= [(x+5) + y][(x+5)^2 - (x+5)y + y^2]$$
$$= (x+y+5)(x^2 + 10x + 25 - xy - 5y + y^2)$$
$$= (x+y+5)(x^2 + 10x - xy - 5y + y^2 + 25)$$

28. $(y-1)^3 + 27x^3$
$$= (y-1)^3 + (3x)^3$$
$$= [(y-1) + 3x][(y-1)^2 - (y-1)(3x) + (3x)^2]$$
$$= (y-1+3x)(y^2 - 2y + 1 - 3xy + 3x + 9x^2)$$

29. Let $y = 5a - 3$. Then

$$(5a-3)^2 - 6(5a-3) + 9 = y^2 - 6y + 9$$
$$= (y-3)(y-3)$$
$$= (y-3)^2$$
$$= [(5a-3)-3]^2$$
$$= (5a-6)^2$$

30. Let $y = 4r + 1$. Then

$$(4r+1)^2 + 8(4r+1) + 16 = y^2 + 8y + 16$$
$$= (y+4)(y+4)$$
$$= (y+4)^2$$
$$= [(4r+1)+4]^2$$
$$= (4r+5)^2$$

31. $7x^2 - 63x = 7x(x-9)$

32. $20x^2 + 23x + 6 = (4x+3)(5x+2)$

33. $ab - 6a + 7b - 42 = a(b-6) + 7(b-6)$
$$= (a+7)(b-6)$$

34. $20x^2 - 220x + 600 = 20(x^2 - 11x + 30)$
$$= 20(x-6)(x-5)$$

35. $x^4 - 1 = (x^2)^2 - 1^2$
$$= (x^2 + 1)(x^2 - 1)$$
$$= (x^2 + 1)(x+1)(x-1)$$

36. $15x^2 - 20x = 5x(3x-4)$

37. $10x^2 - 7x - 33 = (5x - 11)(2x + 3)$

38. $45m^3n^3 - 27m^2n^2 = 9m^2n^2(5mn - 3)$

39. $5a^3b^3 - 50a^3b = 5a^3b(b^2 - 10)$

40. $x^4 + x = x(x^3 + 1)$
$$= x(x^3 + 1^3)$$
$$= x(x + 1)(x^2 - x + 1)$$

41. $16x^2 + 25$ is a prime polynomial.

42. $20x^3 + 20y^3 = 20(x^3 + y^3)$
$$= 20(x + y)(x^2 - xy + y^2)$$

43. $10x^3 - 210x^2 + 1100x = 10x(x^2 - 21x + 110)$
$$= 10x(x - 11)(x - 10)$$

44. $9y^2 - 42y + 49 = (3y)^2 - 2 \cdot 3y \cdot 7 + 7^2$
$$= (3y - 7)^2$$

45. $64a^3b^4 - 27a^3b$
$$= a^3b(64b^3 - 27)$$
$$= a^3b[(4b)^3 - 3^3]$$
$$= a^3b(4b - 3)(16b^2 + 12b + 9)$$

46. $y^4 - 16 = (y^2)^2 - 4^2$
$$= (y^2 + 4)(y^2 - 4)$$
$$= (y^2 + 4)(y + 2)(y - 2)$$

47. $2x^3 - 54 = 2(x^3 - 27)$
$$= 2(x^3 - 3^3)$$
$$= 2(x - 3)(x^2 + 3x + 9)$$

48. $2sr + 10s - r - 5 = 2s(r + 5) - 1(r + 5)$
$$= (2s - 1)(r + 5)$$

49. $3y^5 - 5y^4 + 6y - 10 = y^4(3y - 5) + 2(3y - 5)$
$$= (y^4 + 2)(3y - 5)$$

50. $64a^2 + b^2$ is a prime polynomial.

51. $100z^3 + 100 = 100(z^3 + 1)$
$$= 100(z + 1)(z^2 - z + 1)$$

52. $250x^4 - 16x = 2x(125x^3 - 8)$
$$= 2x[(5x)^3 - 2^3]$$
$$= 2x(5x - 2)(25x^2 + 10x + 4)$$

53. $4b^2 - 36b + 81 = (2b)^2 - 2 \cdot 2b \cdot 9 + 9^2$
$$= (2b - 9)^2$$

54. $2a^5 - a^4 + 6a - 3 = a^4(2a - 1) + 3(2a - 1)$
$$= (a^4 + 3)(2a - 1)$$

55. Let $x = y - 6$. Then
$$(y - 6)^2 + 3(y - 6) + 2 = x^2 + 3x + 2$$
$$= (x + 2)(x + 1)$$
$$= [(y - 6) + 2][(y - 6) + 1]$$
$$= (y - 4)(y - 5)$$

56. Let $x = c + 2$. Then
$$(c + 2)^2 - 6(c + 2) + 5 = x^2 - 6x + 5$$
$$= (x - 5)(x - 1)$$
$$= [(c + 2) - 5][(c + 2) - 1]$$
$$= (c - 3)(c + 1)$$

57. Area $= 3^2 - 4x^2 = 3^2 - (2x)^2 = (3 + 2x)(3 - 2x)$

Section 5.8 Practice Exercises

1. $(x + 8)(x - 5) = 0$
$$x + 8 = 0 \quad \text{or} \quad x - 5 = 0$$
$$x = -8 \quad \text{or} \quad x = 5$$
Both solutions check. The solution set is $\{-8, 5\}$.

2. $3x^2 + 10x - 8 = 0$
$$(3x - 2)(x + 4) = 0$$
$$3x - 2 = 0 \quad \text{or} \quad x + 4 = 0$$
$$3x = 2$$
$$x = \frac{2}{3} \quad \text{or} \quad x = -4$$
Both solutions check. The solution set is $\left\{-4, \frac{2}{3}\right\}$.

3.
$$x(3x+14) = -8$$
$$3x^2 + 14x = -8$$
$$3x^2 + 14x + 8 = 0$$
$$(3x+2)(x+4) = 0$$
$$3x+2 = 0 \quad \text{or} \quad x+4 = 0$$
$$3x = -2$$
$$x = -\frac{2}{3} \quad \text{or} \quad x = -4$$

The solutions are $-\dfrac{2}{3}$ and -4.

4.
$$8(x^2+3)+4 = -8x(x+3)+19$$
$$8x^2 + 24 + 4 = -8x^2 - 24x + 19$$
$$16x^2 + 24x + 9 = 0$$
$$(4x+3)(4x+3) = 0$$
$$4x+3 = 0 \quad \text{or} \quad 4x+3 = 0$$
$$4x = -3 \quad \text{or} \quad 4x = -3$$
$$x = -\frac{3}{4} \quad \text{or} \quad x = -\frac{3}{4}$$

The solution is $-\dfrac{3}{4}$.

5.
$$4x^2 = \frac{15}{2}x + 1$$
$$2(4x^2) = 2\left(\frac{15}{2}x + 1\right)$$
$$8x^2 = 15x + 2$$
$$8x^2 - 15x - 2 = 0$$
$$(8x+1)(x-2) = 0$$
$$8x+1 = 0 \quad \text{or} \quad x-2 = 0$$
$$8x = -1$$
$$x = -\frac{1}{8} \quad \text{or} \quad x = 2$$

The solutions are $-\dfrac{1}{8}$ and 2.

6.
$$x^3 = 2x^2 + 3x$$
$$x^3 - 2x^2 - 3x = 0$$
$$x(x^2 - 2x - 3) = 0$$
$$x(x-3)(x+1) = 0$$
$$x = 0 \quad \text{or} \quad x-3 = 0 \quad \text{or} \quad x+1 = 0$$
$$x = 3 \quad \text{or} \quad x = -1$$

The solutions are -1, 0, and 3.

7.
$$x^3 - 9x = 18 - 2x^2$$
$$x^3 + 2x^2 - 9x - 18 = 0$$
$$x^2(x+2) - 9(x+2) = 0$$
$$(x+2)(x^2 - 9) = 0$$
$$(x+2)(x+3)(x-3) = 0$$
$$x+2 = 0 \quad \text{or} \quad x+3 = 0 \quad \text{or} \quad x-3 = 0$$
$$x = -2 \quad \text{or} \quad x = -3 \quad \text{or} \quad x = 3$$

The solutions are -3, -2, and 3.

8. Let $h = 0$.
$$-16t^2 + 96t = 0$$
$$-16t(t-6) = 0$$
$$-16t = 0 \quad \text{or} \quad t-6 = 0$$
$$t = 0 \quad \text{or} \quad t = 6$$

The rocket will return to the ground in 6 seconds.

9. Let x = first even integer, then
$x + 2$ = next even integer, and
$x + 4$ = third even integer.
$$x^2 + (x+2)^2 = (x+4)^2$$
$$x^2 + x^2 + 4x + 4 = x^2 + 8x + 16$$
$$x^2 - 4x - 12 = 0$$
$$(x-6)(x+2) = 0$$
$$x-6 = 0 \quad \text{or} \quad x+2 = 0$$
$$x = 6 \quad \text{or} \quad x = -2$$

Discard $x = -2$ since length cannot be negative. The legs are $x = 6$ units, $x + 2 = 8$ units, and $x + 4 = 10$ units.

10. The graph of $f(x) = (x-1)(x+3)$ has two x-intercepts, $(1, 0)$ and $(-3, 0)$, so the graph is C.
The graph of $g(x) = x(x+3)(x-2)$ has three x-intercepts, $(0, 0)$, $(-3, 0)$, and $(2, 0)$, so the graph is A.
The graph of $h(x) = (x-3)(x+2)(x-2)$ has three x-intercepts, $(3, 0)$, $(-2, 0)$, and $(2, 0)$, so the graph is B.

Vocabulary, Readiness & Video Check 5.8

1.
$$(x-3)(x+5) = 0$$
$$x-3 = 0 \quad \text{or} \quad x+5 = 0$$
$$x = 3 \quad \text{or} \quad x = -5$$
The solutions are -5, 3.

2. $(y+5)(y+3)=0$

　　$y+5=0$　or　$y+3=0$

　　　$y=-5$　or　　$y=-3$

　The solutions are –5, –3.

3. $(z-3)(z+7)=0$

　　$z-3=0$　or　$z+7=0$

　　　$z=3$　or　　$z=-7$

　The solutions are –7, 3.

4. $(c-2)(c-4)=0$

　　$c-2=0$　or　$c-4=0$

　　　$c=2$　or　　$c=4$

　The solutions are 2, 4.

5. $x(x-9)=0$

　　$x=0$　or　$x-9=0$

　　　　　　　　$x=9$

　The solutions are 0, 9.

6. $w(w+7)=0$

　　$w=0$　or　$w+7=0$

　　　　　　　　$w=-7$

　The solutions are –7, 0.

7. One side of the equation must be a factored polynomial and the other side must be zero.

8. A negative solution makes no sense in the context of our application since we are looking for a positive length.

9. Finding the *x*-intercepts of any equation in two variables means you let $y=0$ and solve for *x*. Doing this with a quadratic equation gives us an equation = 0, which may be solved by factoring.

Exercise Set 5.8

1. $(x+3)(3x-4)=0$

　　$x+3=0$　or　$3x-4=0$

　　$x=-3$　or　　$3x=4$

　　　　　　　　　$x=\dfrac{4}{3}$

　The solutions are $-3, \dfrac{4}{3}$.

3. $3(2x-5)(4x+3)=0$

　　$2x-5=0$　or　$4x+3=0$

　　　$2x=5$　or　　$4x=-3$

　　$x=\dfrac{5}{2}$　or　　$x=-\dfrac{3}{4}$

　The solutions are $-\dfrac{3}{4}, \dfrac{5}{2}$.

5. $x^2+11x+24=0$

　　$(x+8)(x+3)=0$

　　$x+8=0$　or　$x+3=0$

　　　$x=-8$　or　　$x=-3$

　The solutions are –8, –3.

7. $12x^2+5x-2=0$

　　$(4x-1)(3x+2)=0$

　　$4x-1=0$　or　$3x+2=0$

　　　$4x=1$　or　　$3x=-2$

　　$x=\dfrac{1}{4}$　or　　$x=-\dfrac{2}{3}$

　The solutions are $-\dfrac{2}{3}, \dfrac{1}{4}$.

9. $z^2+9=10z$

　　$z^2-10z+9=0$

　　$(z-9)(z-1)=0$

　　$z-9=0$　or　$z-1=0$

　　　$z=9$　or　　$z=1$

　The solutions are 1, 9.

11. $x(5x+2)=3$

　　$5x^2+2x-3=0$

　　$(5x-3)(x+1)=0$

　　$5x-3=0$　or　$x+1=0$

　　　$5x=3$　or　　$x=-1$

　　$x=\dfrac{3}{5}$

　The solutions are $-1, \dfrac{3}{5}$.

13. $x^2-6x=x(8+x)$

　　$x^2-6x=8x+x^2$

　　　$-14x=0$

　　　　$x=0$

　The solution is 0.

15. $\dfrac{z^2}{6} - \dfrac{z}{2} - 3 = 0$

$z^2 - 3z - 18 = 0$

$(z-6)(z+3) = 0$

$z - 6 = 0$ or $z + 3 = 0$

$z = 6$ or $z = -3$

The solutions are -3, 6.

17. $\dfrac{x^2}{2} + \dfrac{x}{20} = \dfrac{1}{10}$

$10x^2 + x = 2$

$10x^2 + x - 2 = 0$

$(5x-2)(2x+1) = 0$

$5x - 2 = 0$ or $2x + 1 = 0$

$5x = 2$ or $2x = -1$

$x = \dfrac{2}{5}$ or $x = -\dfrac{1}{2}$

The solutions are $-\dfrac{1}{2}$, $\dfrac{2}{5}$.

19. $\dfrac{4t^2}{5} = \dfrac{t}{5} + \dfrac{3}{10}$

$8t^2 = 2t + 3$

$8t^2 - 2t - 3 = 0$

$(4t-3)(2t+1) = 0$

$4t - 3 = 0$ or $2t + 1 = 0$

$4t = 3$ or $2t = -1$

$t = \dfrac{3}{4}$ or $t = -\dfrac{1}{2}$

The solutions are $-\dfrac{1}{2}$, $\dfrac{3}{4}$.

21. $(x+2)(x-7)(3x-8) = 0$

$x + 2 = 0$ or $x - 7 = 0$ or $3x - 8 = 0$

$x = -2$ or $x = 7$ or $3x = 8$

$x = \dfrac{8}{3}$

The solutions are -2, 7, $\dfrac{8}{3}$.

23. $(x^2 - 1)(6x+1) = 0$

$(x+1)(x-1)(6x+1) = 0$

$x + 1 = 0$ or $x - 1 = 0$ or $6x + 1 = 0$

$x = -1$ or $x = 1$ or $x = -\dfrac{1}{6}$

The solutions are -1, 1, $-\dfrac{1}{6}$.

25. $y^3 = 9y$

$y^3 - 9y = 0$

$y(y^2 - 9) = 0$

$y(y+3)(y-3) = 0$

$y = 0$ or $y + 3 = 0$ or $y - 3 = 0$

$y = -3$ or $y = 3$

The solutions are -3, 0, 3.

27. $x^3 - x = 2x^2 - 2$

$x^3 - 2x^2 - x + 2 = 0$

$x^2(x-2) - 1(x-2) = 0$

$(x^2 - 1)(x-2) = 0$

$(x+1)(x-1)(x-2) = 0$

$x + 1 = 0$ or $x - 1 = 0$ or $x - 2 = 0$

$x = -1$ or $x = 1$ or $x = 2$

The solutions are -1, 1, 2.

29. $(2x+7)(x-10) = 0$

$2x + 7 = 0$ or $x - 10 = 0$

$2x = -7$ or $x = 10$

$x = -\dfrac{7}{2}$

The solutions are $-\dfrac{7}{2}$, 10.

31. $3x(x-5) = 0$

$3x = 0$ or $x - 5 = 0$

$x = 0$ or $x = 5$

The solutions are 0, 5.

33. $x^2 - 2x - 15 = 0$

$(x-5)(x+3) = 0$

$x - 5 = 0$ or $x + 3 = 0$

$x = 5$ or $x = -3$

The solutions are -3, 5.

35. $12x^2 + 2x - 2 = 0$

$2(6x^2 + x - 1) = 0$

$2(3x-1)(2x+1) = 0$

$3x - 1 = 0$ or $2x + 1 = 0$

$3x = 1$ or $2x = -1$

$x = \dfrac{1}{3}$ or $x = -\dfrac{1}{2}$

The solutions are $-\dfrac{1}{2}$, $\dfrac{1}{3}$.

37.
$$w^2 - 5w = 36$$
$$w^2 - 5w - 36 = 0$$
$$(w - 9)(w + 4) = 0$$
$$w - 9 = 0 \text{ or } w + 4 = 0$$
$$w = 9 \text{ or } \qquad w = -4$$
The solutions are –4, 9.

39. $25x^2 - 40x + 16 = 0$
$$(5x - 4)^2 = 0$$
$$5x - 4 = 0$$
$$5x = 4$$
$$x = \frac{4}{5}$$

The solution is $\frac{4}{5}$.

41.
$$2r^3 + 6r^2 = 20r$$
$$2r^3 + 6r^2 - 20r = 0$$
$$2r(r^2 + 3r - 10) = 0$$
$$2r(r + 5)(r - 2) = 0$$
$$2r = 0 \text{ or } r + 5 = 0 \quad \text{ or } r - 2 = 0$$
$$r = 0 \text{ or } \qquad r = -5 \text{ or } \qquad r = 2$$
The solutions are –5, 0, 2.

43. $z(5z - 4)(z + 3) = 0$
$$z = 0 \quad \text{ or } 5z - 4 = 0 \text{ or } z + 3 = 0$$
$$5z = 4 \text{ or } \qquad z = -3$$
$$z = \frac{4}{5}$$

The solutions are $-3, 0, \frac{4}{5}$.

45. $2z(z + 6) = 2z^2 + 12z - 8$
$$2z^2 + 12z = 2z^2 + 12z - 8$$
$$0 = -8 \text{ False}$$
No solutions exist; $\varnothing$.

47. $(x - 1)(x + 4) = 24$
$$x^2 + 3x - 4 = 24$$
$$x^2 + 3x - 28 = 0$$
$$(x + 7)(x - 4) = 0$$
$$x + 7 = 0 \quad \text{ or } x - 4 = 0$$
$$x = -7 \text{ or } \qquad x = 4$$
The solutions are –7, 4.

49. $\dfrac{x^2}{4} - \dfrac{5}{2}x + 6 = 0$
$$x^2 - 10x + 24 = 0$$
$$(x - 6)(x - 4) = 0$$
$$x - 6 = 0 \text{ or } x - 4 = 0$$
$$x = 6 \text{ or } \qquad x = 4$$
The solutions are 4, 6.

51.
$$y^2 + \frac{1}{4} = -y$$
$$4y^2 + 1 = -4y$$
$$4y^2 + 4y + 1 = 0$$
$$(2y + 1)^2 = 0$$
$$2y + 1 = 0$$
$$2y = -1$$
$$y = -\frac{1}{2}$$

The solution is $-\dfrac{1}{2}$.

53.
$$y^3 + 4y^2 = 9y + 36$$
$$y^3 + 4y^2 - 9y - 36 = 0$$
$$y^2(y + 4) - 9(y + 4) = 0$$
$$(y^2 - 9)(y + 4) = 0$$
$$(y + 3)(y - 3)(y + 4) = 0$$
$$y + 3 = 0 \quad \text{ or } y - 3 = 0 \text{ or } y + 4 = 0$$
$$y = -3 \text{ or } \qquad y = 3 \text{ or } \qquad y = -4$$
The solutions are –4, –3, 3.

55.
$$2x^3 = 50x$$
$$2x^3 - 50x = 0$$
$$2x(x^2 - 25) = 0$$
$$2x(x + 5)(x - 5) = 0$$
$$2x = 0 \text{ or } x + 5 = 0 \quad \text{ or } x - 5 = 0$$
$$x = 0 \text{ or } \qquad x = -5 \text{ or } \qquad x = 5$$
The solutions are –5, 0, 5.

57.
$$x^2 + (x + 1)^2 = 61$$
$$x^2 + x^2 + 2x + 1 = 61$$
$$2x^2 + 2x - 60 = 0$$
$$2(x^2 + x - 30) = 0$$
$$2(x + 6)(x - 5) = 0$$
$$x + 6 = 0 \quad \text{ or } x - 5 = 0$$
$$x = -6 \text{ or } \qquad x = 5$$
The solutions are –6, 5.

59.
$$m^2(3m-2) = m$$
$$3m^3 - 2m^2 = m$$
$$3m^3 - 2m^2 - m = 0$$
$$m(3m^2 - 2m - 1) = 0$$
$$m(3m+1)(m-1) = 0$$
$$m = 0 \quad \text{or} \quad 3m+1 = 0 \quad \text{or} \quad m-1 = 0$$
$$3m = -1 \quad \text{or} \quad m = 1$$
$$m = -\frac{1}{3}$$

The solutions are $-\frac{1}{3}$, 0, 1.

61.
$$3x^2 = -x$$
$$3x^2 + x = 0$$
$$x(3x+1) = 0$$
$$x = 0 \quad \text{or} \quad 3x+1 = 0$$
$$3x = -1$$
$$x = -\frac{1}{3}$$

The solutions are $-\frac{1}{3}$, 0.

63.
$$x(x-3) = x^2 + 5x + 7$$
$$x^2 - 3x = x^2 + 5x + 7$$
$$-8x = 7$$
$$x = -\frac{7}{8}$$

The solution is $-\frac{7}{8}$.

65.
$$3(t-8) + 2t = 7 + t$$
$$3t - 24 + 2t = 7 + t$$
$$5t - 24 = 7 + t$$
$$4t = 31$$
$$t = \frac{31}{4}$$

The solution is $\frac{31}{4}$.

67.
$$-3(x-4) + x = 5(3-x)$$
$$-3x + 12 + x = 15 - 5x$$
$$-2x + 12 = 15 - 5x$$
$$3x = 3$$
$$x = 1$$

The solution is 1.

69. Let n = the one number and
$n + 5$ = the other number.
$$n(n+5) = 66$$
$$n^2 + 5n - 66 = 0$$
$$(n+11)(n-6) = 0$$
$$n + 11 = 0 \quad \text{or} \quad n - 6 = 0$$
$$n = -11 \quad \text{or} \quad n = 6$$
The two solutions are –11 and –6 and 6 and 11.

71. Let d = amount of cable needed. Then from the
Pythagorean theorem, $d^2 = 45^2 + 60^2 = 5625$ so
$d = \sqrt{5625} = 75$ ft.

73.
$$C(x) = x^2 - 15x + 50$$
$$9500 = x^2 - 15x + 50$$
$$0 = x^2 - 15x - 9450$$
$$0 = (x-105)(x+90)$$
$$x - 105 = 0 \quad \text{or} \quad x + 90 = 0$$
$$x = 105 \quad \text{or} \quad x = -90$$
Disregard the negative; 105 units.

75. Let x = one leg of a right triangle and
$x - 3$ = the other leg of the right triangle.
$$15^2 = x^2 + (x-3)^2$$
$$225 = x^2 + x^2 - 6x + 9$$
$$225 = 2x^2 - 6x + 9$$
$$0 = 2x^2 - 6x - 216$$
$$0 = 2(x^2 - 3x - 108)$$
$$0 = 2(x-12)(x+9)$$
$$x - 12 = 0 \quad \text{or} \quad x + 9 = 0$$
$$x = 12 \quad \text{or} \quad x = -9$$
Disregarding the negative solution, we find that
one leg of the right triangle is 12 cm and the
other leg is 9 cm.

77. Note that the outer rectangle has lengths of
$2x + 12$ and $2x + 16$. Thus, the area of the border
is $(2x + 12)(2x + 16) - 12 \cdot 16$. Set this equal to
128 and solve for x.
$$(2x+12)(2x+16) - 12 \cdot 16 = 128$$
$$4x^2 + 56x + 192 - 192 = 128$$
$$4x^2 + 56x = 128$$
$$4x^2 + 56x - 128 = 0$$
$$x^2 + 14x - 32 = 0$$
$$(x+16)(x-2) = 0$$
$$x + 16 = 0 \quad \text{or} \quad x - 2 = 0$$
$$x = -16 \quad \text{or} \quad x = 2$$

Since x must be positive, we see that $x = 2$ inches.

79. The sunglasses will hit the ground when $h(t)$ equals 0.

$$-16t^2 + 1600 = 0$$
$$-16(t^2 - 100) = 0$$
$$-16(t - 10)(t + 10) = 0$$
$$t - 10 = 0 \quad \text{or} \quad t + 10 = 0$$
$$t = 10 \quad \text{or} \qquad t = -10$$

The sunglasses will hit the ground 10 seconds after being dropped.

81. Let the width of the floor $= w$. Then the length is $2w - 3$ and so the area is

$$(2w - 3)w = 90$$
$$2w^2 - 3w - 90 = 0$$
$$(2w - 15)(w + 6) = 0$$
$$2w - 15 = 0 \qquad \text{or} \quad w + 6 = 0$$
$$2w = 15 \qquad \text{or} \qquad w = -6$$
$$w = \frac{15}{2} = 7.5$$

Disregard –6.

$$2w - 3 = 2(7.5) - 3 = 15 - 3 = 12$$

The width is 7.5 ft and the length is 12 ft.

83.
$$0.5x^2 = 50$$
$$0.5x^2 - 50 = 0$$
$$5x^2 - 500 = 0$$
$$x^2 - 100 = 0$$
$$(x + 10)(x - 10) = 0$$
$$x + 10 = 0 \qquad \text{or } x - 10 = 0$$
$$x = -10 \quad \text{or} \qquad x = 10$$

Disregard the negative solution. A 10-inch square tier is needed, provided each person has one serving.

85. The object will hit the ground when $h(t)$ equals 0.

$$-16t^2 + 80t + 576 = 0$$
$$-16(t^2 - 5t - 36) = 0$$
$$-16(t - 9)(t + 4) = 0$$
$$t - 9 = 0 \quad \text{or } t + 4 = 0$$
$$t = 9 \quad \text{or} \qquad t = -4$$

The object will hit the ground 9 seconds after being dropped.

87. E; x-intercepts $(2, 0), (-5, 0)$

89. F; x-intercepts $(0, 0), (-3, 0), (3, 0)$

91. B; x-intercepts $\left(-\dfrac{1}{2}, 0\right), (-4, 0)$

93. $(-3, 0), (0, 2)$; function, because any vertical line will cross only once.

95. $(-4, 0), (0, 2), (4, 0), (0, -2)$; not a function, because a vertical line can be drawn that crosses twice.

97. $(x - 5)(x + 2) = 0$
$$x - 5 = 0 \quad \text{or} \quad x + 2 = 0$$
$$x = 5 \quad \text{or} \qquad x = -2$$

99. $\qquad y(y - 5) = -6$
$$y^2 - 5y + 6 = 0$$
$$(y - 2)(y - 3) = 0$$
$$y - 2 = 0 \quad \text{or} \quad y - 3 = 0$$
$$y = 2 \quad \text{or} \qquad y = 3$$

101. $(x^2 + x - 6)(3x^2 - 14x - 5) = 0$

$$x^2 + x - 6 = 0 \quad \text{or} \quad 3x^2 - 14x - 5 = 0$$
$$(x + 3)(x - 2) = 0 \quad \text{or} \quad (3x + 1)(x - 5) = 0$$
$$x + 3 = 0 \quad \text{or } x - 2 = 0 \text{ or } 3x + 1 = 0$$
$$x = -3 \qquad x = 2 \qquad 3x = -1$$
$$x = -\frac{1}{3}$$

or $x - 5 = 0$
$$x = 5$$

The solutions are $-3, -\dfrac{1}{3}, 2, 5$.

103. $\qquad y = -2x^2 + 58x - 42$
$$338 = -2x^2 + 58x - 42$$
$$0 = -2x^2 + 58x - 380$$
$$0 = -2(x^2 - 29x + 190)$$
$$0 = -2(x - 10)(x - 19)$$
$$x - 10 = 0 \quad \text{or} \quad x - 19 = 0$$
$$x = 10 \quad \text{or} \qquad x = 19$$

The solutions are 10, 19.

105. answers may vary

107. no; answers may vary

109. answers may vary

111. answers may vary

113. answers may vary

Chapter 5 Vocabulary Check

1. A <u>polynomial</u> is a finite sum of terms in which all variables are raised to nonnegative integer powers and no variables appear in any denominator.

2. <u>Factoring</u> is the process of writing a polynomial as a product.

3. <u>Exponents</u> are used to write repeated factors in a more compact form.

4. The <u>degree of a term</u> is the sum of the exponents on the variables contained in the term.

5. A <u>monomial</u> is a polynomial with three terms.

6. If a is not 0, $a^0 = \underline{1}$.

7. A <u>trinomial</u> is a polynomial with three terms.

8. A polynomial equation of degree 2 is also called a <u>quadratic equation</u>.

9. A positive number is written in <u>scientific notation</u> if it is written as the product of a number a, where $1 \le a < 10$, and a power of 10.

10. The <u>degree of a polynomial</u> is the largest degree of all its terms.

11. A <u>binomial</u> is a polynomial with two terms.

12. If a and b are real numbers and $a \cdot b = \underline{0}$, then $a = 0$ and $b = 0$.

Chapter 5 Review

1. $(-2)^2 = (-2)(-2) = 4$

2. $(-3)^4 = (-3)(-3)(-3)(-3) = 81$

3. $-2^2 = -(2 \cdot 2) = -4$

4. $-3^4 = -(3 \cdot 3 \cdot 3 \cdot 3) = -81$

5. $8^0 = 1$

6. $-9^0 = -1$

7. $-4^{-2} = -\dfrac{1}{4^2} = -\dfrac{1}{16}$

8. $(-4)^2 = \dfrac{1}{(-4)^2} = \dfrac{1}{16}$

9. $-xy^2 \cdot y^3 \cdot xy^2 z = -x^{1+1} y^{2+3+2} z = -x^2 y^7 z$

10. $(-4xy)(-3xy^2 b) = (-4)(-3)x^{1+1} y^{1+2} b$
$\qquad = 12x^2 y^3 b$

11. $a^{-14} a^5 = a^{-14+5} = a^{-9} = \dfrac{1}{a^9}$

12. $\dfrac{a^{16}}{a^{17}} = a^{16-17} = a^{-1} = \dfrac{1}{a}$

13. $\dfrac{x^{-7}}{x^4} = x^{-7-4} = x^{-11} = \dfrac{1}{x^{11}}$

14. $\dfrac{9a(a^{-3})}{18a^{15}} = \dfrac{a^{1-3-15}}{2} = \dfrac{a^{-17}}{2} = \dfrac{1}{2a^{17}}$

15. $\dfrac{y^{6p-3}}{y^{6p+2}} = y^{(6p-3)-(6p+2)}$
$\qquad = y^{6p-3-6p-2}$
$\qquad = y^{-5}$
$\qquad = \dfrac{1}{y^5}$

16. $(3x^{2a+b} y^{-3b})^2 = 3^2 x^{2(2a+b)} y^{2(-3b)}$
$\qquad = 9x^{4a+2b} y^{-6b}$ or $\dfrac{9x^{4a+2b}}{y^{6b}}$

17. $36{,}890{,}000 = 3.689 \times 10^7$

18. $0.000362 = 3.62 \times 10^{-4}$

19. $1.678 \times 10^{-6} = 0.000001678$

20. $4.1 \times 10^5 = 410{,}000$

21. $(8^5)^3 = 8^{5 \cdot 3} = 8^{15}$

22. $\left(\dfrac{a}{4}\right)^2 = \dfrac{a^2}{4^2} = \dfrac{a^2}{16}$

23. $(3x)^3 = 3^3 x^3 = 27x^3$

24. $(-4x)^{-2} = \dfrac{1}{(-4x)^2} = \dfrac{1}{(-4)^2 x^2} = \dfrac{1}{16x^2}$

25. $\left(\dfrac{6x}{5}\right)^2 = \dfrac{(6x)^2}{5^2} = \dfrac{36x^2}{25}$

26. $(8^6)^{-3} = 8^{6(-3)} = 8^{-18} = \dfrac{1}{8^{18}}$

27. $\left(\dfrac{4}{3}\right)^{-2} = \dfrac{4^{-2}}{3^{-2}} = \dfrac{3^2}{4^2} = \dfrac{9}{16}$

28. $(-2x^3)^{-3} = \dfrac{1}{(-2x^3)^3}$

$= \dfrac{1}{(-2)^3 (x^3)^3}$

$= \dfrac{1}{-8x^9}$

$= -\dfrac{1}{8x^9}$

29. $\left(\dfrac{8p^6}{4p^4}\right)^{-2} = (2p^2)^{-2} = 2^{-2} p^{-4} = \dfrac{1}{4p^4}$

30. $(-3x^{-2} y^2)^3 = (-3)^3 (x^{-2})^3 (y^2)^3$

$= -27x^{-6} y^6$

$= -\dfrac{27y^6}{x^6}$

31. $\left(\dfrac{x^{-5} y^{-3}}{z^3}\right)^{-5} = \dfrac{x^{25} y^{15}}{z^{-15}} = x^{25} y^{15} z^{15}$

32. $\dfrac{4^{-1} x^3 yz}{x^{-2} yx^4} = \dfrac{x^{3-(-2)-4} z}{4} = \dfrac{x^{3+2-4} z}{4} = \dfrac{xz}{4}$

33. $(5xyz)^{-4} (x^{-2})^{-3} = \dfrac{1}{(5xyz)^4} x^6$

$= \dfrac{x^6}{5^4 x^4 y^4 z^4}$

$= \dfrac{x^2}{625 y^4 z^4}$

34. $\dfrac{2(3yz)^{-3}}{y^{-3}} = \dfrac{2(3)^{-3} y^{-3} z^{-3}}{y^{-3}} = \dfrac{2}{3^3 z^3} = \dfrac{2}{27z^3}$

35. $x^{4a} (3x^{5a})^3 = x^{4a} (3^3 x^{15a})$

$= 27x^{4a+15a}$

$= 27x^{19a}$

36. $\dfrac{4y^{3x-3}}{2y^{2x+4}} = 2y^{(3x-3)-(2x+4)}$

$= 2y^{3x-3-2x-4}$

$= 2y^{x-7}$

37. $\dfrac{(0.00012)(144{,}000)}{0.0003} = \dfrac{(1.2 \times 10^{-4})(1.44 \times 10^5)}{3 \times 10^{-4}}$

$= 0.576 \times 10^5$

$= 5.76 \times 10^4$

38. $\dfrac{(-0.00017)(0.00039)}{3000}$

$= \dfrac{(-1.7 \times 10^{-4})(3.9 \times 10^{-4})}{3 \times 10^3}$

$= -2.21 \times 10^{-4-4-3}$

$= -2.21 \times 10^{-11}$

39. $\dfrac{27x^{-5} y^5}{18x^{-6} y^2} \cdot \dfrac{x^4 y^{-2}}{x^{-2} y^3} = \dfrac{3x^{-5+4} y^{5-2}}{2x^{-6-2} y^{2+3}}$

$= \dfrac{3x^{-1} y^3}{2x^{-8} y^5}$

$= \dfrac{3}{2} x^{-1-(-8)} y^{3-5}$

$= \dfrac{3}{2} x^7 y^{-2}$

$= \dfrac{3x^7}{2y^2}$

40. $\dfrac{3x^5}{y^{-4}} \cdot \dfrac{(3xy^{-3})^{-2}}{(z^{-3})^{-4}} = \dfrac{3x^5 \cdot 3^{-2} x^{-2} y^6}{y^{-4} z^{12}}$

$= \dfrac{3^{1-2} x^{5-2} y^{6-(-4)}}{z^{12}}$

$= \dfrac{3^{-1} x^3 y^{10}}{z^{12}}$

$= \dfrac{x^3 y^{10}}{3z^{12}}$

41. The degree of the polynomial
$x^2 y - 3xy^3 z + 5x + 7y$ is the degree of the term
$-3xy^3 z$ which is 5.

42. $3x + 2$ has degree 1.

43. $4x + 8x - 6x^2 - 6x^2 y = (4+8)x - 6x^2 - 6x^2 y$

$= 12x - 6x^2 - 6x^2 y$

44. $-8xy^3 + 4xy^3 - 3x^3 y = (-8+4)xy^3 - 3x^3 y$

$= -4xy^3 - 3x^3 y$

45. $(3x + 7y) + (4x^2 - 3x + 7) + (y - 1)$

$= 3x + 7y + 4x^2 - 3x + 7 + y - 1$

$= 4x^2 + (3-3)x + (7+1)y + (7-1)$

$= 4x^2 + 8y + 6$

46. $(4x^2 - 6xy + 9y^2) - (8x^2 - 6xy - y^2)$

$= 4x^2 - 6xy + 9y^2 - 8x^2 + 6xy + y^2$

$= (4-8)x^2 + (9+1)y^2$

$= -4x^2 + 10y^2$

47. $(3x^2 - 4b + 28) + (9x^2 - 30) - (4x^2 - 6b + 20)$

$= 3x^2 - 4b + 28 + 9x^2 - 30 - 4x^2 + 6b - 20$

$= (3+9-4)x^2 + (-4+6)b + (28-30-20)$

$= 8x^2 + 2b - 22$

48. $(9xy + 4x^2 + 18) + (7xy - 4x^3 - 9x)$

$= 9xy + 4x^2 + 18 + 7xy - 4x^3 - 9x$

$= -4x^3 + 4x^2 + (9+7)xy - 9x + 18$

$= -4x^3 + 4x^2 + 16xy - 9x + 18$

49. $2x - (3x - 5) = 2x - 3x + 5$

$= (2-3)x + 5$

$= -x + 5$

50. $(3x^2 y - 7xy - 4) + (9x^2 y + x) - (x - 7)$

$= 3x^2 y - 7xy - 4 + 9x^2 y + x - x + 7$

$= (3+9)x^2 y - 7xy + (-4+7)$

$= 12x^2 y - 7xy + 3$

51. $\begin{array}{r} x^2 - 5x + 7 \\ -(x+4) \\ \hline x^2 - 6x + 3 \end{array}$

52. $\begin{array}{r} x^3 + 2xy^2 - y \\ + (x - 4xy^2 - 7) \\ \hline x^3 + x - 2xy^2 - y - 7 \end{array}$

53. $P(6) = 9(6)^2 - 7(6) + 8 = 290$

54. $P(0) = 9(0)^2 - 7(0) + 8 = 8$

55. $P(-2) = 9(-2)^2 - 7(-2) + 8 = 58$

56. $P(-3) = 9(-3)^2 - 7(-3) + 8 = 110$

57. $P(x) + Q(x) = (2x - 1) + (x^2 + 2x - 5)$

$= 2x - 1 + x^2 + 2x - 5$

$= x^2 + 4x - 6$

58. $2 \cdot P(x) - Q(x) = 2(2x - 1) - (x^2 + 2x - 5)$

$= 4x - 2 - x^2 - 2x + 5$

$= -x^2 + 2x + 3$

59. $2(2x^2 y - 6x + 1) + 2(x^2 y + 5)$

$= 4x^2 y - 12x + 2 + 2x^2 y + 10$

$= (6x^2 y - 12x + 12)$ cm

60. $(3x + 11) + (3x + 11) + 3x = 3x + 3x + 3x + 11 + 11$

$= 9x + 22$

The perimeter is $(9x + 22)$ inches.

61. $-6x(4x^2 - 6x + 1) = -24x^3 + 36x^2 - 6x$

62. $-4ab^2(3ab^3 + 7ab + 1)$

$= -4ab^2(3ab^3) - 4ab^2(7ab) - 4ab^2(1)$

$= -12a^2 b^5 - 28a^2 b^3 - 4ab^2$

63. $(x-4)(2x+9) = 2x^2 + 9x - 8x - 36$
$$= 2x^2 + x - 36$$

64. $(-3xa+4b)^2 = (-3xa)^2 + 2(-3xa)(4b) + (4b)^2$
$$= 9x^2 a^2 - 24xab + 16b^2$$

65.
$$9x^2 + 4x + 1$$
$$\underline{\qquad\qquad 4x - 3}$$
$$-27x^2 - 12x - 3$$
$$\underline{36x^3 + 16x^2 \;\; + 4x \qquad}$$
$$36x^3 - 11x^2 \;\; - 8x - 3$$

66. $(2x-1)(x^2+2x-5) = 2x(x^2+2x-5) - 1(x^2+2x-5)$
$$= 2x^3 + 4x^2 - 10x - x^2 - 2x + 5$$
$$= 2x^3 + 3x^2 - 12x + 5$$

67. $(5x-9y)(3x+9y) = 15x^2 + 45xy - 27xy + 81y^2$
$$= 15x^2 + 18xy - 81y^2$$

68. $\left(x - \dfrac{1}{3}\right)\left(x + \dfrac{2}{3}\right) = x^2 + \dfrac{2}{3}x - \dfrac{1}{3}x - \dfrac{1}{3}\left(\dfrac{2}{3}\right)$
$$= x^2 + \dfrac{1}{3}x - \dfrac{2}{9}$$

69. $(x^2+9x+1)^2 = (x^2+9x+1)(x^2+9x+1)$
$$= x^2(x^2+9x+1) + 9x(x^2+9x+1) + 1(x^2+9x+1)$$
$$= x^4 + 9x^3 + x^2 + 9x^3 + 81x^2 + 9x + x^2 + 9x + 1$$
$$= x^4 + 18x^3 + 83x^2 + 18x + 1$$

70. $(m^2+m-2)^2 = (m^2+m-2)(m^2+m-2)$
$$= m^2(m^2+m-2) + m(m^2+m-2) - 2(m^2+m-2)$$
$$= m^4 + m^3 - 2m^2 + m^3 + m^2 - 2m - 2m^2 - 2m + 4$$
$$= m^4 + 2m^3 - 3m^2 - 4m + 4$$

71. $(3x-y)^2 = (3x)^2 - 2(3x)y + y^2$
$$= 9x^2 - 6xy + y^2$$

72. $(4x+9)^2 = (4x)^2 + 2(4x)(9) + 9^2$
$$= 16x^2 + 72x + 81$$

73. $(x+3y)(x-3y) = x^2 - (3y)^2 = x^2 - 9y^2$

74. $[4+(3a-b)][4-(3a-b)]$
$= 4^2 - (3a-b)^2$
$= 16 - [(3a)^2 - 2(3a)b + b^2]$
$= 16 - (9a^2 - 6ab + b^2)$
$= 16 - 9a^2 + 6ab - b^2$

75. Area $= lw$
$= (3y+7z)(3y-7z)$
$= (3y)^2 - (7z)^2$
$= 9y^2 - 49z^2$

The area of the rectangle is $(9y^2 - 49z^2)$ square units.

76. $(11y+9)^2 = (11y)^2 + 2(11y)(9) + 9^2$
$= 121y^2 + 198y + 81$
The area of the square is
$(121y^2 + 198y + 81)$ square units.

77. $4a^b(3a^{b+2} - 7) = 4a^b(3a^{b+2}) + 4a^b(-7)$
$= 12a^{b+b+2} - 28a^b$
$= 12a^{2b+2} - 28a^b$

78. $(3x^a - 4)(3x^a + 4) = (3x^a)^2 - 4^2$
$= 3^2(x^a)^2 - 16$
$= 9x^{2a} - 16$

79. $16x^3 - 24x^2 = 8x^2(2x - 3)$

80. $36y - 24y^2 = 12y(3 - 2y)$

81. $6ab^2 + 8ab - 4a^2b^2 = 2ab(3b + 4 - 2ab)$

82. $14a^2b^2 - 21ab^2 + 7ab = 7ab(2ab - 3b + 1)$

83. $6a(a+3b) - 5(a+3b) = (6a-5)(a+3b)$

84. $4x(x-2y) - 5(x-2y) = (4x-5)(x-2y)$

85. $xy - 6y + 3x - 18 = y(x-6) + 3(x-6)$
$= (x-6)(y+3)$

86. $ab - 8b + 4a - 32 = b(a-8) + 4(a-8)$
$= (a-8)(b+4)$

87. $pq - 3p - 5q + 15 = p(q-3) - 5(q-3)$
$= (q-3)(p-5)$

88. $x^3 - x^2 - 2x + 2 = x^2(x-1) - 2(x-1)$
$= (x-1)(x^2 - 2)$

89. $2x \cdot y - x \cdot x = 2xy - x^2 = x(2y - x)$

90. $9y(4y) - 4(x^2) = 36y^2 - 4x^2$
$= 4(9y^2 - x^2)$
$= 4(3y+x)(3y-x)$

91. $x^2 - 14x - 72 = (x-18)(x+4)$

92. $x^2 + 16x - 80 = (x-4)(x+20)$

93. $2x^2 - 18x + 28 = 2(x^2 - 9x + 14)$
$= 2(x-7)(x-2)$

94. $3x^2 + 33x + 54 = 3(x^2 + 11x + 18)$
$= 3(x+9)(x+2)$

95. $2x^3 - 7x^2 - 9x = x(2x^2 - 7x - 9)$
$= x(2x-9)(x+1)$

96. $3x^3 + 2x^2 - 16x = x(3x^2 + 2x - 16)$
$= x(3x+8)(x-2)$

97. $6x^2 + 17x + 10 = (6x+5)(x+2)$

98. $15x^2 - 91x + 6 = (15x-1)(x-6)$

99. $4x^2 + 2x - 12 = 2(2x^2 + x - 6)$
$= 2(2x-3)(x+2)$

100. $9x^2 - 12x - 12 = 3(3x^2 - 4x - 4)$
$= 3(3x+2)(x-2)$

101. $y^2(x+6)^2 - 2y(x+6)^2 - 3(x+6)^2$
$= (x+6)^2(y^2 - 2y - 3)$
$= (x+6)^2(y-3)(y+1)$

102. Let $y = x + 5$. Then
$$(x+5)^2 + 6(x+5) + 8 = y^2 + 6y + 8$$
$$= (y+4)(y+2)$$
$$= [(x+5)+4][(x+5)+2]$$
$$= (x+9)(x+7)$$

103. $x^4 - 6x^2 - 16 = (x^2 - 8)(x^2 + 2)$

104. $x^4 + 8x^2 - 20 = (x^2 + 10)(x^2 - 2)$

105. $x^2 - 100 = x^2 - 10^2 = (x+10)(x-10)$

106. $x^2 - 81 = x^2 - 9^2 = (x+9)(x-9)$

107. $2x^2 - 32 = 2(x^2 - 16)$
$$= 2(x^2 - 4^2)$$
$$= 2(x+4)(x-4)$$

108. $6x^2 - 54 = 6(x^2 - 9)$
$$= 6(x^2 - 3^2)$$
$$= 6(x+3)(x-3)$$

109. $81 - x^4 = 9^2 - (x^2)^2$
$$= (9 + x^2)(9 - x^2)$$
$$= (9 + x^2)(3 + x)(3 - x)$$

110. $16 - y^4 = 4^2 - (y^2)^2$
$$= (4 + y^2)(4 - y^2)$$
$$= (4 + y^2)(2 + y)(2 - y)$$

111. $(y+2)^2 - 25 = (y+2)^2 - 5^2$
$$= [(y+2)+5][(y+2)-5]$$
$$= (y+7)(y-3)$$

112. $(x-3)^2 - 16 = (x-3)^2 - 4^2$
$$= [(x-3)+4][(x-3)-4]$$
$$= (x+1)(x-7)$$

113. $x^3 + 216 = x^3 + 6^3$
$$= (x+6)(x^2 - 6 \cdot x + 6^2)$$
$$= (x+6)(x^2 - 6x + 36)$$

114. $y^3 + 512 = y^3 + 8^3$
$$= (y+8)(y^2 - 8 \cdot y + 8^2)$$
$$= (y+8)(y^2 - 8y + 64)$$

115. $8 - 27y^3 = 2^3 - (3y)^3$
$$= (2 - 3y)(4 + 2 \cdot 3y + (3y)^2)$$
$$= (2 - 3y)(4 + 6y + 9y^2)$$

116. $1 - 64y^3 = 1^3 - (4y)^3$
$$= (1 - 4y)(1^2 + 1 \cdot 4y + (4y)^2)$$
$$= (1 - 4y)(1 + 4y + 16y^2)$$

117. $6x^4 y + 48xy = 6xy(x^3 + 8)$
$$= 6xy(x^3 + 2^3)$$
$$= 6xy(x+2)(x^2 - 2x + 2^2)$$
$$= 6xy(x+2)(x^2 - 2x + 4)$$

118. $2x^5 + 16x^2 y^3 = 2x^2(x^3 + 8y^3)$
$$= 2x^2(x^3 + (2y)^3)$$
$$= 2x^2(x+2y)(x^2 - x \cdot 2y + (2y)^2)$$
$$= 2x^2(x+2y)(x^2 - 2xy + 4y^2)$$

119. $x^2 - 2x + 1 - y^2 = (x^2 - 2x + 1) - y^2$
$$= (x-1)^2 - y^2$$
$$= [(x-1)+y][(x-1)-y]$$
$$= (x-1+y)(x-1-y)$$

120. $x^2 - 6x + 9 - 4y^2 = (x^2 - 6x + 9) - 4y^2$
$$= (x-3)^2 - (2y)^2$$
$$= [(x-3)+2y][(x-3)-2y]$$
$$= (x-3+2y)(x-3-2y)$$

121. $4x^2 + 12x + 9 = (2x+3)(2x+3)$
$$= (2x+3)^2$$

122. $16a^2 - 40ab + 25b^2 = (4a - 5b)(4a - 5b)$
$$= (4a - 5b)^2$$

123. Volume $= \pi R^2 h - \pi r^2 h$
$$= \pi h(R^2 - r^2)$$
$$= \pi h(R+r)(R-r) \text{ cubic units}$$

124. SA $= 2\pi Rh + 2\pi rh$
$$= 2\pi h(R+r) \text{ square units}$$

125. $(3x-1)(x+7)=0$

$3x-1=0$ or $x+7=0$

$x=\dfrac{1}{3}$ or $\quad x=-7$

The solutions are $-7, \dfrac{1}{3}$.

126. $3(x+5)(8x-3)=0$

$x+5=0$ or $8x-3=0$

$x=-5$ or $\quad x=\dfrac{3}{8}$

The solutions are $-5, \dfrac{3}{8}$.

127. $5x(x-4)(2x-9)=0$

$5x=0$ or $x-4=0$ or $2x-9=0$

$x=0$ or $\quad x=4$ or $\quad x=\dfrac{9}{2}$

The solutions are $0, 4, \dfrac{9}{2}$.

128. $6(x+3)(x-4)(5x+1)=0$

$x+3=0$ or $x-4=0$ or $5x+1=0$

$x=-3$ or $\quad x=4$ or $\quad 5x=-1$

$x=-\dfrac{1}{5}$

The solutions are $-\dfrac{1}{5}, -3, 4$.

129. $2x^2=12x$

$2x^2-12x=0$

$2x(x-6)=0$

$2x=0$ or $x-6=0$

$x=0$ or $\quad x=6$

The solutions are $0, 6$.

130. $4x^3-36x=0$

$4x(x^2-9)=0$

$4x(x+3)(x-3)=0$

$4x=0$ or $x+3=0$ or $x-3=0$

$x=0$ or $\quad x=-3$ or $\quad x=3$

The solutions are $-3, 0, 3$.

131. $(1-x)(3x+2)=-4x$

$3x+2-3x^2-2x=-4x$

$-3x^2+x+2=-4x$

$-3x^2+5x+2=0$

$3x^2-5x-2=0$

$(3x+1)(x-2)=0$

$3x+1=0$ or $x-2=0$

$3x=-1$ or $\quad x=2$

$x=-\dfrac{1}{3}$

The solutions are $-\dfrac{1}{3}, 2$.

132. $2x(x-12)=-40$

$2x^2-24x=-40$

$2x^2-24x+40=0$

$2(x^2-12x+20)=0$

$2(x-10)(x-2)=0$

$x-10=0$ or $x-2=0$

$x=10$ or $\quad x=2$

The solutions are $2, 10$.

133. $3x^2+2x=12-7x$

$3x^2+9x-12=0$

$3(x^2+3x-4)=0$

$3(x+4)(x-1)=0$

$x+4=0$ or $x-1=0$

$x=-4$ or $\quad x=1$

The solutions are $-4, 1$.

134. $2x^2+3x=35$

$2x^2+3x-35=0$

$(2x-7)(x+5)=0$

$2x-7=0$ or $x+5=0$

$2x=7$ or $\quad x=-5$

$x=\dfrac{7}{2}$

The solutions are $-5, \dfrac{7}{2}$.

135.
$$x^3 - 18x = 3x^2$$
$$x^3 - 3x^2 - 18x = 0$$
$$x(x^2 - 3x - 18) = 0$$
$$x(x - 6)(x + 3) = 0$$
$$x = 0 \text{ or } x - 6 = 0 \text{ or } x + 3 = 0$$
$$x = 6 \text{ or } \qquad x = -3$$
The solutions are $-3, 0, 6$.

136.
$$19x^2 - 42x = -x^3$$
$$x^3 + 19x^2 - 42x = 0$$
$$x(x^2 + 19x - 42) = 0$$
$$x(x + 21)(x - 2) = 0$$
$$x = 0 \text{ or } x + 21 = 0 \quad \text{ or } x - 2 = 0$$
$$x = -21 \text{ or } \qquad x = 2$$
The solutions are $-21, 0, 2$.

137.
$$12x = 6x^3 + 6x^2$$
$$-6x^3 - 6x^2 + 12x = 0$$
$$-6x(x^2 + x - 2) = 0$$
$$-6x(x + 2)(x - 1) = 0$$
$$-6x = 0 \text{ or } x + 2 = 0 \quad \text{ or } x - 1 = 0$$
$$x = 0 \text{ or } \qquad x = -2 \text{ or } \qquad x = 1$$
The solutions are $-2, 0, 1$.

138.
$$8x^3 + 10x^2 = 3x$$
$$8x^3 + 10x^2 - 3x = 0$$
$$x(8x^2 + 10x - 3) = 0$$
$$x(4x - 1)(2x + 3) = 0$$
$$x = 0 \text{ or } 4x - 1 = 0 \text{ or } 2x + 3 = 0$$
$$4x = 1 \text{ or } \qquad 2x = -3$$
$$x = \frac{1}{4} \text{ or } \qquad x = -\frac{3}{2}$$
The solutions are $-\dfrac{3}{2}, 0, \dfrac{1}{4}$.

139. Let x = the number. Then
$$x + 2x^2 = 105$$
$$2x^2 + x - 105 = 0$$
$$(2x + 15)(x - 7) = 0$$
$$2x + 15 = 0 \qquad \text{ or } x - 7 = 0$$
$$2x = -15 \text{ or } \qquad x = 7$$
$$x = -\frac{15}{2}$$
The number is $-\dfrac{15}{2}$ or 7.

140. Let x = width; then $2x - 5$ = length.
$$x(2x - 5) = 33$$
$$2x^2 - 5x = 33$$
$$2x^2 - 5x - 33 = 0$$
$$(2x - 11)(x + 3) = 0$$
$$2x - 11 = 0 \quad \text{ or } x + 3 = 0$$
$$2x = 11 \text{ or } \qquad x = -3$$
$$x = \frac{11}{2}$$
Disregard the negative.
Width $= \dfrac{11}{2} = 5\dfrac{1}{2}$ m

Length $= 2\left(\dfrac{11}{2}\right) - 5 = 6$ m

141. $h(t) = -16t^2 + 400$
$$0 = -16t^2 + 400$$
$$0 = -16(t^2 - 25)$$
$$0 = -16(t + 5)(t - 5)$$
$$t + 5 = 0 \quad \text{ or } t - 5 = 0$$
$$t = -5 \text{ or } \qquad t = 5$$
Disregard the negative. The stunt dummy will reach the ground after 5 seconds.

142. $P(t) = -16t^2 + 1053$
$$P(1) = -16(1)^2 + 1053 = -16 + 1053 = 1037$$
$$P(8) = -16(8)^2 + 1053 = -1024 + 1053 = 29$$
After 1 second, the object is at 1037 feet and after 8 seconds, the object is at 29 feet.

143. $(x + 5)(3x^2 - 2x + 1)$
$$= x(3x^2 - 2x + 1) + 5(3x^2 - 2x + 1)$$
$$= 3x^3 - 2x^2 + x + 15x^2 - 10x + 5$$
$$= 3x^3 + 13x^2 - 9x + 5$$

144. $(3x^2 + 4x - 1.2) - (5x^2 - x + 5.7)$
$$= 3x^2 + 4x - 1.2 - 5x^2 + x - 5.7$$
$$= -2x^2 + 5x - 6.9$$

145. $(3x^2 + 4x - 1.2) + (5x^2 - x + 5.7)$
$$= 3x^2 + 4x - 1.2 + 5x^2 - x + 5.7$$
$$= 8x^2 + 3x + 4.5$$

146. $\left(7ab - \dfrac{1}{2}\right)^2 = (7ab)^2 - 2(7ab)\left(\dfrac{1}{2}\right) + \left(\dfrac{1}{2}\right)^2$

$$= 49a^2b^2 - 7ab + \dfrac{1}{4}$$

147. $P(x) = -x^2 + x - 4$

$P(5) = -5^2 + 5 - 4 = -25 + 5 - 4 = -24$

148. $P(x) = -x^2 + x - 4$

$P(-2) = -(-2)^2 + (-2) - 4 = -4 - 2 - 4 = -10$

149. $12y^5 - 6y^4 = 6y^4(2y) + 6y^4(-1) = 6y^4(2y - 1)$

150. $x^2y + 4x^2 - 3y - 12 = x^2(y + 4) - 3(y + 4)$

$$= (y + 4)(x^2 - 3)$$

151. $6x^2 - 34x - 12 = 2(3x^2 - 17x - 6)$

$$= 2(3x + 1)(x - 6)$$

152. $y^2(4x + 3)^2 - 19y(4x + 3)^2 - 20(4x + 3)^2$

$$= (4x + 3)^2(y^2 - 19y - 20)$$

$$= (4x + 3)^2(y - 20)(y + 1)$$

153. $4z^7 - 49z^5 = z^5(4z^2 - 49)$

$$= z^5[(2z)^2 - 7^2]$$

$$= z^5(2z + 7)(2z - 7)$$

154. $5x^4 + 4x^2 - 9 = (x^2 - 1)(5x^2 + 9)$

$$= (x + 1)(x - 1)(5x^2 + 9)$$

155. $\quad\quad 8x^2 = 24x$

$8x^2 - 24x = 0$

$8x(x - 3) = 0$

$8x = 0 \quad$ or $\quad x - 3 = 0$

$x = 0 \quad$ or $\quad\quad x = 3$

156. $\quad\quad x(x - 11) = 26$

$x^2 - 11x = 26$

$x^2 - 11x - 26 = 0$

$(x + 2)(x - 13) = 0$

$x + 2 = 0 \quad$ or $\quad x - 13 = 0$

$x = -2 \quad$ or $\quad\quad x = 13$

Chapter 5 Test

1. $(-9x)^{-2} = \dfrac{1}{(-9x)^2} = \dfrac{1}{81x^2}$

2. $-3xy^{-2}(4xy^2)z = -12x^{1+1}y^{-2+2}z = -12x^2z$

3. $\dfrac{6^{-1}a^2b^{-3}}{3^{-2}a^{-5}b^2} = \dfrac{3^2\,a^{2+5}}{6^1b^{2+3}} = \dfrac{9a^7}{6b^5} = \dfrac{3a^7}{2b^5}$

4. $\left(\dfrac{-xy^{-5}z}{xy^3}\right)^{-5} = \dfrac{-x^{-5}y^{25}z^{-5}}{x^{-5}y^{-15}}$

$$= \dfrac{x^{-5+5}y^{25-(-15)}}{z^5}$$

$$= -\dfrac{y^{40}}{z^5}$$

5. $630,000,000 = 6.3 \times 10^8$

6. $0.01200 = 1.2 \times 10^{-2}$

7. $5 \times 10^{-6} = 0.000005$

8. $\dfrac{(0.0024)(0.00012)}{0.00032} = \dfrac{(2.4 \times 10^{-3})(1.2 \times 10^{-4})}{3.2 \times 10^{-4}}$

$$= \dfrac{(2.4)(1.2)}{3.2} \times 10^{-3+(-4)-(-4)}$$

$$= 0.9 \times 10^{-3}$$

$$= 0.0009$$

9. $(4x^3y - 3x - 4) - (9x^3y + 8x + 5)$

$$= 4x^3y - 3x - 4 - 9x^3y - 8x - 5$$

$$= -5x^3y - 11x - 9$$

10. $-3xy(4x + y) = -3xy(4x) - 3xy(y)$

$$= -12x^2y - 3xy^2$$

11. $(3x + 4)(4x - 7) = 12x^2 - 21x + 16x - 28$

$$= 12x^2 - 5x - 28$$

12. $(5a - 2b)(5a + 2b) = (5a)^2 - (2b)^2 = 25a^2 - 4b^2$

13. $(6m + n)^2 = (6m)^2 + 2(6m)n + n^2$

$$= 36m^2 + 12mn + n^2$$

14.
$$x^2 - 6x + 4$$
$$\times \qquad 2x - 1$$
$$-x^2 + 6x - 4$$
$$2x^3 - 12x^2 + 8x$$
$$2x^3 - 13x^2 + 14x - 4$$

15. $16x^3 y - 12x^2 y^4 = 4x^2 y(4x - 3y^3)$

16. $x^2 - 13x - 30 = (x - 15)(x + 2)$

17. $4y^2 + 20y + 25 = (2y + 5)(2y + 5)$
$$= (2y + 5)^2$$

18. $6x^2 - 15x - 9 = 3(2x^2 - 5x - 3)$
$$= 3(2x + 1)(x - 3)$$

19. $4x^2 - 25 = (2x)^2 - 5^2 = (2x + 5)(2x - 5)$

20. $x^3 + 64 = x^3 + 4^3 = (x + 4)(x^2 - 4x + 16)$

21. $3x^2 y - 27y^3 = 3y(x^2 - 9y^2)$
$$= 3y(x^2 - (3y)^2)$$
$$= 3y(x + 3y)(x - 3y)$$

22. $6x^2 + 24 = 6(x^2 + 4)$

23. $16y^3 - 2 = 2(8y^3 - 1)$
$$= 2((2y)^3 - 1^3)$$
$$= 2(2y - 1)(4y^2 + 2y + 1)$$

24. $x^2 y - 9y - 3x^2 + 27 = y(x^2 - 9) - 3(x^2 - 9)$
$$= (x^2 - 9)(y - 3)$$
$$= (x + 3)(x - 3)(y - 3)$$

25.
$$3n(7n - 20) = 96$$
$$21n^2 - 60n = 96$$
$$21n^2 - 60n - 96 = 0$$
$$3(7n^2 - 20n - 32) = 0$$
$$3(7n + 8)(n - 4) = 0$$
$$7n + 8 = 0 \quad \text{or} \quad n - 4 = 0$$
$$7n = -8 \quad \text{or} \qquad n = 4$$
$$n = -\frac{8}{7}$$

The solutions are $-\dfrac{8}{7}$, 4.

26. $(x + 2)(x - 2) = 5(x + 4)$
$$x^2 - 4 = 5x + 20$$
$$x^2 - 5x - 24 = 0$$
$$(x - 8)(x + 3) = 0$$
$$x - 8 = 0 \quad \text{or} \quad x + 3 = 0$$
$$x = 8 \quad \text{or} \qquad x = -3$$
The solutions are 8, –3.

27.
$$2x^3 + 5x^2 - 8x - 20 = 0$$
$$x^2(2x + 5) - 4(2x + 5) = 0$$
$$(2x + 5)(x^2 - 4) = 0$$
$$(2x + 5)(x + 2)(x - 2) = 0$$
$$2x + 5 = 0 \quad \text{or} \quad x + 2 = 0 \quad \text{or} \quad x - 2 = 0$$
$$2x = -5 \quad \text{or} \qquad x = -2 \quad \text{or} \qquad x = 2$$
$$x = -\frac{5}{2}$$

The solutions are $-\dfrac{5}{2}$, –2, 2.

28. Area $= x^2 - (2y)^2$
$$= (x + 2y)(x - 2y) \text{ square units}$$

29. $h(t) = -16t^2 + 96t + 880$

a. $-16(1)^2 + 96(1) + 880 = -16 + 96 + 880$
$$= 960$$
After 1 second, the height is 960 feet.

b. $-16(5.1)^2 + 96(5.1) + 880$
$$= -416.16 + 489.6 + 880$$
$$= 953.44$$
After 5.1 seconds, the height is 953.44 feet.

c. $-16t^2 + 96t + 880 = -16(t^2 - 6t - 55)$
$$= -16(t - 11)(t + 5)$$

d. $-16(t-11)(t+5) = 0$

$t - 11 = 0$ or $t + 5 = 0$

$t = 11$ or $t = -5$

Disregard the negative. The pebble will hit the ground in 11 seconds.

Chapter 5 Cumulative Review

1. a. $\sqrt[3]{-27} = -3$ since $(-3)^3 = -27$.

   b. $\sqrt[5]{1} = 1$ since $1^5 = 1$.

   c. $\sqrt[4]{16} = 2$ since $2^4 = 16$.

2. a. $\sqrt[3]{64} = 4$ since $4^3 = 64$.

   b. $\sqrt[4]{81} = 3$ since $3^4 = 81$.

   c. $\sqrt[5]{32} = 2$ since $2^5 = 32$.

3. $2(x-3) = 5x - 9$

$2x - 6 = 5x - 9$

$-3x = -3$

$x = 1$

The solution is 1.

4. $0.3y + 2.4 = 0.1y + 4$

$10(0.3y + 2.4) = 10(0.1y + 4)$

$3y + 24 = y + 40$

$2y = 16$

$y = 8$

The solution is 8.

5. $A = 10,000\left(1 + \dfrac{0.05}{4}\right)^{4(3)}$

$= 10,000(1.0125)^{12}$

$= 10,000(1.160754518)$

$= 11,607.54518$

There will be \$11,607.55 in the account.

6. The area of the room is
$2(14 \cdot 8) + 2(18 \cdot 8) = 512$ sq ft . Two coats means
$2 \cdot 512 = 1024$ sq ft of wall needs paint.

$$\frac{1}{400} = \frac{x}{1024}$$

$$1024\left(\frac{1}{400}\right) = 1024\left(\frac{x}{512}\right)$$

$$2.56 = x$$

$$x \approx 3$$

3 gallons of paint are needed.

7. a. $\dfrac{1}{4}x \le \dfrac{3}{8}$

$8\left(\dfrac{1}{4}x\right) \le 8\left(\dfrac{3}{8}\right)$

$2x \le 3$

$x \le \dfrac{3}{2}$

$\left\{x \middle| x \le \dfrac{3}{2}\right\}$ or $\left(-\infty, \dfrac{3}{2}\right]$

b. $-2.3x < 6.9$

$10(-2.3x) < 10(6.9)$

$-23x < 69$

$\dfrac{-23x}{-23} > \dfrac{69}{-23}$

$x > -3$

$\{x \mid x > -3\}$ or $(-3, \infty)$

8. $x + 2 \le \dfrac{1}{4}(x - 7)$

$4(x+2) \le 4\left[\dfrac{1}{4}(x-7)\right]$

$4x + 8 \le x - 7$

$3x \le -15$

$x \le -5$

$\{x \mid x \le -5\}$ or $(-\infty, -5]$

9. $-1 \le \dfrac{2x}{3} + 5 \le 2$

$3(-1) \le 3\left(\dfrac{2x}{3} + 5\right) \le 3(2)$

$-3 \le 2x + 15 \le 6$

$-18 \le 2x \le -9$

$-9 \le x \le -\dfrac{9}{2}$

$\left[-9, -\dfrac{9}{2}\right]$

10. $-\dfrac{1}{3} < \dfrac{3x+1}{6} \le \dfrac{1}{3}$

$6\left(-\dfrac{1}{3}\right) < 6\left(\dfrac{3x+1}{6}\right) \le 6\left(\dfrac{1}{3}\right)$

$-2 < 3x + 1 \le 2$

$-3 < 3x \le 1$

$-1 < x \le \dfrac{1}{3}$

$\left(-1, \dfrac{1}{3}\right]$

11. $|y| = 0$

$y = 0$

The solution is 0.

12. $8 + |4c| = 24$

$|4c| = 16$

$4c = 16$ or $4c = -16$

$c = 4$ or $c = -4$

The solutions are –4, 4.

13. $\left|2x - \dfrac{1}{10}\right| < -13$ is impossible; the solution set is

$\{\ \ \}$ or $\varnothing$.

14. $|5x - 1| + 9 > 5$

$|5x - 1| > -4$ is always true.

$(-\infty, \infty)$

15. $y = \dfrac{1}{3}x$

16. $y = 3x$

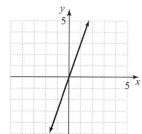

17. $f(x) = \begin{cases} 2x + 3 & \text{if} \quad x \le 0 \\ -x - 1 & \text{if} \quad x > 0 \end{cases}$

$f(2) = -2 - 1 = -3 \colon (2, -3)$

$f(-6) = 2(-6) + 3 = -9 \colon (-6, -9)$

$f(0) = 2(0) + 3 = 3 \colon (0, 3)$

18. $f(x) = 3x^2 + 2x + 3$

$f(-3) = 3(-3)^2 + 2(-3) + 3$

$= 3(9) - 6 + 3$

$= 27 - 6 + 3$

$= 24$

19. $x = 2$

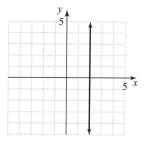

20. $y - 5 = 0$
$\quad\quad y = 5$

21. $y = 2$
This is a horizontal line.
$m = 0$

22. $f(x) = -2x - 3$
$m = -2$

23. $y = 3$

24. $x = -3$

25. $x + \dfrac{1}{2}y \geq -4 \quad\quad$ or $\ y \leq -2$
$\quad\quad \dfrac{1}{2}y \geq -x - 4 \quad$ or $\ y \leq -2$
$\quad\quad\quad\quad y \geq -2x - 8 \ $ or $\ y \leq -2$

26. $y - 3 = 0[x - (-2)]$
$\quad\quad y - 3 = 0$
$\quad\quad\quad\quad y = 3$

27. $\begin{cases} 2x + 4y = -6 & (1) \\ \quad\quad x = 2y - 5 & (2) \end{cases}$
Substitute $2y - 5$ for x in E1.
$2(2y - 5) + 4y = -6$
$\quad 4y - 10 + 4y = -6$
$\quad\quad\quad\quad\quad 8y = 4$
$\quad\quad\quad\quad\quad\quad y = \dfrac{1}{2}$
Substitute $\dfrac{1}{2}$ for y in E2.
$x = 2\left(\dfrac{1}{2}\right) - 5 = 1 - 5 = -4$
The solution is $\left(-4, \dfrac{1}{2}\right)$.

28. $\begin{cases} 4x - 2y = 8 & (1) \\ \quad\quad y = 3x - 6 & (2) \end{cases}$
Substitute $3x - 6$ for y in E1.
$4x - 2(3x - 6) = 8$
$\quad 4x - 6x + 12 = 8$
$\quad\quad\quad\quad -2x = -4$
$\quad\quad\quad\quad\quad\quad x = 2$
Substitute 2 for x in E2.
$y = 3(2) - 6 = 6 - 6 = 0$
The solution is (2, 0).

29. $\begin{cases} 2x + 4y \quad\quad = 1 & (1) \\ 4x \quad\quad - 4z = -1 & (2) \\ \quad\quad y - 4z = -3 & (3) \end{cases}$
Multiply E2 by -1 and add to E3.
$-4x \quad\quad + 4z = 1$
$\underline{\quad\quad\quad y - 4z = -3\quad\quad}$
$-4x + y \quad\quad = -2 \ (4)$
Muliply E1 by 2 and add to E4.
$\quad 4x + 8y = 2$
$\underline{-4x \ + y = -2\quad}$
$\quad\quad\quad 9y = 0$
$\quad\quad\quad\quad y = 0$
Replace y with 0 in E1.
$2x + 4(0) = 1$
$\quad\quad\quad 2x = 1$
$\quad\quad\quad\quad x = \dfrac{1}{2}$
Replace y with 0 in E3.

$$y - 4z = -3$$
$$-4z = -3$$
$$z = \frac{3}{4}$$

The solution is $\left(\frac{1}{2}, 0, \frac{3}{4}\right)$.

30. $\begin{cases} x + y - \dfrac{3}{2}z = \dfrac{1}{2} & (1) \\ -y - 2z = 14 & (2) \\ x - \dfrac{2}{3}y = -\dfrac{1}{3} & (3) \end{cases}$

Multiply E1 by 2 and E3 by 3 to clear fractions.

$\begin{cases} 2x + 2y - 3z = 1 & (1) \\ -y - 2z = 14 & (2) \\ 3x - 2y = -1 & (3) \end{cases}$

Add E1 and E3.
$$5x - 3z = 0 \quad (4)$$
Multiply E2 by 2 and add to E1.
$$-2y - 4z = 28$$
$$\underline{2x + 2y - 3z = 1}$$
$$2x \quad\quad - 7z = 29 \quad (5)$$

Solve the new system $\begin{cases} 5x - 3z = 0 & (4) \\ 2x - 7z = 29 & (5) \end{cases}$.

Multiply E4 by –2, multiply E5 by 5, and add.
$$-10x + 6z = 0$$
$$\underline{10x - 35z = 145}$$
$$-29z = 145$$
$$z = -5$$

Replace z with –5 in E4.
$$5x - 3(-5) = 0$$
$$5x + 15 = 0$$
$$5x = -15$$
$$x = -3$$
Replace z with –5 in E2.
$$-y - 2(-5) = 14$$
$$-y + 10 = 14$$
$$-y = 4$$
$$y = -4$$
The solution is (–3, –4, –5).

31. Let x = the first number and y = the second number.

$\begin{cases} x = y - 4 & (1) \\ 4x = 2y + 6 & (2) \end{cases}$

Multiply E1 by –4 and add to E2.

$$-4x = -4y + 16$$
$$\underline{4x = \quad 2y + 6}$$
$$0 = -2y + 22$$
$$2y = 22$$
$$y = 11$$
Replace y with 11 in E1.
$$x = 11 - 4 = 7$$
The numbers are 7 and 11.

32. Let x = ounces of 20% solution and y = ounces of 60% solution.

$\begin{cases} x + y = 50 & (1) \\ 0.20x + 0.60y = 50(0.30) & (2) \end{cases}$

Multiply E2 by 100 to clear decimals.

$\begin{cases} x + y = 50 & (1) \\ 20x + 60y = 1500 & (2) \end{cases}$

Multiply E1 by –20 and add to E2.
$$-20x - 20y = -1000$$
$$\underline{20x + 60y = 1500}$$
$$40y = 500$$
$$y = \frac{500}{40} = 12.5$$

Replace y with 12.5 in E1.
$$x + 12.5 = 50$$
$$x = 37.5$$
You should mix 37.5 ounces of the 20% solution and 12.5 ounces of the 60% solution.

33. $\begin{cases} 2x - y = 3 \\ 4x - 2y = 5 \end{cases}$

$$\begin{bmatrix} 2 & -1 & | & 3 \\ 4 & -2 & | & 5 \end{bmatrix}$$

Divide R1 by 2.

$$\begin{bmatrix} 1 & -\frac{1}{2} & | & \frac{3}{2} \\ 4 & -2 & | & 5 \end{bmatrix}$$

Multiply R1 by –4 and add to R2.

$$\begin{bmatrix} 1 & -\frac{1}{2} & | & \frac{3}{2} \\ 0 & 0 & | & -1 \end{bmatrix}$$

This corresponds to $\begin{cases} x - \dfrac{1}{2}y = \dfrac{3}{2} \\ 0 = -1 \end{cases}$. The last

equation is impossible. The system is inconsistent. The solution set is $\varnothing$.

34. $\begin{cases} 4y = 8 \\ x + y = 7 \end{cases}$

$\begin{bmatrix} 0 & 4 & | & 8 \\ 1 & 1 & | & 7 \end{bmatrix}$

Interchange R1 and R2.

$\begin{bmatrix} 1 & 1 & | & 7 \\ 0 & 4 & | & 8 \end{bmatrix}$

Divide R2 by 4.

$\begin{bmatrix} 1 & 1 & | & 7 \\ 0 & 1 & | & 2 \end{bmatrix}$

This corresponds to $\begin{cases} x + y = 7 \\ \quad\ y = 2 \end{cases}$.

Replace y with 2 in the equation $x + y = 7$.

$x + 2 = 7$

$\quad x = 5$

The solution is $(5, 2)$.

35. Let x = measure of smallest angle, then $x + 80$ = measure of largest angle, and $x + 10$ = measure of remaining angle.

$x + (x + 80) + (x + 10) = 180$

$\qquad\qquad\quad 3x + 90 = 180$

$\qquad\qquad\qquad\quad 3x = 90$

$\qquad\qquad\qquad\quad\ x = 30$

$x + 80 = 110$

$x + 10 = 40$

The angles measure 30°, 110°, and 40°.

36. $m = \dfrac{1}{2}$, y-intercept $(0, 5)$, $b = 5$

$y = mx + b$

$y = \dfrac{1}{2}x + 5$

$f(x) = \dfrac{1}{2}x + 5$

37. a. $730{,}000 = 7.3 \times 10^5$

   b. $0.00000104 = 1.04 \times 10^{-6}$

38. a. $8{,}250{,}000 = 8.25 \times 10^6$

   b. $0.0000346 = 3.46 \times 10^{-5}$

39. a. $(2x^0 y^{-3})^{-2} = 2^{-2}(1)^{-2}(y^{-3})^{-2} = \dfrac{y^6}{2^2} = \dfrac{y^6}{4}$

   b. $\left(\dfrac{x^{-5}}{x^{-2}}\right)^{-3} = \dfrac{(x^{-5})^{-3}}{(x^{-2})^{-3}} = \dfrac{x^{15}}{x^6} = x^{15-6} = x^9$

   c. $\left(\dfrac{2}{7}\right)^{-2} = \dfrac{2^{-2}}{7^{-2}} = \dfrac{7^2}{2^2} = \dfrac{49}{4}$

   d. $\dfrac{5^{-2}x^{-3}y^{11}}{x^2 y^{-5}} = \dfrac{x^{-3-2}y^{11-(-5)}}{5^2}$

   $\qquad\qquad = \dfrac{x^{-5}y^{16}}{25}$

   $\qquad\qquad = \dfrac{y^{16}}{25x^5}$

40. a. $(4a^{-1}b^0)^{-3} = 4^{-3}(a^{-1})^{-3}(1)^{-3} = \dfrac{a^3}{4^3} = \dfrac{a^3}{64}$

   b. $\left(\dfrac{a^{-6}}{a^{-8}}\right)^{-2} = \dfrac{(a^{-6})^{-2}}{(a^{-8})^{-2}}$

   $\qquad\qquad = \dfrac{a^{12}}{a^{16}}$

   $\qquad\qquad = a^{12-16}$

   $\qquad\qquad = a^{-4}$

   $\qquad\qquad = \dfrac{1}{a^4}$

   c. $\left(\dfrac{2}{3}\right)^{-3} = \dfrac{2^{-3}}{3^{-3}} = \dfrac{3^3}{2^3} = \dfrac{27}{8}$

   d. $\dfrac{3^{-2}a^{-2}b^{12}}{a^4 b^{-5}} = \dfrac{a^{-2-4}b^{12-(-5)}}{3^2}$

   $\qquad\qquad = \dfrac{a^{-6}b^{17}}{9}$

   $\qquad\qquad = \dfrac{b^{17}}{9a^6}$

41. The degree is the degree of the term $x^2 y^2$, which is $2 + 2 = 4$.

42. $(3x^2 - 2x) - (5x^2 + 3x) = 3x^2 - 2x - 5x^2 - 3x$

$\qquad\qquad\qquad\qquad\qquad = -2x^2 - 5x$

43. a. $(2x^3)(5x^6) = 2(5)x^{3+6} = 10x^9$

    b. $(7y^4z^4)(-xy^{11}z^5) = -7xy^{4+11}z^{4+5}$
$$= -7xy^{15}z^9$$

44. a. $(3y^6)(4y^2) = 3(4)y^{6+2} = 12y^8$

    b. $(6a^3b^2)(-a^2bc^4) = -6a^{3+2}b^{2+1}c^4$
$$= -6a^5b^3c^4$$

45. $17x^3y^2 - 34x^4y^2 = 17x^3y^2(1-2x)$

46. $12x^3y - 3xy^3 = 3xy(4x^2 - y^2)$
$$= 3xy((2x)^2 - y^2)$$
$$= 3xy(2x+y)(2x-y)$$

47. $x^2 + 10x + 16 = (x+8)(x+2)$

48. $5a^2 + 14a - 3 = (5a-1)(a+3)$

49. $2x^2 + 9x - 5 = 0$
$$(2x-1)(x+5) = 0$$
$$2x - 1 = 0 \text{ or } x + 5 = 0$$
$$2x = 1 \text{ or } \quad x = -5$$
$$x = \frac{1}{2}$$

The solution is $-5, \dfrac{1}{2}$.

50. $3x^2 - 10x - 8 = 0$
$$(3x+2)(x-4) = 0$$
$$3x + 2 = 0 \quad \text{or } x - 4 = 0$$
$$3x = -2 \text{ or } \quad x = 4$$
$$x = -\frac{2}{3}$$

The solution is $-\dfrac{2}{3}, 4$.

Chapter 6

Section 6.1 Practice Exercises

1. a. The denominator of $f(x)$ is never 0.
Domain: $\{x \mid x$ is a real number$\}$

b. Undefined values when
$x + 3 = 0$, or $x = -3$
Domain: $\{x \mid x$ is a real number and $x \neq -3\}$

c. Undefined values when
$$x^2 - 5x + 6 = 0$$
$$(x - 3)(x - 2) = 0$$
$$x - 3 = 0 \quad \text{or} \quad x - 2 = 0$$
$$x = 3 \quad \text{or} \quad x = 2$$
Domain:
$\{x \mid x$ is a real number and $x \neq 2, x \neq 3\}$

2. a. $x + 3 = 0$
$\qquad x = -3$
Domain: $\{x \mid x$ is a real number and $x \neq -3\}$

b. $3x^2 - 8x + 5 = 0$
$(3x - 5)(x - 1) = 0$
$3x - 5 = 0 \quad \text{or} \quad x - 1 = 0$
$\quad 3x = 5 \quad \text{or} \qquad x = 1$
$\qquad x = \dfrac{5}{3}$
Domain:
$\left\{ x \,\middle|\, x \text{ is a real number and } x \neq 1,\ x \neq \dfrac{5}{3} \right\}$

3. a. $\dfrac{5z^4}{10z^5 - 5z^4} = \dfrac{5z^4 \cdot 1}{5z^4(2z - 1)}$
$\qquad = 1 \cdot \dfrac{1}{2z - 1} = \dfrac{1}{2z - 1}$

b. $\dfrac{5x^2 + 13x + 6}{6x^2 + 7x - 10} = \dfrac{(5x + 3)(x + 2)}{(6x - 5)(x + 2)}$
$\qquad = \dfrac{5x + 3}{6x - 5} \cdot 1$
$\qquad = \dfrac{5x + 3}{6x - 5}$

4. a. $\dfrac{x + 3}{3 + x} = \dfrac{x + 3}{x + 3} = 1$

b. $\dfrac{3 - x}{x - 3} = \dfrac{-1(-3 + x)}{x - 3} = \dfrac{-1(x - 3)}{x - 3} = \dfrac{-1}{1} = -1$

5. $\dfrac{20 - 5x^2}{x^2 + x - 6} = \dfrac{5(4 - x^2)}{(x + 3)(x - 2)}$
$\qquad = \dfrac{5(2 + x)(2 - x)}{(x + 3)(x - 2)}$
$\qquad = \dfrac{5(2 + x) \cdot (-1)(x - 2)}{(x + 3)(x - 2)}$
$\qquad = -\dfrac{5(2 + x)}{x + 3}$

6. a. $\dfrac{x^3 + 64}{4 + x} = \dfrac{(x + 4)(x^2 - 4x + 16)}{x + 4}$
$\qquad = x^2 - 4x + 16$

b. $\dfrac{5z^2 + 10}{z^3 - 3z^2 + 2z - 6} = \dfrac{5(z^2 + 2)}{(z^3 - 3z^2) + (2z - 6)}$
$\qquad = \dfrac{5(z^2 + 2)}{z^2(z - 3) + 2(z - 3)}$
$\qquad = \dfrac{5(z^2 + 2)}{(z - 3)(z^2 + 2)}$
$\qquad = \dfrac{5}{z - 3}$

7. a. $\dfrac{2 + 5n}{3n} \cdot \dfrac{6n + 3}{5n^2 - 3n - 2}$
$\qquad = \dfrac{2 + 5n}{3n} \cdot \dfrac{3(2n + 1)}{(5n + 2)(n - 1)}$
$\qquad = \dfrac{2n + 1}{n(n - 1)}$

b.
$$\frac{x^3 - 8}{-6x + 12} \cdot \frac{6x^2}{x^2 + 2x + 4}$$

$$= \frac{(x-2)(x^2 + 2x + 4)}{-6(x-2)} \cdot \frac{6x^2}{x^2 + 2x + 4}$$

$$= \frac{(x-2)(x^2 + 2x + 4) \cdot 6 \cdot x^2}{-1 \cdot 6(x-2)(x^2 + 2x + 4)}$$

$$= \frac{x^2}{-1}$$

$$= -x^2$$

8. a.
$$\frac{6y^3}{3y^2 - 27} \div \frac{42}{3 - y} = \frac{6y^3}{3y^2 - 27} \cdot \frac{3 - y}{42}$$

$$= \frac{6y^3(3 - y)}{3(y+3)(y-3) \cdot 42}$$

$$= \frac{6y^3 \cdot (-1)(y - 3)}{3(y+3)(y-3) \cdot 6 \cdot 7}$$

$$= -\frac{y^3}{21(y+3)}$$

b.
$$\frac{10x^2 + 23x - 5}{5x^2 - 51x + 10} \div \frac{2x^2 + 9x + 10}{7x^2 - 68x - 20}$$

$$= \frac{10x^2 + 23x - 5}{5x^2 - 51x + 10} \cdot \frac{7x^2 - 68x - 20}{2x^2 + 9x + 10}$$

$$= \frac{(5x-1)(2x+5)}{(5x-1)(x-10)} \cdot \frac{(7x+2)(x-10)}{(2x+5)(x+2)}$$

$$= \frac{7x+2}{x+2}$$

9.
$$\frac{x^2 - 16}{(x-4)^2} \cdot \frac{5x - 20}{3x} \div \frac{x^2 + x - 12}{x}$$

$$= \frac{x^2 - 16}{(x-4)^2} \cdot \frac{5x - 20}{3x} \cdot \frac{x}{x^2 + x - 12}$$

$$= \frac{(x+4)(x-4)}{(x-4)(x-4)} \cdot \frac{5(x-4)}{3x} \cdot \frac{x}{(x+4)(x-3)}$$

$$= \frac{5}{3(x-3)}$$

10. a. $C(100) = \dfrac{3.2(100) + 400}{100} = \dfrac{720}{100} = 7.2$

    $7.20 per tee shirt

b. $C(1000) = \dfrac{3.2(1000) + 400}{1000} = \dfrac{3600}{1000} = 3.6$

    $3.60 per tee shirt

Vocabulary, Readiness & Video Check 6.1

1. A <u>rational</u> expression is an expression that can be written as the quotient $\dfrac{P}{Q}$ of two polynomials P and Q as long as $Q \neq 0$.

2. A rational expression is undefined if the denominator is <u>0</u>.

3. The <u>domain</u> of the rational function $f(x) = \dfrac{2}{x}$ is $\{x | x$ is a real number and $x \neq 0\}$.

4. A rational expression is <u>simplified</u> if the numerator and denominator have no common factors other than 1 or −1.

5. The expression $\dfrac{x^2 + 2}{2 + x^2}$ simplifies to <u>1</u>.

6. The expression $\dfrac{y - z}{z - y}$ simplifies to <u>−1</u>.

7. For a rational expression, $-\dfrac{a}{b} = \dfrac{-a}{b} = \dfrac{a}{-b}$.

8. The statement $\dfrac{a-6}{a+2} = \dfrac{-(a-6)}{-(a+2)} = \dfrac{-a+6}{-a-2}$ is true.

9. Rational expressions are fractions and are therefore undefined if the denominator is zero; the domain of a rational function consists of all real numbers except those for which the rational expression is undefined.

10. If the numerator and denominator of a rational expression are opposites of each other, the expression simplifies to −1.

11. Yes, multiplying and also simplifying rational expressions often require polynomial factoring. Example 6 involves factoring out a GCF and factoring a difference of squares.

12. Division of rational expressions is similar to division of <u>fractions</u>. Therefore, to divide rational expressions, it's the first rational expression times the <u>reciprocal</u> of the second rational expression.

13. Since the domain consists of all allowable x-values and x is defined as the number of years since the book was published, we do not allow x to be negative for the context of this application.

Exercise Set 6.1

1. 4 is never 0, so the domain of $f(x) = \dfrac{5x - 7}{4}$ is

$\{x | x$ is a real number$\}$.

3. $2t = 0$
$t = 0$

The domain of $s(t) = \dfrac{t^2 + 1}{2t}$ is

$\{t | t$ is a real number and $t \neq 0\}$.

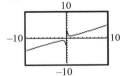

5. $7 - x = 0$
$7 = x$

The domain of $f(x) = \dfrac{3x}{7 - x}$ is

$\{x | x$ is a real number and $x \neq 7\}$.

7. $3x - 1 = 0$
$3x = 1$
$x = \dfrac{1}{3}$

The domain of $f(x) = \dfrac{x}{3x - 1}$ is

$\left\{ x \Big| x \text{ is a real number and } x \neq \dfrac{1}{3} \right\}$.

9. $x^3 + x^2 - 2x = 0$
$x(x^2 + x - 2) = 0$
$x(x + 2)(x - 1) = 0$
$x = 0$ or $x + 2 = 0$ or $x - 1 = 0$
$x = 0$ or $x = -2$ or $x = 1$

The domain of $R(x) = \dfrac{3 + 2x}{x^3 + x^2 - 2x}$ is

$\{x | x$ is a real number and $x \neq -2$, $x \neq 0$, $x \neq 1\}$.

11. $x^2 - 4 = 0$
$(x + 2)(x - 2) = 0$
$x + 2 = 0$ or $x - 2 = 0$
$x = -2$ or $x = 2$

The domain of $C(x) = \dfrac{x + 3}{x^2 - 4}$ is

$\{x | x$ is a real number and $x \neq 2$, $x \neq -2\}$.

13. $\dfrac{8x - 16x^2}{8x} = \dfrac{8x(1 - 2x)}{8x} = 1 - 2x$

15. $\dfrac{x^2 - 9}{3 + x} = \dfrac{(x + 3)(x - 3)}{3 + x}$
$= \dfrac{(x + 3)(x - 3)}{x + 3}$
$= x - 3$

17. $\dfrac{9y - 18}{7y - 14} = \dfrac{9(y - 2)}{7(y - 2)} = \dfrac{9}{7}$

19. $\dfrac{x^2 + 6x - 40}{x + 10} = \dfrac{(x + 10)(x - 4)}{x + 10} = x - 4$

21. $\dfrac{x - 9}{9 - x} = \dfrac{-1(9 - x)}{9 - x} = -1$

23. $\dfrac{x^2-49}{7-x} = \dfrac{(x+7)(x-7)}{7-x}$

$\qquad = \dfrac{(x+7)(x-7)}{-1(-7+x)}$

$\qquad = \dfrac{(x+7)(x-7)}{-1(x-7)}$

$\qquad = -(x+7)$

25. $\dfrac{2x^2-7x-4}{x^2-5x+4} = \dfrac{(2x+1)(x-4)}{(x-1)(x-4)} = \dfrac{2x+1}{x-1}$

27. $\dfrac{x^3-125}{2x-10} = \dfrac{(x-5)(x^2+5x+25)}{2(x-5)}$

$\qquad = \dfrac{x^2+5x+25}{2}$

29. $\dfrac{3x^2-5x-2}{6x^3+2x^2+3x+1} = \dfrac{(3x+1)(x-2)}{2x^2(3x+1)+1(3x+1)}$

$\qquad = \dfrac{(3x+1)(x-2)}{(3x+1)(2x^2+1)}$

$\qquad = \dfrac{x-2}{2x^2+1}$

31. $\dfrac{9x^2-15x+25}{27x^3+125} = \dfrac{9x^2-15x+25}{(3x+5)(9x^2-15x+25)}$

$\qquad = \dfrac{1}{3x+5}$

33. $\dfrac{2x-4}{15} \cdot \dfrac{6}{2-x} = \dfrac{2(x-2)}{3\cdot 5} \cdot \dfrac{2\cdot 3}{-(x-2)}$

$\qquad = \dfrac{2\cdot 2}{5(-1)}$

$\qquad = -\dfrac{4}{5}$

35. $\dfrac{18a-12a^2}{4a^2+4a+1} \cdot \dfrac{4a^2+8a+3}{4a^2-9}$

$\qquad = \dfrac{6a(3-2a)}{(2a+1)(2a+1)} \cdot \dfrac{(2a+3)(2a+1)}{(2a+3)(2a-3)}$

$\qquad = \dfrac{-6a(2a-3)}{(2a+1)(2a+1)} \cdot \dfrac{(2a+3)(2a+1)}{(2a+3)(2a-3)}$

$\qquad = -\dfrac{6a}{2a+1}$

37. $\dfrac{9x+9}{4x+8} \cdot \dfrac{2x+4}{3x^2-3} = \dfrac{9(x+1)}{4(x+2)} \cdot \dfrac{2(x+2)}{3(x^2-1)}$

$\qquad = \dfrac{3\cdot 3(x+1)}{2\cdot 2(x+2)} \cdot \dfrac{2(x+2)}{3(x+1)(x-1)}$

$\qquad = \dfrac{3}{2(x-1)}$

39. $\dfrac{2x^3-16}{6x^2+6x-36} \cdot \dfrac{9x+18}{3x^2+6x+12}$

$\qquad = \dfrac{2(x^3-8)}{6(x^2+x-6)} \cdot \dfrac{9(x+2)}{3(x^2+2x+4)}$

$\qquad = \dfrac{2(x-2)(x^2+2x+4)}{2\cdot 3(x-2)(x+3)} \cdot \dfrac{3\cdot 3(x+2)}{3(x^2+2x+4)}$

$\qquad = \dfrac{x+2}{x+3}$

41. $\dfrac{a^3+a^2b+a+b}{5a^3+5a} \cdot \dfrac{6a^2}{2a^2-2b^2}$

$\qquad = \dfrac{a^2(a+b)+1(a+b)}{5a(a^2+1)} \cdot \dfrac{6a^2}{2(a^2-b^2)}$

$\qquad = \dfrac{(a+b)(a^2+1)}{5a(a^2+1)} \cdot \dfrac{2\cdot 3\cdot a\cdot a}{2(a+b)(a-b)}$

$\qquad = \dfrac{3a}{5(a-b)}$

43. $\dfrac{x^2-6x-16}{2x^2-128} \cdot \dfrac{x^2+16x+64}{3x^2+30x+48}$

$\qquad = \dfrac{(x-8)(x+2)}{2(x^2-64)} \cdot \dfrac{(x+8)(x+8)}{3(x^2+10x+16)}$

$\qquad = \dfrac{(x-8)(x+2)}{2(x+8)(x-8)} \cdot \dfrac{(x+8)(x+8)}{3(x+2)(x+8)}$

$\qquad = \dfrac{1}{2\cdot 3}$

$\qquad = \dfrac{1}{6}$

45. $\dfrac{2x}{5} \div \dfrac{6x+12}{5x+10} = \dfrac{2x}{5} \cdot \dfrac{5x+10}{6x+12}$

$\qquad = \dfrac{2x}{5} \cdot \dfrac{5(x+2)}{6(x+2)}$

$\qquad = \dfrac{2x}{5} \cdot \dfrac{5(x+2)}{2\cdot 3(x+2)}$

$\qquad = \dfrac{x}{3}$

47. $\dfrac{a+b}{ab} \div \dfrac{a^2-b^2}{4a^3b} = \dfrac{a+b}{ab} \cdot \dfrac{4a^3b}{a^2-b^2}$

$= \dfrac{a+b}{ab} \cdot \dfrac{4a^3b}{(a+b)(a-b)}$

$= \dfrac{4a^2}{a-b}$

49. $\dfrac{x^2-6x+9}{x^2-x-6} \div \dfrac{x^2-9}{4}$

$= \dfrac{x^2-6x+9}{x^2-x-6} \cdot \dfrac{4}{x^2-9}$

$= \dfrac{(x-3)^2}{(x-3)(x+2)} \cdot \dfrac{4}{(x+3)(x-3)}$

$= \dfrac{4}{(x+2)(x+3)}$

51. $\dfrac{x^2-6x-16}{2x^2-128} \div \dfrac{x^2+10x+16}{x^2+16x+64}$

$= \dfrac{x^2-6x-16}{2x^2-128} \cdot \dfrac{x^2+16x+64}{x^2+10x+16}$

$= \dfrac{(x-8)(x+2)}{2(x^2-64)} \cdot \dfrac{(x+8)(x+8)}{(x+2)(x+8)}$

$= \dfrac{(x-8)(x+2)}{2(x-8)(x+8)} \cdot \dfrac{(x+8)(x+8)}{(x+2)(x+8)}$

$= \dfrac{1}{2}$

53. $\dfrac{3x-x^2}{x^3-27} \div \dfrac{x}{x^2+3x+9}$

$= \dfrac{3x-x^2}{x^3-27} \cdot \dfrac{x^2+3x+9}{x}$

$= \dfrac{x(3-x)}{(x-3)(x^2+3x+9)} \cdot \dfrac{x^2+3x+9}{x}$

$= \dfrac{-x(x-3)}{(x-3)(x^2+3x+9)} \cdot \dfrac{x^2+3x+9}{x}$

$= -1$

55. $\dfrac{8b+24}{3a+6} \div \dfrac{ab-2b+3a-6}{a^2-4a+4}$

$= \dfrac{8b+24}{3a+6} \cdot \dfrac{a^2-4a+4}{ab-2b+3a-6}$

$= \dfrac{8(b+3)}{3(a+2)} \cdot \dfrac{(a-2)(a-2)}{b(a-2)+3(a-2)}$

$= \dfrac{8(b+3)}{3(a+2)} \cdot \dfrac{(a-2)(a-2)}{(a-2)(b+3)}$

$= \dfrac{8(a-2)}{3(a+2)}$

57. $\dfrac{x^2-9}{4} \cdot \dfrac{x^2-x-6}{x^2-6x+9}$

$= \dfrac{(x+3)(x-3)}{4} \cdot \dfrac{(x+2)(x-3)}{(x-3)^2}$

$= \dfrac{(x+3)(x+2)}{4}$

59. $\dfrac{2x^2-4x-30}{5x^2-40x-75} \div \dfrac{x^2-8x+15}{x^2-6x+9}$

$= \dfrac{2x^2-4x-30}{5x^2-40x-75} \cdot \dfrac{x^2-6x+9}{x^2-8x+15}$

$= \dfrac{2(x^2-2x-15)}{5(x^2-8x-15)} \cdot \dfrac{(x-3)(x-3)}{(x-3)(x-5)}$

$= \dfrac{2(x-5)(x+3)}{5(x^2-8x-15)} \cdot \dfrac{(x-3)(x-3)}{(x-3)(x-5)}$

$= \dfrac{2(x+3)(x-3)}{5(x^2-8x-15)}$

61. $\dfrac{r^3+s^3}{r+s} = \dfrac{(r+s)(r^2-rs+s^2)}{r+s} = r^2-rs+s^2$

63. $\dfrac{4}{x} \div \dfrac{3xy}{x^2} \cdot \dfrac{6x^2}{x^4} = \dfrac{4}{x} \cdot \dfrac{x^2}{3xy} \cdot \dfrac{6x^2}{x^4} = \dfrac{8}{x^2y}$

65. $\dfrac{3x^2-5x-2}{y^2+y-2} \cdot \dfrac{y^2+4y-5}{12x^2+7x+1} \div \dfrac{5x^2-9x-2}{8x^2-2x-1}$

$= \dfrac{3x^2-5x-2}{y^2+y-2} \cdot \dfrac{y^2+4y-5}{12x^2+7x+1} \cdot \dfrac{8x^2-2x-1}{5x^2-9x-2}$

$= \dfrac{(3x+1)(x-2)(y+5)(y-1)(4x+1)(2x-1)}{(y+2)(y-1)(4x+1)(3x+1)(5x+1)(x-2)}$

$= \dfrac{(y+5)(2x-1)}{(y+2)(5x+1)}$

67. $\dfrac{5a^2-20}{3a^2-12a} \div \dfrac{a^3+2a^2}{2a^2-8a} \cdot \dfrac{9a^3+6a^2}{2a^2-4a}$

$= \dfrac{5a^2-20}{3a^2-12a} \cdot \dfrac{2a^2-8a}{a^3+2a^2} \cdot \dfrac{9a^3+6a^2}{2a^2-4a}$

$= \dfrac{5(a^2-4)}{3a(a-4)} \cdot \dfrac{2a(a-4)}{a^2(a+2)} \cdot \dfrac{3a^2(3a+2)}{2a(a-2)}$

$= \dfrac{5(a+2)(a-2)}{3a(a-4)} \cdot \dfrac{2a(a-4)}{a^2(a+2)} \cdot \dfrac{3a^2(3a+2)}{2a(a-2)}$

$= \dfrac{5(3a+2)}{a}$

69. $\dfrac{5x^4+3x^2-2}{x-1} \cdot \dfrac{x+1}{x^4-1}$

$= \dfrac{(5x^2-2)(x^2+1)}{x-1} \cdot \dfrac{x+1}{(x^2+1)(x^2-1)}$

$= \dfrac{(5x^2-2)(x^2+1)}{x-1} \cdot \dfrac{x+1}{(x^2+1)(x+1)(x-1)}$

$= \dfrac{5x^2-2}{(x-1)^2}$

71. $f(x)=\dfrac{x+8}{2x-1}$

$f(2)=\dfrac{2+8}{2(2)-1}=\dfrac{10}{4-1}=\dfrac{10}{3}$

$f(0)=\dfrac{0+8}{2(0)-1}=\dfrac{8}{0-1}=\dfrac{8}{-1}=-8$

$f(-1)=\dfrac{-1+8}{2(-1)-1}=\dfrac{7}{-2-1}=\dfrac{7}{-3}=-\dfrac{7}{3}$

73. $g(x)=\dfrac{x^2+8}{x^3-25x}$

$g(3)=\dfrac{3^2+8}{3^3-25(3)}=\dfrac{9+8}{27-75}=\dfrac{17}{-48}=-\dfrac{17}{48}$

$g(-2)=\dfrac{(-2)^2+8}{(-2)^3-25(-2)}=\dfrac{4+8}{-8+50}=\dfrac{12}{42}=\dfrac{2}{7}$

$g(1)=\dfrac{1^2+8}{1^3-25(1)}=\dfrac{1+8}{1-25}=\dfrac{9}{-24}=-\dfrac{3}{8}$

75. $R(x)=\dfrac{1000x^2}{x^2+4}$

a. $R(1)=\dfrac{1000\cdot1^2}{1^2+4}=\dfrac{1000}{5}=200$

The revenue at the end of the first year is $200 million.

b. $R(2)=\dfrac{1000\cdot2^2}{2^2+4}$

$=\dfrac{1000\cdot4}{4+4}$

$=\dfrac{4000}{8}$

$=500$

The revenue at the end of the second year is $500 million.

c. The revenue during the second year is equal to the revenue at the end of the second year minus the revenue at the end of the first year.

$500-200=300$

The revenue during the second year is $300 million.

d. $x^2+4=0$

$x^2=-4$

This equation has no solutions. Thus there are no values of x that would make the denominator equal to zero and the function undefined. The domain of $R(x)$ is $\{x|x \text{ is a real number}\}$.

77. $\dfrac{4}{5}+\dfrac{3}{5}=\dfrac{7}{5}$

79. $\dfrac{5}{28}-\dfrac{2}{21}=\dfrac{15}{84}-\dfrac{8}{84}=\dfrac{7}{84}=\dfrac{1}{12}$

81. $\dfrac{3}{8}+\dfrac{1}{2}-\dfrac{3}{16}=\dfrac{6}{16}+\dfrac{8}{16}-\dfrac{3}{16}=\dfrac{11}{16}$

83. a. $\dfrac{-x}{5-x} \neq \dfrac{x}{5-x}$

b. $\dfrac{-x}{-5+x}=\dfrac{x}{-(-5+x)}=\dfrac{x}{5-x}$

c. $\dfrac{x}{x-5} = \dfrac{x}{-5+x} \neq \dfrac{x}{5-x}$

d. $\dfrac{-x}{x-5} = \dfrac{x}{-(x-5)} = \dfrac{x}{-x+5} = \dfrac{x}{5-x}$

85. No; answers may vary

87. $A = l \cdot w$

$\qquad = \left(\dfrac{x+2}{x}\right)\left(\dfrac{5x}{x^2-4}\right)$

$\qquad = \dfrac{x+2}{x} \cdot \dfrac{5x}{(x+2)(x-2)}$

$\qquad = \dfrac{5}{x-2}$

The area is $\dfrac{5}{x-2}$ square meters.

89. Since $A = b \cdot h$, $b = \dfrac{A}{h}$.

$b = \dfrac{\dfrac{x^2+x-2}{x^3}}{\dfrac{x^2}{x-1}}$

$\quad = \dfrac{x^2+x-2}{x^3} \cdot \dfrac{x-1}{x^2}$

$\quad = \dfrac{(x+2)(x-1)}{x^3} \cdot \dfrac{x-1}{x^2}$

$\quad = \dfrac{(x+2)(x-1)^2}{x^5}$

The length of the base is $\dfrac{(x+2)(x-1)^2}{x^5}$ feet.

91. answers may vary

93. a. $\dfrac{x+5}{5+x} = \dfrac{x+5}{x+5} = 1$

b. $\dfrac{x-5}{5-x} = \dfrac{x-5}{-(x-5)} = -1$

c. $\dfrac{x+5}{x-5}$ neither

d. $\dfrac{-x-5}{x+5} = \dfrac{-(x+5)}{x+5} = -1$

e. $\dfrac{x-5}{-x+5} = \dfrac{x-5}{-(x-5)} = -1$

f. $\dfrac{-5+x}{x-5} = \dfrac{x-5}{x-5} = 1$

95. answers may vary

97. $f(x) = \dfrac{20x}{100-x}$

x	0	10	30	50	70	90	95	99
y	0	$\dfrac{20}{9}$	$\dfrac{60}{7}$	20	$\dfrac{140}{3}$	180	380	1980

99. $\dfrac{p^x-4}{4-p^x} = \dfrac{p^x-4}{-(-4+p^x)} = \dfrac{p^x-4}{-(p^x-4)} = \dfrac{1}{-1} = -1$

101. $\dfrac{x^n+4}{x^{2n}-16} = \dfrac{x^n+4}{(x^n+4)(x^n-4)} = \dfrac{1}{x^n-4}$

103. $\dfrac{x^{2n}-4}{7x} \cdot \dfrac{14x^3}{x^n-2} = \dfrac{(x^n+2)(x^n-2)}{7x} \cdot \dfrac{14x^3}{x^n-2}$

$\qquad\qquad = 2x^2(x^n+2)$

105. $\dfrac{y^{2n}+9}{10y} \cdot \dfrac{y^n-3}{y^{4n}-81}$

$\quad = \dfrac{y^{2n}+9}{10y} \cdot \dfrac{y^n-3}{(y^{2n}+9)(y^{2n}-9)}$

$\quad = \dfrac{y^{2n}+9}{10y} \cdot \dfrac{y^n-3}{(y^{2n}+9)(y^n+3)(y^n-3)}$

$\quad = \dfrac{1}{10y(y^n+3)}$

107. $\dfrac{y^{2n} - y^n - 2}{2y^n - 4} \div \dfrac{y^{2n} - 1}{1 + y^n}$

$= \dfrac{y^{2n} - y^n - 2}{2y^n - 4} \cdot \dfrac{1 + y^n}{y^{2n} - 1}$

$= \dfrac{(y^n - 2)(y^n + 1)}{2(y^n - 2)} \cdot \dfrac{1 + y^n}{(y^n + 1)(y^n - 1)}$

$= \dfrac{1 + y^n}{2(y^n - 1)}$

Section 6.2 Practice Exercises

1. a. $\dfrac{9}{11z^2} + \dfrac{x}{11z^2} = \dfrac{9 + x}{11z^2}$

b. $\dfrac{x}{8} + \dfrac{5x}{8} = \dfrac{x + 5x}{8} = \dfrac{6x}{8} = \dfrac{3x}{4}$

c. $\dfrac{x^2}{x + 4} - \dfrac{16}{x + 4} = \dfrac{x^2 - 16}{x + 4}$

$= \dfrac{(x + 4)(x - 4)}{x + 4}$

$= x - 4$

d. $\dfrac{z}{2a^2} - \dfrac{z + 3}{2a^2} = \dfrac{z - (z + 3)}{2a^2}$

$= \dfrac{z - z - 3}{2a^2}$

$= \dfrac{-3}{2a^2}$

2. a. $\dfrac{7}{6x^3 y^5}, \dfrac{2}{9x^2 y^4}$

We factor each denominator.

$6x^3 y^5 = 2 \cdot 3 \cdot x^3 \cdot y^5$

$9x^2 y^4 = 3^2 \cdot x^2 \cdot y^4$

$\text{LCD} = 2 \cdot 3^2 \cdot x^3 \cdot y^5 = 18x^3 y^5$

b. $\dfrac{11}{x - 2}, \dfrac{x}{x - 3}$

The denominators $x - 2$ and $x - 3$ do not factor further.

$\text{LCD} = (x - 2)(x - 3)$

c. $\dfrac{b + 2}{b^2 - 16}, \dfrac{8}{b^2 - 8b + 16}, \dfrac{5b}{2b^2 - 5b - 12}$

We factor each denominator.

$b^2 - 16 = (b - 4)(b + 4)$

$b^2 - 8b + 16 = (b - 4)(b - 4)$

$2b^2 - 5b - 12 = (2b + 3)(b - 4)$

$\text{LCD} = (b - 4)^2 (b + 4)(2b + 3)$

d. $\dfrac{y}{y^2 - 9}, \dfrac{3}{12 - 4y}$

We factor each denominator.

$y^2 - 9 = (y - 3)(y + 3)$

$12 - 4y = 4(3 - y) = 4(-1)(y - 3)$

$\text{LCD} = -4(y - 3)(y + 3)$

3. a. The LCD is $5p^4 q$.

$\dfrac{4}{p^3 q} + \dfrac{3}{5p^4 q} = \dfrac{4 \cdot 5p}{p^3 q \cdot 5p} + \dfrac{3}{5p^4 q}$

$= \dfrac{20p}{5p^4 q} + \dfrac{3}{5p^4 q}$

$= \dfrac{20p + 3}{5p^4 q}$

b. The LCD is the product of the two denominators: $(y + 3)(y - 3)$.

$\dfrac{4}{y + 3} + \dfrac{5y}{y - 3}$

$= \dfrac{4 \cdot (y - 3)}{(y + 3) \cdot (y - 3)} + \dfrac{5y \cdot (y + 3)}{(y - 3) \cdot (y + 3)}$

$= \dfrac{4y - 12}{(y + 3)(y - 3)} + \dfrac{5y^2 + 15y}{(y + 3)(y - 3)}$

$= \dfrac{4y - 12 + 5y^2 + 15y}{(y + 3)(y - 3)}$

$= \dfrac{5y^2 + 19y - 12}{(y + 3)(y - 3)}$

c. The LCD is either $z - 5$ or $5 - z$.

$$\frac{3z-18}{z-5} - \frac{3}{5-z} = \frac{3z-18}{z-5} - \frac{3}{-1(z-5)}$$
$$= \frac{3z-18}{z-5} - \frac{-1 \cdot 3}{z-5}$$
$$= \frac{3z-18-(-3)}{z-5}$$
$$= \frac{3z-18+3}{z-5}$$
$$= \frac{3z-15}{z-5}$$
$$= \frac{3(z-5)}{z-5}$$
$$= 3$$

4. $\dfrac{t}{t^2 - 25} - \dfrac{3}{t^2 - 3t - 10} = \dfrac{t}{(t+5)(t-5)} - \dfrac{3}{(t-5)(t+2)}$

The LCD is $(t + 5)(t - 5)(t + 2)$.

$$\frac{t}{(t+5)(t-5)} - \frac{3}{(t-5)(t+2)} = \frac{t \cdot (t+2)}{(t+5)(t-5) \cdot (t+2)} - \frac{3 \cdot (t+5)}{(t-5)(t+2) \cdot (t+5)}$$
$$= \frac{t^2 + 2t}{(t+5)(t-5)(t+2)} - \frac{3t+15}{(t+5)(t-5)(t+2)}$$
$$= \frac{t^2 + 2t - 3t - 15}{(t+5)(t-5)(t+2)}$$
$$= \frac{t^2 - t - 15}{(t+5)(t-5)(t+2)}$$

5. $\dfrac{2x+3}{3x^2 - 5x - 2} + \dfrac{x-6}{6x^2 - 13x - 5} = \dfrac{2x+3}{(3x+1)(x-2)} + \dfrac{x-6}{(3x+1)(2x-5)}$

The LCD is $(3x + 1)(x - 2)(2x - 5)$.

$$= \frac{(2x+3) \cdot (2x-5)}{(3x+1)(x-2) \cdot (2x-5)} + \frac{(x-6) \cdot (x-2)}{(3x+1)(2x-5) \cdot (x-2)}$$
$$= \frac{4x^2 - 4x - 15}{(3x+1)(x-2)(2x-5)} + \frac{x^2 - 8x + 12}{(3x+1)(x-2)(2x-5)}$$
$$= \frac{4x^2 - 4x - 15 + x^2 - 8x + 12}{(3x+1)(x-2)(2x-5)}$$
$$= \frac{5x^2 - 12x - 3}{(3x+1)(x-2)(2x-5)}$$

6. $\dfrac{2}{x-2}+\dfrac{3x}{x^2-x-2}-\dfrac{1}{x+1}=\dfrac{2}{x-2}+\dfrac{3x}{(x-2)(x+1)}-\dfrac{1}{x+1}$

    The LCD is $(x-2)(x-1)$.

$$=\frac{2\cdot(x+1)}{(x-2)\cdot(x+1)}+\frac{3x}{(x-2)(x+1)}-\frac{1\cdot(x-2)}{(x+1)\cdot(x-2)}$$

$$=\frac{2x+2}{(x-2)(x+1)}+\frac{3x}{(x-2)(x+1)}-\frac{x-2}{(x-2)(x+1)}$$

$$=\frac{2x+2+3x-x+2}{(x-2)(x+1)}$$

$$=\frac{4x+4}{(x-2)(x+1)}$$

$$=\frac{4(x+1)}{(x-2)(x+1)}$$

$$=\frac{4}{x-2}$$

Vocabulary, Readiness & Video Check 6.2

1. The denominators must be the same before performing the operations of addition and subtraction (<u>a, b</u>).

2. To perform the operation of division (<u>d</u>), you multiply the first rational expression by the reciprocal of the second rational expression.

3. Numerator times numerator all over denominator times denominator is multiplication (<u>c</u>).

4. The operations of addition and multiplication (<u>a, c</u>) are commutative (order doesn't matter).

5. Addition: $\dfrac{5}{y}+\dfrac{7}{y}=\dfrac{12}{y}$

6. Subtraction: $\dfrac{5}{y}-\dfrac{7}{y}=-\dfrac{2}{y}$

7. Multiplication: $\dfrac{5}{y}\cdot\dfrac{7}{y}=\dfrac{35}{y^2}$

8. Division: $\dfrac{5}{y}\div\dfrac{7}{y}=\dfrac{5}{y}\cdot\dfrac{y}{7}=\dfrac{5}{7}$

9. We need to be sure we subtract the entire second numerator—that is, make sure we "distribute" the subtraction to each term in the second numerator.

10. The LCD contains a factor the greatest number of times it appears in any *one* factored denominator. $(a-b)$ appears at most two times in any one factored denominator.

Copyright © 2014 Pearson Education, Inc.

11. To write an equivalent rational expression, you multiply the <u>numerator</u> of the expression by the exact same thing as the denominator. This is the same as multiplying the original rational expression by <u>one</u>, which doesn't change the <u>value</u> of the original expression.

Exercise Set 6.2

1. $\dfrac{5}{2x} - \dfrac{x+1}{2x} = \dfrac{5-(x+1)}{2x} = \dfrac{-x+4}{2x}$

3. $\dfrac{y+11}{y-2} - \dfrac{y-5}{y-2} = \dfrac{y+11-(y-5)}{y-2} = \dfrac{16}{y-2}$

5. $\dfrac{2}{xz^2} - \dfrac{5}{xz^2} = \dfrac{2-5}{xz^2} = \dfrac{-3}{xz^2} = -\dfrac{3}{xz^2}$

7. $\dfrac{2}{x-2} + \dfrac{x}{x-2} = \dfrac{2+x}{x-2} = \dfrac{x+2}{x-2}$

9. $\dfrac{x^2}{x+2} - \dfrac{4}{x+2} = \dfrac{x^2-4}{x+2}$
$= \dfrac{(x+2)(x-2)}{x+2}$
$= x-2$

11. $\dfrac{2x-6}{x^2+x-6} + \dfrac{3-3x}{x^2+x-6} = \dfrac{2x-6+3-3x}{x^2+x-6}$
$= \dfrac{-x-3}{x^2+x-6}$
$= \dfrac{-1(x+3)}{(x+3)(x-2)}$
$= \dfrac{-1}{x-2}$ or $\dfrac{1}{-(x-2)}$
$= -\dfrac{1}{x-2}$ or $\dfrac{1}{2-x}$

13. $\dfrac{x-5}{2x} - \dfrac{x+5}{2x} = \dfrac{x-5-x-5}{2x} = \dfrac{-10}{2x} = -\dfrac{5}{x}$

15. $7 = 7$
$5x = 5x$
$LCD = 7 \cdot 5x = 35x$

17. $x = x$
$x+1 = x+1$
$LCD = x(x+1)$

19. $x+7 = x+7$
$x-7 = x-7$
$LCD = (x+7)(x-7)$

21. $3x+6 = 3(x+2)$
$2x-4 = 2(x-2)$
$LCD = 3 \cdot 2(x+2)(x-2) = 6(x+2)(x-2)$

23. $a^2 - b^2 = (a+b)(a-b)$
$a^2 - 2ab + b^2 = (a-b)^2$
$LCD = (a+b)(a-b)^2$

25. $x^2 - 9 = (x+3)(x-3)$
$x = x$
$12 - 4x = -4(x-3)$
$LCD = -4x(x+3)(x-3)$

27. $\dfrac{4}{3x} + \dfrac{3}{2x} = \dfrac{4 \cdot 2}{3x(2)} + \dfrac{3 \cdot 3}{2x(3)} = \dfrac{8}{6x} + \dfrac{9}{6x} = \dfrac{17}{6x}$

29. $\dfrac{5}{2y^2} - \dfrac{2}{7y} = \dfrac{5 \cdot 7}{2y^2 \cdot 7} - \dfrac{2 \cdot 2y}{7y \cdot 2y}$
$= \dfrac{35}{14y^2} - \dfrac{4y}{14y^2}$
$= \dfrac{35-4y}{14y^2}$

31. $\dfrac{x-3}{x+4} - \dfrac{x+2}{x-4}$
$= \dfrac{(x-3)(x-4)}{(x+4)(x-4)} - \dfrac{(x+2)(x+4)}{(x-4)(x+4)}$
$= \dfrac{x^2-7x+12}{(x+4)(x-4)} - \dfrac{x^2+6x+8}{(x+4)(x-4)}$
$= \dfrac{x^2-7x+12-x^2-6x-8}{(x+4)(x-4)}$
$= \dfrac{-13x+4}{(x+4)(x-4)}$

33.
$$\frac{1}{x-5} - \frac{19-2x}{(x-5)(x+4)}$$
$$= \frac{1 \cdot (x+4)}{(x-5) \cdot (x+4)} - \frac{19-2x}{(x-5)(x+4)}$$
$$= \frac{x+4+19-2x}{(x-5)(x+4)}$$
$$= \frac{3x-15}{(x-5)(x+4)}$$
$$= \frac{3(x-5)}{(x-5)(x+4)}$$
$$= \frac{3}{x+4}$$

35.
$$\frac{1}{a-b} + \frac{1}{b-a} = \frac{1}{a-b} + \frac{1}{-1(-b+a)}$$
$$= \frac{1}{a-b} + \frac{-1}{a-b}$$
$$= \frac{0}{a-b}$$
$$= 0$$

37.
$$\frac{x+1}{1-x} + \frac{1}{x-1} = \frac{x+1}{-(x-1)} + \frac{1}{x-1}$$
$$= \frac{-(x+1)}{x-1} + \frac{1}{x-1}$$
$$= \frac{-x-1+1}{x-1}$$
$$= \frac{-x}{x-1}$$
$$= -\frac{x}{x-1}$$

39.
$$\frac{5}{x-2} + \frac{x+4}{2-x} = \frac{5}{x-2} + \frac{x+4}{-(-2+x)}$$
$$= \frac{5}{x-2} + \frac{-(x+4)}{x-2}$$
$$= \frac{5-x-4}{x-2}$$
$$= \frac{-x+1}{x-2}$$

41.
$$\frac{y+1}{y^2-6y+8} - \frac{3}{y^2-16}$$
$$= \frac{y+1}{(y-2)(y-4)} - \frac{3}{(y+4)(y-4)}$$
$$= \frac{(y+1)(y+4)}{(y-2)(y-4)(y+4)} - \frac{3(y-2)}{(y-2)(y+4)(y-4)}$$
$$= \frac{(y+1)(y+4)-3(y-2)}{(y-2)(y-4)(y+4)}$$
$$= \frac{y^2+5y+4-3y+6}{(y-2)(y-4)(y+4)}$$
$$= \frac{y^2+2y+10}{(y-2)(y-4)(y+4)}$$

43.
$$\frac{x+4}{3x^2+11x+6} + \frac{x}{2x^2+x-15}$$
$$= \frac{x+4}{(3x+2)(x+3)} + \frac{x}{(2x-5)(x+3)}$$
$$= \frac{(x+4)(2x-5)}{(3x+2)(x+3)(2x-5)} + \frac{x(3x+2)}{(2x-5)(x+3)(3x+2)}$$
$$= \frac{2x^2+3x-20}{(3x+2)(x+3)(2x-5)} + \frac{3x^2+2x}{(3x+2)(x+3)(2x-5)}$$
$$= \frac{2x^2+3x-20+3x^2+2x}{(3x+2)(x+3)(2x-5)}$$
$$= \frac{5x^2+5x-20}{(3x+2)(x+3)(2x-5)}$$
$$= \frac{5(x^2+x-4)}{(3x+2)(x+3)(2x-5)}$$

45.
$$\frac{7}{x^2-x-2} - \frac{x-1}{x^2+4x+3}$$
$$= \frac{7}{(x-2)(x+1)} - \frac{x-1}{(x+3)(x+1)}$$
$$= \frac{7(x+3)}{(x-2)(x+1)(x+3)} - \frac{(x-1)(x-2)}{(x+3)(x+1)(x-2)}$$
$$= \frac{7(x+3)-(x-1)(x-2)}{(x-2)(x+1)(x+3)}$$
$$= \frac{7x+21-x^2+3x-2}{(x-2)(x+1)(x+3)}$$
$$= \frac{-x^2+10x+19}{(x-2)(x+1)(x+3)}$$

47. $\dfrac{x}{x^2 - 8x + 7} - \dfrac{x+2}{2x^2 - 9x - 35}$

$= \dfrac{x}{(x-1)(x-7)} - \dfrac{x+2}{(2x+5)(x-7)}$

$= \dfrac{x(2x+5)}{(x-1)(x-7)(2x+5)} - \dfrac{(x+2)(x-1)}{(2x+5)(x-7)(x-1)}$

$= \dfrac{x(2x+5) - (x+2)(x-1)}{(2x+5)(x-7)(x-1)}$

$= \dfrac{2x^2 + 5x - (x^2 + x - 2)}{(2x+5)(x-7)(x-1)}$

$= \dfrac{2x^2 + 5x - x^2 - x + 2}{(2x+5)(x-7)(x-1)}$

$= \dfrac{x^2 + 4x + 2}{(2x+5)(x-7)(x-1)}$

49. $\dfrac{2}{a^2 + 2a + 1} + \dfrac{3}{a^2 - 1}$

$= \dfrac{2}{(a+1)^2} + \dfrac{3}{(a+1)(a-1)}$

$= \dfrac{2(a-1)}{(a+1)^2(a-1)} + \dfrac{3(a+1)}{(a+1)(a-1)(a+1)}$

$= \dfrac{2(a-1) + 3(a+1)}{(a+1)^2(a-1)}$

$= \dfrac{2a - 2 + 3a + 3}{(a+1)^2(a-1)}$

$= \dfrac{5a+1}{(a+1)^2(a-1)}$

51. $\dfrac{4}{3x^2 y^3} + \dfrac{5}{3x^2 y^3} = \dfrac{9}{3x^2 y^3} = \dfrac{3}{x^2 y^3}$

53. $\dfrac{13x - 5}{2x} - \dfrac{13x + 5}{2x} = \dfrac{13x - 5 - 13x - 5}{2x}$

$= \dfrac{-10}{2x}$

$= -\dfrac{5}{x}$

55. $\dfrac{3}{2x+10} + \dfrac{8}{3x+15} = \dfrac{3}{2(x+5)} + \dfrac{8}{3(x+5)}$

$= \dfrac{3 \cdot 3}{2(x+5) \cdot 3} + \dfrac{8 \cdot 2}{3(x+5) \cdot 2}$

$= \dfrac{9}{6(x+5)} + \dfrac{16}{6(x+5)}$

$= \dfrac{25}{6(x+5)}$

57. $\dfrac{-2}{x^2 - 3x} - \dfrac{1}{x^3 - 3x^2} = \dfrac{-2}{x(x-3)} - \dfrac{1}{x^2(x-3)}$

$= \dfrac{-2x}{x^2(x-3)} - \dfrac{1}{x^2(x-3)}$

$= \dfrac{-2x - 1}{x^2(x-3)}$

59. $\dfrac{ab}{a^2 - b^2} + \dfrac{b}{a+b}$

$= \dfrac{ab}{(a+b)(a-b)} + \dfrac{b}{a+b}$

$= \dfrac{ab}{(a+b)(a-b)} + \dfrac{b(a-b)}{(a+b)(a-b)}$

$= \dfrac{ab}{(a+b)(a-b)} + \dfrac{ab - b^2}{(a+b)(a-b)}$

$= \dfrac{ab + ab - b^2}{(a+b)(a-b)}$

$= \dfrac{2ab - b^2}{(a+b)(a-b)}$

$= \dfrac{b(2a-b)}{(a+b)(a-b)}$

61. $\dfrac{5}{x^2 - 4} - \dfrac{3}{x^2 + 4x + 4}$

$= \dfrac{5}{(x+2)(x-2)} - \dfrac{3}{(x+2)^2}$

$= \dfrac{5(x+2)}{(x+2)(x-2)(x+2)} - \dfrac{3(x-2)}{(x+2)^2(x-2)}$

$= \dfrac{5(x+2) - 3(x-2)}{(x+2)^2(x-2)}$

$= \dfrac{5x + 10 - 3x + 6}{(x+2)^2(x-2)}$

$= \dfrac{2x + 16}{(x+2)^2(x-2)}$

$= \dfrac{2(x+8)}{(x+2)^2(x-2)}$

63. $\dfrac{3x}{2x^2-11x+5}+\dfrac{7}{x^2-2x-15}=\dfrac{3x}{(2x-1)(x-5)}+\dfrac{7}{(x-5)(x+3)}$

$$=\dfrac{3x(x+3)}{(2x-1)(x-5)(x+3)}+\dfrac{7(2x-1)}{(x-5)(x+3)(2x-1)}$$

$$=\dfrac{3x^2+9x}{(2x-1)(x-5)(x+3)}+\dfrac{14x-7}{(2x-1)(x-5)(x+3)}$$

$$=\dfrac{3x^2+9x+14x-7}{(2x-1)(x-5)(x+3)}$$

$$=\dfrac{3x^2+23x-7}{(2x-1)(x-5)(x+3)}$$

65. $\dfrac{2}{x+1}-\dfrac{3x}{3x+3}+\dfrac{1}{2x+2}=\dfrac{2}{x+1}-\dfrac{3x}{3(x+1)}+\dfrac{1}{2(x+1)}$

$$=\dfrac{2}{x+1}-\dfrac{x}{x+1}+\dfrac{1}{2(x+1)}$$

$$=\dfrac{2\cdot2}{2(x+1)}-\dfrac{2\cdot x}{2(x+1)}+\dfrac{1}{2(x+1)}$$

$$=\dfrac{4-2x+1}{2(x+1)}$$

$$=\dfrac{5-2x}{2(x+1)}$$

67. $\dfrac{3}{x+3}+\dfrac{5}{x^2+6x+9}-\dfrac{x}{x^2-9}=\dfrac{3}{x+3}+\dfrac{5}{(x+3)^2}-\dfrac{x}{(x+3)(x-3)}$

$$=\dfrac{3(x+3)(x-3)}{(x+3)(x+3)(x-3)}+\dfrac{5(x-3)}{(x+3)^2(x-3)}-\dfrac{x(x+3)}{(x+3)(x-3)(x+3)}$$

$$=\dfrac{3x^2-27}{(x+3)^2(x-3)}+\dfrac{5x-15}{(x+3)^2(x-3)}-\dfrac{x^2+3x}{(x+3)^2(x-3)}$$

$$=\dfrac{3x^2-27+5x-15-x^2-3x}{(x+3)^2(x-3)}$$

$$=\dfrac{2x^2+2x-42}{(x+3)^2(x-3)}$$

$$=\dfrac{2(x^2+x-21)}{(x+3)^2(x-3)}$$

69. $\dfrac{x}{x^2-9}+\dfrac{3}{x^2-6x+9}-\dfrac{1}{x+3}=\dfrac{x}{(x+3)(x-3)}+\dfrac{3}{(x-3)^2}-\dfrac{1}{x+3}$

$\qquad\qquad\qquad = \dfrac{x(x-3)}{(x+3)(x-3)^2}+\dfrac{3(x+3)}{(x-3)^2(x+3)}-\dfrac{1(x-3)^2}{(x+3)(x-3)^2}$

$\qquad\qquad\qquad = \dfrac{x(x-3)+3(x+3)-(x-3)^2}{(x+3)(x-3)^2}$

$\qquad\qquad\qquad = \dfrac{x^2-3x+3x+9-(x^2-6x+9)}{(x+3)(x-3)^2}$

$\qquad\qquad\qquad = \dfrac{x^2+9-x^2+6x-9}{(x+3)(x-3)^2}$

$\qquad\qquad\qquad = \dfrac{6x}{(x+3)(x-3)^2}$

71. $\left(\dfrac{1}{x}+\dfrac{2}{3}\right)-\left(\dfrac{1}{x}-\dfrac{2}{3}\right)=\left(\dfrac{3}{3x}+\dfrac{2x}{3x}\right)-\left(\dfrac{3}{3x}-\dfrac{2x}{3x}\right)$

$\qquad\qquad\qquad = \left(\dfrac{3+2x}{3x}\right)-\left(\dfrac{3-2x}{3x}\right)$

$\qquad\qquad\qquad = \dfrac{3+2x-3+2x}{3x}$

$\qquad\qquad\qquad = \dfrac{4x}{3x}$

$\qquad\qquad\qquad = \dfrac{4}{3}$

73. $\left(\dfrac{2}{3}-\dfrac{1}{x}\right)\cdot\left(\dfrac{3}{x}+\dfrac{1}{2}\right)=\left(\dfrac{2x}{3x}-\dfrac{3}{3x}\right)\cdot\left(\dfrac{3\cdot2}{2x}+\dfrac{x}{2x}\right)$

$\qquad\qquad\qquad = \left(\dfrac{2x-3}{3x}\right)\cdot\left(\dfrac{6+x}{2x}\right)$

$\qquad\qquad\qquad = \dfrac{(2x-3)(x+6)}{6x^2}\ \text{or}\ \dfrac{2x^2+9x-18}{6x^2}$

75. $\left(\dfrac{2a}{3}\right)^2\div\left(\dfrac{a^2}{a+1}-\dfrac{1}{a+1}\right)=\dfrac{4a^2}{9}\div\dfrac{a^2-1}{a+1}$

$\qquad\qquad\qquad = \dfrac{4a^2}{9}\div\dfrac{(a+1)(a-1)}{a+1}$

$\qquad\qquad\qquad = \dfrac{4a^2}{9}\div(a-1)$

$\qquad\qquad\qquad = \dfrac{4a^2}{9}\cdot\dfrac{1}{a-1}$

$\qquad\qquad\qquad = \dfrac{4a^2}{9(a-1)}$

77. $\left(\dfrac{2x}{3}\right)^2 \div \left(\dfrac{x}{3}\right)^2 = \left(\dfrac{2^2 x^2}{3^2}\right) \div \left(\dfrac{x^2}{3^2}\right)$

$\qquad\qquad\qquad = \dfrac{2^2 x^2}{3^2} \cdot \dfrac{3^2}{x^2}$

$\qquad\qquad\qquad = 2^2$

$\qquad\qquad\qquad = 4$

79. $\left(\dfrac{x}{x+1} - \dfrac{x}{x-1}\right) \div \dfrac{x}{2x+2}$

$= \left(\dfrac{x \cdot (x-1)}{(x+1)(x-1)} - \dfrac{x \cdot (x+1)}{(x-1)(x+1)}\right) \div \dfrac{x}{2(x+1)}$

$= \dfrac{x(x-1) - x(x+1)}{(x+1)(x-1)} \div \dfrac{x}{2(x+1)}$

$= \dfrac{x^2 - x - x^2 - x}{(x+1)(x-1)} \div \dfrac{x}{2(x+1)}$

$= \dfrac{-2x}{(x+1)(x-1)} \cdot \dfrac{2(x+1)}{x}$

$= -\dfrac{4}{x-1}$

81. $\dfrac{4}{x} \cdot \left(\dfrac{2}{x+2} - \dfrac{2}{x-2}\right)$

$= \dfrac{4}{x} \cdot \left(\dfrac{2(x-2)}{(x+2)(x-2)} - \dfrac{2(x+2)}{(x-2)(x+2)}\right)$

$= \dfrac{4}{x} \cdot \left(\dfrac{2x-4}{(x+2)(x-2)} - \dfrac{2x+4}{(x+2)(x-2)}\right)$

$= \dfrac{4}{x} \cdot \left(\dfrac{2x-4-2x-4}{(x+2)(x-2)}\right)$

$= \dfrac{4}{x} \left(\dfrac{-8}{(x+2)(x-2)}\right)$

$= -\dfrac{32}{x(x+2)(x-2)}$

83. $12\left(\dfrac{2}{3} + \dfrac{1}{6}\right) = 12 \cdot \dfrac{2}{3} + 12 \cdot \dfrac{1}{6} = \dfrac{24}{3} + \dfrac{12}{6} = 8 + 2 = 10$

85. $x^2\left(\dfrac{4}{x^2} + 1\right) = x^2 \cdot \dfrac{4}{x^2} + x^2 \cdot 1 = 4 + x^2$

87. $\sqrt{100} = 10$ because $10^2 = 100$.

89. $\sqrt[3]{8} = 2$ because $2^3 = 8$.

91. $\sqrt[4]{81} = 3$ because $3^4 = 81$.

93. $a^2 + b^2 = c^2$

$\qquad 3^2 + 4^2 = c^2$

$\qquad\quad 9 + 16 = c^2$

$\qquad\qquad 25 = c^2$

$\qquad\qquad c = 5$ meters

95. $\dfrac{2x-3}{x^2+1} - \dfrac{x-6}{x^2+1} = \dfrac{2x-3-(x-6)}{x^2+1}$

$\qquad\qquad\qquad\qquad = \dfrac{2x-3-x+6}{x^2+1}$

$\qquad\qquad\qquad\qquad = \dfrac{x+3}{x^2+1}$

97. $P = 4s$

$\quad P = 4 \cdot \left(\dfrac{x}{x+5}\right) = \dfrac{4x}{x+5}$

$\quad A = s^2$

$\quad A = \left(\dfrac{x}{x+5}\right)^2$

$\quad A = \left(\dfrac{x}{x+5}\right)\left(\dfrac{x}{x+5}\right) = \dfrac{x^2}{(x+5)^2}$

The perimeter is $\dfrac{4x}{x+5}$ feet, and the area is

$\dfrac{x^2}{(x+5)^2}$ square feet.

99. answers may vary

101. answers may vary

103. answers may vary

105. $x^{-1} + (2x)^{-1} = \dfrac{1}{x} + \dfrac{1}{2x}$

$\qquad\qquad\qquad = \dfrac{1 \cdot 2}{x \cdot 2} + \dfrac{1}{2x}$

$\qquad\qquad\qquad = \dfrac{2+1}{2x}$

$\qquad\qquad\qquad = \dfrac{3}{2x}$

107. $4x^{-2} - 3x^{-1} = \dfrac{4}{x^2} - \dfrac{3}{x} = \dfrac{4}{x^2} - \dfrac{3x}{x^2} = \dfrac{4-3x}{x^2}$

109.

Section 6.3 Practice Exercises

1. a. $\dfrac{\frac{5k}{36m}}{\frac{15k}{9}} = \dfrac{5k}{36m} \div \dfrac{15k}{9}$

$= \dfrac{5k}{36m} \cdot \dfrac{9}{15k}$

$= \dfrac{5k \cdot 9}{36m \cdot 15k}$

$= \dfrac{1}{12m}$

b. $\dfrac{\frac{8x}{x-4}}{\frac{3}{x+4}} = \dfrac{8x}{x-4} \div \dfrac{3}{x+4}$

$= \dfrac{8x}{x-4} \cdot \dfrac{x+4}{3}$

$= \dfrac{8x(x+4)}{3(x-4)}$

c. $\dfrac{\frac{5}{a}+\frac{b}{a^2}}{\frac{5a}{b^2}+\frac{1}{b}} = \dfrac{\frac{5 \cdot a}{a \cdot a}+\frac{b}{a^2}}{\frac{5a}{b^2}+\frac{1 \cdot b}{b \cdot b}}$

$= \dfrac{\frac{5a+b}{a^2}}{\frac{5a+b}{b^2}}$

$= \dfrac{5a+b}{a^2} \cdot \dfrac{b^2}{5a+b}$

$= \dfrac{b^2(5a+b)}{a^2(5a+b)}$

$= \dfrac{b^2}{a^2}$

2. a. The LCD is $(x-4)(x+4)$.

$\dfrac{\frac{8x}{x-4}}{\frac{3}{x+4}} = \dfrac{\left(\frac{8x}{x-4}\right) \cdot (x-4)(x+4)}{\left(\frac{3}{x+4}\right) \cdot (x-4)(x+4)}$

$= \dfrac{8x(x+4)}{3(x-4)}$

b. The LCD is $a^2 b^2$.

$\dfrac{\frac{b}{a^2}+\frac{1}{a}}{\frac{a}{b^2}+\frac{1}{b}} = \dfrac{\left(\frac{b}{a^2}+\frac{1}{a}\right) \cdot a^2 b^2}{\left(\frac{a}{b^2}+\frac{1}{b}\right) \cdot a^2 b^2}$

$= \dfrac{\frac{b}{a^2} \cdot a^2 b^2 + \frac{1}{a} \cdot a^2 b^2}{\frac{a}{b^2} \cdot a^2 b^2 + \frac{1}{b} \cdot a^2 b^2}$

$= \dfrac{b^3 + ab^2}{a^3 + a^2 b}$

$= \dfrac{b^2(b+a)}{a^2(a+b)}$

$= \dfrac{b^2}{a^2}$

3. $\dfrac{3x^{-1}+x^{-2}y^{-1}}{y^{-2}+xy^{-1}} = \dfrac{\frac{3}{x}+\frac{1}{x^2 y}}{\frac{1}{y^2}+\frac{x}{y}}$

The LCD is $x^2 y^2$.

$= \dfrac{\left(\frac{3}{x}+\frac{1}{x^2 y}\right) \cdot x^2 y^2}{\left(\frac{1}{y^2}+\frac{x}{y}\right) \cdot x^2 y^2}$

$= \dfrac{\frac{3}{x} \cdot x^2 y^2 + \frac{1}{x^2 y} \cdot x^2 y^2}{\frac{1}{y^2} \cdot x^2 y^2 + \frac{x}{y} \cdot x^2 y^2}$

$= \dfrac{3xy^2 + y}{x^2 + x^3 y}$ or $\dfrac{y(3xy+1)}{x^2(1+xy)}$

4. $\dfrac{(3x)^{-1}-2}{5x^{-1}+2} = \dfrac{\frac{1}{3x}-2}{\frac{5}{x}+2}$

$= \dfrac{\left(\frac{1}{3x}-2\right) \cdot 3x}{\left(\frac{5}{x}+2\right) \cdot 3x}$

$= \dfrac{\frac{1}{3x} \cdot 3x - 2 \cdot 3x}{\frac{5}{x} \cdot 3x + 2 \cdot 3x}$

$= \dfrac{1-6x}{15+6x}$ or $\dfrac{1-6x}{3(5+2x)}$

Vocabulary, Readiness & Video Check 6.3

1. $\dfrac{\frac{7}{x}}{\frac{1}{x}+\frac{z}{x}} = \dfrac{x\left(\frac{7}{x}\right)}{x\left(\frac{1}{x}\right)+x\left(\frac{z}{x}\right)} = \dfrac{7}{1+z}$

2. $\dfrac{\frac{x}{4}}{\frac{x^2}{2}+\frac{1}{4}}=\dfrac{4\left(\frac{x}{4}\right)}{4\left(\frac{x^2}{2}\right)+4\left(\frac{1}{4}\right)}=\dfrac{x}{\underline{2x^2+1}}$

3. $x^{-2}=\dfrac{1}{x^2}$

4. $y^{-3}=\dfrac{1}{y^3}$

5. $2x^{-1}=\dfrac{2}{x}$

6. $(2x)^{-1}=\dfrac{1}{2x}$

7. $(9y)^{-1}=\dfrac{1}{9y}$

8. $9y^{-2}=\dfrac{9}{y^2}$

9. a single fraction in the numerator and in the denominator

10. In method 2, you find the LCD of all fractions in both the numerator and the denominator in order to clear the complex fraction of fractions in the numerator and denominator; in method 1, you find the LCD of the fractions only in the numerator and/or only in the denominator in order to get single fractions in the numerator and denominator.

11. Since a negative exponent moves its base from a numerator to a denominator of the expression only, a rational expression containing negative exponents can become a complex fraction when rewritten with positive exponents.

Exercise Set 6.3

1. $\dfrac{\frac{10}{3x}}{\frac{5}{6x}}=\dfrac{10}{3x}\cdot\dfrac{6x}{5}=\dfrac{60x}{15x}=4$

3. $\dfrac{1+\frac{2}{5}}{2+\frac{3}{5}}=\dfrac{5\left(1+\frac{2}{5}\right)}{5\left(2+\frac{3}{5}\right)}=\dfrac{5+2}{10+3}=\dfrac{7}{13}$

5. $\dfrac{\frac{4}{x-1}}{\frac{x}{x-1}}=\dfrac{4}{x-1}\cdot\dfrac{x-1}{x}=\dfrac{4}{x}$

7. $\dfrac{1-\frac{2}{x}}{x+\frac{4}{9x}}=\dfrac{9x\left(1-\frac{2}{x}\right)}{9x\left(x+\frac{4}{9x}\right)}=\dfrac{9x-18}{9x^2+4}=\dfrac{9(x-2)}{9x^2+4}$

9. $\dfrac{\frac{4x^2-y^2}{xy}}{\frac{2}{y}-\frac{1}{x}}=\dfrac{\left(\frac{4x^2-y^2}{xy}\right)\cdot xy}{\left(\frac{2}{y}-\frac{1}{x}\right)\cdot xy}$

$=\dfrac{4x^2-y^2}{2x-y}$

$=\dfrac{(2x-y)(2x+y)}{2x-y}$

$=2x+y$

11. $\dfrac{\frac{x+1}{3}}{\frac{2x-1}{6}}=\dfrac{x+1}{3}\cdot\dfrac{6}{2x-1}=\dfrac{2(x+1)}{2x-1}$

13. $\dfrac{\frac{2}{x}+\frac{3}{x^2}}{\frac{4}{x^2}-\frac{9}{x}}=\dfrac{\left(\frac{2}{x}+\frac{3}{x^2}\right)x^2}{\left(\frac{4}{x^2}-\frac{9}{x}\right)x^2}=\dfrac{2x+3}{4-9x}$

15. $\dfrac{\frac{1}{x}+\frac{2}{x^2}}{x+\frac{8}{x^2}}=\dfrac{x^2\left(\frac{1}{x}+\frac{2}{x^2}\right)}{x^2\left(x+\frac{8}{x^2}\right)}$

$=\dfrac{x+2}{x^3+8}$

$=\dfrac{x+2}{(x+2)(x^2-2x+4)}$

$=\dfrac{1}{x^2-2x+4}$

17. $\dfrac{\frac{4}{5-x}+\frac{5}{x-5}}{\frac{2}{x}+\frac{3}{x-5}}=\dfrac{-\frac{4}{x-5}+\frac{5}{x-5}}{\frac{2(x-5)+3x}{x(x-5)}}$

$=\dfrac{\frac{1}{x-5}}{\frac{2x-10+3x}{x(x-5)}}$

$=\dfrac{1}{x-5}\cdot\dfrac{x(x-5)}{5x-10}$

$=\dfrac{x}{5x-10}$ or $\dfrac{x}{5(x-2)}$

19.
$$\frac{\frac{x+2}{x}-\frac{2}{x-1}}{\frac{x+1}{x}+\frac{x+1}{x-1}}=\frac{\frac{(x+2)(x-1)-2x}{x(x-1)}}{\frac{(x+1)(x-1)+(x+1)(x)}{x(x-1)}}$$

$$=\frac{\frac{x^2+x-2-2x}{x(x-1)}}{\frac{x^2-1+x^2+x}{x(x-1)}}$$

$$=\frac{x^2-x-2}{x(x-1)}\cdot\frac{x(x-1)}{2x^2+x-1}$$

$$=\frac{(x-2)(x+1)}{x(x-1)}\cdot\frac{x(x-1)}{(2x-1)(x+1)}$$

$$=\frac{x-2}{2x-1}$$

21.
$$\frac{\frac{2}{x}+3}{\frac{4}{x^2}-9}=\frac{\left(\frac{2}{x}+3\right)\cdot x^2}{\left(\frac{4}{x^2}-9\right)\cdot x^2}$$

$$=\frac{2x+3x^2}{4-9x^2}$$

$$=\frac{x(2+3x)}{(2+3x)(2-3x)}$$

$$=\frac{x}{2-3x}$$

23.
$$\frac{1-\frac{x}{y}}{\frac{x^2}{y^2}-1}=\frac{\left(1-\frac{x}{y}\right)\cdot y^2}{\left(\frac{x^2}{y^2}-1\right)\cdot y^2}$$

$$=\frac{y^2-xy}{x^2-y^2}$$

$$=\frac{y(y-x)}{(x+y)(x-y)}$$

$$=\frac{-y(x-y)}{(x+y)(x-y)}$$

$$=-\frac{y}{x+y}$$

25.
$$\frac{\frac{-2x}{x-y}}{\frac{y}{x^2}}=\frac{-2x}{x-y}\cdot\frac{x^2}{y}=-\frac{2x^3}{y(x-y)}$$

27.
$$\frac{\frac{2}{x}+\frac{1}{x^2}}{\frac{y}{x^2}}=\frac{\left(\frac{2}{x}+\frac{1}{x^2}\right)x^2}{\left(\frac{y}{x^2}\right)x^2}=\frac{2x+1}{y}$$

29.
$$\frac{\frac{x}{9}-\frac{1}{x}}{1+\frac{3}{x}}=\frac{\left(\frac{x}{9}-\frac{1}{x}\right)\cdot 9x}{\left(1+\frac{3}{x}\right)\cdot 9x}$$

$$=\frac{x^2-9}{9x+27}$$

$$=\frac{(x+3)(x-3)}{9(x+3)}$$

$$=\frac{x-3}{9}$$

31.
$$\frac{\frac{x-1}{x^2-4}}{1+\frac{1}{x-2}}=\frac{\frac{x-1}{x^2-4}}{\frac{x-2+1}{x-2}}=\frac{\frac{x-1}{x^2-4}}{\frac{x-1}{x-2}}$$

$$=\frac{x-1}{x^2-4}\cdot\frac{x-2}{x-1}$$

$$=\frac{x-1}{(x+2)(x-2)}\cdot\frac{x-2}{x-1}$$

$$=\frac{1}{x+2}$$

33.
$$\frac{\frac{2}{x+5}+\frac{4}{x+3}}{\frac{3x+13}{x^2+8x+15}}=\frac{\frac{2}{x+5}+\frac{4}{x+3}}{\frac{3x+13}{(x+5)(x+3)}}$$

$$=\frac{\left(\frac{2}{x+5}+\frac{4}{x+3}\right)(x+5)(x+3)}{\frac{3x+13}{(x+5)(x+3)}(x+5)(x+3)}$$

$$=\frac{2(x+3)+4(x+5)}{3x+13}$$

$$=\frac{2x+6+4x+20}{3x+13}$$

$$=\frac{6x+26}{3x+13}$$

$$=\frac{2(3x+13)}{3x+13}$$

$$=2$$

35.
$$\frac{x^{-1}}{x^{-2}+y^{-2}}=\frac{\frac{1}{x}}{\frac{1}{x^2}+\frac{1}{y^2}}$$

$$=\frac{x^2y^2\left(\frac{1}{x}\right)}{x^2y^2\left(\frac{1}{x^2}+\frac{1}{y^2}\right)}$$

$$=\frac{xy^2}{y^2+x^2}$$

$$=\frac{xy^2}{x^2+y^2}$$

37. $\dfrac{2a^{-1}+3b^{-2}}{a^{-1}-b^{-1}}=\dfrac{\frac{2}{a}+\frac{3}{b^2}}{\frac{1}{a}-\frac{1}{b}}$

$=\dfrac{ab^2\left(\frac{2}{a}+\frac{3}{b^2}\right)}{ab^2\left(\frac{1}{a}-\frac{1}{b}\right)}$

$=\dfrac{2b^2+3a}{b^2-ab}$

$=\dfrac{2b^2+3a}{b(b-a)}$

39. $\dfrac{1}{x-x^{-1}}=\dfrac{1}{x-\frac{1}{x}}$

$=\dfrac{x(1)}{x\left(x-\frac{1}{x}\right)}$

$=\dfrac{x}{x^2-1}$

$=\dfrac{x}{(x+1)(x-1)}$

41. $\dfrac{a^{-1}+1}{a^{-1}-1}=\dfrac{\frac{1}{a}+1}{\frac{1}{a}-1}=\dfrac{a\left(\frac{1}{a}+1\right)}{a\left(\frac{1}{a}-1\right)}=\dfrac{1+a}{1-a}$

43. $\dfrac{3x^{-1}+(2y)^{-1}}{x^{-2}}=\dfrac{\frac{3}{x}+\frac{1}{2y}}{\frac{1}{x^2}}$

$=\dfrac{2x^2y\left(\frac{3}{x}+\frac{1}{2y}\right)}{2x^2y\left(\frac{1}{x^2}\right)}$

$=\dfrac{6xy+x^2}{2y}$

$=\dfrac{x(x+6y)}{2y}$

45. $\dfrac{2a^{-1}+(2a)^{-1}}{a^{-1}+2a^{-2}}=\dfrac{\frac{2}{a}+\frac{1}{2a}}{\frac{1}{a}+\frac{2}{a^2}}$

$=\dfrac{2a^2\left(\frac{2}{a}+\frac{1}{2a}\right)}{2a^2\left(\frac{1}{a}+\frac{2}{a^2}\right)}$

$=\dfrac{4a+a}{2a+4}$

$=\dfrac{5a}{2(a+2)}$

47. $\dfrac{5x^{-1}+2y^{-1}}{x^{-2}y^{-2}}=\dfrac{\frac{5}{x}+\frac{2}{y}}{\frac{1}{x^2y^2}}$

$=\dfrac{x^2y^2\left(\frac{5}{x}+\frac{2}{y}\right)}{x^2y^2\left(\frac{1}{x^2y^2}\right)}$

$=5xy^2+2x^2y$

$=xy(5y+2x)$

49. $\dfrac{5x^{-1}-2y^{-1}}{25x^{-2}-4y^{-2}}=\dfrac{\frac{5}{x}-\frac{2}{y}}{\frac{25}{x^2}-\frac{4}{y^2}}$

$=\dfrac{x^2y^2\left(\frac{5}{x}-\frac{2}{y}\right)}{x^2y^2\left(\frac{25}{x^2}-\frac{4}{y^2}\right)}$

$=\dfrac{5xy^2-2x^2y}{25y^2-4x^2}$

$=\dfrac{xy(5y-2x)}{(5y+2x)(5y-2x)}$

$=\dfrac{xy}{5y+2x}$ or $\dfrac{xy}{2x+5y}$

51. $\dfrac{3x^3y^2}{12x}=\dfrac{3x\cdot x^2y^2}{3x\cdot 4}=\dfrac{x^2y^2}{4}$

53. $\dfrac{144x^5y^5}{-16x^2y}=\dfrac{16x^2y\cdot 9x^3y^4}{16x^2y\cdot(-1)}=-9x^3y^4$

55. $|x-5|=9$

$x-5=-9$ or $x-5=9$

$x=-4$ or $x=14$

The solution set is $\{-4, 14\}$.

57. $\dfrac{\frac{x+1}{9}}{\frac{y-2}{5}}=\dfrac{x+1}{9}\div\dfrac{y-2}{5}=\dfrac{x+1}{9}\cdot\dfrac{5}{y-2}$

Both a and c are equivalent to the original expression.

59. $\dfrac{a}{1-\frac{s}{770}}=\dfrac{770(a)}{770\left(1-\frac{s}{770}\right)}=\dfrac{770a}{770-s}$

61. $\dfrac{\frac{1}{x}}{\frac{3}{y}} = \dfrac{1}{x} \div \dfrac{3}{y} = \dfrac{1}{x} \cdot \dfrac{y}{3}$

Both a and b are equivalent to the original expression.

63. answers may vary

65. $\dfrac{1}{1+(1+x)^{-1}} = \dfrac{1}{1+\frac{1}{1+x}}$

$= \dfrac{(1+x) \cdot 1}{(1+x)\left(1+\frac{1}{1+x}\right)}$

$= \dfrac{1+x}{1+x+1}$

$= \dfrac{1+x}{2+x}$

67. $\dfrac{x}{1-\frac{1}{1+\frac{1}{x}}} = \dfrac{x}{1-\frac{1}{\frac{x+1}{x}}}$

$= \dfrac{x}{1-\frac{x}{x+1}}$

$= \dfrac{(x+1)(x)}{(x+1)\left(1-\frac{x}{x+1}\right)}$

$= \dfrac{x(x+1)}{x+1-x}$

$= \dfrac{x(x+1)}{1}$

$= x(x+1)$

69. $\dfrac{\frac{2}{y^2}-\frac{5}{xy}-\frac{3}{x^2}}{\frac{2}{y^2}+\frac{7}{xy}+\frac{3}{x^2}} = \dfrac{x^2y^2\left(\frac{2}{y^2}-\frac{5}{xy}-\frac{3}{x^2}\right)}{x^2y^2\left(\frac{2}{y^2}+\frac{7}{xy}+\frac{3}{x^2}\right)}$

$= \dfrac{2x^2-5xy-3y^2}{2x^2+7xy+3y^2}$

$= \dfrac{(2x+y)(x-3y)}{(2x+y)(x+3y)}$

$= \dfrac{x-3y}{x+3y}$

71. $\dfrac{3(a+1)^{-1}+4a^{-2}}{(a^3+a^2)^{-1}} = \dfrac{\frac{3}{a+1}+\frac{4}{a^2}}{\frac{1}{a^3+a^2}}$

$= \dfrac{\frac{3a^2+4(a+1)}{a^2(a+1)}}{\frac{1}{a^2(a+1)}}$

$= \dfrac{3a^2+4a+4}{a^2(a+1)} \cdot \dfrac{a^2(a+1)}{1}$

$= 3a^2+4a+4$

73. $f(x) = \dfrac{1}{x}$

a. $f(a+h) = \dfrac{1}{a+h}$

b. $f(a) = \dfrac{1}{a}$

c. $\dfrac{f(a+h)-f(a)}{h} = \dfrac{\frac{1}{a+h}-\frac{1}{a}}{h}$

d. $\dfrac{\frac{1}{a+h}-\frac{1}{a}}{h} = \dfrac{a(a+h)\left(\frac{1}{a+h}-\frac{1}{a}\right)}{a(a+h) \cdot h}$

$= \dfrac{a-(a+h)}{ah(a+h)}$

$= \dfrac{-h}{ah(a+h)}$

$= -\dfrac{1}{a(a+h)}$

75. $f(x) = \dfrac{3}{x+1}$

a. $f(a+h) = \dfrac{3}{a+h+1}$

b. $f(a) = \dfrac{3}{a+1}$

c. $\dfrac{f(a+h)-f(a)}{h} = \dfrac{\frac{3}{a+h+1}-\frac{3}{a+1}}{h}$

d. $\dfrac{\frac{3}{a+h+1} - \frac{3}{a+1}}{h}$

$= \dfrac{\left(\frac{3}{a+h+1} - \frac{3}{a+1}\right) \cdot (a+h+1)(a+1)}{h \cdot (a+h+1)(a+1)}$

$= \dfrac{3(a+1) - 3(a+h+1)}{h(a+h+1)(a+1)}$

$= \dfrac{3a+3 - 3a - 3h - 3}{h(a+h+1)(a+1)}$

$= \dfrac{-3h}{h(a+h+1)(a+1)}$

$= -\dfrac{3}{(a+h+1)(a+1)}$

Section 6.4 Practice Exercises

1. $\dfrac{18a^3 - 12a^2 + 30a}{6a} = \dfrac{18a^3}{6a} - \dfrac{12a^2}{6a} + \dfrac{30a}{6a}$

$\qquad\qquad = 3a^2 - 2a + 5$

2. $\dfrac{5a^3b^4 - 8a^2b^3 + ab^2 - 8b}{ab^2}$

$= \dfrac{5a^3b^4}{ab^2} - \dfrac{8a^2b^3}{ab^2} + \dfrac{ab^2}{ab^2} - \dfrac{8b}{ab^2}$

$= 5a^2b^2 - 8ab + 1 - \dfrac{8}{ab}$

3.
$$
\begin{array}{r}
3x - 2 \\
x+3\overline{\smash{)}3x^2 + 7x - 6} \\
\underline{3x^2 + 9x} \\
-2x - 6 \\
\underline{-2x - 6} \\
0
\end{array}
$$

Answer: $3x - 2$

4.
$$
\begin{array}{r}
3x - 2 \\
2x-1\overline{\smash{)}6x^2 - 7x + 8} \\
\underline{6x^2 - 3x} \\
-4x + 8 \\
\underline{-4x + 2} \\
6
\end{array}
$$

Answer: $3x - 2 + \dfrac{6}{2x - 1}$

5.
$$
\begin{array}{r}
5x^2 - 6x + 8 \\
x+3\overline{\smash{)}5x^3 + 9x^2 - 10x + 30} \\
\underline{5x^3 + 15x^2} \\
-6x^2 - 10x \\
\underline{-6x^2 - 18x} \\
8x + 30 \\
\underline{8x + 24} \\
6
\end{array}
$$

Answer: $5x^2 - 6x + 8 + \dfrac{6}{x + 3}$

6.
$$
\begin{array}{r}
2x^2 + 3x - 2 \\
x^2+0x+1\overline{\smash{)}2x^4 + 3x^3 + 0x^2 - 5x + 2} \\
\underline{2x^4 + 0x^3 + 2x^2} \\
3x^3 - 2x^2 - 5x \\
\underline{3x^3 + 0x^2 + 3x} \\
-2x^2 - 8x + 2 \\
\underline{-2x^2 + 0x - 2} \\
-8x + 4
\end{array}
$$

Answer: $2x^2 + 3x - 2 + \dfrac{-8x + 4}{x^2 + 1}$

7.
$$
\begin{array}{r}
16x^2 + 20x + 25 \\
4x-5\overline{\smash{)}64x^3 + 0x^2 + 0x - 125} \\
\underline{64x^3 - 80x^2} \\
80x^2 + 0x \\
\underline{80x^2 - 100x} \\
100x - 125 \\
\underline{100x - 125} \\
0
\end{array}
$$

Answer: $16x^2 + 20x + 25$

8. Since $x - c = x - 1$, c is 1.

$$
\begin{array}{r|rrrr}
1 & 4 & -3 & 6 & 5 \\
  &   & 4 & 1 & 7 \\
\hline
  & 4 & 1 & 7 & 12
\end{array}
$$

$4x^2 + x + 7 + \dfrac{12}{x - 1}$

9. Since $x - c = x + 3 = x - (-3)$, c is -3.

$$-3 \overline{\smash{\big|}\ \begin{array}{rrrrr} 1 & 3 & -5 & 0 & 12 \\ & -3 & 0 & 15 & -45 \\ \hline 1 & 0 & -5 & 15 & -33 \end{array}}$$

$$x^3 - 5x + 15 - \frac{33}{x+3}$$

10. a. $P(x) = x^3 - 5x - 2$

$$P(2) = 2^3 - 5(2) - 2$$
$$= 8 - 10 - 2$$
$$= -4$$

b. Since $x - c = x - 2$, c is 2.

$$2 \overline{\smash{\big|}\ \begin{array}{rrrr} 1 & 0 & -5 & -2 \\ & 2 & 4 & -2 \\ \hline 1 & 2 & -1 & -4 \end{array}}$$

The remainder is -4.

11.
$$3 \overline{\smash{\big|}\ \begin{array}{rrrrrr} 2 & -18 & 0 & 90 & 59 & 0 \\ & 6 & -36 & -108 & -54 & 15 \\ \hline 2 & -12 & -36 & -18 & 5 & 15 \end{array}}$$

$$P(3) = 15$$

Vocabulary, Readiness & Video Check 6.4

1. the common denominator

2. When the degree of the remainder is less than the degree of the divisor.

3. The last number is the remainder and the other numbers are the coefficients of the variables in the quotient; the degree of the quotient is one less than the degree of the dividend.

4. When the calculations for substituting in c might be tedious, that is, if $P(x)$ is a many-termed, high-degree polynomial and/or of the coefficients are very large numbers, etc.

Exercise Set 6.4

1. $\dfrac{4a^2 + 8a}{2a} = \dfrac{4a^2}{2a} + \dfrac{8a}{2a} = 2a + 4$

3. $\dfrac{12a^5 b^2 + 16a^4 b}{4a^4 b} = \dfrac{12a^5 b^2}{4a^4 b} + \dfrac{16a^4 b}{4a^4 b}$
$$= 3ab + 4$$

5. $\dfrac{4x^2 y^2 + 6xy^2 - 4y^2}{2x^2 y}$

$$= \dfrac{4x^2 y^2}{2x^2 y} + \dfrac{6xy^2}{2x^2 y} - \dfrac{4y^2}{2x^2 y}$$

$$= 2y + \dfrac{3y}{x} - \dfrac{2y}{x^2}$$

7.
$$\begin{array}{r} x+1 \\ x+2 \overline{\smash{\big)}\ x^2 + 3x + 2} \\ \underline{x^2 + 2x} \\ x + 2 \\ \underline{x + 2} \\ 0 \end{array}$$

$$\dfrac{x^2 + 3x + 2}{x + 2} = x + 1$$

9.
$$\begin{array}{r} 2x - 8 \\ x+1 \overline{\smash{\big)}\ 2x^2 - 6x - 8} \\ \underline{2x^2 + 2x} \\ -8x - 8 \\ \underline{-8x - 8} \\ 0 \end{array}$$

$$\dfrac{2x^2 - 6x - 8}{x + 1} = 2x - 8$$

11.
$$\begin{array}{r} x - \dfrac{1}{2} \\ 2x+4 \overline{\smash{\big)}\ 2x^2 + 3x - 2} \\ \underline{2x^2 + 4x} \\ -x - 2 \\ \underline{-x - 2} \\ 0 \end{array}$$

$$\dfrac{2x^2 + 3x - 2}{2x + 4} = x - \dfrac{1}{2}$$

13.

$$\begin{array}{r} 2x^2 - \dfrac{1}{2}x + 5 \\[2pt] 2x+4 \overline{\smash{\big)}\, 4x^3 + 7x^2 + 8x + 20} \\[2pt] \underline{4x^3 + 8x^2} \\[2pt] -x^2 + 8x \\[2pt] \underline{-x^2 - 2x} \\[2pt] 10x + 20 \\[2pt] \underline{10x + 20} \\[2pt] 0 \end{array}$$

$$\frac{4x^3 + 7x^2 + 8x + 20}{2x+4} = 2x^2 - \frac{1}{2}x + 5$$

15.

$$\begin{array}{r} 2x^2 \qquad\;\; -6 \\[2pt] 3x+1 \overline{\smash{\big)}\, 6x^3 + 2x^2 - 18x - 6} \\[2pt] \underline{6x^3 + 2x^2} \\[2pt] -18x - 6 \\[2pt] \underline{-18x - 6} \\[2pt] 0 \end{array}$$

$$\frac{2x^2 + 6x^3 - 18x - 6}{3x+1} = 2x^2 - 6$$

17.

$$\begin{array}{r} 3x^3 + 5x + 4 \\[2pt] x^2 + 0x - 2 \overline{\smash{\big)}\, 3x^5 + 0x^4 - x^3 + 4x^2 - 12x - 8} \\[2pt] \underline{3x^5 + 0x^4 - 6x^3} \\[2pt] 5x^3 + 4x^2 - 12x \\[2pt] \underline{5x^3 + 0x^2 - 10x} \\[2pt] 4x^2 - 2x - 8 \\[2pt] \underline{4x^2 + 0x - 8} \\[2pt] -2x \end{array}$$

$$\frac{3x^5 - x^3 + 4x^2 - 12x - 8}{x^2 - 2} = 3x^3 + 5x + 4 - \frac{2x}{x^2 - 2}$$

19.

$$\begin{array}{r} 2x^3 + \dfrac{9}{2}x^2 + 10x + 21 \\[2pt] x-2 \overline{\smash{\big)}\, 2x^4 + \dfrac{1}{2}x^3 + x^2 + x + 0} \\[2pt] \underline{2x^4 - 4x^3} \\[2pt] \dfrac{9}{2}x^3 + x^2 \\[2pt] \underline{\dfrac{9}{2}x^3 - 9x^2} \\[2pt] 10x^2 + x \\[2pt] \underline{10x^2 - 20x} \\[2pt] 21x + 0 \\[2pt] \underline{21x - 42} \\[2pt] 42 \end{array}$$

$$\frac{2x^4 + \frac{1}{2}x^3 + x^2 + x}{x-2}$$
$$= 2x^3 + \frac{9}{2}x^2 + 10x + 21 + \frac{42}{x-2}$$

21. $x - 5 = x - c$ where $c = 5$.

$$\begin{array}{r|rrr} 5 & 1 & 3 & -40 \\ & & 5 & 40 \\ \hline & 1 & 8 & 0 \end{array}$$

$$\frac{x^2 + 3x - 40}{x-5} = x + 8$$

23. $x + 6 = x - c$ where $c = -6$.

$$\begin{array}{r|rrr} -6 & 1 & 5 & -6 \\ & & -6 & 6 \\ \hline & 1 & -1 & 0 \end{array}$$

$$\frac{x^2 + 5x - 6}{x+6} = x - 1$$

25. $x - 2 = x - c$ where $c = 2$.

$$\begin{array}{r|rrrr} 2 & 1 & -7 & -13 & 5 \\ & & 2 & -10 & -46 \\ \hline & 1 & -5 & -23 & -41 \end{array}$$

$$\frac{x^3 - 7x^2 - 13x + 5}{x-2} = x^2 - 5x - 23 - \frac{41}{x-2}$$

27. $x - 2 = x - c$ where $c = 2$.

$$
\begin{array}{r|rrr}
2 & 4 & 0 & -9 \\
  &   & 8 & 16 \\
\hline
  & 4 & 8 & 7
\end{array}
$$

$$\frac{4x^2 - 9}{x - 2} = 4x + 8 + \frac{7}{x - 2}$$

29. $\dfrac{4x^7 y^4 + 8xy^2 + 4xy^3}{4xy^3}$

$$= \frac{4x^7 y^4}{4xy^3} + \frac{8xy^2}{4xy^3} + \frac{4xy^3}{4xy^3}$$

$$= x^6 y + \frac{2}{y} + 1$$

31.
$$
\require{enclose}
\begin{array}{r}
5x^2 -6 \\
2x - 1 \enclose{longdiv}{10x^3 - 5x^2 - 12x + 1} \\
\underline{10x^3 - 5x^2 } \\
-12x + 1 \\
\underline{-12x + 6} \\
-5
\end{array}
$$

$$\frac{10x^3 - 5x^2 - 12x + 1}{2x - 1} = 5x^2 - 6 - \frac{5}{2x - 1}$$

33. $x - 4$ has the form $x - c$, where $c = 4$, so synthetic division can be used.

$2x^3 - 6x^2 - 4 = 2x^3 - 6x^2 + 0x - 4$

$$
\begin{array}{r|rrrr}
4 & 2 & -6 & 0 & -4 \\
  &   & 8 & 8 & 32 \\
\hline
  & 2 & 2 & 8 & 28
\end{array}
$$

$$\frac{2x^3 - 6x^2 - 4}{x - 4} = 2x^2 + 2x + 8 + \frac{28}{x - 4}$$

35. $x - 5$ has the form $x - c$, where $c = 5$, so synthetic division can be used.

$$
\begin{array}{r|rrrrr}
5 & 2 & -13 & 16 & -9 & 20 \\
  &   & 10 & -15 & 5 & -20 \\
\hline
  & 2 & -3 & 1 & -4 & 0
\end{array}
$$

$$\frac{2x^4 - 13x^3 + 16x^2 - 9x + 20}{x - 5}$$
$$= 2x^3 - 3x^2 + x - 4$$

37. $x + 1$ has the form $x - c$, where $c = -1$, so synthetic division can be used.

$7x^2 - 4x + 12 + 3x^3 = 3x^3 + 7x^2 - 4x + 12$

$$
\begin{array}{r|rrrr}
-1 & 3 & 7 & -4 & 12 \\
   &   & -3 & -4 & 8 \\
\hline
   & 3 & 4 & -8 & 20
\end{array}
$$

$$\frac{7x^2 - 4x + 12 + 3x^3}{x + 1} = 3x^2 + 4x - 8 + \frac{20}{x + 1}$$

39. $x - \dfrac{1}{3}$ has the form $x - c$, where $c = \dfrac{1}{3}$, so synthetic division can be used.

$$
\begin{array}{r|rrrr}
\frac{1}{3} & 3 & 2 & -4 & 1 \\
   &   & 1 & 1 & -1 \\
\hline
   & 3 & 3 & -3 & 0
\end{array}
$$

$$\frac{3x^3 + 2x^2 - 4x + 1}{x - \frac{1}{3}} = 3x^2 + 3x - 3$$

41. $x - 1$ has the form $x - c$, where $c = 1$, so synthetic division can be used.

$x^3 - 1 = x^3 + 0x^2 + 0x - 1$

$$
\begin{array}{r|rrrr}
1 & 1 & 0 & 0 & -1 \\
  &   & 1 & 1 & 1 \\
\hline
  & 1 & 1 & 1 & 0
\end{array}
$$

$$\frac{x^3 - 1}{x - 1} = x^2 + x + 1$$

43. $\dfrac{25xy^2 + 75xyz + 125x^2 yz}{-5x^2 y}$

$$= \frac{25xy^2}{-5x^2 y} + \frac{75xyz}{-5x^2 y} + \frac{125x^2 yz}{-5x^2 y}$$

$$= -\frac{5y}{x} - \frac{15z}{x} - 25z$$

45. $9x^5 + 6x^4 - 6x^2 - 4x$

$= 9x^5 + 6x^4 + 0x^3 - 6x^2 - 4x + 0$

$$
\begin{array}{r}
3x^4 - 2x \\
3x+2\overline{)\,9x^5 + 6x^4 + 0x^3 - 6x^2 - 4x + 0} \\
\underline{9x^5 + 6x^4} \\
0 + 0x^3 - 6x^2 - 4x \\
\underline{-6x^2 - 4x} \\
0
\end{array}
$$

$(9x^5 + 6x^4 - 6x^2 - 4x) \div (3x+2) = 3x^4 - 2x$

47.
$$
\begin{array}{r|rrrr}
1 & 1 & 3 & -7 & 4 \\
 & & 1 & 4 & -3 \\
\hline
 & 1 & 4 & -3 & 1
\end{array}
$$
Thus, $P(1) = 1$.

49.
$$
\begin{array}{r|rrrr}
-3 & 3 & -7 & -2 & 5 \\
 & & -9 & 48 & -138 \\
\hline
 & 3 & -16 & 46 & -133
\end{array}
$$
Thus, $P(-3) = -133$.

51.
$$
\begin{array}{r|rrrrr}
-1 & 4 & 0 & 1 & 0 & -2 \\
 & & -4 & 4 & -5 & 5 \\
\hline
 & 4 & -4 & 5 & -5 & 3
\end{array}
$$
Thus, $P(-1) = 3$.

53.
$$
\begin{array}{r|rrrrr}
\frac{1}{3} & 2 & 0 & -3 & 0 & -2 \\
 & & \frac{2}{3} & \frac{2}{9} & -\frac{25}{27} & -\frac{25}{81} \\
\hline
 & 2 & \frac{2}{3} & -\frac{25}{9} & -\frac{25}{27} & -\frac{187}{181}
\end{array}
$$
Thus, $P\left(\dfrac{1}{3}\right) = -\dfrac{187}{81}$.

55.
$$
\begin{array}{r|rrrrrr}
\frac{1}{2} & 1 & 1 & -1 & 0 & 0 & 3 \\
 & & \frac{1}{2} & \frac{3}{4} & -\frac{1}{8} & -\frac{1}{16} & -\frac{1}{32} \\
\hline
 & 1 & \frac{3}{2} & -\frac{1}{4} & -\frac{1}{8} & -\frac{1}{16} & \frac{95}{32}
\end{array}
$$
Thus, $P\left(\dfrac{1}{2}\right) = \dfrac{95}{32}$.

57. $7x + 2 = x - 3$

$7x - x = -3 - 2$

$6x = -5$

$x = -\dfrac{5}{6}$

The solution is $-\dfrac{5}{6}$.

59. $x^2 = 4x - 4$

$x^2 - 4x + 4 = 0$

$(x-2)^2 = 0$

$x - 2 = 0$

$x = 2$

The solution is 2.

61. $\dfrac{x}{3} - 5 = 13$

$3\left(\dfrac{x}{3} - 5\right) = (13) \cdot 3$

$x - 15 = 39$

$x = 54$

The solution is 54.

63. $x^3 - 1 = x^3 - 1^3 = (x-1)(x^2 + x + 1)$

65. $125z^3 + 8 = (5z)^3 + 2^3$

$\qquad\qquad = (5z+2)(25z^2 - 10z + 4)$

67. $xy + 2x + 3y + 6 = (xy + 2x) + (3y + 6)$

$\qquad\qquad\qquad\quad = x(y+2) + 3(y+2)$

$\qquad\qquad\qquad\quad = (y+2)(x+3)$

69. $x^3 - 9x = x(x^2 - 9) = x(x+3)(x-3)$

71. $(5x^2 - 3x + 2) \div (x+2)$ is a candidate for synthetic division since $x + 2$ is in the form $x - c$, where $c = -2$.

73. $(x^7 - 2) \div (x^5 + 1)$ is not a candidate for synthetic division since $x^5 + 1$ does not have the form $x - c$.

75. The degree of the remainder must be less than that of the divisor, or 3 in this case. The choices are a or d.

77. $\dfrac{3x^4 + 6x^2 - 18}{3} = \dfrac{3x^4}{3} + \dfrac{6x^2}{3} - \dfrac{18}{3}$

$\qquad\qquad\qquad\quad = x^4 + 2x^2 - 6$

The length of each piece is $(x^4 + 2x^2 - 6)$ meters.

79.

$$5x+2\overline{)15x^2-29x-14}$$

quotient: $3x-7$

$$\underline{15x^2+6x}$$
$$-35x-14$$
$$\underline{-35x-14}$$
$$0$$

The width is $(3x-7)$ inches.

81. $A=bh$ so $h=\dfrac{A}{b}=\dfrac{x^4-23x^2+9x-5}{x+5}$

$$\begin{array}{r|rrrrr} -5 & 1 & 0 & -23 & 9 & -5 \\ & & -5 & 25 & -10 & 5 \\ \hline & 1 & -5 & 2 & -1 & 0 \end{array}$$

The height is (x^3-5x^2+2x-1) cm.

83.

$$x-1\overline{)x^4+\tfrac{2}{3}x^3+0x^2+x+0}$$

quotient: $x^3+\dfrac{5}{3}x^2+\dfrac{5}{3}x+\dfrac{8}{3}$

$$\underline{x^4-\ x^3}$$
$$\tfrac{5}{3}x^3-0x^2$$
$$\underline{\tfrac{5}{3}x^3-\tfrac{5}{3}x^2}$$
$$\tfrac{5}{3}x^2+x$$
$$\underline{\tfrac{5}{3}x^2-\tfrac{5}{3}x}$$
$$\tfrac{8}{3}x+0$$
$$\underline{\tfrac{8}{3}x-\tfrac{8}{3}}$$
$$\tfrac{8}{3}$$

Answer: $x^3+\dfrac{5}{3}x^2+\dfrac{5}{3}x+\dfrac{8}{3}+\dfrac{8}{3(x-1)}$

85.

$$2x-1\overline{)3x^4-\ x^3+0x^2-x+\tfrac{1}{2}}$$

quotient: $\dfrac{3}{2}x^3+\dfrac{1}{4}x^2+\dfrac{1}{8}x-\dfrac{7}{16}$

$$\underline{3x^4-\tfrac{3}{2}x^3}$$
$$\tfrac{1}{2}x^3+0x^2$$
$$\underline{\tfrac{1}{2}x^3-\tfrac{1}{4}x^2}$$
$$\tfrac{1}{4}x^2-\ x$$
$$\underline{\tfrac{1}{4}x^2-\tfrac{1}{8}x}$$
$$-\tfrac{7}{8}x+\tfrac{1}{2}$$
$$\underline{-\tfrac{7}{8}x+\tfrac{7}{16}}$$
$$\tfrac{1}{16}$$

Answer: $\dfrac{3}{2}x^3+\dfrac{1}{4}x^2+\dfrac{1}{8}x-\dfrac{7}{16}+\dfrac{1}{16(2x-1)}$

87.

$$5x+10\overline{)5x^4+10x^3-2x^2-4x+0}$$

quotient: $x^3-\dfrac{2}{5}x$

$$\underline{5x^4+10x^3}$$
$$-2x^2-4x$$
$$\underline{-2x^2-4x}$$
$$0$$

Answer: $x^3-\dfrac{2}{5}x$

89. $\dfrac{f(x)}{g(x)}=\dfrac{25x^2-5x+30}{5x}$

$$=\dfrac{25x^2}{5x}-\dfrac{5x}{5x}+\dfrac{30}{5x}$$

$$=5x-1+\dfrac{6}{x}$$

Setting the denominator equal to 0, we get
$5x=0$
$x=0.$

Thus, $x=0$ is not in the domain of $\dfrac{f(x)}{g(x)}$.

91. $\dfrac{f(x)}{g(x)} = \dfrac{7x^4 - 3x^2 + 2}{x - 2}$

$$
\begin{array}{r}
7x^3 + 14x^2 + 25x + 50 \\
x - 2 \overline{)\,7x^4 + 0x^3 - 3x^2 + 0x + 2} \\
\underline{7x^4 - 14x^3} \\
14x^3 - 3x^2 \\
\underline{14x^3 - 28x^2} \\
25x^2 + 0x \\
\underline{25x^2 - 50x} \\
50x + 2 \\
\underline{50x - 100} \\
102
\end{array}
$$

Therefore, $\dfrac{f(x)}{g(x)} = \dfrac{7x^4 - 3x^2 + 2}{x - 2}$

$$= 7x^3 + 14x^2 + 25x + 50 + \dfrac{102}{x - 2}$$

Setting the denominator equal to 0, we get
$x - 2 = 0$, or $x = 2$. Thus, $x = 2$ is not in the

domain of $\dfrac{f(x)}{g(x)}$.

93. answers may vary

95. answers may vary

97.
$$
\begin{array}{r|rrrr}
-3 & 1 & 3 & 4 & 12 \\
   &   & -3 & 0 & -12 \\
\hline
   & 1 & 0 & 4 & 0
\end{array}
$$
Remainder = 0 and
$(x + 3)(x^2 + 4) = x^3 + 3x^2 + 4x + 12$

99. Multiply $(2x^2 + 5x - 6)$ by $(x - 5)$ and add the remainder, 3.

$(2x^2 + 5x - 6)(x - 5) + 3$
$= 2x^3 - 10x^2 + 5x^2 - 25x - 6x + 30 + 3$
$= 2x^3 - 5x^2 - 31x + 33$

101. a. $m(x) = \dfrac{P(x)}{R(x)}$

$$= \dfrac{0.48x^3 + 2.06x^2 + 141x + 9.71}{1011x - 288}$$

b. $m(15)$

$$= \dfrac{0.48(15)^3 + 2.06(15)^2 + 141(15) + 9.71}{1011(15) - 288}$$

$$= \dfrac{4208.21}{14{,}877} \approx 0.28$$

103. $P(c)$ is equal to the remainder when $P(x)$ is divided by $x - c$. Therefore, $P(c) = 0$.

Section 6.5 Practice Exercises

1. The LCD is 8.

$$\dfrac{5x}{4} - \dfrac{3}{2} = \dfrac{7x}{8}$$

$$8\left(\dfrac{5x}{4} - \dfrac{3}{2}\right) = 8\left(\dfrac{7x}{8}\right)$$

$$8 \cdot \dfrac{5x}{4} - 8 \cdot \dfrac{3}{2} = 8 \cdot \dfrac{7x}{8}$$

$$10x - 12 = 7x$$

$$3x = 12$$

$$x = 4$$

2. The LCD of the denominators x, $5x$, and 5 is $5x$.

$$\dfrac{6}{x} - \dfrac{x + 9}{5x} = \dfrac{2}{5}$$

$$5x\left(\dfrac{6}{x} - \dfrac{x + 9}{5x}\right) = 5x\left(\dfrac{2}{5}\right)$$

$$5x \cdot \dfrac{6}{x} - 5x \cdot \dfrac{x + 9}{5x} = 5x \cdot \dfrac{2}{5}$$

$$30 - (x + 9) = 2x$$

$$30 - x - 9 = 2x$$

$$21 = 3x$$

$$7 = x$$

3. The LCD is $x + 3$.

$$\dfrac{x - 5}{x + 3} = \dfrac{2(x - 1)}{x + 3}$$

$$(x + 3) \cdot \dfrac{x - 5}{x + 3} = (x + 3) \cdot \dfrac{2(x - 1)}{x + 3}$$

$$x - 5 = 2(x - 1)$$

$$x - 5 = 2x - 2$$

$$-3 = x$$

The number -3 makes the denominator $x + 3$ equal to 0, so it is not a solution. The solution set is $\{\ \}$ or $\varnothing$.

4. The LCD is $x(5x - 1)$.

$$\frac{5x}{5x-1} + \frac{1}{x} = \frac{1}{5x-1}$$

$$x(5x-1)\cdot\frac{5x}{5x-1} + x(5x-1)\cdot\frac{1}{x} = x(5x-1)\cdot\frac{1}{5x-1}$$

$$x(5x) + (5x-1) = x$$

$$5x^2 + 5x - 1 - x = 0$$

$$5x^2 + 4x - 1 = 0$$

$$(5x-1)(x+1) = 0$$

$$5x-1 = 0 \text{ or } x+1 = 0$$

$$x = \frac{1}{5} \text{ or } \quad x = -1$$

The number $\frac{1}{5}$ makes the denominator $5x-1$ equal 0, so it is not a solution. The solution is -1.

5. $x^2 - 4 = (x+2)(x-2)$

The LCD is $(x + 2)(x - 2)$.

$$\frac{2}{x-2} - \frac{5+2x}{x^2-4} = \frac{x}{x+2}$$

$$(x+2)(x-2)\cdot\frac{2}{x-2} - (x+2)(x-2)\cdot\frac{5+2x}{(x+2)(x-2)}$$

$$= (x+2)(x-2)\cdot\frac{x}{x+2}$$

$$2(x+2) - (5+2x) = x(x-2)$$

$$2x + 4 - 5 - 2x = x^2 - 2x$$

$$x^2 - 2x + 1 = 0$$

$$(x-1)(x-1) = 0$$

$$x - 1 = 0$$

$$x = 1$$

Since 1 does not make any denominator 0, the solution is 1.

6. $2z^2 - z - 6 = (2z+3)(z-2)$

$z^2 - 2z = z(z-2)$

The LCD is $3z(2z + 3)(z - 2)$.

$$\frac{z}{2z^2-z-6} - \frac{1}{3z} = \frac{2}{z^2-2z}$$

$$\frac{z}{(2z+3)(z-2)} - \frac{1}{3z} = \frac{2}{z(z-2)}$$

$$3z(2z+3)(z-2)\cdot\frac{z}{(2z+3)(z-2)} - 3z(2z+3)(z-2)\cdot\frac{1}{3z} = 3z(2z+3)(z-2)\cdot\frac{2}{z(z-2)}$$

$$3z(z) - (2z+3)(z-2) = 2\cdot 3(2z+3)$$

$$3z^2 - (2z^2 - z - 6) = 12z + 18$$

$$3z^2 - 2z^2 + z + 6 = 12z + 18$$

$$z^2 + z + 6 = 12z + 18$$

$$z^2 - 11z - 12 = 0$$

$$(z-12)(z+1) = 0$$

$z - 12 = 0$ or $z + 1 = 0$
$z = 12$ or $z = -1$

Neither 12 nor −1 makes any denominator 0, so they are both solutions. The solutions are 12 and −1.

Vocabulary, Readiness & Video Check 6.5

1. $\dfrac{x}{2} = \dfrac{3x}{5} + \dfrac{x}{6}$ is an equation.

2. $\dfrac{3x}{5} + \dfrac{x}{6}$ is an expression.

3. $\dfrac{x}{x-1} + \dfrac{2x}{x+1}$ is an expression.

4. $\dfrac{x}{x-1} + \dfrac{2x}{x+1} = 5$ is an equation.

5. $\dfrac{y+7}{2} = \dfrac{y+1}{6} + \dfrac{1}{y}$ is an equation.

6. $\dfrac{y+1}{6} + \dfrac{1}{y}$ is an expression.

7. The LCD of $\dfrac{x}{7}, \dfrac{x}{2},$ and $\dfrac{1}{2}$ is <u>14</u>. **c**

8. The LCD of $\dfrac{9}{x+1}, \dfrac{5}{(x+1)^2},$ and $\dfrac{x}{x+1}$ is <u>$(x+1)^2$</u>. **b**

9. The LCD of $\dfrac{7}{x-4}, \dfrac{x}{x^2-16} = \dfrac{x}{(x+4)(x-4)},$ and $\dfrac{1}{x+4}$ is <u>$(x+4)(x-4)$</u>. **a**

10. The LCD of $3 = \dfrac{3}{1}, \dfrac{1}{x-5},$ and $\dfrac{2}{x^2-5x} = \dfrac{2}{x(x-5)}$ is <u>$x(x-5)$</u>. **d**

11. Linear and quadratic equations are solved in very different ways, so you need to determine the next correct move to make.

12. First check to see if a proposed solution makes any denominator of the original equation zero, giving you an undefined rational expression. If so, the solution is an extraneous solution and is not a solution of the equation.

Exercise Set 6.5

1. $\dfrac{x}{2} - \dfrac{x}{3} = 12$
$6\left(\dfrac{x}{2} - \dfrac{x}{3}\right) = 6(12)$
$3x - 2x = 72$
$x = 72$

3. $\dfrac{x}{3} = \dfrac{1}{6} + \dfrac{x}{4}$
$12\left(\dfrac{x}{3}\right) = 12\left(\dfrac{1}{6} + \dfrac{x}{4}\right)$
$4x = 2 + 3x$
$x = 2$

5. $\dfrac{2}{x} + \dfrac{1}{2} = \dfrac{5}{x}$
$2x\left(\dfrac{2}{x} + \dfrac{1}{2}\right) = 2x\left(\dfrac{5}{x}\right)$
$4 + x = 10$
$x = 6$

7. $\dfrac{x^2+1}{x} = \dfrac{5}{x}$
$x\left(\dfrac{x^2+1}{x}\right) = x\left(\dfrac{5}{x}\right)$
$x^2 + 1 = 5$
$x^2 - 4 = 0$
$x + 2 = 0$ or $x - 2 = 0$
$x = -2$ or $x = 2$

9. $\dfrac{x+5}{x+3} = \dfrac{2}{x+3}$
$(x+3)\cdot\dfrac{x+5}{x+3} = (x+3)\cdot\dfrac{2}{x+3}$
$x + 5 = 2$
$x = -3$
which we disregard as extraneous. No solution, or $\varnothing$.

11.
$$\frac{5}{x-2} - \frac{2}{x+4} = -\frac{4}{x^2+2x-8}$$
$$\frac{5}{x-2} - \frac{2}{x+4} = -\frac{4}{(x+4)(x-2)}$$
$$(x-4)(x+2)\left(\frac{5}{x-2} - \frac{2}{x+4}\right) = (x-4)(x+2)\left(-\frac{4}{(x+4)(x-2)}\right)$$
$$5(x+4) - 2(x-2) = -4$$
$$5x+20-2x+4 = -4$$
$$3x+24 = -4$$
$$3x = -28$$
$$x = -\frac{28}{3}$$

13.
$$\frac{1}{x-1} = \frac{2}{x+1}$$
$$(x+1)(x-1)\cdot\frac{1}{x-1} = (x+1)(x-1)\cdot\frac{2}{x+1}$$
$$1(x+1) = 2(x-1)$$
$$x+1 = 2x-2$$
$$-x = -3$$
$$x = 3$$

15.
$$\frac{x^2-23}{2x^2-5x-3} + \frac{2}{x-3} = \frac{-1}{2x+1}$$
$$\frac{x^2-23}{(2x+1)(x-3)} + \frac{2}{x-3} = \frac{-1}{2x+1}$$
$$(2x+1)(x-3)\left(\frac{x^2-23}{(2x+1)(x-3)} + \frac{2}{x-3}\right) = (2x+1)(x-3)\left(\frac{-1}{2x+1}\right)$$
$$(x^2-23) + 2(2x+1) = -1(x-3)$$
$$x^2-23+4x+2 = -x+3$$
$$x^2+5x-24 = 0$$
$$(x+8)(x-3) = 0$$
$$x+8 = 0 \quad \text{or} \quad x-3 = 0$$
$$x = -8 \quad \text{or} \quad x = 3$$

We discard 3 as extraneous.
$$x = -8$$

17.
$$\frac{1}{x-4} - \frac{3x}{x^2-16} = \frac{2}{x+4}$$
$$\frac{1}{x-4} - \frac{3x}{(x+4)(x-4)} = \frac{2}{x+4}$$
$$(x+4)(x-4)\left[\frac{1}{x-4} - \frac{3x}{(x+4)(x-4)}\right] = (x+4)(x-4)\left(\frac{2}{x+4}\right)$$
$$1(x+4) - 3x = 2(x-4)$$
$$x+4-3x = 2x-8$$
$$-2x+4 = 2x-8$$
$$-4x = -12$$
$$x = 3$$

19.
$$\frac{1}{x-4} = \frac{8}{x^2-16}$$
$$\frac{1}{x-4} = \frac{8}{(x+4)(x-4)}$$
$$(x+4)(x-4)\left(\frac{1}{x-4}\right) = (x+4)(x-4)\left(\frac{8}{(x+4)(x-4)}\right)$$
$$1(x+4) = 8$$
$$x+4 = 8$$
$$x = -4$$
which we discard as extraneous. No solution, or $\varnothing$.

21.
$$\frac{1}{x-2} - \frac{2}{x^2-2x} = 1$$
$$\frac{1}{x-2} - \frac{2}{x(x-2)} = 1$$
$$x(x-2)\left[\frac{1}{x-2} - \frac{2}{x(x-2)}\right] = x(x-2) \cdot 1$$
$$x-2 = x(x-2)$$
$$x-2 = x^2 - 2x$$
$$0 = x^2 - 3x + 2$$
$$0 = (x-2)(x-1)$$
$$x-2 = 0 \text{ or } x-1 = 1$$
$$x = 2 \text{ or } \quad x = 1$$
We discard 2 as extraneous.
$$x = 1$$

23.
$$\frac{5}{x} = \frac{20}{12}$$
$$12x\left(\frac{5}{x}\right) = 12x\left(\frac{20}{12}\right)$$
$$60 = 20x$$
$$3 = x$$

25.

$$1 - \frac{4}{a} = 5$$

$$a\left(1 - \frac{4}{a}\right) = a(5)$$

$$a - 4 = 5a$$

$$-4 = 4a$$

$$-1 = a$$

27.

$$\frac{x^2 + 5}{x} - 1 = \frac{5(x+1)}{x}$$

$$x\left(\frac{x^2 + 5}{x} - 1\right) = x\left[\frac{5(x+1)}{x}\right]$$

$$x^2 + 5 - x = 5x + 5$$

$$x^2 - 6x = 0$$

$$x(x - 6) = 0$$

$$x = 0 \ \text{ or } \ x - 6 = 0$$

$$x = 6$$

We discard 0 as extraneous.

$$x = 6$$

29.

$$\frac{1}{2x} - \frac{1}{x+1} = \frac{1}{3x^2 + 3x}$$

$$\frac{1}{2x} - \frac{1}{x+1} = \frac{1}{3x(x+1)}$$

$$6x(x+1)\left(\frac{1}{2x} - \frac{1}{x+1}\right) = 6x(x+1)\left(\frac{1}{3x(x+1)}\right)$$

$$3(x+1) - 6x = 2$$

$$3x + 3 - 6x = 2$$

$$-3x + 3 = 2$$

$$-3x = -1$$

$$x = \frac{1}{3}$$

31.

$$\frac{1}{x} - \frac{x}{25} = 0$$

$$25x\left(\frac{1}{x} - \frac{x}{25}\right) = 25x(0)$$

$$25 - x^2 = 0$$

$$-(x^2 - 25) = 0$$

$$-(x + 5)(x - 5) = 0$$

$$x + 5 = 0 \ \text{ or } \ x - 5 = 0$$

$$x = -5 \ \text{ or } \quad x = 5$$

33.
$$5 - \frac{2}{2y-5} = \frac{3}{2y-5}$$
$$(2y-5)\left(5 - \frac{2}{2y-5}\right) = (2y-5) \cdot \frac{3}{2y-5}$$
$$5(2y-5) - 2 = 3$$
$$10y - 25 - 2 = 3$$
$$10y - 27 = 3$$
$$10y = 30$$
$$y = 3$$

35.
$$\frac{x-1}{x+2} = \frac{2}{3}$$
$$3(x+2)\left(\frac{x-1}{x+2}\right) = 3(x+2)\left(\frac{2}{3}\right)$$
$$3(x-1) = 2(x+2)$$
$$3x - 3 = 2x + 4$$
$$x = 7$$

37.
$$\frac{x+3}{x+2} = \frac{1}{x+2}$$
$$(x+2) \cdot \frac{x+3}{x+2} = (x+2) \cdot \frac{1}{x+2}$$
$$x + 3 = 1$$
$$x = -2$$
which we discard as extraneous. No solution, or $\varnothing$.

39.
$$\frac{1}{a-3} + \frac{2}{a+3} = \frac{1}{a^2-9}$$
$$\frac{1}{a-3} + \frac{2}{a+3} = \frac{1}{(a+3)(a-3)}$$
$$(a+3)(a-3)\left(\frac{1}{a-3} + \frac{2}{a+3}\right) = (a+3)(a-3) \cdot \frac{1}{(a+3)(a-3)}$$
$$1(a+3) + 2(a-3) = 1$$
$$a + 3 + 2a - 6 = 1$$
$$3a - 3 = 1$$
$$3a = 4$$
$$a = \frac{4}{3}$$

41.

$$\frac{64}{x^2 - 16} + 1 = \frac{2x}{x - 4}$$

$$\frac{64}{(x+4)(x-4)} + 1 = \frac{2x}{x - 4}$$

$$(x+4)(x-4)\left[\frac{64}{(x+4)(x-4)} + 1\right] = (x+4)(x-4) \cdot \frac{2x}{x-4}$$

$$64 + 1(x+4)(x-4) = 2x(x+4)$$

$$64 + (x^2 - 16) = 2x^2 + 8x$$

$$x^2 + 48 = 2x^2 + 8x$$

$$0 = x^2 + 8x - 48$$

$$0 = (x+12)(x-4)$$

$$x + 12 = 0 \quad \text{or } x - 4 = 0$$
$$x = -12 \quad \text{or} \quad x = 4$$

We discard 4 as extraneous.

$$x = -12$$

43.

$$\frac{-15}{4y + 1} + 4 = y$$

$$(4y + 1)\left(\frac{-15}{4y + 1} + 4\right) = (4y + 1)y$$

$$-15 + 4(4y + 1) = 4y^2 + y$$

$$-15 + 16y + 4 = 4y^2 + y$$

$$-11 + 16y = 4y^2 + y$$

$$0 = 4y^2 - 15y + 11$$

$$0 = (4y - 11)(y - 1)$$

$$4y - 11 = 0 \quad \text{or } y - 1 = 0$$
$$4y = 11 \quad \text{or} \quad y = 1$$

$$y = \frac{11}{4}$$

45.

$$\frac{28}{x^2 - 9} + \frac{2x}{x - 3} + \frac{6}{x + 3} = 0$$

$$\frac{28}{(x+3)(x-3)} + \frac{2x}{x - 3} + \frac{6}{x + 3} = 0$$

$$(x+3)(x-3)\left[\frac{28}{(x+3)(x-3)} + \frac{2x}{x-3} + \frac{6}{x+3}\right] = (x+3)(x-3) \cdot 0$$

$$28 + 2x(x+3) + 6(x-3) = 0$$

$$28 + 2x^2 + 6x + 6x - 18 = 0$$

$$2x^2 + 12x + 10 = 0$$

$$2(x^2 + 6x + 5) = 0$$

$$2(x+5)(x+1) = 0$$

$$x + 5 = 0 \quad \text{or } x + 1 = 0$$
$$x = -5 \quad \text{or} \quad x = -1$$

47.

$$\frac{x+2}{x^2+7x+10}=\frac{1}{3x+6}-\frac{1}{x+5}$$

$$\frac{x+2}{(x+5)(x+2)}=\frac{1}{3(x+2)}-\frac{1}{x+5}$$

$$3(x+5)(x+2)\cdot\frac{x+2}{(x+5)(x+2)}=3(x+5)(x+2)\left[\frac{1}{3(x+2)}-\frac{1}{x+5}\right]$$

$$3(x+2)=1(x+5)-1\cdot3(x+2)$$

$$3x+6=x+5-3x-6$$

$$3x+6=-2x-1$$

$$5x=-7$$

$$x=-\frac{7}{5}$$

49. Let x = the number.
$$3x+4=19$$
$$3x=15$$
$$x=5$$
The number is 5.

51. Let w = width. Then $w+5=$ length.
$$2l+2w=50$$
$$2(w+5)+2w=50$$
$$2w+10+2w=50$$
$$4w+10=50$$
$$4w=40$$
$$w=10;$$
$$w+5=10+5=15$$
The length is 15 inches and the width is 10 inches.

53. From the graph, 14.8% of state and federal prison inmates are ages 45 to 54.

55. From the graph, the category that shows the highest percent of prison inmates is the 25–29-year-olds.

57. 17.2% of $748,728=0.172(748,728)\approx128,781$
Approximately 128,781 25- to 29-year-old inmates would be expected at the end of 2010.

59. answers may vary

61. $f(x)=3.3+\dfrac{5400}{x}$

$$5.10=3.3+\frac{5400}{x}$$
$$1.8=\frac{5400}{x}$$
$$1.8x=5400$$
$$x=3000$$
3000 game disks

63. $x^{-2} - 19x^{-1} + 48 = 0$

$$\frac{1}{x^2} - \frac{19}{x} + 48 = 0$$

$$x^2 \left(\frac{1}{x^2} - \frac{19}{x} + 48 \right) = x^2 \cdot 0$$

$$1 - 19x + 48x^2 = 0$$

$$48x^2 - 19x + 1 = 0$$

$$(16x - 1)(3x - 1) = 0$$

$$x = \frac{1}{16} \ \text{ or } \ x = \frac{1}{3}$$

65. $p^{-2} + 4p^{-1} - 5 = 0$

$$\frac{1}{p^2} + \frac{4}{p} - 5 = 0$$

$$p^2 \left(\frac{1}{p^2} + \frac{4}{p} - 5 \right) = p^2 \cdot 0$$

$$1 + 4p - 5p^2 = 0$$

$$5p^2 - 4p - 1 = 0$$

$$(5p + 1)(p - 1) = 0$$

$$p = -\frac{1}{5} \ \text{ or } \ p = 1$$

67.

$$\frac{1.4}{x - 2.6} = \frac{-3.5}{x + 7.1}$$

$$(x - 2.6)(x + 7.1) \cdot \frac{1.4}{x - 2.6} = (x - 2.6)(x + 7.1) \cdot \frac{-3.5}{x + 7.1}$$

$$1.4(x + 7.1) = -3.5(x - 2.6)$$

$$1.4x + 9.94 = -3.5x + 9.1$$

$$4.9x = -0.84$$

$$x \approx -0.17$$

69. $\dfrac{10.6}{y} - 14.7 = \dfrac{9.92}{3.2} + 7.6$

$$\frac{10.6}{y} - 14.7 = 10.7$$

$$\frac{10.6}{y} = 25.4$$

$$y \cdot \frac{10.6}{y} = y \cdot 25.4$$

$$10.6 = 25.4y$$

$$0.42 \approx y$$

71. $(x-1)^2 + 3(x-1) + 2 = 0$

Let $u = x - 1$.

$u^2 + 3u + 2 = 0$

$(u+2)(u+1) = 0$

$u = -2$ or $\quad u = -1$

$x - 1 = -2$ or $x - 1 = -1$

$x = -1$ or $\quad x = 0$

73. $\left(\dfrac{3}{x-1}\right)^2 + 2\left(\dfrac{3}{x-1}\right) + 1 = 0$

Let $u = \dfrac{3}{x-1}$.

$u^2 + 2u + 1 = 0$

$(u+1)^2 = 0$

$u + 1 = 0$

$u = -1$

$\dfrac{3}{x-1} = -1$

$3 = -1(x-1)$

$3 = -x + 1$

$x = -2$

75.

77.

Integrated Review

1. $\dfrac{x}{2} = \dfrac{1}{8} + \dfrac{x}{4}$

The LCD is 8.

$8 \cdot \dfrac{x}{2} = 8 \cdot \dfrac{1}{8} + 8 \cdot \dfrac{x}{4}$

$4x = 1 + 2x$

$2x = 1$

$x = \dfrac{1}{2}$

The solution set is $\left\{\dfrac{1}{2}\right\}$.

2. $\dfrac{x}{4} = \dfrac{3}{2} + \dfrac{x}{10}$

The LCD is 20.

$20 \cdot \dfrac{x}{4} = 20 \cdot \dfrac{3}{2} + 20 \cdot \dfrac{x}{10}$

$5x = 30 + 2x$

$3x = 30$

$x = 10$

The solution set is $\{10\}$.

3. $\dfrac{1}{8} + \dfrac{x}{4} = \dfrac{1}{8} + \dfrac{x}{4} \cdot \dfrac{2}{2} = \dfrac{1}{8} + \dfrac{2x}{8} = \dfrac{1+2x}{8}$

4. $\dfrac{3}{2} + \dfrac{x}{10} = \dfrac{5}{5} \cdot \dfrac{3}{2} + \dfrac{x}{10} = \dfrac{15}{10} + \dfrac{x}{10} = \dfrac{15+x}{10}$

5. $\dfrac{4}{x+2} - \dfrac{2}{x-1} = \dfrac{4}{x+2} \cdot \dfrac{x-1}{x-1} - \dfrac{2}{x-1} \cdot \dfrac{x+2}{x+2}$

$= \dfrac{4(x-1)}{(x+2)(x-1)} - \dfrac{2(x+2)}{(x+2)(x+1)}$

$= \dfrac{4x-4-2x-4}{(x+2)(x-1)}$

$= \dfrac{2x-8}{(x+2)(x-1)}$

$= \dfrac{2(x-4)}{(x+2)(x-1)}$

6. $\dfrac{5}{x-2} - \dfrac{10}{x+4} = \dfrac{5}{x-2} \cdot \dfrac{x+4}{x+4} - \dfrac{10}{x+4} \cdot \dfrac{x-2}{x-2}$

$= \dfrac{5(x+4)}{(x-2)(x+4)} - \dfrac{10(x-2)}{(x-2)(x+4)}$

$= \dfrac{5x+20-10x+20}{(x-2)(x+4)}$

$= \dfrac{-5x+40}{(x-2)(x+4)}$

$= \dfrac{-5(x-8)}{(x-2)(x+4)}$ or $-\dfrac{5(x-8)}{(x-2)(x+4)}$

7. $\dfrac{4}{x+2} = \dfrac{2}{x-1}$

The LCD is $(x + 2)(x - 1)$.

$(x+2)(x-1) \cdot \dfrac{4}{x+2} = (x+2)(x-1) \cdot \dfrac{2}{x-1}$

$4(x-1) = 2(x+2)$

$4x - 4 = 2x + 4$

$4x = 2x + 8$

$2x = 8$

$x = 4$

The solution set is $\{4\}$.

8. $\dfrac{5}{x-2} = \dfrac{10}{x+4}$

The LCD is $(x-2)(x+4)$.

$$(x-2)(x+4) \cdot \frac{5}{x-2} = (x-2)(x+4) \cdot \frac{10}{x+4}$$
$$5(x+4) = 10(x-2)$$
$$5x+20 = 10x-20$$
$$5x = 10x-40$$
$$-5x = -40$$
$$x = 8$$

The solution set is $\{8\}$.

9. $x^2 - 4 = (x+2)(x-2)$

The LCD is $(x+2)(x-2)$.

$$\frac{2}{x^2-4} = \frac{1}{x+2} - \frac{3}{x-2}$$
$$(x+2)(x-2) \cdot \frac{2}{x^2-4} = (x+2)(x-2) \cdot \frac{1}{x+2} - (x+2)(x-2) \cdot \frac{3}{x-2}$$
$$2 = (x-2) - 3(x+2)$$
$$2 = x-2-3x-6$$
$$2 = -2x-8$$
$$2x = -10$$
$$x = -5$$

The solution set is $\{-5\}$.

10. $x^2 - 25 = (x+5)(x-5)$

The LCD is $(x+5)(x-5)$.

$$\frac{3}{x^2-25} = \frac{1}{x+5} + \frac{2}{x-5}$$
$$(x+5)(x-5) \cdot \frac{3}{(x+5)(x-5)} = (x+5)(x-5) \cdot \frac{1}{x+5} + (x+5)(x-5) \cdot \frac{2}{x-5}$$
$$3 = (x-5) + 2(x+5)$$
$$3 = x-5+2x+10$$
$$3 = 3x+5$$
$$-2 = 3x$$
$$-\frac{2}{3} = x$$

The solution set is $\left\{ -\dfrac{2}{3} \right\}$.

11. $\dfrac{5}{x^2-3x}+\dfrac{4}{2x-6}=\dfrac{5}{x(x-3)}+\dfrac{4}{2(x-3)}$

$$=\dfrac{5}{x(x-3)}\cdot\dfrac{2}{2}+\dfrac{4}{2(x-3)}\cdot\dfrac{x}{x}$$

$$=\dfrac{10}{2x(x-3)}+\dfrac{4x}{2x(x-3)}$$

$$=\dfrac{4x+10}{2x(x-3)}$$

$$=\dfrac{2(2x+5)}{2x(x-3)}$$

$$=\dfrac{2x+5}{x(x-3)}$$

12. $\dfrac{5}{x^2-3x}\div\dfrac{4}{2x-6}=\dfrac{5}{x^2-3x}\cdot\dfrac{2x-6}{4}$

$$=\dfrac{5}{x(x-3)}\cdot\dfrac{2(x-3)}{4}$$

$$=\dfrac{5}{2x}$$

13. $x^2-1=(x-1)(x+1)$

The LCD is $(x-1)(x+1)$.

$$\dfrac{x-1}{x+1}+\dfrac{x+7}{x-1}=\dfrac{4}{x^2-1}$$

$$(x-1)(x+1)\cdot\dfrac{x-1}{x+1}+(x-1)(x+1)\cdot\dfrac{x+7}{x-1}=(x-1)(x+1)\cdot\dfrac{4}{(x-1)(x+1)}$$

$$(x-1)(x-1)+(x+1)(x+7)=4$$

$$x^2-2x+1+x^2+8x+7=4$$

$$2x^2+6x+8=4$$

$$2x^2+6x+4=0$$

$$2(x^2+3x+2)=0$$

$$2(x+1)(x+2)=0$$

$x+1=0$ or $x+2=0$

$x=-1$ or $x=-2$

The number -1 makes the denominator $x+1$ equal to 0, so it is not a solution. The solution set is $\{-2\}$.

14. $\left(1-\dfrac{y}{x}\right)\div\left(1-\dfrac{x}{y}\right)=\left(\dfrac{x}{x}-\dfrac{y}{x}\right)\div\left(\dfrac{y}{y}-\dfrac{x}{y}\right)$

$$=\left(\dfrac{x-y}{x}\right)\div\left(\dfrac{y-x}{y}\right)$$

$$=\dfrac{x-y}{x}\cdot\dfrac{y}{y-x}$$

$$=\dfrac{x-y}{x}\cdot\dfrac{y}{-(x-y)}$$

$$=-\dfrac{y}{x}$$

15. $\dfrac{a^2-9}{a-6}\cdot\dfrac{a^2-5a-6}{a^2-a-6} = \dfrac{(a+3)(a-3)}{a-6}\cdot\dfrac{(a-6)(a+1)}{(a-3)(a+2)}$

$\qquad\qquad\qquad\qquad = \dfrac{(a+3)(a+1)}{a+2}$

16. $\dfrac{2}{a-6}+\dfrac{3a}{a^2-5a-6}-\dfrac{a}{5a+5} = \dfrac{2}{a-6}+\dfrac{3a}{(a-6)(a+1)}-\dfrac{a}{5(a+1)}$

$\qquad\qquad\qquad\qquad\qquad = \dfrac{2}{a-6}\cdot\dfrac{5(a+1)}{5(a+1)}+\dfrac{3a}{(a-6)(a+1)}\cdot\dfrac{5}{5}-\dfrac{a}{5(a+1)}\cdot\dfrac{a-6}{a-6}$

$\qquad\qquad\qquad\qquad\qquad = \dfrac{10a+10}{5(a+1)(a-6)}+\dfrac{15a}{5(a+1)(a-6)}-\dfrac{a^2-6a}{5(a+1)(a-6)}$

$\qquad\qquad\qquad\qquad\qquad = \dfrac{10a+10+15a-a^2+6a}{5(a+1)(a-6)}$

$\qquad\qquad\qquad\qquad\qquad = \dfrac{-a^2+31a+10}{5(a+1)(a-6)}$

17. $\dfrac{2x+3}{3x-2}=\dfrac{4x+1}{6x+1}$

The LCD is $(3x-2)(6x+1)$.

$(3x-2)(6x+1)\cdot\dfrac{2x+3}{3x-2}=(3x-2)(6x+1)\cdot\dfrac{4x+1}{6x+1}$

$\qquad (6x+1)(2x+3)=(3x-2)(4x+1)$

$12x^2+18x+2x+3=12x^2+3x-8x-2$

$\qquad 12x^2+20x+3=12x^2-5x-2$

$\qquad\qquad 20x+3=-5x-2$

$\qquad\qquad 25x+3=-2$

$\qquad\qquad\quad 25x=-5$

$\qquad\qquad\qquad x=-\dfrac{5}{25}$

$\qquad\qquad\qquad x=-\dfrac{1}{5}$

The solution set is $\left\{-\dfrac{1}{5}\right\}$.

18. The LCD is $2x(4x + 1)$.

$$\frac{5x-3}{2x} = \frac{10x+3}{4x+1}$$

$$2x(4x+1)\cdot\frac{5x-3}{2x} = 2x(4x+1)\cdot\frac{10x+3}{4x+1}$$

$$(4x+1)(5x-3) = 2x(10x+3)$$

$$20x^2 - 12x + 5x - 3 = 20x^2 + 6x$$

$$20x^2 - 7x - 3 = 20x^2 + 6x$$

$$-7x - 3 = 6x$$

$$-3 = 13x$$

$$-\frac{3}{13} = x$$

The solution set is $\left\{-\frac{3}{13}\right\}$.

19. $\dfrac{a}{9a^2-1} + \dfrac{2}{6a-2}$

$$= \frac{a}{(3a-1)(3a+1)} + \frac{2}{2(3a-1)}$$

$$= \frac{a}{(3a-1)(3a+1)}\cdot\frac{2}{2} + \frac{2}{2(3a-1)}\cdot\frac{(3a+1)}{(3a+1)}$$

$$= \frac{2a}{2(3a-1)(3a+1)} + \frac{6a+2}{2(3a-1)(3a+1)}$$

$$= \frac{8a+2}{2(3a-1)(3a+1)}$$

$$= \frac{2(4a+1)}{2(3a-1)(3a+1)}$$

$$= \frac{4a+1}{(3a-1)(3a+1)}$$

20. $\dfrac{3}{4a-8} - \dfrac{a+2}{a^2-2a} = \dfrac{3}{4(a-2)} - \dfrac{a+2}{a(a-2)}$

$$= \frac{3}{4(a-2)}\cdot\frac{a}{a} - \frac{a+2}{a(a-2)}\cdot\frac{4}{4}$$

$$= \frac{3a}{4a(a-2)} - \frac{4(a+2)}{4a(a-2)}$$

$$= \frac{3a}{4a(a-2)} - \frac{4a+8}{4a(a-2)}$$

$$= \frac{3a-4a-8}{4a(a-2)}$$

$$= \frac{-a-8}{4a(a-2)}$$

$$= -\frac{a+8}{4a(a-2)}$$

21. The LCD is x^2.

$$-\frac{3}{x^2} - \frac{1}{x} + 2 = 0$$

$$x^2\cdot-\frac{3}{x^2} - x^2\cdot\frac{1}{x} + x^2\cdot 2 = 0$$

$$-3 - x + 2x^2 = 0$$

$$2x^2 - x - 3 = 0$$

$$(2x-3)(x+1) = 0$$

$$2x-3=0 \quad \text{or} \quad x+1=0$$

$$2x=3 \quad \text{or} \quad x=-1$$

$$x=\frac{3}{2} \quad \text{or} \quad x=-1$$

The solution set is $\left\{-1, \dfrac{3}{2}\right\}$.

22. $\dfrac{x}{2x+6} + \dfrac{5}{x^2-9}$

$$= \frac{x}{2(x+3)} + \frac{5}{(x-3)(x+3)}$$

$$= \frac{x}{2(x+3)}\cdot\frac{x-3}{x-3} + \frac{5}{(x-3)(x+3)}\cdot\frac{2}{2}$$

$$= \frac{x(x-3)}{2(x+3)(x-3)} + \frac{10}{2(x+3)(x-3)}$$

$$= \frac{x^2-3x}{2(x+3)(x-3)} + \frac{10}{2(x+3)(x-3)}$$

$$= \frac{x^2-3x+10}{2(x+3)(x-3)}$$

23. $\dfrac{x-8}{x^2-x-2} + \dfrac{2}{x-2} = \dfrac{x-8}{(x-2)(x+1)} + \dfrac{2}{x-2}$

$$= \frac{x-8}{(x-2)(x+1)} + \frac{2}{x-2}\cdot\frac{x+1}{x+1}$$

$$= \frac{x-8}{(x-2)(x+1)} + \frac{2x+2}{(x-2)(x+1)}$$

$$= \frac{x-8+2x+2}{(x-2)(x+1)}$$

$$= \frac{3x-6}{(x-2)(x+1)}$$

$$= \frac{3(x-2)}{(x-2)(x+1)}$$

$$= \frac{3}{x+1}$$

24. $x^2 - x - 2 = (x-2)(x+1)$

The LCD is $(x-2)(x+1)$.

$$\frac{x-8}{x^2-x-2} + \frac{2}{x-2} = \frac{3}{x+1}$$

$$(x-2)(x+1)\cdot\frac{x-8}{(x-2)(x+1)} + (x-2)(x+1)\cdot\frac{2}{x-2} = (x-2)(x+1)\cdot\frac{3}{x+1}$$

$$(x-8) + 2(x+1) = 3(x-2)$$

$$x-8+2x+2 = 3x-6$$

$$3x-6 = 3x-6$$

$$-6 = -6 \quad \text{True}$$

The solution set is
$\{x | x \text{ is a real number and } x \neq 2,\ x \neq -1\}$.

25. The LCD is a.

$$\frac{3}{a} - 5 = \frac{7}{a} - 1$$

$$a\cdot\frac{3}{a} - a\cdot 5 = a\cdot\frac{7}{a} - a\cdot 1$$

$$3 - 5a = 7 - a$$

$$3 = 7 + 4a$$

$$-4 = 4a$$

$$-1 = a$$

The solution set is $\{-1\}$.

26.

$$\frac{7}{3z-9} + \frac{5}{z} = \frac{7}{3(z-3)} + \frac{5}{z}$$

$$= \frac{7}{3(z-3)}\cdot\frac{z}{z} + \frac{5}{z}\cdot\frac{3(z-3)}{3(z-3)}$$

$$= \frac{7z}{3z(z-3)} + \frac{15(z-3)}{3z(z-3)}$$

$$= \frac{7z+15z-45}{3z(z-3)}$$

$$= \frac{22z-45}{3z(z-3)}$$

27. a. $\dfrac{x}{5} - \dfrac{x}{4} + \dfrac{1}{10}$ is an expression.

b. The first step to simplify this expression is to write each rational expression term so that the denominator is the LCD, 20.

c. $\dfrac{x}{5} - \dfrac{x}{4} + \dfrac{1}{10} = \dfrac{x}{5}\cdot\dfrac{4}{4} - \dfrac{x}{4}\cdot\dfrac{5}{5} + \dfrac{1}{10}\cdot\dfrac{2}{2}$

$$= \frac{4x}{20} - \frac{5x}{20} + \frac{2}{20}$$

$$= \frac{4x-5x+2}{20}$$

$$= \frac{-x+2}{20}$$

28. a. $\dfrac{x}{5} - \dfrac{x}{4} = \dfrac{1}{10}$ is an equation.

b. The first step to solve this equation is to clear the equation of fractions by multiplying each term by the LCD, 20.

c.
$$\dfrac{x}{5} - \dfrac{x}{4} = \dfrac{1}{10}$$
$$20 \cdot \dfrac{x}{5} - 20 \cdot \dfrac{x}{4} = 20 \cdot \dfrac{1}{10}$$
$$4x - 5x = 2$$
$$-x = 2$$
$$x = -2$$
The solution set is $\{-2\}$.

29. $\dfrac{\triangle + \square}{\triangle} = \dfrac{\triangle}{\triangle} + \dfrac{\square}{\triangle} = 1 + \dfrac{\square}{\triangle}$

b is the correct answer.

30. $\dfrac{\triangle}{\square} + \dfrac{\square}{\triangle} = \dfrac{\triangle}{\square} \cdot \dfrac{\triangle}{\triangle} + \dfrac{\square}{\triangle} \cdot \dfrac{\square}{\square}$

$$= \dfrac{\triangle\triangle}{\square\triangle} + \dfrac{\square\square}{\square\triangle}$$

$$= \dfrac{\triangle\triangle + \square\square}{\square\triangle}$$

d is the correct answer.

31. $\dfrac{\triangle}{\square} \cdot \dfrac{\bigcirc}{\square} = \dfrac{\triangle\bigcirc}{\square\square}$

d is the correct answer.

32. $\dfrac{\triangle}{\square} \div \dfrac{\bigcirc}{\triangle} = \dfrac{\triangle}{\square} \cdot \dfrac{\triangle}{\bigcirc} = \dfrac{\triangle\triangle}{\square\bigcirc}$

a is the correct answer.

33. $\dfrac{\frac{\triangle + \square}{\bigcirc}}{\frac{\triangle}{\bigcirc}} = \dfrac{\triangle + \square}{\bigcirc} \div \dfrac{\triangle}{\bigcirc} = \dfrac{\triangle + \square}{\bigcirc} \cdot \dfrac{\bigcirc}{\triangle} = \dfrac{\triangle + \square}{\triangle}$

d is the correct answer.

Section 6.6 Practice Exercises

1. abc is the LCD.
$$\dfrac{1}{a} - \dfrac{1}{b} = \dfrac{1}{c}$$
$$abc\left(\dfrac{1}{a} - \dfrac{1}{b}\right) = abc\left(\dfrac{1}{c}\right)$$
$$abc\left(\dfrac{1}{a}\right) - abc\left(\dfrac{1}{b}\right) = abc\left(\dfrac{1}{c}\right)$$
$$bc - ac = ab$$
$$bc = ab + ac$$
$$bc = a(b + c)$$
$$\dfrac{bc}{b + c} = a$$

2. Let n = the number.
$$\dfrac{3 + n}{11 - n} = \dfrac{5}{2}$$
$$2(11 - n) \cdot \dfrac{3 + n}{11 - n} = 2(11 - n) \cdot \dfrac{5}{2}$$
$$2(3 + n) = 5(11 - n)$$
$$6 + 2n = 55 - 5n$$
$$7n = 49$$
$$n = 7$$
The number is 7.

3. Let x = the number of homes heated by fuel oil.
$$\dfrac{1}{12} = \dfrac{x}{36,000}$$
$$12x = 1 \cdot 36,000$$
$$x = \dfrac{36,000}{12}$$
$$x = 3000$$
3000 homes in the community are heated by fuel oil.

4. Let t = the time it takes them to clean the cages together.

	Time	Part done in one hour
Elissa	3	$\dfrac{1}{3}$
Bill	2	$\dfrac{1}{2}$
Together	t	$\dfrac{1}{t}$

Copyright © 2014 Pearson Education, Inc.

$$\frac{1}{3} + \frac{1}{2} = \frac{1}{t}$$

$$6t\left(\frac{1}{3} + \frac{1}{2}\right) = 6t\left(\frac{1}{t}\right)$$

$$2t + 3t = 6$$

$$5t = 6$$

$$t = \frac{6}{5} \text{ or } 1\frac{1}{5}$$

Elissa and Bill can clean the cages in $1\frac{1}{5}$ hours if they work together.

5. Let x = the speed of the tugboat in still water.

	Distance	Rate	Time $\left(\frac{d}{r}\right)$
Upstream	100	$x - 2$	$\frac{100}{x-2}$
Downstream	100	$x + 2$	$\frac{100}{x+2}$

The time spent traveling upstream is $1\frac{2}{3} = \frac{5}{3}$ times the time spent traveling downstream.

$$\frac{100}{x-2} = \frac{5}{3}\left(\frac{100}{x+2}\right)$$

$$3(x-2)(x+2)\left(\frac{100}{x-2}\right) = 3(x-2)(x+2)\left(\frac{5}{3}\right)\left(\frac{100}{x+2}\right)$$

$$300(x+2) = 500(x-2)$$

$$300x + 600 = 500x - 1000$$

$$1600 = 200x$$

$$8 = x$$

The speed of the tugboat in still water is 8 mph.

Vocabulary, Readiness & Video Check 6.6

1. After we multiply through by the LCD, we factor the specified variable out so that we can divide and get the specified variable alone.

2. $\frac{1}{x}$

3. We write the same units in the numerators and the same units in the denominators.

4. Two or more people (or machines) take different amounts of time to complete a task. When working together, they will complete the task in less time than the fastest person, so your answer will be less than the time of the fastest person.

5. $x + y$ (or $y + x$)

Exercise Set 6.6

1.
$$F = \frac{9}{5}C + 32$$
$$F - 32 = \frac{9}{5}C$$
$$C = \frac{5}{9}(F - 32)$$

3.
$$Q = \frac{A - I}{L}$$
$$QL = L\left(\frac{A - I}{L}\right)$$
$$QL = A - I$$
$$I = A - QL$$

5.
$$\frac{1}{R} = \frac{1}{R_1} + \frac{1}{R_2}$$
$$RR_1R_2 \cdot \frac{1}{R} = RR_1R_2\left(\frac{1}{R_1} + \frac{1}{R_2}\right)$$
$$R_1R_2 = RR_2 + RR_1$$
$$R_1R_2 = R(R_2 + R_1)$$
$$R = \frac{R_1R_2}{R_1 + R_2}$$

7.
$$S = \frac{n(a + L)}{2}$$
$$2S = n(a + L)$$
$$n = \frac{2S}{a + L}$$

9.
$$A = \frac{h(a + b)}{2}$$
$$2A = h(a + b)$$
$$2A = ah + bh$$
$$2A - ah = bh$$
$$b = \frac{2A - ah}{h}$$

11.
$$\frac{P_1 V_1}{T_1} = \frac{P_2 V_2}{T_2}$$
$$T_1 T_2 \cdot \frac{P_1 V_1}{T_1} = T_1 T_2 \cdot \frac{P_2 V_2}{T_2}$$
$$P_1 V_1 T_2 = P_2 V_2 T_1$$
$$T_2 = \frac{P_2 V_2 T_1}{P_1 V_1}$$

13.
$$f = \frac{f_1 f_2}{f_1 + f_2}$$
$$(f_1 + f_2)f = f_1 f_2$$
$$f_1 f + f_2 f = f_1 f_2$$
$$f_1 f = f_1 f_2 - f_2 f$$
$$f_1 f = f_2 (f_1 - f)$$
$$\frac{f_1 f}{f_1 - f} = f_2$$

15.
$$\lambda = \frac{2L}{n}$$
$$n\lambda = 2L$$
$$\frac{n\lambda}{2} = L$$

17.
$$\frac{\theta}{\omega} = \frac{2L}{c}$$
$$c\omega \cdot \frac{\theta}{\omega} = c\omega \cdot \frac{2L}{c}$$
$$c\theta = 2L\omega$$
$$c = \frac{2L\omega}{\theta}$$

19. Let x = the number. Then
$\frac{1}{x}$ = the reciprocal of the number.
$$x + 5\left(\frac{1}{x}\right) = 6$$
$$x + \frac{5}{x} = 6$$
$$x\left(x + \frac{5}{x}\right) = 6x$$
$$x^2 + 5 = 6x$$
$$x^2 - 6x + 5 = 0$$
$$(x - 5)(x - 1) = 0$$
$$x - 5 = 0 \text{ or } x - 1 = 0$$
$$x = 5 \text{ or } \qquad x = 1$$
The number is either 1 or 5.

21. Let x = the number.
$$\frac{12 + x}{41 + 2x} = \frac{1}{3}$$
$$3(12 + x) = 1(41 + 2x)$$
$$36 + 3x = 41 + 2x$$
$$x = 5$$
The number is 5.

23. Let a = amount of water in 3 minutes.
$$\frac{15}{10} = \frac{a}{3}$$
$$10a = 15(3)$$
$$10a = 45$$
$$a = 4.5$$
The camel can drink 4.5 gallons.

25. Let w = the number of women.
$$\frac{13.1}{100} = \frac{w}{42,358}$$
$$13.1(42,358) = 100 \cdot w$$
$$554,889.8 = 100w$$
$$5549 \approx w$$
There are 5549 women.

27. Let x = number of hours needed working together.
$$\frac{1}{26} + \frac{1}{39} = \frac{1}{x}$$
$$78x\left(\frac{1}{26} + \frac{1}{39}\right) = 78x\left(\frac{1}{x}\right)$$
$$3x + 2x = 78$$
$$5x = 78$$
$$x = \frac{78}{5} = 15.6$$
The roofers together would take 15.6 hours.

29. Let x = time to sort the stack working together.
$$\frac{1}{20} + \frac{1}{30} + \frac{1}{60} = \frac{1}{x}$$
$$60x\left(\frac{1}{20} + \frac{1}{30} + \frac{1}{60}\right) = 60x \cdot \frac{1}{x}$$
$$3x + 2x + x = 60$$
$$6x = 60$$
$$x = 10$$
It takes them 10 minutes to sort the mail when all three work together.

31. Let r = speed of the car. Then
$r + 150$ = speed of the plane.

$$t_{\text{plane}} = t_{\text{car}}$$

$$\frac{600}{r+150} = \frac{150}{r}$$

$$600r = 150(r+150)$$

$$600r = 150r + 22,500$$

$$450r = 22,500$$

$$r = \frac{22,500}{450} = 50$$

$$r + 150 = 50 + 150 = 200$$

The speed of the plane was 200 mph.

33. Let r = speed of the boat in still water.

$$t_{\text{downstream}} = t_{\text{upstream}}$$

$$\frac{20}{r+5} = \frac{10}{r-5}$$

$$20(r-5) = 10(r+5)$$

$$20r - 100 = 10r + 50$$

$$10r = 150$$

$$r = 15$$

The speed of the boat in still water is 15 mph.

35. Let x = the first integer. Then
$x + 1$ = the next integer.

$$\frac{1}{x} + \frac{1}{x+1} = -\frac{15}{56}$$

$$56x(x+1)\left[\frac{1}{x} + \frac{1}{x+1}\right] = 56x(x+1)\left(-\frac{15}{56}\right)$$

$$56(x+1) + 56x = -15x(x+1)$$

$$56x + 56 + 56x = -15x^2 - 15x$$

$$112x + 56 = -15x^2 - 15x$$

$$15x^2 + 127x + 56 = 0$$

$$(15x+7)(x+8) = 0$$

$$15x + 7 = 0 \quad \text{or} \quad x + 8 = 0$$

$$15x = -7 \quad \text{or} \qquad x = -8$$

$$x = -\frac{7}{15} \qquad x + 1 = -8 + 1 = -7$$

$-\dfrac{7}{15}$ is not an integer.

The integers are –8 and –7.

37. Let t = time for 2nd hose to fill the pond.

$$\frac{1}{45} + \frac{1}{t} = \frac{1}{20}$$

$$180t\left(\frac{1}{45} + \frac{1}{t}\right) = 180t \cdot \frac{1}{20}$$

$$4t + 180 = 9t$$

$$180 = 5t$$

$$36 = t$$

The second hose will take 36 minutes to fill the pond alone.

39. Let r = the speed of the first train. Then
$r + 15$ = the speed of the 2nd train.

$$d_{\text{train 1}} + d_{\text{train 2}} = 630$$

$$6r + 6(r+15) = 630$$

$$6r + 6r + 90 = 630$$

$$12r = 540$$

$$r = 45$$

$$r + 15 = 45 + 15 = 60$$

The speed of the trains were 45 mph and 60 mph.

41. Let t = time to travel 1 mile.

$$\frac{0.17}{1} = \frac{1}{t}$$

$$0.17t = 1$$

$$100(0.17t) = 100(1)$$

$$17t = 100$$

$$t = \frac{100}{17} \approx 5.882352941$$

It would take 5.9 hours.

43. Let t = time to fill quota working together.

$$\frac{1}{5} + \frac{1}{6} + \frac{1}{7.5} = \frac{1}{t}$$

$$225t\left(\frac{1}{5} + \frac{1}{6} + \frac{1}{7.5}\right) = 225t \cdot \frac{1}{t}$$

$$45t + 37.5t + 30t = 225$$

$$112.5t = 225$$

$$t = 2$$

It would take 2 hours using all three machines.

45. Let r = the speed of plane in still air.

$$t_{\text{with}} = t_{\text{against}}$$

$$\frac{465}{r+20} = \frac{345}{r-20}$$

$$465(r-20) = 345(r+20)$$

$$465r - 9300 = 345r + 6900$$

$$120r = 16,200$$

$$r = 135$$

The planes speed in still air is 135 mph.

47. Let d = the distance of the run.

$$t_{\text{jogger 2}} = \frac{1}{2} + t_{\text{jogger 1}}$$
$$\frac{d}{6} = \frac{1}{2} + \frac{d}{8}$$
$$24\left(\frac{d}{6}\right) = 24\left(\frac{1}{2} + \frac{d}{8}\right)$$
$$4d = 12 + 3d$$
$$d = 12$$

The run was 12 miles.

49. Let n = numerator. Then
$n+1$ = the denominator, and

$\dfrac{n}{n+1}$ = the fraction. Now

$$\frac{n-3}{(n+1)-3} = \frac{4}{5}$$
$$\frac{n-3}{n-2} = \frac{4}{5}$$
$$5(n-3) = 4(n-2)$$
$$5n-15 = 4n-8$$
$$n = 7$$

Thus, $\dfrac{n}{n+1} = \dfrac{7}{7+1} = \dfrac{7}{8}$ is the fraction.

51. Let t = time to move the cans working together.

$$\frac{1}{2} + \frac{1}{6} = \frac{1}{t}$$
$$6t\left(\frac{1}{2} + \frac{1}{6}\right) = 6t \cdot \frac{1}{t}$$
$$3t + t = 6$$
$$4t = 6$$
$$t = \frac{6}{4} = 1\frac{1}{2}$$

It would take them $1\dfrac{1}{2}$ minutes.

53.

distance =	rate	· time	
upstream	3	$x - 6$	$\frac{3}{x-6}$
downstream	9	$x + 6$	$\frac{9}{x+6}$

$$\frac{3}{x-6} = \frac{9}{x+6}$$
$$3(x+6) = 9(x-6)$$
$$3x+18 = 9x-54$$
$$3x+72 = 9x$$
$$72 = 6x$$
$$12 = x$$

Time upstream $= \dfrac{3}{12-6} = \dfrac{3}{6} = \dfrac{1}{2}$ hour.

Total time is $2\left(\dfrac{1}{2}\right) = 1$ hour.

55. Let t = time to complete the job working together.

$$\frac{1}{4} + \frac{1}{5} = \frac{1}{t}$$
$$20t\left(\frac{1}{4} + \frac{1}{5}\right) = 20t \cdot \frac{1}{t}$$
$$5t + 4t = 20$$
$$9t = 20$$
$$t = \frac{20}{9} = 2\frac{2}{9}$$

It would take them $2\dfrac{2}{9}$ hours.

57. Let t = time if they worked together.

$$\frac{1}{3} + \frac{1}{6} = \frac{1}{t}$$
$$6t\left(\frac{1}{3} + \frac{1}{6}\right) = 6t \cdot \frac{1}{t}$$
$$2t + t = 6$$
$$3t = 6$$
$$t = 2$$

It would take them 2 hours.

59. Let m = the number of movies that would be rated PG-13.

$$\frac{9}{20} = \frac{m}{599}$$
$$20m = 9 \cdot 599$$
$$20m = 5391$$
$$m = \frac{5391}{20} = 269.55$$

Approximately 270 movies would be PG-13.

61.
$$\frac{x}{5} = \frac{x+2}{3}$$
$$5(x+2) = 3x$$
$$5x + 10 = 3x$$
$$10 = -2x$$
$$-5 = x$$

63.
$$\frac{x-3}{2} = \frac{x-5}{6}$$
$$6(x-3) = 2(x-5)$$
$$6x - 18 = 2x - 10$$
$$4x = 8$$
$$x = 2$$

65. 5 ft 8 in. = (5 · 12 + 8) in. = 68 in.
$$\frac{705w}{h^2} = 25$$
$$\frac{705w}{68^2} = 25$$
$$\frac{705w}{4624} = 25$$
$$705w = 25(4624)$$
$$705w = 115,600$$
$$w \approx 163.9716312$$
The person should weigh 164 pounds.

67. Use $h = \dfrac{a}{1 - \frac{s}{770}}$ with $a = 329.63$ and $s = 50$.

$$h = \frac{a}{1 - \frac{s}{770}} = \frac{329.63}{1 - \frac{50}{770}} \approx 352.52$$

The pitch that the observer hears is higher than the actual pitch—it is closest to the musical note F.

69.
$$\frac{1}{R} = \frac{1}{R_1} + \frac{1}{R_2}$$
$$\frac{1}{2} = \frac{1}{3} + \frac{1}{R_2}$$
$$6R_2\left(\frac{1}{2}\right) = 6R_2\left(\frac{1}{3} + \frac{1}{R_2}\right)$$
$$3R_2 = 2R_2 + 6$$
$$R_2 = 6$$

6 ohms

71.
$$\frac{1}{R} = \frac{1}{R_1} + \frac{1}{R_2} + \frac{1}{R_3}$$
$$\frac{1}{R} = \frac{1}{5} + \frac{1}{6} + \frac{1}{2}$$
$$30R\left(\frac{1}{R}\right) = 30R\left(\frac{1}{5} + \frac{1}{6} + \frac{1}{2}\right)$$
$$30 = 6R + 5R + 15R$$
$$30 = 26R$$
$$R = \frac{30}{26} = \frac{15}{13}$$

$\dfrac{15}{13}$ ohms

Section 6.7 Practice Exercises

1.
$$y = kx$$
$$20 = k(15)$$
$$\frac{4}{3} = k$$
$$k = \frac{4}{3}; \ y = \frac{4}{3}x$$

2.
$$d = kw$$
$$9 = k(36)$$
$$\frac{1}{4} = k$$
$$d = \frac{1}{4}w$$
$$d = \frac{1}{4}(75)$$
$$d = \frac{75}{4} \text{ inches or } 18\frac{3}{4} \text{ inches}$$

3.
$$b = \frac{k}{a}$$
$$5 = \frac{k}{9}$$
$$k = 45; \ b = \frac{45}{a}$$

4. $P = \dfrac{k}{V}$

$350 = \dfrac{k}{2.8}$

$980 = k$

$P = \dfrac{980}{V}$

$P = \dfrac{980}{1.5}$

$P = 653\dfrac{1}{3}$

The pressure is $653\dfrac{1}{3}$ kilopascals.

5. $A = kap$

6. $y = \dfrac{k}{x^3}$

$\dfrac{1}{2} = \dfrac{k}{2^3}$

$\dfrac{1}{2} = \dfrac{k}{8}$

$4 = k$

$k = 4; \ y = \dfrac{4}{x^3}$

7. $y = \dfrac{kz}{x^3}$

$15 = \dfrac{k \cdot 5}{3^3}$

$81 = k$

$k = 81; \ y = \dfrac{81z}{x^3}$

Vocabulary, Readiness & Video Check 6.7

1. $y = 5x$ represents direct variation.

2. $y = \dfrac{700}{x}$ represents inverse variation.

3. $y = 5xz$ represents joint variation.

4. $y = \dfrac{1}{2}abc$ represents joint variation.

5. $y = \dfrac{9.1}{x}$ represents inverse variation.

6. $y = 2.3x$ represents direct variation.

7. $y = \dfrac{2}{3}x$ represents direct variation.

8. $y = 3.1st$ represents joint variation.

9. linear; slope

10. When x and y vary inversely, the product of x and y is the constant of variation, k.

11. $y = ka^2b^5$

12. A combined variation problem involves combinations of direct, inverse, and/or joint variation.

Exercise Set 6.7

1. $y = kx$

$4 = k(20)$

$k = \dfrac{1}{5}$

$y = \dfrac{1}{5}x$

3. $y = kx$

$6 = k(4)$

$k = \dfrac{3}{2}$

$y = \dfrac{3}{2}x$

5. $y = kx$

$7 = k\left(\dfrac{1}{2}\right)$

$k = 14$

$y = 14x$

7. $y = kx$

$0.2 = k(0.8)$

$k = 0.25$

$y = 0.25x$

9. $W = kr^3$

$1.2 = k \cdot 2^3$

$k = \dfrac{1.2}{8} = 0.15$

$W = 0.15r^3$

$\quad = 0.15(3)^3$

$\quad = 0.15(27)$

$\quad = 4.05$

The ball weighs 4.05 pounds.

11.
$$P = kN$$
$$270,000 = k(460,000)$$
$$k = \frac{27}{46}$$
$$P = \frac{27}{46}N$$
$$P = \frac{27}{46}(319,000) \approx 187,239$$

We expect St. Louis to produce 187,239 tons of pollution.

13.
$$y = \frac{k}{x}$$
$$6 = \frac{k}{5}$$
$$k = 30$$
$$y = \frac{30}{x}$$

15.
$$y = \frac{k}{x}$$
$$100 = \frac{k}{7}$$
$$k = 700$$
$$y = \frac{700}{x}$$

17.
$$y = \frac{k}{x}$$
$$\frac{1}{8} = \frac{k}{16}$$
$$k = 2$$
$$y = \frac{2}{x}$$

19.
$$y = \frac{k}{x}$$
$$0.2 = \frac{k}{0.7}$$
$$k = 0.14$$
$$y = \frac{0.14}{x}$$

21.
$$R = \frac{k}{T}$$
$$45 = \frac{k}{6}$$
$$k = 270$$
$$R = \frac{270}{5} = 54$$

The car's speed is 54 mph.

23.
$$I = \frac{k}{R}$$
$$40 = \frac{k}{270}$$
$$k = 10,800$$
$$I = \frac{10,800}{R} = \frac{10,800}{150} = 72$$

The current is 72 amps.

25.
$$I_1 = \frac{k}{d^2}$$
Replace d by $2d$.
$$I_2 = \frac{k}{(2d)^2} = \frac{k}{4d^2} = \frac{1}{4}I_1$$
Thus, the intensity is divided by 4.

27. $x = kyz$

29. $r = kst^3$

31.
$$y = kx^3$$
$$9 = k(3)^3$$
$$9 = 27k$$
$$k = \frac{1}{3}$$
$$y = \frac{1}{3}x^3$$

33.
$$y = k\sqrt{x}$$
$$0.4 = k\sqrt{4}$$
$$0.4 = 2k$$
$$\frac{0.4}{2} = k$$
$$0.2 = k$$
$$y = 0.2\sqrt{x}$$

35. $y = \dfrac{k}{x^2}$

$0.052 = \dfrac{k}{5^2}$

$k = 1.3$

$y = \dfrac{1.3}{x^2}$

37. $y = kxz^3$

$120 = k(5)(2^3)$

$120 = k(5)(8)$

$120 = 40k$

$3 = k$

$y = 3xz^3$

39. Weight $= \dfrac{kwh^2}{l}$

$12 = \dfrac{k\left(\frac{1}{2}\right)\left(\frac{1}{3}\right)^2}{10}$

$120 = k \cdot \dfrac{1}{2} \cdot \dfrac{1}{9}$

$k = 2160$

Weight $= \dfrac{2160wh^2}{l} = \dfrac{2160\left(\frac{2}{3}\right)\left(\frac{1}{2}\right)^2}{16} = 22.5$

The beam can support 22.5 tons.

41. $V = kr^2h$

$32\pi = k(4)^2(6)$

$32\pi = k(16)(6)$

$32\pi = 96k$

$\dfrac{32\pi}{96} = k$

$\dfrac{\pi}{3} = k$

$V = \dfrac{\pi}{3}r^2h$

$V = \dfrac{\pi}{3}(3)^2(5)$

$V = 15\pi$

The volume is 15π cubic inches.

43. $I = \dfrac{k}{x^2}$

$80 = \dfrac{k}{2^2}$

$k = 320$

$I = \dfrac{320}{x^2}$

$5 = \dfrac{320}{x^2}$

$5x^2 = 320$

$x^2 = 64$

$x = 8$

The source is 8 feet from the light source.

45. y varies directly as x is written as $y = kx$.

47. a varies inversely as b is written as $a = \dfrac{k}{b}$.

49. y varies jointly as x and z is written as $y = kxz$.

51. y varies inversely as x^3 is written as $y = \dfrac{k}{x^3}$.

53. y varies directly as x and inversely as p^2 is written as $y = \dfrac{kx}{p^2}$.

55. $r = 4$ in.

$C = 2\pi r = 2\pi(4) = 8\pi$ in.

$A = \pi r^2 = \pi(4)^2 = 16\pi$ sq in.

57. $r = 9$ cm

$C = 2\pi r = 2\pi(9) = 18\pi$ cm

$A = \pi r^2 = \pi(9)^2 = 81\pi$ sq cm

59. $\sqrt{81} = 9$

61. $\sqrt{1} = 1$

63. $\sqrt{\dfrac{1}{4}} = \dfrac{1}{2}$

65. $\sqrt{\dfrac{4}{9}} = \dfrac{2}{3}$

67. $y = \dfrac{2}{3}x$ is an example of direct variation; a.

69. $y = 9ab$ is an example of joint variation; c.

71. $H_1 = ks^3$

$H_2 = k(2s)^3 = 8(ks^3) = 8H_1$

It is multiplied by 8.

73. $y_1 = kx$

$y_2 = k(2x) = 2(kx) = 2y_1$

It is multiplied by 2.

75.

x	$\frac{1}{4}$	$\frac{1}{2}$	1	2	4
$y = \frac{3}{x}$	12	6	3	$\frac{3}{2}$	$\frac{3}{4}$

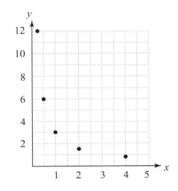

77.

x	$\frac{1}{4}$	$\frac{1}{2}$	1	2	4
$y = \frac{1}{2x}$	2	1	$\frac{1}{2}$	$\frac{1}{4}$	$\frac{1}{8}$

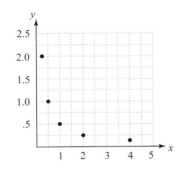

Chapter 6 Vocabulary Check

1. A rational expression whose numerator, denominator, or both contain one or more rational expressions is called a <u>complex fraction</u>.

2. To divide a polynomial by a polynomial other than a monomial, we use <u>long division</u>.

3. In the equation $y = kx$, y varies <u>directly</u> as x.

4. In the equation $y = \dfrac{k}{x}$, y varies <u>inversely</u> as x.

5. The <u>least common denominator</u> of a list of rational expressions is a polynomial of least degree whose factors include the denominator factors in the list.

6. When a polynomial is to be divided by a binomial of the form $x - c$, a shortcut process called <u>synthetic division</u> may be used.

7. In the equation $y = kxz$, y varies <u>jointly</u> as x and z.

8. The expressions $(x - 5)$ and $(5 - x)$ are called <u>opposites</u>.

9. A <u>rational expression</u> is an expression that can be written as the quotient $\dfrac{P}{Q}$ of two polynomials P and Q as long as Q is not 0.

10. Which is an expression and which is an equation? An example of an <u>equation</u> is $\dfrac{2}{x} + \dfrac{2}{x^2} = 7$, and an example of an <u>expression</u> is $\dfrac{2}{x} + \dfrac{5}{x^2}$.

Chapter 6 Review

1. 7 is never 0 so the domain of $f(x) = \dfrac{3 - 5x}{7}$ is $\{x | x \text{ is a real number}\}$.

2. 11 is never 0 so the domain of $g(x) = \dfrac{2x + 4}{11}$ is $\{x | x \text{ is a real number}\}$.

3. $x - 5 = 0$

$x = 5$

The domain of $F(x) = \dfrac{-3x^2}{x - 5}$ is $\{x | x \text{ is a real number and } x \neq 5\}$.

4. $3x - 12 = 0$

$\qquad 3x = 12$

$\qquad x = 4$

The domain of $h(x) = \dfrac{4x}{3x - 12}$ is

$\{x | x \text{ is a real number and } x \neq 4\}$.

5. $x^2 + 8x = 0$

$x(x + 8) = 0$

$x = 0 \quad \text{or} \quad x + 8 = 0$

$x = 0 \quad \text{or} \qquad x = -8$

The domain of $f(x) = \dfrac{x^3 + 2}{x^2 + 8x}$ is

$\{x | x \text{ is a real number and } x \neq 0,\ x \neq -8\}$.

6. $\qquad 3x^2 - 48 = 0$

$\qquad 3(x^2 - 16) = 0$

$3(x + 4)(x - 4) = 0$

$x + 4 = 0 \quad \text{or} \quad x - 4 = 0$

$\qquad x = -4 \quad \text{or} \qquad x = 4$

The domain of $G(x) = \dfrac{20}{3x^2 - 48}$ is

$\{x | x \text{ is a real number and } x \neq -4,\ x \neq 4\}$.

7. $\dfrac{x - 12}{12 - x} = \dfrac{x - 12}{-(x - 12)} = -1$

8. $\dfrac{5x - 15}{25x - 75} = \dfrac{5(x - 3)}{25(x - 3)} = \dfrac{5}{25} = \dfrac{1}{5}$

9. $\dfrac{2x}{2x^2 - 2x} = \dfrac{2x}{2x(x - 1)} = \dfrac{1}{x - 1}$

10. $\dfrac{x + 7}{x^2 - 49} = \dfrac{x + 7}{(x - 7)(x + 7)} = \dfrac{1}{x - 7}$

11. $\dfrac{2x^2 + 4x - 30}{x^2 + x - 20} = \dfrac{2(x^2 + 2x - 15)}{(x + 5)(x - 4)}$

$\qquad = \dfrac{2(x + 5)(x - 3)}{(x + 5)(x - 4)}$

$\qquad = \dfrac{2(x - 3)}{x - 4}$

12. $C(x) = \dfrac{35x + 4200}{x}$

a. $C(50) = \dfrac{35(50) + 4200}{50}$

$\qquad = \dfrac{1750 + 4200}{50}$

$\qquad = \dfrac{5950}{50}$

$\qquad = 119$

The average cost is \$119.

b. $C(100) = \dfrac{35(100) + 4200}{100}$

$\qquad = \dfrac{3500 + 4200}{100}$

$\qquad = \dfrac{7700}{100}$

$\qquad = 77$

The average cost is \$77.

c. It will decrease.

13. $\dfrac{4 - x}{5} \cdot \dfrac{15}{2x - 8} = \dfrac{4 - x}{5} \cdot \dfrac{5 \cdot 3}{2(x - 4)}$

$\qquad = \dfrac{-1(x - 4)}{5} \cdot \dfrac{5 \cdot 3}{2(x - 4)}$

$\qquad = -\dfrac{3}{2}$

14. $\dfrac{x^2 - 6x + 9}{2x^2 - 18} \cdot \dfrac{4x + 12}{5x - 15}$

$\qquad = \dfrac{(x - 3)(x - 3)}{2(x^2 - 9)} \cdot \dfrac{4(x + 3)}{5(x - 3)}$

$\qquad = \dfrac{(x - 3)(x - 3)}{2(x - 3)(x + 3)} \cdot \dfrac{4(x + 3)}{5(x - 3)}$

$\qquad = \dfrac{4}{10}$

$\qquad = \dfrac{2}{5}$

15. $\dfrac{a - 4b}{a^2 + ab} \cdot \dfrac{b^2 - a^2}{8b - 2a}$

$\qquad = \dfrac{a - 4b}{a(a + b)} \cdot \dfrac{(b - a)(b + a)}{2(4b - a)}$

$\qquad = \dfrac{-(4b - a)}{a(a + b)} \cdot -\dfrac{(a - b)(a + b)}{2(4b - a)}$

$\qquad = \dfrac{a - b}{2a}$

16. $\dfrac{x^2-x-12}{2x^2-32} \cdot \dfrac{x^2+8x+16}{3x^2+21x+36}$

$= \dfrac{(x-4)(x+3)}{2(x^2-16)} \cdot \dfrac{(x+4)(x+4)}{3(x^2+7x+12)}$

$= \dfrac{(x-4)(x+3)}{2(x-4)(x+4)} \cdot \dfrac{(x+4)(x+4)}{3(x+4)(x+3)}$

$= \dfrac{1}{2\cdot 3}$

$= \dfrac{1}{6}$

17. $\dfrac{4x+8y}{3} \div \dfrac{5x+10y}{9} = \dfrac{4x+8y}{3} \cdot \dfrac{9}{5x+10y}$

$= \dfrac{4(x+2y)}{3} \cdot \dfrac{3\cdot 3}{5(x+2y)}$

$= \dfrac{12}{5}$

18. $\dfrac{x^2-25}{3} \div \dfrac{x^2-10x+25}{x^2-x-20}$

$= \dfrac{x^2-25}{3} \cdot \dfrac{x^2-x-20}{x^2-10x+25}$

$= \dfrac{(x+5)(x-5)}{3} \cdot \dfrac{(x-5)(x+4)}{(x-5)(x-5)}$

$= \dfrac{(x+5)(x+4)}{3}$

19. $\dfrac{a-4b}{a^2+ab} \div \dfrac{20b-5a}{b^2-a^2}$

$= \dfrac{a-4b}{a^2+ab} \cdot \dfrac{b^2-a^2}{20b-5a}$

$= \dfrac{a-4b}{a(a+b)} \cdot \dfrac{(b+a)(b-a)}{5(4b-a)}$

$= \dfrac{a-4b}{a(a+b)} \cdot \dfrac{-(a+b)(a-b)}{-5(a-4b)}$

$= \dfrac{a-b}{5a}$

20. $\dfrac{3x+3}{x-1} \div \dfrac{x^2-6x-7}{x^2-1} = \dfrac{3x+3}{x-1} \cdot \dfrac{x^2-1}{x^2-6x-7}$

$= \dfrac{3(x+1)}{x-1} \cdot \dfrac{(x+1)(x-1)}{(x-7)(x+1)}$

$= \dfrac{3(x+1)}{x-7}$

21. $\dfrac{2x-x^2}{x^3-8} \div \dfrac{x^2}{x^2+2x+4}$

$= \dfrac{2x-x^2}{x^3-8} \cdot \dfrac{x^2+2x+4}{x^2}$

$= \dfrac{x(2-x)}{(x-2)(x^2+2x+4)} \cdot \dfrac{x^2+2x+4}{x^2}$

$= \dfrac{-x(x-2)}{(x-2)(x^2+2x+4)} \cdot \dfrac{x^2+2x+4}{x^2}$

$= -\dfrac{1}{x}$

22. $\dfrac{5x-15}{3-x} \cdot \dfrac{x+2}{10x+20} \cdot \dfrac{x^2-9}{x^2-x-6}$

$= \dfrac{5(x-3)}{-(x-3)} \cdot \dfrac{x+2}{10(x+2)} \cdot \dfrac{(x+3)(x-3)}{(x-3)(x+2)}$

$= -\dfrac{x+3}{2(x+2)}$

23. $4x^2y^5 = 2\cdot 2x^2y^5$

$10x^2y^4 = 2\cdot 5x^2y^4$

$6y^4 = 2\cdot 3y^4$

The LCD is $2\cdot 2\cdot 5\cdot 3x^2y^5 = 60x^2y^5$.

24. The LCD is $2x(x-2)$.

25. The LCD is $5x(x-5)$.

26. $5x^3 = 5\cdot x^3$

$x^2+3x-28 = (x+7)(x-4)$

$10x^2-30x = 10x(x-3) = 2\cdot 5\cdot x\cdot (x-3)$

The LCD is

$2\cdot 5\cdot x^3\cdot (x-4)(x+7)(x-3)$

$= 10x^3(x-4)(x+7)(x-3)$.

27. $\dfrac{4}{x-4} + \dfrac{x}{x-4} = \dfrac{4+x}{x-4}$

28. $\dfrac{4}{3x^2} + \dfrac{2}{3x^2} = \dfrac{4+2}{3x^2} = \dfrac{6}{3x^2} = \dfrac{2}{x^2}$

29. $\dfrac{1}{x-2} - \dfrac{1}{4-2x} = \dfrac{1}{x-2} - \dfrac{1}{2(2-x)}$

$\qquad\qquad = \dfrac{1}{x-2} - \dfrac{-1}{2(x-2)}$

$\qquad\qquad = \dfrac{1}{x-2} \cdot \dfrac{2}{2} - \dfrac{-1}{2(x-2)}$

$\qquad\qquad = \dfrac{2}{2(x-2)} - \dfrac{-1}{2(x-2)}$

$\qquad\qquad = \dfrac{2+1}{2(x-2)}$

$\qquad\qquad = \dfrac{3}{2(x-2)}$

30. $\dfrac{1}{10-x} + \dfrac{x-1}{x-10} = \dfrac{-1}{x-10} + \dfrac{x-1}{x-10}$

$\qquad\qquad = \dfrac{-1+x-1}{x-10}$

$\qquad\qquad = \dfrac{x-2}{x-10}$

31. $\dfrac{x}{9-x^2} - \dfrac{2}{5x-15} = \dfrac{-x}{x^2-9} - \dfrac{2}{5x-15}$

$\qquad\qquad = \dfrac{-x}{(x-3)(x+3)} - \dfrac{2}{5(x-3)}$

$\qquad\qquad = \dfrac{-x}{(x-3)(x+3)} \cdot \dfrac{5}{5} - \dfrac{2}{5(x-3)} \cdot \dfrac{x+3}{x+3}$

$\qquad\qquad = \dfrac{-5x}{5(x-3)(x+3)} - \dfrac{2(x+3)}{5(x-3)(x+3)}$

$\qquad\qquad = \dfrac{-5x-2(x+3)}{5(x-3)(x+3)}$

$\qquad\qquad = \dfrac{-5x-2x-6}{5(x-3)(x+3)}$

$\qquad\qquad = \dfrac{-7x-6}{5(x-3)(x+3)}$

32. $2x+1 - \dfrac{1}{x-3} = 2x \cdot \dfrac{x-3}{x-3} + 1 \cdot \dfrac{x-3}{x-3} - \dfrac{1}{x-3}$

$\qquad\qquad = \dfrac{2x^2-6x}{x-3} + \dfrac{x-3}{x-3} - \dfrac{1}{x-3}$

$\qquad\qquad = \dfrac{2x^2-6x+x-3-1}{x-3}$

$\qquad\qquad = \dfrac{2x^2-5x-4}{x-3}$

33.
$$\frac{2}{a^2-2a+1}+\frac{3}{a^2-1}=\frac{2}{(a-1)(a-1)}+\frac{3}{(a+1)(a-1)}$$
$$=\frac{2}{(a-1)(a-1)}\cdot\frac{a+1}{a+1}+\frac{3}{(a+1)(a-1)}\cdot\frac{(a-1)}{(a-1)}$$
$$=\frac{2(a+1)}{(a-1)^2(a+1)}+\frac{3(a-1)}{(a-1)^2(a+1)}$$
$$=\frac{2a+2}{(a-1)^2(a+1)}+\frac{3a-3}{(a-1)^2(a+1)}$$
$$=\frac{2a+2+3a-3}{(a-1)^2(a+1)}$$
$$=\frac{5a-1}{(a-1)^2(a+1)}$$

34.
$$\frac{x}{9x^2+12x+16}-\frac{3x+4}{27x^3-64}=\frac{x}{9x^2+12x+16}-\frac{3x+4}{(3x-4)(9x^2+12x+16)}$$
$$=\frac{x}{9x^2+12x+16}\cdot\frac{3x-4}{3x-4}-\frac{3x+4}{(3x-4)(9x^2+12x+16)}$$
$$=\frac{3x^2-4x-3x-4}{(3x-4)(9x^2+12x+16)}$$
$$=\frac{3x^2-7x-4}{(3x-4)(9x^2+12x+16)}$$

35.
$$\frac{2}{x-1}-\frac{3x}{3x-3}+\frac{1}{2x-2}=\frac{2}{x-1}-\frac{3x}{3(x-1)}+\frac{1}{2(x-1)}$$
$$=\frac{2}{x-1}\cdot\frac{6}{6}-\frac{3x}{3(x-1)}\cdot\frac{2}{2}+\frac{1}{2(x-1)}\cdot\frac{3}{3}$$
$$=\frac{12}{6(x-1)}-\frac{6x}{6(x-1)}+\frac{3}{6(x-1)}$$
$$=\frac{12-6x+3}{6(x-1)}$$
$$=\frac{15-6x}{6(x-1)}$$
$$=\frac{3(5-2x)}{6(x-1)}$$
$$=\frac{5-2x}{2(x-1)}$$

36. Perimeter $= \dfrac{1}{x} + \dfrac{1}{x} + \dfrac{1}{x} + \dfrac{2}{x} + \dfrac{2}{x} + \dfrac{3}{2x} + \dfrac{5}{2x}$

$= \dfrac{7}{x} + \dfrac{8}{2x}$

$= \dfrac{7}{x} \cdot \dfrac{2}{2} + \dfrac{8}{2x}$

$= \dfrac{14}{2x} + \dfrac{8}{2x}$

$= \dfrac{14+8}{2x}$

$= \dfrac{22}{2x}$

$= \dfrac{11}{x}$

37. $\dfrac{1-\frac{3x}{4}}{2+\frac{x}{4}} = \dfrac{4\left(1-\frac{3x}{4}\right)}{4\left(2+\frac{x}{4}\right)} = \dfrac{4-3x}{8+x}$

38. $\dfrac{\frac{x^2}{15}}{\frac{x+1}{5x}} = \dfrac{x^2}{15} \div \dfrac{x+1}{5x} = \dfrac{x^2}{15} \cdot \dfrac{5x}{x+1} = \dfrac{x^3}{3(x+1)}$

39. $\dfrac{2-\frac{3}{2x}}{x-\frac{2}{5x}} = \dfrac{10x\left(2-\frac{3}{2x}\right)}{10x\left(x-\frac{2}{5x}\right)}$

$= \dfrac{20x-15}{10x^2-4}$

$= \dfrac{5(4x-3)}{2(5x^2-2)}$

40. $\dfrac{1+\frac{x}{y}}{\frac{x^2}{y^2}-1} = \dfrac{y^2\left(1+\frac{x}{y}\right)}{y^2\left(\frac{x^2}{y^2}-1\right)}$

$= \dfrac{y^2+xy}{x^2-y^2}$

$= \dfrac{y(y+x)}{(x+y)(x-y)}$

$= \dfrac{y}{x-y}$

41. $\dfrac{\frac{5}{x}+\frac{1}{xy}}{\frac{3}{x^2}} = \dfrac{x^2y\left(\frac{5}{x}+\frac{1}{xy}\right)}{x^2y\left(\frac{3}{x^2}\right)} = \dfrac{5xy+x}{3y} = \dfrac{x(5y+1)}{3y}$

42. $\dfrac{\frac{x}{3}-\frac{3}{x}}{1+\frac{3}{x}} = \dfrac{3x\left(\frac{x}{3}-\frac{3}{x}\right)}{3x\left(1+\frac{3}{x}\right)}$

$= \dfrac{x^2-9}{3x+9}$

$= \dfrac{(x+3)(x-3)}{3(x+3)}$

$= \dfrac{x-3}{3}$

43. $\dfrac{\frac{1}{x-1}+1}{\frac{1}{x+1}-1} = \dfrac{(x+1)(x-1)\left(\frac{1}{x-1}+1\right)}{(x+1)(x-1)\left(\frac{1}{x+1}-1\right)}$

$= \dfrac{x+1+(x+1)(x-1)}{x-1-(x+1)(x-1)}$

$= \dfrac{x+1+x^2-1}{x-1-x^2+1}$

$= \dfrac{x+x^2}{x-x^2}$

$= \dfrac{x(1+x)}{x(1-x)}$

$= \dfrac{1+x}{1-x}$

44. $\dfrac{\frac{x-3}{x+3}+\frac{x+3}{x-3}}{\frac{x-3}{x+3}-\frac{x+3}{x-3}} = \dfrac{(x-3)(x+3)\left(\frac{x-3}{x+3}+\frac{x+3}{x-3}\right)}{(x-3)(x+3)\left(\frac{x-3}{x+3}-\frac{x+3}{x-3}\right)}$

$= \dfrac{(x-3)^2+(x+3)^2}{(x-3)^2-(x+3)^2}$

$= \dfrac{x^2-6x+9+x^2+6x+9}{x^2-6x+9-(x^2+6x+9)}$

$= \dfrac{2x^2+18}{x^2-6x+9-x^2-6x-9}$

$= \dfrac{2(x^2+9)}{-12x}$

$= -\dfrac{x^2+9}{6x}$

45. $f(a+h) = \dfrac{3}{a+h}$

46. $f(a) = \dfrac{3}{a}$

47. $\dfrac{f(a+h)-f(a)}{h} = \dfrac{\frac{3}{a+h}-\frac{3}{a}}{h}$

48. $\dfrac{f(a+h)-f(a)}{h} = \dfrac{\dfrac{3}{a+h}-\dfrac{3}{a}}{h}$

$= \dfrac{\left(\dfrac{3}{a+h}-\dfrac{3}{a}\right)\cdot a(a+h)}{h\cdot a(a+h)}$

$= \dfrac{3a-3(a+h)}{h\cdot a(a+h)}$

$= \dfrac{3a-3a-3h}{h\cdot a(a+h)}$

$= \dfrac{-3h}{h\cdot a(a+h)}$

$= -\dfrac{3}{a(a+h)}$

49. $\dfrac{4xy+2x^2-9}{4xy} = \dfrac{4xy}{4xy}+\dfrac{2x^2}{4xy}-\dfrac{9}{4xy}$

$= 1+\dfrac{x}{2y}-\dfrac{9}{4xy}$

50. $\dfrac{12xb^2+16xb^4}{4xb^3} = \dfrac{12xb^2}{4xb^3}+\dfrac{16xb^4}{4xb^3}$

$= \dfrac{3}{b}+4b$

51.

$$
\begin{array}{r}
3x^3+9x^2+2x+6 \\
x-3\overline{)3x^4+0x^3-25x^2+0x-20} \\
\underline{3x^4-9x^3} \\
9x^3-25x^2 \\
\underline{9x^3-27x^2} \\
2x^2+0x \\
\underline{2x^2-6x} \\
6x-20 \\
\underline{6x-18} \\
-2
\end{array}
$$

Answer: $3x^3+9x^2+2x+6-\dfrac{2}{x-3}$

52.

$$
\begin{array}{r}
2x^3-4x^2+7x-9 \\
x+2\overline{)2x^4+0x^3-x^2+5x-12} \\
\underline{2x^4+4x^3} \\
-4x^3-x^2 \\
\underline{-4x^3-8x^2} \\
7x^2+5x \\
\underline{7x^2+14x} \\
-9x-12 \\
\underline{-9x-18} \\
6
\end{array}
$$

Answer: $2x^3-4x^2+7x-9+\dfrac{6}{x+2}$

53.

$$
\begin{array}{r}
x^2-1 \\
2x+3\overline{)2x^3+3x^2-2x+2} \\
\underline{2x^3+3x^2} \\
-2x+2 \\
\underline{-2x-3} \\
5
\end{array}
$$

Answer: $x^2-1+\dfrac{5}{2x+3}$

54.

$$
\begin{array}{r}
3x^2+2x-1 \\
x^2+x+2\overline{)3x^4+5x^3+7x^2+3x-2} \\
\underline{3x^4+3x^3+6x^2} \\
2x^3+x^2+3x \\
\underline{2x^3+2x^2+4x} \\
-x^2-x-2 \\
\underline{-x^2-x-2} \\
0
\end{array}
$$

Answer: $3x^2+2x-1$

55.

$$
\begin{array}{r|rrrr}
2 & 3 & 0 & 12 & -4 \\
  &   & 6 & 12 & 48 \\
\hline
  & 3 & 6 & 24 & 44
\end{array}
$$

Answer: $3x^2+6x+24+\dfrac{44}{x-2}$

56.

$$
\begin{array}{r|rrrrrr}
-1 & 1 & 0 & 0 & 0 & 0 & -1 \\
   &   & -1 & 1 & -1 & 1 & -1 \\
\hline
   & 1 & -1 & 1 & -1 & 1 & -2
\end{array}
$$

Answer: $x^4-x^3+x^2-x+1-\dfrac{2}{x+1}$

57.

$$\begin{array}{r|rrrr} 3 & 1 & 0 & 0 & -81 \\ & & 3 & 9 & 27 \\ \hline & 1 & 3 & 9 & -54 \end{array}$$

Answer: $x^2 + 3x + 9 - \dfrac{54}{x-3}$

58.

$$\begin{array}{r|rrrrr} -2 & 3 & 0 & -2 & 0 & 10 \\ & & -6 & 12 & -20 & 40 \\ \hline & 3 & -6 & 10 & -20 & 50 \end{array}$$

Answer: $3x^3 - 6x^2 + 10x - 20 + \dfrac{50}{x+2}$

59.

$$\begin{array}{r|rrrrrr} 4 & 3 & 0 & 0 & 0 & -9 & 7 \\ & & 12 & 48 & 192 & 768 & 3036 \\ \hline & 3 & 12 & 48 & 192 & 759 & 3043 \end{array}$$

Thus, $P(4) = 3043$.

60.

$$\begin{array}{r|rrrrrr} -5 & 3 & 0 & 0 & 0 & -9 & 7 \\ & & -15 & 75 & -375 & 1875 & -9330 \\ \hline & 3 & -15 & 75 & -375 & 1866 & -9323 \end{array}$$

Thus, $P(-5) = -9323$.

61.

$$\begin{array}{r|rrrrrr} -\frac{1}{2} & 3 & 0 & 0 & 0 & -9 & 7 \\ & & -\frac{3}{2} & \frac{3}{4} & -\frac{3}{8} & \frac{3}{16} & \frac{141}{32} \\ \hline & 3 & -\frac{3}{2} & \frac{3}{4} & -\frac{3}{8} & -\frac{141}{16} & \frac{365}{32} \end{array}$$

Thus, $P\left(-\dfrac{1}{2}\right) = \dfrac{365}{32}$.

62.

$$\begin{array}{r|rrrrr} 3 & 1 & -1 & -6 & -6 & 18 \\ & & 3 & 6 & 0 & -18 \\ \hline & 1 & 2 & 0 & -6 & 0 \end{array}$$

length $= (x^3 + 2x^2 - 6)$ miles

63. The LCD is $3x$.

$$\frac{3}{x} + \frac{1}{3} = \frac{5}{x}$$

$$3x\left(\frac{3}{x} + \frac{1}{3}\right) = 3x\left(\frac{5}{x}\right)$$

$$3x\left(\frac{3}{x}\right) + 3x\left(\frac{1}{3}\right) = 3x\left(\frac{5}{x}\right)$$

$$9 + x = 15$$

$$x = 6$$

The solution is 6.

64. The LCD is $2(5x - 9)$.

$$\frac{2x+3}{5x-9} = \frac{3}{2}$$

$$2(5x-9) \cdot \frac{2x+3}{5x-9} = 2(5x-9) \cdot \frac{3}{2}$$

$$2(2x+3) = (5x-9) \cdot 3$$

$$4x+6 = 15x-27$$

$$6 = 11x-27$$

$$33 = 11x$$

$$3 = x$$

The solution is 3.

65. The LCD is $(x - 2)(x + 2)$.

$$\frac{1}{x-2} - \frac{3x}{x^2-4} = \frac{2}{x+2}$$

$$(x-2)(x+2)\left(\frac{1}{x-2} - \frac{3x}{x^2-4}\right) = (x-2)(x+2)\left(\frac{2}{x+2}\right)$$

$$(x-2)(x+2) \cdot \frac{1}{x-2} - (x-2)(x+2) \cdot \frac{3x}{(x+2)(x-2)} = (x-2)(x+2) \cdot \frac{2}{x+2}$$

$$(x+2) - 3x = 2(x-2)$$

$$-2x+2 = 2x-4$$

$$-4x+2 = -4$$

$$-4x = -6$$

$$x = \frac{-6}{-4} = \frac{3}{2}$$

The solution is $\frac{3}{2}$.

66. The LCD is $7x$.

$$\frac{7}{x} - \frac{x}{7} = 0$$

$$7x\left(\frac{7}{x} - \frac{x}{7}\right) = 7x(0)$$

$$7x\left(\frac{7}{x}\right) - 7x\left(\frac{x}{7}\right) = 0$$

$$49 - x^2 = 0$$

$$(7-x)(7+x) = 0$$

$$7 - x = 0 \quad \text{or} \quad 7 + x = 0$$

$$7 = x \quad \text{or} \quad x = -7$$

The solutions are -7 and 7.

67.
$$\frac{5}{x^2-7x}+\frac{4}{2x-14}=\frac{5}{x(x-7)}+\frac{4}{2(x-7)}$$
$$=\frac{5}{x(x-7)}\cdot\frac{2}{2}+\frac{4}{2(x-7)}\cdot\frac{x}{x}$$
$$=\frac{10}{2x(x-7)}+\frac{4x}{2x(x-7)}$$
$$=\frac{10+4x}{2x(x-7)}$$
$$=\frac{2(5+2x)}{2x(x-7)}$$
$$=\frac{5+2x}{x(x-7)}$$

68.
$$\frac{4}{3-x}-\frac{7}{2x-6}+\frac{5}{x}$$
$$=\frac{-4}{x-3}-\frac{7}{2(x-3)}+\frac{5}{x}$$
$$=\frac{-4}{x-3}\cdot\frac{2x}{2x}-\frac{7}{2(x-3)}\cdot\frac{x}{x}+\frac{5}{x}\cdot\frac{2(x-3)}{2(x-3)}$$
$$=\frac{-8x}{2x(x-3)}-\frac{7x}{2x(x-3)}+\frac{10x-30}{2x(x-3)}$$
$$=\frac{-8x-7x+10x-30}{2x(x-3)}$$
$$=\frac{-5x-30}{2x(x-3)}$$
$$=\frac{-5(x+6)}{2x(x-3)}$$

69. The LCD is x^2.
$$3-\frac{5}{x}-\frac{2}{x^2}=0$$
$$x^2\left(3-\frac{5}{x}-\frac{2}{x^2}\right)=x^2(0)$$
$$x^2\cdot3-x^2\left(\frac{5}{x}\right)-x^2\left(\frac{2}{x^2}\right)=0$$
$$3x^2-5x-2=0$$
$$(3x+1)(x-2)=0$$
$$3x+1=0 \quad\text{or}\quad x-2=0$$
$$3x=-1 \quad\text{or}\quad x=2$$
$$x=-\frac{1}{3} \quad\text{or}\quad x=2$$

The solutions are $-\dfrac{1}{3}$ and 2.

70. The LCD is x^2.
$$2+\frac{15}{x^2}=\frac{13}{x}$$
$$x^2\left(2+\frac{15}{x^2}\right)=x^2\left(\frac{13}{x}\right)$$
$$2x^2+15=13x$$
$$2x^2-13x+15=0$$
$$(2x-3)(x-5)=0$$
$$2x-3=0 \quad\text{or}\quad x-5=0$$
$$x=\frac{3}{2} \quad\text{or}\qquad x=5$$

The solutions are $\dfrac{3}{2}$ and 5.

71.
$$A=\frac{h(a+b)}{2}$$
$$2A=h(a+b)$$
$$\frac{2A}{h}=a+b$$
$$a=\frac{2A}{h}-b$$

72.
$$\frac{1}{R}=\frac{1}{R_1}+\frac{1}{R_2}$$
$$RR_1R_2\left(\frac{1}{R}\right)=RR_1R_2\left(\frac{1}{R_1}\right)+RR_1R_2\left(\frac{1}{R_2}\right)$$
$$R_1R_2=RR_2+RR_1$$
$$R_1R_2-RR_2=RR_1$$
$$R_2(R_1-R)=RR_1$$
$$R_2=\frac{RR_1}{R_1-R}$$

73.
$$I=\frac{E}{R+r}$$
$$I(R+r)=E$$
$$R+r=\frac{E}{I}$$
$$R=\frac{E}{I}-r$$

74.
$$A=P+Prt$$
$$A-P=Prt$$
$$\frac{A-P}{Pt}=r \text{ or } r=\frac{A-P}{Pt}$$

75.

$$\frac{1}{x} = \frac{1}{y} - \frac{1}{z}$$

$$xyz\left(\frac{1}{x}\right) = xyz\left(\frac{1}{y} - \frac{1}{z}\right)$$

$$yz = xz - xy$$

$$yz = x(z - y)$$

$$\frac{yz}{z-y} = \frac{x(z-y)}{z-y}$$

$$\frac{yz}{z-y} = x \text{ or } x = \frac{yz}{z-y}$$

76.

$$H = \frac{kA(T_1 - T_2)}{L}$$

$$HL = kA(T_1 - T_2)$$

$$\frac{HL}{k(T_1 - T_2)} = A \text{ or } A = \frac{HL}{k(T_1 - T_2)}$$

77. Let x = the number.

$$x + 2\left(\frac{1}{x}\right) = 3$$

$$x\left[x + 2\left(\frac{1}{x}\right)\right] = x \cdot 3$$

$$x^2 + 2 = 3x$$

$$x^2 - 3x + 2 = 0$$

$$(x-1)(x-2) = 0$$

$$x - 1 = 0 \quad \text{or} \quad x - 2 = 0$$

$$x = 1 \quad \text{or} \qquad x = 2$$

The numbers are 1 and 2.

78. Let x = the number.

$$\frac{3+x}{7+2x} = \frac{10}{21}$$

$$21(3+x) = 10(7+2x)$$

$$63 + 21x = 70 + 20x$$

$$63 + 1x = 70$$

$$x = 7$$

The number is 7.

79. Let x = amount of time required for all three boys to paint the fence together.

$$\frac{1}{4} + \frac{1}{5} + \frac{1}{6} = \frac{1}{x}$$

$$60x \cdot \frac{1}{4} + 60x \cdot \frac{1}{5} + 60x \cdot \frac{1}{6} = 60x \cdot \frac{1}{x}$$

$$15x + 12x + 10x = 60$$

$$37x = 60$$

$$x = \frac{60}{37} = 1\frac{23}{37}$$

It will take $1\frac{23}{37}$ hours for all three boys to paint the fence together.

80. Let x = amount of time it takes Tom to type the mailing labels when working alone.

$$\frac{1}{6} + \frac{1}{x} = \frac{1}{4}$$

$$12x \cdot \frac{1}{6} + 12x \cdot \frac{1}{x} = 12x \cdot \frac{1}{4}$$

$$2x + 12 = 3x$$

$$12 = x$$

It takes Tom 12 hours to complete the task alone.

81. Let x = the speed of the current.

	distance	=	rate	·	time
Upstream	72		$32 - x$		$\frac{72}{32-x}$
Downstream	120		$32 + x$		$\frac{120}{32+x}$

$$\frac{72}{32-x} = \frac{120}{32+x}$$

$$72(32+x) = 120(32-x)$$

$$2304 + 72x = 3840 - 120x$$

$$72x = 1536 - 120x$$

$$192x = 1536$$

$$x = 8$$

The speed of the current is 8 mph.

82. Let x = the speed of the walker.

	distance	=	rate	·	time
Jogger	14		$x + 3$		$\frac{14}{x+3}$
Walker	8		x		$\frac{8}{x}$

$$\frac{14}{x+3} = \frac{8}{x}$$

$$14x = 8(x+3)$$

$$14x = 8x + 24$$

$$6x = 24$$

$$x = 4$$

The speed of the walker is 4 mph.

83. $A = kB$

$6 = k(14)$

$k = \dfrac{6}{14} = \dfrac{3}{7}$

$A = \dfrac{3}{7}B = \dfrac{3}{7}(21) = 9$

84. $P = \dfrac{K}{V}$

$1250 = \dfrac{K}{2}$

$K = 2500$

$P = \dfrac{2500}{V}$

$800 = \dfrac{2500}{V}$

$800V = 2500$

$V = 3.125$

When the pressure is 800 kilopascals, the volume is 3.125 cubic meters.

85. $\dfrac{22x+8}{11x+4} = \dfrac{2(11x+4)}{11x+4} = 2$

86. $\dfrac{xy-3x+2y-6}{x^2+4x+4} = \dfrac{x(y-3)+2(y-3)}{(x+2)(x+2)}$

$= \dfrac{(x+2)(y-3)}{(x+2)(x+2)}$

$= \dfrac{y-3}{x+2}$

87. $\dfrac{2}{5x} \div \dfrac{4-18x}{6-27x} = \dfrac{2}{5x} \cdot \dfrac{6-27x}{4-18x}$

$= \dfrac{2}{5x} \cdot \dfrac{3(2-9x)}{2(2-9x)}$

$= \dfrac{3}{5x}$

88. $\dfrac{7x+28}{2x+4} \div \dfrac{x^2+2x-8}{x^2-2x-8}$

$= \dfrac{7x+28}{2x+4} \cdot \dfrac{x^2-2x-8}{x^2+2x-8}$

$= \dfrac{7(x+4)}{2(x+2)} \cdot \dfrac{(x-4)(x+2)}{(x+4)(x-2)}$

$= \dfrac{7(x-4)}{2(x-2)}$

89. $\dfrac{5a^2-20}{a^3+2a^2+a+2} \div \dfrac{7a}{a^3+a}$

$= \dfrac{5a^2-20}{a^3+2a^2+a+2} \cdot \dfrac{a^3+a}{7a}$

$= \dfrac{5(a^2-4)}{a^2(a+2)+1(a+2)} \cdot \dfrac{a(a^2+1)}{7a}$

$= \dfrac{5(a+2)(a-2)}{(a+2)(a^2+1)} \cdot \dfrac{a(a^2+1)}{7a}$

$= \dfrac{5(a-2)}{7}$

90. $\dfrac{4a+8}{5a^2-20} \cdot \dfrac{3a^2-6a}{a+3} \div \dfrac{2a^2}{5a+15}$

$= \dfrac{4a+8}{5a^2-20} \cdot \dfrac{3a^2-6a}{a+3} \cdot \dfrac{5a+15}{2a^2}$

$= \dfrac{4(a+2)}{5(a^2-4)} \cdot \dfrac{3a(a-2)}{a+3} \cdot \dfrac{5(a+3)}{2a^2}$

$= \dfrac{4(a+2)}{5(a-2)(a+2)} \cdot \dfrac{3a(a-2)}{a+3} \cdot \dfrac{5(a+3)}{2a^2}$

$= \dfrac{6}{a}$

91. $\dfrac{7}{2x} + \dfrac{5}{6x} = \dfrac{7}{2x} \cdot \dfrac{3}{3} + \dfrac{5}{6x}$

$= \dfrac{21}{6x} + \dfrac{5}{6x}$

$= \dfrac{21+5}{6x}$

$= \dfrac{26}{6x}$

$= \dfrac{13}{3x}$

92. $\dfrac{x-2}{x+1} - \dfrac{x-3}{x-1} = \dfrac{x-2}{x+1} \cdot \dfrac{x-1}{x-1} - \dfrac{x-3}{x-1} \cdot \dfrac{x+1}{x+1}$

$= \dfrac{(x-2)(x-1)}{(x+1)(x-1)} - \dfrac{(x-3)(x+1)}{(x+1)(x-1)}$

$= \dfrac{x^2-3x+2}{(x+1)(x-1)} - \dfrac{x^2-2x-3}{(x+1)(x-1)}$

$= \dfrac{x^2-3x+2-x^2+2x+3}{(x+1)(x-1)}$

$= \dfrac{-x+5}{(x+1)(x-1)}$

93. $\dfrac{2x+1}{x^2+x-6}+\dfrac{2-x}{x^2+x-6}=\dfrac{2x+1+2-x}{x^2+x-6}$

$\qquad\qquad\qquad\qquad\qquad = \dfrac{x+3}{x^2+x-6}$

$\qquad\qquad\qquad\qquad\qquad = \dfrac{x+3}{(x+3)(x-2)}$

$\qquad\qquad\qquad\qquad\qquad = \dfrac{1}{x-2}$

94. $\dfrac{2}{x^2-16}-\dfrac{3x}{x^2+8x+16}+\dfrac{3}{x+4}=\dfrac{2}{(x+4)(x-4)}-\dfrac{3x}{(x+4)^2}+\dfrac{3}{x+4}$

$\qquad\qquad\qquad\qquad\qquad = \dfrac{2}{(x+4)(x-4)}\cdot\dfrac{x+4}{x+4}-\dfrac{3x}{(x+4)^2}\cdot\dfrac{x-4}{x-4}+\dfrac{3}{x+4}\cdot\dfrac{(x+4)(x-4)}{(x+4)(x-4)}$

$\qquad\qquad\qquad\qquad\qquad = \dfrac{2x+8}{(x+4)^2(x-4)}-\dfrac{3x^2-12x}{(x+4)^2(x-4)}+\dfrac{3x^2-48}{(x+4)^2(x-4)}$

$\qquad\qquad\qquad\qquad\qquad = \dfrac{2x+8-3x^2+12x+3x^2-48}{(x+4)^2(x-4)}$

$\qquad\qquad\qquad\qquad\qquad = \dfrac{14x-40}{(x+4)^2(x-4)}$

$\qquad\qquad\qquad\qquad\qquad = \dfrac{2(7x-20)}{(x+4)^2(x-4)}$

95. $\dfrac{\frac{1}{x}-\frac{2}{3x}}{\frac{5}{2x}-\frac{1}{3}}=\dfrac{6x\left(\frac{1}{x}-\frac{2}{3x}\right)}{6x\left(\frac{5}{2x}-\frac{1}{3}\right)}=\dfrac{6-4}{15-2x}=\dfrac{2}{15-2x}$

96. $\dfrac{2}{1-\frac{2}{x}}=\dfrac{x(2)}{x\left(1-\frac{2}{x}\right)}=\dfrac{2x}{x-2}$

97. $\dfrac{\frac{x^2+5x-6}{4x+3}}{\frac{(x+6)^2}{8x+6}}=\dfrac{x^2+5x-6}{4x+3}\div\dfrac{(x+6)^2}{8x+6}$

$\qquad\qquad\qquad = \dfrac{x^2+5x-6}{4x+3}\cdot\dfrac{8x+6}{(x+6)^2}$

$\qquad\qquad\qquad = \dfrac{(x+6)(x-1)}{4x+3}\cdot\dfrac{2(4x+3)}{(x+6)^2}$

$\qquad\qquad\qquad = \dfrac{2(x-1)}{x+6}$

98. $\dfrac{\frac{3}{x-1}-\frac{2}{1-x}}{\frac{2}{x-1}-\frac{2}{x}}=\dfrac{\frac{3}{x-1}-\frac{-2}{x-1}}{\frac{2}{x-1}-\frac{2}{x}}$

$=\dfrac{(x)(x-1)\left(\frac{3}{x-1}-\frac{-2}{x-1}\right)}{(x)(x-1)\left(\frac{2}{x-1}-\frac{2}{x}\right)}$

$=\dfrac{3x+2x}{2x-2(x-1)}$

$=\dfrac{5x}{2x-2x+2}$

$=\dfrac{5x}{2}$

99. $4+\dfrac{8}{x}=8$

The LCD is x.

$x(4)+x\left(\dfrac{8}{x}\right)=x(8)$

$4x+8=8x$

$8=4x$

$x=2$

The solution set is $\{2\}$.

100. $\dfrac{x-2}{x^2-7x+10}=\dfrac{1}{5x-10}-\dfrac{1}{x-5}$

$\dfrac{x-2}{(x-2)(x-5)}=\dfrac{1}{5(x-2)}-\dfrac{1}{x-5}$

The LCD is $5(x-2)(x-5)$.

$5(x-2)(x-5)\cdot\dfrac{x-2}{(x-2)(x-5)}=5(x-2)(x-5)\cdot\dfrac{1}{5(x-2)}-5(x-2)(x-5)\cdot\dfrac{1}{x-5}$

$5(x-2)=(x-5)-5(x-2)$

$5x-10=x-5-5x+10$

$5x-10=-4x+5$

$9x-10=5$

$9x=15$

$x=\dfrac{15}{9}=\dfrac{5}{3}$

The solution set is $\left\{\dfrac{5}{3}\right\}$.

101. Let x be the numerator of a fraction. Then $x+2$ is the denominator.

$\dfrac{x-3}{x+2+5}=\dfrac{2}{3}$

$\dfrac{x-3}{x+7}=\dfrac{2}{3}$

$3(x-3)=2(x+7)$

$3x-9=2x+14$

$x-9=14$

$x=23$

$x + 2 = 23 + 2 = 25$

The fraction is $\dfrac{23}{25}$.

102. Let x be the first even integer and $x + 2$ be the next consecutive even integer.

$$\frac{1}{x} + \frac{1}{x+2} = -\frac{9}{40}$$

$$40x(x+2) \cdot \frac{1}{x} + 40x(x+2) \cdot \frac{1}{x+2} = 40x(x+2) \cdot \frac{-9}{40}$$

$$40(x+2) + 40x = -9x(x+2)$$

$$40x + 80 + 40x = -9x^2 - 18x$$

$$80x + 80 = -9x^2 - 18x$$

$$9x^2 + 98x + 80 = 0$$

$$(9x + 8)(x + 10) = 0$$

$9x + 8 = 0 \qquad$ or $\qquad x + 10 = 0$

$9x = -8 \qquad$ or $\qquad x = -10$

$x = -\dfrac{8}{9} \qquad$ or $\qquad x = -10$

$-\dfrac{8}{9}$ is not an integer.

$x + 2 = -10 + 2 = -8$

The two integers are -10 and -8.

103. Let $x =$ time it takes to empty a full tank if both pipes are open.

$$-\frac{1}{2.5} + \frac{1}{2} = \frac{1}{x}$$

$$(2.5)(2x) \cdot -\frac{1}{2.5} + (2.5)(2x) \cdot \frac{1}{2} = (2.5)(2x) \cdot \frac{1}{x}$$

$$-2x + 2.5x = 5$$

$$0.5x = 5$$

$$x = 10$$

It takes 10 hours to empty a full tank if both pipes are open.

104. Let $x =$ the speed of the car and
$x + 430 =$ the speed of the jet.

	distance =	rate	· time
Car	210	x	$\dfrac{210}{x}$
Jet	1715	$x + 430$	$\dfrac{1715}{x+430}$

$$\frac{210}{x} = \frac{1715}{x+430}$$

$$210(x+430) = 1715x$$

$$210x + 90,300 = 1715x$$

$$90,300 = 1505x$$

$$60 = x$$

$x + 430 = 60 + 430 = 490$

The speed of the jet is 490 mph.

105. One train traveled 382 miles in 6 hours and the second train traveled $382 - 112 = 270$ miles in 6 hours.

Recall $D = rt$ or $r = \dfrac{D}{t}$.

$r = \dfrac{382}{6} = 63\dfrac{2}{3}$ $r = \dfrac{270}{6} = 45$

The speeds of the trains are $63\dfrac{2}{3}$ mph and 45 mph.

106. $A = kr^2$

$36\pi = k(3)^2$

$36\pi = 9k$

$k = 4\pi$

$A = 4\pi r^2$

$A = 4\pi(4)^2 = 4\pi \cdot 16 = 64\pi$

The surface area is 64π square inches.

107. Let p = number of MP3-capable phones.

$\dfrac{3}{4} = \dfrac{p}{43{,}560}$

$4 \cdot p = 3 \cdot 43{,}560$

$4p = 130{,}680$

$p = 32{,}670$

In 2013, it is predicted that there will be 32,670 MP3-capable phones.

108. Let d = number of Internet users who purchase digital music.

$\dfrac{8}{25} = \dfrac{d}{2000}$

$25 \cdot d = 8 \cdot 2000$

$25d = 16{,}000$

$d = 640$

In a group of 2000, it is predicted there would be 640 who purchase digital music.

109.

$$
\require{enclose}
\begin{array}{r}
3x^3 + 13x^2 + 51x + 204 \\[2pt]
x-4 \enclose{longdiv}{3x^4 + x^3 - x^2 + 0x - 2} \\
\underline{3x^4 - 12x^3} \\
13x^3 - x^2 \\
\underline{13x^3 - 52x^2} \\
51x^2 + 0x \\
\underline{51x^2 - 204x} \\
204x - 2 \\
\underline{204x - 816} \\
814
\end{array}
$$

$3x^3 + 13x^2 + 51x + 204 + \dfrac{814}{x-4}$

110. $C = \dfrac{k}{D}$

$12 = \dfrac{k}{8}$

$96 = k$

$C = \dfrac{96}{D} = \dfrac{96}{24} = 4$

Chapter 6 Test

1. $1 - x = 0$

$1 = x$

The domain of $f(x) = \dfrac{5x^2}{1-x}$ is $\{x | x \text{ is a real number and } x \neq 1\}$.

2. $x^2 + 4x + 3 = 0$

$(x+3)(x+1) = 0$

$x + 3 = 0$ or $x + 1 = 0$

$x = -3$ or $x = -1$

The domain of $g(x) = \dfrac{9x^2 - 9}{x^2 + 4x + 3}$ is $\{x | x \text{ is a real number and } x \neq -3, x \neq -1\}$.

3. $\dfrac{7x - 21}{24 - 8x} = \dfrac{7(x-3)}{8(3-x)} = \dfrac{7(x-3)}{-8(x-3)} = -\dfrac{7}{8}$

4. $\dfrac{x^2 - 4x}{x^2 + 5x - 36} = \dfrac{x(x-4)}{(x+9)(x-4)} = \dfrac{x}{x+9}$

5. $\dfrac{x^3-8}{x-2}=\dfrac{x^3-2^3}{x-2}$

$\qquad =\dfrac{(x-2)(x^2+2x+4)}{x-2}$

$\qquad =x^2+2x+4$

6. $\dfrac{2x^3+16}{6x^2+12x}\cdot\dfrac{5}{x^2-2x+4}$

$\quad =\dfrac{2(x^3+8)}{6x(x+2)}\cdot\dfrac{5}{x^2-2x+4}$

$\quad =\dfrac{2(x+2)(x^2-2x+4)}{6x(x+2)}\cdot\dfrac{5}{x^2-2x+4}$

$\quad =\dfrac{5}{3x}$

7. $\dfrac{5}{4x^3}+\dfrac{7}{4x^3}=\dfrac{5+7}{4x^3}=\dfrac{12}{4x^3}=\dfrac{3}{x^3}$

8. $\dfrac{3x^2-12}{x^2+2x-8}\div\dfrac{6x+18}{x+4}$

$\quad =\dfrac{3x^2-12}{x^2+2x-8}\cdot\dfrac{x+4}{6x+18}$

$\quad =\dfrac{3(x^2-4)}{(x+4)(x-2)}\cdot\dfrac{x+4}{6(x+3)}$

$\quad =\dfrac{3(x+2)(x-2)}{(x+4)(x-2)}\cdot\dfrac{(x+4)}{6(x+3)}$

$\quad =\dfrac{x+2}{2(x+3)}$

9. $\dfrac{4x-12}{2x-9}\div\dfrac{3-x}{4x^2-81}\cdot\dfrac{x+3}{5x+15}$

$\quad =\dfrac{4x-12}{2x-9}\cdot\dfrac{4x^2-81}{3-x}\cdot\dfrac{x+3}{5x+15}$

$\quad =\dfrac{4(x-3)}{2x-9}\cdot\dfrac{(2x+9)(2x-9)}{-(x-3)}\cdot\dfrac{x+3}{5(x+3)}$

$\quad =\dfrac{4(2x+9)}{-5}$

$\quad =-\dfrac{4(2x+9)}{5}$

10. $\dfrac{3+2x}{10-x}+\dfrac{13+x}{x-10}=\dfrac{-(3+2x)}{x-10}+\dfrac{13+x}{x-10}$

$\qquad =\dfrac{-3-2x+13+x}{x-10}$

$\qquad =\dfrac{-x+10}{x-10}$

$\qquad =-\dfrac{x-10}{x-10}$

$\qquad =-1$

11. $\dfrac{2x^2+7}{2x^4-18x^2}-\dfrac{6x+7}{2x^4-18x^2}=\dfrac{2x^2+7-6x-7}{2x^4-18x^2}$

$\qquad =\dfrac{2x^2-6x}{2x^4-18x^2}$

$\qquad =\dfrac{2x(x-3)}{2x^2(x^2-9)}$

$\qquad =\dfrac{2x(x-3)}{2x^2(x-3)(x+3)}$

$\qquad =\dfrac{1}{x(x+3)}$

12. $\dfrac{3}{x^2-x-6}+\dfrac{2}{x^2-5x+6}$

$\quad =\dfrac{3}{(x-3)(x+2)}+\dfrac{2}{(x-3)(x-2)}$

$\quad =\dfrac{3}{(x-3)(x+2)}\cdot\dfrac{x-2}{x-2}+\dfrac{2}{(x-3)(x-2)}\cdot\dfrac{x+2}{x+2}$

$\quad =\dfrac{3(x-2)}{(x-3)(x+2)(x-2)}+\dfrac{2(x+2)}{(x-3)(x+2)(x-2)}$

$\quad =\dfrac{3x-6}{(x-3)(x+2)(x-2)}+\dfrac{2x+4}{(x-3)(x+2)(x-2)}$

$\quad =\dfrac{3x-6+2x+4}{(x-3)(x+2)(x-2)}$

$\quad =\dfrac{5x-2}{(x-3)(x+2)(x-2)}$

13. $3x - 21 = 3(x - 7)$

$2x - 14 = 2(x - 7)$

The LCD is $3 \cdot 2(x - 7) = 6(x - 7)$.

$\dfrac{5}{x-7} - \dfrac{2x}{3x-21} + \dfrac{x}{2x-14}$

$= \dfrac{5}{x-7} \cdot \dfrac{6}{6} - \dfrac{2x}{3(x-7)} \cdot \dfrac{2}{2} + \dfrac{x}{2(x-7)} \cdot \dfrac{3}{3}$

$= \dfrac{30}{6(x-7)} - \dfrac{4x}{6(x-7)} + \dfrac{3x}{6(x-7)}$

$= \dfrac{30 - 4x + 3x}{6(x-7)}$

$= \dfrac{30 - x}{6(x-7)}$

14. $\dfrac{3x}{5} \cdot \left(\dfrac{5}{x} - \dfrac{5}{2x} \right) = \dfrac{3x}{5} \cdot \left(\dfrac{5}{x} \cdot \dfrac{2}{2} - \dfrac{5}{2x} \right)$

$= \dfrac{3x}{5} \left(\dfrac{10}{2x} - \dfrac{5}{2x} \right)$

$= \dfrac{3x}{5} \left(\dfrac{10 - 5}{2x} \right)$

$= \dfrac{3x}{5} \left(\dfrac{5}{2x} \right)$

$= \dfrac{3}{2}$

15. $\dfrac{\frac{5}{x} - \frac{7}{3x}}{\frac{9}{8x} - \frac{1}{x}} = \dfrac{24x\left(\frac{5}{x} - \frac{7}{3x}\right)}{24x\left(\frac{9}{8x} - \frac{1}{x}\right)} = \dfrac{120 - 56}{27 - 24} = \dfrac{64}{3}$

16. $\dfrac{\frac{x^2-5x+6}{x+3}}{\frac{x^2-4x+4}{x^2-9}} = \dfrac{x^2 - 5x + 6}{x + 3} \div \dfrac{x^2 - 4x + 4}{x^2 - 9}$

$= \dfrac{x^2 - 5x + 6}{x + 3} \cdot \dfrac{x^2 - 9}{x^2 - 4x + 4}$

$= \dfrac{(x-3)(x-2)}{x+3} \cdot \dfrac{(x+3)(x-3)}{(x-2)(x-2)}$

$= \dfrac{(x-3)^2}{x-2}$

17. $\dfrac{4x^2 y + 9x + 3xz}{3xz} = \dfrac{4x^2 y}{3xz} + \dfrac{9x}{3xz} + \dfrac{3xz}{3xz}$

$= \dfrac{4xy}{3z} + \dfrac{3}{z} + 1$

18.
$$2x+1 \overline{)\begin{array}{l} 2x^2 - x - 2 \\ 4x^3 + 0x^2 - 5x + 0 \end{array}}$$

$\underline{4x^3 + 2x^2}$

$-2x^2 - 5x$

$\underline{-2x^2 - x}$

$-4x + 0$

$\underline{-4x - 2}$

2

Answer: $2x^2 - x - 2 + \dfrac{2}{2x+1}$

19.
$$\begin{array}{r|rrrr} -3 & 4 & -3 & 0 & -1 & -1 \\ & & -12 & 45 & -135 & 408 \\ \hline & 4 & -15 & 45 & -136 & 407 \end{array}$$

Answer: $4x^3 - 15x^2 + 45x - 136 + \dfrac{407}{x+3}$

20.
$$\begin{array}{r|rrrr} -2 & 4 & 0 & 7 & -2 & -5 \\ & & -8 & 16 & -46 & 96 \\ \hline & 4 & -8 & 23 & -48 & 91 \end{array}$$

Thus, $P(-2) = 91$.

21. $\dfrac{5x+3}{3x-7} = \dfrac{19}{7}$

$7(5x + 3) = 19(3x - 7)$

$35x + 21 = 57x - 133$

$154 = 22x$

$7 = x$

22. The LCD is $x - 4$.

$\dfrac{x}{x-4} = 3 - \dfrac{4}{x-4}$

$(x-4)\dfrac{x}{x-4} = (x-4) \cdot 3 - (x-4) \cdot \dfrac{4}{x-4}$

$x = 3(x - 4) - 4$

$x = 3x - 12 - 4$

$x = 3x - 16$

$-2x = -16$

$x = 8$

The solution set is $\{8\}$.

23.
$$\frac{3}{x+2} - \frac{1}{5x} = \frac{2}{5x^2 + 10x}$$
$$\frac{3}{x+2} - \frac{1}{5x} = \frac{2}{5x(x+2)}$$
$$5x(x+2)\left(\frac{3}{x+2} - \frac{1}{5x}\right) = 5x(x+2)\cdot\frac{2}{5x(x+2)}$$
$$3(5x) - 1(x+2) = 2$$
$$15x - x - 2 = 2$$
$$14x = 4$$
$$x = \frac{4}{14} = \frac{2}{7}$$

The solution set is $\left\{\frac{2}{7}\right\}$.

24.
$$\frac{x^2 + 8}{x} - 1 = \frac{2(x+4)}{x}$$
$$x\left(\frac{x^2+8}{x} - 1\right) = x\left(\frac{2(x+4)}{x}\right)$$
$$(x^2 + 8) - x = 2(x+4)$$
$$x^2 - x + 8 = 2x + 8$$
$$x^2 - 3x = 0$$
$$x(x-3) = 0$$
$$x = 0 \text{ or } x - 3 = 0$$
$$x = 3$$

Discard the answer 0 as extraneous.
The solution set is {3}.

25.
$$\frac{x+b}{a} = \frac{4x-7a}{b}$$
$$ab\left(\frac{x+b}{a}\right) = ab\left(\frac{4x-7a}{b}\right)$$
$$b(x+b) = a(4x-7a)$$
$$xb + b^2 = 4ax - 7a^2$$
$$b^2 + 7a^2 = 4ax - xb$$
$$b^2 + 7a^2 = x(4a - b)$$
$$x = \frac{b^2 + 7a^2}{4a - b}$$

26. Let x = the number.
$$(x+1)\cdot\frac{2}{x} = \frac{12}{5}$$
$$\frac{2(x+1)}{x} = \frac{12}{5}$$
$$\frac{2x+2}{x} = \frac{12}{5}$$
$$5(2x+2) = 12x$$
$$10x + 10 = 12x$$
$$10 = 2x$$
$$5 = x$$
The number is 5.

27. Let t = time to weed garden together.

Note that 1 hr and 30 min = $\frac{3}{2}$ hours.

$$\frac{1}{2} + \frac{1}{\frac{3}{2}} = \frac{1}{t}$$
$$\frac{1}{2} + \frac{2}{3} = \frac{1}{t}$$
$$6t\left(\frac{1}{2} + \frac{2}{3}\right) = 6t\left(\frac{1}{t}\right)$$
$$3t + 4t = 6$$
$$7t = 6$$
$$t = \frac{6}{7}$$

It takes them $\frac{6}{7}$ hour.

28. $W = \frac{k}{V}$
$$20 = \frac{k}{12}$$
$$k = 20(12) = 240$$
$$W = \frac{240}{V} = \frac{240}{15} = 16$$

29. $Q = kRS^2$
$$24 = k(3)(4)^2$$
$$24 = 48k$$
$$k = \frac{24}{48} = \frac{1}{2}$$
$$Q = \frac{1}{2}RS^2 = \frac{1}{2}(2)(3)^2 = 9$$

30. $\quad S = k\sqrt{d}$

$\qquad 160 = k\sqrt{400}$

$\qquad 160 = 20k$

$\qquad k = \dfrac{160}{20} = 8$

$\qquad S = 8\sqrt{d}$

$\qquad 128 = 8\sqrt{d}$

$\qquad \sqrt{d} = \dfrac{128}{8}$

$\qquad \sqrt{d} = 16$

$\qquad d = 256$

The height of the cliff is 256 feet.

Chapter 6 Cumulative Review

1. a. $\quad 8x$

   b. $\quad 8x + 3$

   c. $\quad x \div (-7)$ or $\dfrac{x}{-7}$

   d. $\quad 2x - 1\dfrac{6}{10} = 2x - 1.6$

   e. $\quad x - 6$

   f. $\quad 2(4 + x)$

2. a. $\quad x - \dfrac{1}{3}$

   b. $\quad 5x - 6$

   c. $\quad 8x + 3$

   d. $\quad \dfrac{7}{2 - x}$

3. $\quad \dfrac{y}{3} - \dfrac{y}{4} = \dfrac{1}{6}$

$\qquad 12\left(\dfrac{y}{3} - \dfrac{y}{4}\right) = 12\left(\dfrac{1}{6}\right)$

$\qquad 4y - 3y = 2$

$\qquad y = 2$

4. $\qquad \dfrac{x}{7} + \dfrac{x}{5} = \dfrac{12}{5}$

$\qquad 35\left(\dfrac{x}{7} + \dfrac{x}{5}\right) = 35\left(\dfrac{12}{5}\right)$

$\qquad 5x + 7x = 84$

$\qquad 12x = 84$

$\qquad x = 7$

5.

X	Y1
-4	-20
10	-12.22
32	0
70	21.111
100	37.778

Y1=5/9(X-32)

Fahrenheit	−4	10	32	70	100
Celsius	−20	−12.22	0	21.11	37.78

6. Let x = score on final exam.

$\qquad \dfrac{78 + 65 + 82 + 79 + 2x}{6} \geq 78$

$\qquad \dfrac{304 + 2x}{6} \geq 78$

$\qquad 6\left(\dfrac{304 + 2x}{6}\right) \geq 6(78)$

$\qquad 304 + 2x \geq 468$

$\qquad 2x \geq 164$

$\qquad x \geq 82$

The minimum score she can make on her final is 82.

7. $\left|\dfrac{3x + 1}{2}\right| = -2$ is impossible. The solution set is $\varnothing$.

8. $\left|\dfrac{2x - 1}{3}\right| + 6 = 3$

$\qquad \left|\dfrac{2x - 1}{3}\right| = -3$, which is impossible.

The solution set is $\varnothing$.

9. $\left|\dfrac{2(x + 1)}{3}\right| \leq 0$

$\qquad \dfrac{2(x + 1)}{3} = 0$

$\qquad 2(x + 1) = 0$

$\qquad 2x + 2 = 0$

$\qquad 2x = -2$

$\qquad x = -1$

The solution is −1.

10. $\left|\dfrac{3(x-1)}{4}\right| \ge 2$

$\dfrac{3(x-1)}{4} \le -2$ or $\dfrac{3(x-1)}{4} \ge 2$

$\dfrac{3x-3}{4} \le -2$ or $\dfrac{3x-3}{4} \ge 2$

$3x-3 \le -8$ or $3x-3 \ge 8$

$3x \le -5$ or $3x \ge 11$

$x \le -\dfrac{5}{3}$ or $x \ge \dfrac{11}{3}$

$\left(-\infty, -\dfrac{5}{3}\right] \cup \left[\dfrac{11}{3}, \infty\right)$

11. $y = -2x + 3$

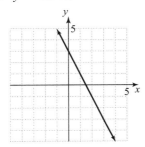

12. $y = -x + 3$

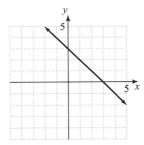

13. a. Function

    b. Not a function

    c. Function

14. $f(x) = -x^2 + 3x - 2$

    a. $f(0) = -(0)^2 + 3(0) - 2 = -2$

    b. $f(-3) = -(-3)^2 + 3(-3) - 2$
$$= -9 - 9 - 2$$
$$= -20$$

c. $f\left(\dfrac{1}{3}\right) = -\left(\dfrac{1}{3}\right)^2 + 3\left(\dfrac{1}{3}\right) - 2$
$$= -\dfrac{1}{9} + 1 - 2$$
$$= -\dfrac{1}{9} + \dfrac{9}{9} - \dfrac{18}{9}$$
$$= \dfrac{-1 + 9 - 18}{9}$$
$$= -\dfrac{10}{9}$$

15. $x - 3y = 6$
Let $y = 0$.
$x - 3(0) = 6$
$x = 6$
Plot $(6, 0)$.
Let $x = 0$.
$0 - 3y = 6$
$-3y = 6$
$y = -2$
Plot $(0, -2)$.

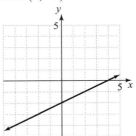

16. $3x - y = 6$
Let $y = 0$.
$3x - 0 = 6$
$3x = 6$
$x = 2$
Plot $(2, 0)$.
Let $x = 0$.
$3(0) - y = 6$
$-y = 6$
$y = -6$
Plot $(0, -6)$.

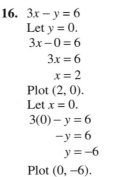

17. $y-(-5)=-3(x-1)$
$\quad\quad\ y+5=-3x+3$
$\quad\quad\quad\quad\ y=-3x-2$

18. $y-3=\dfrac{1}{2}(x-(-1))$

$\quad\quad y-3=\dfrac{1}{2}(x+1)$

$\quad\quad y-3=\dfrac{1}{2}x+\dfrac{1}{2}$

$\quad\quad\quad\ y=\dfrac{1}{2}x+\dfrac{1}{2}+3$

$\quad\quad\quad\ y=\dfrac{1}{2}x+\dfrac{7}{2}$

$\quad\quad f(x)=\dfrac{1}{2}x+\dfrac{7}{2}$

19. $x\ge 1$ and $y\ge 2x-1$

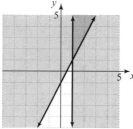

The solution region is the overlap, which has darker shading, along with its boundary.

20. $2x+y\le 4$ or $y>2$

The solution region consists of all shaded regions and the boundary where the boundary line is solid.

21. $\begin{cases}3x-2y=10 & (1)\\ 4x-3y=15 & (2)\end{cases}$

Multiply E1 by -4 and E2 by 3, and add.
$\quad -12x+8y=-40$
$\quad\ \ \underline{12x-9y=45}$
$\quad\quad\quad -y=5$
$\quad\quad\quad\ \ y=-5$
Replace y with -5 in E1.

$3x-2(-5)=10$
$\ 3x+10=10$
$\quad\ \ 3x=0$
$\quad\quad x=0$
The solution is $(0,-5)$.

22. $\begin{cases}-2x+3y=6 & (1)\\ \ 3x-\ y=5 & (2)\end{cases}$

Solve E2 for y:
$y=3x-5$
Replace y with $3x-5$ in E1.
$\quad -2x+3(3x-5)=6$
$\quad\ \ -2x+9x-15=6$
$\quad\quad\quad\quad\quad\ 7x=21$
$\quad\quad\quad\quad\quad\ \ x=3$
Replace x with 3 in the equation $y=3x-5$.
$y=3(3)-5=9-5=4$
The solution is $(3, 4)$.

23. $\begin{cases}2x-4y+8z=2 & (1)\\ -x-3y+\ z=11 & (2)\\ \ x-2y+4z=0 & (3)\end{cases}$

Add E2 and E3.
$-5y+5z=11$ (4)
Multiply E2 by 2 and add to E1.
$\quad -2x-6y+2z=22$
$\quad\ \ \underline{2x-4y+8z=2}$
$\quad\quad\ -10y+10z=24$ (5)
Solve the new system.
$\begin{cases}\ -5y+\ 5z=11 & (4)\\ -10y+10z=24 & (5)\end{cases}$
Multiply E4 by -2 and add to E5.
$\quad\ 10y-10z=-22$
$\quad\underline{-10y+10z=24}$
$\quad\quad\quad\quad\quad 0=2$, which is impossible.
The solution is $\varnothing$.

24. $\begin{cases}\ 2x-2y+4z=6 & (1)\\ -4x-\ y+\ z=-8 & (2)\\ \ 3x-\ y+\ z=6 & (3)\end{cases}$

Multiply E2 by -1 and add to E3.
$\quad 4x+y-z=8$
$\quad\underline{3x-y+z=6}$
$\quad 7x\quad\quad\ =14$
$\quad\ x=2$
Multiply E2 by -2 and add to E1.
$\quad 8x+2y-2z=16$
$\quad\underline{2x-2y+4z=6}$
$\quad 10x\quad\ +2z=22$ or $5x+z=11$

Replace x with 2 in the equation $5x + z = 11$.
$$5(2) + z = 11$$
$$10 + z = 11$$
$$z = 1$$
Replace x with 2 and z with 1 in E3.
$$3(2) - y + 1 = 6$$
$$7 - y = 6$$
$$-y = -1$$
$$y = 1$$
The solution is (2, 1, 1).

25. Let x = measure of the smallest angle,
y = measure of the largest angle and,
z = measure of the remaining angle.
$$\begin{cases} x + y + z = 180 & (1) \\ y = x + 80 & (2) \\ z = x + 10 & (3) \end{cases}$$
Substitute $x + 80$ for y and $x + 10$ for z in E1.
$$x + (x + 80) + (x + 10) = 180$$
$$3x + 90 = 180$$
$$3x = 90$$
$$x = 30$$
Replace x with 30 in E2 and E3.
$$y = 30 + 80 = 110$$
$$z = 30 + 10 = 40$$
The angles measure $30°$, $110°$, and $40°$.

26. Let x = the price of a ream of paper and
y = the price of a box of manila folders.
$$\begin{cases} 3x + 2y = 21.90 & (1) \\ 5x + y = 24.25 & (2) \end{cases}$$
Multiply E2 by -2 and add to E1.
$$\begin{array}{r} -10x - 2y = -48.50 \\ 3x + 2y = 21.90 \\ \hline -7x = -26.60 \\ x = 3.80 \end{array}$$
Replace x with 3.80 in E2.
$$5(3.80) + y = 24.25$$
$$19 + y = 24.25$$
$$y = 5.25$$
A ream of paper cost \$3.80 and a box of manila folders cost \$5.25.

27. $\begin{cases} x + 2y + z = 2 \\ -2x - y + 2z = 5 \\ x + 3y - 2z = -8 \end{cases}$
$$\begin{bmatrix} 1 & 2 & 1 & | & 2 \\ -2 & -1 & 2 & | & 5 \\ 1 & 3 & -2 & | & -8 \end{bmatrix}$$
Multiply R1 by 2 and add to R2.
Multiply R1 by -1 and add to R3.

$$\begin{bmatrix} 1 & 2 & 1 & | & 2 \\ 0 & 3 & 4 & | & 9 \\ 0 & 1 & -3 & | & -10 \end{bmatrix}$$
Interchange R2 and R3.
$$\begin{bmatrix} 1 & 2 & 1 & | & 2 \\ 0 & 1 & -3 & | & -10 \\ 0 & 3 & 4 & | & 9 \end{bmatrix}$$
Multiply R2 by -3 and add to R3.
$$\begin{bmatrix} 1 & 2 & 1 & | & 2 \\ 0 & 1 & -3 & | & -10 \\ 0 & 0 & 13 & | & 39 \end{bmatrix}$$
Divide R3 by 13.
$$\begin{bmatrix} 1 & 2 & 1 & | & 2 \\ 0 & 1 & -3 & | & -10 \\ 0 & 0 & 1 & | & 3 \end{bmatrix}$$
This corresponds to $\begin{cases} x + 2y + z = 2 \\ y - 3z = -10 \\ z = 3. \end{cases}$

$y - 3z = -10$ and so $x + 2(-1) + (3) = 2$
$$y - 3(3) = -10 \qquad\qquad x - 2 + 3 = 2$$
$$y - 9 = -10 \qquad\qquad x + 1 = 2$$
$$y = -1 \qquad\qquad\qquad x = 1$$
The solution is (1, -1, 3).

28. $\begin{cases} x + y + z = 9 \\ 2x - 2y + 3z = 2 \\ -3x + y - z = 1 \end{cases}$
$$\begin{bmatrix} 1 & 1 & 1 & | & 9 \\ 2 & -2 & 3 & | & 2 \\ -3 & 1 & -1 & | & 1 \end{bmatrix}$$
Multiply R1 by -2 and add to R2.
Multiply R1 by 3 and add to R3.
$$\begin{bmatrix} 1 & 1 & 1 & | & 9 \\ 0 & -4 & 1 & | & -16 \\ 0 & 4 & 2 & | & 28 \end{bmatrix}$$
Divide R2 by -4.
$$\begin{bmatrix} 1 & 1 & 1 & | & 9 \\ 0 & 1 & -\frac{1}{4} & | & 4 \\ 0 & 4 & 2 & | & 28 \end{bmatrix}$$
Multiply R2 by -4 and add to R3.
$$\begin{bmatrix} 1 & 1 & 1 & | & 9 \\ 0 & 1 & -\frac{1}{4} & | & 4 \\ 0 & 0 & 3 & | & 12 \end{bmatrix}$$
Divide R3 by 3.
$$\begin{bmatrix} 1 & 1 & 1 & | & 9 \\ 0 & 1 & -\frac{1}{4} & | & 4 \\ 0 & 0 & 1 & | & 4 \end{bmatrix}$$

This corresponds to $\begin{cases} x+y+z=9 \\ y-\dfrac{1}{4}z=4 \\ z=4 \end{cases}$

$y-\dfrac{1}{4}(4)=4$

$y-1=4$ and so

$y=5$

$x+5+4=9$

$x+9=9$

$x=0$

The solution is (0, 5, 4).

29. a. $7^0=1$

　b. $-7^0=-1\cdot 7^0=-1\cdot 1=-1$

　c. $(2x+5)^0=1$

　d. $2x^0=2\cdot x^0=2\cdot 1=2$

30. a. $2^{-2}+3^{-1}=\dfrac{1}{2^2}+\dfrac{1}{3}$

$=\dfrac{1}{4}+\dfrac{1}{3}$

$=\dfrac{3}{12}+\dfrac{4}{12}$

$=\dfrac{7}{12}$

　b. $-6a^0=-6\cdot a^0=-6\cdot 1=-6$

　c. $\dfrac{x^{-5}}{x^{-2}}=x^{-5-(-2)}=x^{-3}=\dfrac{1}{x^3}$

31. a. $x^{-b}(2x^b)^2=\dfrac{2^2(x^b)^2}{x^b}$

$=\dfrac{4x^{2b}}{x^b}$

$=4x^{2b-b}$

$=4x^b$

　b. $\dfrac{(y^{3a})^2}{y^{a-6}}=\dfrac{y^{6a}}{y^{a-6}}=y^{6a-(a-6)}=y^{5a+6}$

32. a. $3x^{4a}(4x^{-a})^2=3x^{4a}\cdot 16x^{-2a}$

$=48x^{4a+(-2a)}$

$=48x^{2a}$

　b. $\dfrac{(y^{4b})^3}{y^{2b-3}}=\dfrac{y^{12b}}{y^{2b-3}}$

$=y^{12b-(2b-3)}$

$=y^{12b-2b+3}$

$=y^{10b+3}$

33. a. $3x^2$ has degree = 2.

　b. $-2^3x^5=-8x^5$ has degree = 5.

　c. y has degree = 1.

　d. $12x^2yz^3$ has degree = 2 + 1 + 3 = 6.

　e. 5 has degree = 0.

34. $(2x^2+8x-3)-(2x-7)$

$=2x^2+8x-3-2x+7$

$=2x^2+6x+4$

35. $[3+(2a+b)]^2$

$=3^2+2(3)(2a+b)+(2a+b)^2$

$=9+6(2a+b)+(4a^2+2(2a)b+b^2)$

$=9+12a+6b+4a^2+4ab+b^2$

36. $[4+(3x-y)]^2$

$=4^2+2(4)(3x-y)+(3x-y)^2$

$=16+8(3x-y)+(9x^2-2(3x)y+y^2)$

$=16+24x-8y+9x^2-6xy+y^2$

37. $ab-6a+2b-12=a(b-6)+2(b-6)$

$=(b-6)(a+2)$

38. $xy+2x-5y-10=x(y+2)-5(y+2)$

$=(y+2)(x-5)$

39. $2n^2-38n+80=2(n^2-19n+40)$

40. $6x^2-x-35=(2x-5)(3x+7)$

41. $x^2+4x+4-y^2=(x^2+4x+4)-y^2$

$=(x+2)^2-y^2$

$=[(x+2)+y][(x+2)-y]$

$=(x+2+y)(x+2-y)$

42. $4x^2 - 4x + 1 - 9y^2$
$= (4x^2 - 4x + 1) - 9y^2$
$= (2x - 1)^2 - (3y)^2$
$= [(2x - 1) + 3y][(2x - 1) - 3y]$
$= (2x - 1 + 3y)(2x - 1 - 3y)$

43. $(x + 2)(x - 6) = 0$
$x + 2 = 0$ or $x - 6 = 0$
$x = -2$ or $x = 6$
The solutions are –2 and 6.

44. $2x(3x + 1)(x - 3) = 0$
$x = 0$ or $3x + 1 = 0$ or $x - 3 = 0$
$3x = -1$ or $x = 3$
$x = -\dfrac{1}{3}$
The solutions are $-\dfrac{1}{3}, 0, 3$.

45. a. $\dfrac{2x^2}{10x^3 - 2x^2} = \dfrac{2x^2}{2x^2(5x - 1)} = \dfrac{1}{5x - 1}$

b. $\dfrac{9x^2 + 13x + 4}{8x^2 + x - 7} = \dfrac{(9x + 4)(x + 1)}{(8x - 7)(x + 1)} = \dfrac{9x + 4}{8x - 7}$

46. a. Domain: $(-\infty, \infty)$; Range: $[-4, \infty)$

b. x-intercepts: (–2, 0), (2, 0)
y-intercept: (0, –4)

c. There is no such point.

d. The point with the least y-value is (0, –4).

e. –2, 2

f. x-values between $x = -2$ and $x = 2$

g. The solutions are –2 and 2.

47. $\dfrac{5k}{k^2 - 4} - \dfrac{2}{k^2 + k - 2}$
$= \dfrac{5k}{(k + 2)(k - 2)} - \dfrac{2}{(k + 2)(k - 1)}$
$= \dfrac{5k(k - 1) - 2(k - 2)}{(k + 2)(k - 2)(k - 1)}$
$= \dfrac{5k^2 - 5k - 2k + 4}{(k + 2)(k - 2)(k - 1)}$
$= \dfrac{5k^2 - 7k + 4}{(k + 2)(k - 2)(k - 1)}$

48. $\dfrac{5a}{a^2 - 4} - \dfrac{3}{2 - a} = \dfrac{5a}{(a + 2)(a - 2)} + \dfrac{3}{a - 2}$
$= \dfrac{5a + 3(a + 2)}{(a + 2)(a - 2)}$
$= \dfrac{5a + 3a + 6}{(a + 2)(a - 2)}$
$= \dfrac{8a + 6}{(a + 2)(a - 2)}$

49. $\dfrac{3}{x} - \dfrac{x + 21}{3x} = \dfrac{5}{3}$
$3x\left(\dfrac{3}{x} - \dfrac{x + 21}{3x}\right) = 3x\left(\dfrac{5}{3}\right)$
$9 - (x + 21) = 5x$
$9 - x - 21 = 5x$
$-x - 12 = 5x$
$-12 = 6x$
$-2 = x$
The solution is –2.

50. $\dfrac{3x - 4}{2x} = -\dfrac{8}{x}$
$x(3x - 4) = -8(2x)$
$3x^2 - 4x = -16x$
$3x^2 + 12x = 0$
$3x(x + 4) = 0$
$3x = 0$ or $x + 4 = 0$
$x = 0$ or $x = -4$
Discard the answer 0 as extraneous. The solution is –4.

Chapter 7

1. a. $\sqrt{49} = 7$ because $7^2 = 49$ and 7 is not negative.

b. $\sqrt{\dfrac{0}{1}} = \sqrt{0} = 0$ because $0^2 = 0$ and 0 is not negative.

c. $\sqrt{\dfrac{16}{81}} = \dfrac{4}{9}$ because $\left(\dfrac{4}{9}\right)^2 = \dfrac{16}{81}$ and $\dfrac{4}{9}$ is not negative.

d. $\sqrt{0.64} = 0.8$ because $(0.8)^2 = 0.64$.

e. $\sqrt{z^8} = z^4$ because $(z^4)^2 = z^8$.

f. $\sqrt{16b^4} = 4b^2$ because $(4b^2)^2 = 16b^4$.

g. $-\sqrt{36} = -6$. The negative in front of the radical indicates the negative square root of 36.

h. $\sqrt{-36}$ is not a real number.

2. $\sqrt{45} \approx 6.708$

Since $36 < 45 < 49$, then $\sqrt{36} < \sqrt{45} < \sqrt{49}$, or $6 < \sqrt{45} < 7$. The approximation is between 6 and 7 and thus is reasonable.

3. a. $\sqrt[3]{-1} = -1$ because $(-1)^3 = -1$.

b. $\sqrt[3]{27} = 3$ because $3^3 = 27$.

c. $\sqrt[3]{\dfrac{27}{64}} = \dfrac{3}{4}$ because $\left(\dfrac{3}{4}\right)^3 = \dfrac{27}{64}$.

d. $\sqrt[3]{x^{12}} = x^4$ because $(x^4)^3 = x^{12}$.

e. $\sqrt[3]{-8x^3} = -2x$ because $(-2x)^3 = -8x^3$.

4. a. $\sqrt[4]{10,000} = 10$ because $10^4 = 10,000$ and 10 is positive.

b. $\sqrt[5]{-1} = -1$ because $(-1)^5 = -1$.

c. $-\sqrt{81} = -9$ because -9 is the opposite of $\sqrt{81}$.

d. $\sqrt[4]{-625}$ is not a real number. There is no real number that, when raised to the fourth power, is -625.

e. $\sqrt[3]{27x^9} = 3x^3$ because $(3x^3)^3 = 27x^9$.

5. a. $\sqrt{(-4)^2} = |-4| = 4$

b. $\sqrt{x^{14}} = \left|x^7\right|$

c. $\sqrt[4]{(x+7)^4} = |x+7|$

d. $\sqrt[3]{(-7)^3} = -7$

e. $\sqrt[5]{(3x-5)^5} = 3x-5$

f. $\sqrt{49x^2} = 7|x|$

g. $\sqrt{x^2 + 16x + 64} = \sqrt{(x+8)^2} = |x+8|$

6. $f(x) = \sqrt{x+5}$, $g(x) = \sqrt[3]{x-3}$

a. $f(11) = \sqrt{11+5} = \sqrt{16} = 4$

b. $f(-1) = \sqrt{-1+5} = \sqrt{4} = 2$

c. $g(11) = \sqrt[3]{11-3} = \sqrt[3]{8} = 2$

d. $g(-6) = \sqrt[3]{-6-3} = \sqrt[3]{-9}$

7. $h(x) = \sqrt{x+2}$

Find the domain.

$x + 2 \geq 0$

$\quad x \geq -2$

The domain of $h(x)$ is $\{x \mid x \geq -2\}$.

x	$h(x) = \sqrt{x+2}$
-2	0
-1	1
1	$\sqrt{1+2} = \sqrt{3} \approx 1.7$
2	2
7	3

8. $f(x) = \sqrt[3]{x} - 4$

The domain is the set of all real numbers.

x	$f(x) = \sqrt[3]{x} - 4$
0	-4
1	-3
-1	-5
6	$\sqrt[3]{6} - 4 \approx 1.8 - 4 = -2.2$
-6	$\sqrt[3]{-6} - 4 \approx -1.8 - 4 = -5.8$
8	-2
-8	-6

Vocabulary, Readiness & Video Check 7.1

1. In the expression $\sqrt[n]{a}$, the n is called the <u>index</u>, the $\sqrt{}$ is called the <u>radical sign</u>, and a is called the <u>radicand</u>.

2. If $\sqrt{a}$ is the positive square root of a, $a \neq 0$, then $-\sqrt{a}$ is the negative square root of a.

3. The square root of a negative number <u>is not</u> a real number.

4. Numbers such as 1, 4, 9, and 25 are called perfect <u>squares</u>, whereas numbers such as 1, 8, 27, and 125 are called perfect <u>cubes</u>.

5. The domain of the function $f(x) = \sqrt{x}$ is <u>$[0, \infty)$</u>.

6. The domain of the function $f(x) = \sqrt[3]{x}$ is <u>$(-\infty, \infty)$</u>.

7. If $f(16) = 4$, the corresponding ordered pair is <u>$(16, 4)$</u>.

8. If $g(-8) = -2$, the corresponding ordered pair is <u>$(-8, -2)$</u>.

9. Divide the index into each exponent in the radicand.

10. Find the nearest perfect squares both less than and greater than the radicand. The square root of the radicand falls between the square roots of these two perfect squares.

11. The square root of a negative number is not a real number, but the cube root of a negative number is a real number.

12. The even root of a negative number is not a real number.

13. For odd roots, there's only one root/answer whether the radicand is positive or negative, so absolute value bars aren't needed.

14. Since the variable x is the radicand of a square root and a square root cannot be negative, then we know that x cannot be negative.

Exercise Set 7.1

1. $\sqrt{100} = 10$ because $10^2 = 100$.

3. $\sqrt{\dfrac{1}{4}} = \dfrac{1}{2}$ because $\left(\dfrac{1}{2}\right)^2 = \dfrac{1}{4}$.

5. $\sqrt{0.0001} = 0.01$ because $(0.01)^2 = 0.0001$.

7. $-\sqrt{36} = -1 \cdot \sqrt{36} = -1 \cdot 6 = -6$

9. $\sqrt{x^{10}} = x^5$ because $(x^5)^2 = x^{10}$.

11. $\sqrt{16y^6} = 4y^3$ because $(4y^3)^2 = 16y^6$.

13. $\sqrt{7} \approx 2.646$
Since $4 < 7 < 9$, then $\sqrt{4} < \sqrt{7} < \sqrt{9}$, or
$2 < \sqrt{7} < 3$. The approximation is between 2 and 3 and thus is reasonable.

15. $\sqrt{38} \approx 6.164$
Since $36 < 38 < 49$, then $\sqrt{36} < \sqrt{38} < \sqrt{49}$, or
$6 < \sqrt{38} < 7$. The approximation is between 6 and 7 and thus is reasonable.

17. $\sqrt{200} \approx 14.142$
Since $196 < 200 < 225$, then
$\sqrt{196} < \sqrt{200} < \sqrt{225}$, or $14 < \sqrt{200} < 15$. The approximation is between 14 and 15 and thus is reasonable.

19. $\sqrt[3]{64} = 4$ because $4^3 = 64$.

21. $\sqrt[3]{\dfrac{1}{8}} = \dfrac{1}{2}$ because $\left(\dfrac{1}{2}\right)^3 = \dfrac{1}{8}$.

23. $\sqrt[3]{-1} = -1$ because $(-1)^3 = -1$.

25. $\sqrt[3]{x^{12}} = x^4$ because $(x^4)^3 = x^{12}$.

27. $\sqrt[3]{-27x^9} = -3x^3$ because $(-3x^3)^3 = -27x^9$.

29. $-\sqrt[4]{16} = -2$ because $2^4 = 16$.

31. $\sqrt[4]{-16}$ is not a real number. There is no real number that, when raised to the fourth power, is -16.

33. $\sqrt[5]{-32} = -2$ because $(-2)^5 = -32$.

35. $\sqrt[5]{x^{20}} = x^4$ because $(x^4)^5 = x^{20}$.

37. $\sqrt[6]{64x^{12}} = 2x^2$ because $(2x^2)^6 = 64x^{12}$.

39. $\sqrt{81x^4} = 9x^2$ because $(9x^2)^2 = 81x^4$.

41. $\sqrt[4]{256x^8} = 4x^2$ because $(4x^2)^4 = 256x^8$.

43. $\sqrt{(-8)^2} = |-8| = 8$

45. $\sqrt[3]{(-8)^3} = -8$

47. $\sqrt{4x^2} = |2x| = 2|x|$

49. $\sqrt[3]{x^3} = x$

51. $\sqrt{(x-5)^2} = |x-5|$

53. $\sqrt{x^2 + 4x + 4} = \sqrt{(x+2)^2} = |x+2|$

55. $-\sqrt{121} = -11$

57. $\sqrt[3]{8x^3} = 2x$

59. $\sqrt{y^{12}} = y^6$

61. $\sqrt{25a^2b^{20}} = 5ab^{10}$

63. $\sqrt[3]{-27x^{12}y^9} = -3x^4y^3$

65. $\sqrt[4]{a^{16}b^4} = a^4b$

67. $\sqrt[5]{-32x^{10}y^5} = -2x^2y$

69. $\sqrt{\dfrac{25}{49}} = \dfrac{5}{7}$

71. $\sqrt{\dfrac{x^{20}}{4y^2}} = \dfrac{x^{10}}{2y}$

73. $-\sqrt[3]{\dfrac{z^{21}}{27x^3}} = -\dfrac{z^7}{3x}$

75. $\sqrt[4]{\dfrac{x^4}{16}} = \dfrac{x}{2}$

77. $f(x) = \sqrt{2x+3}$
$f(0) = \sqrt{2(0)+3} = \sqrt{3}$

79. $g(x) = \sqrt[3]{x-8}$
$g(7) = \sqrt[3]{7-8} = \sqrt[3]{-1} = -1$

81. $g(x) = \sqrt[3]{x-8}$
$g(-19) = \sqrt[3]{-19-8} = \sqrt[3]{-27} = -3$

83. $f(x) = \sqrt{2x+3}$
$f(2) = \sqrt{2(2)+3} = \sqrt{7}$

85. $f(x) = \sqrt{x}+2$
This is the graph of $\sqrt{x}$ moved up 2 units; C.
$x \geq 0$
Domain: $[0, \infty)$

87. $f(x) = \sqrt{x-3}$
This is the graph of $\sqrt{x}$ shifted right 3 units; D.
$x-3 \geq 0$
$x \geq 3$
Domain: $[3, \infty)$

89. $f(x) = \sqrt[3]{x}+1$
This is the graph of $\sqrt[3]{x}$ shifted up 1 unit; A.
Domain: $(-\infty, \infty)$

91. $g(x) = \sqrt[3]{x-1}$
This is the graph of $\sqrt[3]{x}$ shifted right 1 unit; B.
Domain: $(-\infty, \infty)$

93. $(-2x^3y^2)^5 = (-2)^5 x^{3 \cdot 5} y^{2 \cdot 5} = -32x^{15}y^{10}$

95. $(-3x^2y^3z^5)(20x^5y^7) = -3(20)x^{2+5}y^{3+7}z^5$
$\qquad\qquad\qquad\qquad = -60x^7y^{10}z^5$

97. $\dfrac{7x^{-1}y}{14(x^5y^2)^{-2}} = \dfrac{7x^{-1}y}{14x^{-10}y^{-4}} = \dfrac{x^9y^5}{2}$

99. $\sqrt{-17}$ is not a real number.

101. $\sqrt[10]{-17}$ is not a real number.

103. The radical that is not a real number is $\sqrt{-10}$, choice **d**.

105. The radical that simplifies to -3 is $\sqrt[3]{-27}$, choice **d**.

107. $144 < 160 < 169$ so $\sqrt{144} < \sqrt{160} < \sqrt{169}$, or $12 < \sqrt{160} < 13$. Thus $\sqrt{160}$ is between 12 and 13. Therefore, the answer is **b**.

109. $\sqrt{30} \approx 5$, $\sqrt{10} \approx 3$, and $\sqrt{90} \approx 10$ so
$P = \sqrt{30} + \sqrt{10} + \sqrt{90} \approx 5 + 3 + 10 = 18$.
Therefore, the answer is **b**.

111. answers may vary

113. $B = \sqrt{\dfrac{hw}{3131}} = \sqrt{\dfrac{66 \cdot 135}{3131}}$
$\qquad = \sqrt{\dfrac{8910}{3131}}$
$\qquad \approx 1.69$ sq meters

115. $v = \sqrt{\dfrac{2Gm}{r}} = \sqrt{\dfrac{2(6.67 \times 10^{-11})(5.97 \times 10^{24})}{6.37 \times 10^6}}$
$\qquad = \sqrt{\dfrac{2(3.98199 \times 10^{14})}{6.37 \times 10^6}}$
$\qquad = \sqrt{\dfrac{7.96398 \times 10^{14}}{6.37 \times 10^6}}$
$\qquad \approx \sqrt{125,023,233.9}$
$\qquad \approx 11,181$
The escape velocity is 11,181 meters per second.

117. answers may vary

119. $f(x) = \sqrt{x}+2$

Domain: $[0, \infty)$

121. $f(x) = \sqrt[3]{x} + 1$

Domain: $(-\infty, \infty)$

Section 7.2 Practice Exercises

1. a. $36^{1/2} = \sqrt{36} = 6$

b. $1000^{1/3} = \sqrt[3]{1000} = 10$

c. $x^{1/3} = \sqrt[3]{x}$

d. $1^{1/4} = \sqrt[4]{1} = 1$

e. $-64^{1/2} = -\sqrt{64} = -8$

f. $(125x^9)^{1/3} = \sqrt[3]{125x^9} = 5x^3$

g. $3x^{1/4} = 3\sqrt[4]{x}$

2. a. $16^{3/2} = \left(\sqrt{16}\right)^3 = 4^3 = 64$

b. $-1^{3/5} = -\left(\sqrt[5]{1}\right)^3 = -(1)^3 = -1$

c. $-(81)^{3/4} = -\left(\sqrt[4]{81}\right)^3 = -(3)^3 = -27$

d. $\left(\dfrac{1}{25}\right)^{3/2} = \left(\sqrt{\dfrac{1}{25}}\right)^3 = \left(\dfrac{1}{5}\right)^3 = \dfrac{1}{125}$

e. $(3x+2)^{5/9} = \sqrt[9]{(3x+2)^5}$

3. a. $9^{-3/2} = \dfrac{1}{9^{3/2}} = \dfrac{1}{\left(\sqrt{9}\right)^3} = \dfrac{1}{3^3} = \dfrac{1}{27}$

b. $(-64)^{-2/3} = \dfrac{1}{(-64)^{2/3}} = \dfrac{1}{\left(\sqrt[3]{-64}\right)^2} = \dfrac{1}{(-4)^2} = \dfrac{1}{16}$

4. a. $y^{2/3} \cdot y^{8/3} = y^{(2/3+8/3)} = y^{10/3}$

b. $x^{3/5} \cdot x^{1/4} = x^{3/5+1/4} = x^{12/20+5/20} = x^{17/20}$

c. $\dfrac{9^{2/7}}{9^{9/7}} = 9^{2/7-9/7} = 9^{-7/7} = 9^{-1} = \dfrac{1}{9}$

d. $b^{4/9} \cdot b^{-2/9} = b^{4/9+(-2/9)} = b^{2/9}$

e. $\dfrac{\left(3x^{1/4}y^{-2/3}\right)^4}{x^4 y} = \dfrac{3^4 (x^{1/4})^4 (y^{-2/3})^4}{x^4 y}$

$= \dfrac{81xy^{-8/3}}{x^4 y}$

$= 81x^{1-4}y^{-8/3-3/3}$

$= 81x^{-3}y^{-11/3}$

$= \dfrac{81}{x^3 y^{11/3}}$

5. a. $x^{3/5}(x^{1/3} - x^2) = x^{3/5}x^{1/3} - x^{3/5}x^2$

$= x^{(3/5+1/3)} - x^{(3/5+2)}$

$= x^{(9/15+5/15)} - x^{(3/5+10/5)}$

$= x^{14/15} - x^{13/5}$

b. $(x^{1/2} + 6)(x^{1/2} - 2)$

$= x^{2/2} - 2x^{1/2} + 6x^{1/2} - 12$

$= x + 4x^{1/2} - 12$

6. $2x^{-1/5} - 7x^{4/5} = (x^{-1/5})(2) - (x^{-1/5})(7x^{5/5})$

$= x^{-1/5}(2 - 7x)$

7. a. $\sqrt[9]{x^3} = x^{3/9} = x^{1/3} = \sqrt[3]{x}$

b. $\sqrt[4]{36} = 36^{1/4} = (6^2)^{1/4} = 6^{2/4} = 6^{1/2} = \sqrt{6}$

c. $\sqrt[8]{a^4 b^2} = (a^4 b^2)^{1/8}$

$= a^{4/8}b^{2/8}$

$= a^{2/4}b^{1/4}$

$= (a^2 b)^{1/4}$

$= \sqrt[4]{a^2 b}$

8. a. $\sqrt[3]{x} \cdot \sqrt[4]{x} = x^{1/3} \cdot x^{1/4}$
$= x^{1/3+1/4}$
$= x^{4/12+3/12}$
$= x^{7/12}$
$= \sqrt[12]{x^7}$

b. $\dfrac{\sqrt[3]{y}}{\sqrt[5]{y}} = \dfrac{y^{1/3}}{y^{1/5}}$
$= y^{1/3-1/5}$
$= y^{5/15-3/15}$
$= y^{2/15}$
$= \sqrt[15]{y^2}$

c. $\sqrt[3]{5} \cdot \sqrt{3} = 5^{1/3} \cdot 3^{1/2}$
$= 5^{2/6} \cdot 3^{3/6}$
$= (5^2 \cdot 3^3)^{1/6}$
$= \sqrt[6]{5^2 \cdot 3^3}$
$= \sqrt[6]{675}$

Vocabulary, Readiness & Video Check 7.2

1. It is true that $9^{-1/2}$ is a positive number.

2. It is false that $9^{-1/2}$ is a whole number.

3. It is true that $\dfrac{1}{a^{-m/n}} = a^{m/n}$ (where $a^{m/n}$ is a nonzero real number).

4. To simplify $x^{2/3} \cdot x^{1/5}$, <u>add</u> the exponents; a.

5. To simplify $(x^{2/3})^{1/5}$, <u>multiply</u> the exponents; c.

6. To simplify $\dfrac{x^{2/3}}{x^{1/5}}$, <u>subtract</u> the exponents; b.

7. $-\sqrt[5]{3x}$

8. The numerator is the power; the denominator is the index.

9. A negative fractional exponent will move a base from the numerator to the <u>denominator</u> with the fractional exponent becoming <u>positive</u>.

10. If applying the product rule of exponents, you <u>add</u> the exponents. If applying the quotient rule of exponents, you <u>subtract</u> the exponents. If applying the power rule of exponents, you <u>multiply</u> the exponents.

11. Write the radical using an equivalent fractional exponent form, simplify the fraction, then write as a radical again.

Exercise Set 7.2

1. $49^{1/2} = \sqrt{49} = 7$

3. $27^{1/3} = \sqrt[3]{27} = 3$

5. $\left(\dfrac{1}{16}\right)^{1/4} = \sqrt[4]{\dfrac{1}{16}} = \dfrac{1}{2}$

7. $169^{1/2} = \sqrt{169} = 13$

9. $2m^{1/3} = 2\sqrt[3]{m}$

11. $(9x^4)^{1/2} = \sqrt{9x^4} = 3x^2$

13. $(-27)^{1/3} = \sqrt[3]{-27} = -3$

15. $-16^{1/4} = -\sqrt[4]{16} = -2$

17. $16^{3/4} = \left(\sqrt[4]{16}\right)^3 = 2^3 = 8$

19. $(-64)^{2/3} = \left(\sqrt[3]{-64}\right)^2 = (-4)^2 = 16$

21. $(-16)^{3/4} = \left(\sqrt[4]{-16}\right)^3$ is not a real number.

23. $(2x)^{3/5} = \sqrt[5]{(2x)^3}$ or $\left(\sqrt[5]{2x}\right)^3$

25. $(7x+2)^{2/3} = \sqrt[3]{(7x+2)^2}$ or $\left(\sqrt[3]{7x+2}\right)^2$

27. $\left(\dfrac{16}{9}\right)^{3/2} = \left(\sqrt{\dfrac{16}{9}}\right)^3 = \left(\dfrac{4}{3}\right)^3 = \dfrac{64}{27}$

29. $8^{-4/3} = \dfrac{1}{8^{4/3}} = \dfrac{1}{\left(\sqrt[3]{8}\right)^4} = \dfrac{1}{2^4} = \dfrac{1}{16}$

31. $(-64)^{-2/3} = \dfrac{1}{(-64)^{2/3}} = \dfrac{1}{\left(\sqrt[3]{-64}\right)^2} = \dfrac{1}{(-4)^2} = \dfrac{1}{16}$

33. $(-4)^{-3/2} = \dfrac{1}{(-4)^{3/2}} = \dfrac{1}{\left(\sqrt{-4}\right)^3}$ is not a real

number.

35. $x^{-1/4} = \dfrac{1}{x^{1/4}}$

37. $\dfrac{1}{a^{-2/3}} = a^{2/3}$

39. $\dfrac{5}{7x^{-3/4}} = \dfrac{5x^{3/4}}{7}$

41. $a^{2/3}a^{5/3} = a^{2/3+5/3} = a^{7/3}$

43. $x^{-2/5} \cdot x^{7/5} = x^{-\frac{2}{5}+\frac{7}{5}} = x^{5/5} = x$

45. $3^{1/4} \cdot 3^{3/8} = 3^{\frac{1}{4}+\frac{3}{8}} = 3^{\frac{2}{8}+\frac{3}{8}} = 3^{5/8}$

47. $\dfrac{y^{1/3}}{y^{1/6}} = y^{\frac{1}{3}-\frac{1}{6}} = y^{\frac{2}{6}-\frac{1}{6}} = y^{1/6}$

49. $(4u^2)^{3/2} = 4^{3/2}u^{2(3/2)}$

$\qquad = \left(\sqrt{4}\right)^3 u^3$

$\qquad = 2^3 u^3$

$\qquad = 8u^3$

51. $\dfrac{b^{1/2}b^{3/4}}{-b^{1/4}} = -b^{\frac{1}{2}+\frac{3}{4}-\frac{1}{4}} = -b^{\frac{2}{4}+\frac{3}{4}-\frac{1}{4}} = -b^1 = -b$

53. $\dfrac{(x^3)^{1/2}}{x^{7/2}} = \dfrac{x^{3/2}}{x^{7/2}}$

$\qquad = x^{3/2-7/2}$

$\qquad = x^{-2}$

$\qquad = \dfrac{1}{x^2}$

55. $\dfrac{(3x^{1/4})^3}{x^{1/12}} = \dfrac{3^3 x^{3/4}}{x^{1/12}}$

$\qquad = 27x^{\frac{3}{4}-\frac{1}{12}}$

$\qquad = 27x^{\frac{9}{12}-\frac{1}{12}}$

$\qquad = 27x^{8/12}$

$\qquad = 27x^{2/3}$

57. $\dfrac{(y^3z)^{1/6}}{y^{-1/2}z^{1/3}} = \dfrac{y^{3/6}z^{1/6}}{y^{-1/2}z^{1/3}}$

$\qquad = y^{3/6-(-1/2)}z^{1/6-1/3}$

$\qquad = y^{1/2+1/2}z^{1/6-2/6}$

$\qquad = y^1 z^{-1/6}$

$\qquad = \dfrac{y}{z^{1/6}}$

59. $\dfrac{(x^3y^2)^{1/4}}{(x^{-5}y^{-1})^{-1/2}} = \dfrac{x^{3/4}y^{2/4}}{x^{5/2}y^{1/2}}$

$\qquad = x^{\frac{3}{4}-\frac{5}{2}}y^{\frac{2}{4}-\frac{1}{2}}$

$\qquad = x^{\frac{3}{4}-\frac{10}{4}}y^{\frac{1}{2}-\frac{1}{2}}$

$\qquad = x^{-7/4}$

$\qquad = \dfrac{1}{x^{7/4}}$

61. $y^{1/2}(y^{1/2}-y^{2/3}) = y^{1/2}y^{1/2} - y^{1/2}y^{2/3}$

$\qquad = y^{1/2+1/2} - y^{1/2+2/3}$

$\qquad = y^1 - y^{7/6}$

$\qquad = y - y^{7/6}$

63. $x^{2/3}(x-2) = x \cdot x^{2/3} - 2x^{2/3}$

$\qquad = x^{1+2/3} - 2x^{2/3}$

$\qquad = x^{5/3} - 2x^{2/3}$

65. $(2x^{1/3}+3)(2x^{1/3}-3) = (2x^{1/3})^2 - 3^2$

$\qquad = 2^2(x^{1/3})^2 - 9$

$\qquad = 4x^{2/3} - 9$

67. $x^{8/3} + x^{10/3} = x^{8/3}(1) + x^{8/3}(x^{2/3})$

$\qquad = x^{8/3}(1 + x^{2/3})$

69. $x^{2/5} - 3x^{1/5} = x^{1/5}(x^{1/5}) - x^{1/5}(3)$

$\qquad = x^{1/5}(x^{1/5} - 3)$

71. $5x^{-1/3} + x^{2/3} = x^{-1/3}(5) + x^{-1/3}(x^{3/3})$
$$= x^{-1/3}(5 + x)$$

73. $\sqrt[6]{x^3} = x^{3/6} = x^{1/2} = \sqrt{x}$

75. $\sqrt[6]{4} = 4^{1/6} = (2^2)^{1/6} = 2^{1/3} = \sqrt[3]{2}$

77. $\sqrt[4]{16x^2} = (16x^2)^{1/4}$
$$= 16^{1/4}x^{2/4} = 2x^{1/2} = 2\sqrt{x}$$

79. $\sqrt[8]{x^4 y^4} = (x^4 y^4)^{1/8}$
$$= x^{4/8} y^{4/8}$$
$$= x^{1/2} y^{1/2}$$
$$= (xy)^{1/2}$$
$$= \sqrt{xy}$$

81. $\sqrt[12]{a^8 b^4} = a^{8/12} b^{4/12}$
$$= a^{2/3} b^{1/3}$$
$$= (a^2 b)^{1/3}$$
$$= \sqrt[3]{a^2 b}$$

83. $\sqrt[4]{(x+3)^2} = (x+3)^{2/4} = (x+3)^{1/2} = \sqrt{x+3}$

85. $\sqrt[3]{y} \cdot \sqrt[5]{y^2} = y^{1/3} \cdot y^{2/5}$
$$= y^{\frac{1}{3} + \frac{2}{5}}$$
$$= y^{\frac{5}{15} + \frac{6}{15}}$$
$$= y^{11/15}$$
$$= \sqrt[15]{y^{11}}$$

87. $\dfrac{\sqrt[3]{b^2}}{\sqrt[4]{b}} = \dfrac{b^{2/3}}{b^{1/4}} = b^{\frac{2}{3} - \frac{1}{4}} = b^{\frac{8}{12} - \frac{3}{12}} = b^{5/12} = \sqrt[12]{b^5}$

89. $\sqrt[3]{x} \cdot \sqrt[4]{x} \cdot \sqrt[8]{x^3} = x^{1/3} \cdot x^{1/4} \cdot x^{3/8}$
$$= x^{8/24} \cdot x^{6/24} \cdot x^{9/24}$$
$$= x^{23/24}$$
$$= \sqrt[24]{x^{23}}$$

91. $\dfrac{\sqrt[3]{a^2}}{\sqrt[6]{a}} = \dfrac{a^{2/3}}{a^{1/6}}$
$$= a^{\frac{2}{3} - \frac{1}{6}} = a^{\frac{4}{6} - \frac{1}{6}} = a^{3/6} = a^{1/2} = \sqrt{a}$$

93. $\sqrt{3} \cdot \sqrt[3]{4} = 3^{1/2} \cdot 4^{1/3}$
$$= 3^{3/6} \cdot 4^{2/6}$$
$$= (3^3 \cdot 4^2)^{1/6}$$
$$= (27 \cdot 16)^{1/6}$$
$$= (432)^{1/6}$$
$$= \sqrt[6]{432}$$

95. $\sqrt[5]{7} \cdot \sqrt[3]{y} = 7^{1/5} \cdot y^{1/3}$
$$= 7^{3/15} \cdot y^{5/15}$$
$$= (7^3 \cdot y^5)^{1/15}$$
$$= (343y^5)^{1/15}$$
$$= \sqrt[15]{343y^5}$$

97. $\sqrt{5r} \cdot \sqrt[3]{s} = (5r)^{1/2} \cdot s^{1/3}$
$$= (5r)^{3/6} \cdot s^{2/6}$$
$$= [(5r)^3 \cdot s^2]^{1/6}$$
$$= (125r^3 s^2)^{1/6}$$
$$= \sqrt[6]{125r^3 s^2}$$

99. $75 = 25 \cdot 3$ where 25 is a perfect square.

101. $48 = 4 \cdot 12$ or $16 \cdot 3$ where both 4 and 16 are perfect squares.

103. $16 = 8 \cdot 2$ where 8 is a perfect cube.

105. $54 = 27 \cdot 2$ where 27 is a perfect cube.

107. $4^{1/2} = 2$, A

109. $(-4)^{1/2}$ is not a real number, C

111. $-8^{1/3} = -2$, B

113. $B(w) = 70w^{3/4}$
$$B(60) = 70(60)^{3/4}$$
$$\approx 1509$$
The BMR is 1509 calories.

115. 2010 is 15 years after 1995.
$$f(x) = 25x^{23/25}$$
$$f(15) = 25(15)^{23/25}$$
$$\approx 302.0$$
The model predicts 302.0 million subscriptions in 2010.

117. answers may vary

119. $\square \cdot a^{2/3} = a^{3/3}$

$$\square = \frac{a^{3/3}}{a^{2/3}}$$

$$\square = a^{3/3 - 2/3}$$

$$\square = a^{1/3}$$

121.
$$\frac{\square}{x^{-2/5}} = x^{3/5}$$

$$x^{-2/5}\left(\frac{\square}{x^{-2/5}}\right) = x^{3/5} \cdot x^{-2/5}$$

$$\square = x^{3/5 - 2/5}$$

$$\square = x^{1/5}$$

123. $8^{1/4} \approx 1.6818$

125. $18^{3/5} \approx 5.6645$

127. $\dfrac{\sqrt{t}}{\sqrt{u}} = \dfrac{t^{1/2}}{u^{1/2}}$

Section 7.3 Practice Exercises

1. a. $\sqrt{5} \cdot \sqrt{7} = \sqrt{5 \cdot 7} = \sqrt{35}$

b. $\sqrt{13} \cdot \sqrt{z} = \sqrt{13z}$

c. $\sqrt[4]{125} \cdot \sqrt[4]{5} = \sqrt[4]{125 \cdot 5} = \sqrt[4]{625} = 5$

d. $\sqrt[3]{5y} \cdot \sqrt[3]{3x^2} = \sqrt[3]{5y \cdot 3x^2} = \sqrt[3]{15x^2 y}$

e. $\sqrt{\dfrac{5}{m}} \cdot \sqrt{\dfrac{t}{2}} = \sqrt{\dfrac{5}{m} \cdot \dfrac{t}{2}} = \sqrt{\dfrac{5t}{2m}}$

2. a. $\sqrt{\dfrac{36}{49}} = \dfrac{\sqrt{36}}{\sqrt{49}} = \dfrac{6}{7}$

b. $\sqrt{\dfrac{z}{16}} = \dfrac{\sqrt{z}}{\sqrt{16}} = \dfrac{\sqrt{z}}{4}$

c. $\sqrt[3]{\dfrac{125}{8}} = \dfrac{\sqrt[3]{125}}{\sqrt[3]{8}} = \dfrac{5}{2}$

d. $\sqrt[4]{\dfrac{5}{81x^8}} = \dfrac{\sqrt[4]{5}}{\sqrt[4]{81x^8}} = \dfrac{\sqrt[4]{5}}{3x^2}$

3. a. $\sqrt{98} = \sqrt{49 \cdot 2} = \sqrt{49} \cdot \sqrt{2} = 7\sqrt{2}$

b. $\sqrt[3]{54} = \sqrt[3]{27 \cdot 2} = \sqrt[3]{27} \cdot \sqrt[3]{2} = 3\sqrt[3]{2}$

c. The largest perfect square factor of 35 is 1, so $\sqrt{35}$ cannot be simplified further.

d. $\sqrt[4]{243} = \sqrt[4]{81 \cdot 3} = \sqrt[4]{81} \cdot \sqrt[4]{3} = 3\sqrt[4]{3}$

4. a. $\sqrt{36z^7} = \sqrt{36z^6 \cdot z} = \sqrt{36z^6} \cdot \sqrt{z} = 6z^3 \sqrt{z}$

b. $\sqrt[3]{32p^4 q^7} = \sqrt[3]{8 \cdot 4 \cdot p^3 \cdot p \cdot q^6 \cdot q}$
$$= \sqrt[3]{8p^3 q^6 \cdot 4pq}$$
$$= \sqrt[3]{8p^3 q^6} \cdot \sqrt[3]{4pq}$$
$$= 2pq^2 \sqrt[3]{4pq}$$

c. $\sqrt[4]{16x^{15}} = \sqrt[4]{16 \cdot x^{12} \cdot x^3}$
$$= \sqrt[4]{16x^{12}} \cdot \sqrt[4]{x^3}$$
$$= 2x^3 \sqrt[4]{x^3}$$

5. a. $\dfrac{\sqrt{80}}{\sqrt{5}} = \sqrt{\dfrac{80}{5}} = \sqrt{16} = 4$

b. $\dfrac{\sqrt{98z}}{3\sqrt{2}} = \dfrac{1}{3} \cdot \sqrt{\dfrac{98z}{2}}$
$$= \dfrac{1}{3} \cdot \sqrt{49z}$$
$$= \dfrac{1}{3} \cdot \sqrt{49} \cdot \sqrt{z}$$
$$= \dfrac{1}{3} \cdot 7 \cdot \sqrt{z}$$
$$= \dfrac{7}{3}\sqrt{z}$$

c. $\dfrac{5\sqrt[3]{40x^5 y^7}}{\sqrt[3]{5y}} = 5 \cdot \sqrt[3]{\dfrac{40x^5 y^7}{5y}}$
$$= 5 \cdot \sqrt[3]{8x^5 y^6}$$
$$= 5 \cdot \sqrt[3]{8x^3 y^6 \cdot x^2}$$
$$= 5 \cdot \sqrt[3]{8x^3 y^6} \cdot \sqrt[3]{x^2}$$
$$= 5 \cdot 2xy^2 \cdot \sqrt[3]{x^2}$$
$$= 10xy^2 \sqrt[3]{x^2}$$

d. $\dfrac{3\sqrt[5]{64x^9y^8}}{\sqrt[5]{x^{-1}y^2}} = 3 \cdot \sqrt[5]{\dfrac{64x^9y^8}{x^{-1}y^2}}$

$= 3 \cdot \sqrt[5]{64x^{10}y^6}$

$= 3 \cdot \sqrt[5]{32 \cdot x^{10} \cdot y^5 \cdot 2 \cdot y}$

$= 3 \cdot \sqrt[5]{32x^{10}y^5} \cdot \sqrt[5]{2y}$

$= 3 \cdot 2x^2 y \cdot \sqrt[5]{2y}$

$= 6x^2 y\sqrt[5]{2y}$

6. Let $(x_1, y_1) = (-3, 7)$ and $(x_2, y_2) = (-2, 3)$.

$d = \sqrt{(x_2 - x_1)^2 + (y_2 - y_1)^2}$

$= \sqrt{[-2-(-3)]^2 + (3-7)^2}$

$= \sqrt{(1)^2 + (-4)^2}$

$= \sqrt{1+16}$

$= \sqrt{17} \approx 4.123$

The distance between the two points is exactly $\sqrt{17}$ units, or approximately 4.123 units.

7. Let $(x_1, y_1) = (5, -2)$ and $(x_2, y_2) = (8, -6)$.

midpoint $= \left(\dfrac{x_1 + x_2}{2}, \dfrac{y_1 + y_2}{2}\right)$

$= \left(\dfrac{5+8}{2}, \dfrac{-2+(-6)}{2}\right)$

$= \left(\dfrac{13}{2}, \dfrac{-8}{2}\right)$

$= \left(\dfrac{13}{2}, -4\right)$

The midpoint of the segment is $\left(\dfrac{13}{2}, -4\right)$.

Vocabulary, Readiness & Video Check 7.3

1. The <u>midpoint</u> of a line segment is a <u>point</u> exactly halfway between the two endpoints of the line segment.

2. The <u>distance</u> between two points is a distance, measured in units.

3. The <u>distance</u> formula is $d = \sqrt{(x_2 - x_1)^2 + (y_2 - y_1)^2}$.

4. The <u>midpoint</u> formula is $\left(\dfrac{x_1 + x_2}{2}, \dfrac{y_1 + y_2}{2}\right)$.

5. the indexes must be the same

6. If you see that simplifying can be done by separating a fraction radical into separate numerator and denominator radicands or by combining separate numerator and denominator radicands under one radical.

7. The power must be 1. Any even power is a perfect square and will leave no factor in the radicand; any higher odd power can have an even power factored from it, leaving one factor remaining in the radicand.

8. Be careful of signs since you're dealing with subtraction.

9. The *x*-value of the midpoint is the <u>average</u> of the *x*-values of the endpoints and the *y*-value of the midpoint is the <u>average</u> of the *y*-values of the endpoints.

Exercise Set 7.3

1. $\sqrt{7} \cdot \sqrt{2} = \sqrt{7 \cdot 2} = \sqrt{14}$

3. $\sqrt[4]{8} \cdot \sqrt[4]{2} = \sqrt[4]{8 \cdot 2} = \sqrt[4]{16} = 2$

5. $\sqrt[3]{4} \cdot \sqrt[3]{9} = \sqrt[3]{4 \cdot 9} = \sqrt[3]{36}$

7. $\sqrt{2} \cdot \sqrt{3x} = \sqrt{2 \cdot 3x} = \sqrt{6x}$

9. $\sqrt{\dfrac{7}{x}} \cdot \sqrt{\dfrac{2}{y}} = \sqrt{\dfrac{7}{x} \cdot \dfrac{2}{y}} = \sqrt{\dfrac{14}{xy}}$

11. $\sqrt[4]{4x^3} \cdot \sqrt[4]{5} = \sqrt[4]{4x^3 \cdot 5} = \sqrt[4]{20x^3}$

13. $\sqrt{\dfrac{6}{49}} = \dfrac{\sqrt{6}}{\sqrt{49}} = \dfrac{\sqrt{6}}{7}$

15. $\sqrt{\dfrac{2}{49}} = \dfrac{\sqrt{2}}{\sqrt{49}} = \dfrac{\sqrt{2}}{7}$

17. $\sqrt[4]{\dfrac{x^3}{16}} = \dfrac{\sqrt[4]{x^3}}{\sqrt[4]{16}} = \dfrac{\sqrt[4]{x^3}}{2}$

19. $\sqrt[3]{\dfrac{4}{27}} = \dfrac{\sqrt[3]{4}}{\sqrt[3]{27}} = \dfrac{\sqrt[3]{4}}{3}$

21. $\sqrt[4]{\dfrac{8}{x^8}} = \dfrac{\sqrt[4]{8}}{\sqrt[4]{x^8}} = \dfrac{\sqrt[4]{8}}{x^2}$

23. $\sqrt[3]{\dfrac{2x}{81y^{12}}} = \dfrac{\sqrt[3]{2x}}{\sqrt[3]{81y^{12}}}$

$\qquad = \dfrac{\sqrt[3]{2x}}{\sqrt[3]{27y^{12}} \cdot \sqrt[3]{3}}$

$\qquad = \dfrac{\sqrt[3]{2x}}{3y^4\sqrt[3]{3}}$

25. $\sqrt{\dfrac{x^2y}{100}} = \dfrac{\sqrt{x^2y}}{\sqrt{100}} = \dfrac{\sqrt{x^2}\sqrt{y}}{10} = \dfrac{x\sqrt{y}}{10}$

27. $\sqrt{\dfrac{5x^2}{4y^2}} = \dfrac{\sqrt{5x^2}}{\sqrt{4y^2}} = \dfrac{\sqrt{5}\sqrt{x^2}}{2y} = \dfrac{x\sqrt{5}}{2y}$

29. $-\sqrt[3]{\dfrac{z^7}{27x^3}} = -\dfrac{\sqrt[3]{z^7}}{\sqrt[3]{27x^3}} = -\dfrac{\sqrt[3]{z^6 \cdot z}}{3x} = -\dfrac{z^2\sqrt[3]{z}}{3x}$

31. $\sqrt{32} = \sqrt{16 \cdot 2} = \sqrt{16} \cdot \sqrt{2} = 4\sqrt{2}$

33. $\sqrt[3]{192} = \sqrt[3]{64 \cdot 3} = \sqrt[3]{64} \cdot \sqrt[3]{3} = 4\sqrt[3]{3}$

35. $5\sqrt{75} = 5\sqrt{25 \cdot 3} = 5\sqrt{25} \cdot \sqrt{3} = 5(5)\sqrt{3} = 25\sqrt{3}$

37. $\sqrt{24} = \sqrt{4 \cdot 6} = \sqrt{4} \cdot \sqrt{6} = 2\sqrt{6}$

39. $\sqrt{100x^5} = \sqrt{100x^4 \cdot x} = \sqrt{100x^4} \cdot \sqrt{x} = 10x^2\sqrt{x}$

41. $\sqrt[3]{16y^7} = \sqrt[3]{8y^6 \cdot 2y} = \sqrt[3]{8y^6} \cdot \sqrt[3]{2y} = 2y^2\sqrt[3]{2y}$

43. $\sqrt[4]{a^8b^7} = \sqrt[4]{a^8b^4 \cdot b^3} = \sqrt[4]{a^8b^4} \cdot \sqrt[4]{b^3} = a^2b\sqrt[4]{b^3}$

45. $\sqrt{y^5} = \sqrt{y^4 \cdot y} = \sqrt{y^4} \cdot \sqrt{y} = y^2\sqrt{y}$

47. $\sqrt{25a^2b^3} = \sqrt{25a^2b^2 \cdot b}$

$\qquad = \sqrt{25a^2b^2} \cdot \sqrt{b}$

$\qquad = 5ab\sqrt{b}$

49. $\sqrt[5]{-32x^{10}y} = \sqrt[5]{-32x^{10} \cdot y}$

$\qquad = \sqrt[5]{-32x^{10}} \cdot \sqrt[5]{y}$

$\qquad = -2x^2\sqrt[5]{y}$

51. $\sqrt[3]{50x^{14}} = \sqrt[3]{x^{12} \cdot 50x^2}$

$\qquad = \sqrt[3]{x^{12}} \cdot \sqrt[3]{50x^2}$

$\qquad = x^4\sqrt[3]{50x^2}$

53. $-\sqrt{32a^8b^7} = -\sqrt{16a^8b^6 \cdot 2b}$

$\qquad = -\sqrt{16a^8b^6} \cdot \sqrt{2b}$

$\qquad = -4a^4b^3\sqrt{2b}$

55. $\sqrt{9x^7y^9} = \sqrt{9x^6y^8 \cdot xy}$

$\qquad = \sqrt{9x^6y^8} \cdot \sqrt{xy}$

$\qquad = 3x^3y^4\sqrt{xy}$

57. $\sqrt[3]{125r^9s^{12}} = 5r^3s^4$

59. $\sqrt[4]{32x^{12}y^5} = \sqrt[4]{16x^{12}y^4} \cdot \sqrt[4]{2y} = 2x^3y\sqrt[4]{2y}$

61. $\dfrac{\sqrt{14}}{\sqrt{7}} = \sqrt{\dfrac{14}{7}} = \sqrt{2}$

63. $\dfrac{\sqrt[3]{24}}{\sqrt[3]{3}} = \sqrt[3]{\dfrac{24}{3}} = \sqrt[3]{8} = 2$

65. $\dfrac{5\sqrt[4]{48}}{\sqrt[4]{3}} = 5\sqrt[4]{\dfrac{48}{3}} = 5\sqrt[4]{16} = 5(2) = 10$

67. $\dfrac{\sqrt{x^5y^3}}{\sqrt{xy}} = \sqrt{\dfrac{x^5y^3}{xy}} = \sqrt{x^4y^2} = x^2y$

69. $\dfrac{8\sqrt[3]{54m^7}}{\sqrt[3]{2m}} = 8\sqrt[3]{\dfrac{54m^7}{2m}}$

$\qquad = 8\sqrt[3]{27m^6}$

$\qquad = 8(3m^2)$

$\qquad = 24m^2$

71.
$$\frac{3\sqrt{100x^2}}{2\sqrt{2x^{-1}}} = \frac{3}{2}\sqrt{\frac{100x^2}{2x^{-1}}}$$
$$= \frac{3}{2}\sqrt{50x^3}$$
$$= \frac{3}{2}\sqrt{25x^2 \cdot 2x}$$
$$= \frac{3}{2}(5x)\sqrt{2x}$$
$$= \frac{15x}{2}\sqrt{2x}$$

73.
$$\frac{\sqrt[4]{96a^{10}b^3}}{\sqrt[4]{3a^2b^3}} = \sqrt[4]{\frac{96a^{10}b^3}{3a^2b^3}}$$
$$= \sqrt[4]{32a^8}$$
$$= \sqrt[4]{16a^8 \cdot 2}$$
$$= 2a^2\sqrt[4]{2}$$

75.
$$\frac{\sqrt[5]{64x^{10}y^3}}{\sqrt[5]{2x^3y^{-7}}} = \sqrt[5]{\frac{64x^{10}y^3}{2x^3y^{-7}}}$$
$$= \sqrt[5]{32x^7y^{10}}$$
$$= \sqrt[5]{32x^5y^{10}} \cdot \sqrt[5]{x^2}$$
$$= 2xy^2\sqrt[5]{x^2}$$

77. (5, 1), (8, 5)
$$d = \sqrt{(8-5)^2 + (5-1)^2}$$
$$= \sqrt{3^2 + 4^2}$$
$$= \sqrt{9+16}$$
$$= \sqrt{25}$$
$$= 5 \text{ units}$$

79. (−3, 2), (1, −3)
$$d = \sqrt{[1-(-3)]^2 + (-3-2)^2}$$
$$= \sqrt{4^2 + (-5)^2}$$
$$= \sqrt{16+25}$$
$$= \sqrt{41} \approx 6.403 \text{ units}$$

81. (−9, 4), (−8, 1)
$$d = \sqrt{[-8-(-9)]^2 + (1-4)^2}$$
$$= \sqrt{1^2 + (-3)^2}$$
$$= \sqrt{1+9}$$
$$= \sqrt{10} \approx 3.162 \text{ units}$$

83. $\left(0, -\sqrt{2}\right), \left(\sqrt{3}, 0\right)$
$$d = \sqrt{\left(\sqrt{3}-0\right)^2 + \left[0-\left(-\sqrt{2}\right)\right]^2}$$
$$= \sqrt{\left(\sqrt{3}\right)^2 + \left(\sqrt{2}\right)^2}$$
$$= \sqrt{3+2}$$
$$= \sqrt{5} \approx 2.236 \text{ units}$$

85. (1.7, −3.6), (−8.6, 5.7)
$$d = \sqrt{(-8.6-1.7)^2 + [5.7-(-3.6)]^2}$$
$$= \sqrt{(-10.3)^2 + (9.3)^2}$$
$$= \sqrt{192.58} \approx 13.877 \text{ units}$$

87. (6, −8), (2, 4)
$$\left(\frac{6+2}{2}, \frac{-8+4}{2}\right) = \left(\frac{8}{2}, \frac{-4}{2}\right) = (4, -2)$$
The midpoint of the segment is (4, −2).

89. (−2, −1), (−8, 6)
$$\left(\frac{-2+(-8)}{2}, \frac{-1+6}{2}\right) = \left(\frac{-10}{2}, \frac{5}{2}\right) = \left(-5, \frac{5}{2}\right)$$
The midpoint of the segment is $\left(-5, \frac{5}{2}\right)$.

91. (7, 3), (−1, −3)
$$\left(\frac{7+(-1)}{2}, \frac{3+(-3)}{2}\right) = \left(\frac{6}{2}, \frac{0}{2}\right) = (3, 0)$$
The midpoint of the segment is (3, 0).

93. $\left(\frac{1}{2}, \frac{3}{8}\right), \left(-\frac{3}{2}, \frac{5}{8}\right)$
$$\left(\frac{\frac{1}{2}+\left(-\frac{3}{2}\right)}{2}, \frac{\frac{3}{8}+\frac{5}{8}}{2}\right) = \left(\frac{-1}{2}, \frac{1}{2}\right)$$
The midpoint of the segment is $\left(-\frac{1}{2}, \frac{1}{2}\right)$.

95. $\left(\sqrt{2}, 3\sqrt{5}\right), \left(\sqrt{2}, -2\sqrt{5}\right)$
$$\left(\frac{\sqrt{2}+\sqrt{2}}{2}, \frac{3\sqrt{5}+\left(-2\sqrt{5}\right)}{2}\right) = \left(\frac{2\sqrt{2}}{2}, \frac{\sqrt{5}}{2}\right)$$
$$= \left(\sqrt{2}, \frac{\sqrt{5}}{2}\right)$$
The midpoint of the segment is $\left(\sqrt{2}, \frac{\sqrt{5}}{2}\right)$.

97. $(4.6, -3.5), (7.8, -9.8)$
$$\left(\frac{4.6+7.8}{2}, \frac{-3.5+(-9.8)}{2}\right) = \left(\frac{12.4}{2}, \frac{-13.3}{2}\right)$$
$$= (6.2, -6.65)$$
The midpoint of the segment is $(6.2, -6.65)$.

99. $6x + 8x = (6+8)x = 14x$

101. $(2x+3)(x-5) = 2x^2 - 10x + 3x - 15$
$$= 2x^2 - 7x - 15$$

103. $9y^2 - 8y^2 = (9-8)y^2 = 1y^2 = y^2$

105. $-3(x+5) = -3x - 3(5) = -3x - 15$

107. $(x-4)^2 = x^2 - 2(x)(4) + 4^2$
$$= x^2 - 8x + 16$$

109. The statement $\sqrt[n]{a} \cdot \sqrt[n]{b} = \sqrt[n]{ab}$ is <u>true</u>.

111. The statement $\sqrt[3]{7} \cdot \sqrt{11} = \sqrt{77}$ is <u>false</u>.

113. The statement $\dfrac{\sqrt[n]{a}}{\sqrt[n]{b}} = \sqrt[n]{\dfrac{a}{b}}$ is <u>true</u>.

115. $\dfrac{\sqrt[3]{64}}{\sqrt{64}} = \dfrac{4}{8} = \dfrac{1}{2}$

117. $\sqrt[5]{x^{35}} = x^7$

119. $\sqrt[4]{a^{12}b^4c^{20}} = a^3bc^5$

121. $\sqrt[3]{z^{32}} = \sqrt[3]{z^{30} \cdot z^2} = \sqrt[3]{z^{30}} \cdot \sqrt[3]{z^2} = z^{10}\sqrt[3]{z^2}$

123. $\sqrt[7]{q^{17}r^{40}s^7} = \sqrt[7]{q^{14} \cdot q^3 \cdot r^{35} \cdot r^5 \cdot s^7}$
$$= \sqrt[7]{q^{14}r^{35}s^7 \cdot q^3 r^5}$$
$$= q^2 r^5 s \sqrt[7]{q^3 r^5}$$

125. $r = \sqrt{\dfrac{A}{4\pi}} = \sqrt{\dfrac{32.17}{4\pi}} \approx \sqrt{2.56} = 1.6$
The radius of a standard zorb is 1.6 meters.

127. $A = \pi r \sqrt{r^2 + h^2}$

   a. $A = \pi(4)\sqrt{4^2 + 3^2}$
$$= 4\pi\sqrt{16+9}$$
$$= 4\pi\sqrt{25}$$
$$= 4\pi(5)$$
$$= 20\pi$$
The area is 20π square centimeters.

   b. $A = \pi(6.8)\sqrt{(6.8)^2 + (7.2)^2}$
$$= 6.8\pi\sqrt{46.24 + 51.84}$$
$$= 6.8\pi\sqrt{98.08}$$
$$\approx 211.57$$
The area is approximately 211.57 square feet.

Section 7.4 Practice Exercises

1. a. $3\sqrt{17} + 5\sqrt{17} = (3+5)\sqrt{17} = 8\sqrt{17}$

   b. $7\sqrt[3]{5z} - 12\sqrt[3]{5z} = (7-12)\sqrt[3]{5z} = -5\sqrt[3]{5z}$

   c. $3\sqrt{2} + 5\sqrt[3]{2}$
This expression cannot be simplified since $3\sqrt{2}$ and $5\sqrt[3]{2}$ do not contain like radicals.

2. a. $\sqrt{24} + 3\sqrt{54} = \sqrt{4 \cdot 6} + 3\sqrt{9 \cdot 6}$
$$= \sqrt{4} \cdot \sqrt{6} + 3 \cdot \sqrt{9} \cdot \sqrt{6}$$
$$= 2 \cdot \sqrt{6} + 3 \cdot 3 \cdot \sqrt{6}$$
$$= 2\sqrt{6} + 9\sqrt{6}$$
$$= 11\sqrt{6}$$

   b. $\sqrt[3]{24} - 4\sqrt[3]{81} + \sqrt[3]{3}$
$$= \sqrt[3]{8} \cdot \sqrt[3]{3} - 4 \cdot \sqrt[3]{27} \cdot \sqrt[3]{3} + \sqrt[3]{3}$$
$$= 2 \cdot \sqrt[3]{3} - 4 \cdot 3 \cdot \sqrt[3]{3} + \sqrt[3]{3}$$
$$= 2\sqrt[3]{3} - 12\sqrt[3]{3} + \sqrt[3]{3}$$
$$= -9\sqrt[3]{3}$$

   c. $\sqrt{75x} - 3\sqrt{27x} + \sqrt{12x}$
$$= \sqrt{25} \cdot \sqrt{3x} - 3 \cdot \sqrt{9} \cdot \sqrt{3x} + \sqrt{4} \cdot \sqrt{3x}$$
$$= 5 \cdot \sqrt{3x} - 3 \cdot 3 \cdot \sqrt{3x} + 2 \cdot \sqrt{3x}$$
$$= 5\sqrt{3x} - 9\sqrt{3x} + 2\sqrt{3x}$$
$$= -2\sqrt{3x}$$

d. $\sqrt{40} + \sqrt[3]{40} = \sqrt{4} \cdot \sqrt{10} + \sqrt[3]{8} \cdot \sqrt[3]{5}$
$$= 2\sqrt{10} + 2\sqrt[3]{5}$$

e. $\sqrt[3]{81x^4} + \sqrt[3]{3x^4} = \sqrt[3]{27x^3} \cdot \sqrt[3]{3x} + \sqrt[3]{x^3} \cdot \sqrt[3]{3x}$
$$= 3x\sqrt[3]{3x} + x\sqrt[3]{3x}$$
$$= 4x\sqrt[3]{3x}$$

3. a. $\dfrac{\sqrt{28}}{3} - \dfrac{\sqrt{7}}{4} = \dfrac{2\sqrt{7}}{3} - \dfrac{\sqrt{7}}{4}$
$$= \dfrac{2\sqrt{7} \cdot 4}{3 \cdot 4} - \dfrac{\sqrt{7} \cdot 3}{4 \cdot 3}$$
$$= \dfrac{8\sqrt{7}}{12} - \dfrac{3\sqrt{7}}{12}$$
$$= \dfrac{5\sqrt{7}}{12}$$

b. $\sqrt[3]{\dfrac{6y}{64}} + 3\sqrt[3]{6y} = \dfrac{\sqrt[3]{6y}}{\sqrt[3]{64}} + 3\sqrt[3]{6y}$
$$= \dfrac{\sqrt[3]{6y}}{4} + 3\sqrt[3]{6y}$$
$$= \dfrac{\sqrt[3]{6y}}{4} + \dfrac{3\sqrt[3]{6y} \cdot 4}{4}$$
$$= \dfrac{\sqrt[3]{6y}}{4} + \dfrac{12\sqrt[3]{6y}}{4}$$
$$= \dfrac{13\sqrt[3]{6y}}{4}$$

4. a. $\sqrt{5}(2 + \sqrt{15}) = \sqrt{5}(2) + \sqrt{5}(\sqrt{15})$
$$= 2\sqrt{5} + \sqrt{5 \cdot 15}$$
$$= 2\sqrt{5} + \sqrt{5 \cdot 5 \cdot 3}$$
$$= 2\sqrt{5} + 5\sqrt{3}$$

b. $(\sqrt{2} - \sqrt{5})(\sqrt{6} + 2)$
$$= \sqrt{2} \cdot \sqrt{6} + \sqrt{2} \cdot 2 - \sqrt{5} \cdot \sqrt{6} - \sqrt{5} \cdot 2$$
$$= \sqrt{2 \cdot 2 \cdot 3} + 2\sqrt{2} - \sqrt{30} - 2\sqrt{5}$$
$$= 2\sqrt{3} + 2\sqrt{2} - \sqrt{30} - 2\sqrt{5}$$

c. $(3\sqrt{z} - 4)(2\sqrt{z} + 3)$
$$= 3\sqrt{z}(2\sqrt{z}) + 3\sqrt{z}(3) - 4(2\sqrt{z}) - 4(3)$$
$$= 6 \cdot z + 9\sqrt{z} - 8\sqrt{z} - 12$$
$$= 6z + \sqrt{z} - 12$$

d. $(\sqrt{6} - 3)^2 = (\sqrt{6} - 3)(\sqrt{6} - 3)$
$$= \sqrt{6}(\sqrt{6}) - \sqrt{6}(3) - 3(\sqrt{6}) - 3(-3)$$
$$= 6 - 3\sqrt{6} - 3\sqrt{6} + 9$$
$$= 6 - 6\sqrt{6} + 9$$
$$= 15 - 6\sqrt{6}$$

e. $(\sqrt{5x} + 3)(\sqrt{5x} - 3)$
$$= \sqrt{5x} \cdot \sqrt{5x} - 3\sqrt{5x} + 3\sqrt{5x} - 3 \cdot 3$$
$$= 5x - 9$$

f. $(\sqrt{x+2} + 3)^2 = (\sqrt{x+2})^2 + 2 \cdot \sqrt{x+2} \cdot 3 + 3^2$
$$= x + 2 + 6\sqrt{x+2} + 9$$
$$= x + 11 + 6\sqrt{x+2}$$

Vocabulary, Readiness & Video Check 7.4

1. The terms $\sqrt{7}$ and $\sqrt[3]{7}$ are <u>unlike</u> terms.

2. The terms $\sqrt[3]{x^2 y}$ and $\sqrt[3]{yx^2}$ are <u>like</u> terms.

3. The terms $\sqrt[3]{abc}$ and $\sqrt[3]{cba}$ are <u>like</u> terms.

4. The terms $2x\sqrt{5}$ and $2x\sqrt{10}$ are <u>unlike</u> terms.

5. $2\sqrt{3} + 4\sqrt{3} = \underline{6\sqrt{3}}$

6. $5\sqrt{7} + 3\sqrt{7} = \underline{8\sqrt{7}}$

7. $8\sqrt{x} - \sqrt{x} = \underline{7\sqrt{x}}$

8. $3\sqrt{y} - \sqrt{y} = \underline{2\sqrt{y}}$

9. $7\sqrt[3]{x} + \sqrt[3]{x} = \underline{8\sqrt[3]{x}}$

10. $8\sqrt[3]{z} + \sqrt[3]{z} = \underline{9\sqrt[3]{z}}$

11. Sometimes you can't see that there are like radicals until you simplify, so you may incorrectly think you cannot add or subtract if you don't simplify first.

12. The square root of a positive number times the square root of the same positive number is that positive number.

Exercise Set 7.4

1. $\sqrt{8} - \sqrt{32} = \sqrt{4 \cdot 2} - \sqrt{16 \cdot 2}$
$$= \sqrt{4} \cdot \sqrt{2} - \sqrt{16} \cdot \sqrt{2}$$
$$= 2\sqrt{2} - 4\sqrt{2}$$
$$= -2\sqrt{2}$$

3. $2\sqrt{2x^3} + 4x\sqrt{8x} = 2\sqrt{x^2 \cdot 2x} + 4x\sqrt{4 \cdot 2x}$
$$= 2\sqrt{x^2} \cdot \sqrt{2x} + 4x\sqrt{4} \cdot \sqrt{2x}$$
$$= 2x\sqrt{2x} + 4x(2)\sqrt{2x}$$
$$= 2x\sqrt{2x} + 8x\sqrt{2x}$$
$$= 10x\sqrt{2x}$$

5. $2\sqrt{50} - 3\sqrt{125} + \sqrt{98}$
$$= 2\sqrt{25 \cdot 2} - 3\sqrt{25 \cdot 5} + \sqrt{49 \cdot 2}$$
$$= 2\sqrt{25} \cdot \sqrt{2} - 3\sqrt{25} \cdot \sqrt{5} + \sqrt{49} \cdot \sqrt{2}$$
$$= 2(5)\sqrt{2} - 3(5)\sqrt{5} + 7\sqrt{2}$$
$$= 10\sqrt{2} - 15\sqrt{5} + 7\sqrt{2}$$
$$= 17\sqrt{2} - 15\sqrt{5}$$

7. $\sqrt[3]{16x} - \sqrt[3]{54x} = \sqrt[3]{8 \cdot 2x} - \sqrt[3]{27 \cdot 2x}$
$$= \sqrt[3]{8} \cdot \sqrt[3]{2x} - \sqrt[3]{27} \cdot \sqrt[3]{2x}$$
$$= 2\sqrt[3]{2x} - 3\sqrt[3]{2x}$$
$$= -\sqrt[3]{2x}$$

9. $\sqrt{9b^3} - \sqrt{25b^3} + \sqrt{49b^3}$
$$= \sqrt{9b^2 \cdot b} - \sqrt{25b^2 \cdot b} + \sqrt{49b^2 \cdot b}$$
$$= \sqrt{9b^2} \cdot \sqrt{b} - \sqrt{25b^2} \cdot \sqrt{b} + \sqrt{49b^2} \cdot \sqrt{b}$$
$$= 3b\sqrt{b} - 5b\sqrt{b} + 7b\sqrt{b}$$
$$= 5b\sqrt{b}$$

11. $\dfrac{5\sqrt{2}}{3} + \dfrac{2\sqrt{2}}{5} = \dfrac{5\left(5\sqrt{2}\right) + 3\left(2\sqrt{2}\right)}{3(5)}$
$$= \dfrac{25\sqrt{2} + 6\sqrt{2}}{15}$$
$$= \dfrac{31\sqrt{2}}{15}$$

13. $\sqrt[3]{\dfrac{11}{8}} - \dfrac{\sqrt[3]{11}}{6} = \dfrac{\sqrt[3]{11}}{\sqrt[3]{8}} - \dfrac{\sqrt[3]{11}}{6}$
$$= \dfrac{\sqrt[3]{11}}{2} - \dfrac{\sqrt[3]{11}}{6}$$
$$= \dfrac{3\sqrt[3]{11} - \sqrt[3]{11}}{6}$$
$$= \dfrac{2\sqrt[3]{11}}{6}$$
$$= \dfrac{\sqrt[3]{11}}{3}$$

15. $\dfrac{\sqrt{20x}}{9} + \sqrt{\dfrac{5x}{9}} = \dfrac{\sqrt{4 \cdot 5x}}{9} + \dfrac{\sqrt{5x}}{\sqrt{9}}$
$$= \dfrac{2\sqrt{5x}}{9} + \dfrac{\sqrt{5x}}{3}$$
$$= \dfrac{2\sqrt{5x} + 3\sqrt{5x}}{9}$$
$$= \dfrac{5\sqrt{5x}}{9}$$

17. $7\sqrt{9} - 7 + \sqrt{3} = 7(3) - 7 + \sqrt{3}$
$$= 21 - 7 + \sqrt{3}$$
$$= 14 + \sqrt{3}$$

19. $2 + 3\sqrt{y^2} - 6\sqrt{y^2} + 5 = 2 + 3y - 6y + 5$
$$= 7 - 3y$$

21. $3\sqrt{108} - 2\sqrt{18} - 3\sqrt{48}$
$$= 3\sqrt{36 \cdot 3} - 2\sqrt{9 \cdot 2} - 3\sqrt{16 \cdot 3}$$
$$= 3\sqrt{36} \cdot \sqrt{3} - 2\sqrt{9} \cdot \sqrt{2} - 3\sqrt{16} \cdot \sqrt{3}$$
$$= 3(6)\sqrt{3} - 2(3)\sqrt{2} - 3(4)\sqrt{3}$$
$$= 18\sqrt{3} - 6\sqrt{2} - 12\sqrt{3}$$
$$= 6\sqrt{3} - 6\sqrt{2}$$

23. $-5\sqrt[3]{625} + \sqrt[3]{40} = -5\sqrt[3]{125 \cdot 5} + \sqrt[3]{8 \cdot 5}$
$$= -5(5)\sqrt[3]{5} + 2\sqrt[3]{5}$$
$$= -25\sqrt[3]{5} + 2\sqrt[3]{5}$$
$$= -23\sqrt[3]{5}$$

25. $a^3\sqrt{9ab^3} - \sqrt{25a^7b^3} + \sqrt{16a^7b^3}$
$$= a^3\sqrt{9b^2 \cdot ab} - \sqrt{25a^6b^2 \cdot ab} + \sqrt{16a^6b^2 \cdot ab}$$
$$= a^3 \cdot 3b\sqrt{ab} - 5a^3b\sqrt{ab} + 4a^3b\sqrt{ab}$$
$$= 3a^3b\sqrt{ab} - 5a^3b\sqrt{ab} + 4a^3b\sqrt{ab}$$
$$= 2a^3b\sqrt{ab}$$

27. $5y\sqrt{8y} + 2\sqrt{50y^3} = 5y\sqrt{4 \cdot 2y} + 2\sqrt{25y^2 \cdot 2y}$
$$= 5y(2)\sqrt{2y} + 2(5y)\sqrt{2y}$$
$$= 10y\sqrt{2y} + 10y\sqrt{2y}$$
$$= 20y\sqrt{2y}$$

29. $\sqrt[3]{54xy^3} - 5\sqrt[3]{2xy^3} + y\sqrt[3]{128x}$
$$= \sqrt[3]{27y^3 \cdot 2x} - 5\sqrt[3]{y^3 \cdot 2x} + y\sqrt[3]{64 \cdot 2x}$$
$$= 3y\sqrt[3]{2x} - 5y\sqrt[3]{2x} + 4y\sqrt[3]{2x}$$
$$= 2y\sqrt[3]{2x}$$

31. $6\sqrt[3]{11} + 8\sqrt{11} - 12\sqrt{11} = 6\sqrt[3]{11} - 4\sqrt{11}$

33. $-2\sqrt[4]{x^7} + 3\sqrt[4]{16x^7} - x\sqrt[4]{x^3}$
$$= -2\sqrt[4]{x^4 \cdot x^3} + 3\sqrt[4]{16x^4 \cdot x^3} - x\sqrt[4]{x^3}$$
$$= -2x\sqrt[4]{x^3} + 3(2x)\sqrt[4]{x^3} - x\sqrt[4]{x^3}$$
$$= -2x\sqrt[4]{x^3} + 6x\sqrt[4]{x^3} - x\sqrt[4]{x^3}$$
$$= 3x\sqrt[4]{x^3}$$

35. $\dfrac{4\sqrt{3}}{3} - \dfrac{\sqrt{12}}{3} = \dfrac{4\sqrt{3}}{3} - \dfrac{\sqrt{4 \cdot 3}}{3}$
$$= \dfrac{4\sqrt{3} - 2\sqrt{3}}{3}$$
$$= \dfrac{2\sqrt{3}}{3}$$

37. $\dfrac{\sqrt[3]{8x^4}}{7} + \dfrac{3x\sqrt[3]{x}}{7} = \dfrac{\sqrt[3]{8x^3 \cdot x}}{7} + \dfrac{3x\sqrt[3]{x}}{7}$
$$= \dfrac{2x\sqrt[3]{x} + 3x\sqrt[3]{x}}{7}$$
$$= \dfrac{5x\sqrt[3]{x}}{7}$$

39. $\sqrt{\dfrac{28}{x^2}} + \sqrt{\dfrac{7}{4x^2}} = \dfrac{\sqrt{28}}{\sqrt{x^2}} + \dfrac{\sqrt{7}}{\sqrt{4x^2}}$
$$= \dfrac{\sqrt{4 \cdot 7}}{x} + \dfrac{\sqrt{7}}{2x}$$
$$= \dfrac{2\sqrt{7}}{x} + \dfrac{\sqrt{7}}{2x}$$
$$= \dfrac{2(2\sqrt{7}) + \sqrt{7}}{2x}$$
$$= \dfrac{4\sqrt{7} + \sqrt{7}}{2x}$$
$$= \dfrac{5\sqrt{7}}{2x}$$

41. $\sqrt[3]{\dfrac{16}{27}} - \dfrac{\sqrt[3]{54}}{6} = \dfrac{\sqrt[3]{8 \cdot 2}}{\sqrt[3]{27}} - \dfrac{\sqrt[3]{27 \cdot 2}}{6}$
$$= \dfrac{2\sqrt[3]{2}}{3} - \dfrac{3\sqrt[3]{2}}{6}$$
$$= \dfrac{2(2\sqrt[3]{2}) - 3\sqrt[3]{2}}{6}$$
$$= \dfrac{4\sqrt[3]{2} - 3\sqrt[3]{2}}{6}$$
$$= \dfrac{\sqrt[3]{2}}{6}$$

43. $-\dfrac{\sqrt[3]{2x^4}}{9} + \sqrt[3]{\dfrac{250x^4}{27}} = -\dfrac{\sqrt[3]{x^3 \cdot 2x}}{9} + \dfrac{\sqrt[3]{125x^3 \cdot 2x}}{\sqrt[3]{27}}$
$$= \dfrac{-x\sqrt[3]{2x}}{9} + \dfrac{5x\sqrt[3]{2x}}{3}$$
$$= \dfrac{-x\sqrt[3]{2x} + 3(5x\sqrt[3]{2x})}{9}$$
$$= \dfrac{-x\sqrt[3]{2x} + 15x\sqrt[3]{2x}}{9}$$
$$= \dfrac{14x\sqrt[3]{2x}}{9}$$

45. $P = 2\sqrt{12} + \sqrt{12} + 2\sqrt{27} + 3\sqrt{3}$
$$= 2\sqrt{4 \cdot 3} + \sqrt{4 \cdot 3} + 2\sqrt{9 \cdot 3} + 3\sqrt{3}$$
$$= 2(2)\sqrt{3} + 2\sqrt{3} + 2(3)\sqrt{3} + 3\sqrt{3}$$
$$= 4\sqrt{3} + 2\sqrt{3} + 6\sqrt{3} + 3\sqrt{3}$$
$$= 15\sqrt{3} \text{ inches}$$

47. $\sqrt{7}\left(\sqrt{5} + \sqrt{3}\right) = \sqrt{7}\sqrt{5} + \sqrt{7}\sqrt{3}$
$$= \sqrt{35} + \sqrt{21}$$

49. $\left(\sqrt{5}-\sqrt{2}\right)^2 = \left(\sqrt{5}\right)^2 - 2\sqrt{5}\sqrt{2} + \left(\sqrt{2}\right)^2$
$$= 5 - 2\sqrt{10} + 2$$
$$= 7 - 2\sqrt{10}$$

51. $\sqrt{3x}\left(\sqrt{3}-\sqrt{x}\right) = \sqrt{3x}\sqrt{3} - \sqrt{3x}\sqrt{x}$
$$= \sqrt{9x} - \sqrt{3x^2}$$
$$= 3\sqrt{x} - x\sqrt{3}$$

53. $\left(2\sqrt{x}-5\right)\left(3\sqrt{x}+1\right)$
$$= 2\sqrt{x}\left(3\sqrt{x}\right) + 2\sqrt{x}\cdot 1 - 5\left(3\sqrt{x}\right) - 5(1)$$
$$= 6x + 2\sqrt{x} - 15\sqrt{x} - 5$$
$$= 6x - 13\sqrt{x} - 5$$

55. $\left(\sqrt[3]{a}-4\right)\left(\sqrt[3]{a}+5\right)$
$$= \sqrt[3]{a}\left(\sqrt[3]{a}\right) + \sqrt[3]{a}\cdot 5 - 4\sqrt[3]{a} - 4(5)$$
$$= \sqrt[3]{a^2} + 5\sqrt[3]{a} - 4\sqrt[3]{a} - 20$$
$$= \sqrt[3]{a^2} + \sqrt[3]{a} - 20$$

57. $6\left(\sqrt{2}-2\right) = 6\sqrt{2} - 6(2) = 6\sqrt{2} - 12$

59. $\sqrt{2}\left(\sqrt{2}+x\sqrt{6}\right) = \sqrt{2}\sqrt{2} + \sqrt{2}\left(x\sqrt{6}\right)$
$$= 2 + x\sqrt{12}$$
$$= 2 + x\sqrt{4\cdot 3}$$
$$= 2 + 2x\sqrt{3}$$

61. $\left(2\sqrt{7}+3\sqrt{5}\right)\left(\sqrt{7}-2\sqrt{5}\right)$
$$= 2\sqrt{7}\sqrt{7} + 2\sqrt{7}\left(-2\sqrt{5}\right) + 3\sqrt{5}\sqrt{7} + 3\sqrt{5}\left(-2\sqrt{5}\right)$$
$$= 2(7) - 4\sqrt{35} + 3\sqrt{35} - 6(5)$$
$$= 14 - \sqrt{35} - 30$$
$$= -16 - \sqrt{35}$$

63. $\left(\sqrt{x}-y\right)\left(\sqrt{x}+y\right) = \left(\sqrt{x}\right)^2 - y^2 = x - y^2$

65. $\left(\sqrt{3}+x\right)^2 = \left(\sqrt{3}\right)^2 + 2\sqrt{3}\cdot x + x^2$
$$= 3 + 2x\sqrt{3} + x^2$$

67. $\left(\sqrt{5x}-2\sqrt{3x}\right)\left(\sqrt{5x}-3\sqrt{3x}\right)$
$$= \left(\sqrt{5x}\right)^2 - \sqrt{5x}\left(3\sqrt{3x}\right) - 2\sqrt{3x}\left(\sqrt{5x}\right)$$
$$\qquad\qquad - 2\sqrt{3x}\left(-3\sqrt{3x}\right)$$
$$= 5x - 3x\sqrt{15} - 2x\sqrt{15} + 6\cdot 3x$$
$$= 23x - 5x\sqrt{15}$$

69. $\left(\sqrt[3]{4}+2\right)\left(\sqrt[3]{2}-1\right)$
$$= \sqrt[3]{4}\left(\sqrt[3]{2}\right) + \sqrt[3]{4}\cdot(-1) + 2\sqrt[3]{2} + 2(-1)$$
$$= \sqrt[3]{8} - \sqrt[3]{4} + 2\sqrt[3]{2} - 2$$
$$= 2 - \sqrt[3]{4} + 2\sqrt[3]{2} - 2$$
$$= 2\sqrt[3]{2} - \sqrt[3]{4}$$

71. $\left(\sqrt[3]{x}+1\right)\left(\sqrt[3]{x^2}-\sqrt[3]{x}+1\right)$
$$= \sqrt[3]{x}\left(\sqrt[3]{x^2}\right) - \sqrt[3]{x}\left(\sqrt[3]{x}\right) + \sqrt[3]{x}(1)$$
$$\quad + 1\left(\sqrt[3]{x^2}\right) - 1\left(\sqrt[3]{x}\right) + 1(1)$$
$$= \sqrt[3]{x^3} - \sqrt[3]{x^2} + \sqrt[3]{x} + \sqrt[3]{x^2} - \sqrt[3]{x} + 1$$
$$= x + 1$$

73. $\left(\sqrt{x-1}+5\right)^2 = \left(\sqrt{x-1}\right)^2 + 2\sqrt{x-1}\cdot 5 + 5^2$
$$= (x-1) + 10\sqrt{x-1} + 25$$
$$= x + 10\sqrt{x-1} + 24$$

75. $\left(\sqrt{2x+5}-1\right)^2 = \left(\sqrt{2x+5}\right)^2 - 2\sqrt{2x+5}\cdot 1 + 1^2$
$$= (2x+5) - 2\sqrt{2x+5} + 1$$
$$= 2x - 2\sqrt{2x+5} + 6$$

77. $\dfrac{2x-14}{2} = \dfrac{2(x-7)}{2} = x - 7$

79. $\dfrac{7x-7y}{x^2-y^2} = \dfrac{7(x-y)}{(x+y)(x-y)} = \dfrac{7}{x+y}$

81. $\dfrac{6a^2b-9ab}{3ab} = \dfrac{3ab(2a-3)}{3ab} = 2a - 3$

83. $\dfrac{-4+2\sqrt{3}}{6} = \dfrac{2\left(-2+\sqrt{3}\right)}{6} = \dfrac{-2+\sqrt{3}}{3}$

85. $P = 2l + 2w$

$= 2\left(3\sqrt{20}\right) + 2\left(\sqrt{125}\right)$

$= 6\sqrt{4 \cdot 5} + 2\sqrt{25 \cdot 5}$

$= 6(2)\sqrt{5} + 2(5)\sqrt{5}$

$= 12\sqrt{5} + 10\sqrt{5}$

$= 22\sqrt{5}$ feet

$A = lw$

$= \left(3\sqrt{20}\right)\left(\sqrt{125}\right)$

$= 3\sqrt{4 \cdot 5}\sqrt{25 \cdot 5}$

$= 3(2)\sqrt{5} \cdot 5\sqrt{5}$

$= 30 \cdot 5$

$= 150$ square feet

87. a. $\sqrt{3} + \sqrt{3} = 2\sqrt{3}$

    b. $\sqrt{3} \cdot \sqrt{3} = \sqrt{9} = 3$

    c. answers may vary

89. $\left(\sqrt{2} + \sqrt{3} - 1\right)^2$

$= \left[\left(\sqrt{2} + \sqrt{3}\right) - 1\right]^2$

$= \left(\sqrt{2} + \sqrt{3}\right)^2 - 2\left(\sqrt{2} + \sqrt{3}\right) + 1^2$

$= \left(\sqrt{2}\right)^2 + 2\sqrt{2}\sqrt{3} + \left(\sqrt{3}\right)^2 - 2\sqrt{2} - 2\sqrt{3} + 1$

$= 2 + 2\sqrt{6} + 3 - 2\sqrt{2} - 2\sqrt{3} + 1$

$= 6 + 2\sqrt{6} - 2\sqrt{2} - 2\sqrt{3}$

91. answers may vary

Section 7.5 Practice Exercises

1. a. $\dfrac{5}{\sqrt{3}} = \dfrac{5 \cdot \sqrt{3}}{\sqrt{3} \cdot \sqrt{3}} = \dfrac{5\sqrt{3}}{3}$

    b. $\dfrac{3\sqrt{25}}{\sqrt{4x}} = \dfrac{3(5)}{2\sqrt{x}} = \dfrac{15}{2\sqrt{x}} = \dfrac{15 \cdot \sqrt{x}}{2\sqrt{x} \cdot \sqrt{x}} = \dfrac{15\sqrt{x}}{2x}$

    c. $\sqrt[3]{\dfrac{2}{9}} = \dfrac{\sqrt[3]{2}}{\sqrt[3]{9}} = \dfrac{\sqrt[3]{2} \cdot \sqrt[3]{3}}{\sqrt[3]{3^2} \cdot \sqrt[3]{3}} = \dfrac{\sqrt[3]{6}}{3}$

2. $\sqrt{\dfrac{3z}{5y}} = \dfrac{\sqrt{3z}}{\sqrt{5y}} = \dfrac{\sqrt{3z} \cdot \sqrt{5y}}{\sqrt{5y} \cdot \sqrt{5y}} = \dfrac{\sqrt{15yz}}{5y}$

3. $\dfrac{\sqrt[3]{z^2}}{\sqrt[3]{27x^4}} = \dfrac{\sqrt[3]{z^2}}{\sqrt[3]{27x^3} \cdot \sqrt[3]{x}}$

$= \dfrac{\sqrt[3]{z^2}}{3x\sqrt[3]{x}}$

$= \dfrac{\sqrt[3]{z^2} \cdot \sqrt[3]{x^2}}{3x\sqrt[3]{x} \cdot \sqrt[3]{x^2}}$

$= \dfrac{\sqrt[3]{z^2 x^2}}{3x\sqrt[3]{x^3}}$

$= \dfrac{\sqrt[3]{x^2 z^2}}{3x^2}$

4. a. $\dfrac{5}{3\sqrt{5} + 2} = \dfrac{5\left(3\sqrt{5} - 2\right)}{\left(3\sqrt{5} + 2\right)\left(3\sqrt{5} - 2\right)}$

$= \dfrac{5\left(3\sqrt{5} - 2\right)}{\left(3\sqrt{5}\right)^2 - 2^2}$

$= \dfrac{5\left(3\sqrt{5} - 2\right)}{45 - 4}$

$= \dfrac{5\left(3\sqrt{5} - 2\right)}{41}$

    b. $\dfrac{\sqrt{2} + 5}{\sqrt{3} - \sqrt{5}} = \dfrac{\left(\sqrt{2} + 5\right)\left(\sqrt{3} + \sqrt{5}\right)}{\left(\sqrt{3} - \sqrt{5}\right)\left(\sqrt{3} + \sqrt{5}\right)}$

$= \dfrac{\sqrt{2}\sqrt{3} + \sqrt{2}\sqrt{5} + 5\sqrt{3} + 5\sqrt{5}}{\left(\sqrt{3}\right)^2 - \left(\sqrt{5}\right)^2}$

$= \dfrac{\sqrt{6} + \sqrt{10} + 5\sqrt{3} + 5\sqrt{5}}{3 - 5}$

$= \dfrac{\sqrt{6} + \sqrt{10} + 5\sqrt{3} + 5\sqrt{5}}{-2}$

    c. $\dfrac{3\sqrt{x}}{2\sqrt{x} + \sqrt{y}} = \dfrac{3\sqrt{x}\left(2\sqrt{x} - \sqrt{y}\right)}{\left(2\sqrt{x} + \sqrt{y}\right)\left(2\sqrt{x} - \sqrt{y}\right)}$

$= \dfrac{6\sqrt{x^2} - 3\sqrt{xy}}{\left(2\sqrt{x}\right)^2 - \left(\sqrt{y}\right)^2}$

$= \dfrac{6x - 3\sqrt{xy}}{4x - y}$

5. $\dfrac{\sqrt{32}}{\sqrt{80}} = \dfrac{\sqrt{16 \cdot 2}}{\sqrt{16 \cdot 5}} = \dfrac{4\sqrt{2}}{4\sqrt{5}} = \dfrac{\sqrt{2}}{\sqrt{5}} = \dfrac{\sqrt{2} \cdot \sqrt{2}}{\sqrt{5} \cdot \sqrt{2}} = \dfrac{2}{\sqrt{10}}$

6. $\dfrac{\sqrt[3]{5b}}{\sqrt[3]{2a}} = \dfrac{\sqrt[3]{5b} \cdot \sqrt[3]{25b^2}}{\sqrt[3]{2a} \cdot \sqrt[3]{25b^2}} = \dfrac{\sqrt[3]{125b^3}}{\sqrt[3]{50ab^2}} = \dfrac{5b}{\sqrt[3]{50ab^2}}$

7. $\dfrac{\sqrt{x}-3}{4} = \dfrac{\left(\sqrt{x}-3\right)\left(\sqrt{x}+3\right)}{4\left(\sqrt{x}+3\right)}$

$\qquad\qquad = \dfrac{\left(\sqrt{x}\right)^2 - (3)^2}{4\left(\sqrt{x}+3\right)}$

$\qquad\qquad = \dfrac{x-9}{4\left(\sqrt{x}+3\right)}$

Vocabulary, Readiness & Video Check 7.5

1. The <u>conjugate</u> of $a + b$ is $a - b$.

2. The process of writing an equivalent expression, but without a radical in the denominator is called <u>rationalizing the denominator</u>.

3. The process of writing an equivalent expression, but without a radical in the numerator is called <u>rationalizing the numerator</u>.

4. To rationalize the denominator of $\dfrac{5}{\sqrt{3}}$, we multiply by $\dfrac{\sqrt{3}}{\sqrt{3}}$.

5. To write an equivalent expression without a radical in the denominator.

6. Using the FOIL order to multiply, the Outer product and the Inner product are opposites and they will subtract out.

7. No, except for the fact you're working with numerators, the process is the same.

Exercise Set 7.5

1. $\dfrac{\sqrt{2}}{\sqrt{7}} = \dfrac{\sqrt{2}\cdot\sqrt{7}}{\sqrt{7}\cdot\sqrt{7}} = \dfrac{\sqrt{14}}{\sqrt{49}} = \dfrac{\sqrt{14}}{7}$

3. $\sqrt{\dfrac{1}{5}} = \dfrac{\sqrt{1}}{\sqrt{5}} = \dfrac{1\cdot\sqrt{5}}{\sqrt{5}\cdot\sqrt{5}} = \dfrac{\sqrt{5}}{5}$

5. $\sqrt{\dfrac{4}{x}} = \dfrac{\sqrt{4}}{\sqrt{x}} = \dfrac{2\cdot\sqrt{x}}{\sqrt{x}\cdot\sqrt{x}} = \dfrac{2\sqrt{x}}{\sqrt{x^2}} = \dfrac{2\sqrt{x}}{x}$

7. $\dfrac{4}{\sqrt[3]{3}} = \dfrac{4\cdot\sqrt[3]{9}}{\sqrt[3]{3}\cdot\sqrt[3]{9}} = \dfrac{4\sqrt[3]{9}}{\sqrt[3]{27}} = \dfrac{4\sqrt[3]{9}}{3}$

9. $\dfrac{3}{\sqrt{8x}} = \dfrac{3\cdot\sqrt{2x}}{\sqrt{8x}\cdot\sqrt{2x}} = \dfrac{3\sqrt{2x}}{\sqrt{16x^2}} = \dfrac{3\sqrt{2x}}{4x}$

11. $\dfrac{3}{\sqrt[3]{4x^2}} = \dfrac{3\cdot\sqrt[3]{2x}}{\sqrt[3]{4x^2}\cdot\sqrt[3]{2x}} = \dfrac{3\sqrt[3]{2x}}{\sqrt[3]{8x^3}} = \dfrac{3\sqrt[3]{2x}}{2x}$

13. $\dfrac{9}{\sqrt{3a}} = \dfrac{9\cdot\sqrt{3a}}{\sqrt{3a}\cdot\sqrt{3a}} = \dfrac{9\sqrt{3a}}{3a} = \dfrac{3\sqrt{3a}}{a}$

15. $\dfrac{3}{\sqrt[3]{2}} = \dfrac{3\cdot\sqrt[3]{4}}{\sqrt[3]{2}\cdot\sqrt[3]{4}} = \dfrac{3\sqrt[3]{4}}{\sqrt[3]{8}} = \dfrac{3\sqrt[3]{4}}{2}$

17. $\dfrac{2\sqrt{3}}{\sqrt{7}} = \dfrac{2\sqrt{3}\cdot\sqrt{7}}{\sqrt{7}\cdot\sqrt{7}} = \dfrac{2\sqrt{21}}{\sqrt{49}} = \dfrac{2\sqrt{21}}{7}$

19. $\sqrt{\dfrac{2x}{5y}} = \dfrac{\sqrt{2x}}{\sqrt{5y}} = \dfrac{\sqrt{2x}\cdot\sqrt{5y}}{\sqrt{5y}\cdot\sqrt{5y}} = \dfrac{\sqrt{10xy}}{5y}$

21. $\sqrt[3]{\dfrac{3}{5}} = \dfrac{\sqrt[3]{3}}{\sqrt[3]{5}}\cdot\dfrac{\sqrt[3]{25}}{\sqrt[3]{25}} = \dfrac{\sqrt[3]{75}}{5}$

23. $\sqrt{\dfrac{3x}{50}} = \dfrac{\sqrt{3x}}{\sqrt{50}}$

$\qquad\qquad = \dfrac{\sqrt{3x}}{5\sqrt{2}}$

$\qquad\qquad = \dfrac{\sqrt{3x}\cdot\sqrt{2}}{5\sqrt{2}\cdot\sqrt{2}}$

$\qquad\qquad = \dfrac{\sqrt{6x}}{5\cdot 2}$

$\qquad\qquad = \dfrac{\sqrt{6x}}{10}$

25. $\dfrac{1}{\sqrt{12z}} = \dfrac{1}{\sqrt{4\cdot 3z}} = \dfrac{1}{2\sqrt{3z}}\cdot\dfrac{\sqrt{3z}}{\sqrt{3z}} = \dfrac{\sqrt{3z}}{6z}$

27. $\dfrac{\sqrt[3]{2y^2}}{\sqrt[3]{9x^2}} = \dfrac{\sqrt[3]{2y^2}\cdot\sqrt[3]{3x}}{\sqrt[3]{9x^2}\cdot\sqrt[3]{3x}} = \dfrac{\sqrt[3]{6xy^2}}{3x}$

29. $\sqrt[4]{\dfrac{81}{8}} = \dfrac{\sqrt[4]{81}}{\sqrt[4]{8}} = \dfrac{3\cdot\sqrt[4]{2}}{\sqrt[4]{8}\cdot\sqrt[4]{2}} = \dfrac{3\sqrt[4]{2}}{\sqrt[4]{16}} = \dfrac{3\sqrt[4]{2}}{2}$

31. $\sqrt[4]{\dfrac{16}{9x^7}} = \dfrac{\sqrt[4]{16}}{\sqrt[4]{9x^7}} = \dfrac{2\cdot\sqrt[4]{9x}}{\sqrt[4]{9x^7}\cdot\sqrt[4]{9x}} = \dfrac{2\sqrt[4]{9x}}{\sqrt[4]{81x^8}} = \dfrac{2\sqrt[4]{9x}}{3x^2}$

33. $\dfrac{5a}{\sqrt[5]{8a^9b^{11}}} = \dfrac{5a\cdot\sqrt[5]{4ab^4}}{\sqrt[5]{8a^9b^{11}}\cdot\sqrt[5]{4ab^4}}$

$= \dfrac{5a\sqrt[5]{4ab^4}}{\sqrt[5]{32a^{10}b^{15}}}$

$= \dfrac{5a\sqrt[5]{4ab^4}}{2a^2b^3}$

$= \dfrac{5\sqrt[5]{4ab^4}}{2ab^3}$

35. The conjugate of $\sqrt{2}+x$ is $\sqrt{2}-x$.

37. The conjugate of $5-\sqrt{a}$ is $5+\sqrt{a}$.

39. The conjugate of $-7\sqrt{5}+8\sqrt{x}$ is $-7\sqrt{5}-8\sqrt{x}$.

41. $\dfrac{6}{2-\sqrt{7}} = \dfrac{6(2+\sqrt{7})}{(2-\sqrt{7})(2+\sqrt{7})}$

$= \dfrac{6(2+\sqrt{7})}{2^2-(\sqrt{7})^2}$

$= \dfrac{6(2+\sqrt{7})}{4-7}$

$= \dfrac{6(2+\sqrt{7})}{-3}$

$= -2(2+\sqrt{7})$

43. $\dfrac{-7}{\sqrt{x}-3} = \dfrac{-7(\sqrt{x}+3)}{(\sqrt{x}-3)(\sqrt{x}+3)}$

$= \dfrac{-7(\sqrt{x}+3)}{(\sqrt{x})^2-(3)^2}$

$= \dfrac{-7(\sqrt{x}+3)}{x-9}$ or $\dfrac{7(\sqrt{x}+3)}{9-x}$

45. $\dfrac{\sqrt{2}-\sqrt{3}}{\sqrt{2}+\sqrt{3}} = \dfrac{(\sqrt{2}-\sqrt{3})(\sqrt{2}-\sqrt{3})}{(\sqrt{2}+\sqrt{3})(\sqrt{2}-\sqrt{3})}$

$= \dfrac{(\sqrt{2})^2-2\sqrt{2}\sqrt{3}+(\sqrt{3})^2}{(\sqrt{2})^2-(\sqrt{3})^2}$

$= \dfrac{2-2\sqrt{6}+3}{2-3}$

$= \dfrac{5-2\sqrt{6}}{-1}$

$= -5+2\sqrt{6}$

47. $\dfrac{\sqrt{a}+1}{2\sqrt{a}-\sqrt{b}}$

$= \dfrac{(\sqrt{a}+1)(2\sqrt{a}+\sqrt{b})}{(2\sqrt{a}-\sqrt{b})(2\sqrt{a}+\sqrt{b})}$

$= \dfrac{\sqrt{a}\cdot 2\sqrt{a}+\sqrt{a}\sqrt{b}+1\cdot 2\sqrt{a}+1\cdot\sqrt{b}}{(2\sqrt{a})^2-(\sqrt{b})^2}$

$= \dfrac{2a+\sqrt{ab}+2\sqrt{a}+\sqrt{b}}{4a-b}$

49. $\dfrac{8}{1+\sqrt{10}} = \dfrac{8(1-\sqrt{10})}{(1+\sqrt{10})(1-\sqrt{10})}$

$= \dfrac{8(1-\sqrt{10})}{1^2-(\sqrt{10})^2}$

$= \dfrac{8(1-\sqrt{10})}{1-10}$

$= -\dfrac{8(1-\sqrt{10})}{9}$

51.
$$\frac{\sqrt{x}}{\sqrt{x}+\sqrt{y}} = \frac{\sqrt{x}\left(\sqrt{x}-\sqrt{y}\right)}{\left(\sqrt{x}+\sqrt{y}\right)\left(\sqrt{x}-\sqrt{y}\right)}$$
$$= \frac{\sqrt{x}\left(\sqrt{x}-\sqrt{y}\right)}{\left(\sqrt{x}\right)^2-\left(\sqrt{y}\right)^2}$$
$$= \frac{\sqrt{x}\left(\sqrt{x}-\sqrt{y}\right)}{x-y}$$
$$= \frac{\sqrt{x}\sqrt{x}-\sqrt{x}\sqrt{y}}{x-y}$$
$$= \frac{x-\sqrt{xy}}{x-y}$$

53.
$$\frac{2\sqrt{3}+\sqrt{6}}{4\sqrt{3}-\sqrt{6}} = \frac{\left(2\sqrt{3}+\sqrt{6}\right)\left(4\sqrt{3}+\sqrt{6}\right)}{\left(4\sqrt{3}-\sqrt{6}\right)\left(4\sqrt{3}+\sqrt{6}\right)}$$
$$= \frac{8\cdot3+2\sqrt{18}+4\sqrt{18}+6}{\left(4\sqrt{3}\right)^2-\left(\sqrt{6}\right)^2}$$
$$= \frac{30+6\sqrt{18}}{16\cdot3-6}$$
$$= \frac{30+6(3)\sqrt{2}}{42}$$
$$= \frac{30+18\sqrt{2}}{42}$$
$$= \frac{6\left(5+3\sqrt{2}\right)}{42}$$
$$= \frac{5+3\sqrt{2}}{7}$$

55.
$$\sqrt{\frac{5}{3}} = \frac{\sqrt{5}}{\sqrt{3}} = \frac{\sqrt{5}\cdot\sqrt{5}}{\sqrt{3}\cdot\sqrt{5}} = \frac{\sqrt{25}}{\sqrt{15}} = \frac{5}{\sqrt{15}}$$

57.
$$\sqrt{\frac{18}{5}} = \frac{\sqrt{18}}{\sqrt{5}}$$
$$= \frac{\sqrt{9}\cdot\sqrt{2}}{\sqrt{5}}$$
$$= \frac{3\sqrt{2}}{\sqrt{5}}$$
$$= \frac{3\sqrt{2}\cdot\sqrt{2}}{\sqrt{5}\cdot\sqrt{2}}$$
$$= \frac{3\cdot2}{\sqrt{10}}$$
$$= \frac{6}{\sqrt{10}}$$

59.
$$\frac{\sqrt{4x}}{7} = \frac{2\sqrt{x}}{7} = \frac{2\sqrt{x}\cdot\sqrt{x}}{7\cdot\sqrt{x}} = \frac{2\sqrt{x^2}}{7\sqrt{x}} = \frac{2x}{7\sqrt{x}}$$

61.
$$\frac{\sqrt[3]{5y^2}}{\sqrt[3]{4x}} = \frac{\sqrt[3]{5y^2}\cdot\sqrt[3]{5^2y}}{\sqrt[3]{4x}\cdot\sqrt[3]{5^2y}} = \frac{\sqrt[3]{5^3y^3}}{\sqrt[3]{100xy}} = \frac{5y}{\sqrt[3]{100xy}}$$

63.
$$\sqrt{\frac{2}{5}} = \frac{\sqrt{2}}{\sqrt{5}} = \frac{\sqrt{2}\cdot\sqrt{2}}{\sqrt{5}\cdot\sqrt{2}} = \frac{\sqrt{4}}{\sqrt{10}} = \frac{2}{\sqrt{10}}$$

65.
$$\frac{\sqrt{2x}}{11} = \frac{\sqrt{2x}\cdot\sqrt{2x}}{11\cdot\sqrt{2x}} = \frac{\sqrt{4x^2}}{11\sqrt{2x}} = \frac{2x}{11\sqrt{2x}}$$

67.
$$\sqrt[3]{\frac{7}{8}} = \frac{\sqrt[3]{7}}{\sqrt[3]{8}}$$
$$= \frac{\sqrt[3]{7}}{2}$$
$$= \frac{\sqrt[3]{7}\cdot\sqrt[3]{7^2}}{2\cdot\sqrt[3]{7^2}}$$
$$= \frac{\sqrt[3]{7^3}}{2\sqrt[3]{49}}$$
$$= \frac{7}{2\sqrt[3]{49}}$$

69.
$$\frac{\sqrt[3]{3x^5}}{10} = \frac{\sqrt[3]{x^3\cdot3x^2}}{10}$$
$$= \frac{x\sqrt[3]{3x^2}}{10}$$
$$= \frac{x\sqrt[3]{3x^2}\cdot\sqrt[3]{3^2x}}{10\cdot\sqrt[3]{3^2x}}$$
$$= \frac{x\sqrt[3]{3^3x^3}}{10\sqrt[3]{9x}}$$
$$= \frac{x\cdot3x}{10\sqrt[3]{9x}}$$
$$= \frac{3x^2}{10\sqrt[3]{9x}}$$

71. $\sqrt{\dfrac{18x^4y^6}{3z}} = \dfrac{\sqrt{18x^4y^6}}{\sqrt{3z}}$

$= \dfrac{\sqrt{9x^4y^6 \cdot 2}}{\sqrt{3z}}$

$= \dfrac{3x^2y^3\sqrt{2}}{\sqrt{3z}}$

$= \dfrac{3x^2y^3\sqrt{2}\cdot\sqrt{2}}{\sqrt{3z}\cdot\sqrt{2}}$

$= \dfrac{3x^2y^3\cdot 2}{\sqrt{6z}}$

$= \dfrac{6x^2y^3}{\sqrt{6z}}$

73. $\dfrac{2-\sqrt{11}}{6} = \dfrac{\left(2-\sqrt{11}\right)\left(2+\sqrt{11}\right)}{6\left(2+\sqrt{11}\right)}$

$= \dfrac{4-11}{12+6\sqrt{11}}$

$= \dfrac{-7}{12+6\sqrt{11}}$

75. $\dfrac{2-\sqrt{7}}{-5} = \dfrac{\left(2-\sqrt{7}\right)\left(2+\sqrt{7}\right)}{-5\left(2+\sqrt{7}\right)}$

$= \dfrac{4-7}{-5\left(2+\sqrt{7}\right)}$

$= \dfrac{-3}{-5\left(2+\sqrt{7}\right)}$

$= \dfrac{3}{5\left(2+\sqrt{7}\right)}$

$= \dfrac{3}{10+5\sqrt{7}}$

77. $\dfrac{\sqrt{x}+3}{\sqrt{x}} = \dfrac{\left(\sqrt{x}+3\right)\left(\sqrt{x}-3\right)}{\sqrt{x}\left(\sqrt{x}-3\right)}$

$= \dfrac{\sqrt{x^2}-9}{\sqrt{x^2}-3\sqrt{x}}$

$= \dfrac{x-9}{x-3\sqrt{x}}$

79. $\dfrac{\sqrt{2}-1}{\sqrt{2}+1} = \dfrac{\left(\sqrt{2}-1\right)\left(\sqrt{2}+1\right)}{\left(\sqrt{2}+1\right)\left(\sqrt{2}+1\right)}$

$= \dfrac{\sqrt{4}-1}{\sqrt{4}+2\sqrt{2}+1}$

$= \dfrac{2-1}{2+2\sqrt{2}+1}$

$= \dfrac{1}{3+2\sqrt{2}}$

81. $\dfrac{\sqrt{x}+1}{\sqrt{x}-1} = \dfrac{\left(\sqrt{x}+1\right)\left(\sqrt{x}-1\right)}{\left(\sqrt{x}-1\right)\left(\sqrt{x}-1\right)}$

$= \dfrac{\sqrt{x^2}-1}{\sqrt{x^2}-2\sqrt{x}+1}$

$= \dfrac{x-1}{x-2\sqrt{x}+1}$

83. $2x-7 = 3(x-4)$

$2x-7 = 3x-12$

$-x-7 = -12$

$-x = -5$

$x = 5$

The solution is 5.

85. $(x-6)(2x+1) = 0$

$x-6 = 0$ or $2x+1 = 0$

$x = 6$ or $\quad 2x = -1$

$\qquad\qquad\qquad x = -\dfrac{1}{2}$

The solutions are $-\dfrac{1}{2}, 6$.

87. $\qquad x^2-8x = -12$

$x^2-8x+12 = 0$

$(x-6)(x-2) = 0$

$x-6 = 0$ or $x-2 = 0$

$x = 6$ or $\qquad x = 2$

The solutions are 2, 6.

89. $r = \sqrt{\dfrac{A}{4\pi}}$

$= \dfrac{\sqrt{A}}{\sqrt{4\pi}}$

$= \dfrac{\sqrt{A}}{2\sqrt{\pi}}$

$= \dfrac{\sqrt{A} \cdot \sqrt{\pi}}{2\sqrt{\pi} \cdot \sqrt{\pi}}$

$= \dfrac{\sqrt{A\pi}}{2\pi}$

91. a. $\dfrac{\sqrt{5y^3} \cdot \sqrt{12x^3}}{\sqrt{12x^3} \cdot \sqrt{12x^3}} = \dfrac{\sqrt{60x^3y^3}}{\sqrt{(12x^3)^2}}$

$= \dfrac{2xy\sqrt{15xy}}{12x^3}$

$= \dfrac{y\sqrt{15xy}}{6x^2}$

b. $\dfrac{\sqrt{5y^3} \cdot \sqrt{3x}}{\sqrt{12x^3} \cdot \sqrt{3x}} = \dfrac{\sqrt{15xy^3}}{\sqrt{36x^4}} = \dfrac{y\sqrt{15xy}}{6x^2}$

c. answers may vary

93. $\dfrac{9}{\sqrt[3]{5}} = \dfrac{9}{\sqrt[3]{5}} \cdot \dfrac{\sqrt[3]{25}}{\sqrt[3]{25}} = \dfrac{9\sqrt[3]{25}}{\sqrt[3]{125}} = \dfrac{9\sqrt[3]{25}}{5}$

The smallest number is $\sqrt[3]{25}$.

95. answers may vary

97. answers may vary

Integrated Review

1. $\sqrt{81} = 9$ because $9^2 = 81$.

2. $\sqrt[3]{-8} = -2$ because $(-2)^3 = -8$.

3. $\sqrt[4]{\dfrac{1}{16}} = \dfrac{1}{2}$ because $\left(\dfrac{1}{2}\right)^4 = \dfrac{1}{16}$.

4. $\sqrt{x^6} = x^3$ because $(x^3)^2 = x^6$.

5. $\sqrt[3]{y^9} = y^3$ because $(y^3)^3 = y^9$.

6. $\sqrt{4y^{10}} = 2y^5$ because $(2y^5)^2 = 4y^{10}$.

7. $\sqrt[5]{-32y^5} = -2y$ because $(-2y)^5 = -32y^5$.

8. $\sqrt[4]{81b^{12}} = 3b^3$ because $(3b^3)^4 = 81b^{12}$.

9. $36^{1/2} = \sqrt{36} = 6$

10. $(3y)^{1/4} = \sqrt[4]{3y}$

11. $64^{-2/3} = \dfrac{1}{\left(\sqrt[3]{64}\right)^2} = \dfrac{1}{4^2} = \dfrac{1}{16}$

12. $(x+1)^{3/5} = \sqrt[5]{(x+1)^3}$

13. $y^{-1/6} \cdot y^{7/6} = y^{-\frac{1}{6}+\frac{7}{6}} = y^{6/6} = y$

14. $\dfrac{(2x^{1/3})^4}{x^{5/6}} = 16x^{4/3}x^{-5/6}$

$= 16x^{\frac{8}{6}-\frac{5}{6}}$

$= 16x^{3/6}$

$= 16x^{1/2}$

15. $\dfrac{x^{1/4}x^{3/4}}{x^{-1/4}} = x^{\frac{1}{4}+\frac{3}{4}+\frac{1}{4}} = x^{5/4}$

16. $4^{1/3} \cdot 4^{2/5} = 4^{\frac{1}{3}+\frac{2}{5}} = 4^{\frac{5}{15}+\frac{6}{15}} = 4^{11/15}$

17. $\sqrt[3]{8x^6} = (8x^6)^{1/3} = (2^3x^6)^{1/3} = 2^{3/3}x^{6/3} = 2x^2$

18. $\sqrt[12]{a^9b^6} = (a^9b^6)^{1/12}$

$= a^{9/12}b^{6/12}$

$= a^{3/4}b^{1/2}$

$= a^{3/4}b^{2/4}$

$= (a^3b^2)^{1/4}$

$= \sqrt[4]{a^3b^2}$

19. $\sqrt[4]{x} \cdot \sqrt{x} = x^{1/4} \cdot x^{1/2} = x^{\frac{1}{4}+\frac{2}{4}} = x^{3/4} = \sqrt[4]{x^3}$

20. $\sqrt{5} \cdot \sqrt[3]{2} = 5^{1/2} \cdot 2^{1/3}$

$= 5^{3/6} \cdot 2^{2/6}$

$= (5^3 \cdot 2^2)^{1/6}$

$= \sqrt[6]{5^3 \cdot 2^2}$

$= \sqrt[6]{500}$

21. $\sqrt{40} = \sqrt{4}\sqrt{10} = 2\sqrt{10}$

22. $\sqrt[4]{16x^7y^{10}} = \sqrt[4]{16x^4y^8}\sqrt[4]{x^3y^2} = 2xy^2\sqrt[4]{x^3y^2}$

23. $\sqrt[3]{54x^4} = \sqrt[3]{27x^3}\sqrt[3]{2x} = 3x\sqrt[3]{2x}$

24. $\sqrt[5]{-64b^{10}} = \sqrt[5]{-32b^{10}}\sqrt[5]{2} = -2b^2\sqrt[5]{2}$

25. $\sqrt{5}\cdot\sqrt{x} = \sqrt{5x}$

26. $\sqrt[3]{8x}\cdot\sqrt[3]{8x^2} = \sqrt[3]{64x^3} = 4x$

27. $\dfrac{\sqrt{98y^6}}{\sqrt{2y}} = \sqrt{\dfrac{98y^6}{2y}}$

$\qquad = \sqrt{49y^5}$

$\qquad = \sqrt{49y^4}\cdot\sqrt{y}$

$\qquad = 7y^2\sqrt{y}$

28. $\dfrac{\sqrt[4]{48a^9b^3}}{\sqrt[4]{ab^3}} = \sqrt[4]{\dfrac{48a^9b^3}{ab^3}}$

$\qquad = \sqrt[4]{48a^8}$

$\qquad = \sqrt[4]{16a^8}\cdot\sqrt[4]{3}$

$\qquad = 2a^2\sqrt[4]{3}$

29. $\sqrt{20} - \sqrt{75} + 5\sqrt{7} = \sqrt{4}\sqrt{5} - \sqrt{25}\sqrt{3} + 5\sqrt{7}$

$\qquad\qquad = 2\sqrt{5} - 5\sqrt{3} + 5\sqrt{7}$

30. $\sqrt[3]{54y^4} - y\sqrt[3]{16y} = \sqrt[3]{27y^3}\sqrt[3]{2y} - y\sqrt[3]{8}\sqrt[3]{2y}$

$\qquad\qquad = 3y\sqrt[3]{2y} - 2y\sqrt[3]{2y}$

$\qquad\qquad = y\sqrt[3]{2y}$

31. $\sqrt{3}\left(\sqrt{5} - \sqrt{2}\right) = \sqrt{3}\sqrt{5} - \sqrt{3}\sqrt{2} = \sqrt{15} - \sqrt{6}$

32. $\left(\sqrt{7} + \sqrt{3}\right)^2 = \left(\sqrt{7}\right)^2 + 2\sqrt{7}\sqrt{3} + \left(\sqrt{3}\right)^2$

$\qquad\qquad = 7 + 2\sqrt{21} + 3$

$\qquad\qquad = 10 + 2\sqrt{21}$

33. $\left(2x - \sqrt{5}\right)\left(2x + \sqrt{5}\right) = (2x)^2 - \left(\sqrt{5}\right)^2$

$\qquad\qquad\qquad = 4x^2 - 5$

34. $\left(\sqrt{x+1} - 1\right)^2 = \left(\sqrt{x+1}\right)^2 - 2\left(\sqrt{x+1}\right) + 1^2$

$\qquad\qquad = x + 1 - 2\sqrt{x+1} + 1$

$\qquad\qquad = x + 2 - 2\sqrt{x+1}$

35. $\sqrt{\dfrac{7}{3}} = \dfrac{\sqrt{7}}{\sqrt{3}} = \dfrac{\sqrt{7}}{\sqrt{3}}\cdot\dfrac{\sqrt{3}}{\sqrt{3}} = \dfrac{\sqrt{21}}{3}$

36. $\dfrac{5}{\sqrt[3]{2x^2}} = \dfrac{5}{\sqrt[3]{2x^2}}\cdot\dfrac{\sqrt[3]{4x}}{\sqrt[3]{4x}} = \dfrac{5\sqrt[3]{4x}}{\sqrt[3]{8x^3}} = \dfrac{5\sqrt[3]{4x}}{2x}$

37. $\dfrac{\sqrt{3} - \sqrt{7}}{2\sqrt{3} + \sqrt{7}}$

$\qquad = \dfrac{\sqrt{3} - \sqrt{7}}{2\sqrt{3} + \sqrt{7}}\cdot\dfrac{\left(2\sqrt{3} - \sqrt{7}\right)}{\left(2\sqrt{3} - \sqrt{7}\right)}$

$\qquad = \dfrac{\sqrt{3}\left(2\sqrt{3}\right) - \sqrt{3}\sqrt{7} - \sqrt{7}\left(2\sqrt{3}\right) + \sqrt{7}\sqrt{7}}{\left(2\sqrt{3}\right)^2 - \left(\sqrt{7}\right)^2}$

$\qquad = \dfrac{6 - \sqrt{21} - 2\sqrt{21} + 7}{12 - 7}$

$\qquad = \dfrac{13 - 3\sqrt{21}}{5}$

38. $\sqrt{\dfrac{7}{3}} = \dfrac{\sqrt{7}}{\sqrt{3}} = \dfrac{\sqrt{7}}{\sqrt{3}}\cdot\dfrac{\sqrt{7}}{\sqrt{7}} = \dfrac{7}{\sqrt{21}}$

39. $\sqrt[3]{\dfrac{9y}{11}} = \dfrac{\sqrt[3]{9y}}{\sqrt[3]{11}} = \dfrac{\sqrt[3]{9y}}{\sqrt[3]{11}}\cdot\dfrac{\sqrt[3]{3y^2}}{\sqrt[3]{3y^2}} = \dfrac{\sqrt[3]{27y^3}}{\sqrt[3]{31y^2}} = \dfrac{3y}{\sqrt[3]{33y^2}}$

40. $\dfrac{\sqrt{x} - 2}{\sqrt{x}} = \dfrac{\sqrt{x} - 2}{\sqrt{x}}\cdot\dfrac{\sqrt{x} + 2}{\sqrt{x} + 2}$

$\qquad = \dfrac{\left(\sqrt{x}\right)^2 - 2^2}{\sqrt{x}\sqrt{x} + 2\sqrt{x}}$

$\qquad = \dfrac{x - 4}{x + 2\sqrt{x}}$

Section 7.6 Practice Exercises

1. $\sqrt{3x-5} = 7$

$\left(\sqrt{3x-5}\right)^2 = 7^2$

$3x - 5 = 49$

$3x = 54$

$x = 18$

The solution is 18.

2. $\sqrt{16x-3} - 4x = 0$

Graph $y_1 = \sqrt{16x-3} - 4x$ in a decimal window.

The solutions are $\dfrac{1}{4}$ and $\dfrac{3}{4}$.

3. $\sqrt[3]{x-2} + 1 = 3$

Graph $y_1 = \sqrt[3]{x-2} + 1$ and $y_2 = 3$.

Check:

$\sqrt[3]{x-2} + 1 = 3$

$\sqrt[3]{10-2} + 1 \overset{?}{=} 3$

$\sqrt[3]{8} + 1 \overset{?}{=} 3$

$2 + 1 = 3$

The solution is 10.

4. $\sqrt{16+x} = x-4$

$\left(\sqrt{16+x}\right)^2 = (x-4)^2$

$16 + x = x^2 - 8x + 16$

$x^2 - 9x = 0$

$x(x-9) = 0$

$x = 0 \text{ or } x - 9 = 0$

$x = 9$

Check 0:

$\sqrt{16+x} = x-4$

$\sqrt{16+0} \overset{?}{=} 0-4$

$\sqrt{16} \overset{?}{=} -4$

$4 \ne -4$

Check 9:

$\sqrt{16+x} = x-4$

$\sqrt{16+9} \overset{?}{=} 9-4$

$\sqrt{25} \overset{?}{=} 5$

$5 = 5$

0 does not check, so the only solution is 9.

5. Graph $y_1 = \sqrt{8x+1} + \sqrt{3x}$ and $y_2 = 2.$

Now solve algebraically.

$\sqrt{8x+1} + \sqrt{3x} = 2$

$\sqrt{8x+1} = 2 - \sqrt{3x}$

$\left(\sqrt{8x+1}\right)^2 = \left(2 - \sqrt{3x}\right)^2$

$8x + 1 = 4 - 4\sqrt{3x} + 3x$

$4\sqrt{3x} = 3 - 5x$

$\left(4\sqrt{3x}\right)^2 = (3-5x)^2$

$16(3x) = 9 - 30x + 25x^2$

$25x^2 - 78x + 9 = 0$

$(25x-3)(x-3) = 0$

$25x - 3 = 0 \text{ or } x - 3 = 0$

$x = \dfrac{3}{25} \text{ or } x = 3$

From the graph, 3 does not check, so the only

solution is $\dfrac{3}{25}.$

6. $a^2 + b^2 = c^2$

$a^2 + 6^2 = 12^2$

$a^2 + 36 = 144$

$a^2 = 108$

$a = \pm\sqrt{108} = \pm\sqrt{36 \cdot 3} = \pm 6\sqrt{3}$

Since a is a length, we will use the positive value only. The unknown leg is $6\sqrt{3}$ meters long.

7. Consider the base of the tank, and the plastic divider in the diagonal. Use the Pythagorean theorem to find l.

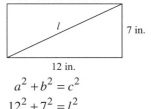

$$a^2 + b^2 = c^2$$
$$12^2 + 7^2 = l^2$$
$$144 + 49 = l^2$$
$$l^2 = 193$$
$$l = \pm\sqrt{193}$$

We will use the positive value because l represents length. The divider must be $\sqrt{193} \approx 13.89$ inches long.

Vocabulary, Readiness & Video Check 7.6

1. A proposed solution that is not a solution of the original equation is called an <u>extraneous solution</u>.

2. The Pythagorean Theorem states that $a^2 + b^2 = c^2$ where a and b are the lengths of the <u>legs</u> of a <u>right</u> triangle and c is the length of the <u>hypotenuse</u>.

3. The square of $x - 5$, or $(x-5)^2 = \underline{x^2 - 10x + 25}$.

4. The square of $4 - \sqrt{7x}$, or $\left(4 - \sqrt{7x}\right)^2 = \underline{16 - 8\sqrt{7x} + 7x}$.

5. Applying the power rule can result in an equation with more solutions than the original equation, so you need to check all proposed solutions in the original equation.

6. The Pythagorean theorem works for a right triangle only and the side opposite the right angle is the hypotenuse, which is c in the formula $a^2 + b^2 = c^2$.

7. Our answer is either a positive square root of a value or a negative square root of a value. We're looking for a length, which must be positive, so our answer must be the positive square root.

Exercise Set 7.6

1. $\sqrt{2x} = 4$
$$\left(\sqrt{2x}\right)^2 = 4^2$$
$$2x = 16$$
$$x = 8$$
The solution is 8.

3. $\sqrt{x-3} = 2$
$$\left(\sqrt{x-3}\right)^2 = 2^2$$
$$x - 3 = 4$$
$$x = 7$$
The solution is 7.

5. $\sqrt{2x} = -4$
No solution since a principle square root does not yield a negative number.

7. $\sqrt{4x-3} - 5 = 0$
$$\sqrt{4x-3} = 5$$
$$\left(\sqrt{4x-3}\right)^2 = 5^2$$
$$4x - 3 = 25$$
$$4x = 28$$
$$x = 7$$
The solution is 7.

9. $\sqrt{2x-3} - 2 = 1$
$$\sqrt{2x-3} = 3$$
$$\left(\sqrt{2x-3}\right)^2 = 3^2$$
$$2x - 3 = 9$$
$$2x = 12$$
$$x = 6$$
The solution is 6.

11. $\sqrt[3]{6x} = -3$
$$\left(\sqrt[3]{6x}\right)^3 = (-3)^3$$
$$6x = -27$$
$$x = -\frac{27}{6} = -\frac{9}{2}$$
The solution is $-\frac{9}{2}$.

13. $\sqrt[3]{x-2}-3=0$

$\sqrt[3]{x-2}=3$

$\left(\sqrt[3]{x-2}\right)^3=3^3$

$x-2=27$

$x=29$

The solution is 29.

15. $\sqrt{13-x}=x-1$

$\left(\sqrt{13-x}\right)^2=(x-1)^2$

$13-x=x^2-2x+1$

$0=x^2-x-12$

$0=(x-4)(x+3)$

$x-4=0$ or $x+3=0$

$x=4$ or $\qquad x=-3$

We discard –3 as extraneous. The solution is 4.

17. $x-\sqrt{4-3x}=-8$

$x+8=\sqrt{4-3x}$

$(x+8)^2=\left(\sqrt{4-3x}\right)^2$

$x^2+16x+64=4-3x$

$x^2+19x+60=0$

$(x+4)(x+15)=0$

$x+4=0$ or $x+15=0$

$x=-4$ or $\qquad x=-15$

We discard –15 as extraneous. The solution is –4.

19. $\sqrt{y+5}=2-\sqrt{y-4}$

$\left(\sqrt{y+5}\right)^2=\left(2-\sqrt{y-4}\right)^2$

$y+5=4-4\sqrt{y-4}+(y-4)$

$y+5=y-4\sqrt{y-4}$

$5=-4\sqrt{y-4}$

$5^2=\left(-4\sqrt{y-4}\right)^2$

$25=16(y-4)$

$25=16y-64$

$89=16y$

$\dfrac{89}{16}=y$

We discard $\dfrac{89}{16}$ as extraneous. There is no solution.

21. $\sqrt{x-3}+\sqrt{x+2}=5$

$\sqrt{x-3}=5-\sqrt{x+2}$

$\left(\sqrt{x-3}\right)^2=\left(5-\sqrt{x+2}\right)^2$

$x-3=25-10\sqrt{x+2}+(x+2)$

$x-3=27-10\sqrt{x+2}+x$

$-30=-10\sqrt{x+2}$

$3=\sqrt{x+2}$

$3^2=\left(\sqrt{x+2}\right)^2$

$9=x+2$

$7=x$

The solution is 7.

23. $\sqrt{3x-2}=5$

$\left(\sqrt{3x-2}\right)^2=5^2$

$3x-2=25$

$3x=27$

$x=9$

The solution is 9.

25. $-\sqrt{2x}+4=-6$

$10=\sqrt{2x}$

$10^2=\left(\sqrt{2x}\right)^2$

$100=2x$

$50=x$

The solution is 50.

27. $\sqrt{3x+1}+2=0$

$\sqrt{3x+1}=-2$

No solution since a principle square root does not yield a negative number.

29. $\sqrt[4]{4x+1}-2=0$

$\sqrt[4]{4x+1}=2$

$\left(\sqrt[4]{4x+1}\right)^4=2^4$

$4x+1=16$

$4x=15$

$x=\dfrac{15}{4}$

The solution is $\dfrac{15}{4}$.

31.
$$\sqrt{4x-3} = 7$$
$$\left(\sqrt{4x-3}\right)^2 = 7^2$$
$$4x-3 = 49$$
$$4x = 52$$
$$x = 13$$
The solution is 13.

33.
$$\sqrt[3]{6x-3} - 3 = 0$$
$$\sqrt[3]{6x-3} = 3$$
$$\left(\sqrt[3]{6x-3}\right)^3 = 3^3$$
$$6x-3 = 27$$
$$6x = 30$$
$$x = 5$$
The solution is 5.

35.
$$\sqrt[3]{2x-3} - 2 = -5$$
$$\sqrt[3]{2x-3} = -3$$
$$\left(\sqrt[3]{2x-3}\right)^3 = (-3)^3$$
$$2x-3 = -27$$
$$2x = -24$$
$$x = -12$$
The solution is -12.

37.
$$\sqrt{x+4} = \sqrt{2x-5}$$
$$\left(\sqrt{x+4}\right)^2 = \left(\sqrt{2x-5}\right)^2$$
$$x+4 = 2x-5$$
$$-x = -9$$
$$x = 9$$
The solution is 9.

39.
$$x - \sqrt{1-x} = -5$$
$$x + 5 = \sqrt{1-x}$$
$$(x+5)^2 = \left(\sqrt{1-x}\right)^2$$
$$x^2 + 10x + 25 = 1 - x$$
$$x^2 + 11x + 24 = 0$$
$$(x+8)(x+3) = 0$$
$$x+8 = 0 \quad \text{or} \quad x+3 = 0$$
$$x = -8 \quad \text{or} \qquad x = -3$$
We discard -8 as extraneous. The solution is -3.

41.
$$\sqrt[3]{-6x-1} = \sqrt[3]{-2x-5}$$
$$\left(\sqrt[3]{-6x-1}\right)^3 = \left(\sqrt[3]{-2x-5}\right)^3$$
$$-6x-1 = -2x-5$$
$$-4x = -4$$
$$x = 1$$
The solution is 1.

43.
$$\sqrt{5x-1} - \sqrt{x+2} = 3$$
$$\sqrt{5x-1} = \sqrt{x} + 1$$
$$\left(\sqrt{5x-1}\right)^2 = \left(\sqrt{x}+1\right)^2$$
$$5x-1 = x + 2\sqrt{x} + 1$$
$$4x-2 = 2\sqrt{x}$$
$$2x-1 = \sqrt{x}$$
$$(2x-1)^2 = \left(\sqrt{x}\right)^2$$
$$4x^2 - 4x + 1 = x$$
$$4x^2 - 5x + 1 = 0$$
$$(4x-1)(x-1) = 0$$
$$4x-1 = 0 \quad \text{or} \quad x-1 = 0$$
$$4x = 1 \quad \text{or} \qquad x = 1$$
$$x = \frac{1}{4}$$
We discard $\frac{1}{4}$ as extraneous. The solution is 1.

45.
$$\sqrt{2x-1} = \sqrt{1-2x}$$
$$\left(\sqrt{2x-1}\right)^2 = \left(\sqrt{1-2x}\right)^2$$
$$2x-1 = 1-2x$$
$$4x = 2$$
$$x = \frac{2}{4} = \frac{1}{2}$$
The solution is $\frac{1}{2}$.

47. $\sqrt{3x+4}-1=\sqrt{2x+1}$

$\sqrt{3x+4}=\sqrt{2x+1}+1$

$\left(\sqrt{3x+4}\right)^2=\left(\sqrt{2x+1}+1\right)^2$

$3x+4=(2x+1)+2\sqrt{2x+1}+1$

$3x+4=2x+2+2\sqrt{2x+1}$

$x+2=2\sqrt{2x+1}$

$(x+2)^2=\left(2\sqrt{2x+1}\right)^2$

$x^2+4x+4=4(2x+1)$

$x^2+4x+4=8x+4$

$x^2-4x=0$

$x(x-4)=0$

$x=0$ or $x-4=0$

$\qquad\qquad x=4$

The solutions are 0 and 4.

49. $\sqrt{y+3}-\sqrt{y-3}=1$

$\sqrt{y+3}=1+\sqrt{y-3}$

$\left(\sqrt{y+3}\right)^2=\left(1+\sqrt{y-3}\right)^2$

$y+3=1+2\sqrt{y-3}+(y-3)$

$y+3=-2+2\sqrt{y-3}+y$

$5=2\sqrt{y-3}$

$(5)^2=\left(2\sqrt{y-3}\right)^2$

$25=4(y-3)$

$25=4y-12$

$37=4y$

$\dfrac{37}{4}=y$

The solution is $\dfrac{37}{4}$.

51. Let c = length of the hypotenuse.

$6^2+3^2=c^2$

$36+9=c^2$

$45=c^2$

$\sqrt{45}=\sqrt{c^2}$

$\sqrt{9\cdot5}=c$

$3\sqrt{5}=c$ so $c=3\sqrt{5}$ feet

53. Let b = length of the unknown leg.

$3^2+b^2=7^2$

$9+b^2=49$

$b^2=40$

$\sqrt{b^2}=\sqrt{40}$

$b=\sqrt{4\cdot10}$

$b=2\sqrt{10}$ meters

55. Let b = length of the unknown leg.

$9^2+b^2=\left(11\sqrt{5}\right)^2$

$81+b^2=121\cdot5$

$81+b^2=605$

$b^2=524$

$\sqrt{b^2}=\sqrt{524}$

$b=\sqrt{4\cdot131}$

$b=2\sqrt{131}\approx22.9$ meters

57. Let c = length of the hypotenuse.

$7^2+(7.2)^2=c^2$

$49+51.84=c^2$

$100.84=c^2$

$100.84=\sqrt{c^2}$

$c=\sqrt{100.84}\approx10.0$ mm

59. Let x = amount of cable needed.

$15^2+8^2=x^2$

$225+64=x^2$

$289=x^2$

$\sqrt{289}=\sqrt{x^2}$

$17=x$

Thus, 17 feet of cable is needed.

61. Let c = length of the ladder.

$12^2+5^2=c^2$

$144+25=c^2$

$169=c^2$

$\sqrt{169}=\sqrt{c^2}$

$13=c$

A 13-foot ladder is needed.

63.
$$r = \sqrt{\frac{A}{4\pi}}$$
$$1080 = \sqrt{\frac{A}{4\pi}}$$
$$(1080)^2 = \left(\sqrt{\frac{A}{4\pi}}\right)^2$$
$$1,166,400 = \frac{A}{4\pi}$$
$$14,657,415 \approx A$$
The surface area is approximately 14,657,415 square miles.

65.
$$v = \sqrt{2gh}$$
$$80 = \sqrt{2(32)h}$$
$$(80)^2 = \left(\sqrt{64h}\right)^2$$
$$6400 = 64h$$
$$100 = h$$
The object fell 100 feet.

67.
$$S = 2\sqrt{I} - 9$$
$$11 = 2\sqrt{I} - 9$$
$$20 = 2\sqrt{I}$$
$$10 = \sqrt{I}$$
$$10^2 = \left(\sqrt{I}\right)^2$$
$$100 = I$$
The estimated IQ is 100.

69.
$$P = 2\pi\sqrt{\frac{l}{32}}$$
$$= 2\pi\sqrt{\frac{2}{32}}$$
$$= 2\pi\sqrt{\frac{1}{16}}$$
$$= 2\pi\left(\frac{1}{4}\right)$$
$$= \frac{\pi}{2} \text{ sec} \approx 1.57 \text{ sec}$$

71.
$$P = 2\pi\sqrt{\frac{l}{32}}$$
$$4 = 2\pi\sqrt{\frac{l}{32}}$$
$$\frac{4}{2\pi} = \sqrt{\frac{l}{32}}$$
$$\left(\frac{2}{\pi}\right)^2 = \left(\sqrt{\frac{l}{32}}\right)^2$$
$$\frac{4}{\pi^2} = \frac{l}{32}$$
$$l = 32\left(\frac{4}{\pi^2}\right) \approx 12.97 \text{ feet}$$

73. answers may vary

75.
$$s = \frac{1}{2}(6+10+14) = \frac{1}{2}(30) = 15$$
$$A = \sqrt{s(s-a)(s-b)(s-c)}$$
$$= \sqrt{15(15-6)(15-10)(15-14)}$$
$$= \sqrt{15(9)(5)(1)}$$
$$= \sqrt{675}$$
$$= \sqrt{225 \cdot 3}$$
$$= 15\sqrt{3} \text{ sq mi} \approx 25.98 \text{ sq mi.}$$

77. answers may vary

79.
$$D(h) = 111.7\sqrt{h}$$
$$80 = 111.7\sqrt{h}$$
$$\frac{80}{111.7} = \sqrt{h}$$
$$\left(\frac{80}{111.7}\right)^2 = \left(\sqrt{h}\right)^2$$
$$0.5129483389 \approx h$$
$$h \approx 0.51 \text{ km}$$

81. Function; no vertical line intersects the graph more than one time.

83. Function; no vertical line intersects the graph more than one time.

85. Not a function; the y-axis is an example of a vertical line that intersects the graph more than one time.

87.
$$\frac{\frac{x}{6}}{\frac{2x}{3}+\frac{1}{2}} = \frac{\left(\frac{x}{6}\right)6}{\left(\frac{2x}{3}+\frac{1}{2}\right)6} = \frac{x}{4x+3}$$

89. $\dfrac{\frac{z}{5}+\frac{1}{10}}{\frac{z}{20}-\frac{z}{5}} = \dfrac{\left(\frac{z}{5}+\frac{1}{10}\right)20}{\left(\frac{z}{20}-\frac{z}{5}\right)20}$

$\qquad = \dfrac{4z+2}{z-4z}$

$\qquad = \dfrac{4z+2}{-3z}$

$\qquad = -\dfrac{4z+2}{3z}$

91. $\sqrt{5x-1}+4=7$

$\qquad \sqrt{5x-1}=3$

$\qquad \left(\sqrt{5x-1}\right)^2 = 3^2$

$\qquad 5x-1=9$

$\qquad 5x=10$

$\qquad x=2$

93. $\sqrt{\sqrt{x+3}+\sqrt{x}} = \sqrt{3}$

$\qquad \left(\sqrt{\sqrt{x+3}+\sqrt{x}}\right)^2 = \left(\sqrt{3}\right)^2$

$\qquad \sqrt{x+3}+\sqrt{x} = 3$

$\qquad \sqrt{x+3} = 3-\sqrt{x}$

$\qquad \left(\sqrt{x+3}\right)^2 = \left(3-\sqrt{x}\right)^2$

$\qquad x+3 = 9-6\sqrt{x}+x$

$\qquad -6 = -6\sqrt{x}$

$\qquad (-6)^2 = \left(-6\sqrt{x}\right)^2$

$\qquad 36 = 36x$

$\qquad 1 = x$

95. a.   answers may vary

  b.   answers may vary

97. $3\sqrt{x^2-8x} = x^2-8x$

Let $t = x^2-8x$. Then

$\qquad 3\sqrt{t} = t$

$\qquad \left(3\sqrt{t}\right)^2 = t^2$

$\qquad 9t = t^2$

$\qquad 0 = t^2-9t$

$\qquad 0 = t(t-9)$

$\qquad t=0 \quad \text{or} \quad t=9$

Replace t with x^2-8x.

$x^2-8x=0 \qquad \text{or} \qquad x^2-8x=9$

$x(x-8)=0 \qquad\qquad x^2-8x-9=0$

$x=0 \quad \text{or} \quad x=8 \qquad (x-9)(x+1)=0$

$\qquad\qquad\qquad\qquad\qquad x=9 \ \text{or} \ x=-1$

The solutions are -1, 0, 8, and 9.

99. $7-(x^2-3x) = \sqrt{(x^2-3x)+5}$

Let $t = x^2-3x$. Then

$\qquad 7-t = \sqrt{t+5}$

$\qquad (7-t)^2 = \left(\sqrt{t+5}\right)^2$

$\qquad 49-14t+t^2 = t+5$

$\qquad t^2-15t+44 = 0$

$\qquad (t-11)(t-4) = 0$

$\qquad t=11 \ \text{or} \ t=4$

Replace t with x^2-3x.

$x^2-3x=11 \qquad \text{or} \qquad x^2-3x=4$

$x^2-3x-11=0 \qquad\qquad x^2-3x-4=0$

Can't factor $\qquad\qquad (x-4)(x+1)=0$

$\qquad\qquad\qquad\qquad\qquad x=4 \ \text{or} \ x=-1$

The solutions are -1 and 4.

Section 7.7 Practice Exercises

1. a. $\sqrt{-4} = \sqrt{-1\cdot 4} = \sqrt{-1}\cdot\sqrt{4} = i\cdot 2$, or $2i$

  b. $\sqrt{-7} = \sqrt{-1(7)} = \sqrt{-1}\cdot\sqrt{7} = i\sqrt{7}$

  c. $-\sqrt{-18} = -\sqrt{-1\cdot 18}$

$\qquad\qquad = -\sqrt{-1}\cdot\sqrt{9\cdot 2}$

$\qquad\qquad = -i\cdot 3\sqrt{2}$

$\qquad\qquad = -3i\sqrt{2}$

2. a. $\sqrt{-5}\cdot\sqrt{-6} = i\sqrt{5}\left(i\sqrt{6}\right)$

$\qquad\qquad = i^2\sqrt{30}$

$\qquad\qquad = -1\sqrt{30}$

$\qquad\qquad = -\sqrt{30}$

  b. $\sqrt{-9}\cdot\sqrt{-1} = 3i\cdot i = 3i^2 = 3(-1) = -3$

  c. $\sqrt{125}\cdot\sqrt{-5} = 5\sqrt{5}\left(i\sqrt{5}\right)$

$\qquad\qquad = 5i\left(\sqrt{5}\sqrt{5}\right)$

$\qquad\qquad = 5i(5)$

$\qquad\qquad = 25i$

d. $\dfrac{\sqrt{-27}}{\sqrt{3}} = \dfrac{i\sqrt{27}}{\sqrt{3}} = i\sqrt{9} = 3i$

3. a. $(3-5i)+(-4+i) = (3-4)+(-5+1)i$
$\qquad\qquad = -1-4i$

b. $4i-(3-i) = 4i-3+i$
$\qquad\qquad = -3+(4+1)i$
$\qquad\qquad = -3+5i$

c. $(-5-2i)-(-8) = -5-2i+8$
$\qquad\qquad = (-5+8)-2i$
$\qquad\qquad = 3-2i$

4. a. $-4i \cdot 5i = -20i^2 = -20(-1) = 20 = 20+0i$

b. $5i(2+i) = 5i \cdot 2 + 5i \cdot i$
$\qquad\qquad = 10i + 5i^2$
$\qquad\qquad = 10i + 5(-1)$
$\qquad\qquad = 10i - 5$
$\qquad\qquad = -5+10i$

c. $(2+3i)(6-i) = 2(6)-2(i)+3i(6)-3i(i)$
$\qquad\qquad = 12-2i+18i-3i^2$
$\qquad\qquad = 12+16i-3(-1)$
$\qquad\qquad = 12+16i+3$
$\qquad\qquad = 15+16i$

d. $(3-i)^2 = (3-i)(3-i)$
$\qquad\qquad = 3(3)-3(i)-3(i)+i^2$
$\qquad\qquad = 9-6i+(-1)$
$\qquad\qquad = 8-6i$

e. $(9+2i)(9-2i) = 9(9)-9(2i)+2i(9)-2i(2i)$
$\qquad\qquad = 81-18i+18i-4i^2$
$\qquad\qquad = 81-4(-1)$
$\qquad\qquad = 81+4$
$\qquad\qquad = 85$
$\qquad\qquad = 85+0i$

5. a. $\dfrac{4-i}{3+i} = \dfrac{(4-i)(3-i)}{(3+i)(3-i)}$
$\qquad = \dfrac{4(3)-4(i)-3(i)+i^2}{3^2-i^2}$
$\qquad = \dfrac{12-7i-1}{9+1}$
$\qquad = \dfrac{11-7i}{10}$
$\qquad = \dfrac{11}{10}-\dfrac{7i}{10}$ or $\dfrac{11}{10}-\dfrac{7}{10}i$

b. $\dfrac{5}{2i} = \dfrac{5(-2i)}{2i(-2i)}$
$\qquad = \dfrac{-10i}{-4i^2}$
$\qquad = \dfrac{-10i}{-4(-1)}$
$\qquad = \dfrac{-10i}{4}$
$\qquad = \dfrac{-5i}{2}$
$\qquad = 0-\dfrac{5i}{2}$ or $0-\dfrac{5}{2}i$

6. a. $i^9 = i^4 \cdot i^4 \cdot i = 1 \cdot 1 \cdot i = i$

b. $i^{16} = (i^4)^4 = 1^4 = 1$

c. $i^{34} = i^{32} \cdot i^2 = (i^4)^8 \cdot i^2 = 1^8(-1) = -1$

d. $i^{-24} = \dfrac{1}{i^{24}} = \dfrac{1}{(i^4)^6} = \dfrac{1}{(1)^6} = \dfrac{1}{1} = 1$

Vocabulary, Readiness & Video Check 7.7

1. A <u>complex</u> number is one that can be written in the form $a + bi$ where a and b are real numbers.

2. In the complex number system, i denotes the <u>imaginary unit</u>.

3. $i^2 = \underline{-1}$

4. $i = \underline{\sqrt{-1}}$

5. A complex number, $a + bi$, is a <u>real</u> number if $b = 0$.

6. A complex number, $a + bi$, is a <u>pure imaginary</u> number if $a = 0$ and $b \ne 0$.

7. The product rule for radicals; you need to first simplify each separate radical and have nonnegative radicands before applying the product rule.

8. combining like terms; i is *not* a variable, but a constant, $\sqrt{-1}$

9. The fact that $i^2 = -1$.

10. using conjugates to rationalize denominators with two terms

11. $i, i^2 = -1, i^3 = -i, i^4 = 1$

Exercise Set 7.7

1. $\sqrt{-81} = \sqrt{-1 \cdot 81} = \sqrt{-1}\sqrt{81} = 9i$

3. $\sqrt{-7} = \sqrt{-1 \cdot 7} = \sqrt{-1}\sqrt{7} = i\sqrt{7}$

5. $-\sqrt{16} = -4$

7. $\sqrt{-64} = \sqrt{-1 \cdot 64} = \sqrt{-1}\sqrt{64} = 8i$

9. $\sqrt{-24} = \sqrt{-1 \cdot 24} = \sqrt{-1}\sqrt{4 \cdot 6} = i \cdot 2\sqrt{6} = 2i\sqrt{6}$

11. $-\sqrt{-36} = -\sqrt{-1 \cdot 36} = -\sqrt{-1}\sqrt{36} = -i \cdot 6 = -6i$

13. $8\sqrt{-63} = 8\sqrt{-1 \cdot 63}$
$= 8\sqrt{-1}\sqrt{9 \cdot 7}$
$= 8i \cdot 3\sqrt{7}$
$= 24i\sqrt{7}$

15. $-\sqrt{54} = -\sqrt{9 \cdot 6} = -3\sqrt{6} = -3\sqrt{6} + 0i$

17. $\sqrt{-2} \cdot \sqrt{-7} = i\sqrt{2} \cdot i\sqrt{7}$
$= i^2\sqrt{14}$
$= (-1)\sqrt{14}$
$= -\sqrt{14}$

19. $\sqrt{-5} \cdot \sqrt{-10} = i\sqrt{5} \cdot i\sqrt{10}$
$= i^2\sqrt{50}$
$= (-1)\sqrt{25 \cdot 2}$
$= -5\sqrt{2}$

21. $\sqrt{16} \cdot \sqrt{-1} = 4i$

23. $\dfrac{\sqrt{-9}}{\sqrt{3}} = \dfrac{i\sqrt{9}}{\sqrt{3}} = i\sqrt{\dfrac{9}{3}} = i\sqrt{3}$

25. $\dfrac{\sqrt{-80}}{\sqrt{-10}} = \dfrac{i\sqrt{80}}{i\sqrt{10}} = \sqrt{\dfrac{80}{10}} = \sqrt{8} = \sqrt{4 \cdot 2} = 2\sqrt{2}$

27. $(4 - 7i) + (2 + 3i) = (4 + 2) + (-7 + 3)i$
$= 6 + (-4)i$
$= 6 - 4i$

29. $(6 + 5i) - (8 - i) = 6 + 5i - 8 + i$
$= (6 - 8) + (5 + 1)i$
$= -2 + 6i$

31. $6 - (8 + 4i) = 6 - 8 - 4i$
$= (6 - 8) - 4i$
$= -2 - 4i$

33. $-10i \cdot -4i = 40i^2 = 40(-1) = -40 = -40 + 0i$

35. $6i(2 - 3i) = 12i - 18i^2$
$= 12i - 18(-1)$
$= 18 + 12i$

37. $\left(\sqrt{3} + 2i\right)\left(\sqrt{3} - 2i\right)$
$= \sqrt{3} \cdot \sqrt{3} - \sqrt{3} \cdot 2i + \sqrt{3} \cdot 2i - 4i^2$
$= 3 - 4(-1) + 0i$
$= 3 + 4 + 0i$
$= 7 + 0i$

39. $(4 - 2i)^2 = (4 - 2i)(4 - 2i)$
$= 16 - 4 \cdot 2i - 4 \cdot 2i + 4i^2$
$= 16 - 8i - 8i + 4(-1)$
$= 16 - 16i - 4$
$= 12 - 16i$

41. $\dfrac{4}{i} = \dfrac{4(-i)}{i(-i)} = \dfrac{-4i}{-i^2} = \dfrac{-4i}{-(-1)} = -4i = 0 - 4i$

43. $\dfrac{7}{4+3i} = \dfrac{7(4-3i)}{(4+3i)(4-3i)}$

$\qquad = \dfrac{28-21i}{4^2 - 9i^2}$

$\qquad = \dfrac{28-21i}{16+9}$

$\qquad = \dfrac{28-21i}{25}$

$\qquad = \dfrac{28}{25} - \dfrac{21}{25}i$

45. $\dfrac{3+5i}{1+i} = \dfrac{(3+5i)(1-i)}{(1+i)(1-i)}$

$\qquad = \dfrac{3-3i+5i-5i^2}{1^2 - i^2}$

$\qquad = \dfrac{3+2i+5}{1+1}$

$\qquad = \dfrac{8+2i}{2}$

$\qquad = \dfrac{8}{2} + \dfrac{2}{2}i$

$\qquad = 4+i$

47. $\dfrac{5-i}{3-2i} = \dfrac{(5-i)(3+2i)}{(3-2i)(3+2i)}$

$\qquad = \dfrac{15+10i-3i-2i^2}{3^2 - 4i^2}$

$\qquad = \dfrac{15+7i+2}{9+4}$

$\qquad = \dfrac{17+7i}{13}$

$\qquad = \dfrac{17}{13} + \dfrac{7}{13}i$

49. $(7i)(-9i) = -63i^2 = -63(-1) = 63 = 63+0i$

51. $(6-3i) - (4-2i) = 6-3i-4+2i = 2-i$

53. $-3i(-1+9i) = 3i - 27i^2$

$\qquad = 3i - 27(-1)$

$\qquad = 27+3i$

55. $\dfrac{4-5i}{2i} = \dfrac{4-5i}{2i} \cdot \dfrac{-2i}{-2i}$

$\qquad = \dfrac{-8i + 10i^2}{-4i^2}$

$\qquad = \dfrac{-10-8i}{4}$

$\qquad = \dfrac{-10}{4} - \dfrac{8}{4}i$

$\qquad = -\dfrac{5}{2} - 2i$

57. $(4+i)(5+2i) = 20+8i+5i+2i^2$

$\qquad = 20+13i+2(-1)$

$\qquad = 20+13i-2$

$\qquad = 18+13i$

59. $(6-2i)(3+i) = 18+6i-6i-2i^2$

$\qquad = 18+2+0i$

$\qquad = 20+0i$

61. $(8-3i) + (2+3i) = 8-3i+2+3i = 10+0i$

63. $(1-i)(1+i) = 1+i-i-i^2 = 1+1+0i = 2+0i$

65. $\dfrac{16+15i}{-3i} = \dfrac{(16+15i)(3i)}{-3i(3i)}$

$\qquad = \dfrac{48i+45i^2}{-9i^2}$

$\qquad = \dfrac{-45+48i}{9}$

$\qquad = \dfrac{-45}{9} + \dfrac{48}{9}i$

$\qquad = -5 + \dfrac{16}{3}i$

67. $(9+8i)^2 = 9^2 + 2(9)(8i) + (8i)^2$

$\qquad = 81+144i+64i^2$

$\qquad = 81+144i-64$

$\qquad = 17+144i$

69. $\dfrac{2}{3+i} = \dfrac{2(3-i)}{(3+i)(3-i)}$

$= \dfrac{6-2i}{3^2-i^2}$

$= \dfrac{6-2i}{9+1}$

$= \dfrac{6-2i}{10}$

$= \dfrac{6}{10} - \dfrac{2}{10}i$

$= \dfrac{3}{5} - \dfrac{1}{5}i$

71. $(5-6i)-4i = 5-6i-4i = 5-10i$

73. $\dfrac{2-3i}{2+i} = \dfrac{(2-3i)(2-i)}{(2+i)(2-i)}$

$= \dfrac{4-2i-6i+3i^2}{2^2-i^2}$

$= \dfrac{4-8i-3}{4+1}$

$= \dfrac{1-8i}{5}$

$= \dfrac{1}{5} - \dfrac{8}{5}i$

75. $(2+4i)+(6-5i) = 2+4i+6-5i = 8-i$

77. $\left(\sqrt{6}+i\right)\left(\sqrt{6}-i\right) = \left(\sqrt{6}\right)^2 - i^2$

$= 6-(-1)$

$= 6+1$

$= 7$

$= 7+0i$

79. $4(2-i)^2 = 4(2^2 - 2\cdot 2i + i^2)$

$= 4(4-4i-1)$

$= 4(3-4i)$

$= 12-16i$

81. $i^8 = (i^4)^2 = 1^2 = 1$

83. $i^{21} = i^{20}\cdot i = (i^4)^5 \cdot i = 1^5 \cdot i = i$

85. $i^{11} = i^8 \cdot i^3 = (i^4)^2 \cdot i^3 = 1^2 \cdot (-i) = -i$

87. $i^{-6} = \dfrac{1}{i^6} = \dfrac{1}{i^4 \cdot i^2} = \dfrac{1}{1\cdot(-1)} = -1$

89. $(2i)^6 = 2^6 i^6 = 64i^4 \cdot i^2 = 64(1)(-1) = -64$

91. $(-3i)^5 = (-3)^5 i^5 = -243i^4 \cdot i = -243(1)i = -243i$

93. $x+50° + 90° = 180°$

$x+140° = 180°$

$x = 40°$

95. $\underline{1|}\ \begin{array}{rrrr} 1 & -6 & 3 & -4 \\ & 1 & -5 & -2 \\ \hline 1 & -5 & -2 & -6 \end{array}$

Answer: $x^2 - 5x - 2 - \dfrac{6}{x-1}$

97. 5 people

99. $5+9 = 14$ people

101. $\dfrac{5 \text{ people}}{30 \text{ people}} = \dfrac{1}{6} \approx 0.1666$

About 16.7% of the people reported an average checking balance of $201 to $300.

103. $i^3 - i^4 = -i - 1 = -1 - i$

105. $i^6 + i^8 = i^4 \cdot i^2 + (i^4)^2$

$= 1(-1) + 1^2$

$= -1 + 1$

$= 0$

$= 0 + 0i$

107. $2 + \sqrt{-9} = 2 + i\sqrt{9} = 2 + 3i$

109. $\dfrac{6+\sqrt{-18}}{3} = \dfrac{6 + i\sqrt{9\cdot 2}}{3}$

$= \dfrac{6 + 3i\sqrt{2}}{3}$

$= \dfrac{6}{3} + \dfrac{3\sqrt{2}}{3}i$

$= 2 + i\sqrt{2}$

111.

$$\frac{5-\sqrt{-75}}{10} = \frac{5-i\sqrt{25\cdot3}}{10}$$

$$= \frac{5-5i\sqrt{3}}{10}$$

$$= \frac{5}{10} - \frac{5\sqrt{3}}{10}i$$

$$= \frac{1}{2} - \frac{\sqrt{3}}{2}i$$

113. answers may vary

115. $\left(8-\sqrt{-3}\right) - \left(2+\sqrt{-12}\right)$

$$= 8 - i\sqrt{3} - 2 - 2i\sqrt{3}$$

$$= 6 - 3i\sqrt{3}$$

117. $x^2 + 4 = 0$

$$(2i)^2 + 4 = 0$$

$$4i^2 + 4 = 0$$

$$4(-1) + 4 = 0$$

     $-4 + 4 = 0$, which is true.

Yes, $2i$ is a solution.

Chapter 7 Vocabulary Check

1. The <u>conjugate</u> of $\sqrt{3}+2$ is $\sqrt{3}-2$.

2. The <u>principal square root</u> of a nonnegative number a is written as $\sqrt{a}$.

3. The process of writing a radical expression as an equivalent expression but without a radical in the denominator is called <u>rationalizing</u> the denominator.

4. The <u>imaginary unit</u>, written i, is the number whose square is -1.

5. The <u>cube root</u> of a number is written as $\sqrt[3]{a}$.

6. In the notation $\sqrt[n]{a}$, n is called the <u>index</u> and a is called the <u>radicand</u>.

7. Radicals with the same index and the same radicand are called <u>like radicals</u>.

8. A <u>complex number</u> is a number that can be written in the form $a + bi$, where a and b are real numbers.

9. The <u>distance</u> formula is
$$d = \sqrt{(x_2-x_1)^2 + (y_2-y_1)^2}.$$

10. The <u>midpoint</u> formula is $\left(\dfrac{x_1+x_2}{2}, \dfrac{y_1+y_2}{2}\right)$.

Chapter 7 Review

1. $\sqrt{81} = 9$ because $9^2 = 81$.

2. $\sqrt[4]{81} = 3$ because $3^4 = 81$.

3. $\sqrt[3]{-8} = -2$ because $(-2)^3 = -8$.

4. $\sqrt[4]{-16}$ is not a real number.

5. $-\sqrt{\dfrac{1}{49}} = -\dfrac{1}{7}$ because $\left(\dfrac{1}{7}\right)^2 = \dfrac{1}{49}$.

6. $\sqrt{x^{64}} = x^{32}$ because $(x^{32})^2 = x^{32\cdot2} = x^{64}$.

7. $-\sqrt{36} = -6$ because $6^2 = 36$.

8. $\sqrt[3]{64} = 4$ because $4^3 = 64$.

9. $\sqrt[3]{-a^6b^9} = \sqrt[3]{-1}\sqrt[3]{a^6}\sqrt[3]{b^9}$
$$= -1a^2b^3$$
$$= -a^2b^3$$

10. $\sqrt{16a^4b^{12}} = \sqrt{16}\sqrt{a^4}\sqrt{b^{12}} = 4a^2b^6$

11. $\sqrt[5]{32a^5b^{10}} = \sqrt[5]{32}\sqrt[5]{a^5}\sqrt[5]{b^{10}} = 2ab^2$

12. $\sqrt[5]{-32x^{15}y^{20}} = \sqrt[5]{-32}\sqrt[5]{x^{15}}\sqrt[5]{y^{20}} = -2x^3y^4$

13. $\sqrt{\dfrac{x^{12}}{36y^2}} = \dfrac{\sqrt{x^{12}}}{\sqrt{36y^2}} = \dfrac{x^6}{6y}$

14. $\sqrt[3]{\dfrac{27y^3}{z^{12}}} = \dfrac{\sqrt[3]{27y^3}}{\sqrt[3]{z^{12}}} = \dfrac{3y}{z^4}$

15. $\sqrt{(-x)^2} = |-x|$

16. $\sqrt[4]{(x^2-4)^4} = |x^2-4|$

17. $\sqrt[3]{(-27)^3} = -27$

18. $\sqrt[5]{(-5)^5} = -5$

19. $-\sqrt[5]{x^5} = -x$

20. $-\sqrt[3]{x^3} = -x$

21. $\sqrt[4]{16(2y+z)^4} = 2|2y+z|$

22. $\sqrt{25(x-y)^2} = 5|x-y|$

23. $\sqrt[5]{y^5} = y$

24. $\sqrt[6]{x^6} = |x|$

25. a. $f(x) = \sqrt{x} + 3$
   $f(0) = \sqrt{0} + 3 = 0 + 3 = 3$
   $f(9) = \sqrt{9} + 3 = 3 + 3 = 6$

   b. $f(x) = \sqrt{x} + 3$
   $x \ge 0$
   Domain: $[0, \infty)$

   c.

x	0	1	4	9
$f(x)$	3	4	5	6

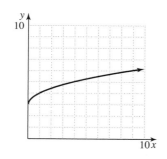

26. a. $g(x) = \sqrt[3]{x-3}$
   $g(11) = \sqrt[3]{11-3} = \sqrt[3]{8} = 2$
   $g(20) = \sqrt[3]{20-3} = \sqrt[3]{17}$

   b. $g(x) = \sqrt[3]{x-3}$
   Domain: $(-\infty, \infty)$

   c.

x	-5	2	3	4	11
$g(x)$	-2	-1	0	1	2

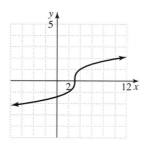

27. $\left(\dfrac{1}{81}\right)^{1/4} = \dfrac{\sqrt[4]{1}}{\sqrt[4]{81}} = \dfrac{1}{3}$

28. $\left(-\dfrac{1}{27}\right)^{1/3} = \dfrac{\sqrt[3]{-1}}{\sqrt[3]{27}} = -\dfrac{1}{3}$

29. $(-27)^{-1/3} = \dfrac{1}{\sqrt[3]{-27}} = -\dfrac{1}{3}$

30. $(-64)^{-1/3} = \dfrac{1}{\sqrt[3]{-64}} = -\dfrac{1}{4}$

31. $-9^{3/2} = -\left(\sqrt{9}\right)^3 = -(3)^3 = -27$

32. $64^{-1/3} = \dfrac{1}{\sqrt[3]{64}} = \dfrac{1}{4}$

33. $(-25)^{5/2} = \left(\sqrt{-25}\right)^5$ is not a real number.

34. $\left(\dfrac{25}{49}\right)^{-3/2} = \dfrac{1}{\left(\frac{25}{49}\right)^{3/2}}$

$= \left(\dfrac{49}{25}\right)^{3/2}$

$= \left(\dfrac{\sqrt{49}}{\sqrt{25}}\right)^{3}$

$= \left(\dfrac{7}{5}\right)^{3}$

$= \dfrac{343}{125}$

35. $\left(\dfrac{8}{27}\right)^{-2/3} = \dfrac{1}{\left(\frac{8}{27}\right)^{2/3}}$

$= \left(\dfrac{27}{8}\right)^{2/3}$

$= \left(\dfrac{\sqrt[3]{27}}{\sqrt[3]{8}}\right)^{2}$

$= \left(\dfrac{3}{2}\right)^{2}$

$= \dfrac{9}{4}$

36. $\left(-\dfrac{1}{36}\right)^{-1/4}$ is not a real number.

37. $\sqrt[3]{x^2} = x^{2/3}$

38. $\sqrt[5]{5x^2 y^3} = 5^{1/5} x^{2/5} y^{3/5}$

39. $y^{4/5} = \sqrt[5]{y^4}$

40. $5(xy^2 z^5)^{1/3} = 5\sqrt[3]{xy^2 z^5}$

41. $(x+2)^{-1/3} = \dfrac{1}{\sqrt[3]{x+2}}$

42. $(x+2y)^{-1/2} = \dfrac{1}{\sqrt{x+2y}}$

43. $a^{1/3} a^{4/3} a^{1/2} = a^{2/6} a^{8/6} a^{3/6} = a^{13/6}$

44. $\dfrac{b^{1/3}}{b^{4/3}} = b^{\frac{1}{3}-\frac{4}{3}} = b^{-3/3} = b^{-1} = \dfrac{1}{b}$

45. $(a^{1/2} a^{-2})^3 = a^{3/2} a^{-6} = a^{\frac{3}{2}-\frac{12}{2}} = a^{-9/2} = \dfrac{1}{a^{9/2}}$

46. $(x^{-3} y^6)^{1/3} = x^{-3/3} y^{6/3} = x^{-1} y^2 = \dfrac{y^2}{x}$

47. $\left(\dfrac{b^{3/4}}{a^{-1/2}}\right)^8 = \dfrac{b^{24/4}}{a^{-8/2}} = \dfrac{b^6}{a^{-4}} = a^4 b^6$

48. $\dfrac{x^{1/4} x^{-1/2}}{x^{2/3}} = x^{\frac{1}{4}-\frac{1}{2}-\frac{2}{3}} = x^{\frac{3}{12}-\frac{6}{12}-\frac{8}{12}} = x^{-\frac{11}{12}} = \dfrac{1}{x^{11/12}}$

49. $\left(\dfrac{49c^{5/3}}{a^{-1/4} b^{5/6}}\right)^{-1} = \left(\dfrac{a^{-1/4} b^{5/6}}{49c^{5/3}}\right) = \dfrac{b^{5/6}}{49a^{1/4} c^{5/3}}$

50. $a^{-1/4}(a^{5/4} - a^{9/4}) = a^{-\frac{1}{4}+\frac{5}{4}} - a^{-\frac{1}{4}+\frac{9}{4}}$

$= a^{\frac{4}{4}} - a^{\frac{8}{4}}$

$= a - a^2$

51. $\sqrt{20} \approx 4.472$

52. $\sqrt[3]{-39} \approx -3.391$

53. $\sqrt[4]{726} \approx 5.191$

54. $56^{1/3} \approx 3.826$

55. $-78^{3/4} \approx -26.246$

56. $105^{-2/3} \approx 0.045$

57. $\sqrt[3]{2} \cdot \sqrt{7} = 2^{1/3} \cdot 7^{1/2}$

$= 2^{2/6} \cdot 7^{3/6}$

$= \sqrt[6]{2^2 \cdot 7^3}$

$= \sqrt[6]{1372}$

58. $\sqrt[3]{3} \cdot \sqrt[4]{x} = 3^{1/3} x^{1/4}$

$= 3^{4/12} \cdot x^{3/12}$

$= \sqrt[12]{3^4 x^3}$

$= \sqrt[12]{81x^3}$

59. $\sqrt{3} \cdot \sqrt{8} = \sqrt{24} = \sqrt{4}\sqrt{6} = 2\sqrt{6}$

60. $\sqrt[3]{7y} \cdot \sqrt[3]{x^2 z} = \sqrt[3]{7x^2 yz}$

61. $\dfrac{\sqrt{44x^3}}{\sqrt{11x}} = \sqrt{\dfrac{44x^3}{11x}} = \sqrt{4x^2} = 2x$

62. $\dfrac{\sqrt[4]{a^6 b^{13}}}{\sqrt[4]{a^2 b}} = \sqrt[4]{\dfrac{a^6 b^{13}}{a^2 b}} = \sqrt[4]{a^4 b^{12}} = ab^3$

63. $\sqrt{60} = \sqrt{4}\sqrt{15} = 2\sqrt{15}$

64. $-\sqrt{75} = -\sqrt{25}\sqrt{3} = -5\sqrt{3}$

65. $\sqrt[3]{162} = \sqrt[3]{27}\sqrt[3]{6} = 3\sqrt[3]{6}$

66. $\sqrt[3]{-32} = \sqrt[3]{-8}\sqrt[3]{4} = -2\sqrt[3]{4}$

67. $\sqrt{36x^7} = \sqrt{36x^6}\sqrt{x} = 6x^3 \sqrt{x}$

68. $\sqrt[3]{24a^5 b^7} = \sqrt[3]{8a^3 b^6}\sqrt[3]{3a^2 b^1} = 2ab^2 \sqrt[3]{3a^2 b}$

69. $\sqrt{\dfrac{p^{17}}{121}} = \dfrac{\sqrt{p^{16}}\sqrt{p}}{\sqrt{121}} = \dfrac{p^8 \sqrt{p}}{11}$

70. $\sqrt[3]{\dfrac{y^5}{27x^6}} = \dfrac{\sqrt[3]{y^3}\sqrt[3]{y^2}}{\sqrt[3]{27x^6}} = \dfrac{y\sqrt[3]{y^2}}{3x^2}$

71. $\sqrt[4]{\dfrac{xy^6}{81}} = \dfrac{\sqrt[4]{y^4}\sqrt[4]{xy^2}}{\sqrt[4]{81}} = \dfrac{y\sqrt[4]{xy^2}}{3}$

72. $\sqrt{\dfrac{2x^3}{49y^4}} = \dfrac{\sqrt{x^2}\sqrt{2x}}{\sqrt{49y^4}} = \dfrac{x\sqrt{2x}}{7y^2}$

73. $r = \sqrt{\dfrac{A}{\pi}}$

$\quad = \sqrt{\dfrac{25}{\pi}}$

$\quad = \dfrac{\sqrt{25}}{\sqrt{\pi}}$

$\quad = \dfrac{5}{\sqrt{\pi}}$ meters or $\dfrac{5\sqrt{\pi}}{\pi}$ meters

74. $r = \sqrt{\dfrac{A}{\pi}} = \sqrt{\dfrac{104}{\pi}} = 5.75$ inches

75. $(x_1,\ y_1) = (-6,\ 3),\ (x_2,\ y_2) = (8,\ 4)$

$d = \sqrt{(x_2 - x_1)^2 + (y_2 - y_1)^2}$

$\quad = \sqrt{(8+6)^2 + (4-3)^2}$

$\quad = \sqrt{196+1}$

$\quad = \sqrt{197}$ units ≈ 14.036 units

76. $(x_1,\ y_1) = (-4,\ -6),\ (x_2,\ y_2) = (-1,\ 5)$

$d = \sqrt{(x_2 - x_1)^2 + (y_2 - y_1)^2}$

$\quad = \sqrt{(-1+4)^2 + (5+6)^2}$

$\quad = \sqrt{9+121}$

$\quad = \sqrt{130}$ units ≈ 11.402 units

77. $(x_1,\ y_1) = (-1,\ 5),\ (x_2,\ y_2) = (2,\ -3)$

$d = \sqrt{(x_2 - x_1)^2 + (y_2 - y_1)^2}$

$\quad = \sqrt{(2+1)^2 + (-3-5)^2}$

$\quad = \sqrt{9+64}$

$\quad = \sqrt{73}$ units ≈ 8.544 units

78. $(x_1,\ y_1) = \left(-\sqrt{2},\ 0\right),\ (x_2,\ y_2) = \left(0,\ -4\sqrt{6}\right)$

$d = \sqrt{(x_2 - x_1)^2 + (y_2 - y_1)^2}$

$\quad = \sqrt{\left(0+\sqrt{2}\right)^2 + \left(-4\sqrt{6}-0\right)^2}$

$\quad = \sqrt{2+96}$

$\quad = \sqrt{98}$

$\quad = 7\sqrt{2}$ units ≈ 9.899 units

79. $(x_1,\ y_1) = \left(-\sqrt{5},\ -\sqrt{11}\right),$

$\quad (x_2,\ y_2) = \left(-\sqrt{5},\ -3\sqrt{11}\right)$

$d = \sqrt{(x_2 - x_1)^2 + (y_2 - y_1)^2}$

$\quad = \sqrt{\left(-\sqrt{5}+\sqrt{5}\right)^2 + \left(-3\sqrt{11}+\sqrt{11}\right)^2}$

$\quad = \sqrt{0+44}$

$\quad = \sqrt{44}$

$\quad = 2\sqrt{11}$ units ≈ 6.633 units

80. $(x_1, y_1) = (7.4, -8.6), (x_2, y_2) = (-1.2, 5.6)$

$$d = \sqrt{(-1.2 - 7.4)^2 + (5.6 + 8.6)^2}$$
$$= \sqrt{(-8.6)^2 + (14.2)^2}$$
$$= \sqrt{73.96 + 201.64}$$
$$= \sqrt{275.6} \text{ units} \approx 16.601 \text{ units}$$

81. $(x_1, y_1) = (2, 6), (x_2, y_2) = (-12, 4)$

$$\text{midpoint} = \left(\frac{x_1 + x_2}{2}, \frac{y_1 + y_2}{2} \right)$$
$$= \left(\frac{2 - 12}{2}, \frac{6 + 4}{2} \right)$$
$$= \left(\frac{-10}{2}, \frac{10}{2} \right)$$
$$= (-5, 5)$$

82. $(x_1, y_1) = (-6, -5), (x_2, y_2) = (-9, 7)$

$$\text{midpoint} = \left(\frac{x_1 + x_2}{2}, \frac{y_1 + y_2}{2} \right)$$
$$= \left(\frac{-6 - 9}{2}, \frac{-5 + 7}{2} \right)$$
$$= \left(-\frac{15}{2}, 1 \right)$$

83. $(x_1, y_1) = (4, -6), (x_2, y_2) = (-15, 2)$

$$\text{midpoint} = \left(\frac{x_1 + x_2}{2}, \frac{y_1 + y_2}{2} \right)$$
$$= \left(\frac{4 - 15}{2}, \frac{-6 + 2}{2} \right)$$
$$= \left(-\frac{11}{2}, -2 \right)$$

84. $(x_1, y_1) = \left(0, -\frac{3}{8} \right), (x_2, y_2) = \left(\frac{1}{10}, 0 \right)$

$$\text{midpoint} = \left(\frac{x_1 + x_2}{2}, \frac{y_1 + y_2}{2} \right)$$
$$= \left(\frac{0 + \frac{1}{10}}{2}, \frac{-\frac{3}{8} + 0}{2} \right)$$
$$= \left(\frac{1}{20}, -\frac{3}{16} \right)$$

85. $(x_1, y_1) = \left(\frac{3}{4}, -\frac{1}{7} \right), (x_2, y_2) = \left(-\frac{1}{4}, -\frac{3}{7} \right)$

$$\text{midpoint} = \left(\frac{\frac{3}{4} - \frac{1}{4}}{2}, \frac{-\frac{1}{7} - \frac{3}{7}}{2} \right)$$
$$= \left(\frac{\frac{1}{2}}{2}, \frac{-\frac{11}{7}}{2} \right)$$
$$= \left(\frac{1}{4}, -\frac{2}{7} \right)$$

86. $(x_1, y_1) = \left(\sqrt{3}, -2\sqrt{6} \right), (x_2, y_2) = \left(\sqrt{3}, -4\sqrt{6} \right)$

$$\text{midpoint} = \left(\frac{x_1 + x_2}{2}, \frac{y_1 + y_2}{2} \right)$$
$$= \left(\frac{\sqrt{3} + \sqrt{3}}{2}, \frac{-2\sqrt{6} - 4\sqrt{6}}{2} \right)$$
$$= \left(\frac{2\sqrt{3}}{2}, \frac{-6\sqrt{6}}{2} \right)$$
$$= \left(\sqrt{3}, -3\sqrt{6} \right)$$

87. $\sqrt{20} + \sqrt{45} - 7\sqrt{5} = 2\sqrt{5} + 3\sqrt{5} - 7\sqrt{5} = -2\sqrt{5}$

88. $x\sqrt{75x} - \sqrt{27x^3} = 5x\sqrt{3x} - 3x\sqrt{3x} = 2x\sqrt{3x}$

89. $\sqrt[3]{128} + \sqrt[3]{250} = 4\sqrt[3]{2} + 5\sqrt[3]{2} = 9\sqrt[3]{2}$

90. $3\sqrt[4]{32a^5} - a\sqrt[4]{162a} = 6a\sqrt[4]{2a} - 3a\sqrt[4]{2a} = 3a\sqrt[4]{2a}$

91. $\dfrac{5}{\sqrt{4}} + \dfrac{\sqrt{3}}{3} = \dfrac{5}{2} + \dfrac{\sqrt{3}}{3} = \dfrac{15}{6} + \dfrac{2\sqrt{3}}{6} = \dfrac{15 + 2\sqrt{3}}{6}$

92. $\sqrt{\dfrac{8}{x^2}} - \sqrt{\dfrac{50}{16x^2}} = \dfrac{2\sqrt{2}}{x} - \dfrac{5\sqrt{2}}{4x}$
$$= \frac{8\sqrt{2} - 5\sqrt{2}}{4x}$$
$$= \frac{3\sqrt{2}}{4x}$$

93. $2\sqrt{50} - 3\sqrt{125} + \sqrt{98} = 10\sqrt{2} - 15\sqrt{5} + 7\sqrt{2}$
$$= 17\sqrt{2} - 15\sqrt{5}$$

94. $2a\sqrt[4]{32b^5} - 3b\sqrt[4]{162a^4b} + \sqrt[4]{2a^4b^5}$
$$= 4ab\sqrt[4]{2b} - 9ab\sqrt[4]{2b} + ab\sqrt[4]{2b}$$
$$= -4ab\sqrt[4]{2b}$$

95. $\sqrt{3}\left(\sqrt{27}-\sqrt{3}\right)=\sqrt{3}\sqrt{27}-\sqrt{3}\sqrt{3}$
$=\sqrt{81}-\sqrt{9}$
$=9-3$
$=6$

96. $\left(\sqrt{x}-3\right)^2=\left(\sqrt{x}\right)^2-(2)(3)\sqrt{x}+9=x-6\sqrt{x}+9$

97. $\left(\sqrt{5}-5\right)\left(2\sqrt{5}+2\right)$
$=\sqrt{5}\left(2\sqrt{5}\right)+2\sqrt{5}-10\sqrt{5}-10$
$=10-8\sqrt{5}-10$
$=-8\sqrt{5}$

98. $\left(2\sqrt{x}-3\sqrt{y}\right)\left(2\sqrt{x}+3\sqrt{y}\right)=\left(2\sqrt{x}\right)^2-\left(3\sqrt{y}\right)^2$
$=4x-9y$

99. $\left(\sqrt{a}+3\right)\left(\sqrt{a}-3\right)=\left(\sqrt{a}\right)^2-(3)^2=a-9$

100. $\left(\sqrt[3]{a}+2\right)^2=\left(\sqrt[3]{a}\right)^2+2(2)\left(\sqrt[3]{a}\right)+2^2$
$=\sqrt[3]{a^2}+4\sqrt[3]{a}+4$

101. $\left(\sqrt[3]{5x}+9\right)\left(\sqrt[3]{5x}-9\right)=\left(\sqrt[3]{5x}\right)^2-9^2$
$=\sqrt[3]{25x^2}-81$

102. $\left(\sqrt[3]{a}+4\right)\left(\sqrt[3]{a^2}-4\sqrt[3]{a}+16\right)$
$=\sqrt[3]{a}\sqrt[3]{a^2}-\sqrt[3]{a}\left(4\sqrt[3]{a}\right)+16\sqrt[3]{a}+4\sqrt[3]{a^2}-16\sqrt[3]{a}+64$
$=a-4\sqrt[3]{a^2}+4\sqrt[3]{a^2}+64$
$=a+64$

103. $\dfrac{3}{\sqrt{7}}=\dfrac{3}{\sqrt{7}}\cdot\dfrac{\sqrt{7}}{\sqrt{7}}=\dfrac{3\sqrt{7}}{7}$

104. $\sqrt{\dfrac{x}{12}}=\dfrac{\sqrt{x}}{\sqrt{12}}=\dfrac{\sqrt{x}}{2\sqrt{3}}\cdot\dfrac{\sqrt{3}}{\sqrt{3}}=\dfrac{\sqrt{3x}}{6}$

105. $\dfrac{5}{\sqrt[3]{4}}=\dfrac{5}{\sqrt[3]{4}}\cdot\dfrac{\sqrt[3]{2}}{\sqrt[3]{2}}=\dfrac{5\sqrt[3]{2}}{2}$

106. $\sqrt{\dfrac{24x^5}{3y}}=\dfrac{\sqrt{24x^5}}{\sqrt{3y}}$
$=\dfrac{2x^2\sqrt{6x}}{\sqrt{3y}}\cdot\dfrac{\sqrt{3y}}{\sqrt{3y}}$
$=\dfrac{2x^2\sqrt{9xy}}{3y}$
$=\dfrac{6x^2\sqrt{2xy}}{3y}$
$=\dfrac{2x^2\sqrt{2xy}}{y}$

107. $\sqrt[3]{\dfrac{15x^6y^7}{z^2}}=\dfrac{\sqrt[3]{15x^6y^7}}{\sqrt[3]{z^2}}$
$=\dfrac{x^2y^2\sqrt[3]{15y}}{\sqrt[3]{z^2}}\cdot\dfrac{\sqrt[3]{z}}{\sqrt[3]{z}}$
$=\dfrac{x^2y^2\sqrt[3]{15yz}}{z}$

108. $\sqrt[4]{\dfrac{81}{8x^{10}}}=\dfrac{\sqrt[4]{81}}{\sqrt[4]{8x^{10}}}=\dfrac{3}{x^2\sqrt[4]{8x^2}}\cdot\dfrac{\sqrt[4]{2x^2}}{\sqrt[4]{2x^2}}=\dfrac{3\sqrt[4]{2x^2}}{2x^3}$

109. $\dfrac{3}{\sqrt{y}-2}=\dfrac{3}{\sqrt{y}-2}\cdot\dfrac{\sqrt{y}+2}{\sqrt{y}+2}=\dfrac{3\sqrt{y}+6}{y-4}$

110. $\dfrac{\sqrt{2}-\sqrt{3}}{\sqrt{2}+\sqrt{3}}=\dfrac{\sqrt{2}-\sqrt{3}}{\sqrt{2}+\sqrt{3}}\cdot\dfrac{\sqrt{2}-\sqrt{3}}{\sqrt{2}-\sqrt{3}}$
$=\dfrac{2-2\sqrt{6}+3}{2-3}$
$=-5+2\sqrt{6}$

111. $\dfrac{\sqrt{11}}{3}=\dfrac{\sqrt{11}}{3}\cdot\dfrac{\sqrt{11}}{\sqrt{11}}=\dfrac{11}{3\sqrt{11}}$

112. $\sqrt{\dfrac{18}{y}}=\dfrac{\sqrt{18}}{\sqrt{y}}=\dfrac{3\sqrt{2}}{\sqrt{y}}\cdot\dfrac{\sqrt{2}}{\sqrt{2}}=\dfrac{6}{\sqrt{2y}}$

113. $\dfrac{\sqrt[3]{9}}{7}=\dfrac{\sqrt[3]{9}}{7}\cdot\dfrac{\sqrt[3]{3}}{\sqrt[3]{3}}=\dfrac{3}{7\sqrt[3]{3}}$

114. $\sqrt{\dfrac{24x^5}{3y^2}} = \dfrac{\sqrt{24x^5}}{\sqrt{3y^2}}$

$\qquad = \dfrac{2x^2\sqrt{6x}}{y\sqrt{3}} \cdot \dfrac{\sqrt{6x}}{\sqrt{6x}}$

$\qquad = \dfrac{12x^3}{3y\sqrt{2x}}$

$\qquad = \dfrac{4x^3}{y\sqrt{2x}}$

115. $\sqrt[3]{\dfrac{xy^2}{10z}} = \dfrac{\sqrt[3]{xy^2}}{\sqrt[3]{10z}} = \dfrac{\sqrt[3]{xy^2}}{\sqrt[3]{10z}} \cdot \dfrac{\sqrt[3]{x^2y}}{\sqrt[3]{x^2y}} = \dfrac{xy}{\sqrt[3]{10x^2yz}}$

116. $\dfrac{\sqrt{x}+5}{-3} = \dfrac{\sqrt{x}+5}{-3} \cdot \dfrac{\sqrt{x}-5}{\sqrt{x}-5} = \dfrac{x-25}{-3\sqrt{x}+15}$

117. $\sqrt{y-7} = 5$

$\qquad y - 7 = 25$

$\qquad\qquad y = 32$

The solution is 32.

118. $\sqrt{2x}+10 = 4$

$\qquad \sqrt{2x} = -6$

No solution since a principal square root does not yield a negative number. The solution set is $\varnothing$.

119. $\sqrt[3]{2x-6} = 4$

$\qquad \left(\sqrt[3]{2x-6}\right)^3 = 4^3$

$\qquad\qquad 2x-6 = 64$

$\qquad\qquad\quad 2x = 70$

$\qquad\qquad\quad\ x = 35$

The solution is 35.

120. $\sqrt{x+6} = \sqrt{x+2}$

$\qquad \left(\sqrt{x+6}\right)^2 = \left(\sqrt{x+2}\right)^2$

$\qquad\qquad x+6 = x+2$

$\qquad\qquad\quad 6 = 2 \quad$ False

The solution set is $\varnothing$.

121. $2x - 5\sqrt{x} = 3$

$\qquad 2x - 3 = 5\sqrt{x}$

$\qquad 4x^2 - 12x + 9 = 25x$

$\qquad 4x^2 - 37x + 9 = 0$

$\qquad (4x-1)(x-9) = 0$

$\qquad 4x - 1 = 0 \quad$ or $\quad x - 9 = 0$

$\qquad\quad 4x = 1 \qquad\qquad\quad x = 9$

$\qquad\qquad x = \dfrac{1}{4}$

We discard the $\dfrac{1}{4}$ as extraneous, leaving $x = 9$ as the only solution. The solution is 9.

122. $\sqrt{x+9} = 2 + \sqrt{x-7}$

$\qquad x+9 = 4 + (2)2\sqrt{x-7} + x - 7$

$\qquad 12 = 4\sqrt{x-7}$

$\qquad 3 = \sqrt{x-7}$

$\qquad 9 = x - 7$

$\qquad 16 = x$

The solution is 16.

123. $a^2 + b^2 = c^2$

$\qquad 3^2 + 3^2 = c^2$

$\qquad 9 + 9 = c^2$

$\qquad 18 = c^2$

$\qquad \sqrt{18} = c$

$\qquad 3\sqrt{2} = c$

The unknown length is $3\sqrt{2}$ centimeters.

124. $a^2 + b^2 = c^2$

$\qquad 7^2 + \left(8\sqrt{3}\right)^2 = c^2$

$\qquad 49 + 192 = c^2$

$\qquad 241 = c^2$

$\qquad \sqrt{241} = c$

The unknown length is $\sqrt{241}$ feet.

125. $a^2 + b^2 = c^2$

$\qquad a^2 + 40^2 = 65^2$

$\qquad a^2 + 1600 = 4225$

$\qquad a^2 = 2625$

$\qquad a = 51.2$

The width of the pond is 51.2 feet.

126.
$$a^2 + b^2 = c^2$$
$$3^2 + 3^2 = c^2$$
$$9 + 9 = c^2$$
$$18 = c^2$$
$$\sqrt{18} = c$$
$3\sqrt{2}$ or $4.24 = c$
The length is 4.24 feet.

127. $\sqrt{-8} = \sqrt{-1} \cdot \sqrt{8} = 2i\sqrt{2} = 0 + 2i\sqrt{2}$

128. $-\sqrt{-6} = -\sqrt{-1} \cdot \sqrt{6} = -i\sqrt{6} = 0 - i\sqrt{6}$

129.
$$\sqrt{-4} + \sqrt{-16} = \sqrt{-1}\sqrt{4} + \sqrt{-1}\sqrt{16}$$
$$= 2i + 4i$$
$$= 6i$$
$$= 0 + 6i$$

130.
$$\sqrt{-2} \cdot \sqrt{-5} = \sqrt{-1} \cdot \sqrt{2} \cdot \sqrt{-1} \cdot \sqrt{5}$$
$$= i\sqrt{2} \cdot i\sqrt{5}$$
$$= i^2\sqrt{10}$$
$$= -\sqrt{10}$$
$$= -\sqrt{10} + 0i$$

131. $(12 - 6i) + (3 + 2i) = (12 + 3) + (-6 + 2)i = 15 - 4i$

132.
$$(-8 - 7i) - (5 - 4i) = (-8 - 5) + [-7 - (-4)]i$$
$$= -13 - 3i$$

133. $(2i)^6 = 2^6 \cdot i^6 = 64 \cdot (-1) = -64$

134. $(3i)^4 = 3^4 \cdot i^4 = 81 \cdot 1 = 81$

135. $-3i(6 - 4i) = -18i + 12(-1) = -12 - 18i$

136. $(3 + 2i)(1 + i) = 3 + 3i + 2i - 2 = 1 + 5i$

137.
$$(2 - 3i)^2 = 2^2 - 2(2)(3i) + (3i)^2$$
$$= 4 - 12i - 9$$
$$= -5 - 12i$$

138. $\left(\sqrt{6} - 9i\right)\left(\sqrt{6} + 9i\right) = \left(\sqrt{6}\right)^2 - (9i)^2 = 6 + 81 = 87$

139. $\dfrac{2 + 3i}{2i} = \dfrac{2 + 3i}{2i} \cdot \dfrac{-2i}{-2i} = \dfrac{-4i + 6}{4} = \dfrac{3}{2} - i$

140. $\dfrac{1 + i}{-3i} = \dfrac{1 + i}{-3i} \cdot \dfrac{3i}{3i} = \dfrac{3i - 3}{9} = \dfrac{-1 + i}{3} = -\dfrac{1}{3} + \dfrac{1}{3}i$

141. $\sqrt[3]{x^3} = x$

142. $\sqrt{(x + 2)^2} = |x + 2|$

143. $-\sqrt{100} = -10$

144. $\sqrt[3]{-x^{12}y^3} = -x^4 y$

145. $\sqrt[4]{\dfrac{y^{20}}{16x^{12}}} = \dfrac{\sqrt[4]{y^{20}}}{\sqrt[4]{16x^{12}}} = \dfrac{y^5}{2x^3}$

146. $9^{1/2} = \sqrt{9} = 3$

147. $64^{-1/2} = \dfrac{1}{64^{1/2}} = \dfrac{1}{\sqrt{64}} = \dfrac{1}{8}$

148. $\left(\dfrac{27}{64}\right)^{-2/3} = \left(\dfrac{64}{27}\right)^{2/3} = \left(\sqrt[3]{\dfrac{64}{27}}\right)^2 = \left(\dfrac{4}{3}\right)^2 = \dfrac{16}{9}$

149.
$$\dfrac{(x^{2/3}x^{-3})^3}{x^{-1/2}} = \dfrac{x^{6/3}x^{-9}}{x^{-1/2}}$$
$$= x^{2 - 9 + \frac{1}{2}}$$
$$= x^{-13/2}$$
$$= \dfrac{1}{x^{13/2}}$$

150. $\sqrt{200x^9} = 10x^4\sqrt{2x}$

151. $\sqrt{\dfrac{3n^3}{121m^{10}}} = \dfrac{\sqrt{3n^3}}{\sqrt{121m^{10}}} = \dfrac{n\sqrt{3n}}{11m^5}$

152.
$$3\sqrt{20} - 7x\sqrt[3]{40} + 3\sqrt[3]{5x^3}$$
$$= 3\sqrt{4}\sqrt{5} - 7x\sqrt[3]{8}\sqrt[3]{54} + 3\sqrt[3]{x^3}\sqrt[3]{5}$$
$$= 6\sqrt{5} - 14x\sqrt[3]{5} + 3x\sqrt[3]{5}$$
$$= 6\sqrt{5} - 11x\sqrt[3]{5}$$

153.
$$\left(2\sqrt{x} - 5\right)^2 = \left(2\sqrt{x}\right)^2 - 2(5)\left(2\sqrt{x}\right) + 5^2$$
$$= 4x - 20\sqrt{x} + 25$$

154. $(x_1, y_1) = (-3, 5), (x_2, y_2) = (-8, 9)$

$$d = \sqrt{(x_2 - x_1)^2 + (y_2 - y_1)^2}$$
$$= \sqrt{(-8+3)^2 + (9-5)^2}$$
$$= \sqrt{(-5)^2 + (4)^2}$$
$$= \sqrt{25+16}$$
$$= \sqrt{41}$$

The distance is $\sqrt{41}$ units.

155. $(x_1, y_1) = (-3, 8), (x_2, y_2) = (11, 24)$

$$\text{midpoint} = \left(\frac{x_1 + x_2}{2}, \frac{y_1 + y_2}{2} \right)$$
$$= \left(\frac{-3+11}{2}, \frac{8+24}{2} \right)$$
$$= \left(\frac{8}{2}, \frac{32}{2} \right)$$
$$= (4, 16)$$

156. $\dfrac{7}{\sqrt{13}} = \dfrac{7}{\sqrt{13}} \cdot \dfrac{\sqrt{13}}{\sqrt{13}} = \dfrac{7\sqrt{13}}{13}$

157. $\dfrac{2}{\sqrt{x}+3} = \dfrac{2}{\sqrt{x}+3} \cdot \dfrac{\sqrt{x}-3}{\sqrt{x}-3} = \dfrac{2\sqrt{x}-6}{x-9}$

158. $\sqrt{x}+2 = x$

$$\sqrt{x} = x-2$$
$$\left(\sqrt{x}\right)^2 = (x-2)^2$$
$$x = x^2 - 4x + 4$$
$$0 = x^2 - 5x + 4$$
$$0 = (x-4)(x-1)$$
$$x-4 = 0 \quad \text{or} \quad x-1 = 0$$
$$x = 4 \qquad\qquad x = 1$$

Discard the extraneous solution $x = 1$. The solution is 4.

159. $\sqrt{2x-1}+2 = x$

$$\sqrt{2x-1} = x-2$$
$$\left(\sqrt{2x-1}\right)^2 = (x-2)^2$$
$$2x-1 = x^2 - 4x + 4$$
$$0 = x^2 - 6x + 5$$
$$0 = (x-5)(x-1)$$
$$x-5 = 0 \quad \text{or} \quad x-1 = 0$$
$$x = 5 \quad \text{or} \qquad x = 1$$

Discard the extraneous solution $x = 1$. The solution is 5.

Chapter 7 Test

1. $\sqrt{216} = \sqrt{36 \cdot 6} = 6\sqrt{6}$

2. $-\sqrt[4]{x^{64}} = -x^{16}$

3. $\left(\dfrac{1}{125} \right)^{1/3} = \dfrac{1}{125^{1/3}} = \dfrac{1}{\sqrt[3]{125}} = \dfrac{1}{5}$

4. $\left(\dfrac{1}{125} \right)^{-1/3} = \dfrac{1}{\left(\frac{1}{125}\right)^{1/3}} = \dfrac{1}{\frac{1}{5}} = 5$

5. $\left(\dfrac{8x^3}{27} \right)^{2/3} = \dfrac{(8x^3)^{2/3}}{27^{2/3}}$

$$= \dfrac{\left(\sqrt[3]{8x^3}\right)^2}{\left(\sqrt[3]{27}\right)^2}$$
$$= \dfrac{(2x)^2}{3^2}$$
$$= \dfrac{4x^2}{9}$$

6. $\sqrt[3]{-a^{18}b^9} = \sqrt[3]{-1a^{18}b^9} = (-1)a^6b^3 = -a^6b^3$

7. $\left(\dfrac{64c^{4/3}}{a^{-2/3}b^{5/6}} \right)^{1/2} = \left(\dfrac{64a^{2/3}c^{4/3}}{b^{5/6}} \right)^{1/2}$

$$= \dfrac{64^{1/2}(a^{2/3})^{1/2}(c^{4/3})^{1/2}}{(b^{5/6})^{1/2}}$$
$$= \dfrac{\sqrt{64}a^{1/3}c^{2/3}}{b^{5/12}}$$
$$= \dfrac{8a^{1/3}c^{2/3}}{b^{5/12}}$$

8. $a^{-2/3}(a^{5/4} - a^3) = a^{-2/3}a^{5/4} - a^{-2/3}a^3$

$$= a^{-\frac{2}{3}+\frac{5}{4}} - a^{-\frac{2}{3}+3}$$
$$= a^{-\frac{8}{12}+\frac{15}{12}} - a^{-\frac{2}{3}+\frac{9}{3}}$$
$$= a^{7/12} - a^{7/3}$$

9. $\sqrt[4]{(4xy)^4} = |4xy| = 4|xy|$

10. $\sqrt[3]{(-27)^3} = -27$

11. $\sqrt{\dfrac{9}{y}} = \dfrac{\sqrt{9}}{\sqrt{y}} = \dfrac{3}{\sqrt{y}} = \dfrac{3 \cdot \sqrt{y}}{\sqrt{y} \cdot \sqrt{y}} = \dfrac{3\sqrt{y}}{y}$

12. $\dfrac{4 - \sqrt{x}}{4 + 2\sqrt{x}} = \dfrac{4 - \sqrt{x}}{2\left(2 + \sqrt{x}\right)}$

$= \dfrac{\left(4 - \sqrt{x}\right)\left(2 - \sqrt{x}\right)}{2\left(2 + \sqrt{x}\right)\left(2 - \sqrt{x}\right)}$

$= \dfrac{8 - 4\sqrt{x} - 2\sqrt{x} + x}{2\left[2^2 - \left(\sqrt{x}\right)^2\right]}$

$= \dfrac{8 - 6\sqrt{x} + x}{2(4 - x)} \text{ or } \dfrac{8 - 6\sqrt{x} + x}{8 - 2x}$

13. $\dfrac{\sqrt[3]{ab}}{\sqrt[3]{ab^2}} = \sqrt[3]{\dfrac{ab}{ab^2}}$

$= \sqrt[3]{\dfrac{1}{b}}$

$= \dfrac{1}{\sqrt[3]{b}}$

$= \dfrac{1 \cdot \sqrt[3]{b^2}}{\sqrt[3]{b} \cdot \sqrt[3]{b^2}}$

$= \dfrac{\sqrt[3]{b^2}}{b}$

14. $\dfrac{\sqrt{6} + x}{8} = \dfrac{\left(\sqrt{6} + x\right)\left(\sqrt{6} - x\right)}{8\left(\sqrt{6} - x\right)}$

$= \dfrac{\left(\sqrt{6}\right)^2 - x^2}{8\left(\sqrt{6} - x\right)}$

$= \dfrac{6 - x^2}{8\left(\sqrt{6} - x\right)}$

15. $\sqrt{125x^3} - 3\sqrt{20x^3} = \sqrt{25x^2 \cdot 5x} - 3\sqrt{4x^2 \cdot 5x}$

$= 5x\sqrt{5x} - 3 \cdot 2x\sqrt{5x}$

$= 5x\sqrt{5x} - 6x\sqrt{5x}$

$= -x\sqrt{5x}$

16. $\sqrt{3}\left(\sqrt{16} - \sqrt{2}\right) = \sqrt{3}\left(4 - \sqrt{2}\right)$

$= 4\sqrt{3} - \sqrt{3}\sqrt{2}$

$= 4\sqrt{3} - \sqrt{6}$

17. $\left(\sqrt{x} + 1\right)^2 = \left(\sqrt{x}\right)^2 + 2\sqrt{x} + 1^2$

$= x + 2\sqrt{x} + 1$

18. $\left(\sqrt{2} - 4\right)\left(\sqrt{3} + 1\right) = \sqrt{2}\sqrt{3} + 1 \cdot \sqrt{2} - 4\sqrt{3} - 4$

$= \sqrt{6} + \sqrt{2} - 4\sqrt{3} - 4$

19. $\left(\sqrt{5} + 5\right)\left(\sqrt{5} - 5\right) = \left(\sqrt{5}\right)^2 - 5^2$

$= 5 - 25$

$= -20$

20. $\sqrt{561} \approx 23.685$

21. $386^{-2/3} \approx 0.019$

22.

$$x = \sqrt{x - 2} + 2$$
$$x - 2 = \sqrt{x - 2}$$
$$(x - 2)^2 = \left(\sqrt{x - 2}\right)^2$$
$$x^2 - 4x + 4 = x - 2$$
$$x^2 - 5x + 6 = 0$$
$$(x - 2)(x - 3) = 0$$
$$x = 2 \text{ or } x = 3$$

The solutions are 2 and 3.

23. $\sqrt{x^2 - 7} + 3 = 0$

$\sqrt{x^2 - 7} = -3$

No solution exists since the principle square root of a number is not negative.

24.

$$\sqrt[3]{x + 5} = \sqrt[3]{2x - 1}$$
$$\left(\sqrt[3]{x + 5}\right)^3 = \left(\sqrt[3]{2x - 1}\right)^3$$
$$x + 5 = 2x - 1$$
$$-x = -6$$
$$x = 6$$

The solution is 6.

25. $\sqrt{-2} = i\sqrt{2} = 0 + i\sqrt{2}$

26. $-\sqrt{-8} = -i\sqrt{4 \cdot 2} = -2i\sqrt{2} = 0 - 2i\sqrt{2}$

27. $(12 - 6i) - (12 - 3i) = 12 - 6i - 12 + 3i = 0 - 3i$

28. $(6-2i)(6+2i) = 6^2 - (2i)^2$
$$= 36 - 4i^2$$
$$= 36 + 4$$
$$= 40$$
$$= 40 + 0i$$

29. $(4+3i)^2 = 4^2 + 2 \cdot 4 \cdot 3i + (3i)^2$
$$= 16 + 24i + 9i^2$$
$$= 16 + 24i - 9$$
$$= 7 + 24i$$

30. $\dfrac{1+4i}{1-i} = \dfrac{(1+4i)(1+i)}{(1-i)(1+i)}$
$$= \dfrac{1+i+4i+4i^2}{1^2 - i^2}$$
$$= \dfrac{1+5i-4}{1-(-1)}$$
$$= \dfrac{-3+5i}{2}$$
$$= -\dfrac{3}{2} + \dfrac{5}{2}i$$

31. $x^2 + x^2 = 5^2$
$$2x^2 = 25$$
$$x^2 = \dfrac{25}{2}$$
$$\sqrt{x^2} = \sqrt{\dfrac{25}{2}}$$
$$x = \dfrac{5}{\sqrt{2}} = \dfrac{5 \cdot \sqrt{2}}{\sqrt{2} \cdot \sqrt{2}} = \dfrac{5\sqrt{2}}{2}$$

32. $g(x) = \sqrt{x+2}$
$$x + 2 \geq 0$$
$$x \geq -2$$
Domain: $[-2, \infty)$

x	-2	-1	2	7
$g(x)$	0	1	2	3

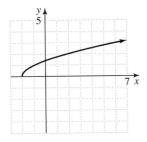

33. $(x_1, y_1) = (-6, 3), (x_2, y_2) = (-8, -7)$
$$d = \sqrt{(-8-(-6))^2 + (-7-3)^2}$$
$$= \sqrt{(-2)^2 + (-10)^2}$$
$$= \sqrt{4+100}$$
$$= \sqrt{104}$$
$$= \sqrt{4 \cdot 26}$$
$$= 2\sqrt{26}$$
The distance is $2\sqrt{26}$ units.

34. $(x_1, y_1) = \left(-2\sqrt{5}, \sqrt{10}\right)$,
$(x_2, y_2) = \left(-\sqrt{5}, 4\sqrt{10}\right)$
$$d = \sqrt{(x_2 - x_1)^2 + (y_2 - y_1)^2}$$
$$= \sqrt{\left(-\sqrt{5} + 2\sqrt{5}\right)^2 + \left(4\sqrt{10} - \sqrt{10}\right)^2}$$
$$= \sqrt{\left(\sqrt{5}\right)^2 + \left(3\sqrt{10}\right)^2}$$
$$= \sqrt{5+90}$$
$$= \sqrt{95}$$
The distance is $\sqrt{95}$ units.

35. $(x_1, y_1) = (-2, -5), (x_2, y_2) = (-6, 12)$
$$\text{midpoint} = \left(\dfrac{x_1 + x_2}{2}, \dfrac{y_1 + y_2}{2}\right)$$
$$= \left(\dfrac{-2-6}{2}, \dfrac{-5+12}{2}\right)$$
$$= \left(-\dfrac{8}{2}, \dfrac{7}{2}\right)$$
$$= \left(-4, \dfrac{7}{2}\right)$$

36. $(x_1, y_1) = \left(-\dfrac{2}{3}, -\dfrac{1}{5}\right), (x_2, y_2) = \left(-\dfrac{1}{3}, \dfrac{4}{5}\right)$
$$\text{midpoint} = \left(\dfrac{x_1 + x_2}{2}, \dfrac{y_1 + y_2}{2}\right)$$
$$= \left(\dfrac{-\frac{2}{3} - \frac{1}{3}}{2}, \dfrac{-\frac{1}{5} + \frac{4}{5}}{2}\right)$$
$$= \left(\dfrac{-\frac{3}{3}}{2}, \dfrac{\frac{3}{5}}{2}\right)$$
$$= \left(-\dfrac{1}{2}, \dfrac{3}{10}\right)$$

37. $V(r) = \sqrt{2.5r}$

$V(300) = \sqrt{2.5(300)} = \sqrt{750} \approx 27$ mph

38. $V(r) = \sqrt{2.5r}$

$30 = \sqrt{2.5r}$

$30^2 = \left(\sqrt{2.5r}\right)^2$

$900 = 2.5r$

$r = \dfrac{900}{2.5} = 360$ feet

Chapter 7 Cumulative Review

1. a. $3xy - 2xy + 5 - 7 + xy = 2xy - 2$

   b. $7x^2 + 3 - 5(x^2 - 4) = 7x^2 + 3 - 5x^2 + 20$
   $= 2x^2 + 23$

   c. $(2.1x - 5.6) - (-x - 5.3) = 2.1x - 5.6 + x + 5.3$
   $= 3.1x - 0.3$

   d. $\dfrac{1}{2}(4a - 6b) - \dfrac{1}{3}(9a + 12b - 1) + \dfrac{1}{4}$
   $= 2a - 3b - 3a - 4b + \dfrac{1}{3} + \dfrac{1}{4}$
   $= -a - 7b + \dfrac{4}{12} + \dfrac{3}{12}$
   $= -a - 7b + \dfrac{7}{12}$

2. a. $2(x - 3) + (5x + 3) = 2x - 6 + 5x + 3$
   $= 7x - 3$

   b. $4(3x + 2) - 3(5x - 1) = 12x + 8 - 15x + 3$
   $= -3x + 11$

   c. $7x + 2(x - 7) - 3x = 7x + 2x - 14 - 3x$
   $= 6x - 14$

3. $\dfrac{x + 5}{2} + \dfrac{1}{2} = 2x - \dfrac{x - 3}{8}$

$8\left(\dfrac{x + 5}{2} + \dfrac{1}{2}\right) = 8\left(2x - \dfrac{x - 3}{8}\right)$

$4(x + 5) + 4 = 16x - (x - 3)$

$4x + 20 + 4 = 16x - x + 3$

$4x + 24 = 15x + 3$

$-11x = -21$

$x = \dfrac{21}{11}$

4. $\dfrac{a - 1}{2} + a = 2 - \dfrac{2a + 7}{8}$

$8\left(\dfrac{a - 1}{2} + a\right) = 8\left(2 - \dfrac{2a + 7}{8}\right)$

$4(a - 1) + 8a = 16 - (2a + 7)$

$4a - 4 + 8a = 16 - 2a - 7$

$12a - 4 = 9 - 2a$

$14a = 13$

$a = \dfrac{13}{14}$

5. Let $x =$ the sales needed.

$600 + 0.20x > 1500$

$0.20x > 900$

$x > \dfrac{900}{0.20}$

$x > 4500$

The salesperson needs sales of at least \$4500.

6. Let $r =$ their average speed.

$t_{\text{going}} + t_{\text{returning}} = 4.5$ hrs

$\dfrac{121.5}{r} + \dfrac{121.5}{r} = 4.5$

$\dfrac{243}{r} = 4.5$

$243 = 4.5r$

$r = \dfrac{243}{4.5} = 54$

Their average speed was 54 mph.

7. $2|x| + 25 = 23$

$2|x| = -2$

$|x| = -1$, which is impossible.

There is no solution, or the solution set is $\varnothing$.

8. $|3x - 2| + 5 = 5$

$|3x - 2| = 0$

$3x - 2 = 0$

$3x = 2$

$x = \dfrac{2}{3}$

Copyright © 2014 Pearson Education, Inc.

9. $\left|\dfrac{x}{3}-1\right|-2\ge 0$

$$\left|\dfrac{x}{3}-1\right|\ge 2$$

$$\dfrac{x}{3}-1\le -2 \ \text{ or } \ \dfrac{x}{3}-1\ge 2$$

$$\dfrac{x}{3}\le -1 \ \text{ or } \ \ \dfrac{x}{3}\ge 3$$

$$x\le -3 \ \text{ or } \ \ \ \ x\ge 9$$

$$(-\infty,\,-3]\cup[9,\,\infty)$$

10. $\left|\dfrac{x}{2}-1\right|\le 0$

$$\dfrac{x}{2}-1=0$$

$$\dfrac{x}{2}=1$$

$$x=2$$

11. $y=|x|$

12. $y=|x-2|$

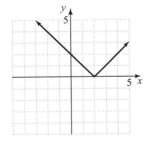

13. a. Domain: $\{2,\,0,\,3\}$
Range: $\{3,\,4,\,-1\}$

    b. Domain: $\{-4,\,-3,\,-2,\,-1,\,0,\,1,\,2,\,3\}$
Range: $\{1\}$

    c. Domain: $\{$Lubbock, Colorado Springs, Omaha, Yonkers, Sacramento$\}$
Range: $\{370,\,404,\,445,\,206,\,197\}$

14. a. Domain: $(-\infty,\,0]$, Range: $(-\infty,\,\infty)$
not a function

    b. Domain: $(-\infty,\,\infty)$, Range: $(-\infty,\,\infty)$
function

    c. Domain: $(-\infty,\,-2]\cup[2,\,\infty)$
Range: $(-\infty,\,\infty)$
not a function

15. $y=-3$
This is a horizontal line passing through $(0,\,-3)$.

16. $f(x)=-2$
This is a horizontal line passing through $(0,\,-2)$.

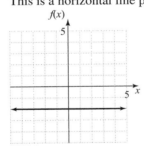

17. $x=-5$ is a vertical line. The slope is undefined.

18. $y=-3$ is a horizontal line. The slope is 0.

19. $\begin{cases} -\dfrac{x}{6}+\dfrac{y}{2}=\dfrac{1}{2} \\[2mm] \dfrac{x}{3}-\dfrac{y}{6}=-\dfrac{3}{4} \end{cases}$ or $\begin{cases} -x+3y=3 & (1) \\ 4x-2y=-9 & (2) \end{cases}$

Solve equation (1) for x.

$-x+3y=3$

$3y-3=x$

Replace x with $3y-3$ in equation (2).

$4(3y-3)-2y=-9$

$12y-12-2y=-9$

$10y-12=-9$

$10y=3$

$y=\dfrac{3}{10}$

Substitute $\frac{3}{10}$ for y in $x = 3y - 3$.

$x = 3\left(\frac{3}{10}\right) - 3 = \frac{9}{10} - \frac{30}{10} = -\frac{21}{10}$

The solution is $\left(-\frac{21}{10}, \frac{3}{10}\right)$.

20. $\begin{cases} \dfrac{x}{6} - \dfrac{y}{2} = 1 \\ \dfrac{x}{3} - \dfrac{y}{4} = 2 \end{cases}$ or $\begin{cases} x - 3y = 6 & (1) \\ 4x - 3y = 24 & (2) \end{cases}$

Solve equation (1) for x.
$x - 3y = 6$
$\quad x = 3y + 6$

Replace x with $3y + 6$ in equation (2).
$4(3y + 6) - 3y = 24$
$12y + 24 - 3y = 24$
$\quad\quad 9y + 24 = 24$
$\quad\quad\quad\quad 9y = 0$
$\quad\quad\quad\quad\ y = 0$

Substitute 0 for y in $x = 3y + 6$.
$x = 3(0) + 6 = 0 + 6 = 6$
The solution is (6, 0).

21. a. $2^2 \cdot 2^5 = 2^{2+5} = 2^7$

b. $x^7 x^3 = x^{7+3} = x^{10}$

c. $y \cdot y^2 \cdot y^4 = y^{1+2+4} = y^7$

22. Let x = number of tee-shirts and
y = number of shorts.
$\begin{cases} x + y = 9 & (1) \\ 3.50x + 4.25y = 33.75 & (2) \end{cases}$

Solve equation (1) for y.
$x + y = 9$
$\quad\ y = 9 - x$

Substitute $9 - x$ for y in equation (2).
$3.50x + 4.25(9 - x) = 33.75$
$3.50x + 38.25 - 4.25x = 33.75$
$\quad\quad -0.75x + 38.25 = 33.75$
$\quad\quad\quad\quad\quad -0.75x = -4.5$
$\quad\quad\quad\quad\quad\quad\quad x = \dfrac{-4.5}{-0.75} = 6$

Replace x with 6 in $y = 9 - x$.
$y = 9 - 6 = 3$
Nana bought 6 shirts and 3 shorts.

23. $\dfrac{2000 \times 0.000021}{700} = \dfrac{(2 \times 10^3) \times (2.1 \times 10^{-5})}{7 \times 10^2}$

$\quad\quad = \dfrac{2(2.1)}{7} \times 10^{3+(-5)-2}$

$\quad\quad = 0.6 \times 10^{-4}$

$\quad\quad = (6 \times 10^{-1}) \times 10^{-4}$

$\quad\quad = 6 \times 10^{-5}$

24. $\dfrac{0.0000035 \times 4000}{0.28} = \dfrac{(3.5 \times 10^{-6}) \times (4 \times 10^3)}{2.8 \times 10^{-1}}$

$\quad\quad = \dfrac{3.5 \times 4}{2.8} \times 10^{-6+3-(-1)}$

$\quad\quad = 5 \times 10^{-2}$

25. $P(x) = 3x^2 - 2x - 5$

a. $P(1) = 3(1)^2 - 2(1) - 5$
$\quad\quad = 3(1) - 2(1) - 5$
$\quad\quad = 3 - 2 - 5$
$\quad\quad = -4$

b. $P(-2) = 3(-2)^2 - 2(-2) - 5$
$\quad\quad = 3(4) - (-4) - 5$
$\quad\quad = 12 + 4 - 5$
$\quad\quad = 11$

26. $(5x^2 - 3x + 6) + (4x^2 + 5x - 3) - (2x - 5)$
$= 5x^2 - 3x + 6 + 4x^2 + 5x - 3 - 2x + 5$
$= 9x^2 + 8$

27. a. $(x + 3)(2x + 5) = 2x^2 + 5x + 6x + 15$
$\quad\quad\quad\quad\quad\quad\quad\ = 2x^2 + 11x + 15$

b. $(2x - 3)(5x^2 - 6x + 7)$
$\quad = 10x^3 - 12x^2 + 14x - 15x^2 + 18x - 21$
$\quad = 10x^3 - 27x^2 + 32x - 21$

28. a. $(y - 2)(3y + 4) = 3y^2 + 4y - 6y - 8$
$\quad\quad\quad\quad\quad\quad\quad = 3y^2 - 2y - 8$

b. $(3y - 1)(2y^2 + 3y - 1)$
$\quad = 6y^3 + 9y^2 - 3y - 2y^2 - 3y + 1$
$\quad = 6y^3 + 7y^2 - 6y + 1$

29. $20x^3 y = 2^2 \cdot 5x^3 y$

$10x^2 y^2 = 2 \cdot 5x^2 y^2$

$35x^3 = 5 \cdot 7x^3$

$\text{GCF} = 5x^2$

30. $x^3 - x^2 + 4x - 4 = (x^3 - x^2) + (4x - 4)$

$\qquad\qquad = x^2(x-1) + 4(x-1)$

$\qquad\qquad = (x-1)(x^2 + 4)$

31. a. $\dfrac{x^3 + 8}{2 + x} = \dfrac{x^3 + 2^3}{x + 2}$

$\qquad = \dfrac{(x+2)(x^2 - 2x + 4)}{x + 2}$

$\qquad = x^2 - 2x + 4$

b. $\dfrac{2y^2 + 2}{y^3 - 5y^2 + y - 5} = \dfrac{2(y^2 + 1)}{y^2(y-5) + 1(y-5)}$

$\qquad\qquad = \dfrac{2(y^2 + 1)}{(y-5)(y^2 + 1)}$

$\qquad\qquad = \dfrac{2}{y - 5}$

32. a. $\dfrac{a^3 - 8}{2 - a} = \dfrac{a^3 - 2^3}{2 - a}$

$\qquad = \dfrac{(a-2)(a^2 + 2a + 4)}{-1(a-2)}$

$\qquad = -1(a^2 + 2a + 4)$

$\qquad = -a^2 - 2a - 4$

b. $\dfrac{3a^2 - 3}{a^3 + 5a^2 - a - 5} = \dfrac{3(a^2 - 1)}{a^2(a+5) - 1(a+5)}$

$\qquad\qquad = \dfrac{3(a^2 - 1)}{(a+5)(a^2 - 1)}$

$\qquad\qquad = \dfrac{3}{a + 5}$

33. a. $\dfrac{2}{x^2 y} + \dfrac{5}{3x^3 y} = \dfrac{2 \cdot 3x}{x^2 y \cdot 3x} + \dfrac{5}{3x^3 y}$

$\qquad\qquad = \dfrac{6x + 5}{3x^3 y}$

b. $\dfrac{3x}{x+2} + \dfrac{2x}{x-2} = \dfrac{3x(x-2) + 2x(x+2)}{(x+2)(x-2)}$

$\qquad\qquad = \dfrac{3x^2 - 6x + 2x^2 + 4x}{(x+2)(x-2)}$

$\qquad\qquad = \dfrac{5x^2 - 2x}{(x+2)(x-2)}$

c. $\dfrac{2x-6}{x-1} - \dfrac{4}{1-x} = \dfrac{2x-6}{x-1} + \dfrac{4}{x-1}$

$\qquad\qquad = \dfrac{2x-2}{x-1}$

$\qquad\qquad = \dfrac{2(x-1)}{x-1}$

$\qquad\qquad = 2$

34. a. $\dfrac{3}{xy^2} - \dfrac{2}{3x^2 y} = \dfrac{3 \cdot 3x}{xy^2 \cdot 3x} - \dfrac{2 \cdot y}{3x^2 y \cdot y}$

$\qquad\qquad = \dfrac{9x - 2y}{3x^2 y^2}$

b. $\dfrac{5x}{x+3} - \dfrac{2x}{x-3} = \dfrac{5x(x-3) - 2x(x+3)}{(x+3)(x-3)}$

$\qquad\qquad = \dfrac{5x^2 - 15x - 2x^2 - 6x}{(x+3)(x-3)}$

$\qquad\qquad = \dfrac{3x^2 - 21x}{(x+3)(x-3)}$

$\qquad\qquad \text{or } \dfrac{3x(x-7)}{(x+3)(x-3)}$

c. $\dfrac{x}{x-2} - \dfrac{5}{2-x} = \dfrac{x}{x-2} + \dfrac{5}{x-2} = \dfrac{x+5}{x-2}$

35. a. $\dfrac{\frac{5x}{x+2}}{\frac{10}{x-2}} = \dfrac{5x}{x+2} \cdot \dfrac{x-2}{10} = \dfrac{x(x-2)}{2(x+2)}$

b. $\dfrac{\frac{x}{y^2} + \frac{1}{y}}{\frac{y}{x^2} + \frac{1}{x}} = \dfrac{\left(\frac{x}{y^2} + \frac{1}{y}\right) x^2 y^2}{\left(\frac{y}{x^2} + \frac{1}{x}\right) x^2 y^2}$

$\qquad\qquad = \dfrac{x^3 + x^2 y}{y^3 + xy^2}$

$\qquad\qquad = \dfrac{x^2(x + y)}{y^2(y + x)}$

$\qquad\qquad = \dfrac{x^2}{y^2}$

36. a. $\dfrac{\frac{y-2}{16}}{\frac{2y+3}{12}} = \dfrac{y-2}{16}\cdot\dfrac{12}{2y+3} = \dfrac{3(y-2)}{4(2y+3)}$

b. $\dfrac{\frac{x}{16}-\frac{1}{x}}{1-\frac{4}{x}} = \dfrac{\left(\frac{x}{16}-\frac{1}{x}\right)16x}{\left(1-\frac{4}{x}\right)16x}$

$\qquad = \dfrac{x^2-16}{16x-64}$

$\qquad = \dfrac{(x+4)(x-4)}{16(x-4)}$

$\qquad = \dfrac{x+4}{16}$

37. $\dfrac{10x^3-5x^2+20x}{5x} = \dfrac{10x^3}{5x}-\dfrac{5x^2}{5x}+\dfrac{20x}{5x}$

$\qquad\qquad\qquad\qquad = 2x^2-x+4$

38.
$$
\begin{array}{r}
x^2 \qquad\; +3 \\
x-2\overline{\smash{\big)}\,x^3-2x^2+3x-6} \\
\underline{x^3-2x^2} \\
3x-6 \\
\underline{3x-6} \\
0
\end{array}
$$

Answer: x^2+3

39.
$$
\begin{array}{r|rrrr}
3 & 2 & -1 & -13 & 1 \\
 & & 6 & 15 & 6 \\
\hline
 & 2 & 5 & 2 & 7
\end{array}
$$

Answer: $2x^2+5x+2+\dfrac{7}{x-3}$

40.
$$
\begin{array}{r|rrrr}
3 & 4 & -12 & -1 & 12 \\
 & & 12 & 0 & -3 \\
\hline
 & 4 & 0 & -1 & 9
\end{array}
$$

Answer: $4y^2-1+\dfrac{9}{y-3}$

41. $\dfrac{x+6}{x-2} = \dfrac{2(x+2)}{x-2}$

$(x-2)\left(\dfrac{x+6}{x-2}\right) = (x-2)\left[\dfrac{2(x+2)}{x-2}\right]$

$\qquad\qquad x+6 = 2(x+2)$

$\qquad\qquad x+6 = 2x+4$

$\qquad\qquad -x = -2$

$\qquad\qquad\quad x = 2$

which we discard as extraneous. There is no solution.

42.

$$\frac{28}{9-a^2} = \frac{2a}{a-3} + \frac{6}{a+3}$$

$$\frac{28}{-(a^2-9)} = \frac{2a}{a-3} + \frac{6}{a+3}$$

$$\frac{-28}{(a+3)(a-3)} = \frac{2a}{a-3} + \frac{6}{a+3}$$

$$(a+3)(a-3) \cdot \frac{-28}{(a+3)(a-3)} = (a+3)(a-3) \cdot \left(\frac{2a}{a-3} + \frac{6}{a+3}\right)$$

$$-28 = 2a(a+3) + 6(a-3)$$

$$-28 = 2a^2 + 6a + 6a - 18$$

$$0 = 2a^2 + 12a + 10$$

$$0 = 2(a^2 + 6a + 5)$$

$$0 = 2(a+5)(a+1)$$

$a = -5$ or $a = -1$

The solutions are -5 and -1.

43.

$$\frac{1}{x} + \frac{1}{y} = \frac{1}{z}$$

$$xyz\left(\frac{1}{x} + \frac{1}{y}\right) = xyz\left(\frac{1}{z}\right)$$

$$yz + xz = xy$$

$$yz = xy - xz$$

$$yz = x(y - z)$$

$$x = \frac{yz}{y - z}$$

44.

$$A = \frac{h(a+b)}{2}$$

$$2A = h(a+b)$$

$$2A = ah + bh$$

$$2A - bh = ah$$

$$\frac{2A - bh}{h} = a$$

45. $u = \dfrac{k}{w}$

$$3 = \frac{k}{5}$$

$$k = 3(5) = 15$$

$$u = \frac{15}{w}$$

46. $y = kx$

$$0.51 = k(3)$$

$$k = \frac{0.51}{3} = 0.17$$

$$y = 0.17x$$

47. a. $16^{-3/4} = \dfrac{1}{16^{3/4}} = \dfrac{1}{\left(\sqrt[4]{16}\right)^3} = \dfrac{1}{(2)^3} = \dfrac{1}{8}$

b. $(-27)^{-2/3} = \dfrac{1}{(-27)^{2/3}}$

$= \dfrac{1}{\left(\sqrt[3]{-27}\right)^2}$

$= \dfrac{1}{(-3)^2}$

$= \dfrac{1}{9}$

48. a. $81^{-3/4} = \dfrac{1}{81^{3/4}} = \dfrac{1}{\left(\sqrt[4]{81}\right)^3} = \dfrac{1}{(3)^3} = \dfrac{1}{27}$

b. $(-125)^{-2/3} = \dfrac{1}{(-125)^{2/3}}$

$= \dfrac{1}{\left(\sqrt[3]{-125}\right)^2}$

$= \dfrac{1}{(-5)^2}$

$= \dfrac{1}{25}$

49. $\dfrac{\sqrt{x}+2}{5} = \dfrac{\left(\sqrt{x}+2\right)\left(\sqrt{x}-2\right)}{5\left(\sqrt{x}-2\right)}$

$= \dfrac{\left(\sqrt{x}\right)^2 - 2^2}{5\left(\sqrt{x}-2\right)}$

$= \dfrac{x-4}{5\left(\sqrt{x}-2\right)}$

50. a. $\sqrt{36a^3} - \sqrt{144a^3} + \sqrt{4a^3}$

$= \sqrt{36a^2 \cdot a} - \sqrt{144a^2 \cdot a} + \sqrt{4a^2 \cdot a}$

$= 6a\sqrt{a} - 12a\sqrt{a} + 2a\sqrt{a}$

$= -4a\sqrt{a}$

b. $\sqrt[3]{128ab^3} - 3\sqrt[3]{2ab^3} + b\sqrt[3]{16a}$

$= \sqrt[3]{64b^3 \cdot 2a} - 3\sqrt[3]{b^3 \cdot 2a} + b\sqrt[3]{8 \cdot 2a}$

$= 4b\sqrt[3]{2a} - 3b\sqrt[3]{2a} + 2b\sqrt[3]{2a}$

$= 3b\sqrt[3]{2a}$

c. $\dfrac{\sqrt[3]{81}}{10} + \sqrt[3]{\dfrac{192}{125}} = \dfrac{\sqrt[3]{27 \cdot 3}}{10} + \dfrac{\sqrt[3]{192}}{\sqrt[3]{125}}$

$= \dfrac{3\sqrt[3]{3}}{10} + \dfrac{\sqrt[3]{64 \cdot 3}}{5}$

$= \dfrac{3\sqrt[3]{3}}{10} + \dfrac{4\sqrt[3]{3}}{5}$

$= \dfrac{3\sqrt[3]{3}}{10} + \dfrac{4\sqrt[3]{3} \cdot 2}{5 \cdot 2} = \dfrac{11\sqrt[3]{3}}{10}$

Chapter 8

1. $x^2 = 32$

$x = \pm\sqrt{32}$

$x = \pm 4\sqrt{2}$

Check:

Let $x = 4\sqrt{2}$.   Let $x = -4\sqrt{2}$.

$x^2 = 32$ $x^2 = 32$

$\left(4\sqrt{2}\right)^2 \overset{?}{=} 32$ $\left(-4\sqrt{2}\right)^2 \overset{?}{=} 32$

$16 \cdot 2 \overset{?}{=} 32$ $16 \cdot 2 \overset{?}{=} 32$

$32 = 32$ True $32 = 32$ True

The solutions are $4\sqrt{2}$ and $-4\sqrt{2}$, or the

solution set is $\left\{-4\sqrt{2}, 4\sqrt{2}\right\}$.

2. First we get the squared variable alone on one side of the equation.

$5x^2 - 50 = 0$

$5x^2 = 50$

$x^2 = 10$

$x = \pm\sqrt{10}$

The solutions are $\sqrt{10}$ and $-\sqrt{10}$, or the

solution set is $\left\{-\sqrt{10}, \sqrt{10}\right\}$.

3. $(x+3)^2 = 20$

$x + 3 = \pm\sqrt{20}$

$x + 3 = \pm 2\sqrt{5}$

$x = -3 \pm 2\sqrt{5}$

Check:

$(x+3)^2 = 20$

$\left(-3 + 2\sqrt{5} + 3\right)^2 \overset{?}{=} 20$

$\left(2\sqrt{5}\right)^2 \overset{?}{=} 20$

$4 \cdot 5 \overset{?}{=} 20$

$20 = 20$ True

$(x+3)^2 = 20$

$\left(-3 - 2\sqrt{5} + 3\right)^2 \overset{?}{=} 20$

$\left(-2\sqrt{5}\right)^2 \overset{?}{=} 20$

$4 \cdot 5 \overset{?}{=} 20$

$20 = 20$ True

The solutions are $-3 + 2\sqrt{5}$ and $-3 - 2\sqrt{5}$.

4. $(5x-2)^2 = -9$

$5x - 2 = \pm\sqrt{-9}$

$5x - 2 = \pm 3i$

$5x = 2 \pm 3i$

$x = \dfrac{2 \pm 3i}{5}$

The solutions are $\dfrac{2+3i}{5}$ and $\dfrac{2-3i}{5}$.

5. $b^2 + 4b = 3$

Add the square of half the coefficient of b to both sides.

$b^2 + 4b + \left(\dfrac{4}{2}\right)^2 = 3 + \left(\dfrac{4}{2}\right)^2$

$b^2 + 4b + 4 = 7$

$(b+2)^2 = 7$

$b + 2 = \pm\sqrt{7}$

$b = -2 \pm\sqrt{7}$

The solutions are $-2 + \sqrt{7}$ and $-2 - \sqrt{7}$.

6. $p^2 - 3p + 1 = 0$

Subtract 1 from both sides.

$p^2 - 3p = -1$

Add the square of half the coefficient of p to both sides.

$p^2 - 3p + \left(\dfrac{-3}{2}\right)^2 = -1 + \left(\dfrac{-3}{2}\right)^2$

$p^2 - 3p + \dfrac{9}{4} = -1 + \dfrac{9}{4} = \dfrac{5}{4}$

$\left(p - \dfrac{3}{2}\right)^2 = \dfrac{5}{4}$

$p - \dfrac{3}{2} = \pm\dfrac{\sqrt{5}}{2}$

$p = \dfrac{3 \pm \sqrt{5}}{2}$

The solutions are $\dfrac{3 + \sqrt{5}}{2}$ and $\dfrac{3 - \sqrt{5}}{2}$.

7. $3x^2 - 12x + 1 = 0$

Divide both sides by 3.

$3x^2 - 12x + 1 = 0$

$x^2 - 4x + \dfrac{1}{3} = 0$

$x^2 - 4x = -\dfrac{1}{3}$

Find the square of half of -4.

$\left(\dfrac{-4}{2}\right)^2 = (-2)^2 = 4$

Add 4 to both sides of the equation.

$x^2 - 4x + 4 = -\dfrac{1}{3} + 4$

$(x - 2)^2 = -\dfrac{1}{3} + \dfrac{12}{3} = \dfrac{11}{3}$

$x - 2 = \pm\sqrt{\dfrac{11}{3}} = \pm\dfrac{\sqrt{33}}{3}$

$x = \dfrac{6}{3} \pm \dfrac{\sqrt{33}}{3} = \dfrac{6 \pm \sqrt{33}}{3}$

The solutions are $\dfrac{6 + \sqrt{33}}{3}$ and $\dfrac{6 - \sqrt{33}}{3}$.

8. $2x^2 - 5x + 7 = 0$

$2x^2 - 5x = -7$

$x^2 - \dfrac{5}{2}x = -\dfrac{7}{2}$

Since $\dfrac{1}{2}\left(-\dfrac{5}{2}\right) = -\dfrac{5}{4}$ and $\left(-\dfrac{5}{4}\right)^2 = \dfrac{25}{16}$, we add

$\dfrac{25}{16}$ to both sides of the equation.

$x^2 - \dfrac{5}{2}x + \dfrac{25}{16} = -\dfrac{7}{2} + \dfrac{25}{16}$

$\left(x - \dfrac{5}{4}\right)^2 = -\dfrac{56}{16} + \dfrac{25}{16} = -\dfrac{31}{16}$

$x - \dfrac{5}{4} = \pm\sqrt{-\dfrac{31}{16}}$

$x = \dfrac{5}{4} \pm \dfrac{i\sqrt{31}}{4} = \dfrac{5 \pm i\sqrt{31}}{4}$

The solutions are $\dfrac{5 + i\sqrt{31}}{4}$ and $\dfrac{5 - i\sqrt{31}}{4}$.

9. $A = P(1 + r)^t$; $A = 5618$, $P = 5000$, $t = 2$

$A = P(1 + r)^t$

$5618 = 5000(1 + r)^2$

$1.1236 = (1 + r)^2$

$\pm\sqrt{1.1236} = 1 + r$

$-1 \pm 1.06 = r$

$0.06 = r$ or $-2.06 = r$

The rate cannot be negative, so we reject -2.06.

Check: $A = 5000(1 + 0.06)^2$

$= 5000(1.06)^2$

$= 5000 \cdot 1.1236$

$= 5618$

The interest rate is 6% compounded annually.

Vocabulary, Readiness & Video Check 8.1

1. By the square root property, if b is a real number, and $a^2 = b$, then $a = \underline{\pm\sqrt{b}}$.

2. A <u>quadratic</u> equation can be written in the form $ax^2 + bx + c = 0$, $a \neq 0$.

3. The process of writing a quadratic equation so that one side is a perfect square trinomial is called <u>completing the square</u>.

4. A perfect square trinomial is one that can be factored as a <u>binomial</u> squared.

5. To solve $x^2 + 6x = 10$ by completing the square, add $\underline{9}$ to both sides.

6. To solve $x^2 + bx = c$ by completing the square, add $\underline{\left(\dfrac{b}{2}\right)^2}$ to both sides.

7. We need a quantity shown squared by itself on one side of the equation. The only quantity squared is x, so divide both sides by 2 before applying the square root property.

8. The coefficient of y^2 is 3. To use the completing the square method, the coefficient of the squared variable must be 1, so we first divide through by 3.

9. We're looking for an interest rate so a negative value does not make sense.

Exercise Set 8.1

1. $x^2 = 16$
$$x = \pm\sqrt{16}$$
$$x = \pm 4$$
The solutions are -4 and 4.

3. $x^2 - 7 = 0$
$$x^2 = 7$$
$$x = \pm\sqrt{7}$$
The solutions are $\sqrt{7}$ and $-\sqrt{7}$.

5. $x^2 = 18$
$$x = \pm\sqrt{18}$$
$$x = \pm\sqrt{9 \cdot 2}$$
$$x = \pm 3\sqrt{2}$$
The solutions are $3\sqrt{2}$ and $-3\sqrt{2}$.

7. $3z^2 - 30 = 0$
$$3z^2 = 30$$
$$z^2 = 10$$
$$z = \pm\sqrt{10}$$
The solutions are $\sqrt{10}$ and $-\sqrt{10}$.

9. $(x+5)^2 = 9$
$$x + 5 = \pm\sqrt{9}$$
$$x + 5 = \pm 3$$
$$x = -5 \pm 3$$
$$x = -8 \text{ or } x = -2$$
The solutions are -8 and -2.

11. $(z-6)^2 = 18$
$$z - 6 = \pm\sqrt{18}$$
$$z - 6 = \pm 3\sqrt{2}$$
$$z = 6 \pm 3\sqrt{2}$$
The solutions are $6 + 3\sqrt{2}$ and $6 - 3\sqrt{2}$.

13. $(2x-3)^2 = 8$
$$2x - 3 = \pm\sqrt{8}$$
$$2x - 3 = \pm 2\sqrt{2}$$
$$2x = 3 \pm 2\sqrt{2}$$
$$x = \frac{3 \pm 2\sqrt{2}}{2}$$
The solutions are $\dfrac{3 + 2\sqrt{2}}{2}$ and $\dfrac{3 - 2\sqrt{2}}{2}$.

15. $x^2 + 9 = 0$
$$x^2 = -9$$
$$x = \pm\sqrt{-9}$$
$$x = \pm 3i$$
The solutions are $3i$ and $-3i$.

17. $x^2 - 6 = 0$
$$x^2 = 6$$
$$x = \pm\sqrt{6}$$
The solutions are $\sqrt{6}$ and $-\sqrt{6}$.

19. $2z^2 + 16 = 0$
$$2z^2 = -16$$
$$z^2 = -8$$
$$z = \pm\sqrt{-8}$$
$$z = \pm i\sqrt{8}$$
$$z = \pm 2i\sqrt{2}$$
The solutions are $2i\sqrt{2}$ and $-2i\sqrt{2}$.

21. $(3x-1)^2 = -16$

$$3x-1 = \pm\sqrt{-16}$$
$$3x-1 = \pm 4i$$
$$3x = 1 \pm 4i$$
$$x = \frac{1 \pm 4i}{3}$$

The solutions are $\dfrac{1+4i}{3}$ and $\dfrac{1-4i}{3}$.

23. $(z+7)^2 = 5$

$$z+7 = \pm\sqrt{5}$$
$$z = -7 \pm\sqrt{5}$$

The solutions are $-7+\sqrt{5}$ and $-7-\sqrt{5}$.

25. $(x+3)^2 + 8 = 0$

$$(x+3)^2 = -8$$
$$x+3 = \pm\sqrt{-8}$$
$$x+3 = \pm i\sqrt{8}$$
$$x+3 = \pm 2i\sqrt{2}$$
$$x = -3 \pm 2i\sqrt{2}$$

The solutions are $-3+2i\sqrt{2}$ and $-3-2i\sqrt{2}$.

27. $x^2 + 16x + \left(\dfrac{16}{2}\right)^2 = x^2 + 16x + 64$

$$= (x+8)^2$$

29. $z^2 - 12z + \left(\dfrac{-12}{2}\right)^2 = z^2 - 12z + 36$

$$= (z-6)^2$$

31. $p^2 + 9p + \left(\dfrac{9}{2}\right)^2 = p^2 + 9p + \dfrac{81}{4}$

$$= \left(p + \dfrac{9}{2}\right)^2$$

33. $x^2 + x + \left(\dfrac{1}{2}\right)^2 = x^2 + 16x + \dfrac{1}{4}$

$$= \left(x + \dfrac{1}{2}\right)^2$$

35. $\qquad x^2 + 8x = -15$

$$x^2 + 8x + \left(\dfrac{8}{2}\right)^2 = -15 + 16$$
$$x^2 + 8x + 16 = 1$$
$$(x+4)^2 = 1$$
$$x+4 = \pm\sqrt{1}$$
$$x = -4 \pm 1$$
$$x = -5 \text{ or } x = -3$$

The solutions are -5 and -3.

37. $\qquad x^2 + 6x + 2 = 0$

$$x^2 + 6x = -2$$
$$x^2 + 6x + \left(\dfrac{6}{2}\right)^2 = -2 + 9$$
$$x^2 + 6x + 9 = 7$$
$$(x+3)^2 = 7$$
$$x+3 = \pm\sqrt{7}$$
$$x = -3 \pm\sqrt{7}$$

The solutions are $-3+\sqrt{7}$ and $-3-\sqrt{7}$.

39. $\qquad x^2 + x - 1 = 0$

$$x^2 + x = 1$$
$$x^2 + x + \left(\dfrac{1}{2}\right)^2 = 1 + \dfrac{1}{4}$$
$$x^2 + x + \dfrac{1}{4} = \dfrac{5}{4}$$
$$\left(x + \dfrac{1}{2}\right)^2 = \dfrac{5}{4}$$
$$x + \dfrac{1}{2} = \pm\sqrt{\dfrac{5}{4}}$$
$$x = -\dfrac{1}{2} \pm \dfrac{\sqrt{5}}{2} = \dfrac{-1 \pm\sqrt{5}}{2}$$

The solutions are $\dfrac{-1+\sqrt{5}}{2}$ and $\dfrac{-1-\sqrt{5}}{2}$.

41.
$$x^2 + 2x - 5 = 0$$
$$x^2 + 2x = 5$$
$$x^2 + 2x + \left(\frac{2}{2}\right)^2 = 5 + 1$$
$$x^2 + 2x + 1 = 6$$
$$(x+1)^2 = 6$$
$$x + 1 = \pm\sqrt{6}$$
$$x = -1 \pm \sqrt{6}$$
The solutions are $-1 + \sqrt{6}$ and $-1 - \sqrt{6}$.

43.
$$y^2 + y - 7 = 0$$
$$y^2 + y = 7$$
$$y^2 + y + \left(\frac{1}{2}\right)^2 = 7 + \frac{1}{4}$$
$$y^2 + y + \frac{1}{4} = \frac{29}{4}$$
$$\left(y + \frac{1}{2}\right)^2 = \frac{29}{4}$$
$$y + \frac{1}{2} = \pm\sqrt{\frac{29}{4}}$$
$$y = -\frac{1}{2} \pm \frac{\sqrt{29}}{2} = \frac{-1 \pm \sqrt{29}}{2}$$
The solutions are $\dfrac{-1 + \sqrt{29}}{2}$ and $\dfrac{-1 - \sqrt{29}}{2}$.

45.
$$3p^2 - 12p + 2 = 0$$
$$p^2 - 4p + \frac{2}{3} = 0$$
$$p^2 - 4p = -\frac{2}{3}$$
$$p^2 - 4p + \left(\frac{-4}{2}\right)^2 = -\frac{2}{3} + 4$$
$$(p-2)^2 = \frac{10}{3}$$
$$p - 2 = \pm\sqrt{\frac{10}{3}}$$
$$p - 2 = \pm\frac{\sqrt{10} \cdot \sqrt{3}}{\sqrt{3} \cdot \sqrt{3}}$$
$$p - 2 = \pm\frac{\sqrt{30}}{3}$$
$$p = 2 \pm \frac{\sqrt{30}}{3} = \frac{6 \pm \sqrt{30}}{3}$$
The solutions are $\dfrac{6 + \sqrt{30}}{3}$ and $\dfrac{6 - \sqrt{30}}{3}$.

47.
$$4y^2 - 2 = 12y$$
$$4y^2 - 12y - 2 = 0$$
$$y^2 - 3y - \frac{1}{2} = 0$$
$$y^2 - 3y = \frac{1}{2}$$
$$y^2 - 3y + \left(\frac{-3}{2}\right)^2 = \frac{1}{2} + \frac{9}{4}$$
$$y^2 - 3y + \frac{9}{4} = \frac{11}{4}$$
$$\left(y - \frac{3}{2}\right)^2 = \frac{11}{4}$$
$$y - \frac{3}{2} = \pm\sqrt{\frac{11}{4}}$$
$$y = \frac{3}{2} \pm \frac{\sqrt{11}}{2} = \frac{3 \pm \sqrt{11}}{2}$$
The solutions are $\dfrac{3 + \sqrt{11}}{2}$ and $\dfrac{3 - \sqrt{11}}{2}$.

49.
$$2x^2 + 7x = 4$$
$$x^2 + \frac{7}{2}x = 2$$
$$x^2 + \frac{7}{2}x + \left(\frac{\frac{7}{2}}{2}\right)^2 = 2 + \frac{49}{16}$$
$$x^2 + \frac{7}{2}x + \frac{49}{16} = \frac{81}{16}$$
$$\left(x + \frac{7}{4}\right)^2 = \frac{81}{16}$$
$$x + \frac{7}{4} = \pm\sqrt{\frac{81}{16}}$$
$$x = -\frac{7}{4} \pm \frac{9}{4} = \frac{-7 \pm 9}{4}$$
$$x = -4 \text{ or } \frac{1}{2}$$
The solutions are -4 and $\dfrac{1}{2}$.

51.
$$x^2 + 8x + 1 = 0$$
$$x^2 + 8x = -1$$
$$x^2 + 8x + \left(\frac{8}{2}\right)^2 = -1 + 16$$
$$x^2 + 8x + 16 = 15$$
$$(x+4)^2 = 15$$
$$x + 4 = \pm\sqrt{15}$$
$$x = -4 \pm \sqrt{15}$$
The solutions are $-4 + \sqrt{15}$ and $-4 - \sqrt{15}$.

53.
$$3y^2 + 6y - 4 = 0$$
$$y^2 + 2y - \frac{4}{3} = 0$$
$$y^2 + 2y = \frac{4}{3}$$
$$y^2 + 2y + \left(\frac{2}{2}\right)^2 = \frac{4}{3} + 1$$
$$y^2 + 2y + 1 = \frac{7}{3}$$
$$(y+1)^2 = \frac{7}{3}$$
$$y + 1 = \pm\sqrt{\frac{7}{3}}$$
$$y + 1 = \pm\frac{\sqrt{7} \cdot \sqrt{3}}{\sqrt{3} \cdot \sqrt{3}}$$
$$y + 1 = \pm\frac{\sqrt{21}}{3}$$
$$y = -1 \pm \frac{\sqrt{21}}{3} = \frac{-3 \pm \sqrt{21}}{3}$$
The solutions are $\dfrac{-3 + \sqrt{21}}{3}$ and $\dfrac{-3 - \sqrt{21}}{3}$.

55.
$$2x^2 - 3x - 5 = 0$$
$$x^2 - \frac{3}{2}x - \frac{5}{2} = 0$$
$$x^2 - \frac{3}{2}x = \frac{5}{2}$$
$$x^2 - \frac{3}{2}x + \left(\frac{\frac{3}{2}}{2}\right)^2 = \frac{5}{2} + \frac{9}{16}$$
$$x^2 - \frac{3}{2}x + \frac{9}{16} = \frac{49}{16}$$
$$\left(x - \frac{3}{4}\right)^2 = \frac{49}{16}$$
$$x - \frac{3}{4} = \pm\sqrt{\frac{49}{16}}$$
$$x = \frac{3}{4} \pm \frac{7}{4} = \frac{3 \pm 7}{4}$$
$$x = -1 \text{ or } \frac{5}{2}$$
The solutions are -1 and $\dfrac{5}{2}$.

57.
$$y^2 + 2y + 2 = 0$$
$$y^2 + 2y = -2$$
$$y^2 + 2y + \left(\frac{2}{2}\right)^2 = -2 + 1$$
$$y^2 + 2y + 1 = -1$$
$$(y+1)^2 = -1$$
$$y + 1 = \pm\sqrt{-1}$$
$$y = -1 \pm i$$
The solutions are $-1 + i$ and $-1 - i$.

59.
$$y^2 + 6y - 8 = 0$$
$$y^2 + 6y = 8$$
$$y^2 + 6y + \left(\frac{6}{2}\right)^2 = 8 + 9$$
$$y^2 + 6y + 9 = 17$$
$$(y+3)^2 = 17$$
$$y + 3 = \pm\sqrt{17}$$
$$y = -3 \pm \sqrt{17}$$
The solutions are $-3 + \sqrt{17}$ and $-3 - \sqrt{17}$.

61.
$$2a^2 + 8a = -12$$
$$a^2 + 4a = -6$$
$$a^2 + 4a + \left(\frac{4}{2}\right)^2 = -6 + 4$$
$$a^2 + 4a + 4 = -2$$
$$(a+2)^2 = -2$$
$$a + 2 = \pm\sqrt{-2}$$
$$a + 2 = \pm i\sqrt{2}$$
$$a = -2 \pm i\sqrt{2}$$

The solutions are $-2 + i\sqrt{2}$ and $-2 - i\sqrt{2}$.

63.
$$5x^2 + 15x - 1 = 0$$
$$x^2 + 3x - \frac{1}{5} = 0$$
$$x^2 + 3x = \frac{1}{5}$$
$$x^2 + 3x + \left(\frac{3}{2}\right)^2 = \frac{1}{5} + \frac{9}{4}$$
$$x^2 + 3x + \frac{9}{4} = \frac{49}{20}$$
$$\left(x + \frac{3}{2}\right)^2 = \frac{49}{20}$$
$$x + \frac{3}{2} = \pm\sqrt{\frac{49}{20}}$$
$$x + \frac{3}{2} = \pm\frac{7}{\sqrt{20}}$$
$$x + \frac{3}{2} = \pm\frac{7}{2\sqrt{5}}$$
$$x + \frac{3}{2} = \pm\frac{7 \cdot \sqrt{5}}{2\sqrt{5} \cdot \sqrt{5}}$$
$$x + \frac{3}{2} = \pm\frac{7\sqrt{5}}{10}$$
$$x = -\frac{3}{2} \pm \frac{7\sqrt{5}}{10} = \frac{-15 \pm 7\sqrt{5}}{10}$$

The solutions are $\dfrac{-15 + 7\sqrt{5}}{10}$ and $\dfrac{-15 - 7\sqrt{5}}{10}$.

65.
$$2x^2 - x + 6 = 0$$
$$x^2 - \frac{1}{2}x + 3 = 0$$
$$x^2 - \frac{1}{2}x = -3$$
$$x^2 - \frac{1}{2}x + \left(\frac{-\frac{1}{2}}{2}\right)^2 = -3 + \frac{1}{16}$$
$$x^2 - \frac{1}{2}x + \frac{1}{16} = -\frac{47}{16}$$
$$\left(x - \frac{1}{4}\right)^2 = -\frac{47}{16}$$
$$x - \frac{1}{4} = \pm\sqrt{-\frac{47}{16}}$$
$$x - \frac{1}{4} = \pm i\frac{\sqrt{47}}{4}$$
$$x = \frac{1}{4} \pm i\frac{\sqrt{47}}{4} = \frac{1 \pm i\sqrt{47}}{4}$$

The solutions are $\dfrac{1 + i\sqrt{47}}{4}$ and $\dfrac{1 - i\sqrt{47}}{4}$.

67.
$$x^2 + 10x + 28 = 0$$
$$x^2 + 10x = -28$$
$$x^2 + 10x + \left(\frac{10}{2}\right)^2 = -28 + 25$$
$$(x+5)^2 = -3$$
$$x + 5 = \pm\sqrt{-3}$$
$$x = -5 \pm i\sqrt{3}$$

The solutions are $-5 + i\sqrt{3}$ and $-5 - i\sqrt{3}$.

69.
$$z^2 + 3z - 4 = 0$$
$$z^2 + 3z = 4$$
$$z^2 + 3z + \left(\frac{3}{2}\right)^2 = 4 + \frac{9}{4}$$
$$z^2 + 3z + \frac{9}{4} = \frac{25}{4}$$
$$\left(z + \frac{3}{2}\right)^2 = \frac{25}{4}$$
$$z + \frac{3}{2} = \pm\sqrt{\frac{25}{4}}$$
$$z = -\frac{3}{2} \pm \frac{5}{2} = \frac{-3 \pm 5}{2}$$
$$z = -4 \text{ or } 1$$

The solutions are -4 and 1.

71.
$$2x^2 - 4x = -3$$
$$x^2 - 2x = -\frac{3}{2}$$
$$x^2 - 2x + \left(\frac{-2}{2}\right)^2 = -\frac{3}{2} + 1$$
$$x^2 - 2x + 1 = -\frac{1}{2}$$
$$(x-1)^2 = -\frac{1}{2}$$
$$x - 1 = \pm\sqrt{-\frac{1}{2}}$$
$$x - 1 = \pm i\frac{1}{\sqrt{2}}$$
$$x - 1 = \pm i\frac{1 \cdot \sqrt{2}}{\sqrt{2} \cdot \sqrt{2}}$$
$$x - 1 = \pm i\frac{\sqrt{2}}{2}$$
$$x = 1 \pm i\frac{\sqrt{2}}{2} = \frac{2 \pm i\sqrt{2}}{2}$$

The solutions are $\dfrac{2 + i\sqrt{2}}{2}$ and $\dfrac{2 - i\sqrt{2}}{2}$.

73.
$$3x^2 + 3x = 5$$
$$x^2 + x = \frac{5}{3}$$
$$x^2 + x + \left(\frac{1}{2}\right)^2 = \frac{5}{3} + \frac{1}{4}$$
$$x^2 + x + \frac{1}{4} = \frac{23}{12}$$
$$\left(x + \frac{1}{2}\right)^2 = \frac{23}{12}$$
$$x + \frac{1}{2} = \pm\sqrt{\frac{23}{12}}$$
$$x + \frac{1}{2} = \pm\frac{\sqrt{23}}{2\sqrt{3}}$$
$$x + \frac{1}{2} = \pm\frac{\sqrt{23} \cdot \sqrt{3}}{2\sqrt{3} \cdot \sqrt{3}}$$
$$x + \frac{1}{2} = \pm\frac{\sqrt{69}}{6}$$
$$x = -\frac{1}{2} \pm \frac{\sqrt{69}}{6} = \frac{-3 \pm \sqrt{69}}{6}$$

The solutions are $\dfrac{-3 + \sqrt{69}}{6}$ and $\dfrac{-3 - \sqrt{69}}{6}$.

75. The graph has two *x*-intercepts. Thus, the equation has two real number solutions.

77. The graph has no *x*-intercepts. Thus, the equation has no real number solutions.

79.
$$A = P(1+r)^t$$
$$4320 = 3000(1+r)^2$$
$$\frac{4320}{3000} = (1+r)^2$$
$$1.44 = (1+r)^2$$
$$\pm\sqrt{1.44} = 1 + r$$
$$\pm 1.2 = 1 + r$$
$$-1 \pm 1.2 = r$$
$$-2.2 = r \text{ or } 0.2 = r$$
Rate cannot be negative, so the rate is $r = 0.2 = 20\%$.

81.
$$A = P(1+r)^t$$
$$16,224 = 15,000(1+r)^2$$
$$\frac{16,224}{15,000} = (1+r)^2$$
$$\pm\sqrt{1.0816} = 1 + r$$
$$\pm 1.04 = 1 + r$$
$$-1 \pm 1.04 = r$$
$$0.04 = r \text{ or } -2.04 = r$$
Rate cannot be negative, so the rate is $r = 0.04$, or 4%.

83.
$$s(t) = 16t^2$$
$$1483 = 16t^2$$
$$t^2 = \frac{1483}{16}$$
$$t = \pm\sqrt{\frac{1483}{16}}$$
$$t \approx 9.63 \text{ or } -9.63 \text{ (disregard)}$$
It would take 9.63 seconds.

85.
$$s(t) = 16t^2$$
$$1100 = 16t^2$$
$$\frac{1100}{16} = t^2$$
$$\frac{275}{4} = t^2$$
$$\pm\sqrt{\frac{275}{4}} = t^2$$
$$\pm\frac{5\sqrt{11}}{2} = t$$

The time cannot be negative, so $-\dfrac{5\sqrt{11}}{2}$ is rejected.

It takes the object $\dfrac{5\sqrt{11}}{2} \approx 8.29$ seconds to fall from the top to the base of the dam.

87. Let x be the length of one side.

$$x^2 = 225$$
$$x = \pm\sqrt{225}$$
$$x = \pm 15$$

The length cannot be negative, so -15 is rejected. The dimensions are 15 feet by 15 feet.

89. $a^2 + b^2 = c^2$

$$x^2 + x^2 = 20^2$$
$$2x^2 = 400$$
$$x^2 = 200$$
$$x = \pm\sqrt{200}$$
$$x = \pm 10\sqrt{2}$$

The length cannot be negative, so $-10\sqrt{2}$ is rejected. The length of each leg is $10\sqrt{2}$ centimeters.

91. $\dfrac{1}{2} - \sqrt{\dfrac{9}{4}} = \dfrac{1}{2} - \dfrac{3}{2} = -\dfrac{2}{2} = -1$

93. $\dfrac{6 + 4\sqrt{5}}{2} = \dfrac{6}{2} + \dfrac{4\sqrt{5}}{2} = 3 + 2\sqrt{5}$

95. $\dfrac{3 - 9\sqrt{2}}{6} = \dfrac{3}{6} - \dfrac{9\sqrt{2}}{6} = \dfrac{1}{2} - \dfrac{3\sqrt{2}}{2} = \dfrac{1 - 3\sqrt{2}}{2}$

97. $\sqrt{b^2 - 4ac} = \sqrt{(4)^2 - 4(2)(-1)}$
$$= \sqrt{16 + 8}$$
$$= \sqrt{24}$$
$$= 2\sqrt{6}$$

99. $\sqrt{b^2 - 4ac} = \sqrt{(-1)^2 - 4(3)(-2)}$
$$= \sqrt{1 + 24}$$
$$= \sqrt{25}$$
$$= 5$$

101. The solutions of $(x+1)^2 = -1$ are complex, but not real numbers.

103. The solutions of $3z^2 = 10$ are real numbers.

105. The solutions of $(2y-5)^2 + 7 = 3$ are complex, but not real numbers.

107. $x^2 + \underline{\quad} + 16$

$$\left(\frac{b}{2}\right)^2 = 16$$
$$\frac{b}{2} = \pm\sqrt{16}$$
$$\frac{b}{2} = \pm 4$$
$$b = \pm 8$$

Answer: $\pm 8x$

109. $z^2 + \underline{\quad} + \dfrac{25}{4}$

$$\left(\frac{b}{2}\right)^2 = \frac{25}{4}$$
$$\frac{b}{2} = \pm\sqrt{\frac{25}{4}}$$
$$\frac{b}{2} = \pm\frac{5}{2}$$
$$b = \pm 5$$

Answer: $\pm 5z$

111. answers may vary

113. compound interest is preferable; answers may vary

115. $p = -x^2 + 47$

$$11 = -x^2 + 47$$
$$x^2 = 36$$
$$x = \pm\sqrt{36}$$
$$x = \pm 6$$

Demand cannot be negative. Therefore, the demand is 6 thousand scissors.

Section 8.2 Practice Exercises

1. $3x^2 - 5x - 2 = 0$

$a = 3, b = -5, c = -2$

$x = \dfrac{-b \pm \sqrt{b^2 - 4ac}}{2a}$

$= \dfrac{-(-5) \pm \sqrt{(-5)^2 - 4(3)(-2)}}{2(3)}$

$= \dfrac{5 \pm \sqrt{25 + 24}}{6}$

$= \dfrac{5 \pm \sqrt{49}}{6}$

$= \dfrac{5 \pm 7}{6}$

$x = \dfrac{5 + 7}{6} = \dfrac{12}{6} = 2$ or $x = \dfrac{5 - 7}{6} = \dfrac{-2}{6} = -\dfrac{1}{3}$

The solutions are $-\dfrac{1}{3}$ and 2, or the solution set

is $\left\{ -\dfrac{1}{3}, 2 \right\}$.

2. $3x^2 - 8x = 2$

Write in standard form.

$3x^2 - 8x - 2 = 0$

$a = 3, b = -8, c = -2$

$x = \dfrac{-b \pm \sqrt{b^2 - 4ac}}{2a}$

$= \dfrac{-(-8) \pm \sqrt{(-8)^2 - 4(3)(-2)}}{2(3)}$

$= \dfrac{8 \pm \sqrt{64 + 24}}{6}$

$= \dfrac{8 \pm \sqrt{88}}{6}$

$= \dfrac{8 \pm 2\sqrt{22}}{6}$

$= \dfrac{4 \pm \sqrt{22}}{3}$

The solutions are $\dfrac{4 + \sqrt{22}}{3}$ and $\dfrac{4 - \sqrt{22}}{3}$, or the

solution set is $\left\{ \dfrac{4 + \sqrt{22}}{3}, \dfrac{4 - \sqrt{22}}{3} \right\}$.

3. $\dfrac{1}{8}x^2 - \dfrac{1}{4}x - 2 = 0$

Multiply both sides of the equation by 8.

$8\left(\dfrac{1}{8}x^2 - \dfrac{1}{4}x - 2 \right) = 8 \cdot 0$

$x^2 - 2x - 16 = 0$

Substitute $a = 1$, $b = -2$, and $c = -16$ into the quadratic formula and simplify.

$x = \dfrac{-(-2) \pm \sqrt{(-2)^2 - 4(1)(-16)}}{2(1)}$

$= \dfrac{2 \pm \sqrt{4 + 64}}{2}$

$= \dfrac{2 \pm \sqrt{68}}{2}$

$= \dfrac{2 \pm 2\sqrt{17}}{2}$

$= 1 \pm \sqrt{17}$

The solutions are $1 + \sqrt{17}$ or $1 - \sqrt{17}$.

4. $x = -2x^2 - 2$

The equation in standard form is

$2x^2 + x + 2 = 0$. Thus, let $a = 2$, $b = 1$, and $c = 2$ in the quadratic formula.

$x = \dfrac{-1 \pm \sqrt{1^2 - 4(2)(2)}}{2(2)}$

$= \dfrac{-1 \pm \sqrt{1 - 16}}{4}$

$= \dfrac{-1 \pm \sqrt{-15}}{4}$

$= \dfrac{-1 \pm i\sqrt{15}}{4}$

The solutions are $\dfrac{-1 + i\sqrt{15}}{4}$ and $\dfrac{-1 - i\sqrt{15}}{4}$.

5. **a.** $x^2 - 6x + 9 = 0$

In $x^2 - 6x + 9$, $a = 1$, $b = -6$, and $c = 9$. Thus,

$b^2 - 4ac = (-6)^2 - 4(1)(9) = 36 - 36 = 0$

Since $b^2 - 4ac = 0$, this equation has one real solution.

b. $x^2 - 3x - 1 = 0$

In this equation, $a = 1$, $b = -3$, and $c = -1$.

$b^2 - 4ac = (-3)^2 - 4(1)(-1) = 9 + 4 = 13 > 0$

Since $b^2 - 4ac$ is positive, this equation has two real solutions.

c. $7x^2 + 11 = 0$

In this equation, $a = 7$, $b = 0$, and $c = 11$.

$b^2 - 4ac = 0^2 - 4(7)(11) = -308 < 0$

Since $b^2 - 4ac$ is negative, this equation has two complex but not real solutions.

6. By the Pythagorean theorem, we have

$$x^2 + (x+3)^2 = 15^2$$
$$x^2 + x^2 + 6x + 9 = 225$$
$$2x^2 + 6x - 216 = 0$$
$$x^2 + 3x - 108 = 0$$

Here, $a = 1$, $b = 3$, and $c = -108$. By the quadratic formula,

$$x = \frac{-3 \pm \sqrt{3^2 - 4(1)(-108)}}{2(1)}$$
$$= \frac{-3 \pm \sqrt{9 + 432}}{2}$$
$$= \frac{-3 \pm \sqrt{441}}{2}$$
$$= \frac{-3 \pm 21}{2}$$

$$x = \frac{-3 + 21}{2} = \frac{18}{2} = 9 \text{ or}$$
$$x = \frac{-3 - 21}{2} = \frac{-24}{2} = -12$$

The length can't be negative, so reject -12. The distance along the sidewalk is
$x + (x + 3) = 2x + 3 = 2(9) + 3 = 18 + 3 = 21$ feet
A person can save $21 - 15 = 6$ feet by cutting across the lawn.

7. $h = -16t^2 + 20t + 45$

At the ground, $h = 0$.

$$0 = -16t^2 + 20t + 45$$

Here, $a = -16$, $b = 20$, and $c = 45$. By the quadratic formula,

$$t = \frac{-20 \pm \sqrt{20^2 - 4(-16)(45)}}{2(-16)}$$
$$= \frac{-20 \pm \sqrt{400 + 2880}}{-32}$$
$$= \frac{-20 \pm \sqrt{3280}}{-32}$$
$$= \frac{20 \pm \sqrt{16 \cdot 205}}{32}$$
$$= \frac{20 \pm 4\sqrt{205}}{32}$$
$$= \frac{5 \pm \sqrt{205}}{8}$$

$$t = \frac{5 + \sqrt{205}}{8} \approx 2.4 \text{ or } t = \frac{5 - \sqrt{205}}{8} \approx -1.2$$

Since the time won't be negative, we reject -1.2. The rocket will strike the ground 2.4 seconds after launch.

Vocabulary, Readiness & Video Check 8.2

1. The quadratic formula is $x = \dfrac{-b \pm \sqrt{b^2 - 4ac}}{2a}$.

2. For $2x^2 + x + 1 = 0$, if $a = 2$, then $b = \underline{1}$ and $c = \underline{1}$.

3. For $5x^2 - 5x - 7 = 0$, if $a = 5$, then $b = \underline{-5}$ and $c = \underline{-7}$.

4. For $7x^2 - 4 = 0$, if $a = 7$, then $b = \underline{0}$ and $c = \underline{-4}$.

5. For $x^2 + 9 = 0$, if $c = 9$, then $a = \underline{1}$ and $b = \underline{0}$.

6. The correct simplified form of $\dfrac{5 \pm 10\sqrt{2}}{5}$ is $1 \pm 2\sqrt{2}$. The answer is **c**.

7. a. Yes, in order to make sure we have correct values for a, b, and c.

    b. No; clearing fractions makes the work less tedious, but it's not a necessary step.

8. The discriminant is the <u>radicand</u> in the quadratic formula and can be used to find the number and type of solutions of a quadratic equation without <u>solving</u> the equation. To use the discriminant, the quadratic equation needs to be written in <u>standard</u> form.

9. With applications, we need to make sure we answer the question(s) asked. Here we're asked how much distance is saved, so once the dimensions of the triangle are known, further calculations are needed to answer this question and solve the problem.

Exercise Set 8.2

1. $m^2 + 5m - 6 = 0$
$a = 1, b = 5, c = -6$

$m = \dfrac{-5 \pm \sqrt{(5)^2 - 4(1)(-6)}}{2(1)}$

$= \dfrac{-5 \pm \sqrt{25 + 24}}{2}$

$= \dfrac{-5 \pm \sqrt{49}}{2}$

$= \dfrac{-5 \pm 7}{2}$

$= -6$ or 1

The solutions are -6 and 1.

3. $\quad 2y = 5y^2 - 3$
$5y^2 - 2y - 3 = 0$
$a = 5, b = -2, c = -3$

$y = \dfrac{2 \pm \sqrt{(-2)^2 - 4(5)(-3)}}{2(5)}$

$= \dfrac{2 \pm \sqrt{4 + 60}}{10}$

$= \dfrac{2 \pm \sqrt{64}}{10}$

$= \dfrac{2 \pm 8}{10}$

$= -\dfrac{3}{5}$ or 1

The solutions are $-\dfrac{3}{5}$ and 1.

5. $x^2 - 6x + 9 = 0$
$a = 1, b = -6, c = 9$

$x = \dfrac{6 \pm \sqrt{(-6)^2 - 4(1)(9)}}{2(1)}$

$= \dfrac{6 \pm \sqrt{36 - 36}}{2}$

$= \dfrac{6 \pm \sqrt{0}}{2}$

$= \dfrac{6}{2}$

$= 3$

The solution is 3.

7. $x^2 + 7x + 4 = 0$
$a = 1, b = 7, c = 4$

$x = \dfrac{-7 \pm \sqrt{(7)^2 - 4(1)(4)}}{2(1)}$

$= \dfrac{-7 \pm \sqrt{49 - 16}}{2}$

$= \dfrac{-7 \pm \sqrt{33}}{2}$

The solutions are $\dfrac{-7 + \sqrt{33}}{2}$ and $\dfrac{-7 - \sqrt{33}}{2}$.

9. $\quad 8m^2 - 2m = 7$
$8m^2 - 2m - 7 = 0$
$a = 8, b = -2, c = -7$

$m = \dfrac{2 \pm \sqrt{(-2)^2 - 4(8)(-7)}}{2(8)}$

$= \dfrac{2 \pm \sqrt{4 + 224}}{16}$

$= \dfrac{2 \pm \sqrt{228}}{16}$

$= \dfrac{2 \pm \sqrt{4 \cdot 57}}{16}$

$= \dfrac{2 \pm 2\sqrt{57}}{16}$

$= \dfrac{1 \pm \sqrt{57}}{8}$

The solutions are $\dfrac{1 + \sqrt{57}}{8}$ and $\dfrac{1 - \sqrt{57}}{8}$.

11. $3m^2 - 7m = 3$

$3m^2 - 7m - 3 = 0$

$a = 3, b = -7, c = -3$

$m = \dfrac{7 \pm \sqrt{(-7)^2 - 4(3)(-3)}}{2(3)}$

$= \dfrac{7 \pm \sqrt{49 + 36}}{6}$

$= \dfrac{7 \pm \sqrt{85}}{6}$

The solutions are $\dfrac{7 + \sqrt{85}}{6}$ and $\dfrac{7 - \sqrt{85}}{6}$.

13. $\dfrac{1}{2}x^2 - x - 1 = 0$

$x^2 - 2x - 2 = 0$

$a = 1, b = -2, c = -2$

$x = \dfrac{2 \pm \sqrt{(-2)^2 - 4(1)(-2)}}{2(1)}$

$= \dfrac{2 \pm \sqrt{4 + 8}}{2}$

$= \dfrac{2 \pm \sqrt{12}}{2}$

$= \dfrac{2 \pm 2\sqrt{3}}{2}$

$= 1 \pm \sqrt{3}$

The solutions are $1 + \sqrt{3}$ and $1 - \sqrt{3}$.

15. $\dfrac{2}{5}y^2 + \dfrac{1}{5}y = \dfrac{3}{5}$

$2y^2 + y - 3 = 0$

$a = 2, b = 1, c = -3$

$y = \dfrac{-1 \pm \sqrt{(1)^2 - 4(2)(-3)}}{2(2)}$

$= \dfrac{-1 \pm \sqrt{1 + 24}}{4}$

$= \dfrac{-1 \pm \sqrt{25}}{4}$

$= \dfrac{-1 \pm 5}{4}$

$= -\dfrac{3}{2}$ or 1

The solutions are $-\dfrac{3}{2}$ and 1.

17. $\dfrac{1}{3}y^2 = y + \dfrac{1}{6}$

$\dfrac{1}{3}y^2 - y - \dfrac{1}{6} = 0$

$2y^2 - 6y - 1 = 0$

$a = 2, b = -6, c = -1$

$y = \dfrac{6 \pm \sqrt{(-6)^2 - 4(2)(-1)}}{2(2)}$

$= \dfrac{6 \pm \sqrt{36 + 8}}{4}$

$= \dfrac{6 \pm \sqrt{44}}{4}$

$= \dfrac{6 \pm 2\sqrt{11}}{4}$

$= \dfrac{3 \pm \sqrt{11}}{2}$

The solutions are $\dfrac{3 + \sqrt{11}}{2}$ and $\dfrac{3 - \sqrt{11}}{2}$.

19. $x^2 + 5x = -2$

$x^2 + 5x + 2 = 0$

$a = 1, b = 5, c = 2$

$x = \dfrac{-5 \pm \sqrt{(5)^2 - 4(1)(2)}}{2(1)}$

$= \dfrac{-5 \pm \sqrt{25 - 8}}{2}$

$= \dfrac{-5 \pm \sqrt{17}}{2}$

The solutions are $\dfrac{-5 + \sqrt{17}}{2}$ and $\dfrac{-5 - \sqrt{17}}{2}$.

21.
$$(m+2)(2m-6) = 5(m-1) - 12$$
$$2m^2 - 6m + 4m - 12 = 5m - 5 - 12$$
$$2m^2 - 7m + 5 = 0$$
$$a = 2, b = -7, c = 5$$
$$m = \frac{7 \pm \sqrt{(-7)^2 - 4(2)(5)}}{2(2)}$$
$$= \frac{7 \pm \sqrt{49 - 40}}{4}$$
$$= \frac{7 \pm \sqrt{9}}{4}$$
$$= \frac{7 \pm 3}{4}$$
$$= 1 \text{ or } \frac{5}{2}$$

The solutions are 1 and $\frac{5}{2}$.

23. $x^2 + 6x + 13 = 0$
$$a = 1, b = 6, c = 13$$
$$x = \frac{-6 \pm \sqrt{(6)^2 - 4(1)(13)}}{2(1)}$$
$$= \frac{-6 \pm \sqrt{36 - 52}}{2}$$
$$= \frac{-6 \pm \sqrt{-16}}{2}$$
$$= \frac{-6 \pm 4i}{2}$$
$$= -3 \pm 2i$$
The solutions are $-3 + 2i$ and $-3 - 2i$.

25. $(x+5)(x-1) = 2$
$$x^2 + 4x - 5 = 2$$
$$x^2 + 4x - 7 = 0$$
$$a = 1, b = 4, c = -7$$
$$x = \frac{-4 \pm \sqrt{(4)^2 - 4(1)(-7)}}{2(1)}$$
$$= \frac{-4 \pm \sqrt{16 + 28}}{2}$$
$$= \frac{-4 \pm \sqrt{44}}{2}$$
$$= \frac{-4 \pm 2\sqrt{11}}{2}$$
$$= -2 \pm \sqrt{11}$$
The solutions are $-2 + \sqrt{11}$ and $-2 - \sqrt{11}$.

27.
$$6 = -4x^2 + 3x$$
$$4x^2 - 3x + 6 = 0$$
$$a = 4, b = -3, c = 6$$
$$x = \frac{3 \pm \sqrt{(-3)^2 - 4(4)(6)}}{2(4)}$$
$$= \frac{3 \pm \sqrt{9 - 96}}{8}$$
$$= \frac{3 \pm \sqrt{-87}}{8}$$
$$= \frac{3 \pm i\sqrt{87}}{8}$$

The solutions are $\frac{3 + i\sqrt{87}}{8}$ and $\frac{3 - i\sqrt{87}}{8}$.

29.
$$\frac{x^2}{3} - x = \frac{5}{3}$$
$$x^2 - 3x = 5$$
$$x^2 - 3x - 5 = 0$$
$$a = 1, b = -3, c = -5$$
$$x = \frac{3 \pm \sqrt{(-3)^2 - 4(1)(-5)}}{2(1)}$$
$$= \frac{3 \pm \sqrt{9 + 20}}{2}$$
$$= \frac{3 \pm \sqrt{29}}{2}$$

The solutions are $\frac{3 + \sqrt{29}}{2}$ and $\frac{3 - \sqrt{29}}{2}$.

31. $10y^2 + 10y + 3 = 0$
$$a = 10, b = 10, c = 3$$
$$y = \frac{-10 \pm \sqrt{(10)^2 - 4(10)(3)}}{2(10)}$$
$$= \frac{-10 \pm \sqrt{100 - 120}}{20}$$
$$= \frac{-10 \pm \sqrt{-20}}{20}$$
$$= \frac{-10 \pm i\sqrt{4 \cdot 5}}{20}$$
$$= \frac{-10 \pm 2i\sqrt{5}}{20}$$
$$= \frac{-5 \pm i\sqrt{5}}{10}$$

The solutions are $\frac{-5 + i\sqrt{5}}{10}$ and $\frac{-5 - i\sqrt{5}}{10}$.

33.
$$x(6x + 2) = 3$$
$$x(6x + 2) - 3 = 0$$
$$6x^2 + 2x - 3 = 0$$
$$a = 6, b = 2, c = -3$$
$$x = \frac{-2 \pm \sqrt{(2)^2 - 4(6)(-3)}}{2(6)}$$
$$= \frac{-2 \pm \sqrt{4 + 72}}{12}$$
$$= \frac{-2 \pm \sqrt{76}}{12}$$
$$= \frac{-2 \pm \sqrt{4 \cdot 19}}{12}$$
$$= \frac{-2 \pm 2\sqrt{19}}{12}$$
$$= \frac{-1 \pm \sqrt{19}}{6}$$

The solutions are $\dfrac{-1 + \sqrt{19}}{6}$ and $\dfrac{-1 - \sqrt{19}}{6}$.

35.
$$\frac{2}{5}y^2 + \frac{1}{5}y + \frac{3}{5} = 0$$
$$2y^2 + y + 3 = 0$$
$$a = 2, b = 1, c = 3$$
$$y = \frac{-1 \pm \sqrt{(1)^2 - 4(2)(3)}}{2(2)}$$
$$= \frac{-1 \pm \sqrt{1 - 24}}{4}$$
$$= \frac{-1 \pm \sqrt{-23}}{4}$$
$$= \frac{-1 \pm i\sqrt{23}}{4}$$

The solutions are $\dfrac{-1 + i\sqrt{23}}{4}$ and $\dfrac{-1 - i\sqrt{23}}{4}$.

37.
$$\frac{1}{2}y^2 = y - \frac{1}{2}$$
$$y^2 = 2y - 1$$
$$y^2 - 2y + 1 = 0$$
$$a = 1, b = -2, c = 1$$
$$y = \frac{2 \pm \sqrt{(-2)^2 - 4(1)(1)}}{2(1)}$$
$$= \frac{2 \pm \sqrt{4 - 4}}{2}$$
$$= \frac{2 \pm \sqrt{0}}{2}$$
$$= \frac{2}{2}$$
$$= 1$$

The solution is 1.

39.
$$(n - 2)^2 = 2n$$
$$n^2 - 4n + 4 = 2n$$
$$n^2 - 6n + 4 = 0$$
$$a = 1, b = -6, c = 4$$
$$n = \frac{6 \pm \sqrt{(-6)^2 - 4(1)(4)}}{2(1)}$$
$$= \frac{6 \pm \sqrt{36 - 16}}{2}$$
$$= \frac{6 \pm \sqrt{20}}{2}$$
$$= \frac{6 \pm 2\sqrt{5}}{2}$$
$$= 3 \pm \sqrt{5}$$

The solutions are $3 + \sqrt{5}$ and $3 - \sqrt{5}$.

41.
$$x^2 - 5 = 0$$
$$a = 1, b = 0, c = -5$$
$$b^2 - 4ac = 0^2 - 4(1)(-5) = 20 > 0$$
Therefore, there are two real solutions.

43.
$$4x^2 + 12x = -9$$
$$4x^2 - 12x + 9 = 0$$
$$a = 4, b = -12, c = 9$$
$$b^2 - 4ac = (-12)^2 - 4(4)(9)$$
$$= 144 - 144$$
$$= 0$$
Therefore, there is one real solution.

45.
$$3x = -2x^2 + 7$$
$$2x^2 + 3x - 7 = 0$$
$$a = 2, b = 3, c = -7$$
$$b^2 - 4ac = 3^2 - 4(2)(-7)$$
$$= 9 + 56$$
$$= 65 > 0$$
Therefore, there are two real solutions.

47.
$$6 = 4x - 5x^2$$
$$5x^2 - 4x + 6 = 0$$
$$a = 5, b = -4, c = 6$$
$$b^2 - 4ac = (-4)^2 - 4(5)(6)$$
$$= 16 - 120$$
$$= -104 < 0$$
Therefore, there are two complex but not real solutions.

49. a. The graph has two *x*-intercepts. Thus, the related equation has two real number solutions.

b. The graph has one *x*-intercept. Thus, the related equation has one real number solution.

51.
$$(x+8)^2 + x^2 = 36^2$$
$$(x^2 + 16x + 64) + x^2 = 1296$$
$$2x^2 + 16x - 1232 = 0$$
$$a = 2, b = 16, c = -1232$$
$$x = \frac{-16 \pm \sqrt{(16)^2 - 4(2)(-1232)}}{2(2)}$$
$$= \frac{-16 \pm \sqrt{10,112}}{4}$$
$$x \approx 21 \text{ or } x \approx -29 \text{ (disregard)}$$
$$x + (x+8) = 21 + 21 + 8 = 50$$
$$50 - 36 = 14$$
They save about 14 feet of walking distance.

53. Let *x* = length of leg. Then
x + 2 = length of hypotenuse.
$$x^2 + x^2 = (x+2)^2$$
$$2x^2 = x^2 + 4x + 4$$
$$x^2 - 4x - 4 = 0$$
$$a = 1, b = -4, c = -4$$

$$x = \frac{4 \pm \sqrt{(-4)^2 - 4(1)(-4)}}{2(1)}$$
$$= \frac{4 \pm \sqrt{32}}{2}$$
$$= \frac{4 \pm 4\sqrt{2}}{2}$$
$$= 2 \pm 2\sqrt{2} \text{ (disregard the negative)}$$
$$= 2 + 2\sqrt{2}$$
The sides measure $\left(2 + 2\sqrt{2}\right)$ cm, $\left(2 + 2\sqrt{2}\right)$ cm, and $\left(4 + 2\sqrt{2}\right)$ cm.

55. Let *x* = width; then *x* + 10 = length.
Area = length · width
$$400 = (x+10)x$$
$$0 = x^2 + 10x - 400$$
$$a = 1, b = 10, c = -400$$
$$x = \frac{-10 \pm \sqrt{(10)^2 - 4(1)(-400)}}{2(1)}$$
$$= \frac{-10 \pm \sqrt{1700}}{2}$$
$$= \frac{-10 \pm 10\sqrt{17}}{2}$$
$$= -5 \pm 5\sqrt{17}$$
Disregard the negative length. The width is $\left(-5 + 5\sqrt{17}\right)$ ft and the length is $\left(5 + 5\sqrt{17}\right)$ ft.

57. a. Let *x* = length.
$$x^2 + x^2 = 100^2$$
$$2x^2 - 10,000 = 0$$
$$a = 2, b = 0, c = -10,000$$
$$x = \frac{0 \pm \sqrt{(0)^2 - 4(2)(-10,000)}}{2(2)}$$
$$= \frac{\pm\sqrt{80,000}}{4}$$
$$= \frac{\pm 200\sqrt{2}}{4}$$
$$= \pm 50\sqrt{2}$$
Disregard the negative length. The side measures $50\sqrt{2}$ meters.

b. $\text{Area} = s^2$

$= \left(50\sqrt{2}\right)^2$

$= 2500(2)$

$= 5000$

The area is 5000 square meters.

59. Let $w = $ width; then $w + 1.1 = $ height.

$\text{Area} = \text{length} \cdot \text{width}$

$1439.9 = (w + 1.1)w$

$0 = w^2 + 1.1w - 1439.9$

$a = 1, b = 1.1, c = -1439.9$

$w = \dfrac{-1.1 \pm \sqrt{(1.1)^2 - 4(1)(-1439.9)}}{2(1)}$

$= \dfrac{-1.1 \pm \sqrt{5760.81}}{2}$

$= 37.4 \text{ or } -38.5 \text{ (disregard)}$

Its width is 37.4 ft and its height is 38.5 ft.

61. Let $h = $ height. Then $2h + 4 = $ base.

$\text{Area} = \dfrac{1}{2} \text{base} \cdot \text{height}$

$42 = \dfrac{1}{2}(2h + 4)h$

$42 = h^2 + 2h$

$0 = h^2 + 2h - 42$

$a = 1, b = 2, c = -42$

$h = \dfrac{-2 \pm \sqrt{(2)^2 - 4(1)(-42)}}{2(1)}$

$= \dfrac{-2 \pm \sqrt{172}}{2}$

$= \dfrac{-2 \pm 2\sqrt{43}}{2}$

$= -1 \pm \sqrt{43} \text{ (disregard the negative)}$

$\text{base} = 2\left(-1 + \sqrt{43}\right) + 4 = 2 + 2\sqrt{43}$

Height: $\left(-1 + \sqrt{43}\right)$ cm

Base: $\left(2 + 2\sqrt{43}\right)$ cm

63. $h = -16t^2 + 20t + 1100$

$0 = -16t^2 + 20t + 1100$

$a = -16, b = 20, c = 1100$

$t = \dfrac{-20 \pm \sqrt{(20)^2 - 4(-16)(1100)}}{2(-16)}$

$= \dfrac{-20 \pm \sqrt{70,800}}{-32}$

$\approx 8.9 \text{ or } -7.7 \text{ (disregard)}$

It will take about 8.9 seconds.

65. $h = -16t^2 - 20t + 180$

$0 = -16t^2 - 20t + 180$

$a = -16, b = -20, c = 180$

$t = \dfrac{20 \pm \sqrt{(-20)^2 - 4(-16)(180)}}{2(-16)}$

$= \dfrac{20 \pm \sqrt{11,920}}{-32}$

$\approx 2.8 \text{ or } -4.0 \text{ (disregard)}$

It will take about 2.8 seconds.

67. $\sqrt{5x - 2} = 3$

$\left(\sqrt{5x - 2}\right)^2 = 3^2$

$5x - 2 = 9$

$5x = 11$

$x = \dfrac{11}{5}$

69. $\dfrac{1}{x} + \dfrac{2}{5} = \dfrac{7}{x}$

$5x\left(\dfrac{1}{x} + \dfrac{2}{5}\right) = 5x\left(\dfrac{7}{x}\right)$

$5 + 2x = 35$

$2x = 30$

$x = 15$

71. $x^4 + x^2 - 20 = (x^2 + 5)(x^2 - 4)$

$= (x^2 + 5)(x + 2)(x - 2)$

73. $z^4 - 13z^2 + 36 = (z^2 - 9)(z^2 - 4)$

$= (z + 3)(z - 3)(z + 2)(z - 2)$

75. $x^2 = -10$

$x^2 + 10 = 0$

$a = 1, b = 0, c = 10$

The correct substitution is **b**.

77. $m^2 + 5m - 6 = 0$

$(m+6)(m-1) = 0$

$m + 6 = 0$ or $m - 1 = 0$

$m = -6$ or $m = 1$

The results are the same. answers may vary.

79. $2x^2 - 6x + 3 = 0$

$a = 2, b = -6, c = 3$

$x = \dfrac{6 \pm \sqrt{(-6)^2 - 4(2)(3)}}{2(2)}$

$= \dfrac{6 \pm \sqrt{12}}{4}$

≈ 0.6 or 2.4

81. From Sunday to Monday

83. Wednesday

85. $f(x) = 3x^2 - 18x + 56$

$f(4) = 3(4)^2 - 18(4) + 56 = 32$

This answers appears to agree with the graph.

87. a. 2010 is 10 years after 2000, so $x = 10$.

$f(x) = 22x^2 + 274x + 15,628$

$f(10) = 22(10)^2 + 274(10) + 15,628$

$= 20,568$

The number of college students in the United States in 2010 was 20,568 thousand students.

b. $24,500 = 22x^2 + 274x + 15,628$

$0 = 22x^2 + 274x - 8872$

$0 = 2(11x^2 + 137x - 4436)$

$a = 11, b = 137, c = -4436$

$x = \dfrac{-b \pm \sqrt{b^2 - 4ac}}{2a}$

$x = \dfrac{-137 \pm \sqrt{137^2 - 4(11)(-4436)}}{2(11)}$

$x \approx \dfrac{-137 \pm 462.55}{22}$

$x \approx 15$ (reject a negative time because we are not concerned with the past)

The model predicts 24,500 thousand students 15 years after 2000, or in 2015.

89. a. 2004 is 4 years after 2000, so $x = 4$.

$y = -10x^2 + 193x + 8464$

$y = -10(4)^2 + 193(4) + 8464 = 9076$

The average total daily supply in 2004 was 9076 thousand barrels per day.

b. $9325 = -10x^2 + 193x + 8464$

$10x^2 - 193x + 861 = 0$

$a = 10, b = -193, c = 861$

$x = \dfrac{-b \pm \sqrt{b^2 - 4ac}}{2a}$

$x = \dfrac{193 \pm \sqrt{(-193)^2 - 4(10)(861)}}{2(10)}$

$= \dfrac{193 \pm \sqrt{2809}}{20}$

$= \dfrac{193 \pm 53}{20}$

$x = 12.3$ or $x = 7$

The first time the supply reaches this level is 7 years after 2000, or in 2007. The supply was 9325 thousand barrels per day in 2007.

c. 12.3 years after 2000, or in 2012, the supply will be 9325 thousand barrels per day.

91. $\dfrac{-b + \sqrt{b^2 - 4ac}}{2a} + \dfrac{-b - \sqrt{b^2 - 4ac}}{2a}$

$= \dfrac{-b + \sqrt{b^2 - 4ac} - b - \sqrt{b^2 - 4ac}}{2a}$

$= \dfrac{-2b}{2a}$

$= -\dfrac{b}{a}$

93. $3x^2 - \sqrt{12}x + 1 = 0$

$a = 3, b = -\sqrt{12}, c = 1$

$$x = \frac{\sqrt{12} \pm \sqrt{\left(-\sqrt{12}\right)^2 - 4(3)(1)}}{2(3)}$$

$$= \frac{\sqrt{12} \pm \sqrt{12 - 12}}{6}$$

$$= \frac{\sqrt{4 \cdot 3} \pm \sqrt{0}}{6}$$

$$= \frac{2\sqrt{3}}{6}$$

$$= \frac{\sqrt{3}}{3}$$

The solution is $\dfrac{\sqrt{3}}{3}$.

95. $x^2 + \sqrt{2}x + 1 = 0$

$a = 1, b = \sqrt{2}, c = 1$

$$x = \frac{-\sqrt{2} \pm \sqrt{\left(\sqrt{2}\right)^2 - 4(1)(1)}}{2(1)}$$

$$= \frac{-\sqrt{2} \pm \sqrt{2 - 4}}{2}$$

$$= \frac{-\sqrt{2} \pm \sqrt{-2}}{2}$$

$$= \frac{-\sqrt{2} \pm i\sqrt{2}}{2}$$

The solutions are $\dfrac{-\sqrt{2} + i\sqrt{2}}{2}$ and $\dfrac{-\sqrt{2} - i\sqrt{2}}{2}$.

97. $2x^2 - \sqrt{3}x - 1 = 0$

$a = 2, b = -\sqrt{3}, c = -1$

$$x = \frac{\sqrt{3} \pm \sqrt{\left(-\sqrt{3}\right)^2 - 4(2)(-1)}}{2(2)}$$

$$= \frac{\sqrt{3} \pm \sqrt{3 + 8}}{4}$$

$$= \frac{\sqrt{3} \pm \sqrt{11}}{4}$$

The solutions are $\dfrac{\sqrt{3} + \sqrt{11}}{4}$ and $\dfrac{\sqrt{3} - \sqrt{11}}{4}$.

Section 8.3 Practice Exercises

1. $x - \sqrt{x+1} - 5 = 0$

Get the radical alone on one side of the equation. Then square both sides.

$$x - \sqrt{x+1} - 5 = 0$$

$$x - 5 = \sqrt{x+1}$$

$$(x-5)^2 = x+1$$

$$x^2 - 10x + 25 = x+1$$

$$x^2 - 11x + 24 = 0$$

$$(x-8)(x-3) = 0$$

$$x - 8 = 0 \quad \text{or} \quad x - 3 = 0$$

$$x = 8 \quad \text{or} \quad x = 3$$

Check:

Let $x = 3$.

$$x - \sqrt{x+1} - 5 = 0$$

$$3 - \sqrt{3+1} - 5 \stackrel{?}{=} 0$$

$$-2 - \sqrt{4} \stackrel{?}{=} 0$$

$$-2 - 2 \stackrel{?}{=} 0$$

$$-4 = 0 \quad \text{False}$$

Let $x = 8$.

$$x - \sqrt{x+1} - 5 = 0$$

$$8 - \sqrt{8+1} - 5 \stackrel{?}{=} 0$$

$$3 - \sqrt{9} \stackrel{?}{=} 0$$

$$3 - 3 \stackrel{?}{=} 0$$

$$0 = 0 \quad \text{True}$$

The solution is 8 or the solution set is $\{8\}$.

2. $\dfrac{5x}{x+1} - \dfrac{x+4}{x} = \dfrac{3}{x(x+1)}$

x cannot be either −1 or 0, because these values cause denominators to equal zero. Multiply both sides of the equation by $x(x + 1)$.

$$x(x+1)\left(\frac{5x}{x+1}\right) - x(x+1)\left(\frac{x+4}{x}\right) = x(x+1)\left[\frac{3}{x(x+1)}\right]$$

$$5x^2 - (x+1)(x+4) = 3$$

$$5x^2 - x^2 - 5x - 4 = 3$$

$$4x^2 - 5x - 7 = 0$$

Use the quadratic formula with $a = 4$, $b = -5$, and $c = -7$.

$$x = \frac{-(-5)\pm\sqrt{(-5)^2 - 4(4)(-7)}}{2(4)} = \frac{5\pm\sqrt{25+112}}{8} = \frac{5\pm\sqrt{137}}{8}$$

Neither proposed solution will make denominators 0. The solutions are $\dfrac{5+\sqrt{137}}{8}$ and $\dfrac{5-\sqrt{137}}{8}$ or the solution

set is $\left\{ \dfrac{5+\sqrt{137}}{8}, \dfrac{5-\sqrt{137}}{8} \right\}$.

3.
$$p^4 - 7p^2 - 144 = 0$$
$$(p^2 + 9)(p^2 - 16) = 0$$
$$(p^2 + 9)(p + 4)(p - 4) = 0$$

$p^2 + 9 = 0$ or $p + 4 = 0$ or $p - 4 = 0$

$\qquad p^2 = -9 \qquad\qquad\qquad p = -4 \qquad\qquad p = 4$

$\qquad p = \pm\sqrt{-9}$

$\qquad p = \pm 3i$

The solutions are 4, −4, $3i$, and $-3i$.

4. $(x+2)^2 - 2(x+2) - 3 = 0$

Let $y = x + 2$.

$$y^2 - 2y - 3 = 0$$
$$(y-3)(y+1) = 0$$

$y - 3 = 0$ or $y + 1 = 0$

$\quad y = 3 \qquad\qquad y = -1$

Substitute $x + 2$ for y.

$x + 2 = 3$ or $x + 2 = -1$

$\quad x = 1 \qquad\qquad x = -3$

Both 1 and −3 check. The solutions are 1 and −3.

5. $x^{2/3} - 5x^{1/3} + 4 = 0$

Let $m = x^{1/3}$.

$$m^2 - 5m + 4 = 0$$
$$(m-4)(m-1) = 0$$

$m - 4 = 0$ or $m - 1 = 0$

$\quad m = 4 \qquad\qquad m = 1$

Since $m = x^{1/3}$, we have

$$x^{1/3} = 4 \qquad \text{or} \qquad x^{1/3} = 1$$
$$x = 4^3 = 64 \qquad\qquad x = 1^3 = 1$$

Both 64 and 1 check. The solutions are 64 and 1.

6. Let $x =$ the time in hours it takes Steve to groom all the dogs. Then, $x - 1 =$ the time it takes Katy to groom all the dogs.

The part of the job completed in one hour by Steve is $\dfrac{1}{x}$, and the part completed by Katy in one hour is $\dfrac{1}{x-1}$. In one hour, $\dfrac{1}{4}$ of the job is completed. We have,

$$\frac{1}{x} + \frac{1}{x-1} = \frac{1}{4}$$

$$4x(x-1)\left(\frac{1}{x}\right) + 4x(x-1)\left(\frac{1}{x-1}\right) = 4x(x-1)\left(\frac{1}{4}\right)$$

$$4(x-1) + 4x = x(x-1)$$

$$4x - 4 + 4x = x^2 - x$$

$$0 = x^2 - 9x + 4$$

Use the quadratic formula with $a = 1$, $b = -9$, and $c = 4$.

$$x = \frac{-(-9) \pm \sqrt{(-9)^2 - 4(1)(4)}}{2(1)}$$

$$x = \frac{9 \pm \sqrt{81 - 16}}{2} = \frac{9 \pm \sqrt{65}}{2}$$

$$x \approx 8.53 \quad \text{or} \quad x \approx 0.47$$

Since $x - 1 = 0.47 - 1 = -0.53 < 0$, representing negative time worked, we reject 0.47. It takes Steve $\dfrac{9 + \sqrt{65}}{2} \approx 8.5$ hours and Katy $\dfrac{9 + \sqrt{65}}{2} - 1 = \dfrac{7 + \sqrt{65}}{2} \approx 7.5$ hours to groom all the dogs when working alone.

7. Let $x =$ the speed driven to Shanghai. Then $x + 50 =$ the speed driven to Ningbo.

	distance =	*rate* ·	*time*
To Shanghai	36	x	$\dfrac{36}{x}$
To Ningbo	36	$x + 50$	$\dfrac{36}{x+50}$

The total travel time was 1.3 hours, so

$$\frac{36}{x} + \frac{36}{x+50} = 1.3$$

$$x(x+50)\left(\frac{36}{x}\right) + x(x+50)\left(\frac{36}{x+50}\right) = 1.3x(x+50)$$

$$36(x+50) + 36x = 1.3x^2 + 65x$$

$$36x + 1800 + 36x = 1.3x^2 + 65x$$

$$0 = 1.3x^2 - 7x - 1800$$

Use the quadratic formula with $a = 1.3$, $b = -7$, and $c = -1800$.

$$x = \frac{-(-7) \pm \sqrt{(-7)^2 - 4(1.3)(-1800)}}{2(1.3)}$$

$$= \frac{7 \pm \sqrt{9409}}{2.6}$$

$$x = \frac{7 + \sqrt{9409}}{2.6} = 40 \quad \text{or}$$

$$x = \frac{7 - \sqrt{9409}}{2.6} \approx -34.6$$

The speed is not negative, so reject −34.6. The speed to Shanghai was 40 km/hr and to Ningbo it was 40 + 50 = 90 km/hr.

Vocabulary, Readiness & Video Check 8.3

1. The values we get for the substituted variable are *not* our final answers. Remember to always substitute back to the original variable and solve for it if necessary.

2. The rational equation simplifies to a quadratic equation once you multiply through by the LCD to rid the equation of fractions.

Exercise Set 8.3

1.
$$2x = \sqrt{10 + 3x}$$
$$4x^2 = 10 + 3x$$
$$4x^2 - 3x - 10 = 0$$
$$(4x + 5)(x - 2) = 0$$
$$4x + 5 = 0 \quad \text{or} \quad x - 2 = 0$$
$$x = -\frac{5}{4} \quad \text{or} \quad x = 2$$

Discard $-\frac{5}{4}$. The solution is 2.

3.
$$x - 2\sqrt{x} = 8$$
$$x - 8 = 2\sqrt{x}$$
$$(x - 8)^2 = \left(2\sqrt{x}\right)^2$$
$$x^2 - 16x + 64 = 4x$$
$$x^2 - 20x + 64 = 0$$
$$(x - 16)(x - 4) = 0$$
$$x - 16 = 0 \quad \text{or} \quad x - 4 = 0$$
$$x = 16 \quad \text{or} \quad x = 4 \text{ (discard)}$$

The solution is 16.

5.
$$\sqrt{9x} = x + 2$$
$$\left(\sqrt{9x}\right)^2 = (x + 2)^2$$
$$9x = x^2 + 4x + 4$$
$$0 = x^2 - 5x + 4$$
$$0 = (x - 4)(x - 1)$$
$$x - 4 = 0 \quad \text{or} \quad x - 1 = 0$$
$$x = 4 \quad \text{or} \quad x = 1$$

The solutions are 1 and 4.

7. $\dfrac{2}{x} + \dfrac{3}{x - 1} = 1$

Multiply each term by $x(x - 1)$.
$$2(x - 1) + 3x = x(x - 1)$$
$$2x - 2 + 3x = x^2 - x$$
$$0 = x^2 - 6x + 2$$
$$x = \frac{6 \pm \sqrt{(-6)^2 - 4(1)(2)}}{2(1)}$$
$$= \frac{6 \pm \sqrt{28}}{2}$$
$$= \frac{6 \pm 2\sqrt{7}}{2} = 3 \pm \sqrt{7}$$

The solutions are $3 + \sqrt{7}$ and $3 - \sqrt{7}$.

9. $\dfrac{3}{x} + \dfrac{4}{x + 2} = 2$

Multiply each term by $x(x + 2)$.
$$3(x + 2) + 4x = 2x(x + 2)$$
$$3x + 6 + 4x = 2x^2 + 4x$$
$$0 = 2x^2 - 3x - 6$$
$$x = \frac{3 \pm \sqrt{(-3)^2 - 4(2)(-6)}}{2(2)}$$
$$= \frac{3 \pm \sqrt{57}}{4}$$

The solutions are $\dfrac{3 + \sqrt{57}}{4}$ and $\dfrac{3 - \sqrt{57}}{4}$.

11.
$$\frac{7}{x^2 - 5x + 6} = \frac{2x}{x - 3} - \frac{x}{x - 2}$$
$$\frac{7}{(x - 3)(x - 2)} = \frac{2x}{x - 3} - \frac{x}{x - 2}$$

Multiply each term by $(x - 3)(x - 2)$.
$$7 = 2x(x - 2) - x(x - 3)$$
$$7 = 2x^2 - 4x - x^2 + 3x$$
$$0 = x^2 - x - 7$$

$$x = \frac{1 \pm \sqrt{(-1)^2 - 4(1)(-7)}}{2(1)}$$

$$= \frac{1 \pm \sqrt{29}}{2}$$

The solutions are $\dfrac{1+\sqrt{29}}{2}$ and $\dfrac{2-\sqrt{29}}{2}$.

13.
$$p^4 - 16 = 0$$
$$(p^2 - 4)(p^2 + 4) = 0$$
$$(p+2)(p-2)(p^2+4) = 0$$
$$p+2 = 0 \quad \text{or } p-2 = 0 \text{ or } p^2 + 4 = 0$$
$$p = -2 \text{ or } \quad p = 2 \text{ or } \quad p^2 = -4$$
$$p = \pm\sqrt{-4}$$
$$p = \pm 2i$$

The solutions are –2, 2, –2*i*, and 2*i*.

15.
$$4x^4 + 11x^2 = 3$$
$$4x^4 + 11x^2 - 3 = 0$$
$$(4x^2 - 1)(x^2 + 3) = 0$$
$$(2x+1)(2x-1)(x^2+3) = 0$$
$$2x+1 = 0 \quad \text{or } 2x-1 = 0 \text{ or } x^2 + 3 = 0$$
$$x = -\frac{1}{2} \text{ or } \quad x = \frac{1}{2} \text{ or } \quad x^2 = -3$$
$$x = \pm\sqrt{-3}$$
$$x = \pm i\sqrt{3}$$

The solutions are $-\dfrac{1}{2}, \dfrac{1}{2}, -i\sqrt{3}$, and $i\sqrt{3}$.

17.
$$z^4 - 13z^2 + 36 = 0$$
$$(z^2 - 9)(z^2 - 4) = 0$$
$$(z+3)(z-3)(z+2)(z-2) = 0$$
$$z = -3, z = 3, z = -2, z = 2$$
The solutions are –3, 3, –2, and 2.

19. $x^{2/3} - 3x^{1/3} - 10 = 0$

Let $y = x^{1/3}$. Then $y^2 = x^{2/3}$ and
$$y^2 - 3y - 10 = 0$$
$$(y-5)(y+2) = 0$$
$$y-5 = 0 \quad \text{or } y+2 = 0$$
$$y = 5 \quad \text{or } \quad y = -2$$
$$x^{1/3} = 5 \quad \text{or } \quad x^{1/3} = -2$$
$$x = 125 \text{ or } \quad x = -8$$
The solutions are –8 and 125.

21. $(5n+1)^2 + 2(5n+1) - 3 = 0$

Let $y = 5n + 1$. Then $y^2 = (5n+1)^2$ and
$$y^2 + 2y - 3 = 0$$
$$(y+3)(y-1) = 0$$
$$y+3 = 0 \quad \text{or } y-1 = 0$$
$$y = -3 \text{ or } \quad y = 1$$
$$5n+1 = -3 \text{ or } 5n+1 = 1$$
$$5n = -4 \text{ or } \quad 5n = 0$$
$$n = -\frac{4}{5} \text{ or } \quad n = 0$$

The solutions are $-\dfrac{4}{5}$ and 0.

23. $2x^{2/3} - 5x^{1/3} = 3$

Let $y = x^{1/3}$. Then $y^2 = x^{2/3}$ and
$$2y^2 - 5y = 3$$
$$2y^2 - 5y - 3 = 0$$
$$(2y+1)(y-3) = 0$$
$$2y+1 = 0 \quad \text{or } y-3 = 0$$
$$y = -\frac{1}{2} \text{ or } \quad y = 3$$
$$x^{1/3} = -\frac{1}{2} \text{ or } \quad x^{1/3} = 3$$
$$x = -\frac{1}{8} \text{ or } \quad x = 27$$

The solutions are $-\dfrac{1}{8}$ and 27.

25.
$$1 + \frac{2}{3t-2} = \frac{8}{(3t-2)^2}$$
$$(3t-2)^2 + 2(3t-2) = 8$$
$$(3t-2)^2 + 2(3t-2) - 8 = 0$$

Let $y = 3t - 2$. Then $y^2 = (3t-2)^2$ and
$$y^2 + 2y - 8 = 0$$
$$(y+4)(y-2) = 0$$
$$y+4 = 0 \quad \text{or } y-2 = 0$$
$$y = -4 \text{ or } \quad y = 2$$
$$3t-2 = -4 \text{ or } 3t-2 = 2$$
$$3t = -2 \text{ or } \quad 3t = 4$$
$$t = -\frac{2}{3} \text{ or } \quad t = \frac{4}{3}$$

The solutions are $-\dfrac{2}{3}$ and $\dfrac{4}{3}$.

27. $20x^{2/3} - 6x^{1/3} - 2 = 0$

Let $y = x^{1/3}$. Then $y^2 = x^{2/3}$ and

$20y^2 - 6y - 2 = 0$

$2(10y^2 - 3y - 1) = 0$

$2(5y + 1)(2y - 1) = 0$

$5y + 1 = 0 \quad$ or $\quad 2y - 1 = 0$

$y = -\dfrac{1}{5} \quad$ or $\quad y = \dfrac{1}{2}$

$x^{1/3} = -\dfrac{1}{5} \quad$ or $\quad x^{1/3} = \dfrac{1}{2}$

$x = -\dfrac{1}{125} \quad$ or $\quad x = \dfrac{1}{8}$

The solutions are $\dfrac{1}{8}$ and $-\dfrac{1}{125}$.

29. $a^4 - 5a^2 + 6 = 0$

$(a^2 - 3)(a^2 - 2) = 0$

$a^2 - 3 = 0 \quad$ or $a^2 - 2 = 0$

$a^2 = 3 \quad$ or $\quad a^2 = 2$

$a = \pm\sqrt{3} \quad$ or $\quad a = \pm\sqrt{2}$

The solutions are $-\sqrt{3}, \sqrt{3}, -\sqrt{2}$, and $\sqrt{2}$.

31. $\dfrac{2x}{x-2} + \dfrac{x}{x+3} = -\dfrac{5}{x+3}$

Multiply each term by $(x + 3)(x - 2)$.

$2x(x + 3) + x(x - 2) = -5(x - 2)$

$2x^2 + 6x + x^2 - 2x = -5x + 10$

$3x^2 + 9x - 10 = 0$

$x = \dfrac{-9 \pm \sqrt{(9)^2 - 4(3)(-10)}}{2(3)}$

$= \dfrac{-9 \pm \sqrt{201}}{6}$

The solutions are $\dfrac{-9 + \sqrt{201}}{6}$ and $\dfrac{-9 - \sqrt{201}}{6}$.

33. $(p + 2)^2 = 9(p + 2) - 20$

$(p + 2)^2 - 9(p + 2) + 20 = 0$

Let $x = p + 2$. Then $x^2 = (p + 2)^2$ and

$x^2 - 9x + 20 = 0$

$(x - 5)(x - 4) = 0$

$x = 5 \quad$ or $\quad x = 4$

$p + 2 = 5 \quad$ or $p + 2 = 4$

$p = 3 \quad$ or $\quad p = 2$

The solutions are 2 and 3.

35. $2x = \sqrt{11x + 3}$

$(2x)^2 = \left(\sqrt{11x + 3}\right)^2$

$4x^2 = 11x + 3$

$4x^2 - 11x - 3 = 0$

$(4x + 1)(x - 3) = 0$

$x = -\dfrac{1}{4}$ (discard) or $x = 3$

The solution is 3.

37. $x^{2/3} - 8x^{1/3} + 15 = 0$

Let $y = x^{1/3}$. Then $y^2 = x^{2/3}$ and

$y^2 - 8y + 15 = 0$

$(y - 5)(y - 3) = 0$

$y = 5 \quad$ or $\quad y = 3$

$x^{1/3} = 5 \quad$ or $\quad x^{1/3} = 3$

$x = 125 \quad$ or $\quad x = 27$

The solutions are 27 and 125.

39. $y^3 + 9y - y^2 - 9 = 0$

$y(y^2 + 9) - 1(y^2 + 9) = 0$

$(y^2 + 9)(y - 1) = 0$

$y^2 + 9 = 0 \quad$ or $y - 1 = 0$

$y^2 = -9 \quad$ or $\quad y = 1$

$y = \pm\sqrt{-9}$

$y = \pm 3i$

The solutions are 1, $-3i$, and $3i$.

41. $2x^{2/3} + 3x^{1/3} - 2 = 0$

Let $y = x^{1/3}$. Then $y^2 = x^{2/3}$ and

$2y^2 + 3y - 2 = 0$

$(2y - 1)(y + 2) = 0$

$y = \dfrac{1}{2} \quad$ or $\quad y = -2$

$x^{1/3} = \dfrac{1}{2} \quad$ or $x^{1/3} = -2$

$x = \dfrac{1}{8} \quad$ or $\quad x = -8$

The solutions are -8 and $\dfrac{1}{8}$.

43. $x^{-2} - x^{-1} - 6 = 0$

Let $y = x^{-1}$. Then $y^2 = x^{-2}$ and

$y^2 - y - 6 = 0$

$(y - 3)(y + 2) = 0$

$\quad y = 3$ or $\quad y = -2$

$\quad x^{-1} = 3$ or $\quad x^{-1} = -2$

$\quad \dfrac{1}{x} = 3$ or $\quad \dfrac{1}{x} = -2$

$\quad x = \dfrac{1}{3}$ or $\quad x = -\dfrac{1}{2}$

The solutions are $-\dfrac{1}{2}$ and $\dfrac{1}{3}$.

45. $\quad x - \sqrt{x} = 2$

$\quad x - 2 = \sqrt{x}$

$\quad (x - 2)^2 = x$

$\quad x^2 - 4x + 4 = x$

$\quad x^2 - 5x + 4 = 0$

$\quad (x - 4)(x - 1) = 0$

$x = 4$ or $x = 1$ (discard)

The solution is 4.

47. $\quad \dfrac{x}{x-1} + \dfrac{1}{x+1} = \dfrac{2}{x^2 - 1}$

$\quad \dfrac{x}{x-1} + \dfrac{1}{x+1} = \dfrac{2}{(x+1)(x-1)}$

$\quad x(x+1) + (x-1) = 2$

$\quad x^2 + x + x - 1 = 2$

$\quad x^2 + 2x - 3 = 0$

$\quad (x+3)(x-1) = 0$

$x = -3$ or $x = 1$ (discard)

The solution is -3.

49. $\quad p^4 - p^2 - 20 = 0$

$\quad (p^2 - 5)(p^2 + 4) = 0$

$\quad p^2 - 5 = 0 \quad$ or $p^2 + 4 = 0$

$\quad p^2 = 5 \quad$ or $\quad p^2 = -4$

$\quad p = \pm\sqrt{5} \quad$ or $\quad p = \pm 2i$

The solutions are $-\sqrt{5}, \sqrt{5}, -2i,$ and $2i$.

51. $(x + 3)(x^2 - 3x + 9) = 0$

$x + 3 = 0 \quad$ or $x^2 - 3x + 9 = 0$

$x = -3$ or

$x = \dfrac{3 \pm \sqrt{(-3)^2 - 4(1)(9)}}{2(1)}$

$\quad = \dfrac{3 \pm \sqrt{-27}}{2}$

$\quad = \dfrac{3 \pm 3i\sqrt{3}}{2}$

The solutions are -3, $\dfrac{3 + 3i\sqrt{3}}{2}$, and $\dfrac{3 - 3i\sqrt{3}}{2}$.

53. $\qquad 1 = \dfrac{4}{x-7} + \dfrac{5}{(x-7)^2}$

$(x - 7)^2 - 4(x - 7) - 5 = 0$

Let $y = x - 7$. Then $y^2 = (x - 7)^2$ and

$\quad y^2 - 4y - 5 = 0$

$\quad (y - 5)(y + 1) = 0$

$\quad y = 5 \quad$ or $\quad y = -1$

$\quad x - 7 = 5 \quad$ or $x - 7 = -1$

$\quad x = 12$ or $\quad x = 6$

The solutions are 6 and 12.

55. $\qquad 27y^4 + 15y^2 = 2$

$\qquad 27y^4 + 15y^2 - 2 = 0$

$\qquad (9y^2 - 1)(3y^2 + 2) = 0$

$(3y + 1)(3y - 1)(3y^2 + 2) = 0$

$y = -\dfrac{1}{3} \quad$ or $\; y = \dfrac{1}{3} \quad$ or $\; y^2 = -\dfrac{2}{3}$

$\qquad\qquad\qquad\qquad y = \pm\sqrt{-\dfrac{2}{3}}$

$\qquad\qquad\qquad\qquad y = \pm\dfrac{i\sqrt{6}}{3}$

The solutions are $-\dfrac{1}{3}, \dfrac{1}{3}, -\dfrac{i\sqrt{6}}{3},$ and $\dfrac{i\sqrt{6}}{3}$.

57. $x - \sqrt{19 - 2x} - 2 = 0$

$$x - 2 = \sqrt{19 - 2x}$$
$$(x-2)^2 = \left(\sqrt{19 - 2x}\right)^2$$
$$x^2 - 4x + 4 = 19 - 2x$$
$$x^2 - 2x - 15 = 0$$
$$(x-5)(x+3) = 0$$
$$x - 5 = 0 \quad \text{or} \quad x + 3 = 0$$
$$x = 5 \quad \text{or} \quad x = -3$$

Reject $x = -3$ as an extraneous solution. The solution is 5.

59. Let x be the rate to Tucson. Then $x + 11$ is the rate returning home.

	distance	=	rate	·	time
To Tucson	330		x		$\frac{320}{x}$
Return trip	330		$x + 11$		$\frac{330}{x-11}$

The LCD is $x(x + 11)$.

$$\frac{330}{x} = \frac{330}{x+11} + 1$$
$$\frac{330}{x} - \frac{330}{x+11} = 1$$
$$x(x+11)\left(\frac{330}{x}\right) - x(x+11)\left(\frac{330}{x+11}\right) = x(x+11)(1)$$
$$330(x+11) - 330x = x(x+11)$$
$$330x + 3630 - 330x = x^2 + 11x$$
$$0 = x^2 + 11x - 3630$$
$$0 = (x+66)(x-55)$$

$$x + 66 = 0 \quad \text{or} \quad x - 55 = 0$$
$$x = -66 \qquad\qquad x = 55$$

Reject -66 since rate cannot be negative. The speed to Tucson is 55 mph and the speed returning is $x + 11 = 55 + 11 = 66$ mph.

61. Let x = speed on the first part. Then $x - 1$ = speed on the second part.

$$d = rt \implies t = \frac{d}{r}$$

$$t_{\text{on first part}} + t_{\text{on second part}} = 1\frac{3}{5}$$
$$\frac{3}{x} + \frac{4}{x-1} = \frac{8}{5}$$
$$3 \cdot 5(x-1) + 4 \cdot 5x = 8x(x-1)$$
$$15x - 15 + 20x = 8x^2 - 8x$$
$$0 = 8x^2 - 43x + 15$$
$$0 = (8x-3)(x-5)$$

$8x-3=0$ or $x-5=0$

$x=\dfrac{3}{8}$ or $x=5$

$x-1=4$

Discard $\dfrac{3}{8}$. Her speeds were 5 mph then 4 mph.

63. Let x = time for hose alone. Then
$x-1$ = time for the inlet pipe alone.

$\dfrac{1}{x}+\dfrac{1}{x-1}=\dfrac{1}{8}$

$8(x-1)+8x=x(x-1)$

$8x-8+8x=x^2-x$

$0=x^2-17x+8$

$x=\dfrac{17\pm\sqrt{(-17)^2-4(1)(8)}}{2(1)}$

$=\dfrac{17\pm\sqrt{257}}{2}$

$x\approx0.5$ (discard)   or $x\approx16.5$

$x-1\approx15.5$

Hose: 16.5 hrs; Inlet pipe: 15.5 hrs

65. Let x = time for son alone. Then
$x-1$ = time for dad alone.

$\dfrac{1}{x}+\dfrac{1}{x-1}=\dfrac{1}{4}$

$4(x-1)+4x=x(x-1)$

$4x-4+4x=x^2-x$

$0=x^2-9x+4$

$x=\dfrac{9\pm\sqrt{(-9)^2-4(1)(4)}}{2(1)}$

$=\dfrac{9\pm\sqrt{65}}{2}$

≈0.5 (discard) or 8.5

It takes his son about 8.5 hours.

67. Let x = the number.

$x(x-4)=96$

$x^2-4x-96=0$

$(x-12)(x+8)=0$

$x=12$ or $x=-8$

The number is 12 or -8.

69. a.   length $=x-3-3=x-6=(x-6)$ in.

     b. $V=lwh$

          $300=(x-6)(x-6)\cdot3$

c. $300=3(x-6)^2$

       $100=x^2-12x+36$

         $0=x^2-12x-64$

         $0=(x-16)(x+4)$

    $x=16$ or $x=-4$ (discard)

The sheet is 16 in. by 16 in.

Check: $V=3(x-6)(x-6)$

           $=3(16-6)(16-6)$

           $=3(10)(10)$

           $=300$ cubic in.

71. Let x = length of the side of the square.

Area $=x^2$

$920=x^2$

$\sqrt{920}=x$

Adding another radial line to a different corner
would yield a right triangle with legs r and
hypotenuse x.

$r^2+r^2=x^2$

$2r^2=\left(\sqrt{920}\right)^2$

$2r^2=920$

$r^2=460$

$r=\pm\sqrt{460}=\pm21.4476$

Disregard the negative. The smallest radius
would be 22 feet.

73. $\dfrac{5x}{3}+2\le7$

         $\dfrac{5x}{3}\le5$

        $5x\le15$

          $x\le3$

   $(-\infty,3]$

75. $\dfrac{y-1}{15}>-\dfrac{2}{5}$

   $15\left(\dfrac{y-1}{15}\right)>15\left(-\dfrac{2}{5}\right)$

         $y-1>-6$

           $y>-5$

   $(-5,\infty)$

77. Domain: $\{x\,|\,x$ is a real number$\}$ or $(-\infty,\infty)$

Range: $\{y\,|\,y$ is a real number$\}$ or $(-\infty,\infty)$

It is a function.

79. Domain: $\{x \mid x \text{ is a real number}\}$ or $(-\infty, \infty)$

Range: $\{y \mid y \geq -1\}$ or $[-1, \infty)$

It is a function.

81. $5y^3 - 45y - 5y^2 - 45 = 0$

$5y(y^2 + 9) - 5(y^2 + 9) = 0$

$(5y - 5)(y^2 + 9) = 0$

$5(y - 1)(y^2 + 9) = 0$

$y - 1 = 0$ or $y^2 + 9 = 0$

$\quad y = 1 \qquad\qquad y^2 = -9$

$\qquad\qquad\qquad\quad y = \pm\sqrt{-9} = \pm 3i$

The solutions are 1, $3i$, and $-3i$.

83. Let $u = x^{-1}$, then $x^{-2} = u^2$.

$3x^{-2} - 3x^{-1} - 18 = 0$

$3u^2 - 3u - 18 = 0$

$3(u^2 - u - 6) = 0$

$3(u - 3)(u + 2) = 0$

$u - 3 = 0$ or $u + 2 = 0$

$\quad u = 3 \qquad\qquad u = -2$

$\quad x^{-1} = 3 \qquad\quad x^{-1} = -2$

$\quad x = \dfrac{1}{3} \qquad\qquad x = -\dfrac{1}{2}$

The solutions are $\dfrac{1}{3}$ and $-\dfrac{1}{2}$.

85. $\qquad\qquad 2x^3 = -54$

$\qquad\qquad\qquad x^3 = -27$

$\qquad\qquad x^3 + 27 = 0$

$(x + 3)(x^2 - 3x + 9) = 0$

$x + 3 = 0$ or $x^2 - 3x + 9 = 0$

$\quad x = -3$

$x = \dfrac{-(-3) \pm \sqrt{(-3)^2 - 4(1)(9)}}{2(1)}$

$x = \dfrac{3 \pm \sqrt{9 - 36}}{2} = \dfrac{3 \pm \sqrt{-27}}{2}$

$x = \dfrac{3 \pm 3i\sqrt{3}}{2}$

The solutions are -3, $\dfrac{3 + 3i\sqrt{3}}{2}$, and $\dfrac{3 - 3i\sqrt{3}}{2}$.

87. answers may vary

89. a. Let x be Voeckler's average speed. Using $d = r \cdot t$ or $t = \dfrac{d}{r}$, Voeckler's time for that stage was

$\dfrac{119,000}{x}$ seconds. Contador's average speed is then $(x - 0.0034)$ meters per second and the time for his stage

was $\dfrac{119,000}{x - 0.0034}$ seconds, which is also $\dfrac{119,000}{x} + 3$.

The LCD is $x(x - 0.0034)$.

$$\frac{119,000}{x} + 3 = \frac{119,000}{x - 0.0034}$$

$$x(x - 0.0034)\left(\frac{119,000}{x}\right) + x(x - 0.0034)(3) = x(x - 0.0034)\left(\frac{119,000}{x - 0.0034}\right)$$

$$119,000(x - 0.0034) + 3x(x - 0.0034) = x(119,000)$$

$$119,000x - 404.6 + 3x^2 - 0.0102x = 119,000x$$

$$3x^2 - 0.0102x - 404.6 = 0$$

$a = 3,\ b = -0.0102,\ c = -404.6$

$$x = \frac{-b \pm \sqrt{b^2 - 4ac}}{2a}$$

$$x = \frac{-(-0.0102) \pm \sqrt{(-0.0102)^2 - 4(3)(-404.6)}}{2(3)}$$

$x \approx -11.612$ or $x \approx 11.615$

Since the speed cannot be negative, reject -11.612.

Voeckler's average speed was approximately 11.615 meters per second.

b. Contador's average speed was $11.615 - 0.0034 \approx 11.612$ meters per second.

c. $\dfrac{11.615 \text{ m}}{1 \text{ sec}} \cdot \dfrac{3600 \text{ sec}}{\text{hr}} \cdot \dfrac{1 \text{ km}}{1000 \text{ m}} \cdot \dfrac{0.62 \text{ mi}}{1 \text{ km}} \approx 25.925$ mi/hr

Voeckler's average speed was about 25.925 miles per hour.

Integrated Review

1. $x^2 - 10 = 0$

$\quad x^2 = 10$

$\quad x = \pm\sqrt{10}$

The solutions are $\sqrt{10}$ and $-\sqrt{10}$.

2. $x^2 - 14 = 0$

$\quad x^2 = 14$

$\quad x = \pm\sqrt{14}$

The solutions are $\sqrt{14}$ and $-\sqrt{14}$.

3. $(x - 1)^2 = 8$

$\quad x - 1 = \pm\sqrt{8}$

$\quad x - 1 = \pm 2\sqrt{2}$

$\quad x = 1 \pm 2\sqrt{2}$

The solutions are $1 + 2\sqrt{2}$ and $1 - 2\sqrt{2}$.

4. $(x+5)^2 = 12$

$\quad x+5 = \pm\sqrt{12}$

$\quad x+5 = \pm 2\sqrt{3}$

$\quad\quad x = -5 \pm 2\sqrt{3}$

The solutions are $-5+2\sqrt{3}$ and $-5-2\sqrt{3}$.

5. $\quad x^2 + 2x - 12 = 0$

$\quad x^2 + 2x + \left(\dfrac{2}{2}\right)^2 = 12 + 1$

$\quad\quad x^2 + 2x + 1 = 13$

$\quad\quad\quad (x+1)^2 = 13$

$\quad\quad\quad x + 1 = \pm\sqrt{13}$

$\quad\quad\quad\quad x = -1 \pm\sqrt{13}$

The solutions are $-1+\sqrt{13}$ and $-1-\sqrt{13}$.

6. $\quad\quad x^2 - 12x + 11 = 0$

$\quad x^2 - 12x + \left(\dfrac{-12}{2}\right)^2 = -11 + 36$

$\quad\quad x^2 - 12x + 36 = 25$

$\quad\quad\quad (x-6)^2 = \pm\sqrt{25}$

$\quad\quad\quad\quad x - 6 = \pm 5$

$\quad\quad\quad\quad\quad x = 6 \pm 5$

$\quad\quad\quad\quad\quad x = 1 \text{ or } x = 11$

The solutions are 1 and 11.

7.

$\quad\quad 3x^2 + 3x = 5$

$\quad\quad x^2 + x = \dfrac{5}{3}$

$\quad x^2 + x + \left(\dfrac{1}{2}\right)^2 = \dfrac{5}{3} + \dfrac{1}{4}$

$\quad x^2 + x + \dfrac{1}{4} = \dfrac{23}{12}$

$\quad\quad \left(x+\dfrac{1}{2}\right)^2 = \dfrac{23}{12}$

$\quad\quad x + \dfrac{1}{2} = \pm\sqrt{\dfrac{23}{12}}$

$\quad\quad x + \dfrac{1}{2} = \pm\dfrac{\sqrt{23}}{2\sqrt{3}}$

$\quad\quad x + \dfrac{1}{2} = \pm\dfrac{\sqrt{23}\cdot\sqrt{3}}{2\sqrt{3}\cdot\sqrt{3}}$

$\quad\quad x + \dfrac{1}{2} = \pm\dfrac{\sqrt{69}}{6}$

$\quad\quad x = -\dfrac{1}{2} \pm \dfrac{\sqrt{69}}{6} = \dfrac{-3 \pm\sqrt{69}}{6}$

The solutions are $\dfrac{-3+\sqrt{69}}{6}$ and $\dfrac{-3-\sqrt{69}}{6}$.

8. $\quad\quad 16y^2 + 16y = 1$

$\quad\quad y^2 + y = \dfrac{1}{16}$

$\quad y^2 + y + \left(\dfrac{1}{2}\right)^2 = \dfrac{1}{16} + \dfrac{1}{4}$

$\quad\quad y^2 + y + \dfrac{1}{4} = \dfrac{5}{16}$

$\quad\quad \left(y+\dfrac{1}{2}\right)^2 = \dfrac{5}{16}$

$\quad\quad y + \dfrac{1}{2} = \pm\sqrt{\dfrac{5}{16}}$

$\quad\quad y + \dfrac{1}{2} = \pm\dfrac{\sqrt{5}}{4}$

$\quad\quad y = -\dfrac{1}{2} \pm \dfrac{\sqrt{5}}{4} = \dfrac{-2 \pm\sqrt{5}}{4}$

The solutions are $\dfrac{-2+\sqrt{5}}{4}$ and $\dfrac{-2-\sqrt{5}}{4}$.

9. $2x^2 - 4x + 1 = 0$

$a = 2, b = -4, c = 1$

$x = \dfrac{4 \pm \sqrt{(-4)^2 - 4(2)(1)}}{2(2)}$

$= \dfrac{4 \pm \sqrt{8}}{4}$

$= \dfrac{4 \pm 2\sqrt{2}}{4} = \dfrac{2 \pm \sqrt{2}}{2}$

The solutions are $\dfrac{2 + \sqrt{2}}{2}$ and $\dfrac{2 - \sqrt{2}}{2}$.

10. $\dfrac{1}{2}x^2 + 3x + 2 = 0$

$x^2 + 6x + 4 = 0$

$a = 1, b = 6, c = 4$

$x = \dfrac{-6 \pm \sqrt{(6)^2 - 4(1)(4)}}{2(1)}$

$= \dfrac{-6 \pm \sqrt{20}}{2}$

$= \dfrac{-6 \pm 2\sqrt{5}}{2} = -3 \pm \sqrt{5}$

The solutions are $-3 + \sqrt{5}$ and $-3 - \sqrt{5}$.

11. $x^2 + 4x = -7$

$x^2 + 4x + 7 = 0$

$a = 1, b = 4, c = 7$

$x = \dfrac{-4 \pm \sqrt{(4)^2 - 4(1)(7)}}{2(1)}$

$= \dfrac{-4 \pm \sqrt{-12}}{2}$

$= \dfrac{-4 \pm i\sqrt{4 \cdot 3}}{2}$

$= \dfrac{-4 \pm 2i\sqrt{3}}{2} = -2 \pm i\sqrt{3}$

The solutions are $-2 + i\sqrt{3}$ and $-2 - i\sqrt{3}$.

12. $x^2 + x = -3$

$x^2 + x + 3 = 0$

$a = 1, b = 1, c = 3$

$x = \dfrac{-1 \pm \sqrt{(1)^2 - 4(1)(3)}}{2(1)}$

$= \dfrac{-1 \pm \sqrt{-11}}{2}$

$= \dfrac{-1 \pm i\sqrt{11}}{2}$

The solutions are $\dfrac{-1 + i\sqrt{11}}{2}$ and $\dfrac{-1 - i\sqrt{11}}{2}$.

13. $x^2 + 3x + 6 = 0$

$a = 1, b = 3, c = 6$

$x = \dfrac{-3 \pm \sqrt{(3)^2 - 4(1)(6)}}{2(1)}$

$= \dfrac{-3 \pm \sqrt{-15}}{2}$

$= \dfrac{-3 \pm i\sqrt{15}}{2}$

The solutions are $\dfrac{-3 + i\sqrt{15}}{2}$ and $\dfrac{-3 - i\sqrt{15}}{2}$.

14. $2x^2 + 18 = 0$

$2x^2 = -18$

$x^2 = -9$

$x = \pm\sqrt{-9}$

$x = \pm 3i$

The solutions are $3i$ and $-3i$.

15. $x^2 + 17x = 0$

$x(x + 17) = 0$

$x = 0$ or $x + 17 = 0$

$\qquad\qquad\qquad x = -17$

$x = 0, -17$

The solutions are 0 and -17.

16. $4x^2 - 2x - 3 = 0$

$a = 4, b = -2, c = -3$

$$x = \frac{2 \pm \sqrt{(-2)^2 - 4(4)(-3)}}{2(4)}$$

$$= \frac{2 \pm \sqrt{52}}{8}$$

$$= \frac{2 \pm 2\sqrt{13}}{8}$$

$$= \frac{1 \pm \sqrt{13}}{4}$$

The solutions are $\dfrac{1 + \sqrt{13}}{4}$ and $\dfrac{1 - \sqrt{13}}{4}$.

17. $(x - 2)^2 = 27$

$x - 2 = \pm\sqrt{27}$

$x - 2 = \pm 3\sqrt{3}$

$x = 2 \pm 3\sqrt{3}$

The solutions are $2 + 3\sqrt{3}$ and $2 - 3\sqrt{3}$.

18. $\dfrac{1}{2}x^2 - 2x + \dfrac{1}{2} = 0$

$x^2 - 4x + 1 = 0$

$x^2 - 4x + \left(\dfrac{-4}{2}\right)^2 = -1 + 4$

$x^2 - 4x + 4 = 3$

$(x - 2)^2 = 3$

$x - 2 = \pm\sqrt{3}$

$x = 2 \pm \sqrt{3}$

The solutions are $2 + \sqrt{3}$ and $2 - \sqrt{3}$.

19. $3x^2 + 2x = 8$

$3x^2 + 2x - 8 = 0$

$(3x - 4)(x + 2) = 0$

$3x - 4 = 0$ or $x + 2 = 0$

$x = \dfrac{4}{3}$ or $x = -2$

The solutions are $\dfrac{4}{3}$ and -2.

20. $2x^2 = -5x - 1$

$2x^2 + 5x + 1 = 0$

$a = 2, b = 5, c = 1$

$$x = \frac{-5 \pm \sqrt{(5)^2 - 4(2)(1)}}{2(2)}$$

$$= \frac{-5 \pm \sqrt{17}}{4}$$

The solutions are $\dfrac{-5 + \sqrt{17}}{4}$ and $\dfrac{-5 - \sqrt{17}}{4}$.

21. $x(x - 2) = 5$

$x^2 - 2x = 5$

$x^2 - 2x + \left(\dfrac{-2}{2}\right)^2 = 5 + 1$

$x^2 - 2x + 1 = 6$

$(x - 1)^2 = 6$

$x - 1 = \pm\sqrt{6}$

$x = 1 \pm \sqrt{6}$

The solutions are $1 + \sqrt{6}$ and $1 - \sqrt{6}$.

22. $x^2 - 31 = 0$

$x^2 = 31$

$x = \pm\sqrt{31}$

The solutions are $\sqrt{31}$ and $-\sqrt{31}$.

23. $5x^2 - 55 = 0$

$5x^2 = 55$

$x^2 = 11$

$x = \pm\sqrt{11}$

The solutions are $\sqrt{11}$ and $-\sqrt{11}$.

24. $5x^2 + 55 = 0$

$5x^2 = -55$

$x^2 = -11$

$x = \pm\sqrt{-11}$

$x = \pm i\sqrt{11}$

The solutions are $i\sqrt{11}$ and $-i\sqrt{11}$.

25.
$$x(x+5) = 66$$
$$x^2 + 5x = 66$$
$$x^2 + 5x - 66 = 0$$
$$(x+11)(x-6) = 0$$
$$x + 11 = 0 \quad \text{or} \quad x - 6 = 0$$
$$x = -11 \quad \text{or} \quad x = 6$$
The solutions are -11 and 6.

26. $5x^2 + 6x - 2 = 0$
$a = 5, b = 6, c = -2$
$$x = \frac{-6 \pm \sqrt{(6)^2 - 4(5)(-2)}}{2(5)}$$
$$= \frac{-6 \pm \sqrt{76}}{10}$$
$$= \frac{-6 \pm \sqrt{4 \cdot 19}}{10}$$
$$= \frac{-6 \pm 2\sqrt{19}}{10}$$
$$= \frac{-3 \pm \sqrt{19}}{5}$$
The solutions are $\dfrac{-3 + \sqrt{19}}{5}$ and $\dfrac{-3 - \sqrt{19}}{5}$.

27.
$$2x^2 + 3x = 1$$
$$2x^2 + 3x - 1 = 0$$
$$a = 2, b = 3, c = -1$$
$$x = \frac{-3 \pm \sqrt{(3)^2 - 4(2)(-1)}}{2(2)}$$
$$= \frac{-3 \pm \sqrt{17}}{4}$$
The solutions are $\dfrac{-3 + \sqrt{17}}{4}$ and $\dfrac{-3 - \sqrt{17}}{4}$.

28. $x - \sqrt{13 - 3x} - 3 = 0$
$$x - 3 = \sqrt{13 - 3x}$$
$$(x-3)^2 = \left(\sqrt{13 - 3x}\right)^2$$
$$x^2 - 6x + 9 = 13 - 3x$$
$$x^2 - 3x - 4 = 0$$
$$(x-4)(x+1) = 0$$
$$x - 4 = 0 \quad \text{or} \quad x + 1 = 0$$
$$x = 4 \qquad\qquad x = -1$$
The value -1 does not check, so the solution is 4.

29. The LCD is $x(x-2)$.

$$\frac{5x}{x-2} - \frac{x+1}{x} = \frac{3}{x(x-2)}$$

$$x(x-2)\left(\frac{5x}{x-2}\right) - x(x-2)\left(\frac{x+1}{x}\right) = x(x-2)\left[\frac{3}{x(x-2)}\right]$$

$$x(5x) - (x-2)(x+1) = 3$$

$$5x^2 - (x^2 - x - 2) = 3$$

$$5x^2 - x^2 + x + 2 = 3$$

$$4x^2 + x - 1 = 0$$

$a = 4$, $b = 1$, $c = -1$

$$x = \frac{-b \pm \sqrt{b^2 - 4ac}}{2a}$$

$$x = \frac{-1 \pm \sqrt{1^2 - 4(4)(-1)}}{2(4)}$$

$$= \frac{-1 \pm \sqrt{1 + 16}}{8}$$

$$= \frac{-1 \pm \sqrt{17}}{8}$$

The solutions are $\dfrac{-1 - \sqrt{17}}{8}$ and $\dfrac{-1 + \sqrt{17}}{8}$.

30. $a^2 + b^2 = c^2$

$$x^2 + x^2 = 20^2$$

$$2x^2 = 400$$

$$x^2 = 200$$

$$x = \pm\sqrt{200}$$

$$= \pm 10\sqrt{2} \approx 14.1421$$

Disregard the negative. A side of the room is $10\sqrt{2}$ feet ≈ 14.1 feet.

31. Let x = time for Jack alone. Then
$x - 2$ = time for Lucy alone.

$$\frac{1}{x} + \frac{1}{x-2} = \frac{1}{4}$$

$$4(x-2) + 4x = x(x-2)$$

$$4x - 8 + 4x = x^2 - 2x$$

$$0 = x^2 - 10x + 8$$

$$x = \frac{10 \pm \sqrt{(-10)^2 - 4(1)(8)}}{2(1)}$$

$$= \frac{10 \pm \sqrt{68}}{2}$$

$$\approx 9.1 \text{ or } 0.9 \text{ (disregard)}$$

$$x - 2 = 9.1 - 2 = 7.1$$

It would take Jack 9.1 hours and Lucy 7.1 hours.

32. Let x = initial speed on treadmill. Then $x + 1$ = speed increased.

$$t_{\text{initial}} + t_{\text{increased}} = \frac{4}{3}$$

$$\frac{5}{x} + \frac{2}{x+1} = \frac{4}{3}$$

$$5 \cdot 3(x+1) + 2 \cdot 3x = 4x(x+1)$$

$$15x + 15 + 6x = 4x^2 + 4x$$

$$0 = 4x^2 - 17x - 15$$

$$0 = (4x + 3)(x - 5)$$

$x = -\dfrac{4}{3}$ (disregard) or $x = 5$

$x + 1 = 5 + 1 = 6$

Initial speed: 5 mph

Increased speed: 6 mph

Section 8.4 Practice Exercises

1. $(x - 4)(x + 3) > 0$

Solve the related equation, $(x - 4)(x + 3) = 0$.

$(x - 4)(x + 3) = 0$

$x - 4 = 0$ or $x + 3 = 0$

$\quad x = 4 \qquad\qquad x = -3$

Test points in the three regions separated by $x = 4$ and $x = -3$.

Region	Test Point	$(x - 4)(x + 3) > 0$ Result
A: $(-\infty, -3)$	-4	$(-8)(-1) > 0$ True
B: $(-3, 4)$	0	$(-4)(3) > 0$ False
C: $(4, \infty)$	5	$(1)(8) > 0$ True

The points in regions A and C satisfy the inequality. The numbers 4 and -3 are not included in the solution since the inequality symbol is $>$. The solution set is $(-\infty, -3) \cup (4, \infty)$.

2. $x^2 - 8x \le 0$

Solve the related equation, $x^2 - 8x = 0$.

$x^2 - 8x = 0$

$x(x - 8) = 0$

$x = 0$ or $x - 8 = 0$

$\qquad\qquad\qquad x = 8$

The numbers 0 and 8 separate the number line into three regions, A, B, and C. Test a point in each region.

Region	Test Point	$x^2 - 8x \le 0$ Result
A: $(-\infty, 0]$	-1	$1 + 8 \le 0$ False
B: $[0, 8]$	1	$1 - 8 \le 0$ True
C: $[8, \infty)$	9	$81 - 72 \le 0$ False

Values in region B satisfy the inequality. The numbers 0 and 8 are included in the solution since the inequality symbol is $\le$. The solution set is $[0, 8]$.

3. The solutions of $x^2 - 8x \ge 0$ are where the graph is on or above the x-axis.

The solution set is $(-\infty, 0] \cup [8, \infty)$.

4. The solutions of $(x + 3)(x - 2)(x + 1) \le 0$ are where the graph is on or below the x-axis.

The solution set is $(-\infty, -3] \cup [-1, 2]$.

5. In standard form, the inequality is $x^2 + 3x + 7 < 0$.

Since the entire graph lies above the x-axis, the inequality has no solution or $\varnothing$.

6. $\dfrac{x - 5}{x + 4} \le 0$

$x + 4 = 0$

$\qquad x = -4$

$x = -4$ makes the denominator zero. Solve the related equation $\dfrac{x - 5}{x + 4} = 0$.

$$\frac{x-5}{x+4}=0$$
$$x-5=0$$
$$x=5$$

Test points in the three regions separated by $x=-4$ and $x=5$.

Region	Test Point	$\dfrac{x-5}{x+4}\le 0$ Result
A: $(-\infty,-4)$	-5	$\dfrac{-10}{-1}\le 0$ False
B: $(-4, 5]$	0	$\dfrac{-5}{4}\le 0$ True
C: $[5, \infty)$	6	$\dfrac{1}{10}\le 0$ False

The solution set is $(-4, 5]$. The interval includes 5 because 5 satisfies the original inequality. This interval does not include -4, because -4 would make the denominator zero.

7. $\dfrac{7}{x+3}<5$

$$x+3=0$$
$$x=-3$$

$x=-3$ makes the denominator zero.

Solve $\dfrac{7}{x+3}=5$.

$$(x+3)\left(\frac{7}{x+3}\right)=5(x+3)$$
$$7=5x+15$$
$$-8=5x$$
$$-\frac{8}{5}=x$$

We use these two solutions to divide the number line into three regions and choose test points.

Region	Test Point	$\dfrac{7}{x+3}<5$ Result
A: $(-\infty,-3)$	-4	$\dfrac{7}{-1}<5$ True
B: $\left(-3,-\dfrac{8}{5}\right)$	-2	$\dfrac{7}{1}<5$ False
C: $\left(-\dfrac{8}{5},\infty\right)$	0	$\dfrac{7}{3}<5$ True

The solution set is $(-\infty,-3)\cup\left(-\dfrac{8}{5},\infty\right)$.

Vocabulary, Readiness & Video Check 8.4

1. $[-7, 3)$

2. $(-1, 5]$

3. $(-\infty, 0]$

4. $(-\infty, -8]$

5. $(-\infty, -12)\cup[-10, \infty)$

6. $(-\infty, -3]\cup(4, \infty)$

7. We use the solutions to the related equation to divide the number line into regions that either entirely are or entirely are not solution regions; the solutions to the related equation are solutions to the inequality only if the inequality symbol is $\le$ or $\ge$.

8. The solution set cannot include values that make the denominator zero.

Exercise Set 8.4

1. $(x+1)(x+5) > 0$
$\quad x+1 = 0 \quad$ or $\quad x+5 = 0$
$\qquad x = -1 \quad$ or $\qquad x = -5$

Region	Test Point	$(x+1)(x+5) > 0$ Result
$A: (-\infty, -5)$	-6	$(-5)(-1) > 0$ True
$B: (-5, -1)$	-2	$(-1)(3) > 0$ False
$C: (-1, \infty)$	0	$(1)(5) > 0$ True

Solution: $(-\infty, -5) \cup (-1, \infty)$

3. $(x-3)(x+4) \le 0$
$\quad x-3 = 0 \quad$ or $\quad x+4 = 0$
$\qquad x = 3 \quad$ or $\qquad x = -4$

Region	Test Point	$(x-3)(x+4) \le 0$ Result
$A: (-\infty, -4]$	-5	$(-8)(-1) \le 0$ False
$B: [-4, 3]$	0	$(-3)(4) \le 0$ True
$C: [3, \infty)$	4	$(1)(8) \le 0$ False

Solution: $[-4, 3]$

5. $x^2 - 7x + 10 \le 0$
$\quad (x-5)(x-2) \le 0$
$\quad x-5 = 0 \quad$ or $\quad x-2 = 0$
$\qquad x = 5 \quad$ or $\qquad x = 2$

Region	Test Point	$(x-5)(x-2) \le 0$ Result
$A: (-\infty, 2]$	0	$(-5)(-2) \le 0$ False
$B: [2, 5]$	3	$(-2)(1) \le 0$ True
$C: [5, \infty)$	6	$(1)(4) \le 0$ False

Solution: $[2, 5]$

7. $3x^2 + 16 < -5$
$\quad 3x^2 + 16x + 5 < 0$
$\quad (3x+1)(x+5) < 0$
$\quad 3x+1 = 0 \quad$ or $\quad x+5 = 0$
$\qquad x = -\dfrac{1}{3} \quad$ or $\qquad x = -5$

Region	Test Point	$(3x+1)(x+5) < 0$ Result
$A: (-\infty, -5)$	-6	$(-17)(-1) < 0$ False
$B: \left(-5, -\dfrac{1}{3}\right)$	-1	$(-2)(4) < 0$ True
$C: \left(-\dfrac{1}{3}, \infty\right)$	0	$(1)(5) < 0$ False

Solution: $\left(-5, -\dfrac{1}{3}\right)$

9. $(x-6)(x-4)(x-2) > 0$
$\quad x-6 = 0 \quad$ or $\quad x-4 = 0 \quad$ or $\quad x-2 = 0$
$\qquad x = 6 \quad$ or $\qquad x = 4 \quad$ or $\qquad x = 2$

Region	Test Point	$(x-6)(x-4)(x-2) > 0$ Result
$A: (-\infty, 2)$	0	$(-6)(-4)(-2) > 0$ False
$B: (2, 4)$	3	$(-3)(-1)(1) > 0$ True
$C: (4, 6)$	5	$(-1)(1)(3) > 0$ False
$D: (6, \infty)$	7	$(1)(3)(5) > 0$ True

Solution: $(2, 4) \cup (6, \infty)$

11. $x(x-1)(x+4) \leq 0$
$$x = 0 \quad \text{or} \quad x - 1 = 0 \quad \text{or} \quad x + 4 = 0$$
$$x = 1 \quad \text{or} \quad x = -4$$

Region	Test Point	$x(x-1)(x+4) \leq 0$ Result
A: $(-\infty, -4]$	-5	$-5(-6)(-1) \leq 0$ True
B: $[-4, 0]$	-1	$-1(-2)(3) \leq 0$ False
C: $[0, 1]$	$\dfrac{1}{2}$	$\dfrac{1}{2}\left(-\dfrac{1}{2}\right)\left(\dfrac{9}{2}\right) \leq 0$ True
D: $[1, \infty)$	2	$2(1)(6) \leq 0$ False

Solution: $(-\infty, -4] \cup [0, 1]$

13. The solutions of the inequality are where the graph is on or below the x-axis. The solution set is $[-3, -2] \cup [2, 3]$.

15. $\dfrac{x+7}{x-2} < 0$
$$x + 7 = 0 \quad \text{or} \quad x - 2 = 0$$
$$x = -7 \quad \text{or} \quad x = 2$$

Region	Test Point	$\dfrac{x+7}{x-2} < 0$ False
A: $(-\infty, -7)$	-8	$\dfrac{-1}{-10} < 0$ False
B: $(-7, 2)$	0	$\dfrac{7}{-2} < 0$ True
C: $(2, \infty)$	3	$\dfrac{10}{1} < 0$ False

Solution: $(-7, 2)$

17. $\dfrac{5}{x+1} > 0$
$$x + 1 = 0$$
$$x = -1$$

Region	Test Point	$\dfrac{5}{x+1} > 0$ Result
A: $(-\infty, -1)$	-2	$\dfrac{5}{-1} > 0$ False
B: $(-1, \infty)$	0	$\dfrac{5}{1} > 0$ True

Solution: $(-1, \infty)$

19. From the graph, the solution set is $(-\infty, -1] \cup (4, \infty)$.

21. $\dfrac{3}{x-2} < 4$

The denominator is equal to 0 when $x - 2 = 0$, or $x = 2$.
$$\dfrac{3}{x-2} = 4$$
$$3 = 4x - 8$$
$$11 = 4x$$
$$\dfrac{11}{4} = x$$

Region	Test Point	$\dfrac{3}{x-2} < 4$ Result
A: $(-\infty, 2)$	0	$\dfrac{3}{-2} < 4$ True
B: $\left(2, \dfrac{11}{4}\right)$	$\dfrac{5}{2}$	$\dfrac{3}{\frac{1}{2}} = 6 < 4$ False
C: $\left(\dfrac{11}{4}, \infty\right)$	4	$\dfrac{3}{2} < 4$ True

Solution: $(-\infty, 2) \cup \left(\dfrac{11}{4}, \infty\right)$

23. $\dfrac{x^2+6}{5x} \geq 1$

The denominator is equal to 0 when $5x = 0$, or $x = 0$.

$$\dfrac{x^2+6}{5x} = 1$$
$$x^2+6 = 5x$$
$$x^2-5x+6 = 0$$
$$(x-2)(x-3) = 0$$
$$x-2 = 0 \quad \text{or} \quad x-3 = 0$$
$$x = 2 \quad \text{or} \qquad x = 3$$

Region	Test Point	$\dfrac{x^2+6}{5x} \geq 1$ Result
A: $(-\infty, 0)$	-1	$\dfrac{7}{-5} \geq 1$ False
B: $(0, 2]$	1	$\dfrac{7}{5} \geq 1$ True
C: $[2, 3]$	$\dfrac{5}{2}$	$\dfrac{\frac{49}{4}}{\frac{25}{2}} = \dfrac{49}{50} \geq 1$ False
D: $[3, \infty)$	4	$\dfrac{22}{20} \geq 1$ True

Solution: $(0, 2] \cup [3, \infty)$

25. $\dfrac{x+2}{x-3} < 1$

$x - 3 = 0$

$x = 3$ makes the denominator 0.

$$\dfrac{x+2}{x-3} = 1$$
$$x+2 = x-3$$
$$2 = -3 \quad \text{False}$$

The related equation has no solution.

Region	Test Point Value	$\dfrac{x+2}{x-3} < 1$ Result
$(-\infty, 3)$	0	$\dfrac{2}{-3} < 1$ True
$(3, \infty)$	4	$\dfrac{6}{1} < 1$ False

Solution: $(-\infty, 3)$

27. $(x-8)(x+7) > 0$

$$x-8 = 0 \quad \text{or} \quad x+7 = 0$$
$$x = 8 \quad \text{or} \qquad x = -7$$

Region	Test Point	$(x-8)(x+7) > 0$ Result
A: $(-\infty, -7)$	-8	$(-16)(-1) > 0$ True
B: $(-7, 8)$	0	$(-8)(7) > 0$ False
C: $(8, \infty)$	9	$(1)(16) > 0$ True

Solution: $(-\infty, -7) \cup (8, \infty)$

29. $(2x-3)(4x+5) \leq 0$

$$2x-3 = 0 \quad \text{or} \quad 4x+5 = 0$$
$$x = \dfrac{3}{2} \quad \text{or} \qquad x = -\dfrac{5}{4}$$

Region	Test Point	$(2x-3)(4x+5) \leq 0$ Result
A: $\left(-\infty, -\dfrac{5}{4}\right]$	-2	$(-7)(-3) \leq 0$ False
B: $\left[-\dfrac{5}{4}, \dfrac{3}{2}\right]$	0	$(-3)(5) \leq 0$ True
C: $\left[\dfrac{3}{2}, \infty\right)$	2	$(1)(13) \leq 0$ False

Solution: $\left[-\dfrac{5}{4}, \dfrac{3}{2}\right]$

31. $x^2 > x$

$x^2 - x > 0$

$x(x-1) > 0$

$x = 0$ or $x - 1 = 0$

$x = 1$

Region	Test Point	$x(x-1) > 0$ Result
$A: (-\infty, 0)$	-1	$-1(-2) > 0$ True
$B: (0, 1)$	$\dfrac{1}{2}$	$\dfrac{1}{2}\left(-\dfrac{1}{2}\right) > 0$ False
$C: (1, \infty)$	2	$2(1) > 0$ True

Solution: $(-\infty, 0) \cup (1, \infty)$

33. $(2x - 8)(x + 4)(x - 6) \le 0$

$2x - 8 = 0$ or $x + 4 = 0$ or $x - 6 = 0$

$x = 4$ or $x = -4$ or $x = 6$

Region	Test Point	$(2x - 8)(x + 4)(x - 6) \le 0$ Result
$A: (-\infty, -4]$	-5	$(-18)(-1)(-11) \le 0$ True
$B: [-4, 4]$	0	$(-8)(4)(-6) \le 0$ False
$C: [4, 6]$	5	$(2)(9)(-1) \le 0$ True
$D: [6, \infty)$	7	$(6)(11)(1) \le 0$ False

Solution: $(-\infty, -4] \cup [4, 6]$

35.
$$6x^2 - 5x \geq 6$$
$$6x^2 - 5x - 6 \geq 0$$
$$(3x+2)(2x-3) \geq 0$$
$$3x+2=0 \quad \text{or} \quad 2x-3=0$$
$$x = -\frac{2}{3} \quad \text{or} \qquad x = \frac{3}{2}$$

Region	Test Point	$(3x+2)(2x-3) \geq 0$ Result
A: $\left(-\infty, -\dfrac{2}{3}\right]$	-1	$(-1)(-5) \geq 0$ True
B: $\left[-\dfrac{2}{3}, \dfrac{3}{2}\right]$	0	$(2)(-3) \geq 0$ False
C: $\left[\dfrac{3}{2}, \infty\right)$	2	$(8)(1) \geq 0$ True

Solution: $\left(-\infty, -\dfrac{2}{3}\right] \cup \left[\dfrac{3}{2}, \infty\right)$

37.
$$4x^3 + 16x^2 - 9x - 36 > 0$$
$$4x^2(x+4) - 9(x+4) > 0$$
$$(x+4)(4x^2 - 9) > 0$$
$$(x+4)(2x+3)(2x-3) > 0$$
$$x+4=0 \quad \text{or} \quad 2x+3=0 \quad \text{or} \quad 2x-3=0$$
$$x = -4 \quad \text{or} \qquad x = -\frac{3}{2} \quad \text{or} \qquad x = \frac{3}{2}$$

Region	Test Point	$(x+4)(2x+3)(2x-3) > 0$
A: $(-\infty, -4)$	-5	$(-1)(-7)(-13) > 0$ False
B: $\left(-4, -\dfrac{3}{2}\right)$	-3	$(1)(-3)(-9) > 0$ True
C: $\left(-\dfrac{3}{2}, \dfrac{3}{2}\right)$	0	$(4)(3)(-3) > 0$ False
D: $\left(\dfrac{3}{2}, \infty\right)$	4	$(8)(11)(5) > 0$ True

Solution: $\left(-4, -\dfrac{3}{2}\right) \cup \left(\dfrac{3}{2}, \infty\right)$

39.
$$x^4 - 26x^2 + 25 \geq 0$$
$$(x^2 - 25)(x^2 - 1) \geq 0$$
$$(x+5)(x-5)(x+1)(x-1) \geq 0$$
$$x = -5 \quad \text{or} \quad x = 5 \quad \text{or} \quad x = -1 \quad \text{or} \quad x = 1$$

Region	Test Point	$(x + 5)(x - 5)(x + 1)(x - 1) \geq 0$ Result
$A: (-\infty, -5]$	-6	$(-1)(-11)(-5)(-7) \geq 0$ True
$B: [-5, -1]$	-2	$(3)(-7)(-1)(-3) \geq 0$ False
$C: [-1, 1]$	0	$(5)(-5)(1)(-1) \geq 0$ True
$D: [1, 5]$	2	$(7)(-3)(3)(1) \geq 0$ False
$E: [5, \infty)$	6	$(11)(1)(7)(5) \geq 0$ True

Solution: $(-\infty, -5] \cup [-1, 1] \cup [5, \infty)$

41. $(2x - 7)(3x + 5) > 0$
$$2x - 7 = 0 \quad \text{or} \quad 3x + 5 = 0$$
$$x = \frac{7}{2} \quad \text{or} \quad x = -\frac{5}{3}$$

Region	Test Point	$(2x - 7)(3x + 5) > 0$
$A: \left(-\infty, -\dfrac{5}{3}\right)$	-2	$(-11)(-1) > 0$ True
$B: \left(-\dfrac{5}{3}, \dfrac{7}{2}\right)$	0	$(-7)(5) > 0$ False
$C: \left(\dfrac{7}{2}, \infty\right)$	4	$(1)(17) > 0$ True

Solution: $\left(-\infty, -\dfrac{5}{3}\right) \cup \left(\dfrac{7}{2}, \infty\right)$

43. $\dfrac{x}{x-10} < 0$

$x = 0$ or $x - 10 = 0$

$\qquad\qquad\qquad x = 10$

Region	Test Point	$\dfrac{x}{x-10} < 0$ Result
$A: (-\infty, 0)$	-1	$\dfrac{-1}{-11} < 0$ False
$B: (0, 10)$	5	$\dfrac{5}{-5} < 0$ True
$C: (10, \infty)$	11	$\dfrac{11}{1} < 0$ False

Solution: $(0, 10)$

45. $\dfrac{x-5}{x+4} \geq 0$

$x - 5 = 0$ or $x + 4 = 0$

$\quad x = 5$ or $\qquad x = -4$

Region	Test Point	$\dfrac{x-5}{x+4} \geq 0$ Result
$A: (-\infty, -4)$	-5	$\dfrac{-10}{-1} \geq 0$ True
$B: (-4, 5]$	0	$\dfrac{-5}{4} \geq 0$ False
$C: [5, \infty)$	6	$\dfrac{1}{10} \geq 0$ True

Solution: $(-\infty, -4) \cup [5, \infty)$

47. $\dfrac{x(x+6)}{(x-7)(x+1)} \geq 0$

$x = 0$ or $x + 6 = 0$ or $x - 7 = 0$ or $x + 1 = 0$

$\qquad\qquad x = -6$ or $\qquad x = 7$ or $\qquad x = -1$

Region	Test Point	$\dfrac{x(x+6)}{(x-7)(x+1)} \geq 0$ Result
A: $(-\infty, -6]$	-7	$\dfrac{-7(-1)}{(-14)(-6)} \geq 0$ True
B: $[-6, -1)$	-3	$\dfrac{-3(3)}{(-10)(-2)} \geq 0$ False
C: $(-1, 0]$	$-\dfrac{1}{2}$	$\dfrac{-\frac{1}{2}\left(\frac{11}{2}\right)}{\left(-\frac{15}{2}\right)\left(\frac{1}{2}\right)} \geq 0$ True
D: $[0, 7)$	2	$\dfrac{2(8)}{(-5)(3)} \geq 0$ False
E: $(7, \infty)$	8	$\dfrac{8(14)}{(1)(9)} \geq 0$ True

Solution: $(-\infty, -6] \cup (-1, 0] \cup (7, \infty)$

49. $\dfrac{-1}{x-1} > -1$

The denominator is equal to 0 when $x - 1 = 0$, or $x = 1$.

$$\frac{-1}{x-1} = -1$$
$$-1 = -1(x-1)$$
$$-1 = -x + 1$$
$$x = 2$$

Region	Test Point	$\dfrac{-1}{x-1} > -1$ Result
A: $(-\infty, 1)$	0	$\dfrac{-1}{-1} > -1$ True
B: $(1, 2)$	$\dfrac{3}{2}$	$\dfrac{-1}{\frac{1}{2}} = -2 > -1$ False
C: $(2, \infty)$	3	$\dfrac{-1}{2} > -1$ True

Solution: $(-\infty, 1) \cup (2, \infty)$

51. $\dfrac{x}{x+4} \leq 2$

The denominator is equal to 0 when $x + 4 = 0$, or $x = -4$.

$$\frac{x}{x+4} = 2$$
$$x = 2x + 8$$
$$-x = 8$$
$$x = -8$$

Region	Test Point	$\dfrac{x}{x+4} \leq 2$ Result
A: $(-\infty, -8]$	-9	$\dfrac{-9}{-5} \leq 2$ True
B: $[-8, -4)$	-6	$\dfrac{-6}{-2} \leq 2$ False
C: $(-4, \infty)$	0	$\dfrac{0}{4} \leq 2$ True

Solution: $(-\infty, -8] \cup (-4, \infty)$

53. $\dfrac{z}{z-5} \geq 2z$

The denominator is equal to 0 when $z - 5 = 0$, or $z = 5$.

$$\dfrac{z}{z-5} = 2z$$
$$z = 2z(z-5)$$
$$z = 2z^2 - 10z$$
$$0 = 2z^2 - 11z$$
$$0 = z(2z-11)$$
$$z = 0 \quad \text{or} \quad 2z - 11 = 0$$
$$z = \dfrac{11}{2}$$

Region	Test Point	$\dfrac{z}{z-5} \geq 2z$ Result
A: $(-\infty, 0]$	-1	$\dfrac{-1}{-6} \geq -2$ True
B: $[0, 5)$	1	$\dfrac{1}{-4} \geq 2$ False
C: $\left(5, \dfrac{11}{2}\right]$	$\dfrac{21}{4}$	$\dfrac{\left(\frac{21}{4}\right)}{\left(\frac{1}{4}\right)} \geq \dfrac{21}{2}$ $21 \geq \dfrac{21}{2}$ True
D: $\left[\dfrac{11}{2}, \infty\right)$	6	$\dfrac{6}{1} \geq 12$ False

Solution: $(-\infty, 0] \cup \left(5, \dfrac{11}{2}\right]$

55. $\dfrac{(x+1)^2}{5x} > 0$

The denominator is equal to 0 when $5x = 0$, or $x = 0$.

$$\dfrac{(x+1)^2}{5x} = 0$$
$$(x+1)^2 = 0$$
$$x + 1 = 0$$
$$x = -1$$

Region	Test Point	$\dfrac{(x+1)^2}{5x} > 0$ Result
A: $(-\infty, -1)$	-2	$\dfrac{1}{-10} > 0$ False
B: $(-1, 0)$	$-\dfrac{1}{2}$	$\dfrac{\left(\frac{1}{4}\right)}{\left(-\frac{5}{2}\right)} > 0$ False
C: $(0, \infty)$	1	$\dfrac{4}{5} > 0$ True

Solution: $(0, \infty)$

57. $g(x) = |x| + 2$

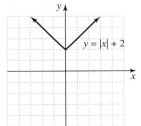

59. $F(x) = |x| - 1$

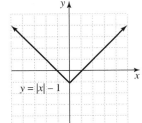

61. $F(x) = x^2 - 3$

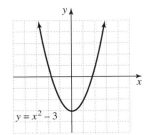

63. $H(x) = x^2 + 1$

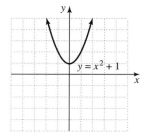

65. answers may vary

67. Let x = the number. Then

$\dfrac{1}{x}$ = the reciprocal of the number.

$$x - \frac{1}{x} < 0$$

$$\frac{x^2 - 1}{x} < 0$$

$$\frac{(x+1)(x-1)}{x} < 0$$

$x + 1 = 0$ or $x - 1 = 0$ or $x = 0$

$x = -1$ or $x = 1$

Region	Test Point	$\dfrac{(x+1)(x-1)}{x} < 0$ Result
$A: (-\infty, -1)$	-2	$\dfrac{(-1)(-3)}{-2} < 0$ True
$B: (-1, 0)$	$-\dfrac{1}{2}$	$\dfrac{\left(\frac{1}{2}\right)\left(-\frac{3}{2}\right)}{\left(-\frac{1}{2}\right)} < 0$ False
$C: (0, 1)$	$\dfrac{1}{2}$	$\dfrac{\left(\frac{3}{2}\right)\left(-\frac{1}{2}\right)}{\left(\frac{1}{2}\right)} < 0$ True
$D: (1, \infty)$	2	$\dfrac{(3)(1)}{2} < 0$ False

$(-\infty, -1) \cup (0, 1)$ or any number less than -1 or between 0 and 1 satisfies the conditions.

69. $P(x) = -2x^2 + 26x - 44$

$$-2x^2 + 26x - 44 > 0$$
$$-2(x^2 + 13x - 22) > 0$$
$$-2(x - 11)(x - 2) > 0$$
$$x - 11 = 0 \quad \text{or} \quad x - 2 = 0$$
$$x = 11 \quad \text{or} \qquad x = 2$$

Region	Test Point	$-2(x-11)(x-2) > 0$ Result
$A: (0, 2)$	1	$-2(-10)(-3) > 0$ False
$B: (2, 11)$	3	$-2(-8)(1) > 0$ True
$C: (11, \infty)$	12	$-2(1)(10) > 0$ False

The company makes a profit when x is between 2 and 11.

71.

73.

Section 8.5 Practice Exercises

1. a. $f(x) = x^2 - 5$

The graph of $f(x)$ is obtained by shifting the graph of $y = x^2$ downward 5 units.

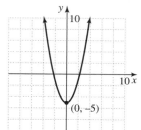

b. $g(x) = x^2 + 3$

The graph of $g(x)$ is obtained by shifting the graph of $y = x^2$ upward 3 units.

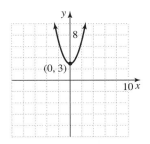

2. a. $G(x) = (x+4)^2$

The graph of $G(x)$ is obtained by shifting the graph of $y = x^2$ to the left 4 units.

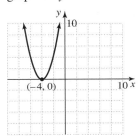

b. $H(x) = (x-7)^2$

The graph of $H(x)$ is obtained by shifting the graph of $y = x^2$ to the right 7 units.

3. $f(x) = (x+2)^2 + 2$

The graph of $f(x)$ is the graph of $y = x^2$ shifted 2 units to the left and 2 units upward. The vertex is then $(-2, 2)$, and the axis of symmetry is $x = -2$.

x	$f(x) = (x+2)^2 + 2$
-4	6
-3	3
-1	3
0	6

4. $f(x) = -\dfrac{1}{2}x^2$

Because $a = -\dfrac{1}{2}$, a negative value, this parabola opens downward. Since $\left|-\dfrac{1}{2}\right| = \dfrac{1}{2} < 1$, the parabola is wider than the graph of $y = x^2$. The vertex is $(0, 0)$, and the axis of symmetry is the y-axis.

x	$f(x) = -\dfrac{1}{2}x^2$
-2	-2
-1	$-\dfrac{1}{2}$
0	0
1	$-\dfrac{1}{2}$
2	-2

5. $h(x) = \frac{1}{3}(x-4)^2 - 3$

This graph is the same as $y = x^2$ shifted 4 units to the right and 3 units downward, and it is wider because a is $\frac{1}{3}$. The vertex is $(4, -3)$, and the axis of symmetry is $x = 4$.

x	$h(x) = \frac{1}{3}(x-4)^2 - 3$
2	$-\frac{5}{3}$
3	$-\frac{8}{3}$
4	-3
5	$-\frac{8}{3}$
6	$-\frac{5}{3}$

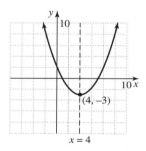

Vocabulary, Readiness & Video Check 8.5

1. A <u>quadratic</u> function is one that can be written in the form $f(x) = ax^2 + bx + c, \ a \neq 0$.

2. The graph of a quadratic function is a <u>parabola</u> opening <u>upward</u> or <u>downward</u>.

3. If $a > 0$, the graph of the quadratic function opens <u>upward</u>.

4. If $a < 0$, the graph of the quadratic function opens <u>downward</u>.

5. The vertex of a parabola is the <u>lowest</u> point if $a > 0$.

6. The vertex of a parabola is the <u>highest</u> point if $a < 0$.

7. $f(x) = x^2$; vertex: $(0, 0)$

8. $f(x) = -5x^2$; vertex: $(0, 0)$

9. $g(x) = (x-2)^2$; vertex: $(2, 0)$

10. $g(x) = (x+5)^2$; vertex: $(-5, 0)$

11. $f(x) = 2x^2 + 3$; vertex: $(0, 3)$

12. $h(x) = x^2 - 1$; vertex: $(0, -1)$

13. $g(x) = (x+1)^2 + 5$; vertex: $(-1, 5)$

14. $h(x) = (x-10)^2 - 7$; vertex: $(10, -7)$

15. Graphs of the form $f(x) = x^2 + k$ shift up or down the y-axis k units from $y = x^2$; the y-intercept.

16. Graphs of the form $f(x) = (x-h)^2$ shift right or left on the x-axis h units from $y = x^2$; the x-intercept.

17. The vertex, (h, k) and the axis of symmetry, $x = h$; the basic shape of $y = x^2$ does not change.

18. whether the graph is wider or narrower than $y = x^2$

19. the coordinates of the vertex, whether the graph opens upward or downward, whether the graph is narrower or wider than $y = x^2$, and the graph's axis of symmetry

Exercise Set 8.5

1. $f(x) = x^2 - 1$

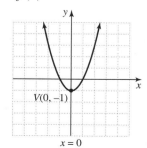

3. $h(x) = x^2 + 5$

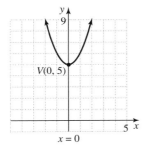

$V(0, 5)$

$x = 0$

5. $g(x) = x^2 + 7$

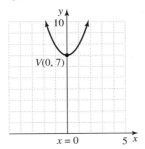

$V(0, 7)$

$x = 0$

7. $f(x) = (x - 5)^2$

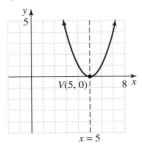

$V(5, 0)$

$x = 5$

9. $h(x) = (x + 2)^2$

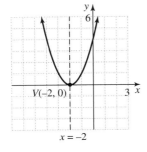

$V(-2, 0)$

$x = -2$

11. $G(x) = (x + 3)^2$

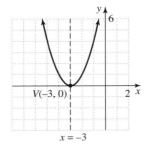

$V(-3, 0)$

$x = -3$

13. $f(x) = (x - 2)^2 + 5$

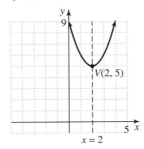

$V(2, 5)$

$x = 2$

15. $h(x) = (x + 1)^2 + 4$

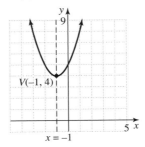

$V(-1, 4)$

$x = -1$

17. $g(x) = (x + 2)^2 - 5$

$V(-2, -5)$

$x = -2$

     457

19. $H(x) = 2x^2$

21. $h(x) = \dfrac{1}{3}x^2$

23. $g(x) = -x^2$

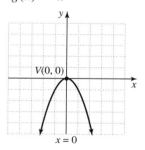

25. $f(x) = 2(x-1)^2 + 3$

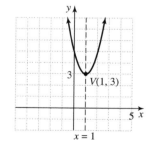

27. $h(x) = -3(x+3)^2 + 1$

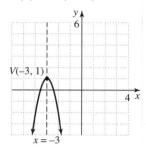

29. $H(x) = \dfrac{1}{2}(x-6)^2 - 3$

31. $f(x) = -(x-2)^2$

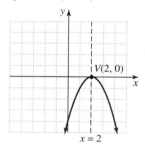

33. $F(x) = -x^2 + 4$

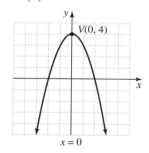

35. $F(x) = 2x^2 - 5$

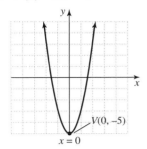

37. $h(x) = (x-6)^2 + 4$

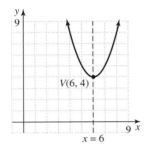

39. $F(x) = \left(x + \dfrac{1}{2}\right)^2 - 2$

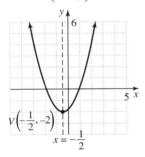

41. $F(x) = \dfrac{3}{2}(x+7)^2 + 1$

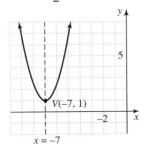

43. $f(x) = \dfrac{1}{4}x^2 - 9$

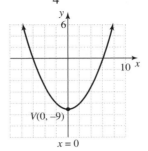

45. $G(x) = 5\left(x + \dfrac{1}{2}\right)^2$

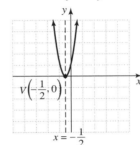

47. $h(x) = -(x-1)^2 - 1$

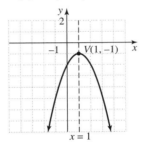

49. $g(x) = \sqrt{3}(x+5)^2 + \dfrac{3}{4}$

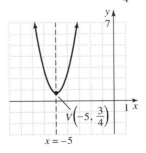

51. $h(x) = 10(x+4)^2 - 6$

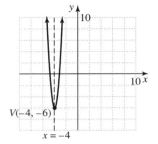

53. $f(x) = -2(x-4)^2 + 5$

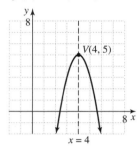

55. $x^2 + 8x$

$$\left[\frac{1}{2}(8)\right]^2 = (4)^2 = 16$$

$$x^2 + 8x + 16$$

57. $z^2 - 16z$

$$\left[\frac{1}{2}(-16)\right]^2 = (-8)^2 = 64$$

$$z^2 - 16z + 64$$

59. $y^2 + y$

$$\left[\frac{1}{2}(1)\right]^2 = \left(\frac{1}{2}\right)^2 = \frac{1}{4}$$

$$y^2 + y + \frac{1}{4}$$

61.
$$x^2 + 4x = 12$$
$$x^2 + 4x + \left(\frac{4}{2}\right)^2 = 12 + 4$$
$$x^2 + 4x + 4 = 16$$
$$(x+2)^2 = 16$$
$$x + 2 = \pm\sqrt{16}$$
$$x + 2 = \pm 4$$
$$x = -2 \pm 4$$
$$x = -6 \ \text{or} \ 2$$

63.
$$z^2 + 10z - 1 = 0$$
$$z^2 + 10z = 1$$
$$z^2 + 10z + \left(\frac{10}{2}\right)^2 = 1 + 25$$
$$z^2 + 10z + 25 = 26$$
$$(z+5)^2 = 26$$
$$z + 5 = \pm\sqrt{26}$$
$$z = -5 \pm \sqrt{26}$$

65.
$$z^2 - 8z = 2$$
$$z^2 - 8z + \left(\frac{-8}{2}\right)^2 = 2 + 16$$
$$z^2 - 8z + 16 = 18$$
$$(z-4)^2 = 18$$
$$z - 4 = \pm\sqrt{18}$$
$$z - 4 = \pm 3\sqrt{2}$$
$$z = 4 \pm 3\sqrt{2}$$

67. $f(x) = -213(x-0.1)^2 + 3.6$

$a = -213 < 0$, so $f(x)$ opens downward.
The vertex is (0.1, 3.6). The correct answer is **c**.

69. $f(x) = 5(x-2)^2 + 3$

71. $f(x) = 5[x-(-3)]^2 + 6$
$$= 5(x+3)^2 + 6$$

73. $y = f(x) + 1$

75. $y = f(x-3)$

77. $y = f(x+2)+2$

79. $f(x) = 2158x^2 - 10,339x + 6731$

a. $x = 2010 - 2000 = 10$
$$f(10) = 2158(10)^2 - 10,339(10) + 6731$$
$$= 119,141$$
There were approximately 119,141 million text messages sent.

b. $x = 2014 - 2000 = 14$
$$f(14) = 2158(14)^2 - 10,339(14) + 6731$$
$$= 284,953$$
The model predicts approximately 284,953 million text messages sent in 2014.

Section 8.6 Practice Exercises

1. $g(x) = x^2 - 2x - 3$

Write in the form $y = (x-h)^2 + k$ by completing the square.
$$y = x^2 - 2x - 3$$
$$y + 3 = x^2 - 2x$$
$$y + 3 + \left(\frac{-2}{2}\right)^2 = x^2 - 2x + \left(\frac{-2}{2}\right)^2$$
$$y + 4 = x^2 - 2x + 1$$
$$y = (x-1)^2 - 4$$
The vertex is at $(1, -4)$.
Let $g(x) = 0$.
$$0 = x^2 - 2x - 3$$
$$0 = (x-3)(x+1)$$
$$x - 3 = 0 \quad \text{or} \quad x + 1 = 0$$
$$x = 3 \qquad\qquad x = -1$$
The x-intercepts are $(3, 0)$ and $(-1, 0)$.
Let $x = 0$.
$$g(0) = 0^2 - 2(0) - 3 = -3$$
The y-intercept is $(0, -3)$.

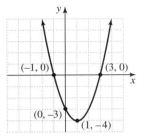

2. Graph $y_1 = x^2 - x - 6$ in a standard window. Since the parabola opens upward, use the minimum feature.

The vertex is $(0.5, -6.25)$ and the axis of symmetry is the line $x = 0.5$.

3. $g(x) = 4x^2 + 4x + 3$

Replace $g(x)$ with y and complete the square to write the equation in the form $y = a(x-h)^2 + k$.
$$y = 4x^2 + 4x + 3$$
$$y - 3 = 4x^2 + 4x = 4(x^2 + x)$$
$$y - 3 + 4\left(\frac{1}{2}\right)^2 = 4\left[x^2 + x + \left(\frac{1}{2}\right)^2\right]$$
$$y - 3 + 1 = 4\left(x^2 + x + \frac{1}{4}\right)$$
$$y = 4\left(x + \frac{1}{2}\right)^2 + 2$$
$a = 4$, $h = -\dfrac{1}{2}$, and $k = 2$.
The parabola opens upward with vertex $\left(-\dfrac{1}{2}, 2\right)$, and has an axis of symmetry $x = -\dfrac{1}{2}$.
Let $x = 0$.
$$g(0) = 4(0)^2 + 4(0) + 3 = 3$$
The y-intercept is $(0, 3)$. There are no x-intercepts.

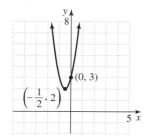

4. $g(x) = x^2 - 2x - 3$

$a = 1$, $b = -2$, and $c = -3$

$\dfrac{-b}{2a} = \dfrac{-(-2)}{2(1)} = \dfrac{2}{2} = 1$

The x-value of the vertex is 1.

$g(1) = 1^2 - 2(1) - 3 = 1 - 2 - 3 = -4$

The vertex is (1, −4).

5.

For $g(x) = -x^2 + 5x + 6$, $a = -1$, $b = 5$, and $c = 6$.

$x = \dfrac{-b}{2a} = \dfrac{-5}{2(-1)} = \dfrac{5}{2}$

$g\left(\dfrac{5}{2}\right) = \dfrac{49}{4}$

The vertex is $\left(\dfrac{5}{2}, \dfrac{49}{4}\right)$.

$g(0) = 6$, so the y-intercept is (0, 6).
Let y or $g(x) = 0$.

$g(x) = -x^2 + 5x + 6$

$0 = -x^2 + 5x + 6$

$0 = -(x^2 - 5x - 6)$

$0 = -(x + 1)(x - 6)$

$x + 1 = 0 \quad \text{or} \quad x - 6 = 0$

$x = -1 \quad \text{or} \qquad x = 6$

The x-intercepts are (−1, 0) and (6, 0).

6. $h(t) = -16t^2 + 24t$

Find the vertex of $h(t)$ to find its maximum value.

$a = -16$, $b = 24$, and $c = 0$

$\dfrac{-b}{2a} = \dfrac{-24}{2(-16)} = \dfrac{3}{4}$

The t-value of the vertex is $\dfrac{3}{4}$.

$h\left(\dfrac{3}{4}\right) = -16\left(\dfrac{3}{4}\right)^2 + 24\left(\dfrac{3}{4}\right)$

$\qquad = -16\left(\dfrac{9}{16}\right) + 18$

$\qquad = -9 + 18$

$\qquad = 9$

The vertex is $\left(\dfrac{3}{4}, 9\right)$. Thus, the ball reaches its

maximum height of 9 feet in $\dfrac{3}{4}$ second.

Vocabulary, Readiness & Video Check 8.6

1. If a quadratic function is in the form $f(x) = a(x - h)^2 + k$, the vertex of its graph is <u>(h, k)</u>.

2. The graph of $f(x) = ax^2 + bx + c$, $a \neq 0$, is a parabola whose vertex has x-value $\dfrac{-b}{2a}$.

3. We can immediately identify the vertex (h, k), whether the parabola opens upward or downward, and know its axis of symmetry; completing the square.

4. This information tells us whether or not the graph has x-intercepts. For example, if the vertex is in quadrant III or IV and the parabola opens downward, then there aren't any x-intercepts and there's no need to go through the steps to locate any.

5. the vertex

Exercise Set 8.6

	Parabola Opens	Vertex Location	Number of x-intercept(s)	Number of y-intercept(s)
1.	up	Q I	0	1
3.	down	Q II	2	1
5.	up	x-axis	1	1
7.	down	Q III	0	
9.	up	Q IV	2	

11. $f(x) = x^2 + 8x + 7$

$-\dfrac{b}{2a} = \dfrac{-8}{2(1)} = -4$ and

$f(-4) = (-4)^2 + 8(-4) + 7$
$= 16 - 32 + 7$
$= -9$

Thus, the vertex is $(-4, -9)$.

13. $f(x) = -x^2 + 10x + 5$

$-\dfrac{b}{2a} = \dfrac{-10}{2(-1)} = 5$ and

$f(5) = -(5)^2 + 10(5) + 5$
$= -25 + 50 + 5$
$= 30$

Thus, the vertex is $(5, 30)$.

15. $f(x) = 5x^2 - 10x + 3$

$-\dfrac{b}{2a} = \dfrac{-(-10)}{2(5)} = 1$ and

$f(1) = 5(1)^2 - 10(1) + 3$
$= 5 - 10 + 3$
$= -2$

Thus, the vertex is $(1, -2)$.

17. $f(x) = -x^2 + x + 1$

$-\dfrac{b}{2a} = \dfrac{-1}{2(-1)} = \dfrac{1}{2}$ and

$f\left(\dfrac{1}{2}\right) = -\left(\dfrac{1}{2}\right)^2 + \left(\dfrac{1}{2}\right) + 1$

$= -\dfrac{1}{4} + \dfrac{1}{2} + 1$

$= \dfrac{5}{4}$

Thus, the vertex is $\left(\dfrac{1}{2}, \dfrac{5}{4}\right)$.

19. $f(x) = x^2 - 4x + 3$

$-\dfrac{b}{2a} = \dfrac{-(-4)}{2(1)} = 2$ and

$f(2) = (2)^2 - 4(2) + 3 = -1$

The vertex is (2, –1), so the graph is D.

21. $f(x) = x^2 - 2x - 3$

$-\dfrac{b}{2a} = \dfrac{-(-2)}{2(1)} = 1$ and

$f(1) = (1)^2 - 2(1) - 3 = -4$

The vertex is (1, –4), so the graph is B.

23. $f(x) = x^2 + 4x - 5$

$-\dfrac{b}{2a} = \dfrac{-4}{2(1)} = -2$ and

$f(-2) = (-2)^2 + 4(-2) - 5 = -9$

Thus, the vertex is (–2, –9).
The graph opens upward ($a = 1 > 0$).

$x^2 + 4x - 5 = 0$

$(x + 5)(x - 1) = 0$

$x + 5 = 0$ or $x - 1 = 0$

$x = -5$ or $x = 1$

x-intercepts: (–5, 0) and (1, 0).

$f(0) = -5$, so the y-intercept is (0, –5).

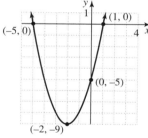

25. $f(x) = -x^2 + 2x - 1$

$-\dfrac{b}{2a} = \dfrac{-2}{2(-1)} = 1$ and

$f(1) = -(1)^2 + 2(1) - 1 = 0$

Thus, the vertex is (1, 0).
The graph opens downward ($a = -1 < 0$).

$-x^2 + 2x - 1 = 0$

$x^2 - 2x + 1 = 0$

$(x - 1)^2 = 0$

$x - 1 = 0$

$x = 1$

x-intercept: (1, 0).

$f(0) = -1$, so the y-intercept is (0, –1).

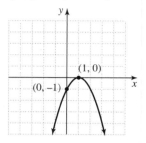

27. $f(x) = x^2 - 4$

$-\dfrac{b}{2a} = \dfrac{-0}{2(1)} = 0$ and

$f(0) = (0)^2 - 4 = -4$

Thus, the vertex is (0, –4).
The graph opens upward ($a = 1 > 0$).

$x^2 - 4 = 0$

$(x + 2)(x - 2) = 0$

$x + 2 = 0$ or $x - 2 = 0$

$x = -2$ or $x = 2$

x-intercepts: (–2, 0) and (2, 0).

$f(0) = -4$, so the y-intercept is (0, –4).

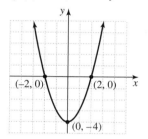

29. $f(x) = 4x^2 + 4x - 3$

$-\dfrac{b}{2a} = -\dfrac{4}{2(4)} = -\dfrac{1}{2}$ and

$f\left(-\dfrac{1}{2}\right) = 4\left(-\dfrac{1}{2}\right)^2 + 4\left(-\dfrac{1}{2}\right) - 3 = -4$

Thus, the vertex is $\left(-\dfrac{1}{2}, -4\right)$.

The graph opens upward ($a = 4 > 0$).

$4x^2 + 4x - 3 = 0$

$(2x + 3)(2x - 1) = 0$

$2x + 3 = 0$ or $2x - 1 = 0$

$x = -\dfrac{3}{2}$ or $x = \dfrac{1}{2}$

x-intercepts: $\left(-\dfrac{3}{2}, 0\right)$ and $\left(\dfrac{1}{2}, 0\right)$.

$f(0) = -3$, so the y-intercept is $(0, -3)$.

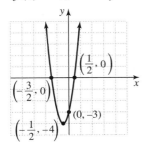

31. $f(x) = \dfrac{1}{2}x^2 + 4x + \dfrac{15}{2}$

$-\dfrac{b}{2a} = -\dfrac{4}{2\left(\frac{1}{2}\right)} = -\dfrac{4}{1} = -4$ and

$f(-4) = \dfrac{1}{2}(-4)^2 + 4(-4) + \dfrac{15}{2} = -\dfrac{1}{2}.$

Thus, the vertex is $\left(-4, -\dfrac{1}{2}\right)$.

The graph opens upward $\left(a = \dfrac{1}{2} > 0\right)$.

$\dfrac{1}{2}x^2 + 4x + \dfrac{15}{2} = 0$

$2\left(\dfrac{1}{2}x^2 + 4x + \dfrac{15}{2}\right) = 2(0)$

$x^2 + 8x + 15 = 0$

$(x+5)(x+3) = 0$

$x + 5 = 0$ or $x + 3 = 0$

$x = -5$ or $x = -3$

x-intercepts: $(-5, 0)$ and $(-3, 0)$.

$f(0) = \dfrac{15}{2}$, so the y-intercept is $\left(0, \dfrac{15}{2}\right)$.

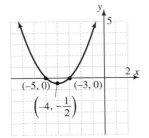

33. $f(x) = x^2 - 6x + 5$

$y = x^2 - 6x + 5$

$y - 5 = x^2 - 6x$

$y - 5 + 9 = x^2 - 6x + 9$

$y + 4 = (x-3)^2$

$y = (x-3)^2 - 4$

$f(x) = (x-3)^2 - 4$

Thus, the vertex is $(3, -4)$.

The graph opens upward $(a = 1 > 0)$.

$x^2 - 6x + 5 = 0$

$(x-5)(x-1) = 0$

$x = 5$ or $x = 1$

x-intercepts: $(5, 0)$ and $(1, 0)$.

$f(0) = 5$, so the y-intercept is $(0, 5)$.

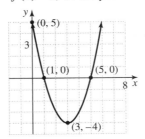

35. $f(x) = x^2 - 4x + 5$

$y = x^2 - 4x + 5$

$y - 5 = x^2 - 4x$

$y - 5 + 4 = x^2 - 4x + 4$

$y - 1 = (x-2)^2$

$y = (x-2)^2 + 1$

$f(x) = (x-2)^2 + 1$

Thus, the vertex is $(2, 1)$.

The graph opens upward $(a = 1 > 0)$.

$x^2 - 4x + 5 = 0$

$x = \dfrac{4 \pm \sqrt{(-4)^2 - 4(1)(5)}}{2(1)} = \dfrac{4 \pm \sqrt{-4}}{2}$

which give non-real solutions.

Hence, there are no x-intercepts.

$f(0) = 5$, so the y-intercept is $(0, 5)$.

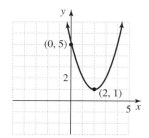

37.
$$f(x) = 2x^2 + 4x + 5$$
$$y = 2x^2 + 4x + 5$$
$$y - 5 = 2(x^2 + 2x)$$
$$y - 5 + 2(1) = 2(x^2 + 2x + 1)$$
$$y - 3 = 2(x+1)^2$$
$$y = 2(x+1)^2 + 3$$
$$f(x) = 2(x+1)^2 + 3$$
Thus, the vertex is (–1, 3).
The graph opens upward ($a = 2 > 0$).
$$2x^2 + 4x + 5 = 0$$
$$x = \frac{-4 \pm \sqrt{(4)^2 - 4(2)(5)}}{2(2)} = \frac{-4 \pm \sqrt{-24}}{4}$$
which give non-real solutions.
Hence, there are no *x*-intercepts.
$f(0) = 5$, so the *y*-intercept is (0, 5).

39.
$$f(x) = -2x^2 + 12x$$
$$y = -2(x^2 - 6x)$$
$$y + [-2(9)] = -2(x^2 - 6x + 9)$$
$$y - 18 = -2(x-3)^2$$
$$y = -2(x-3)^2 + 18$$
$$f(x) = -2(x-3)^2 + 18$$
Thus, the vertex is (3, 18).
The graph opens downward ($a = -2 < 0$).
$$-2x^2 + 12x = 0$$
$$-2x(x - 6) = 0$$
$$x = 0 \text{ or } x - 6 = 0$$
$$x = 6$$
x-intercepts: (0, 0) and (6, 0)

$f(0) = 0$, so the *y*-intercept is (0, 0).

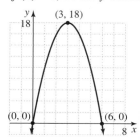

41. $f(x) = x^2 + 1$
$$x = -\frac{b}{2a} = -\frac{0}{2(1)} = 0$$
$$f(0) = (0)^2 + 1 = 1$$
Thus, the vertex is (0, 1).
The graph opens upward ($a = 1 > 0$).
$$x^2 + 1 = 0$$
$$x^2 = -1$$
which give non-real solutions.
Hence, there are no *x*-intercepts.
$f(0) = 1$, so the *y*-intercept is (0, 1).

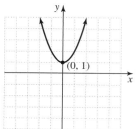

43.
$$f(x) = x^2 - 2x - 15$$
$$y = x^2 - 2x - 15$$
$$y + 15 = x^2 - 2x$$
$$y + 15 + 1 = x^2 - 2x + 1$$
$$y + 16 = (x-1)^2$$
$$y = (x-1)^2 - 16$$
$$f(x) = (x-1)^2 - 16$$
Thus, the vertex is (1, –16).
The graph opens upward ($a = 1 > 0$).
$$x^2 - 2x - 15 = 0$$
$$(x - 5)(x + 3) = 0$$
$$x = 5 \text{ or } x = -3$$
x-intercepts: (–3, 0) and (5, 0).
$f(0) = -15$ so the *y*-intercept is (0, –15).

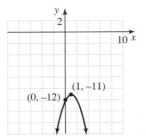

45. $f(x) = -5x^2 + 5x$

$x = -\dfrac{b}{2a} = \dfrac{-5}{2(-5)} = \dfrac{1}{2}$ and

$f\left(\dfrac{1}{2}\right) = -5\left(\dfrac{1}{2}\right)^2 + 5\left(\dfrac{1}{2}\right) = -\dfrac{5}{4} + \dfrac{5}{2} = \dfrac{5}{4}$

Thus, the vertex is $\left(\dfrac{1}{2}, \dfrac{5}{4}\right)$.

The graph opens downward $(a = -5 < 0)$.

$-5x^2 + 5x = 0$

$-5x(x-1) = 0$

$x = 0$ or $x - 1 = 0$

$\qquad\qquad x = 1$

x-intercepts: $(0, 0)$ and $(1, 0)$

$f(0) = 0$, so the y-intercept is $(0, 0)$.

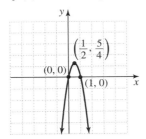

47. $f(x) = -x^2 + 2x - 12$

$x = -\dfrac{b}{2a} = \dfrac{-2}{2(-1)} = 1$ and

$f(1) = -(1)^2 + 2(1) - 12 = -11$

Thus, the vertex is $(1, -11)$.

The graph opens downward $(a = -1 < 0)$.

$-x^2 + 2x - 12 = 0$

$x^2 - 2x + 12 = 0$

$x = \dfrac{2 \pm \sqrt{(-2)^2 - 4(1)(12)}}{2(1)} = \dfrac{2 \pm \sqrt{-44}}{2}$

which yields non-real solutions.

Hence, there are no x-intercepts.

$f(0) = -12$ so the y-intercept is $(0, -12)$.

49. $f(x) = 3x^2 - 12x + 15$

$x = -\dfrac{b}{2a} = \dfrac{-(-12)}{2(3)} = \dfrac{12}{6} = 2$ and

$f(2) = 3(2)^2 - 12(2) + 15$

$\qquad = 12 - 24 + 15$

$\qquad = 3$

Thus, the vertex is $(2, 3)$.

The graph opens upward $(a = 3 > 0)$.

$3x^2 - 12x + 15 = 0$

$x^2 - 4x + 5 = 0$

$x = \dfrac{4 \pm \sqrt{(-4)^2 - 4(1)(5)}}{2(1)} = \dfrac{4 \pm \sqrt{-4}}{2}$

which yields non-real solutions.

Hence, there are no x-intercepts.

$f(0) = 15$, so the y-intercept is $(0, 15)$.

51. $f(x) = x^2 + x - 6$

$x = -\dfrac{b}{2a} = \dfrac{-1}{2(1)} = -\dfrac{1}{2}$ and

$f\left(-\dfrac{1}{2}\right) = \left(-\dfrac{1}{2}\right)^2 + \left(-\dfrac{1}{2}\right) - 6$

$\qquad = \dfrac{1}{4} - \dfrac{1}{2} - 6$

$\qquad = -\dfrac{25}{4}$

Thus, the vertex is $\left(-\dfrac{1}{2}, -\dfrac{25}{4}\right)$.

The graph opens upward $(a = 1 > 0)$.

$$x^2 + x - 6 = 0$$
$$(x+3)(x-2) = 0$$
$$x = -3 \quad \text{or} \quad x = 2$$
x-intercepts: $(-3, 0)$ and $(2, 0)$.
$f(0) = -6$ so the y-intercept is $(0, -6)$.

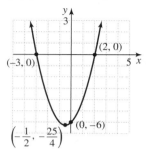

53. $f(x) = -2x^2 - 3x + 35$
$$x = -\frac{b}{2a} = \frac{-(-3)}{2(-2)} = -\frac{3}{4} \quad \text{and}$$
$$f\left(-\frac{3}{4}\right) = -2\left(-\frac{3}{4}\right)^2 - 3\left(-\frac{3}{4}\right) + 35$$
$$= -\frac{9}{8} + \frac{9}{4} + 35$$
$$= \frac{289}{8}$$

Thus, the vertex is $\left(-\frac{3}{4}, \frac{289}{8}\right)$.
The graph opens downward ($a = -2 < 0$).
$$-2x^2 - 3x + 35 = 0$$
$$2x^2 + 3x - 35 = 0$$
$$(2x-7)(x+5) = 0$$
$$2x - 7 = 0 \quad \text{or} \quad x + 5 = 0$$
$$x = \frac{7}{2} \quad \text{or} \quad x = -5$$

x-intercepts: $(-5, 0)$ and $\left(\frac{7}{2}, 0\right)$.

$f(0) = 35$ so the y-intercept is $(0, 35)$.

55. $h(t) = -16t^2 + 96t$
$$t = -\frac{b}{2a} = \frac{-96}{2(-16)} = \frac{96}{32} = 3 \quad \text{and}$$
$$h(3) = -16(3)^2 + 96(3)$$
$$= -144 + 288$$
$$= 144$$
The maximum height is 144 feet.

57. $C(x) = 2x^2 - 800x + 92,000$

a. $x = -\frac{b}{2a} = \frac{-(-800)}{2(2)} = 200$
200 bicycles are needed to minimize the cost.

b. $C(200) = 2(200)^2 - 800(200) + 92,000$
$$= 12,000$$
The minimum cost is \$12,000.

59. Let $x =$ one number. Then
$60 - x =$ the other number.
$$f(x) = x(60 - x)$$
$$= 60x - x^2$$
$$= -x^2 + 60x$$
The maximum will occur at the vertex.
$$x = -\frac{b}{2a} = \frac{-60}{2(-1)} = 30$$
$60 - x = 60 - 30 = 30$
The numbers are 30 and 30.

61. Let $x =$ one number. Then
$10 + x =$ the other number.
$$f(x) = x(10 + x)$$
$$= 10x + x^2$$
$$= x^2 + 10x$$
The minimum will occur at the vertex.
$$x = -\frac{b}{2a} = \frac{-10}{2(1)} = -5$$
$10 + x = 10 + (-5) = 5$
The numbers are -5 and 5.

63. Let $x =$ width. Then $40 - x =$ the length.
Area = length · width
$$A(x) = (40 - x)x$$
$$= 40x - x^2$$
$$= -x^2 + 40x$$
The maximum will occur at the vertex.

$$x = -\frac{b}{2a} = \frac{-40}{2(-1)} = 20$$

$$40 - x = 40 - 20 = 20$$

The maximum area will occur when the length and width are 20 units each.

65. $f(x) = x^2 + 2$

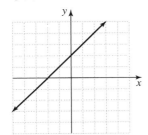

67. $g(x) = x + 2$

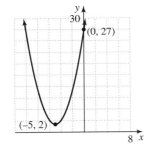

69. $f(x) = (x+5)^2 + 2$

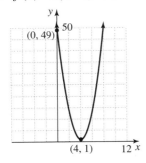

71. $f(x) = 3(x-4)^2 + 1$

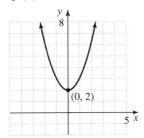

73. $f(x) = -(x-4)^2 + \frac{3}{2}$

75. $f(x) = 2x^2 - 5$

Since $a = 2 > 0$, the graph opens upward; thus, $f(x)$ has a minimum value.

77. $F(x) = 3 - \frac{1}{2}x^2$

Since $a = -\frac{1}{2} < 0$, the graph opens downward; thus, $F(x)$ has a maximum value.

79. $f(x) = x^2 + 10x + 15$

$$x = -\frac{b}{2a} = \frac{-10}{2(1)} = -5 \text{ and}$$

$$f(-5) = (-5)^2 + 10(-5) + 15 = -10$$

Thus, the vertex is $(-5, -10)$.

The graph opens upward ($a = 1 > 0$).

$f(0) = 15$ so the y-intercept is $(0, 15)$.

$$x^2 + 10x + 15 = 0$$

$$x = \frac{-10 \pm \sqrt{(10)^2 - 4(1)(15)}}{2(1)}$$

$$= \frac{-10 \pm \sqrt{40}}{2}$$

$$\approx -8.2 \text{ or } -1.8$$

The x-intercepts are approximately $(-8.2, 0)$ and $(-1.8, 0)$.

81. $f(x) = 3x^2 - 6x + 7$

$x = -\dfrac{b}{2a} = \dfrac{-(-6)}{2(3)} = 1$ and

$f(1) = 3(1)^2 - 6(1) + 7 = 4$

Thus, the vertex is $(1, 4)$.

The graph opens upward $(a = 3 > 0)$.

$f(0) = 7$ so the y-intercept is $(0, 7)$.

$3x^2 - 6x + 7 = 0$

$x = \dfrac{6 \pm \sqrt{(-6)^2 - 4(3)(7)}}{2(3)} = \dfrac{6 \pm \sqrt{-48}}{6}$

which yields non-real solutions.

Hence, there are no x-intercepts.

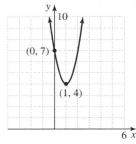

83. $f(x) = 2.3x^2 - 6.1x + 3.2$

$x = \dfrac{-(-6.1)}{2(2.3)} \approx 1.33$

$f(1.33) \approx -0.84$

minimum ≈ -0.84

Alternative solution:

85. $f(x) = -1.9x^2 + 5.6x - 2.7$

$x = \dfrac{-5.6}{2(-1.9)} \approx 1.47$

$f(1.47) \approx 1.43$

maximum ≈ 1.43

Alternate solution:

87. a.   The function will have a maximum value; answers may vary.

b. $c(x) = -0.4x^2 + 21x + 35$

$a = -0.4$, $b = 21$, $c = 35$

$-\dfrac{b}{2a} = -\dfrac{21}{2(-0.4)} = 26.25$

The maximum will be reached 26.25 years after 2009 or in $2009 + 26 = 2035$.

c. $c(26.25) = -0.4(26.25)^2 + 21(26.25) + 35$

≈ 310.6

The maximum number of Wi-Fi-enabled cell phones is predicted to be about 311 million.

Section 8.7 Practice Exercises

1. a.

b.

The linear regression equation is $y = 0.874x - 52.438$. Store this in Y_1.

c.

The quadratic regression equation is

$y = -0.004x^2 + 1.656x - 87.469$. Store this in Y_2.

d.   From the R^2 values, the quadratic equation is the better fit.

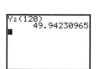

The cost of a first-class stamp is predicted to be 50¢ in 2020.

2. a.

The linear regression equation is
$y = 0.054x + 22.951$.

b.

The quadratic regression equation is

$y = 0.002x^2 + 0.026x + 23.016$.

c.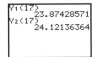

The linear model predicts 23.9 MPG in 2017, while the quadratic model predicts 24.1 MPG.

d. answers may vary

3. a.

From the graph, a linear model appears to fit the data.

b.

The linear regression equation is
$y = 7.767x + 363.2$.

c.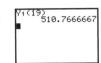

Truck sales of 511 are predicted for 2019.

4. a.

A quadratic model appears to fit the data.

b.

The quadratic regression equation is

$y = -2.484x^2 + 91.205x - 502.769$.

c.

The expected number of broadband connections in 2020 is 327.8 million.

5. a.

Either a linear or quadratic model could fit the data, but comparing the R^2 values, the quadratic model is better.

The quadratic regression model is

$y = 0.003x^2 + 0.207x + 5.343$.

b.

The average ticket price in 2019 is predicted to be $10.28.

Vocabulary, Readiness & Video Check 8.7

1. Create a scatter plot of the data. If the data points appear to form a line, a linear regression would be best; if they appear to form a parabola, a quadratic regression equation would be best.

Exercise Set 8.7

1. The data appears to be linear.

3. The data appears to be neither linear nor quadratic.

5. The data appears to be quadratic.

7. The data points show some curvature, so a quadratic model is indicated.

9.

   The data do not appear to be either linear or quadratic.

11.

   The linear model is $y = 7.3x + 559.8$.

   The quadratic model is

   $y = -1.107x^2 + 20.586x + 528.8$.

   The linear model predicts earnings of $684; the quadratic model predicts $559.

13.

   The quadratic regression equation is

   $y = 1815.714x^2 - 27,563.057x + 108,947.429$.

   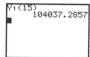

   The number of digital cinema screens in 2015 is predicted to be 104,037.

15. answers may vary

17. a.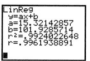

   The linear regression equation is
   $y = 15.321x + 101.929$.

   b.

   The quadratic regression equation is

   $y = -0.190x^2 + 17.607x + 98.119$.

   c.

   The linear model predicts 378 thousand cell sites in the U.S. in 2018; the quadratic model predicts 353 thousand.

   d. answers may vary

19. a.

   Based on the graph, it appears that a quadratic model would best fit the data.

   b.

   The quadratic model is

   $y = -13.010x^2 + 1727.626x - 37,489.948$.

   c.

   If the shoes are sold for $80, the projected profit is about $17,457. If the shoes are sold for $60, the projected profit is about $19,332.

   d. If the shoes are sold for $105, the profit would be $1500. This is not a good selling price; answers may vary

e.

The shoes should be sold for about $66 to achieve the maximum profit.

21.

The data appears to be linear so we will fit a linear model.

The linear model of best fit is
$y = 82.509x - 51,908.333$.

23. The model from Exercise 21 is
$y = 82.509x - 51,908.333$.
Evaluate this model when $x = 2400$.
$y = 82.509(2400) - 51,908.333$
$ = 198,021.6 - 51,908.333$
$ = 146,113.267$

The linear model predicts that a house with 2400 square feet will cost about $146,113.

25.

The quadratic regression equation is
$y = 77.512x^2 - 2265.226x + 18,625.643$.

Use a table of values to get the predicted revenue.

x (year)	Actual Revenue	Predicted Revenue	Difference between Actual and Predicted
2008	5540	5465	75
2009	4432	4517	−85
2010	3656	3725	−69
2011	3108	3087	21
2012	2673	2605	68
2013	2325	2277	48
2014	2046	2105	−59

27.

Based on the graph, it appears that a quadratic model would best fit the data.

The quadratic model of best fit is
$y = 0.75x^2 - 10.774x + 86.655$.

29. Since the area of a square is equal to the square of the length of one side, a quadratic model would best describe the relationship.

31. The cost of a number of calculators is directly related to how many are purchased. Therefore, a linear model would best describe the relationship.

33. The average temperature in Cleveland (for the months of March through October) will tend to rise as summer approaches, peak sometime during summer, and then begin to decline as autumn approaches. This indicates the data will have some curvature, so a quadratic model would best describe this relationship.

35. The number of bacteria will generally be small at first when the person is coming down with the flu, increase as the flu worsens, and decrease as the person recovers. This indicates some curvature to the graph, so a quadratic model would describe the relationship better than a linear model.

37. $(x+3)(x-5) = 0$

$x + 3 = 0$ or $x - 5 = 0$

$x = -3$ $x = 5$

The solutions are -3 and 5.

39. $2x^2 - 7x - 15 = 0$

$(2x + 3)(x - 5) = 0$

$2x + 3 = 0$ or $x - 5 = 0$

$2x = -3$ $x = 5$

$x = -\dfrac{3}{2}$

The solutions are $-\dfrac{3}{2}$ and 5.

41. $3(x - 4) + 2 = 5(x - 6)$

$3x - 12 + 2 = 5x - 30$

$3x - 10 = 5x - 30$

$-2x = -20$

$x = 10$

The solution is 10.

43. $f(x) = x^3 + 3x^2 - 5x - 8$

To find the y-intercept, we evaluate the function when $x = 0$.

$f(0) = (0)^3 + 3(0)^2 - 5(0) - 8 = -8$

The y-intercept is $(0, -8)$.

45. $g(x) = x^2 - 3x + 5$

To find the y-intercept, we evaluate the function when $x = 0$.

$g(0) = (0)^2 - 3(0) + 5 = 5$

The y-intercept is $(0, 5)$.

Chapter 8 Vocabulary Check

1. The <u>discriminant</u> helps us find the number and type of solutions of a quadratic equation.

2. If $a^2 = b$, then $a = \underline{\pm\sqrt{b}}$.

3. The graph of $f(x) = ax^2 + bx + c$, where a is not 0, is a parabola whose vertex has x-value $\underline{\dfrac{-b}{2a}}$.

4. A <u>quadratic inequality</u> is an inequality that can be written so that one side is a quadratic expression and the other side is 0.

5. The process of writing a quadratic equation so that one side is a perfect square trinomial is called <u>completing the square</u>.

6. The graph of $f(x) = x^2 + k$ has vertex <u>(0, k)</u>.

7. The graph of $f(x) = (x - h)^2$ has vertex <u>(h, 0)</u>.

8. The graph of $f(x) = (x - h)^2 + k$ has vertex <u>(h, k)</u>.

9. The formula $x = \dfrac{-b \pm \sqrt{b^2 - 4ac}}{2a}$ is called the <u>quadratic formula</u>.

10. A <u>quadratic</u> equation is one that can be written in the form $ax^2 + bx + c = 0$ where a, b, and c are real numbers and a is not 0.

Chapter 8 Review

1. $x^2 - 15x + 14 = 0$

$(x - 14)(x - 1) = 0$

$x - 14 = 0$ or $x - 1 = 0$

$x = 14$ or $x = 1$

The solutions are 1 and 14.

2. $7a^2 = 29a + 30$

$7a^2 - 29a - 30 = 0$

$(7a + 6)(a - 5) = 0$

$7a + 6 = 0$ or $a - 5 = 0$

$7a = -6$ or $a = 5$

$a = -\dfrac{6}{7}$

The solutions are $-\dfrac{6}{7}$ and 5.

3. $4m^2 = 196$

$m^2 = 49$

$m = \pm\sqrt{49}$

$m = \pm 7$

The solutions are -7 and 7.

4. $(5x - 2)^2 = 2$

$$5x - 2 = \pm\sqrt{2}$$

$$5x = 2 \pm \sqrt{2}$$

$$x = \frac{2 \pm \sqrt{2}}{5}$$

The solutions are $\dfrac{2 + \sqrt{2}}{5}$ and $\dfrac{2 - \sqrt{2}}{5}$.

5. $z^2 + 3z + 1 = 0$

$$z^2 + 3z = -1$$

$$z^2 + 3z + \left(\frac{3}{2}\right)^2 = -1 + \frac{9}{4}$$

$$\left(z + \frac{3}{2}\right)^2 = \frac{5}{4}$$

$$z + \frac{3}{2} = \pm\sqrt{\frac{5}{4}}$$

$$z + \frac{3}{2} = \pm\frac{\sqrt{5}}{2}$$

$$z = -\frac{3}{2} \pm \frac{\sqrt{5}}{2} = \frac{-3 \pm \sqrt{5}}{2}$$

The solutions are $\dfrac{-3 + \sqrt{5}}{2}$ and $\dfrac{-3 - \sqrt{5}}{2}$.

6. $(2x + 1)^2 = x$

$$4x^2 + 4x + 1 = x$$

$$4x^2 + 3x = -1$$

$$x^2 + \frac{3}{4}x = -\frac{1}{4}$$

$$x^2 + \frac{3}{4}x + \left(\frac{\frac{3}{4}}{2}\right)^2 = -\frac{1}{4} + \frac{9}{64}$$

$$\left(x + \frac{3}{8}\right)^2 = -\frac{7}{64}$$

$$x + \frac{3}{8} = \pm\sqrt{-\frac{7}{64}}$$

$$x + \frac{3}{8} = \pm\frac{i\sqrt{7}}{8}$$

$$x = -\frac{3}{8} \pm \frac{i\sqrt{7}}{8} = \frac{-3 \pm i\sqrt{7}}{8}$$

The solutions are $\dfrac{-3 + i\sqrt{7}}{8}$ and $\dfrac{-3 - i\sqrt{7}}{8}$.

7. $A = P(1 + r)^2$

$$2717 = 2500(1 + r)^2$$

$$\frac{2717}{2500} = (1 + r)^2$$

$$(1 + r)^2 = 1.0868$$

$$1 + r = \pm\sqrt{1.0868}$$

$$1 + r = \pm 1.0425$$

$$r = -1 \pm 1.0425$$

$$= 0.0425 \text{ or } -2.0425 \text{ (disregard)}$$

The interest rate is 4.25%.

8. Let x = distance traveled.

$$a^2 + b^2 = c^2$$

$$x^2 + x^2 = (150)^2$$

$$2x^2 = 22,500$$

$$x^2 = 11,250$$

$$x = \pm 75\sqrt{2} \approx \pm 106.1$$

Disregard the negative. The ships each traveled $75\sqrt{2} \approx 106.1$ miles.

9. Two complex but not real solutions exist.

10. Two real solutions exist.

11. Two real solutions exist.

12. One real solution exists.

13. $x^2 - 16x + 64 = 0$

$a = 1, b = -16, c = 64$

$$x = \frac{16 \pm \sqrt{(-16)^2 - 4(1)(64)}}{2(1)}$$

$$= \frac{16 \pm \sqrt{256 - 256}}{2}$$

$$= \frac{16 \pm \sqrt{0}}{2}$$

$$= 8$$

The solution is 8.

14. $x^2 + 5x = 0$

$a = 1, b = 5, c = 0$

$x = \dfrac{-5 \pm \sqrt{(5)^2 - 4(1)(0)}}{2(1)}$

$= \dfrac{-5 \pm \sqrt{25}}{2}$

$= \dfrac{-5 \pm 5}{2}$

$= 0 \text{ or } -5$

The solutions are -5 and 0.

15. $2x^2 + 3x = 5$

$2x^2 + 3x - 5 = 0$

$a = 2, b = 3, c = -5$

$x = \dfrac{-3 \pm \sqrt{(3)^2 - 4(2)(-5)}}{2(2)}$

$= \dfrac{-3 \pm \sqrt{49}}{4}$

$= \dfrac{-3 \pm 7}{4}$

$= 1 \text{ or } -\dfrac{5}{2}$

The solutions are $-\dfrac{5}{2}$ and 1.

16. $9x^2 + 4 = 2x$

$9x^2 - 2x + 4 = 0$

$x = \dfrac{2 \pm \sqrt{(-2)^2 - 4(9)(4)}}{2(9)}$

$= \dfrac{2 \pm \sqrt{-140}}{18}$

$= \dfrac{2 \pm i\sqrt{4 \cdot 35}}{18}$

$= \dfrac{2 \pm 2i\sqrt{35}}{18}$

$= \dfrac{1 \pm i\sqrt{35}}{9}$

The solutions are $\dfrac{1 + i\sqrt{35}}{9}$ and $\dfrac{1 - i\sqrt{35}}{9}$.

17. $6x^2 + 7 = 5x$

$6x^2 - 5x + 7 = 0$

$a = 6, b = -5, c = 7$

$x = \dfrac{5 \pm \sqrt{(-5)^2 - 4(6)(7)}}{2(6)}$

$= \dfrac{5 \pm \sqrt{25 - 168}}{12}$

$= \dfrac{5 \pm \sqrt{-143}}{12}$

$= \dfrac{5 \pm i\sqrt{143}}{12}$

The solutions are $\dfrac{5 + i\sqrt{143}}{12}$ and $\dfrac{5 - i\sqrt{143}}{12}$.

18. $(2x - 3)^2 = x$

$4x^2 - 12x + 9 - x = 0$

$4x^2 - 13x + 9 = 0$

$a = 4, b = -13, c = 9$

$x = \dfrac{13 \pm \sqrt{(-13)^2 - 4(4)(9)}}{2(4)}$

$= \dfrac{13 \pm \sqrt{169 - 144}}{8}$

$= \dfrac{13 \pm \sqrt{25}}{8}$

$= \dfrac{13 \pm 5}{8}$

$= \dfrac{9}{4} \text{ or } 1$

The solutions are 1 and $\dfrac{9}{4}$.

19. $d(t) = -16t^2 + 30t + 6$

a. $d(1) = -16(1)^2 + 30(1) + 6$

$\qquad = -16 + 30 + 6$

$\qquad = 20 \text{ feet}$

b. $-16t^2 + 30t + 6 = 0$

$$8t^2 - 15t - 3 = 0$$
$$a = 8, b = -15, c = -3$$
$$t = \frac{15 \pm \sqrt{(-15)^2 - 4(8)(-3)}}{2(8)}$$
$$= \frac{15 \pm \sqrt{225 + 96}}{16}$$
$$= \frac{15 \pm \sqrt{321}}{16}$$

Disregarding the negative, we have

$$t = \frac{15 + \sqrt{321}}{16} \text{ seconds}$$
$$\approx 2.1 \text{ seconds.}$$

20. Let $x =$ length of the legs. Then
$x + 6 =$ length of the hypotenuse.

$$x^2 + x^2 = (x+6)^2$$
$$2x^2 = x^2 + 12x + 36$$
$$x^2 - 12x - 36 = 0$$
$$a = 1, b = -12, c = -36$$
$$x = \frac{12 \pm \sqrt{(-12)^2 - 4(1)(-36)}}{2(1)}$$
$$= \frac{12 \pm \sqrt{144 + 144}}{2}$$
$$= \frac{12 \pm \sqrt{144 \cdot 2}}{2}$$
$$= \frac{12 \pm 12\sqrt{2}}{2}$$
$$= 6 \pm 6\sqrt{2}$$

Disregard the negative. The length of each leg is
$\left(6 + 6\sqrt{2}\right)$ cm.

21.
$$x^3 = 27$$
$$x^3 - 27 = 0$$
$$(x-3)(x^2 + 3x + 9) = 0$$
$$x - 3 = 0 \text{ or } x^2 + 3x + 9 = 0$$
$$x = 3 \qquad a = 1, b = 3, c = 9$$
$$x = \frac{-3 \pm \sqrt{(3)^2 - 4(1)(9)}}{2(1)}$$
$$= \frac{-3 \pm \sqrt{9 - 36}}{2}$$
$$= \frac{-3 \pm \sqrt{-27}}{2}$$
$$= \frac{-3 \pm 3i\sqrt{3}}{2}$$

The solutions are 3, $\dfrac{-3 + 3i\sqrt{3}}{2}$, and $\dfrac{-3 - 3i\sqrt{3}}{2}$.

22.
$$y^3 = -64$$
$$y^3 + 64 = 0$$
$$(y+4)(y^2 - 4y + 16) = 0$$
$$y + 4 = 0 \text{ or } y^2 - 4y + 16 = 0$$
$$y = -4 \qquad a = 1, b = -4, c = 16$$
$$y = \frac{4 \pm \sqrt{(-4)^2 - 4(1)(16)}}{2(1)}$$
$$= \frac{4 \pm \sqrt{16 - 64}}{2}$$
$$= \frac{4 \pm \sqrt{-48}}{2}$$
$$= \frac{4 \pm 4i\sqrt{3}}{2}$$
$$= 2 \pm 2i\sqrt{3}$$

The solutions are -4, $2 + 2i\sqrt{3}$, and $2 - 2i\sqrt{3}$.

23.
$$\frac{5}{x}+\frac{6}{x-2}=3$$
$$x(x-2)\left(\frac{5}{x}+\frac{6}{x-2}\right)=3x(x-2)$$
$$5(x-2)+6x=3x^2-6x$$
$$5x-10+6x=3x^2-6x$$
$$0=3x^2-17x+10$$
$$0=(3x-2)(x-5)$$
$$3x-2=0 \ \text{ or } \ x-5=0$$
$$x=\frac{2}{3} \ \text{ or } \quad x=5$$

The solutions are $\frac{2}{3}$ and 5.

24.
$$x^4-21x^2-100=0$$
$$(x^2-25)(x^2+4)=0$$
$$(x+5)(x-5)(x^2+4)=0$$
$$x+5=0 \ \text{ or } x-5=0 \text{ or } x^2+4=0$$
$$x=-5 \text{ or } \quad x=5 \text{ or } \quad x^2=-4$$
$$x=\pm 2i$$

The solutions are –5, 5 –2i, and 2i.

25. $x^{2/3}-6x^{1/3}+5=0$

Let $y=x^{1/3}$. Then $y^2=x^{2/3}$ and
$$y^2-6y+5=0$$
$$(y-5)(y-1)=0$$
$$y-5=0 \ \text{ or } y-1=0$$
$$y=5 \ \text{ or } \quad y=1$$
$$x^{1/3}=5 \ \text{ or } \ x^{1/3}=1$$
$$x=125 \text{ or } \quad x=1$$

The solutions are 1 and 125.

26.
$$5(x+3)^2-19(x+3)=4$$
$$5(x+3)^2-19(x+3)-4=0$$

Let $y=x+3$. Then $y^2=(x+3)^2$ and

$$5y^2-19y-4=0$$
$$(5y+1)(y-4)=0$$
$$5y+1=0 \quad \text{or } y-4=0$$
$$y=-\frac{1}{5} \ \text{ or } \qquad y=4$$
$$x+3=-\frac{1}{5} \ \text{ or } \ x+3=4$$
$$x=-\frac{16}{5} \ \text{ or } \qquad x=1$$

The solutions are $-\frac{16}{5}$ and 1.

27.
$$a^6-a^2=a^4-1$$
$$a^6-a^4-a^2+1=0$$
$$a^4(a^2-1)-1(a^2-1)=0$$
$$(a^2-1)(a^4-1)=0$$
$$(a+1)(a-1)(a^2+1)(a^2-1)=0$$
$$(a+1)(a-1)(a^2+1)(a+1)(a-1)=0$$
$$(a+1)^2(a-1)^2(a^2+1)=0$$
$$(a+1)^2=0 \ \text{ or } (a-1)^2=0 \text{ or } a^2+1=0$$
$$a+1=0 \ \text{ or } \quad a-1=0 \text{ or } \quad a^2=-1$$
$$a=-1 \text{ or } \qquad a=1 \text{ or } \qquad a=\pm i$$

The solutions are –1, 1, –i, and i.

28. $y^{-2}+y^{-1}=20$
$$\frac{1}{y^2}+\frac{1}{y}=20$$
$$1+y=20y^2$$
$$0=20y^2-y-1$$
$$0=(5y+1)(4y-1)$$
$$5y+1=0 \quad \text{or } 4y-1=0$$
$$y=-\frac{1}{5} \ \text{ or } \qquad y=\frac{1}{4}$$

The solutions are $-\frac{1}{5}$ and $\frac{1}{4}$.

29. Let x = time for Jerome alone. Then $x-1$ = time for Tim alone.
$$\frac{1}{x}+\frac{1}{x-1}=\frac{1}{5}$$
$$5(x-1)+5x=x(x-1)$$
$$5x-5+5x=x^2-x$$
$$0=x^2-11x+5$$

$a = 1, b = -11, c = 5$

$$x = \frac{11 \pm \sqrt{(-11)^2 - 4(1)(5)}}{2(1)}$$

$$= \frac{11 \pm \sqrt{101}}{2}$$

$$\approx 0.475 \text{ (disregard) or } 10.525$$

Jerome: 10.5 hours
Tim: 9.5 hours

30. Let x = the number; then

$\dfrac{1}{x}$ = the reciprocal of the number.

$$x - \frac{1}{x} = -\frac{24}{5}$$

$$5x\left(x - \frac{1}{x}\right) = 5x\left(-\frac{24}{5}\right)$$

$$5x^2 - 5 = -24x$$

$$5x^2 + 24x - 5 = 0$$

$$(5x - 1)(x + 5) = 0$$

$$5x - 1 = 0 \text{ or } x + 5 = 0$$

$$x = \frac{1}{5} \text{ or } \quad x = -5$$

Disregard the positive value as extraneous. The number is -5.

31.
$$2x^2 - 50 \le 0$$

$$2(x^2 - 25) \le 0$$

$$2(x + 5)(x - 5) \le 0$$

$$x + 5 = 0 \text{ or } x - 5 = 0$$

$$x = -5 \text{ or } \quad x = 5$$

Region	Test Point	$2(x + 5)(x - 5) \le 0$ Result
$A: (-\infty, -5]$	-6	$2(-1)(-11) \le 0$ False
$B: [-5, 5]$	0	$2(5)(-5) \le 0$ True
$C: [5, \infty)$	6	$2(11)(1) \le 0$ False

Solution: $[-5, 5]$

32.
$$\frac{1}{4}x^2 < \frac{1}{16}$$

$$x^2 < \frac{1}{4}$$

$$x^2 - \frac{1}{4} < 0$$

$$\left(x + \frac{1}{2}\right)\left(x - \frac{1}{2}\right) < 0$$

$$x + \frac{1}{2} = 0 \quad \text{or } x - \frac{1}{2} = 0$$

$$x = -\frac{1}{2} \quad \text{or} \quad x = \frac{1}{2}$$

Region	Test Point	$\left(x + \frac{1}{2}\right)\left(x - \frac{1}{2}\right) < 0$ Result
$A: \left(-\infty, -\frac{1}{2}\right)$	-1	$\left(-\frac{1}{2}\right)\left(-\frac{3}{2}\right) < 0$ False
$B: \left(-\frac{1}{2}, \frac{1}{2}\right)$	0	$\left(\frac{1}{2}\right)\left(-\frac{1}{2}\right) < 0$ True
$C: \left(\frac{1}{2}, \infty\right)$	1	$\left(\frac{3}{2}\right)\left(\frac{1}{2}\right) < 0$ False

Solution: $\left(-\dfrac{1}{2}, \dfrac{1}{2}\right)$

33.
$$(x^2 - 4)(x^2 - 25) \le 0$$
$$(x + 2)(x - 2)(x + 5)(x - 5) \le 0$$

$x + 2 = 0$ or $x - 2 = 0$ or $x + 5 = 0$ or $x - 5 = 0$

    $x = -2$ or $x = 2$ or $x = -5$ or $x = 5$

Region	Test Point	$(x^2 - 4)(x^2 - 25) \le 0$ Result
A: $(-\infty, -5)$	−6	$352 \le 0$ False
B: $(-5, -2)$	−3	$-80 \le 0$ True
C: $(-2, 2)$	0	$100 \le 0$ False
D: $(2, 5)$	3	$-80 \le 0$ True
E: $(5, \infty)$	6	$352 \le 0$ False

Solution: $[-5, -2] \cup [2, 5]$

34.
$$(x^2 - 16)(x^2 - 1) > 0$$
$$(x + 4)(x - 4)(x + 1)(x - 1) > 0$$

$x + 4 = 0$ or $x - 4 = 0$ or $x + 1 = 0$ or $x - 1 = 0$

    $x = -4$ or $x = 4$ or $x = -1$ or $x = 1$

Region	Test Point	$(x + 4)(x - 4)(x + 1)(x - 1) > 0$ Result
A: $(-\infty, -4)$	−5	$(-1)(-9)(-4)(-6) > 0$ True
B: $(-4, -1)$	−2	$(2)(-6)(-1)(-3) > 0$ False
C: $(-1, 1)$	0	$(4)(-4)(1)(-1) > 0$ True
D: $(1, 4)$	2	$(6)(-2)(3)(1) > 0$ False
E: $(4, \infty)$	5	$(9)(1)(6)(4) > 0$ True

Solution: $(-\infty, -4) \cup (-1, 1) \cup (4, \infty)$

35. $\dfrac{x-5}{x-6} < 0$

$x-5=0 \ \text{ or } \ x-6=0$

$\qquad x=5 \ \text{ or } \qquad x=6$

Region	Test Point	$\dfrac{x-5}{x-6}<0$ Result
A: $(-\infty, 5)$	0	$\dfrac{-5}{-6}<0$ False
B: $(5, 6)$	$\dfrac{11}{2}$	$\dfrac{\frac{1}{2}}{-\frac{1}{2}}<0$ True
C: $(6, \infty)$	7	$\dfrac{2}{1}<0$ False

Solution: $(5, 6)$

36. $\dfrac{(4x+3)(x-5)}{x(x+6)} > 0$

$4x+3=0,\ x-5=0,\ x=0,\ \text{or } x+6=0$

$x=-\dfrac{3}{4},\ x=5,\ x=0,\ \text{or } x=-6$

Region	Test Point	$\dfrac{(4x+3)(x-5)}{x(x+6)}>0$ Result
A: $(-\infty, -6)$	-7	$\dfrac{(-25)(-12)}{-7(-1)}>0$ True
B: $\left(-6, -\dfrac{3}{4}\right)$	-3	$\dfrac{(-9)(-8)}{-3(3)}>0$ False
C: $\left(-\dfrac{3}{4}, 0\right)$	$-\dfrac{1}{2}$	$\dfrac{(1)\left(-\frac{11}{2}\right)}{-\frac{1}{2}\left(\frac{11}{2}\right)}>0$ True
D: $(0, 5)$	1	$\dfrac{(7)(-4)}{1(7)}>0$ False
E: $(5, \infty)$	6	$\dfrac{(27)(1)}{6(12)}>0$ True

Solution: $(-\infty, -6)\cup\left(-\dfrac{3}{4}, 0\right)\cup(5, \infty)$

37. $(x+5)(x-6)(x+2)\le 0$

$x+5=0 \ \text{ or } \ x-6=0 \ \text{ or } \ x+2=0$

$x=-5 \ \text{ or } \qquad x=6 \ \text{ or } \qquad x=-2$

Region	Test Point	$(x+5)(x-6)(x+2)\le 0$ Result
A: $(-\infty, -5]$	-6	$(-1)(-12)(-4)\le 0$ True
B: $[-5, -2]$	-3	$(2)(-9)(-1)\le 0$ False
C: $[-2, 6]$	0	$(5)(-6)(2)\le 0$ True
D: $[6, \infty)$	7	$(12)(1)(9)\le 0$ False

Solution: $(-\infty, -5]\cup[-2, 6]$

38. $x^3 + 3x^2 - 25x - 75 > 0$

$x^2(x+3) - 25(x+3) > 0$

$\qquad (x+3)(x^2 - 25) > 0$

$\qquad (x+3)(x+5)(x-5) > 0$

$x+3 = 0 \quad$ or $\quad x+5 = 0 \quad$ or $\quad x-5 = 0$

$x = -3 \quad$ or $\qquad x = -5 \quad$ or $\qquad x = 5$

Region	Test Point	$(x + 3)(x + 5)(x - 5) > 0$ Result
A: $(-\infty, -5)$	-6	$(-3)(-1)(-11) > 0$ False
B: $(-5, -3)$	-4	$(-1)(1)(-9) > 0$ True
C: $(-3, 5)$	0	$(3)(5)(-5) > 0$ False
D: $(5, \infty)$	6	$(9)(11)(1) > 0$ True

Solution: $(-5, -3) \cup (5, \infty)$

39. $\dfrac{x^2 + 4}{3x} \le 1$

The denominator equals 0 when $3x = 0$, or $x = 0$.

$$\frac{x^2 + 4}{3x} = 1$$

$$x^2 + 4 = 3x$$

$$x^2 - 3x + 4 = 0$$

$$x = \frac{3 \pm \sqrt{(-3)^2 - 4(1)(4)}}{2(1)} = \frac{3 \pm \sqrt{-7}}{2}$$

which yields non-real solutions.

Region	Test Point	$\dfrac{x^2 + 4}{3x} \le 1$ Result
A: $(-\infty, 0)$	-1	$\dfrac{5}{-3} \le 1$ True
B: $(0, \infty)$	1	$\dfrac{5}{3} \le 1$ False

Solution: $(\infty, 0)$

40. $\dfrac{3}{x-2} > 2$

The denominator is equal to 0 when $x - 2 = 0$, or $x = 2$.

$$\frac{3}{x-2} = 2$$

$$3 = 2(x-2)$$

$$3 = 2x - 4$$

$$7 = 2x$$

$$\frac{7}{2} = x$$

Region	Test Point	$\dfrac{3}{x-2} > 2$ Result
A: $(-\infty, 2)$	0	$\dfrac{3}{-2} > 2$ False
B: $\left(2, \dfrac{7}{2}\right)$	3	$\dfrac{3}{1} > 2$ True
C: $\left(\dfrac{7}{2}, \infty\right)$	5	$\dfrac{3}{3} > 2$ False

Solution: $\left(2, \dfrac{7}{2}\right)$

41. $f(x) = x^2 - 4$

Vertex: $(0, -4)$

Axis of symmetry: $x = 0$

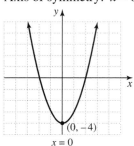

42. $g(x) = x^2 + 7$

Vertex: (0, 7)
Axis of symmetry: $x = 0$

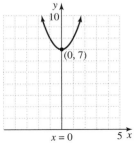

43. $H(x) = 2x^2$

Vertex: (0, 0)
Axis of symmetry: $x = 0$

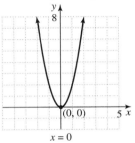

44. $h(x) = -\dfrac{1}{3}x^2$

Vertex: (0, 0)
Axis of symmetry: $x = 0$

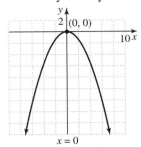

45. $F(x) = (x-1)^2$

Vertex: (1, 0)
Axis of symmetry: $x = 1$

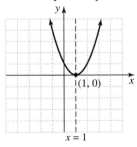

46. $G(x) = (x+5)^2$

Vertex: (–5, 0)
Axis of symmetry: $x = -5$

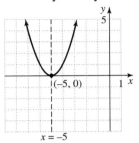

47. $f(x) = (x-4)^2 - 2$

Vertex: (4, –2)
Axis of symmetry: $x = 4$

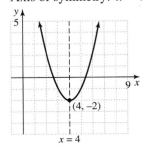

48. $f(x) = -3(x-1)^2 + 1$

Vertex: (1, 1)
Axis of symmetry: $x = 1$

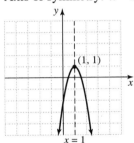

49. $f(x) = x^2 + 10x + 25$

$$x = -\frac{b}{2a} = \frac{-10}{2(1)} = -5$$

$$f(-5) = (-5)^2 + 10(-5) + 25 = 0$$

Vertex: $(-5, 0)$

$$x^2 + 10x + 25 = 0$$

$$(x+5)^2 = 0$$

$$x + 5 = 0$$

$$x = -5$$

x-intercept: $(-5, 0)$

$f(0) = 25$ so the y-intercept is $(0, 25)$.

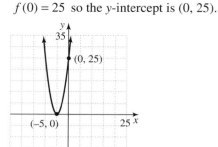

50. $f(x) = -x^2 + 6x - 9$

$$x = -\frac{b}{2a} = \frac{-6}{2(-1)} = 3$$

$$f(3) = -(3)^2 + 6(3) - 9 = 0$$

Vertex: $(3, 0)$

x-intercept: $(3, 0)$

$f(0) = -9$

y-intercept: $(0, -9)$

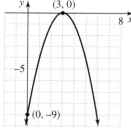

51. $f(x) = 4x^2 - 1$

$$x = -\frac{b}{2a} = \frac{-0}{2(4)} = 0$$

$$f(0) = 4(0)^2 - 1 = -1$$

Vertex: $(0, -1)$

$$4x^2 - 1 = 0$$

$$(2x+1)(2x-1) = 0$$

$$x = -\frac{1}{2} \quad \text{or} \quad x = \frac{1}{2}$$

x-intercepts: $\left(-\frac{1}{2}, 0\right), \left(\frac{1}{2}, 0\right)$

$f(0) = -1$

y-intercept: $(0, -1)$

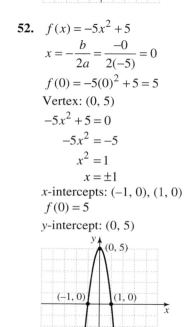

52. $f(x) = -5x^2 + 5$

$$x = -\frac{b}{2a} = \frac{-0}{2(-5)} = 0$$

$$f(0) = -5(0)^2 + 5 = 5$$

Vertex: $(0, 5)$

$$-5x^2 + 5 = 0$$

$$-5x^2 = -5$$

$$x^2 = 1$$

$$x = \pm 1$$

x-intercepts: $(-1, 0), (1, 0)$

$f(0) = 5$

y-intercept: $(0, 5)$

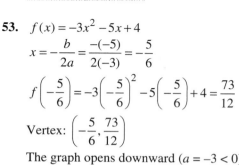

53. $f(x) = -3x^2 - 5x + 4$

$$x = -\frac{b}{2a} = \frac{-(-5)}{2(-3)} = -\frac{5}{6}$$

$$f\left(-\frac{5}{6}\right) = -3\left(-\frac{5}{6}\right)^2 - 5\left(-\frac{5}{6}\right) + 4 = \frac{73}{12}$$

Vertex: $\left(-\frac{5}{6}, \frac{73}{12}\right)$

The graph opens downward ($a = -3 < 0$).

$f(0) = 4 \Rightarrow y$-intercept: $(0, 4)$

$$-3x^2 - 5x + 4 = 0$$

$$x = \frac{5 \pm \sqrt{(-5)^2 - 4(-3)(4)}}{2(-3)}$$

$$= \frac{5 \pm \sqrt{73}}{-6}$$

$$\approx -2.2573 \text{ or } 0.5907$$

x-intercepts: (–2.3, 0), (0.6, 0)

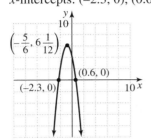

54. $h(t) = -16t^2 + 120t + 300$

    a.
    $$350 = -16t^2 + 120t + 300$$
    $$16t^2 - 120t + 50 = 0$$
    $$8t^2 - 60t + 25 = 0$$
    $$a = 8, b = -60, c = 25$$
    $$t = \frac{60 \pm \sqrt{(-60)^2 - 4(8)(25)}}{2(8)}$$
    $$= \frac{60 \pm \sqrt{2800}}{16}$$
    $$\approx 0.4 \text{ second and } 7.1 \text{ seconds}$$

    b. The object will be at 350 feet on the way up and on the way down.

55. Let *x* = one number; then
    420 – *x* = the other number.
    Let $f(x)$ represent their product.
    $$f(x) = x(420 - x)$$
    $$= 420x - x^2$$
    $$= -x^2 + 420x$$
    $$x = -\frac{b}{2a} = \frac{-420}{2(-1)} = 210;$$
    $$420 - x = 420 - 210 = 210$$
    Therefore, the numbers are both 210.

56. $y = a(x - h)^2 + k$

    vertex (–3, 7) with $a = -\frac{7}{9}$ gives

    $$y = -\frac{7}{9}(x + 3)^2 + 7.$$

57. **a.**

    The linear regression equation is
    $y = 2.753x + 223.950$.

    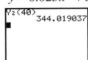

    The population of the U.S. is predicted to be 334 million in 2020.

    b.

    The quadratic regression equation is
    $y = 0.025x^2 + 1.932x + 227.01$.

    The population of the U.S. is predicted to be 344 million in 2020.

58. **a.**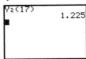

    The linear regression equation is
    $y = 0.47x + 0.11$.

    The sales of digital music online in 2017 is predicted to be $8.1 billion.

    b.

    The quadratic regression equation is
    $y = -0.125x^2 + 2.845x - 11.015$.

    The sales of digital music online in 2017 is predicted to be $1.225 billion.

59. $x^2 - x - 30 = 0$
$(x+5)(x-6) = 0$
$x+5 = 0$ or $x-6 = 0$
$x = -5$ or $x = 6$
The solutions are –5 and 6.

60. $(9n+1)^2 = 9$
$9n+1 = \pm\sqrt{9}$
$9n+1 = \pm 3$
$9n = -1 \pm 3$
$n = \dfrac{-1 \pm 3}{9} = \dfrac{2}{9}, -\dfrac{4}{9}$
The solutions are $-\dfrac{4}{9}$ and $\dfrac{2}{9}$.

61. $x^2 + x + 7 = 0$
$x^2 + x = -7$
$x^2 + x + \left(\dfrac{1}{2}\right)^2 = -7 + \dfrac{1}{4}$
$\left(x + \dfrac{1}{2}\right)^2 = -\dfrac{27}{4}$
$x + \dfrac{1}{2} = \pm\sqrt{-\dfrac{27}{4}}$
$x + \dfrac{1}{2} = \pm\dfrac{i\sqrt{9 \cdot 3}}{2}$
$x + \dfrac{1}{2} = \pm\dfrac{3i\sqrt{3}}{2}$
$x = -\dfrac{1}{2} \pm \dfrac{3i\sqrt{3}}{2} = \dfrac{-1 \pm 3i\sqrt{3}}{2}$
The solutions are $\dfrac{-1 + 3i\sqrt{3}}{2}$ and $\dfrac{-1 - 3i\sqrt{3}}{2}$.

62. $(3x-4)^2 = 10x$
$9x^2 - 24x + 16 = 10x$
$9x^2 - 34x = -16$
$x^2 - \dfrac{34}{9}x = -\dfrac{16}{9}$
$x^2 - \dfrac{34}{9}x + \left(\dfrac{-\dfrac{34}{9}}{2}\right)^2 = -\dfrac{16}{9} + \dfrac{289}{81}$
$\left(x - \dfrac{17}{9}\right)^2 = \dfrac{145}{81}$
$x - \dfrac{17}{9} = \pm\sqrt{\dfrac{145}{81}}$
$x - \dfrac{17}{9} = \pm\dfrac{\sqrt{145}}{9}$
$x = \dfrac{17 \pm \sqrt{145}}{9}$
The solutions are $\dfrac{17 + \sqrt{145}}{9}$ and $\dfrac{17 - \sqrt{145}}{9}$.

63. $x^2 + 11 = 0$
$a = 1, b = 0, c = 11$
$x = \dfrac{0 \pm \sqrt{(0)^2 - 4(1)(11)}}{2(1)}$
$= \dfrac{\pm\sqrt{-44}}{2}$
$= \dfrac{\pm 2i\sqrt{11}}{2}$
$= \pm i\sqrt{11}$
The solutions are $-i\sqrt{11}$ and $i\sqrt{11}$.

64. $x^2 + 7 = 0$
$x^2 = -7$
$\sqrt{x^2} = \pm\sqrt{-7}$
$x = \pm i\sqrt{7}$
The solutions are $-i\sqrt{7}$ and $i\sqrt{7}$.

65.
$$(5a-2)^2 - a = 0$$
$$25a^2 - 20a + 4 - a = 0$$
$$25a^2 - 21a + 4 = 0$$
$$a = \frac{21 \pm \sqrt{(-21)^2 - 4(25)(4)}}{2(25)}$$
$$= \frac{21 \pm \sqrt{441 - 400}}{50}$$
$$= \frac{21 \pm \sqrt{41}}{50}$$

The solutions are $\dfrac{21+\sqrt{41}}{50}$ and $\dfrac{21-\sqrt{41}}{50}$.

66.
$$\frac{7}{8} = \frac{8}{x^2}$$
$$7x^2 = 64$$
$$x^2 = \frac{64}{7}$$
$$x = \pm\sqrt{\frac{64}{7}}$$
$$x = \pm\frac{8}{\sqrt{7}} = \pm\frac{8 \cdot \sqrt{7}}{\sqrt{7} \cdot \sqrt{7}} = \pm\frac{8\sqrt{7}}{7}$$

The solutions are $-\dfrac{8\sqrt{7}}{7}$ and $\dfrac{8\sqrt{7}}{7}$.

67.
$$x^{2/3} - 6x^{1/3} = -8$$
$$x^{2/3} - 6x^{1/3} + 8 = 0$$
Let $y = x^{1/3}$. Then $y^2 = x^{2/3}$ and
$$y^2 - 6y + 8 = 0$$
$$(y-4)(y-2) = 0$$
$$y - 4 = 0 \quad \text{or} \quad y - 2 = 0$$
$$y = 4 \quad \text{or} \quad y = 2$$
$$x^{1/3} = 4 \quad \text{or} \quad x^{1/3} = 2$$
$$x = 64 \quad \text{or} \quad x = 8$$
The solutions are 8 and 64.

68.
$$(2x-3)(4x+5) \geq 0$$
$$2x - 3 = 0 \quad \text{or} \quad 4x + 5 = 0$$
$$x = \frac{3}{2} \quad \text{or} \quad x = -\frac{5}{4}$$

Region	Test Point	$(2x-3)(4x+5) \geq 0$ Result
$A: \left(-\infty, -\dfrac{5}{4}\right]$	-2	$(-7)(-3) \geq 0$ True
$B: \left[-\dfrac{5}{4}, \dfrac{3}{2}\right]$	0	$(-3)(5) \geq 0$ False
$C: \left[\dfrac{3}{2}, \infty\right)$	3	$(3)(17) \geq 0$ True

Solution: $\left(-\infty, -\dfrac{5}{4}\right] \cup \left[\dfrac{3}{2}, \infty\right)$

69.
$$\frac{x(x+5)}{4x-3} \geq 0$$
$$x = 0 \quad \text{or} \quad x + 5 = 0 \quad \text{or} \quad 4x - 3 = 0$$
$$x = -5 \quad \text{or} \quad x = \frac{3}{4}$$

Region	Test Point	$\dfrac{x(x+5)}{4x-3} \geq 0$ Result
$A: (-\infty, -5]$	-6	$\dfrac{-6(-1)}{-27} \geq 0$ False
$B: [-5, 0]$	-1	$\dfrac{-1(4)}{-7} \geq 0$ True
$C: \left[0, \dfrac{3}{4}\right)$	$\dfrac{1}{2}$	$\dfrac{\frac{1}{2}\left(\frac{11}{2}\right)}{-1} \geq 0$ False
$D: \left(\dfrac{3}{4}, \infty\right)$	1	$\dfrac{1(6)}{1} \geq 0$ True

Solution: $[-5, 0] \cup \left(\dfrac{3}{4}, \infty\right)$

70. $\dfrac{3}{x-2} > 2$

The denominator is equal to 0 when $x - 2 = 0$, or $x = 2$.

$$\dfrac{3}{x-2} = 2$$
$$3 = 2(x-2)$$
$$3 = 2x - 4$$
$$7 = 2x$$
$$\dfrac{7}{2} = x$$

Region	Test Point	$\dfrac{3}{x-2} > 2$ Result
$A: (-\infty, 2)$	0	$\dfrac{3}{-2} > 2$ False
$B: \left(2, \dfrac{7}{2}\right)$	3	$\dfrac{3}{1} > 2$ True
$C: \left(\dfrac{7}{2}, \infty\right)$	5	$\dfrac{3}{3} > 2$ False

Solution: $\left(2, \dfrac{7}{2}\right)$

71. $y = -32x^2 + 1733x + 76{,}362$

a. $x = 2015 - 2000 = 15$

$$y = -32(15)^2 + 1733(15) + 76{,}362$$
$$= 95{,}175$$

The passenger traffic will be approximately 95,157 thousand.

b. Let $y = 99{,}000$.

$$99{,}000 = -32x^2 + 1733x + 76{,}362$$
$$0 = -32x^2 + 1733x - 22{,}638$$

$a = -32,\ b = 1733,\ c = -22{,}638$

$$x = \dfrac{-b \pm \sqrt{b^2 - 4ac}}{2a}$$

$$x = \dfrac{-1733 \pm \sqrt{1733^2 - 4(-32)(-22{,}638)}}{2(-32)}$$
$$= \dfrac{-1733 \pm \sqrt{105{,}625}}{-64}$$
$$= \dfrac{-1733 \pm 325}{-64}$$
$$= 22 \text{ or approximately } 32$$

Choosing the first root, $x = 22$, we see that there will first be 99,000 thousand passengers in 2022.

Chapter 8 Test

1.
$$5x^2 - 2x = 7$$
$$5x^2 - 2x - 7 = 0$$
$$(5x - 7)(x + 1) = 0$$
$$5x - 7 = 0 \quad \text{or} \quad x + 1 = 0$$
$$x = \dfrac{7}{5} \quad \text{or} \quad x = -1$$

The solutions are -1 and $\dfrac{7}{5}$.

2. $(x+1)^2 = 10$
$$x + 1 = \pm\sqrt{10}$$
$$x = -1 \pm \sqrt{10}$$

The solutions are $-1 + \sqrt{10}$ and $-1 - \sqrt{10}$.

3. $m^2 - m + 8 = 0$
$a = 1,\ b = -1,\ c = 8$

$$m = \dfrac{1 \pm \sqrt{(-1)^2 - 4(1)(8)}}{2(1)}$$
$$= \dfrac{1 \pm \sqrt{1 - 32}}{2}$$
$$= \dfrac{1 \pm \sqrt{-31}}{2}$$
$$= \dfrac{1 \pm i\sqrt{31}}{2}$$

The solutions are $\dfrac{1 + i\sqrt{31}}{2}$ and $\dfrac{1 - i\sqrt{31}}{2}$.

4. $u^2 - 6u + 2 = 0$

$a = 1, b = -6, c = 2$

$u = \dfrac{-(-6) \pm \sqrt{(-6)^2 - 4(1)(2)}}{2(1)}$

$ = \dfrac{6 \pm \sqrt{36 - 8}}{2}$

$ = \dfrac{6 \pm \sqrt{28}}{2}$

$ = \dfrac{6 \pm 2\sqrt{7}}{2}$

$ = 3 \pm \sqrt{7}$

The solutions are $3 + \sqrt{7}$ and $3 - \sqrt{7}$.

5. $7x^2 + 8x + 1 = 0$

$(7x + 1)(x + 1) = 0$

$7x + 1 = 0 \quad$ or $\quad x + 1 = 0$

$7x = -1 \qquad\qquad x = -1$

$x = -\dfrac{1}{7}$

The solutions are $-\dfrac{1}{7}$ and -1.

6. $y^2 - 3y = 5$

$y^2 - 3y - 5 = 0$

$a = 1, b = -3, c = -5$

$y = \dfrac{3 \pm \sqrt{(-3)^2 - 4(1)(-5)}}{2(1)}$

$ = \dfrac{3 \pm \sqrt{9 + 20}}{2}$

$ = \dfrac{3 \pm \sqrt{29}}{2}$

The solutions are $\dfrac{3 + \sqrt{29}}{2}$ and $\dfrac{3 - \sqrt{29}}{2}$.

7. $\dfrac{4}{x+2} + \dfrac{2x}{x-2} = \dfrac{6}{x^2 - 4}$

$\dfrac{4}{x+2} + \dfrac{2x}{x-2} = \dfrac{6}{(x+2)(x-2)}$

$4(x-2) + 2x(x+2) = 6$

$4x - 8 + 2x^2 + 4x = 6$

$2x^2 + 8x - 14 = 0$

$x^2 + 4x - 7 = 0$

$a = 1, b = 4, c = -7$

$x = \dfrac{-4 \pm \sqrt{(4)^2 - 4(1)(-7)}}{2(1)}$

$ = \dfrac{-4 \pm \sqrt{16 + 28}}{2}$

$ = \dfrac{-4 \pm \sqrt{44}}{2}$

$ = \dfrac{-4 \pm 2\sqrt{11}}{2}$

$ = -2 \pm \sqrt{11}$

The solutions are $-2 + \sqrt{11}$ and $-2 - \sqrt{11}$.

8. $x^5 + 3x^4 = x + 3$

$x^5 + 3x^4 - x - 3 = 0$

$x^4(x+3) - 1(x+3) = 0$

$(x+3)(x^4 - 1) = 0$

$(x+3)(x^2 + 1)(x^2 - 1) = 0$

$x + 3 = 0 \quad$ or $\; x^2 + 1 = 0 \quad$ or $\; x^2 - 1 = 0$

$\quad x = -3$ or $\qquad x^2 = -1$ or $\qquad x^2 = 1$

$\qquad\qquad\qquad\quad x = \pm i$ or $\qquad x = \pm 1$

The solutions are $-3, -1, 1, -i,$ and i.

9. $x^6 + 1 = x^4 + x^2$

$x^6 - x^4 - x^2 + 1 = 0$

$x^4(x^2 - 1) - (x^2 - 1) = 0$

$(x^4 - 1)(x^2 - 1) = 0$

$(x^2 + 1)(x^2 - 1)(x+1)(x-1) = 0$

$(x^2 + 1)(x+1)^2(x-1)^2 = 0$

$x^2 + 1 = 0 \quad$ or $\; x + 1 = 0 \quad$ or $\; x - 1 = 0$

$\quad x^2 = -1 \qquad\qquad x = -1 \qquad\qquad x = 1$

$\quad x = \pm i$

The solutions are $-i, i, -1,$ and 1.

10. $(x+1)^2 - 15(x+1) + 56 = 0$

Let $y = x + 1$. Then $y^2 = (x+1)^2$ and

$y^2 - 15y + 56 = 0$

$(y - 8)(y - 7) = 0$

$\quad y = 8 \quad$ or $\qquad y = 7$

$x + 1 = 8 \quad$ or $\; x + 1 = 7$

$\quad x = 7 \quad$ or $\qquad x = 6$

The solutions are 6 and 7.

11.
$$x^2 - 6x = -2$$
$$x^2 - 6x + \left(\frac{-6}{2}\right)^2 = -2 + 9$$
$$x^2 - 6x + 9 = 7$$
$$(x-3)^2 = 7$$
$$x - 3 = \pm\sqrt{7}$$
$$x = 3 \pm \sqrt{7}$$
The solutions are $3 + \sqrt{7}$ and $3 - \sqrt{7}$.

12.
$$2a^2 + 5 = 4a$$
$$2a^2 - 4a = -5$$
$$a^2 - 2a = -\frac{5}{2}$$
$$a^2 - 2a + \left(\frac{-2}{2}\right)^2 = -\frac{5}{2} + 1$$
$$a^2 - 2a + 1 = -\frac{3}{2}$$
$$(a-1)^2 = -\frac{3}{2}$$
$$a - 1 = \pm\sqrt{-\frac{3}{2}} = \pm\frac{i\sqrt{3}}{\sqrt{2}}$$
$$a - 1 = \pm\frac{i\sqrt{6}}{2}$$
$$a = 1 \pm \frac{i\sqrt{6}}{2} \quad \text{or} \quad \frac{2 \pm i\sqrt{6}}{2}$$
The solutions are $\dfrac{2 + i\sqrt{6}}{2}$ and $\dfrac{2 - i\sqrt{6}}{2}$.

13.
$$2x^2 - 7x > 15$$
$$2x^2 - 7x - 15 > 0$$
$$(2x+3)(x-5) > 0$$
$$2x + 3 = 0 \quad \text{or} \quad x - 5 = 0$$
$$x = -\frac{3}{2} \quad \text{or} \quad x = 5$$

Region	Test Point	$(2x+3)(x-5) > 0$ Result
A: $\left(-\infty, -\frac{3}{2}\right)$	-2	$(-1)(-7) > 0$ True
B: $\left(-\frac{3}{2}, 5\right)$	0	$(3)(-5) > 0$ False
C: $(5, \infty)$	6	$(15)(1) > 0$ True

Solution: $\left(-\infty, -\frac{3}{2}\right) \cup (5, \infty)$

Copyright © 2014 Pearson Education, Inc.

14. $(x^2 - 16)(x^2 - 25) \geq 0$
   $(x + 4)(x - 4)(x + 5)(x - 5) \geq 0$
   $x + 4 = 0$ or $x - 4 = 0$ or $x + 5 = 0$ or $x - 5 = 0$
   $x = -4$ or $x = 4$ or $x = -5$ or $x = 5$

Region	Test Point	$(x + 4)(x - 4)(x + 5)(x - 5) \geq 0$ Result
A: $(-\infty, -5]$	-6	$(-2)(-10)(-1)(-11) \geq 0$ True
B: $[-5, -4]$	$-\dfrac{9}{2}$	$\left(-\dfrac{1}{2}\right)\left(-\dfrac{17}{2}\right)\left(\dfrac{1}{2}\right)\left(-\dfrac{19}{2}\right) \geq 0$ False
C: $[-4, 4]$	0	$(4)(-4)(5)(-5) \geq 0$ True
D: $[4, 5]$	$\dfrac{9}{2}$	$\left(\dfrac{17}{2}\right)\left(\dfrac{1}{2}\right)\left(\dfrac{19}{2}\right)\left(-\dfrac{1}{2}\right) \geq 0$ False
E: $[5, \infty)$	6	$(10)(2)(11)(1) \geq 0$ True

Solution: $(-\infty, -5] \cup [-4, 4] \cup [5, \infty)$

15. $\dfrac{5}{x+3} < 1$

The denominator is equal to 0 when $x + 3 = 0$, or $x = -3$.

$\dfrac{5}{x+3} = 1$
   $5 = x + 3$ so $x = 2$

Region	Test Point	$\dfrac{5}{x+3} < 1$ Result
A: $(-\infty, -3)$	-4	$\dfrac{5}{-1} < 1$ True
B: $(-3, 2)$	0	$\dfrac{5}{3} < 1$ False
C: $(2, \infty)$	3	$\dfrac{5}{6} < 1$ True

Solution: $(-\infty, -3) \cup (2, \infty)$

16. $\dfrac{7x-14}{x^2-9} \le 0$

$\dfrac{7(x-2)}{(x+3)(x-3)} \le 0$

$x-2=0$ or $x+3=0$ or $x-3=0$
    $x=2$ or $x=-3$ or $x=3$

Region	Test Point	$\dfrac{7(x-2)}{(x+3)(x-3)} \le 0$ Result
A: $(-\infty,-3)$	-4	$\dfrac{7(-6)}{(-1)(-7)} \le 0$ True
B: $(-3, 2]$	0	$\dfrac{7(-2)}{(3)(-3)} \le 0$ False
C: $[2, 3)$	$\dfrac{5}{2}$	$\dfrac{7\left(\frac{1}{2}\right)}{\left(\frac{11}{2}\right)\left(-\frac{1}{2}\right)} \le 0$ True
D: $(3, \infty)$	4	$\dfrac{7(2)}{(7)(1)} \le 0$ False

Solution: $(-\infty,-3) \cup [2, 3)$

17. $f(x) = 3x^2$
Vertex: $(0, 0)$

18. $G(x) = -2(x-1)^2 + 5$

Vertex: (1, 5)

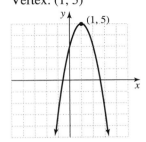

19. $h(x) = x^2 - 4x + 4$

$x = -\dfrac{b}{2a} = \dfrac{-(-4)}{2(1)} = 2$

$h(2) = (2)^2 - 4(2) + 4 = 0$

Vertex: (2, 0)

$h(0) = 4 \Rightarrow$ *y*-intercept: (0, 4)

x-intercept: (2, 0)

20. $F(x) = 2x^2 - 8x + 9$

$x = -\dfrac{b}{2a} = \dfrac{-(-8)}{2(2)} = 2$

$F(2) = 2(2)^2 - 8(2) + 9 = 1$

Vertex: (2, 1)

$F(0) = 9 \Rightarrow$ *y*-intercept: (0, 9)

$2x^2 - 8x + 9 = 0$

$a = 2, b = -8, c = 9$

$x = \dfrac{8 \pm \sqrt{(-8)^2 - 4(2)(9)}}{2(2)}$

$= \dfrac{8 \pm \sqrt{-8}}{4}$

which yields non-real solutons.

Therefore, there are no *x*-intercepts.

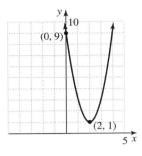

21. Let t = time for Sandy alone. Then
$t - 2$ = time for Dave alone.

$\dfrac{1}{t} + \dfrac{1}{t-2} = \dfrac{1}{4}$

$4(t-2) + 4t = t(t-2)$

$4t - 8 + 4t = t^2 - 2t$

$0 = t^2 - 10t + 8$

$a = 1, b = -10, c = 8$

$t = \dfrac{10 \pm \sqrt{(-10)^2 - 4(1)(8)}}{2(1)}$

$= \dfrac{10 \pm \sqrt{68}}{2}$

$= \dfrac{10 \pm 2\sqrt{17}}{2}$

$= 5 \pm \sqrt{17}$

≈ 9.12 or 0.88 (discard)

It takes her about 9.12 hours.

22. $s(t) = -16t^2 + 32t + 256$

a. $t = -\dfrac{b}{2a} = \dfrac{-32}{2(-16)} = 1$

$s(1) = -16(1)^2 + 32(1) + 256 = 272$

Vertex: (1, 272)

The maximum height is 272 feet.

b. $-16t^2 + 32t + 256 = 0$

$$t^2 - 2t - 16 = 0$$

$$a = 1, b = -2, c = -16$$

$$t = \frac{2 \pm \sqrt{(-2)^2 - 4(1)(-16)}}{2(1)}$$

$$= \frac{2 \pm \sqrt{68}}{2}$$

$$= \frac{2 \pm 2\sqrt{17}}{2}$$

$$= 1 \pm \sqrt{17}$$

$$\approx -3.12 \text{ and } 5.12$$

Disregard the negative. The stone will hit the water in about 5.12 seconds.

23.

$$a^2 + b^2 = c^2$$

$$x^2 + (x+8)^2 = (20)^2$$

$$x^2 + (x^2 + 16x + 64) = 400$$

$$2x^2 + 16x - 336 = 0$$

$$x^2 + 8x - 168 = 0$$

$$a = 1, b = 8, c = -168$$

$$x = \frac{-8 \pm \sqrt{(8)^2 - 4(1)(-168)}}{2(1)}$$

$$= \frac{-8 \pm \sqrt{736}}{2}$$

$$\approx -17.565 \text{ or } 9.565$$

Disregard the negative.

$$x \approx 9.6$$

$$x + 8 \approx 9.6 + 8 = 17.6$$

$$17.6 + 9.6 = 27.2$$

$$27.2 - 20 = 7.2$$

They would save about 7.2 feet.

24. a. Let $x =$ the number of years since 1990.

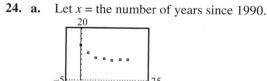

A quadratic model appears to fit the data.

b.

The quadratic regression equation is

$$y = 0.016x^2 - 0.417x + 16.581.$$

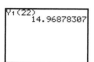

The birth rate for 2012 is predicted to be 14.97 births per 1000 people.

Chapter 8 Cumulative Review

1. a. $5 + y \geq 7$

b. $11 \neq z$

c. $20 < 5 - 2x$

2. $|3x - 2| = -5$ which is impossible. Thus, there is no solution, or $\varnothing$.

3. $m = \dfrac{5-3}{2-0} = \dfrac{2}{2} = 1$

Plot the given points and draw a line through them.

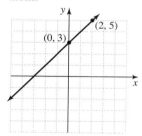

4. $\begin{cases} -6x + y = 5 & (1) \\ 4x - 2y = 6 & (2) \end{cases}$

Multiply E1 by 2 and add to E2.

$$-12x + 2y = 10$$
$$\underline{4x - 2y = 6}$$
$$-8x \quad\quad = 16$$
$$x = -2$$

Replace x with -2 in E1.

$$-6(-2) + y = 5$$
$$12 + y = 5$$
$$y = -7$$

The solution is $(-2, -7)$.

5. $\begin{cases} x - 5y = -12 & (1) \\ -x + y = 4 & (2) \end{cases}$

Add E1 and E2.

$$-4y = -8$$
$$y = 2$$

Replace y with 2 in E1.

$$x - 5(2) = -12$$
$$x - 10 = -12$$
$$x = -2$$

The solution is $(-2, 2)$.

6. a. $(a^{-2}bc^3)^{-3} = (a^{-2})^{-3}b^{-3}(c^3)^{-3}$
$$= a^6 b^{-3} c^{-9}$$
$$= \frac{a^6}{b^3 c^9}$$

b. $\left(\dfrac{a^{-4}b^2}{c^3}\right)^{-2} = \dfrac{(a^{-4})^{-2}(b^2)^{-2}}{(c^3)^{-2}}$
$$= \frac{a^8 b^{-4}}{c^{-6}}$$
$$= \frac{a^8 c^6}{b^4}$$

c. $\left(\dfrac{3a^8 b^2}{12a^5 b^5}\right)^{-2} = \left(\dfrac{a^3}{4b^3}\right)^{-2}$
$$= \frac{(a^3)^{-2}}{4^{-2}(b^3)^{-2}}$$
$$= \frac{4^2 a^{-6}}{b^{-6}}$$
$$= \frac{16 b^6}{a^6}$$

7. a. $(2x-7)(3x-4) = 6x^2 - 8x - 21x + 28$
$$= 6x^2 - 29x + 28$$

b. $(3x^2 + y)(5x^2 - 2y)$
$$= 15x^4 - 6x^2 y + 5x^2 y - 2y^2$$
$$= 15x^4 - x^2 y - 2y^2$$

8. a. $(4a-3)(7a-2) = 28a^2 - 8a - 21a + 6$
$$= 28a^2 - 29a + 6$$

b. $(2a+b)(3a-5b)$
$$= 6a^2 - 10ab + 3ab - 5b^2$$
$$= 6a^2 - 7ab - 5b^2$$

9. a. $8x^2 + 4 = 4(2x^2 + 1)$

b. $5y - 2z^4$ is a prime polynomial.

c. $6x^2 - 3x^3 + 12x^4 = 3x^2(2 - x + 4x^2)$

10. a. $9x^3 + 27x^2 - 15x = 3x(3x^2 + 9x - 5)$

b. $2x(3y-2) - 5(3y-2)$
$$= (3y-2)(2x-5)$$

c. $2xy + 6x - y - 3 = 2x(y+3) - 1(y+3)$
$$= (y+3)(2x-1)$$

11. $x^2 - 12x + 35 = (x-5)(x-7)$

12. $x^2 - 2x - 48 = (x+6)(x-8)$

13. $12a^2 x - 12abx + 3b^2 x = 3x(4a^2 - 4ab + b^2)$
$$= 3x(2a-b)(2a-b)$$
$$= 3x(2a-b)^2$$

14. $2ax^2 - 12axy + 18ay^2 = 2a(x^2 - 6xy + 9y^2)$
$$= 2a(x-3y)(x-3y)$$
$$= 2a(x-3y)^2$$

15. $3(x^2 + 4) + 5 = -6(x^2 + 2x) + 13$
$$3x^2 + 12 + 5 = -6x^2 - 12x + 13$$
$$9x^2 + 12x + 4 = 0$$
$$(3x+2)^2 = 0$$
$$3x + 2 = 0$$
$$3x = -2$$
$$x = -\frac{2}{3}$$

The solutions is $-\dfrac{2}{3}$.

16. $2(a^2 + 2) - 8 = -2a(a-2) - 5$
$$2a^2 + 4 - 8 = -2a^2 + 4a - 5$$
$$4a^2 - 4a + 1 = 0$$
$$(2a-1)^2 = 0$$
$$2a - 1 = 0$$
$$2a = 1$$
$$a = \frac{1}{2}$$

The solution is $\dfrac{1}{2}$.

17.
$$x^3 = 4x$$
$$x^3 - 4x = 0$$
$$x(x^2 - 4) = 0$$
$$x(x+2)(x-2) = 0$$
$$x = 0 \text{ or } x+2 = 0 \quad \text{or } x-2 = 0$$
$$x = -2 \text{ or } \quad x = 2$$
The solutions are -2, 0, and 2.

18. $f(x) = x^2 + x - 12$
$$x = -\frac{b}{2a} = \frac{-1}{2(1)} = -\frac{1}{2}$$
$$f\left(-\frac{1}{2}\right) = \left(-\frac{1}{2}\right)^2 + \left(-\frac{1}{2}\right) - 12$$
$$= \frac{1}{4} - \frac{1}{2} - 12$$
$$= -\frac{49}{4}$$
Vertex: $\left(-\frac{1}{2}, -\frac{49}{4}\right)$
$$x^2 + x - 12 = 0$$
$$(x+4)(x-3) = 0$$
$$x + 4 = 0 \quad \text{or} \quad x - 3 = 0$$
$$x = -4 \qquad x = 3$$
x-intercepts: $(-4, 0)$, $(3, 0)$
$$f(0) = 0^2 + 0 - 12 = -12$$
y-intercept: $(0, -12)$

19. $\dfrac{2x^2}{10x^3 - 2x^2} = \dfrac{2x^2}{2x^2(5x-1)} = \dfrac{1}{5x-1}$

20. $\dfrac{x^2 - 4x + 4}{2-x} = \dfrac{(x-2)^2}{-(x-2)} = \dfrac{x-2}{-1} = 2 - x$

21. $\dfrac{2x-1}{2x^2 - 9x - 5} + \dfrac{x+3}{6x^2 - x - 2}$
$$= \frac{2x-1}{(2x+1)(x-5)} + \frac{x+3}{(2x+1)(3x-2)}$$
$$= \frac{(2x-1)(3x-2) + (x+3)(x-5)}{(2x+1)(x-5)(3x-2)}$$
$$= \frac{(6x^2 - 4x - 3x + 2) + (x^2 - 5x + 3x - 15)}{(2x+1)(x-5)(3x-2)}$$
$$= \frac{7x^2 - 9x - 13}{(2x+1)(x-5)(3x-2)}$$

22. $\dfrac{a+1}{a^2 - 6a + 8} - \dfrac{3}{16 - a^2}$
$$= \frac{a+1}{(a-4)(a-2)} - \frac{3}{(4+a)(4-a)}$$
$$= \frac{a+1}{(a-4)(a-2)} + \frac{3}{(4+a)(a-4)}$$
$$= \frac{(a+1)(a+4) + 3(a-2)}{(a-4)(a-2)(a+4)}$$
$$= \frac{(a^2 + 4a + a + 4) + 3a - 6}{(a-4)(a-2)(a+4)}$$
$$= \frac{a^2 + 8a - 2}{(a-4)(a-2)(a+4)}$$

23. $\dfrac{x^{-1} + 2xy^{-1}}{x^{-2} - x^{-2}y^{-1}} = \dfrac{\dfrac{1}{x} + \dfrac{2x}{y}}{\dfrac{1}{x^2} - \dfrac{1}{x^2 y}}$
$$= \frac{\left(\dfrac{1}{x} + \dfrac{2x}{y}\right)x^2 y}{\left(\dfrac{1}{x^2} - \dfrac{1}{x^2 y}\right)x^2 y}$$
$$= \frac{xy + 2x^3}{y - 1}$$

24. $\dfrac{(2a)^{-1} + b^{-1}}{a^{-1} + (2b)^{-1}} = \dfrac{\dfrac{1}{2a} + \dfrac{1}{b}}{\dfrac{1}{a} + \dfrac{1}{2b}}$
$$= \frac{\left(\dfrac{1}{2a} + \dfrac{1}{b}\right)2ab}{\left(\dfrac{1}{a} + \dfrac{1}{2b}\right)2ab}$$
$$= \frac{b + 2a}{2b + a}$$
$$= \frac{2a + b}{a + 2b}$$

25. $\dfrac{3x^5 y^2 - 15x^3 y - x^2 y - 6x}{x^2 y}$
$$= \frac{3x^5 y^2}{x^2 y} - \frac{15x^3 y}{x^2 y} - \frac{x^2 y}{x^2 y} - \frac{6x}{x^2 y}$$
$$= 3x^3 y - 15x - 1 - \frac{6}{xy}$$

26.

$$x+3 \overline{)\begin{array}{l} x^2-6x+8 \\ x^3-3x^2-10x+24 \end{array}}$$

$$\underline{x^3+3x^2}$$
$$-6x^2-10x$$
$$\underline{-6x^2-18x}$$
$$8x+24$$
$$\underline{8x+24}$$
$$0$$

Answer: x^2-6x+8

27. $P(x)=2x^3-4x^2+5$

    a. $P(2)=2(2)^3-4(2)^2+5$
$$=2(8)-4(4)+5$$
$$=16-16+5$$
$$=5$$

    b.

$$\begin{array}{r|rrrr} 2 & 2 & -4 & 0 & 5 \\ & & 4 & 0 & 0 \\ \hline & 2 & 0 & 0 & 5 \end{array}$$

    Thus, $P(2)=5$.

28. $P(x)=4x^3-2x^2+3$

    a. $P(-2)=4(-2)^3-2(-2)^2+3$
$$=4(-8)-2(4)+3$$
$$=-32-8+3$$
$$=-37$$

    b.

$$\begin{array}{r|rrrr} -2 & 4 & -2 & 0 & 3 \\ & & -8 & 20 & -40 \\ \hline & 4 & -10 & 20 & -37 \end{array}$$

    Thus, $P(-2)=-37$.

29.

$$\frac{4x}{5}+\frac{3}{2}=\frac{3x}{10}$$
$$10\left(\frac{4x}{5}+\frac{3}{2}\right)=10\left(\frac{3x}{10}\right)$$
$$2(4x)+5(3)=3x$$
$$8x+15=3x$$
$$5x=-15$$
$$x=-3$$

The solution is -3.

30.

$$\frac{x+3}{x^2+5x+6}=\frac{3}{2x+4}-\frac{1}{x+3}$$
$$\frac{x+3}{(x+3)(x+2)}=\frac{3}{2(x+2)}-\frac{1}{x+3}$$
$$2(x+3)=3(x+3)-2(x+2)$$
$$2x+6=3x+9-2x-4$$
$$2x+6=x+5$$
$$x=-1$$

31. Let $x=$ the number.

$$\frac{9-x}{19+x}=\frac{1}{3}$$
$$3(9-x)=1(19+x)$$
$$27-3x=19+x$$
$$-4x=-8$$
$$x=2$$

The number is 2.

32. Let $t=$ time to roof the house together.

$$\frac{1}{24}+\frac{1}{40}=\frac{1}{t}$$
$$120t\left(\frac{1}{24}+\frac{1}{40}\right)=120t\left(\frac{1}{t}\right)$$
$$5t+3t=120$$
$$8t=120$$
$$t=\frac{120}{8}=15$$

It would take them 15 hours to roof the house working together.

33. $y=kx$
$$5=k(30)$$
$$k=\frac{5}{30}=\frac{1}{6} \quad \text{and} \quad y=\frac{1}{6}x$$

34. $y=\dfrac{k}{x}$
$$8=\frac{k}{14}$$
$$k=8(14)=112 \quad \text{and} \quad y=\frac{112}{x}$$

35. a. $\sqrt{(-3)^2}=|-3|=3$

    b. $\sqrt{x^2}=|x|$

    c. $\sqrt[4]{(x-2)^4}=|x-2|$

    d. $\sqrt[3]{(-5)^3}=-5$

e. $\sqrt[5]{(2x-7)^5} = 2x - 7$

f. $\sqrt{25x^2} = \sqrt{25} \cdot \sqrt{x^2} = 5|x|$

g. $\sqrt{x^2 + 2x + 1} = \sqrt{(x+1)^2} = |x+1|$

36. a. $\sqrt{(-2)^2} = |-2| = 2$

b. $\sqrt{y^2} = |y|$

c. $\sqrt[4]{(a-3)^4} = |a-3|$

d. $\sqrt[3]{(-6)^3} = -6$

e. $\sqrt[5]{(3x-1)^5} = 3x - 1$

37. a. $\sqrt[8]{x^4} = x^{4/8} = x^{1/2} = \sqrt{x}$

b. $\sqrt[6]{25} = (25)^{1/6}$
$= (5^2)^{1/6} = 5^{2/6} = 5^{1/3} = \sqrt[3]{5}$

c. $\sqrt[4]{r^2 s^6} = (r^2 s^6)^{1/4}$
$= r^{2/4} s^{6/4}$
$= r^{1/2} s^{3/2}$
$= (rs^3)^{1/2} = \sqrt{rs^3}$

38. a. $\sqrt[4]{5^2} = 5^{2/4} = 5^{1/2} = \sqrt{5}$

b. $\sqrt[12]{x^3} = x^{3/12} = x^{1/4} = \sqrt[4]{x}$

c. $\sqrt[6]{x^2 y^4} = (x^2 y^4)^{1/6}$
$= x^{2/6} y^{4/6}$
$= x^{1/3} y^{2/3}$
$= (xy^2)^{1/3} = \sqrt[3]{xy^2}$

39. a. $\sqrt{25x^3} = \sqrt{25x^2 \cdot x} = 5x\sqrt{x}$

b. $\sqrt[3]{54x^6 y^8} = \sqrt[3]{27x^6 y^6 \cdot 2y^2}$
$= 3x^2 y^2 \sqrt[3]{2y^2}$

c. $\sqrt[4]{81z^{11}} = \sqrt[4]{81z^8 \cdot z^3} = 3z^2 \sqrt[4]{z^3}$

40. a. $\sqrt{64a^5} = \sqrt{64a^4 \cdot a} = 8a^2 \sqrt{a}$

b. $\sqrt[3]{24a^7 b^9} = \sqrt[3]{8a^6 b^9 \cdot 3a} = 2a^2 b^3 \sqrt[3]{3a}$

c. $\sqrt[4]{48x^9} = \sqrt[4]{16x^8 \cdot 3x} = 2x^2 \sqrt[4]{3x}$

41. a. $\dfrac{2}{\sqrt{5}} = \dfrac{2 \cdot \sqrt{5}}{\sqrt{5} \cdot \sqrt{5}} = \dfrac{2\sqrt{5}}{5}$

b. $\dfrac{2\sqrt{16}}{\sqrt{9x}} = \dfrac{2 \cdot 4}{3\sqrt{x}} = \dfrac{8 \cdot \sqrt{x}}{3\sqrt{x} \cdot \sqrt{x}} = \dfrac{8\sqrt{x}}{3x}$

c. $\sqrt[3]{\dfrac{1}{2}} = \dfrac{\sqrt[3]{1}}{\sqrt[3]{2}} = \dfrac{1}{\sqrt[3]{2}} = \dfrac{1 \cdot \sqrt[3]{2^2}}{\sqrt[3]{2} \cdot \sqrt[3]{2^2}} = \dfrac{\sqrt[3]{4}}{2}$

42. a. $\left(\sqrt{3} - 4\right)\left(2\sqrt{3} + 2\right)$
$= \sqrt{3} \cdot 2\sqrt{3} + 2\sqrt{3} - 4 \cdot 2\sqrt{3} - 4 \cdot 2$
$= 2(3) + 2\sqrt{3} - 8\sqrt{3} - 8$
$= 6 - 6\sqrt{3} - 8$
$= -2 - 6\sqrt{3}$

b. $\left(\sqrt{5} - x\right)^2 = \left(\sqrt{5}\right)^2 - 2 \cdot \sqrt{5} \cdot x + x^2$
$= 5 - 2x\sqrt{5} + x^2$

c. $\left(\sqrt{a} + b\right)\left(\sqrt{a} - b\right) = \left(\sqrt{a}\right)^2 - b^2$
$= a - b^2$

43. $\sqrt{2x+5} + \sqrt{2x} = 3$
$\sqrt{2x+5} = 3 - \sqrt{2x}$
$\left(\sqrt{2x+5}\right)^2 = \left(3 - \sqrt{2x}\right)^2$
$2x + 5 = 9 - 6\sqrt{2x} + 2x$
$-4 = -6\sqrt{2x}$
$(-4)^2 = \left(-6\sqrt{2x}\right)^2$
$16 = 36(2x)$
$16 = 72x$
$x = \dfrac{16}{72} = \dfrac{2}{9}$

The solution is $\dfrac{2}{9}$.

44.

$$\sqrt{x-2} = \sqrt{4x+1} - 3$$

$$\left(\sqrt{x-2}\right)^2 = \left(\sqrt{4x+1} - 3\right)^2$$

$$x - 2 = (4x+1) - 6\sqrt{4x+1} + 9$$

$$6\sqrt{4x+1} = 3x + 12$$

$$2\sqrt{4x+1} = x + 4$$

$$\left(2\sqrt{4x+1}\right)^2 = (x+4)^2$$

$$4(4x+1) = x^2 + 8x + 16$$

$$16x + 4 = x^2 + 8x + 16$$

$$0 = x^2 - 8x + 12$$

$$0 = (x-6)(x-2)$$

$$x - 6 = 0 \text{ or } x - 2 = 0$$

$$x = 6 \text{ or } \quad x = 2$$

The solutions are 2 and 6.

45. a.

$$\frac{2+i}{1-i} = \frac{(2+i)\cdot(1+i)}{(1-i)\cdot(1+i)}$$

$$= \frac{2 + 2i + 1i + i^2}{1^2 - i^2}$$

$$= \frac{2 + 3i - 1}{1 + 1}$$

$$= \frac{1+3i}{2} \text{ or } \frac{1}{2} + \frac{3}{2}i$$

b.

$$\frac{7}{3i} = \frac{7\cdot(-3i)}{3i\cdot(-3i)} = \frac{-21i}{-9i^2} = \frac{-21i}{9} = -\frac{7}{3}i$$

46. a.

$$3i(5-2i) = 15i - 6i^2$$

$$= 15i + 6$$

$$= 6 + 15i$$

b.

$$(6-5i)^2 = 6^2 - 2(6)(5i) + (5i)^2$$

$$= 36 - 60i + 25i^2$$

$$= 36 - 60i - 25$$

$$= 11 - 60i$$

c.

$$\left(\sqrt{3}+2i\right)\left(\sqrt{3}-2i\right) = \left(\sqrt{3}\right)^2 - (2i)^2$$

$$= 3 - 4i^2$$

$$= 3 + 4$$

$$= 7$$

47.

$$(x+1)^2 = 12$$

$$x + 1 = \pm\sqrt{12}$$

$$x + 1 = \pm 2\sqrt{3}$$

$$x = -1 \pm 2\sqrt{3}$$

The solutions are $-1 + 2\sqrt{3}$ and $-1 - 2\sqrt{3}$.

48.

$$(y-1)^2 = 24$$

$$y - 1 = \pm\sqrt{24}$$

$$y - 1 = \pm 2\sqrt{6}$$

$$y = 1 \pm 2\sqrt{6}$$

The solutions are $1 + 2\sqrt{6}$ and $1 - 2\sqrt{6}$.

49. $x - \sqrt{x} - 6 = 0$

Let $y = \sqrt{x}$. Then $y^2 = x$ and
$$y^2 - y - 6 = 0$$
$$(y-3)(y+2) = 0$$

$$y - 3 = 0 \text{ or } y + 2 = 0$$

$$y = 3 \text{ or } \quad y = -2$$

$$\sqrt{x} = 3 \text{ or } \sqrt{x} = -2 \text{ (can't happen)}$$

$$x = 9$$

The solution is 9.

50.

$$m^2 = 4m + 8$$

$$m^2 - 4m - 8 = 0$$

$$a = 1, b = -4, c = -8$$

$$x = \frac{4 \pm \sqrt{(-4)^2 - 4(1)(-8)}}{2(1)}$$

$$= \frac{4 \pm \sqrt{16 + 32}}{2}$$

$$= \frac{4 \pm \sqrt{48}}{2}$$

$$= \frac{4 \pm 4\sqrt{3}}{2}$$

$$= 2 \pm 2\sqrt{3}$$

The solutions are $2 + 2\sqrt{3}$ and $2 - 2\sqrt{3}$.

Chapter 9

Section 9.1 Practice Exercises

1. $f(x) = x + 2;\ g(x) = 3x + 5$

   a. $(f + g)(x) = f(x) + g(x)$
$$= (x + 2) + (3x + 5)$$
$$= 4x + 7$$

   b. $(f - g)(x) = f(x) - g(x)$
$$= (x + 2) - (3x + 5)$$
$$= x + 2 - 3x - 5$$
$$= -2x - 3$$

   c. $(f \cdot g)(x) = f(x) \cdot g(x)$
$$= (x + 2)(3x + 5)$$
$$= 3x^2 + 6x + 5x + 10$$
$$= 3x^2 + 11x + 10$$

   d. $\left(\dfrac{f}{g}\right)(x) = \dfrac{f(x)}{g(x)} = \dfrac{x + 2}{3x + 5}$, where $x \neq -\dfrac{5}{3}$.

2. $f(x) = x^2 + 1;\ g(x) = 3x - 5$

   a. $(f \circ g)(4) = f(g(4)) = f(7) = 50$
$$(g \circ f)(4) = g(f(4)) = g(17) = 46$$

   b. $(f \circ g)(x) = f(g(x))$
$$= f(3x - 5)$$
$$= (3x - 5)^2 + 1$$
$$= 9x^2 - 30x + 26$$
$$(g \circ f)(x) = g(f(x))$$
$$= g(x^2 + 1)$$
$$= 3(x^2 + 1) - 5$$
$$= 3x^2 - 2$$

3. $f(x) = x^2 + 5;\ g(x) = x + 3$

   a. $(f \circ g)(x) = f(g(x))$
$$= f(x + 3)$$
$$= (x + 3)^2 + 5$$
$$= x^2 + 6x + 14$$

   b. $(g \circ f)(x) = g(f(x))$
$$= g(x^2 + 5)$$
$$= (x^2 + 5) + 3$$
$$= x^2 + 8$$

4. $f(x) = 3x;\ g(x) = x - 4;\ h(x) = |x|$

   a. $F(x) = |x - 4|$
$$F(x) = (h \circ g)(x)$$
$$= h(g(x))$$
$$= h(x - 4)$$
$$= |x - 4|$$

   b. $G(x) = 3x - 4$
$$G(x) = (g \circ f)(x)$$
$$= g(f(x))$$
$$= g(3x)$$
$$= 3x - 4$$

Vocabulary, Readiness & Video Check 9.1

1. $(f \circ g)(x) = f(g(x));\ $ C

2. $(f \cdot g)(x) = f(x) \cdot g(x);\ $ E

3. $(f - g)(x) = f(x) - g(x);\ $ F

4. $(g \circ f)(x) = g(f(x));\ $ A

5. $\left(\dfrac{f}{g}\right)(x) = \dfrac{f(x)}{g(x)},\ g(x) \neq 0;\ $ D

6. $(f + g)(x) = f(x) + g(x);\ $ B

7. You can find $(f + g)(x)$ and then find $(f + g)(2)$ or you can find $f(2)$ and $g(2)$ and then add those results.

8. Yes, sometimes they can be equal.

Exercise Set 9.1

1. a. $(f + g)(x) = (x - 7) + (2x + 1) = 3x - 6$

   b. $(f - g)(x) = (x - 7) - (2x + 1)$
$$= x - 7 - 2x - 1$$
$$= -x - 8$$

   c. $(f \cdot g)(x) = (x - 7)(2x + 1) = 2x^2 - 13x - 7$

d. $\left(\dfrac{f}{g}\right)(x) = \dfrac{x-7}{2x+1}$, where $x \neq -\dfrac{1}{2}$.

3. a. $(f+g)(x) = (x^2+1) + 5x = x^2 + 5x + 1$

b. $(f-g)(x) = (x^2+1) - 5x = x^2 - 5x + 1$

c. $(f \cdot g)(x) = (x^2+1)(5x) = 5x^3 + 5x$

d. $\left(\dfrac{f}{g}\right)(x) = \dfrac{x^2+1}{5x}$, where $x \neq 0$

5. a. $(f+g)(x) = \sqrt[3]{x} + x + 5$

b. $(f-g)(x) = \sqrt[3]{x} - (x+5) = \sqrt{x} - x - 5$

c. $(f \cdot g)(x) = \sqrt[3]{x}(x+5)$
$= x\sqrt[3]{x} + 5\sqrt[3]{x}$

d. $\left(\dfrac{f}{g}\right)(x) = \dfrac{\sqrt[3]{x}}{x+5}$; where $x \neq -5$.

7. a. $(f+g)(x) = -3x + 5x^2$ or $5x^2 - 3x$

b. $(f-g)(x) = -3x - 5x^2$ or $-5x^2 - 3x$

c. $(f \cdot g)(x) = (-3x)(5x^2) = -15x^3$

d. $\left(\dfrac{f}{g}\right)(x) = \dfrac{-3x}{5x^2}$
$= -\dfrac{3}{5x}$, where $x \neq 0$.

9. $(f \circ g)(2) = f(g(2))$
$= f(-4)$
$= (-4)^2 - 6(-4) + 2$
$= 16 + 24 + 2$
$= 42$

11. $(g \circ f)(-1) = g(f(-1))$
$= g(9)$
$= -2(9)$
$= -18$

13. $(g \circ h)(0) = g(h(0))$
$= g(0)$
$= -2(0)$
$= 0$

15. $(f \circ g)(x) = f(g(x))$
$= f(5x)$
$= (5x)^2 + 1$
$= 25x^2 + 1$
$(g \circ f)(x) = g(f(x))$
$= g(x^2 + 1)$
$= 5(x^2 + 1)$
$= 5x^2 + 5$

17. $(f \circ g)(x) = f(g(x))$
$= f(x+7)$
$= 2(x+7) - 3$
$= 2x + 14 - 3$
$= 2x + 11$
$(g \circ f)(x) = g(f(x))$
$= g(2x - 3)$
$= (2x - 3) + 7$
$= 2x + 4$

19. $(f \circ g)(x) = f(g(x))$
$= f(-2x)$
$= (-2x)^3 + (-2x) - 2$
$= -8x^3 - 2x - 2$
$(g \circ f)(x) = g(f(x))$
$= g(x^3 + x - 2)$
$= -2(x^3 + x - 2)$
$= -2x^3 - 2x + 4$

21. $(f \circ g)(x) = f(g(x))$
$= f(10x - 3)$
$= |10x - 3|$
$(g \circ f)(x) = g(f(x)) = g(|x|) = 10|x| - 3$

23. $(f \circ g)(x) = f(g(x)) = f(-5x+2) = \sqrt{-5x+2}$
$(g \circ f)(x) = g(f(x)) = g(\sqrt{x}) = -5\sqrt{x} + 2$

25. $H(x) = (g \circ h)(x)$
$= g(h(x))$
$= g(x^2 + 2)$
$= \sqrt{x^2 + 2}$

27. $F(x) = (h \circ f)(x)$
$= h(f(x))$
$= h(3x)$
$= (3x)^2 + 2$
$= 9x^2 + 2$

29. $G(x) = (f \circ g)(x)$
$= f(g(x))$
$= f\left(\sqrt{x}\right)$
$= 3\sqrt{x}$

31. answers may vary; for example, $g(x) = x + 2$ and $f(x) = x^2$

33. answers may vary; for example, $g(x) = x + 5$ and $f(x) = \sqrt{x} + 2$

35. answers may vary; for example, $g(x) = 2x - 3$ and $f(x) = \dfrac{1}{x}$

37. $x = y + 2$
$y = x - 2$

39. $x = 3y$
$y = \dfrac{x}{3}$

41. $x = -2y - 7$
$2y = -x - 7$
$y = -\dfrac{x + 7}{2}$

43. $(f + g)(2) = f(2) + g(2) = 7 + (-1) = 6$

45. $(f \circ g)(2) = f(g(2)) = f(-1) = 4$

47. $(f \cdot g)(7) = f(7) \cdot g(7) = 1 \cdot 4 = 4$

49. $\left(\dfrac{f}{g}\right)(-1) = \dfrac{f(-1)}{g(-1)} = \dfrac{4}{-4} = -1$

51. answers may vary.

53. Profit is equal to the revenue minus the cost; $P(x) = R(x) - C(x)$

Section 9.2 Practice Exercises

1. a. $f = \{(4, -3), (3, -4), (2, 7), (5, 0)\}$
f is one-to-one since each y-value corresponds to only one x-value.

   b. $g = \{(8, 4), (-2, 0), (6, 4), (2, 6)\}$
g is not one-to-one because the y-value 4 in (8, 4) and (6, 4) corresponds to two different x-values.

c. $h = \{(2, 4), (1, 3), (4, 6), (-2, 4)\}$
h is not one-to-one because the y-value 4 in (2, 4) and (−2, 4) corresponds to two different x-values.

d.

Year	1950	1963	1968	1975	1997	2008
Federal Minimum Wage	$0.75	$1.25	$1.60	$2.10	$5.15	$6.55

This function is one-to-one because each wage corresponds to only one year.

e.   The function represented by the graph is not one-to-one because the y-value 2 in (2, 2) and (3, 2) corresponds to two different x-values.

f.   The function represented by the diagram is not one-to-one because the score 509 corresponds to two different states.

2. Graphs **a**, **b**, and **c** all pass the vertical line test, so only these graphs are functions. But, of these, only **b** and **c** pass the horizontal line test, so only **b** and **c** are graphs of one-to-one functions.

3. $f = \{(3, 4), (-2, 0), (2, 8), (6, 6)\}$
Switching the coordinates of each ordered pair gives $f^{-1} = \{(4, 3), (0, -2), (8, 2), (6, 6)\}$

4. $f(x) = 6 - x$
Replace $f(x)$ with y.
$y = 6 - x$
Interchange x and y.
$x = 6 - y$
Solve for y.
$x = 6 - y$
$y = 6 - x$
Replace y with $f^{-1}(x)$.

$f^{-1}(x) = 6 - x$

5. $f(x) = 5x + 2$
Replace $f(x)$ with y.
$y = 5x + 2$
Interchange x and y.
$x = 5y + 2$
Solve for y.
$x = 5y + 2$
$x - 2 = 5y$
$\dfrac{x - 2}{5} = y$
Replace y with $f^{-1}(x)$.

$f^{-1}(x) = \dfrac{x - 2}{5}$

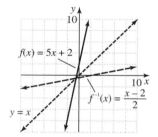

$f(x) = 5x + 2$

$f^{-1}(x) = \dfrac{x-2}{2}$

$y = x$

6. a.

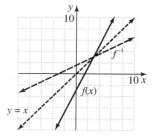

f^{-1}

$f(x)$

$y = x$

b.

f

f^{-1}

$y = x$

7. $f(x) = 4x - 1;\ f^{-1}(x) = \dfrac{x+1}{4}$

$$(f \circ f^{-1})(x) = f(f^{-1}(x))$$

$$= f\left(\dfrac{x+1}{4}\right)$$

$$= 4\left(\dfrac{x+1}{4}\right) - 1$$

$$= x + 1 - 1$$

$$= x$$

$$(f^{-1} \circ f)(x) = f^{-1}(f(x))$$

$$= f^{-1}(4x - 1)$$

$$= \dfrac{(4x - 1) + 1}{4}$$

$$= \dfrac{4x - 1 + 1}{4}$$

$$= \dfrac{4x}{4}$$

$$= x$$

Since $f \circ f^{-1} = x$ and $f^{-1} \circ f = x$, if

$f(x) = 4x - 1,\ f^{-1}(x) = \dfrac{x+1}{4}$.

Vocabulary, Readiness & Video Check 9.2

1. If $f(2) = 11$, the corresponding ordered pair is <u>(2, 11)</u>.

2. If (7, 3) is an ordered pair solution of $f(x)$, and $f(x)$ has an inverse, then an ordered pair solution of $f^{-1}(x)$ is <u>(3, 7)</u>.

3. The symbol f^{-1} means <u>the inverse of f</u>.

4. True or false: The function notation $f^{-1}(x)$ means $\dfrac{1}{f(x)}$. <u>false</u>

5. To tell whether a graph is the graph of a function, use the <u>vertical</u> line test.

6. To tell whether the graph of a function is also a one-to-one function, use the <u>horizontal</u> line test.

7. The graphs of f and f^{-1} are symmetric about the line <u>$y = x$</u>.

8. Two functions are inverse of each other if $(f \circ f^{-1})(x) = \underline{x}$ and $(f^{-1} \circ f)(x) = \underline{x}$.

9. Every function must have each x-value correspond to only one y-value. A one-to-one function must also have each y-value correspond to only one x-value.

10. No, a graph must pass the vertical line test to even be a function—a graph must pass *both* the vertical and horizontal line tests to be a one-to-one function.

11. Yes; by the definition of an inverse function.

12. The definition of inverse function tells us that f^{-1} consists of the ordered pairs (y, x) when (x, y) belongs to f. So it makes sense that switching x and y in the equations would result in switching the x and y values in the ordered pairs.

13. Once you know some points of the original equation or graph, you can switch the x's and y's of these points to find points that satisfy the inverse and then graph it. You can also check that the two graphs (the original and the inverse) are symmetric about the line $y = x$.

14. You must show that $f(f^{-1}(x))$ and $f^{-1}(f(x))$ both equal x in order to prove they are inverses of each other.

Exercise Set 9.2

1. $f = \{(-1,-1),(1,1),(0,2),(2,0)\}$ is a one-to-one function.

$f^{-1} = \{(-1,-1),(1,1),(2,0),(0,2)\}$

3. $h = \{(10,10)\}$ is a one-to-one function.

$h^{-1} = \{(10,10)\}$

5. $f = \{(11,12),(4,3),(3,4),(6,6)\}$ is a one-to-one function.

$f^{-1} = \{(12,11),(3,4),(4,3),(6,6)\}$

7. This function is not one-to-one because the months April and August both have the output 8.1; also, May and June both have the output 8.2.

9. This function is one-to-one.

Rank in Population (input)	1	47	16	25	36	7
State (output)	California	Alaska	Indiana	Louisiana	New Mexico	Ohio

11. $f(x) = x^3 + 2$

    a. $f(1) = 1^3 + 2 = 3$

    b. $f^{-1}(3) = 1$

13. $f(x) = x^3 + 2$

    a. $f(-1) = (-1)^3 + 2 = 1$

    b. $f^{-1}(1) = -1$

15. The graph represents a one-to-one function because it passes the horizontal line test.

17. The graph does not represent a one-to-one function because it does not pass the horizontal line test.

19. The graph represents a one-to-one function because it passes the horizontal line test.

21. The graph does not represent a one-to-one function because it does not pass the horizontal line test.

23. $f(x) = x + 4$
$y = x + 4$
$x = y + 4$
$y = x - 4$
$f^{-1}(x) = x - 4$

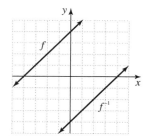

25. $f(x) = 2x - 3$
$y = 2x - 3$
$x = 2y - 3$
$2y = x + 3$
$y = \dfrac{x+3}{2}$
$f^{-1}(x) = \dfrac{x+3}{2}$

27. $f(x) = \dfrac{1}{2}x - 1$
$y = \dfrac{1}{2}x - 1$
$x = \dfrac{1}{2}y - 1$
$\dfrac{1}{2}y = x + 1$
$y = 2x + 2$
$f^{-1}(x) = 2x + 2$

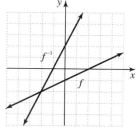

29. $f(x) = x^3$
$y = x^3$
$x = y^3$
$y = \sqrt[3]{x}$
$f^{-1}(x) = \sqrt[3]{x}$

31. $f(x) = 5x + 2$
$y = 5x + 2$
$x = 5y + 2$
$5y = x - 2$
$y = \dfrac{x-2}{5}$
$f^{-1}(x) = \dfrac{x-2}{5}$

33. $f(x) = \dfrac{x-2}{5}$
$y = \dfrac{x-2}{5}$
$x = \dfrac{y-2}{5}$
$5x = y - 2$
$y = 5x + 2$
$f^{-1}(x) = 5x + 2$

35. $f(x) = \sqrt[3]{x}$
$y = \sqrt[3]{x}$
$x = \sqrt[3]{y}$
$x^3 = y$
$f^{-1}(x) = x^3$

37.
$$f(x) = \frac{5}{3x+1}$$
$$y = \frac{5}{3x+1}$$
$$x = \frac{5}{3y+1}$$
$$3y+1 = \frac{5}{x}$$
$$3y = \frac{5}{x} - 1$$
$$3y = \frac{5-x}{x}$$
$$y = \frac{5-x}{3x}$$
$$f^{-1}(x) = \frac{5-x}{3x}$$

39.
$$f(x) = (x+2)^3$$
$$y = (x+2)^3$$
$$x = (y+2)^3$$
$$\sqrt[3]{x} = y+2$$
$$\sqrt[3]{x} - 2 = y$$
$$f^{-1}(x) = \sqrt[3]{x} - 2$$

41.

43.

45.

47.
$$(f \circ f^{-1})(x) = f(f^{-1}(x))$$
$$= f\left(\frac{x-1}{2}\right)$$
$$= 2\left(\frac{x-1}{2}\right) + 1$$
$$= x - 1 + 1$$
$$= x$$
$$(f^{-1} \circ f)(x) = f^{-1}(f(x))$$
$$= f^{-1}(2x+1)$$
$$= \frac{(2x+1)-1}{2}$$
$$= \frac{2x}{2}$$
$$= x$$

49.
$$(f \circ f^{-1})(x) = f(f^{-1}(x))$$
$$= f\left(\sqrt[3]{x-6}\right)$$
$$= \left(\sqrt[3]{x-6}\right)^3 + 6$$
$$= x - 6 + 6$$
$$= x$$
$$(f^{-1} \circ f)(x) = f^{-1}(f(x))$$
$$= f^{-1}(x^3 + 6)$$
$$= \sqrt[3]{(x^3 + 6) - 6}$$
$$= \sqrt[3]{x^3}$$
$$= x$$

51. $25^{1/2} = \sqrt{25} = 5$

53. $16^{3/4} = \left(\sqrt[4]{16}\right)^3 = 2^3 = 8$

55. $9^{-3/2} = \dfrac{1}{9^{3/2}} = \dfrac{1}{\left(\sqrt{9}\right)^3} = \dfrac{1}{3^3} = \dfrac{1}{27}$

57.
$$f(x) = 3^x$$
$$f(2) = 3^2 = 9$$

59. $f(x) = 3^x$

$$f\left(\frac{1}{2}\right) = 3^{1/2} \approx 1.73$$

61. $f(2) = 9$

    a. $(2, 9)$

    b. $(9, 2)$

63. a. $\left(-2, \frac{1}{4}\right), \left(-1, \frac{1}{2}\right), (0,1), (1,2), (2,5)$

    b. $\left(\frac{1}{4}, -2\right), \left(\frac{1}{2}, -1\right), (1,0), (2,1), (5,2)$

    c, d.

65. answers may vary.

67. $f(x) = 3x + 1$

$$y = 3x + 1$$
$$x = 3y + 1$$
$$x - 1 = 3y$$
$$y = \frac{x-1}{3}$$
$$f^{-1}(x) = \frac{x-1}{3}$$

69. $f(x) = \sqrt[3]{x+1}$
$$y = \sqrt[3]{x+1}$$
$$x = \sqrt[3]{y+1}$$
$$x^3 = y+1$$
$$y = x^3 - 1$$
$$f^{-1}(x) = x^3 - 1$$

Section 9.3 Practice Exercises

1. $f(x) = 2^x$

x	0	1	2	3	−1	−2
$f(x)$	1	2	4	8	$\frac{1}{2}$	$\frac{1}{4}$

$g(x) = 7^x$

x	0	1	2	3	−1	−2
$g(x)$	1	7	49	343	$\frac{1}{7}$	$\frac{1}{49}$

2. $f(x) = \left(\frac{1}{3}\right)^x$

x	0	1	2	3	−1	−2
$f(x)$	1	$\frac{1}{3}$	$\frac{1}{9}$	$\frac{1}{27}$	3	9

$g(x) = \left(\frac{1}{5}\right)^x$

x	0	1	2	3	−1	−2
$g(x)$	1	$\frac{1}{5}$	$\frac{1}{25}$	$\frac{1}{125}$	5	25

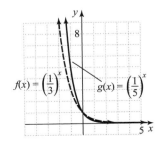

3. $f(x) = 2^{x-3}$

	x	6	5	4	3	2	1	0
$f(x) = 2^{x-3}$	$f(x)$	8	4	2	1	$\frac{1}{2}$	$\frac{1}{4}$	$\frac{1}{8}$

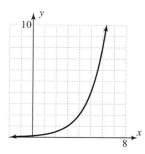

4. a. $3^x = 9$

Write 9 as a power of 3, $9 = 3^2$.

$3^x = 3^2$, thus, $x = 2$.

b. $8^x = 16$

Write 8 and 16 as powers of 2.

$8 = 2^3$ and $16 = 2^4$.

$$8^x = 16$$
$$(2^3)^x = 2^4$$
$$2^{3x} = 2^4$$
$$3x = 4$$
$$x = \frac{4}{3}$$

c. $125^x = 25^{x-2}$

Write 125 and 25 as powers of 5.

$125 = 5^3$ and $25 = 5^2$.

$$125^x = 25^{x-2}$$
$$(5^3)^x = (5^2)^{x-2}$$
$$5^{3x} = 5^{2x-4}$$
$$3x = 2x - 4$$
$$x = -4$$

d. Graph $y_1 = 7^x$ and $y_2 = 21$.

The solution is ≈ 1.565.

5. $P = \$3000$, $r = 7\% = 0.07$, $n = 2$, and $t = 4$.

$$A = P\left(1 + \frac{r}{n}\right)^{nt}$$

$$A = 3000\left(1 + \frac{0.07}{2}\right)^{2(4)}$$

$$= 3000(1.035)^8$$

$$\approx 3950.43$$

Thus, the amount owed is approximately $3950.43.

6. a. $p(n) = 100(2.7)^{-0.05n}$, $n = 2$

$$p(2) = 100(2.7)^{-0.05(2)}$$

$$= 100(2.7)^{-0.1}$$

$$\approx 90.54$$

Thus, approximately 90.54% of the light passes through.

b. $p(n) = 100(2.7)^{-0.05n}$, $n = 10$

$$p(10) = 100(2.7)^{-0.05(10)}$$

$$= 100(2.7)^{-0.5}$$

$$\approx 60.86$$

Thus, approximately 60.86% of the light passes through.

7.

a. The expected percent after 4 days is 67.21%.

b. The expected percent after 35 days is 3.09%.

Vocabulary, Readiness & Video Check 9.3

1. A function such as $f(x) = 2^x$ is an <u>exponential</u> function; **C.**

2. If $7^x = 7^y$, then <u>$x = y$</u>; **B.**

3. Yes, the graph passes the vertical line test.

4. Yes, the function passes both the vertical and horizontal line tests.

5. The function has no *x*-intercept.

6. The function has a *y*-intercept of <u>(0, 1)</u>.

7. The domain of this function, in interval notation, is <u>$(-\infty, \infty)$</u>.

8. The range of this function, in interval notation, is <u>$(0, \infty)$</u>.

9. In a polynomial function, the base is the variable and the exponent is the constant; in an exponential function, the base is the constant and the exponent is the variable.

10. Rewrite the equation so the bases are the same.

11. $y = 30(0.996)^{101} \approx 20.0$ lb

Exercise Set 9.3

1. $y = 5^x$

3. $y = 2^x + 1$

5. $y = \left(\frac{1}{4}\right)^x$

7. $y = \left(\dfrac{1}{2}\right)^x - 2$

9. $y = -2^x$

11. $y = -\left(\dfrac{1}{4}\right)^x$

13. $f(x) = 2^{x+1}$

15. $f(x) = 4^{x-2}$

17. C

19. B

21. $3^x = 27$

$3^x = 3^3$

$x = 3$

The solution is 3.

23. $16^x = 8$

$(2^4)^x = 2^3$

$2^{4x} = 2^3$

$4x = 3$

$x = \dfrac{3}{4}$

The solution is $\dfrac{3}{4}$.

25. $32^{2x-3} = 2$

$(2^5)^{2x-3} = 2^1$

$2^{10x-15} = 2^1$

$10x - 15 = 1$

$10x = 16$

$x = \dfrac{8}{5}$

The solution is $\dfrac{8}{5}$.

27. $\dfrac{1}{4} = 2^{3x}$

$2^{-2} = 2^{3x}$

$3x = -2$

$x = -\dfrac{2}{3}$

The solution is $-\dfrac{2}{3}$.

29. $5^x = 625$

$5^x = 5^4$

$x = 4$

The solution is 4.

31. $4^x = 8$

$(2^2)^x = 2^3$

$2^{2x} = 2^3$

$2x = 3$

$x = \dfrac{3}{2}$

The solution is $\dfrac{3}{2}$.

33. $27^{x+1} = 9$

$(3^3)^{x+1} = 3^2$

$3^{3x+3} = 3^2$

$3x + 3 = 2$

$3x = -1$

$x = -\dfrac{1}{3}$

The solution is $-\dfrac{1}{3}$.

35. $81^{x-1} = 27^{2x}$

$(3^4)^{x-1} = (3^3)^{2x}$

$3^{4x-4} = 3^{6x}$

$4x - 4 = 6x$

$-4 = 2x$

$x = -2$

The solution is -2.

37. $y = 30(0.996)^x$

$y = 30(0.996)^{50} \approx 24.6$

There will be about 24.6 pounds left after 50 days.

39. a. 2007 is 4 years after 2003, so $x = 4$.

$y = 8.6(1.03)^x$

$y = 8.6(1.03)^4 \approx 9.7$

The total cheese production in the United States was about 9.7 billion pounds in 2007.

b. 2015 is 12 years after 2003, so $x = 12$.

$y = 8.6(1.03)^{12} \approx 12.3$

The total cheese production is predicted to be 12.3 billion pounds in 2015.

41. $y = 140,242(1.083)^x$

a. $x = 2004 - 2000 = 4$

$y = 140,242(1.083)^4 \approx 192,927$

192,927 American students studied abroad in 2004.

b. $x = 2015 - 2000 = 15$

$y = 140,242(1.083)^{15} \approx 463,772$

463,722 American students would be studying abroad in 2015.

43. $x = 12$ corresponds to 2012.

$y = 120.48(1.093)^x$

$y = 120.48(1.093)^{12} \approx 350.2$

There were 350.2 million cell phone subscribers in 2012.

45. $y = 200,000(1.08)^x$, $x = 13$

$y = 200,000(1.08)^{13} \approx 544,000$

There will be approximately 544,000 mosquitoes on May 25.

47. $A = P\left(1 + \dfrac{r}{n}\right)^{nt}$

$t = 3$, $P = 6000$, $r = 0.08$, and $n = 12$

$A = 6000\left(1 + \dfrac{0.08}{12}\right)^{12(3)}$

$ = 6000\left(1 + \dfrac{0.08}{12}\right)^{36}$

$ \approx 7621.42$

Erica would owe \$7621.42 after 3 years.

49. $A = P\left(1 + \dfrac{r}{n}\right)^{nt}$

$P = 2000$

$r = 0.06$, $n = 2$, and $t = 12$

$A = 2000\left(1 + \dfrac{0.06}{2}\right)^{2(12)}$

$ = 2000(1.03)^{24}$

$ \approx 4065.59$

Janina has approximately \$4065.59 in her savings account.

51. $5x - 2 = 18$

$\quad 5x = 20$

$\quad\quad x = 4$

The solution is 4.

53. $3x - 4 = 3(x+1)$

$\quad 3x - 4 = 3x + 3$

$\quad\quad -4 = 3$

This is a false statement. The solution set is $\varnothing$.

55. $\quad\quad x^2 + 6 = 5x$

$\quad x^2 - 5x + 6 = 0$

$\quad (x-2)(x-3) = 0$

$\quad x = 2 \quad$ or $\quad x = 3$

The solutions are 2 and 3.

57. $2^x = 8$

$\quad 2^3 = 8$

$\quad\, x = 3$

59. $5^x = \dfrac{1}{5}$

$\quad 5^{-1} = \dfrac{1}{5}$

$\quad\quad x = -1$

61. Since there are no variables in the exponent in $f(x) = 1.5x^2$, it is not an exponential function.

63. Since there are no variables in the exponent in $h(x) = \left(\dfrac{1}{2}x\right)^2$, it is not an exponential function.

65. $f(x) = 2^{-x}$

$\quad f(1) = 2^{-1} = \dfrac{1}{2}$

This is graph C.

67. $f(x) = 4^{-x}$

$\quad f(1) = 4^{-1} = \dfrac{1}{4}$

This is graph D.

69. answers may vary

71. $y = \left|3^x\right|$

73. $y = 3^{|x|}$

75.

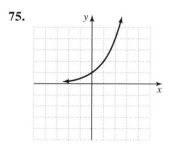

The graphs are the same, since $\left(\dfrac{1}{2}\right)^{-x} = 2^x$.

77. $y = 30(0.996)^x$

$\quad y = 30(0.996)^{100} \approx 20.09$

After 100 days, the estimate is 20.09 pounds left.

Section 9.4 Practice Exercises

1. $C = 25,000, \ r = 0.12, \ x = 2015 - 2000 = 15$

$\quad y = C(1+r)^x$

$\quad y = 25,000(1+0.12)^{15}$

$\quad\quad = 25,000(1.12)^{15}$

$\quad\quad \approx 136,839$

In 2015, the predicted population is 136,839.

2. $C = 800$, $r = 0.30$, $x = 9$

   $y = C(1-r)^x$

   $y = 800(1-0.30)^9 = 800(0.70)^9 \approx 32$

   After 9 rounds, there are 32 players remaining.

3. $C = 500$, $r = 0.50$, $x = \dfrac{51}{15} = 3.4$

   $y = C(1-r)^x$

   $y = 500(1-0.50)^{3.4} = 500(0.50)^{3.4} \approx 47.4$

   In 51 years, 47.4 grams of DDT remain.

Vocabulary, Readiness & Video Check 9.4

1. For Example 1, the growth rate is given as 5% per year. Since this is "per year," the number of time intervals is the "number of years," or 8.

2. The number of employees is decreasing and not increasing. It is exponential decay because the decrease is the same percent per year.

3. time intervals = years/half-life; the decay rate is 50% or $\dfrac{1}{2}$ because half-life is the amount of time it takes half of a substance to decay.

Exercise Set 9.4

	Original Amount	Growth Rate Per Year	Number of Years, x	Final Amount after x Years of Growth
1.	305	5%	8	$y = 305(1+0.05)^8 \approx 451$
3.	2000	11%	41	$y = 2000(1+0.11)^{41} \approx 144,302$
5.	17	29%	28	$y = 17(1+0.29)^{28} \approx 21,231$

	Original Amount	Decay Rate per Year	Number of Years, x	Final Amount after x Years of Decay
7.	305	5%	8	$y = 305(1-0.05)^8 \approx 202$
9.	10,000	12%	15	$y = 10,000(1-0.12)^{15} \approx 1470$
11.	207,000	32%	25	$y = 207,000(1-0.32)^{25} \approx 13$

13. $C = 500,000$, $r = 0.03$, $x = 12$

    $y = C(1+r)^x = 500,000(1+0.03)^{12} \approx 712,880$

    In 12 years, the population is predicted to be 712,880.

15. $C = 640$, $r = 0.05$, $x = 10$

    $y = C(1-r)^x = 640(1-0.05)^{10} \approx 383$

    In 10 years, the number of employees is predicted to be 383.

17. $C = 260$, $r = 0.025$, $x = 10$

$y = C(1+r)^x = 260(1+0.025)^{10} \approx 333$

In 10 years, the predicted number of bison is 333.

19. $C = 5$, $r = 0.15$, $x = 10$

$y = C(1-r)^x = 5(1-0.15)^{10} \approx 1$

After 10 seconds, there will be 1 gram of the isotope.

	Original Amount	Half-Life (in years)	Number of Years	Time intervals, x	Final Amount after x Time Intervals	Is amount reasonable?
21. a.	40	7	14	$\frac{14}{7} = 2$	$40(1-0.5)^2 = 10$	yes
b.	40	7	11	$\frac{11}{7} \approx 1.6$	$40(1-0.5)^{1.6} \approx 13.2$	yes
23.	21	152	500	$\frac{500}{152} \approx 3.3$	$21(1-0.5)^{3.3} \approx 2.1$	yes

25. $C = 30$, $r = 0.5$, $x = \dfrac{250}{96}$

$y = 30(1-0.5)^{250/96}$

$y \approx 4.9$

In 250 years, the predicted amount is 4.9 grams.

27. $2^x = 8$

$2^x = 2^3$

$x = 3$

29. $5^x = \dfrac{1}{5}$

$5^x = 5^{-1}$

$x = -1$

31. no; answers may vary.

Section 9.5 Practice Exercises

1. a. $\log_3 81 = 4$ means $3^4 = 81$.

b. $\log_5 \dfrac{1}{5} = -1$ means $5^{-1} = \dfrac{1}{5}$.

c. $\log_7 \sqrt{7} = \dfrac{1}{2}$ means $7^{1/2} = \sqrt{7}$.

d. $\log_{13} y = 4$ means $13^4 = y$.

2. a. $4^3 = 64$ means $\log_4 64 = 3$.

   b. $6^{1/3} = \sqrt[3]{6}$ means $\log_6 \sqrt[3]{6} = \dfrac{1}{3}$.

   c. $5^{-3} = \dfrac{1}{125}$ means $\log_5 \dfrac{1}{125} = -3$.

   d. $\pi^7 = z$ means $\log_\pi z = 7$.

3. a. $\log_3 9 = 2$ because $3^2 = 9$.

   b. $\log_2 \dfrac{1}{8} = -3$ because $2^{-3} = \dfrac{1}{8}$.

   c. $\log_{49} 7 = \dfrac{1}{2}$ because $49^{1/2} = 7$.

4. a. $\log_5 \dfrac{1}{25} = x$

     $\log_5 \dfrac{1}{25} = x$ means $5^x = \dfrac{1}{25}$. Solve $5^x = \dfrac{1}{25}$.

     $5^x = \dfrac{1}{25}$

     $5^x = 5^{-2}$

     Since the bases are the same, by the uniqueness of b^x, we have that $x = -2$. The solution is -2 or the solution set is $\{-2\}$.

   b. $\log_x 8 = 3$
     $x^3 = 8$
     $x^3 = 2^3$
     $x = 2$

   c. $\log_6 x = 2$
     $6^2 = x$
     $36 = x$

   d. $\log_{13} 1 = x$
     $13^x = 1$
     $13^x = 13^0$
     $x = 0$

   e. $\log_h 1 = x$
     $h^x = 1$
     $h^x = h^0$
     $x = 0$

5. a. From Property 2, $\log_5 5^4 = 4$.

   b. From Property 2, $\log_9 9^{-2} = -2$.

   c. From Property 3, $6^{\log_6 5} = 5$.

   d. From Property 3, $7^{\log_7 4} = 4$.

6. $y = \log_7 x$ means that $7^y = x$. Find some ordered pair solutions that satisfy $7^y = x$.

$x = 7^y$	y
1	0
7	1
$\frac{1}{7}$	-1
$\frac{1}{49}$	-2

7. $y = \log_{1/4} x$ means that $\left(\dfrac{1}{4}\right)^y = x$. Find some ordered-pair solutions that satisfy $\left(\dfrac{1}{4}\right)^y = x$.

$x = \left(\frac{1}{4}\right)^y$	y
1	0
$\frac{1}{4}$	1
4	-1
16	-2

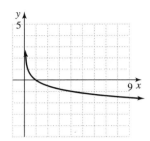

Vocabulary, Readiness & Video Check 9.5

1. A function such as $y = \log_2 x$ is a <u>logarithmic</u> function; **B**.

2. If $y = \log_2 x$, then $\underline{2^y = x}$; **C**.

3. Yes, the function passes both the horizontal- and vertical-line tests.

4. The function has an x-intercept of $\underline{(1, 0)}$.

5. The function has no y-intercept.

6. The domain of this function, in interval notation, is $\underline{(0, \infty)}$.

7. The range of this function, in interval notation, is $\underline{(-\infty, \infty)}$.

8. Logarithms are exponents.

9. First write the equation as an equivalent exponential equation. Then solve.

10. The exponential is solved for x, and y is the exponent in the equation. Since the exponential equation is solved for x, it is easier to choose a y-value and simplify the expression containing y, which is then the x-value.

Exercise Set 9.5

1. $\log_6 36 = 2$
   $6^2 = 36$

3. $\log_3 \dfrac{1}{27} = -3$
   $3^{-3} = \dfrac{1}{27}$

5. $\log_{10} 1000 = 3$
   $10^3 = 1000$

7. $\log_9 x = 4$
   $9^4 = x$

9. $\log_\pi \dfrac{1}{\pi^2} = -2$
   $\pi^{-2} = \dfrac{1}{\pi^2}$

11. $\log_7 \sqrt{7} = \dfrac{1}{2}$
    $7^{1/2} = \sqrt{7}$

13. $\log_{0.7} 0.343 = 3$
    $0.7^3 = 0.343$

15. $\log_3 \dfrac{1}{81} = -4$
    $3^{-4} = \dfrac{1}{81}$

17. $2^4 = 16$
    $\log_2 16 = 4$

19. $10^2 = 100$
    $\log_{10} 100 = 2$

21. $\pi^3 = x$
    $\log_\pi x = 3$

23. $10^{-1} = \dfrac{1}{10}$
    $\log_{10} \dfrac{1}{10} = -1$

25. $4^{-2} = \dfrac{1}{16}$
    $\log_4 \dfrac{1}{16} = -2$

27. $5^{1/2} = \sqrt{5}$
    $\log_5 \sqrt{5} = \dfrac{1}{2}$

29. $\log_2 8 = 3$ since $2^3 = 8$.

31. $\log_3 \dfrac{1}{9} = -2$ since $3^{-2} = \dfrac{1}{9}$.

33. $\log_{25} 5 = \dfrac{1}{2}$ since $25^{1/2} = 5$.

35. $\log_{1/2} 2 = -1$ since $\left(\dfrac{1}{2}\right)^{-1} = 2$.

37. $\log_6 1 = 0$ since $6^0 = 1$.

39. $\log_{10} 100 = \log_{10} 10^2 = 2$

41. $\log_3 81 = \log_3 3^4 = 4$

43. $\log_4 \dfrac{1}{64} = \log_4 4^{-3} = -3$

45. $\log_3 9 = x$
$$3^x = 9$$
$$3^x = 3^2$$
$$x = 2$$

47. $\log_3 x = 4$
$$x = 3^4 = 81$$

49. $\log_x 49 = 2$
$$x^2 = 49$$
$$x = \pm 7$$
We discard the negative base.
$$x = 7$$

51. $\log_2 \dfrac{1}{8} = x$
$$2^x = \dfrac{1}{8}$$
$$2^x = 2^{-3}$$
$$x = -3$$

53. $\log_3 \dfrac{1}{27} = x$
$$\dfrac{1}{27} = 3^x$$
$$3^{-3} = 3^x$$
$$-3 = x$$

55. $\log_8 x = \dfrac{1}{3}$
$$x = 8^{1/3} = 2$$

57. $\log_4 16 = x$
$$4^x = 16$$
$$4^x = 4^2$$
$$x = 2$$

59. $\log_{3/4} x = 3$
$$\left(\dfrac{3}{4}\right)^3 = x$$
$$\dfrac{27}{64} = x$$

61. $\log_x 100 = 2$
$$x^2 = 100$$
$$x = \pm 10$$
We discard the negative base.
$$x = 10$$

63. $\log_2 2^4 = x$
$$2^x = 2^4$$
$$x = 4$$

65. $3^{\log_3 5} = x$
$$5 = x$$

67. $\log_x \dfrac{1}{7} = \dfrac{1}{2}$
$$x^{1/2} = \dfrac{1}{7}$$
$$x = \dfrac{1}{49}$$

69. $\log_5 5^3 = 3$

71. $2^{\log_2 3} = 3$

73. $\log_9 9 = 1$

75. $\log_8 (8)^{-1} = -1$

77. $y = \log_3 x$

$y = 0$:

$\log_3 x = 0$

$x = 3^0 = 1$

(1, 0) is the only *x*-intercept. No *y*-intercept exists.

79. $f(x) = \log_{1/4} x$

$y = 0$: $\quad 0 = \log_{1/4} x$

$x = \left(\dfrac{1}{4}\right)^0 = 1$

(1, 0) is the *x*-intercept. No *y*-intercept exists.

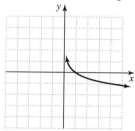

81. $f(x) = \log_5 x$

$y = 0$: $\quad 0 = \log_5 x$

$x = 5^0 = 1$

(1, 0) is the *x*-intercept. No *y*-intercept exists.

83. $f(x) = \log_{1/16} x$

$y = 0$:

$0 = \log_{1/6} x$

$x = \left(\dfrac{1}{6}\right)^0 = 1$

(1, 0) is the *x*-intercept. No *y*-intercept exists.

85. $\dfrac{x+3}{3+x} = \dfrac{x+3}{x+3} = 1$

87. $\dfrac{x^2 - 8x + 16}{2x - 8} = \dfrac{(x-4)^2}{2(x-4)} = \dfrac{x-4}{2}$

89. $\dfrac{2}{x} + \dfrac{3}{x^2} = \dfrac{2x}{x^2} + \dfrac{3}{x^2} = \dfrac{2x+3}{x^2}$

91. $\dfrac{3x}{x+3} + \dfrac{9}{x+3} = \dfrac{3x+9}{x+3} = \dfrac{3(x+3)}{x+3} = 3$

93. $f(x) = \log_5 x;\ f^{-1}(x) = g(x) = 5^x$

   a. (2, 25) implies $g(2) = 25$.

   b. Since $f^{-1}(x) = g(x)$, (25, 2) is a solution of $f(x)$.

   c. (25, 2) implies $f(25) = 2$.

95. answers may vary

97. $\log_7 (5x - 2) = 1$

$5x - 2 = 7^1$

$5x = 9$

$x = \dfrac{9}{5}$

99. $\log_3 \left(\log_5 125\right) = \log_3(3) = 1$

101. $y = 4^x$; $y = \log_4 x$

103. $y = \left(\dfrac{1}{3}\right)^x$; $y = \log_{1/3} x$

105. answers may vary

107. $\log_{10}(1-k) = \dfrac{-0.3}{H}$, $H = 8$

$$\log_{10}(1-k) = \dfrac{-0.3}{8} = -0.0375$$

$$1 - k = 10^{-0.0375}$$

$$1 - 10^{-0.0375} = k$$

$$k \approx 0.0827$$

The rate of decay is 0.0827.

Section 9.6 Practice Exercises

1. a. $\log_8 5 + \log_8 3 = \log_8(5 \cdot 3) = \log_8 15$

   b. $\log_2 \dfrac{1}{3} + \log_2 18 = \log_2\left(\dfrac{1}{3} \cdot 18\right) = \log_2 6$

   c. $\log_5(x-1) + \log_5(x+1) = \log_5[(x+1)(x+1)]$
$$= \log_5(x^2 - 1)$$

2. a. $\log_5 18 - \log_5 6 = \log_5 \dfrac{18}{6} = \log_5 3$

   b. $\log_6 x - \log_6 3 = \log_6 \dfrac{x}{3}$

c. $\log_4(x^2+1) - \log_4(x^2+3) = \log_4 \dfrac{x^2+1}{x^2+3}$

3. a. $\log_7 x^8 = 8\log_7 x$

   b. $\log_5 \sqrt[4]{7} = \log_5 7^{1/4} = \dfrac{1}{4}\log_5 7$

4. a. $2\log_5 4 + 5\log_5 2 = \log_5 4^2 + \log_5 2^5$
$$= \log_5 16 + \log_5 32$$
$$= \log_5(16 \cdot 32)$$
$$= \log_5 512$$

   b. $2\log_8 x - \log_8(x+3) = \log_8 x^2 - \log_8(x+3)$
$$= \log_8 \dfrac{x^2}{x+3}$$

   c. $\log_7 12 + \log_7 5 - \log_7 4$
$$= \log_7(12 \cdot 5) - \log_7 4$$
$$= \log_7 60 - \log_7 4$$
$$= \log_7 \dfrac{60}{4}$$
$$= \log_7 15$$

5. a. $\log_5 \dfrac{4 \cdot 3}{7} = \log_5(4 \cdot 3) - \log_5 7$
$$= \log_5 4 + \log_5 3 - \log_5 7$$

   b. $\log_4 \dfrac{a^2}{b^5} = \log_4 a^2 - \log_4 b^5$
$$= 2\log_4 a - 5\log_4 b$$

6. $\log_b 5 = 0.83$ and $\log_b 3 = 0.56$

   a. $\log_b 15 = \log_b(3 \cdot 5)$
$$= \log_b 3 + \log_b 5$$
$$= 0.56 + 0.83$$
$$= 1.39$$

   b. $\log_b 25 = \log_b 5^2 = 2\log_b 5 = 2(0.83) = 1.66$

   c. $\log_b \sqrt{3} = \log_b 3^{1/2}$
$$= \dfrac{1}{2}\log_b 3$$
$$= \dfrac{1}{2}(0.56)$$
$$= 0.28$$

Vocabulary, Readiness & Video Check 9.6

1. $\log_b 12 + \log_b 3 = \log_b (12 \cdot 3) = \log_b \underline{36}$; **a.**

2. $\log_b 12 - \log_b 3 = \log_b \dfrac{12}{3} = \log_b \underline{4}$; **c.**

3. $7\log_b 2 = \underline{\log_b 2^7}$; **b.**

4. $\log_b 1 = \underline{0}$; **c.**

5. $b^{\log_b x} = \underline{x}$; **a.**

6. $\log_5 5^2 = \underline{2}$; **b.**

7. No, the product property says the logarithm of a product can be written as a sum of logarithms—the expression in Example 2 is a logarithm of a sum.

8. The bases must be the same.

9. Since $\dfrac{1}{x} = x^{-1}$, this gives us $\log_2 x^{-1}$. Using the power property, we get $-1\log_2 x$ or $-\log_2 x$.

10. From writing logarithms as equivalent exponents and then using the rules for exponents.

Exercise Set 9.6

1. $\log_5 2 + \log_5 7 = \log_5 (2 \cdot 7) = \log_5 14$

3. $\log_4 9 + \log_4 x = \log_4 9x$

5. $\log_6 x + \log_6 (x+1) = \log_6 [x(x+1)]$
$$= \log_6 (x^2 + x)$$

7. $\log_{10} 5 + \log_{10} 2 + \log_{10} (x^2 + 2)$
$$= \log_{10} \left[5 \cdot 2 \left(x^2 + 2 \right) \right]$$
$$= \log_{10} \left(10x^2 + 20 \right)$$

9. $\log_5 12 - \log_5 4 = \log_5 \dfrac{12}{4} = \log_5 3$

11. $\log_3 8 - \log_3 2 = \log_3 \dfrac{8}{2} = \log_3 4$

13. $\log_2 x - \log_2 y = \log_2 \dfrac{x}{y}$

15. $\log_2 (x^2 + 6) - \log(x^2 + 1) = \log_2 \dfrac{x^2 + 6}{x^2 + 1}$

17. $\log_3 x^2 = 2\log_3 x$

19. $\log_4 5^{-1} = -\log_4 5$

21. $\log_5 \sqrt{y} = \log_5 y^{1/2} = \dfrac{1}{2}\log_5 y$

23. $\log_2 5 + \log_2 x^3 = \log_2 5x^3$

25. $3\log_4 2 + \log_4 6 = \log_4 2^3 + \log_4 6$
$$= \log_4 8 + \log_4 6$$
$$= \log_4 (8 \cdot 6)$$
$$= \log_4 48$$

27. $3\log_5 x + 6\log_5 z = \log_5 x^3 + \log_5 z^6$
$$= \log_5 x^3 z^6$$

29. $\log_4 2 + \log_4 10 - \log_4 5 = \log_4 (2 \cdot 10) - \log_4 5$
$$= \log_4 \dfrac{20}{5}$$
$$= \log_4 4$$
$$= 1$$

31. $\log_7 6 + \log_7 3 - \log_7 4 = \log_7 (6 \cdot 3) - \log_7 4$
$$= \log_7 \dfrac{18}{4}$$
$$= \log_7 \dfrac{9}{2}$$

33. $\log_{10} x - \log_{10} (x+1) + \log_{10} (x^2 - 2)$
$$= \log_{10} \dfrac{x}{x+1} + \log_{10} (x^2 - 2)$$
$$= \log_{10} \dfrac{x(x^2 - 2)}{x+1}$$
$$= \log_{10} \dfrac{x^3 - 2x}{x+1}$$

35. $3\log_2 x + \dfrac{1}{2}\log_2 x - 2\log_2(x+1)$

$= \log_2 x^3 + \log_2 x^{1/2} - \log_2(x+1)^2$

$= \log_2(x^3 \cdot x^{1/2}) - \log_2(x+1)^2$

$= \log_2 x^{7/2} - \log_2(x+1)^2$

$= \log_2 \dfrac{x^{7/2}}{(x+1)^2}$

37. $2\log_8 x - \dfrac{2}{3}\log_8 x + 4\log_8 x = \left(2 - \dfrac{2}{3} + 4\right)\log_8 x$

$= \dfrac{16}{3}\log_8 x$

$= \log_8 x^{16/3}$

39. $\log_3 \dfrac{4y}{5} = \log_3 4y - \log_3 5$

$= \log_3 4 + \log_3 y - \log_3 5$

41. $\log_4 \dfrac{5}{9z} = \log_4 5 - \log_4 9z$

$= \log_4 5 - (\log_4 9 + \log_4 z)$

$= \log_4 5 - \log_4 9 - \log_4 z$

43. $\log_2 \dfrac{x^3}{y} = \log_2 x^3 - \log_2 y$

$= 3\log_2 x - \log_2 y$

45. $\log_b \sqrt{7x} = \log_b(7x)^{1/2}$

$= \dfrac{1}{2}\log_b(7x)$

$= \dfrac{1}{2}\left[\log_b 7 + \log_b x\right]$

$= \dfrac{1}{2}\log_b 7 + \dfrac{1}{2}\log_b x$

47. $\log_6 x^4 y^5 = \log_6 x^4 + \log_6 y^5$

$= 4\log_6 x + 5\log_6 y$

49. $\log_5 x^3(x+1) = \log_5 x^3 + \log_5(x+1)$

$= 3\log_5 x + \log_5(x+1)$

51. $\log_6 \dfrac{x^2}{x+3} = \log_6 x^2 - \log_6(x+3)$

$= 2\log_6 x - \log_6(x+3)$

53. $\log_b 15 = \log_b(5 \cdot 3)$

$= \log_b 5 + \log_b 3$

$= 0.7 + 0.5$

$= 1.2$

55. $\log_b \dfrac{5}{3} = \log_b 5 - \log_b 3 = 0.7 - 0.5 = 0.2$

57. $\log_b \sqrt{5} = \log_b 5^{1/2} = \dfrac{1}{2}\log_b 5 = \dfrac{1}{2}(0.7) = 0.35$

59. $\log_b 8 = \log_b 2^3 = 3\log_b 2 = 3(0.43) = 1.29$

61. $\log_b \dfrac{3}{9} = \log_b 3 - \log_b 9$

$= \log_b 3 - \log_b 3^2$

$= \log_b 3 - 2\log_b 3$

$= -\log_b 3$

$= -0.68$

63. $\log_b \sqrt{\dfrac{2}{3}} = \log_b \left(\dfrac{2}{3}\right)^{1/2}$

$= \dfrac{1}{2}\log_b \dfrac{2}{3}$

$= \dfrac{1}{2}\left(\log_b 2 - \log_b 3\right)$

$= \dfrac{1}{2}(0.43 - 0.68)$

$= \dfrac{1}{2}(-0.25)$

$= -0.125$

65. $y = 10^x$ and $y = \log_{10} x$

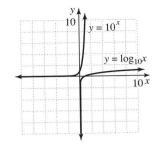

67. $\log_{10} 100 = \log_{10} 10^2 = 2$

69. $\log_7 7^2 = 2$

71. $\log_3 \dfrac{14}{11} = \log_3 14 - \log_3 11;$ **b**

73. $\log_2 x^3 = 3\log_2 x$ is true.

75. $\dfrac{\log_7 10}{\log_7 5} = \log_7 2$ is false.

77. $\dfrac{\log_7 x}{\log_7 y} = (\log_7 x) - (\log_7 y)$ is false.

79. $\log_b 8$ equals $\log_b 8 + \log_b 1$ because $\log_b 1 = 0$.

Integrated Review

1. $(f+g)(x) = x - 6 + x^2 + 1 = x^2 + x - 5$

2. $(f-g)(x) = x - 6 - (x^2 + 1) = -x^2 + x - 7$

3. $(f \cdot g)(x) = (x-6)(x^2+1) = x^3 - 6x^2 + x - 6$

4. $\left(\dfrac{f}{g}\right)(x) = \dfrac{x-6}{x^2+1}$

5. $(f \circ g)(x) = f(g(x)) = f(3x-1) = \sqrt{3x-1}$

6. $(g \circ f)(x) = g(f(x)) = g\left(\sqrt{x}\right) = 3\sqrt{x} - 1$

7. one-to-one; inverse:
$\{(6,-2),(8,4),(-6,2),(3,3)\}$

8. not one-to-one

9. not one-to-one

10. one-to-one

11. not one-to-one

12. $f(x) = 3x$
$y = 3x$

$x = 3y$

$y = \dfrac{x}{3}$

$f^{-1}(x) = \dfrac{x}{3}$

13. $f(x) = x + 4$
$y = x + 4$

$x = y + 4$
$y = x - 4$
$f^{-1}(x) = x - 4$

14. $f(x) = 5x - 1$
$y = 5x - 1$

$x = 5y - 1$
$5y = x + 1$
$y = \dfrac{x+1}{5}$
$f^{-1}(x) = \dfrac{x+1}{5}$

15. $f(x) = 3x + 2$
$y = 3x + 2$

$x = 3y + 2$
$3y = x - 2$
$y = \dfrac{x-2}{3}$
$f^{-1}(x) = \dfrac{x-2}{3}$

16. $y = \left(\dfrac{1}{2}\right)^x$

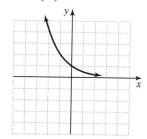

17. $y = 2^x + 1$

18. $y = \log_3 x$

19. $y = \log_{1/3} x$

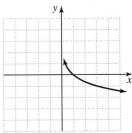

20. $2^x = 8$

$2^x = 2^3$

$x = 3$

The solution is 3.

21. $9 = 3^{x-5}$

$3^2 = 3^{x-5}$

$2 = x - 5$

$7 = x$

The solution is 7.

22. $4^{x-1} = 8^{x+2}$

$(2^2)^{x-1} = (2^3)^{x+2}$

$2^{2x-2} = 2^{3x+6}$

$2x - 2 = 3x + 6$

$-8 = x$

The solution is -8.

23. $25^x = 125^{x-1}$

$(5^2)^x = (5^3)^{x-1}$

$5^{2x} = 5^{3x-3}$

$2x = 3x - 3$

$3 = x$

The solution is 3.

24. $\log_4 16 = x$

$4^x = 16$

$4^x = 4^2$

$x = 2$

The solution is 2.

25. $\log_{49} 7 = x$

$49^x = 7$

$(7^2)^x = 7$

$7^{2x} = 7$

$2x = 1$

$x = \dfrac{1}{2}$

The solution is $\dfrac{1}{2}$.

26. $\log_2 x = 5$

$2^5 = x$

$32 = x$

The solution is 32.

27. $\log_x 64 = 3$

$x^3 = 64$

$x^3 = 4^3$

$x = 4$

The solution is 4.

28. $\log_x \dfrac{1}{125} = -3$

$x^{-3} = \dfrac{1}{125}$

$x^{-3} = 5^{-3}$

$x = 5$

The solution is 5.

29. $\log_3 x = -2$

$3^{-2} = x$

$x = \dfrac{1}{3^2} = \dfrac{1}{9}$

The solution is $\dfrac{1}{9}$.

30. $5\log_2 x = \log_2 x^5$

31. $x\log_2 5 = \log_2 5^x$

32. $3\log_5 x - 5\log_5 y = \log_5 x^3 - \log_5 y^5 = \log_5 \dfrac{x^3}{y^5}$

33. $9\log_5 x + 3\log_5 y = \log_5 x^9 + \log_5 y^3$
$$= \log_5 x^9 y^3$$

34. $\log_2 x + \log_2(x-3) - \log_2(x^2+4)$
$$= \log_2 [x(x-3)] - \log_2(x^2+4)$$
$$= \log_2(x^2-3x) - \log_2(x^2+4)$$
$$= \log_2 \frac{x^2-3x}{x^2+4}$$

35. $\log_3 y - \log_3(y+2) + \log_3(y^3+11)$
$$= \log_3 \frac{y}{y+2} + \log_3(y^3+11)$$
$$= \log_3 \frac{y(y^3+11)}{y+2}$$
$$= \log_3 \frac{y^4+11y}{y+2}$$

36. $\log_7 \dfrac{9x^2}{y} = \log_7 9x^2 - \log_7 y$
$$= \log_7 9 + \log_7 x^2 - \log_7 y$$
$$= \log_7 9 + 2\log_7 x - \log_7 y$$

37. $\log_6 \dfrac{5y}{z^2} = \log_6 5y - \log_6 z^2$
$$= \log_6 5 + \log_6 y - 2\log_6 z$$

38. $C = 100{,}000, \ r = 6\% = 0.06, \ x = 17 - 1 = 16$
$$y = C(1+r)^x$$
$$= 100{,}000(1+0.06)^{16}$$
$$\approx 254{,}000$$
There will be approximately 254,000 mosquitoes on April 17.

Section 9.7 Practice Exercises

1. To four decimal places, $\log 15 \approx 1.1761$.

2. a. $\log \dfrac{1}{100} = \log 10^{-2} = 2$

   b. $\log 100{,}000 = \log 10^5 = 5$

   c. $\log \sqrt[5]{10} = \log 10^{1/5} = \dfrac{1}{5}$

   d. $\log 0.001 = \log 10^{-3} = -3$

3. $\log x = 3.4$
$$x = 10^{3.4}$$
$$x \approx 2511.8864$$

4. $a = 450$ micrometers
$T = 4.2$ seconds
$B = 3.6$
$$R = \log\left(\frac{a}{T}\right) + B$$
$$= \log\left(\frac{450}{4.2}\right) + 3.6$$
$$\approx 2.0 + 3.6$$
$$= 5.6$$
The earthquake had a magnitude of 5.6 on the Richter scale.

5. To four decimal places, $\ln 13 \approx 2.5649$.

6. a. $\ln e^4 = 4$

   b. $\ln \sqrt[3]{e} = \ln e^{1/3} = \dfrac{1}{3}$

7. $\ln 5x = 8$
$$e^8 = 5x$$
$$\frac{e^8}{5} = x$$
$$x = \frac{1}{5}e^8 \approx 596.1916$$

8. $P = \$2400$
$r = 6\% = 0.06$
$t = 4$ years
$$A = Pe^{rt} = 2400e^{0.06(4)} = 2400e^{0.24} \approx 3051.00$$
The total amount of money owed is $3051.00.

9. $\log_8 5 = \dfrac{\log 5}{\log 8} \approx \dfrac{0.6989700043}{0.903089987} \approx 0.773976$
To four decimal places, $\log_8 5 \approx 0.7740$.

Vocabulary, Readiness & Video Check 9.7

1. The base of $\log 7$ is $\underline{10}$; **c.**

2. The base of $\ln 7$ is $\underline{e}$; **a.**

3. $\log_{10} 10^7 = \underline{7}$; **b.**

4. $\log_7 1 = \underline{0}$; **d.**

5. $\log_e e^5 = \underline{5}$; **b.**

6. $\ln e^5 = \underline{5}$; **b.**

7. $\log_2 7 = \dfrac{\log 7}{\log 2}$ or $\dfrac{\ln 7}{\ln 2}$; **a** and **b.**

8. 10

9. The understood base of a common logarithm is 10. If you're finding the common logarithm of a known power of 10, then the common logarithm is the known power of 10.

10. e

11. $\log_b b^x = x$

12. $\dfrac{\ln 4}{\ln 6}$ or $\dfrac{\log 4}{\log 6}$; also $\dfrac{\ln 4}{\ln 6} = \dfrac{\log 4}{\log 6}$

Exercise Set 9.7

1. $\log 8 \approx 0.9031$

3. $\log 2.31 \approx 0.3636$

5. $\ln 2 \approx 0.6931$

7. $\ln 0.0716 \approx -2.6367$

9. $\log 12.6 \approx 1.1004$

11. $\ln 5 \approx 1.6094$

13. $\log 41.5 \approx 1.6180$

15. $\log 100 = \log 10^2 = 2$

17. $\log \dfrac{1}{1000} = \log 10^{-3} = -3$

19. $\ln e^2 = 2$

21. $\ln \sqrt[4]{e} = \ln e^{1/4} = \dfrac{1}{4}$

23. $\log 10^3 = 3$

25. $\ln e^{-7} = -7$

27. $\log 0.0001 = \log 10^{-4} = -4$

29. $\ln \sqrt{e} = \ln e^{1/2} = \dfrac{1}{2}$

31. $\ln 2x = 7$
$$2x = e^7$$
$$x = \frac{1}{2} e^7 \approx 548.3166$$

33. $\log x = 1.3$
$$x = 10^{1.3} \approx 19.9526$$

35. $\log 2x = 1.1$
$$2x = 10^{1.1}$$
$$x = \frac{10^{1.1}}{2} \approx 6.2946$$

37. $\ln x = 1.4$
$$x = e^{1.4} \approx 4.0552$$

39. $\ln(3x - 4) = 2.3$
$$3x - 4 = e^{2.3}$$
$$3x = 4 + e^{2.3}$$
$$x = \frac{4 + e^{2.3}}{3} \approx 4.6581$$

41. $\log x = 2.3$
$$x = 10^{2.3} \approx 199.5262$$

43. $\ln x = -2.3$
$$x = e^{-2.3} \approx 0.1003$$

45. $\log(2x + 1) = -0.5$
$$2x + 1 = 10^{-0.5}$$
$$2x = 10^{-0.5} - 1$$
$$x = \frac{10^{-0.5} - 1}{2} \approx -0.3419$$

47. $\ln 4x = 0.18$
$$4x = e^{0.18}$$
$$x = \frac{e^{0.18}}{4} \approx 0.2993$$

49. $\log_2 3 = \dfrac{\log 3}{\log 2} \approx 1.5850$

51. $\log_{1/2} 5 = \dfrac{\ln 5}{\ln\left(\frac{1}{2}\right)} \approx -2.3219$

53. $\log_4 9 = \dfrac{\ln 9}{\ln 4} \approx 1.5850$

55. $\log_3\left(\dfrac{1}{6}\right) = \dfrac{\log\left(\frac{1}{6}\right)}{\log 3} \approx -1.6309$

57. $\log_8 6 = \dfrac{\log 6}{\log 8} \approx 0.8617$

59. $R = \log\left(\dfrac{a}{T}\right) + B,\ a = 200,\ T = 1.6$

$B = 2.1$

$R = \log\left(\dfrac{200}{1.6}\right) + 2.1 \approx 4.2$

The earthquake measures 4.2 on the Richter scale.

61. $R = \log\left(\dfrac{a}{T}\right) + B,\ a = 400,\ T = 2.6$

$B = 3.1$

$R = \log\left(\dfrac{400}{2.6}\right) + 3.1 \approx 5.3$

The earthquake measures 5.3 on the Richter scale.

63. $A = Pe^{rt},\ t = 12,\ P = 1400,\ r = 0.08$

$A = 1400e^{(0.08)12} = 1400e^{0.96} \approx 3656.38$
Dana has \$3656.38 after 12 years.

65. $A = Pe^{rt},\ t = 4,\ P = 2000,\ r = 0.06$

$A = 2000e^{(0.06)4} = 2000e^{0.24} \approx 2542.50$
Barbara owes \$2542.50 at the end of 4 years.

67. $6x - 3(2 - 5x) = 6$

$6x - 6 + 15x = 6$

$21x = 12$

$x = \dfrac{12}{21} = \dfrac{4}{7}$

The solution is $\dfrac{4}{7}$.

69. $2x + 3y = 6x$

$3y = 4x$

$x = \dfrac{3y}{4}$

71. $x^2 + 7x = -6$

$x^2 + 7x + 6 = 0$

$(x + 6)(x + 1) = 0$

$x + 6 = 0$ or $x + 1 = 0$

$x = -6$ or $x = -1$

The solutions are -6 and -1.

73. $\begin{cases} x + 2y = -4 \\ 3x - y = 9 \end{cases}$

Multiply the second equation by 2, then add.

$\begin{array}{r} x + 2y = -4 \\ 6x - 2y = 18 \\ \hline 7x \quad\ = 14 \\ x = 2 \end{array}$

Replace x with 2 in the first equation.

$x + 2y = -4$

$2 + 2y = -4$

$2y = -6$

$y = -3$

The solution is $(2, -3)$.

75. answers may vary

77. $\ln 50$ is larger. answers may vary

79. $f(x) = e^x$

81. $f(x) = e^{-3x}$

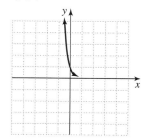

83. $f(x) = e^x + 2$

85. $f(x) = e^{x-1}$

87. $f(x) = 3e^x$

89. $f(x) = \ln x$

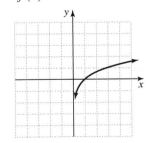

91. $f(x) = -2\log x$

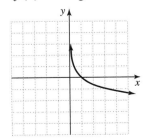

93. $f(x) = \log(x + 2)$

95. $f(x) = \ln x - 3$

97. $f(x) = e^x$

      $f(x) = e^x + 2$

      $f(x) = e^x - 3$

answers may vary

Section 9.8 Practice Exercises

1. $5^x = 9$

     $\log 5^x = \log 9$

     $x \log 5 = \log 9$

         $x = \dfrac{\log 9}{\log 5} \approx 1.3652$

The solution is $\dfrac{\log 9}{\log 5}$, or approximately 1.3652.

2. $\log_2(x-1) = 5$

$$2^5 = x-1$$
$$32 = x-1$$
$$33 = x$$

Check: $\log_2(x-1) = 5$
$$\log_2(33-1) \stackrel{?}{=} 5$$
$$\log_2 32 \stackrel{?}{=} 5$$
$$2^5 = 32 \quad \text{True}$$

The solution is 33.

3. $\log_5 x + \log_5(x+4) = 1$

$$\log_5 x(x+4) = 1$$
$$\log_5(x^2+4x) = 1$$
$$5^1 = x^2 + 4x$$
$$0 = x^2 + 4x - 5$$
$$0 = (x+5)(x-1)$$
$$x+5 = 0 \quad \text{or} \quad x-1 = 0$$
$$x = -5 \qquad\qquad x = 1$$

Since $\log_5(-5)$ is undefined, -5 is rejected. The solution is 1.

4. $\log(x+3) - \log x = 1$

$$\log\frac{x+3}{x} = 1$$
$$10^1 = \frac{x+3}{x}$$
$$10x = x+3$$
$$9x = 3$$
$$x = \frac{1}{3}$$

The solution is $\frac{1}{3}$.

5. $y_0 = 60; \ t = 3$

$$y = y_0 e^{0.916t}$$
$$y = 60e^{0.916(3)} = 60e^{2.748} \approx 937$$

The population will be approximately 937 rabbits.

6. $P = \$3000; \ r = 7\% = 0.07; \ n = 12;$
$A = 2P = \$6000$

$$A = P\left(1+\frac{r}{n}\right)^{nt}$$
$$6000 = 3000\left(1+\frac{0.07}{12}\right)^{12t}$$
$$2 = \left(1+\frac{0.07}{12}\right)^{12t}$$
$$\log 2 = \log\left(1+\frac{0.07}{12}\right)^{12t}$$
$$\log 2 = 12t\log\left(1+\frac{0.07}{12}\right)$$
$$\frac{\log 2}{12\log\left(1+\frac{0.07}{12}\right)} = t$$
$$9.9 \approx t$$

It takes nearly 10 years to double.

7. $y_1 = 5000\left(1+\frac{0.05}{4}\right)^{4x}$, $y_2 = 6000$

It takes 3.67 years, or 3 years and 8 months.

Vocabulary, Readiness & Video Check 9.8

1. $\ln(4x-2) = \ln 3$ is the same as $\log_e(4x-2) = \log_e 3$. Therefore, from the logarithm property of equality, we know that $4x - 2 = 3$.

2. Substituting 8 in the original equation gives us the logarithm of a negative number, which does not exist—we can only take the logarithm of a positive number.

3. $2000 = 1000\left(1+\frac{0.07}{12}\right)^{12 \cdot t}$

$$t \approx 9.9$$

As long as the interest rate and compounding are the same, it takes any amount of money the same time to double.

Exercise Set 9.8

1. $3^x = 6$

$$\log 3^x = \log 6$$
$$x \log 3 = \log 6$$
$$x = \frac{\log 6}{\log 3} \approx 1.6309$$

3. $3^{2x} = 3.8$

$$\log 3^{2x} = \log 3.8$$
$$2x \log 3 = \log 3.8$$
$$x = \frac{\log 3.8}{2 \log 3} \approx 0.6076$$

5. $2^{x-3} = 5$

$$\log 2^{x-3} = \log 5$$
$$(x-3) \log 2 = \log 5$$
$$x \log 2 - 3 \log 2 = \log 5$$
$$x \log 2 = 3 \log 2 + \log 5$$
$$x = \frac{3 \log 2 + \log 5}{\log 2}$$
or
$$x = 3 + \frac{\log 5}{\log 2}$$
$$x \approx 5.3219$$

7. $9^x = 5$

$$\log 9^x = \log 5$$
$$x \log 9 = \log 5$$
$$x = \frac{\log 5}{\log 9} \approx 0.7325$$

9. $4^{x+7} = 3$

$$\log 4^{x+7} = \log 3$$
$$(x+7) \log 4 = \log 3$$
$$x \log 4 + 7 \log 4 = \log 3$$
$$x \log 4 = \log 3 - 7 \log 4$$
$$x = \frac{\log 3 - 7 \log 4}{\log 4}$$
or
$$x = \frac{\log 3}{\log 4} - 7$$
$$x \approx -6.2075$$

11. $\log_2 (x+5) = 4$

$$x + 5 = 2^4$$
$$x + 5 = 16$$
$$x = 11$$

13. $\log_4 2 + \log_4 x = 0$

$$\log_4 (2x) = 0$$
$$2x = 4^0$$
$$2x = 1$$
$$x = \frac{1}{2}$$

15. $\log_2 6 - \log_2 x = 3$

$$\log_2 \left(\frac{6}{x}\right) = 3$$
$$\frac{6}{x} = 2^3$$
$$\frac{6}{x} = 8$$
$$8x = 6$$
$$x = \frac{3}{4}$$

17. $\log_6 (x^2 - x) = 1$

$$6^1 = x^2 - x$$
$$0 = x^2 - x - 6$$
$$0 = (x-3)(x+2)$$
$$x = 3 \text{ or } x = -2$$

19. $\log_4 x + \log_4 (x+6) = 2$

$$\log_4 x(x+6) = 2$$
$$x(x+6) = 4^2$$
$$x^2 + 6x = 16$$
$$x^2 + 6x - 16 = 0$$
$$(x+8)(x-2) = 0$$
$$x = -8 \quad \text{or} \quad x = 2$$

We discard -8 as extraneous, the solution is 2.

21. $\log_5(x+3) - \log_5 x = 2$

$$\log_5\left(\frac{x+3}{x}\right) = 2$$

$$\frac{x+3}{x} = 5^2$$

$$\frac{x+3}{x} = 25$$

$$x+3 = 25x$$

$$3 = 24x$$

$$x = \frac{1}{8}$$

23. $7^{3x-4} = 11$

$$\log 7^{3x-4} = \log 11$$

$$(3x-4)\log 7 = \log 11$$

$$3x\log 7 - 4\log 7 = \log 11$$

$$3x\log 7 = 4\log 7 + \log 11$$

$$x = \frac{4\log 7 + \log 11}{3\log 7}$$

or

$$x = \frac{1}{3}\left(4 + \frac{\log 11}{\log 7}\right)$$

$$x \approx 1.7441$$

25. $\log_4(x^2 - 3x) = 1$

$$x^2 - 3x = 4$$

$$x^2 - 3x - 4 = 0$$

$$(x-4)(x+1) = 0$$

$$x = 4 \quad \text{or} \quad x = -1$$

27. $e^{6x} = 5$

$$\ln e^{6x} = \ln 5$$

$$6x = \ln 5$$

$$x = \frac{\ln 5}{6} \approx 0.2682$$

29. $\log_3 x^2 = 4$

$$x^2 = 3^4$$

$$x^2 = 81$$

$$x = \pm 9$$

31. $\ln 5 + \ln x = 0$

$$\ln(5x) = 0$$

$$e^0 = 5x$$

$$1 = 5x$$

$$\frac{1}{5} = x$$

33. $3\log x - \log x^2 = 2$

$$3\log x - 2\log x = 2$$

$$\log x = 2$$

$$x = 10^2$$

$$x = 100$$

35. $\log_4 x - \log_4(2x-3) = 3$

$$\log_4\left(\frac{x}{2x-3}\right) = 3$$

$$\frac{x}{2x-3} = 4^3$$

$$x = 64(2x-3)$$

$$x = 128x - 192$$

$$192 = 127x$$

$$x = \frac{192}{127}$$

37. $\log_2 x + \log_2(3x+1) = 1$

$$\log_2 x(3x+1) = 1$$

$$x(3x+1) = 2$$

$$3x^2 + x - 2 = 0$$

$$(3x-2)(x+1) = 0$$

$$3x-2 = 0 \quad \text{or} \quad x+1 = 0$$

$$x = \frac{2}{3} \quad \text{or} \quad x = -1$$

We discard -1 as extraneous, the solution is $\frac{2}{3}$.

39. $\log_2 x + \log_2(x+5) = 1$

$$\log_2 x(x+5) = 1$$

$$x(x+5) = 2$$

$$x^2 + 5x - 2 = 0$$

$$a = 1, \, b = 5, \, c = -2$$

$$x = \frac{-5 \pm \sqrt{5^2 - 4(1)(-2)}}{2(1)}$$

$$x = \frac{-5 \pm \sqrt{33}}{2}$$

Discard $\dfrac{-5 - \sqrt{33}}{2}$, the solution is $\dfrac{-5 + \sqrt{33}}{2}$.

41. Let $y_0 = 83$ and $t = 5$.

$$y = y_0 e^{0.043t}$$

$$y = 83e^{0.043(5)} = 83e^{0.215} \approx 103$$

The population is estimated to be 103 wolves in 5 years.

43. Let $y_0 = 11,488$ and $t = 15$.

$$y = y_0 e^{-0.0277t}$$

$$y = 11,488 e^{-0.0277(15)} = 11,488 e^{-0.4155} \approx 7582$$

The population of the Cook Islands is predicted to be 7582 in 2025.

45. $A = P\left(1 + \dfrac{r}{n}\right)^{nt}$, $P = 600$,

$A = 2(600) = 1200$, $r = 0.07$, $n = 12$

$$1200 = 600\left(1 + \frac{0.07}{12}\right)^{12t}$$

$$2 = \left(1 + \frac{0.07}{12}\right)^{12t}$$

$$\log 2 = \log\left(1 + \frac{0.07}{12}\right)^{12t}$$

$$\log 2 = 12t \log\left(1 + \frac{0.07}{12}\right)$$

$$\frac{\log 2}{12 \log\left(1 + \frac{0.07}{12}\right)} = t$$

$$9.9 \approx t$$

It takes approximately 9.9 years for the $600 to double.

47. $A = P\left(1 + \dfrac{r}{n}\right)^{nt}$, $P = 1200$,

$A = P + I = 1200 + 200 = 1400$

$r = 0.09$, $n = 4$

$$1400 = 1200\left(1 + \frac{0.09}{4}\right)^{4t}$$

$$\frac{7}{6} = (1.0225)^{4t}$$

$$\log \frac{7}{6} = \log 1.0225^{4t}$$

$$\log \frac{7}{6} = 4t \log 1.0225$$

$$t = \frac{\log \frac{7}{6}}{4 \log 1.0225}$$

$$t \approx 1.7$$

It would take the investment approximately 1.7 years to earn $200.

49. $A = P\left(1 + \dfrac{r}{n}\right)^{nt}$, $P = 1000$

$A = 2(1000) = 2000$, $r = 0.08$, $n = 2$

$$2000 = 1000\left(1 + \frac{0.08}{2}\right)^{2t}$$

$$2 = (1.04)^{2t}$$

$$\log 2 = \log 1.04^{2t}$$

$$\log 2 = 2t \log 1.04$$

$$t = \frac{\log 2}{2 \log 1.04}$$

$$t \approx 8.8$$

It takes 8.8 years to double.

51. $w = 0.00185h^{2.67}$, and $h = 35$

$$w = 0.00185(35)^{2.67} \approx 24.5$$

The expected weight of a boy 35 inches tall is 24.5 pounds.

53. $w = 0.00185h^{2.67}$, and $w = 85$

$$85 = 0.00185h^{2.67}$$

$$\frac{85}{0.00185} = h^{2.67}$$

$$h = \left(\frac{85}{0.00185}\right)^{1/2.67} \approx 55.7$$

The expected height of the boy is 55.7 inches.

55. $P = 14.7e^{-0.21x}$, $x = 1$

$$P = 14.7e^{-0.21(1)}$$

$$= 14.7e^{-0.21}$$

$$\approx 11.9$$

The average atmospheric pressure in Denver is approximately 11.9 pounds per square inch.

57. $P = 14.7e^{-0.21x}$, $P = 7.5$

$$7.5 = 14.7e^{-0.21x}$$

$$\frac{7.5}{14.7} = e^{-0.21x}$$

$$-0.21x = \ln\left(\frac{7.5}{14.7}\right)$$

$$x = -\frac{1}{0.21} \ln\left(\frac{7.5}{14.7}\right) \approx 3.2$$

The elevation of the jet is approximately 3.2 miles.

59. $t = \dfrac{1}{c}\ln\left(\dfrac{A}{A-N}\right)$

$t = \dfrac{1}{0.09}\ln\left(\dfrac{75}{75-50}\right)$

$t = \dfrac{1}{0.09}\ln(3)$

$t \approx 12.21$

It will take 12 weeks.

61. $t = \dfrac{1}{c}\ln\left(\dfrac{A}{A-N}\right)$

$t = \dfrac{1}{0.07}\ln\left(\dfrac{210}{210-150}\right)$

$t = \dfrac{1}{0.07}\ln(3.5)$

$t \approx 17.9$

It will take 18 weeks.

63. $\dfrac{x^2 - y + 2z}{3x} = \dfrac{(-2)^2 - 0 + 2(3)}{3(-2)}$

$= \dfrac{4+6}{-6}$

$= \dfrac{10}{-6}$

$= -\dfrac{5}{3}$

65. $\dfrac{3z - 4x + y}{x + 2z} = \dfrac{3(3) - 4(-2) + 0}{-2 + 2(3)} = \dfrac{9+8}{-2+6} = \dfrac{17}{4}$

67. $f(x) = 5x + 2$

$y = 5x + 2$

$x = 5y + 2$

$\dfrac{x-2}{5} = y$

$f^{-1}(x) = \dfrac{x-2}{5}$

69. $y = y_0 e^{-0.003t}$

$y = 9{,}500{,}000,\ y_0 = 9{,}939{,}000$

$9{,}500{,}000 = 9{,}939{,}000 e^{-0.003t}$

$\dfrac{9{,}500{,}000}{9{,}939{,}000} = e^{-0.003t}$

$\ln\dfrac{9500}{9939} = -0.003t$

$t = \dfrac{\ln\dfrac{9500}{9939}}{-0.003} \approx 15.1$

The population of Michigan will be 9,500,000 after approximately 15 years.

71. answers may vary

73. a.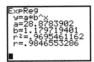

The regression equation is

$y = 28.878(1.18)^x$.

b.

In 2009, spending was approximately $127.81 billion.

c.

Spending is predicted to be $406.47 billion in 2016.

75. a.

The regression equation is

$y = 64.204(1.045)^x$.

b.

In 2009, there were approximately 849 million vehicles on the road.

c.

In 2020, there are predicted to be 1374 million vehicles on the road.

77. a.

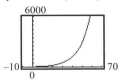

The regression equation is

$$y = 36.720(1.092)^x.$$

b.

Health spending was approximately $3515.2 billion in 2012.

c.

Health spending is predicted to be $4189.4 billion in 2014.

79. $Y_1 = e^{0.3x}$, $Y_2 = 8$

$x \approx 6.93$

81. $Y_1 = 2\log(-5.6x + 1.3) + x + 1$, $Y_2 = 0$

$x \approx -3.68, 0.19$

Chapter 9 Vocabulary Check

1. For a one-to-one function, we can find its <u>inverse</u> function by switching the coordinates of the ordered pairs of the function.

2. The <u>composition</u> of functions *f* and *g* is $(f \circ g)(x) = f(g(x))$.

3. A function of the form $f(x) = b^x$ is called an <u>exponential</u> function if $b > 0$, *b* is not 1, and *x* is a real number.

4. The graphs of *f* and f^{-1} are <u>symmetric</u> about the line $y = x$.

5. <u>Natural</u> logarithms are logarithms to base *e*.

6. <u>Common</u> logarithms are logarithms to base 10.

7. To see whether a graph is the graph of a one-to-one function, apply the <u>vertical</u> line test to see whether it is a function, and then apply the <u>horizontal</u> line test to see whether it is a one-to-one function.

8. A <u>logarithmic</u> function is a function that can be defined by $f(x) = \log_b x$ where *x* is a positive real number, *b* is a constant positive real number, and *b* is not 1.

9. <u>Half-life</u> is the amount of time it takes for half of the amount of a substance to decay.

10. A quantity that grows or decays by the same percent at regular time periods is said to have <u>exponential</u> growth or decay.

Chapter 9 Review

1. $(f + g)(x) = f(x) + g(x)$
$$= (x - 5) + (2x + 1)$$
$$= x - 5 + 2x + 1$$
$$= 3x - 4$$

2. $(f - g)(x) = f(x) - g(x)$
$$= (x - 5) - (2x + 1)$$
$$= x - 5 - 2x - 1$$
$$= -x - 6$$

3. $(f \cdot g)(x) = f(x) \cdot g(x)$
$$= (x - 5)(2x + 1)$$
$$= 2x^2 + x - 10x - 5$$
$$= 2x^2 - 9x - 5$$

4. $\left(\dfrac{g}{f}\right)(x) = \dfrac{g(x)}{f(x)} = \dfrac{2x+1}{x-5}, x \neq 5$

5. $(f \circ g)(x) = f(g(x))$
$$= f(x + 1)$$
$$= (x + 1)^2 - 2$$
$$= x^2 + 2x - 1$$

6. $(g \circ f)(x) = g(f(x))$
$$= g(x^2 - 2)$$
$$= x^2 - 2 + 1$$
$$= x^2 - 1$$

7. $(h \circ g)(2) = h(g(2)) = h(3) = 3^3 - 3^2 = 18$

8. $(f \circ f)(x) = f(f(x))$
$$= f(x^2 - 2)$$
$$= (x^2 - 2)^2 - 2$$
$$= x^4 - 4x^2 + 4 - 2$$
$$= x^4 - 4x^2 + 2$$

9. $(f \circ g)(-1) = f(g(-1)) = f(0) = 0^2 - 2 = -2$

10. $(h \circ h)(2) = h(h(2)) = h(4) = 4^3 - 4^2 = 48$

11. The function is one-to-one.
$$h^{-1} = \{(14, -9), (8, 6), (12, -11), (15, 15)\}$$

12. The function is not one-to-one.

13. The function is one-to-one.

Rank in Housing Starts for a recent year (Input)	4	3	1	2
U.S. Region (Output)	Northeast	Midwest	South	West

14. The function is not one-to-one.

15. $f(x) = \sqrt{x + 2}$

 a. $f(7) = \sqrt{7 + 2} = \sqrt{9} = 3$

 b. $f^{-1}(3) = 7$

16. $f(x) = \sqrt{x + 2}$

 a. $f(-1) = \sqrt{-1 + 2} = \sqrt{1} = 1$

 b. $f^{-1}(1) = -1$

17. The graph does not represent a one-to-one function.

18. The graph does not represent a one-to-one function.

19. The graph does not represent a one-to-one function.

20. The graph represents a one-to-one function.

21. $f(x) = x - 9$
$$y = x - 9$$
$$x = y - 9$$
$$y = x + 9$$
$$f^{-1}(x) = x + 9$$

22. $f(x) = x + 8$
$$y = x + 8$$
$$x = y + 8$$
$$y = x - 8$$
$$f^{-1}(x) = x - 8$$

23. $f(x) = 6x + 11$

$y = 6x + 11$

$x = 6y + 11$

$6y = x - 11$

$y = \dfrac{x - 11}{6}$

$f^{-1}(x) = \dfrac{x - 11}{6}$

24. $f(x) = 12x$

$y = 12x$

$x = 12y$

$y = \dfrac{x}{12}$

$f^{-1}(x) = \dfrac{x}{12}$

25. $f(x) = x^3 - 5$

$y = x^3 - 5$

$x = y^3 - 5$

$y^3 = x + 5$

$y = \sqrt[3]{x + 5}$

$f^{-1}(x) = \sqrt[3]{x + 5}$

26. $f(x) = \sqrt[3]{x + 2}$

$y = \sqrt[3]{x + 2}$

$x = \sqrt[3]{y + 2}$

$x^3 = y + 2$

$y = x^3 - 2$

$f^{-1}(x) = x^3 - 2$

27. $g(x) = \dfrac{12x - 7}{6}$

$y = \dfrac{12x - 7}{6}$

$x = \dfrac{12y - 7}{6}$

$6x = 12y - 7$

$12y = 6x + 7$

$y = \dfrac{6x + 7}{12}$

$g^{-1}(x) = \dfrac{6x + 7}{12}$

28. $r(x) = \dfrac{13x - 5}{2}$

$y = \dfrac{13x - 5}{2}$

$x = \dfrac{13y - 5}{2}$

$2x = 13y - 5$

$2x + 5 = 13y$

$y = \dfrac{2x + 5}{13}$

$r^{-1}(x) = \dfrac{2x + 5}{13}$

29. $f(x) = -2x + 3$

$y = -2x + 3$

$x = -2y + 3$

$x - 3 = -2y$

$-\dfrac{x - 3}{2} = y$

$f^{-1}(x) = -\dfrac{x - 3}{2}$

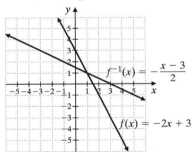

30. $f(x) = 5x - 5$

$y = 5x - 5$

$x = 5y - 5$

$x + 5 = 5y$

$\dfrac{x + 5}{5} = y$

$f^{-1}(x) = \dfrac{x + 5}{5}$

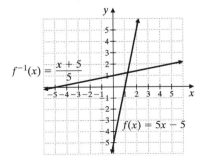

31. $4^x = 64$
$4^x = 4^3$
$x = 3$

32. $3^x = \dfrac{1}{9}$
$3^x = 3^{-2}$
$x = -2$

33. $2^{3x} = \dfrac{1}{16}$
$2^{3x} = 2^{-4}$
$3x = -4$
$x = -\dfrac{4}{3}$

34. $5^{2x} = 125$
$5^{2x} = 5^3$
$2x = 3$
$x = \dfrac{3}{2}$

35. $9^{x+1} = 243$
$(3^2)^{x+1} = 3^5$
$3^{2x+2} = 3^5$
$2x + 2 = 5$
$2x = 3$
$x = \dfrac{3}{2}$

36. $8^{3x-2} = 4$
$(2^3)^{3x-2} = 2^2$
$2^{9x-6} = 2^2$
$9x - 6 = 2$
$9x = 8$
$x = \dfrac{8}{9}$

37. $y = 3^x$

38. $y = \left(\dfrac{1}{3}\right)^x$

39. $y = 2^{x-4}$

40. $y = 2^x + 4$

41. $A = P\left(1 + \dfrac{r}{n}\right)^{nt}$

$A = 1600\left(1 + \dfrac{0.09}{2}\right)^{(2)(7)}$

$A \approx 2963.11$
The amount accrued is \$2963.11.

42. $A = P\left(1 + \dfrac{r}{n}\right)^{nt}$

$A = 800\left(1 + \dfrac{0.07}{4}\right)^{(4)(5)}$

$A \approx 1131.82$
The certificate is worth \$1131.82 at the end of 5 years.

43. $C = 79{,}087, r = 0.044, x = 2020 - 2000 = 20$

$y = C(1+r)^x$

$y = 79{,}087(1+0.044)^{20}$

$y \approx 187{,}118$

The predicted population in 2020 is 187,118.

44. $C = 287{,}370, r = 0.042, x = 2018 - 2000 = 18$

$y = C(1+r)^x$

$y = 287{,}370(1+0.042)^{18}$

$y \approx 602{,}643$

The predicted population in 2018 is 602,643.

45. $C = 1024, r = 0.5, x = 7$

$y = C(1-r)^x$

$y = 1024(1-0.5)^7$

$y = 8$

After 7 rounds there will be 8 players.

46. $C = 1280, r = 0.11, x = 6$

$y = C(1-r)^x$

$y = 1280(1-0.11)^6$

$y \approx 636$

The predicted bear population in 6 years is 636.

47. $\quad 49 = 7^2$

$\log_7 49 = 2$

48. $\quad 2^{-4} = \dfrac{1}{16}$

$\log_2 \dfrac{1}{16} = -4$

49. $\log_{1/2} 16 = -4$

$\left(\dfrac{1}{2}\right)^{-4} = 16$

50. $\log_{0.4} 0.064 = 3$

$\quad 0.4^3 = 0.064$

51. $\log_4 x = -3$

$\quad x = 4^{-3} = \dfrac{1}{64}$

52. $\log_3 x = 2$

$\quad x = 3^2 = 9$

53. $\log_3 1 = x$

$3^x = 1$

$3^x = 3^0$

$x = 0$

54. $\log_4 64 = x$

$4^x = 64$

$4^x = 4^3$

$x = 3$

55. $\log_4 4^5 = x$

$x = 5$

56. $\log_7 7^{-2} = x$

$x = -2$

57. $5^{\log_5 4} = x$

$x = 4$

58. $2^{\log_2 9} = x$

$9 = x$

59. $\log_2 (3x - 1) = 4$

$3x - 1 = 2^4$

$3x - 1 = 16$

$3x = 17$

$x = \dfrac{17}{3}$

60. $\log_3 (2x + 5) = 2$

$2x + 5 = 3^2$

$2x + 5 = 9$

$2x = 4$

$x = 2$

61. $\log_4 (x^2 - 3x) = 1$

$x^2 - 3x = 4$

$x^2 - 3x - 4 = 0$

$(x+1)(x-4) = 0$

$x = -1 \text{ or } x = 4$

62. $\log_8 (x^2 + 7x) = 1$

$x^2 + 7x = 8$

$x^2 + 7x - 8 = 0$

$(x+8)(x-1) = 0$

$x = -8 \text{ or } x = 1$

63. $y = 2^x$ and $y = \log_2 x$

64. $y = \left(\dfrac{1}{2}\right)^x$ and $y = \log_{1/2} x$

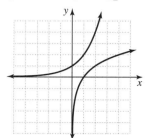

65. $\log_3 8 + \log_3 4 = \log_3 (8 \cdot 4) = \log_3 32$

66. $\log_2 6 + \log_2 3 = \log_2 (6 \cdot 3) = \log_2 18$

67. $\log_7 15 - \log_7 20 = \log_7 \dfrac{15}{20} = \log_7 \dfrac{3}{4}$

68. $\log 18 - \log 12 = \log \dfrac{18}{12} = \log \dfrac{3}{2}$

69. $\log_{11} 8 + \log_{11} 3 - \log_{11} 6 = \log_{11} \dfrac{(8)(3)}{6}$
$\qquad\qquad = \log_{11} 4$

70. $\log_5 14 + \log_5 3 - \log_5 21$
$\quad = \log_5 (14 \cdot 3) - \log_5 21$
$\quad = \log_5 \dfrac{42}{21}$
$\quad = \log_5 2$

71. $2 \log_5 x - 2 \log_5 (x+1) + \log_5 x$
$\quad = \log_5 x^2 - \log_5 (x+1)^2 + \log_5 x$
$\quad = \log_5 \dfrac{(x^2)(x)}{(x+1)^2}$
$\quad = \log_5 \dfrac{x^3}{(x+1)^2}$

72. $4 \log_3 x - \log_3 x + \log_3 (x+2)$
$\quad = 3 \log_3 x + \log_3 (x+2)$
$\quad = \log_3 x^3 + \log_3 (x+2)$
$\quad = \log_3 \left[x^3 (x+2) \right]$
$\quad = \log_3 (x^4 + 2x^3)$

73. $\log_3 \dfrac{x^3}{x+2} = \log_3 x^3 - \log_3 (x+2)$
$\qquad\qquad = 3 \log_3 x - \log_3 (x+2)$

74. $\log_4 \dfrac{x+5}{x^2} = \log_4 (x+5) - \log_4 x^2$
$\qquad\qquad = \log_4 (x+5) - 2 \log_4 x$

75. $\log_2 \dfrac{3x^2 y}{z} = \log_2 (3x^2 y) - \log_2 z$
$\qquad\qquad = \log_2 3 + \log_2 x^2 + \log_2 y - \log_2 z$
$\qquad\qquad = \log_2 3 + 2 \log_2 x + \log_2 y - \log_2 z$

76. $\log_7 \dfrac{yz^3}{x} = \log_7 (yz^3) - \log_7 x$
$\qquad\qquad = \log_7 y + \log_7 z^3 - \log_7 x$
$\qquad\qquad = \log_7 y + 3 \log_7 z - \log_7 x$

77. $\log_b 50 = \log_b (5)(5)(2)$
$\qquad\quad = \log_b (5) + \log_b (5) + \log_b (2)$
$\qquad\quad = 0.83 + 0.83 + 0.36$
$\qquad\quad = 2.02$

78. $\log_b \dfrac{4}{5} = \log_b 4 - \log_b 5$
$\qquad\quad = \log_b 2^2 - \log_b 5$
$\qquad\quad = 2 \log_b 2 - \log_b 5$
$\qquad\quad = 2(0.36) - 0.83$
$\qquad\quad = 0.72 - 0.83$
$\qquad\quad = -0.11$

79. $\log 3.6 \approx 0.5563$

80. $\log 0.15 \approx -0.8239$

81. $\ln 1.25 \approx 0.2231$

82. $\ln 4.63 \approx 1.5326$

83. $\log 1000 = \log 10^3 = 3$

84. $\log \dfrac{1}{10} = \log 10^{-1} = -1$

85. $\ln \dfrac{1}{e} = \ln e^{-1} = -1$

86. $\ln e^4 = 4$

87. $\ln(2x) = 2$

$\qquad 2x = e^2$

$\qquad x = \dfrac{e^2}{2}$

88. $\ln(3x) = 1.6$

$\qquad 3x = e^{1.6}$

$\qquad x = \dfrac{e^{1.6}}{3}$

89. $\ln(2x - 3) = -1$

$\qquad 2x - 3 = e^{-1}$

$\qquad x = \dfrac{e^{-1} + 3}{2}$

90. $\ln(3x + 1) = 2$

$\qquad 3x + 1 = e^2$

$\qquad 3x = e^2 - 1$

$\qquad x = \dfrac{e^2 - 1}{3}$

91. $\qquad \ln \dfrac{I}{I_0} = -kx$

$\qquad \ln \dfrac{0.03 I_0}{I_0} = -2.1x$

$\qquad \ln 0.03 = -2.1x$

$\qquad \dfrac{\ln 0.03}{-2.1} = x$

$\qquad\qquad x \approx 1.67$

The depth is 1.67 millimeters.

92. $\qquad \ln \dfrac{I}{I_0} = -kx$

$\qquad \ln \dfrac{0.02 I_0}{I_0} = -3.2x$

$\qquad \ln 0.02 = -3.2x$

$\qquad \dfrac{\ln 0.02}{-3.2} = x$

$\qquad\qquad x \approx 1.22$

2% of the original radioactivity will penetrate at a depth of approximately 1.22 millimeters.

93. $\log_5 1.6 = \dfrac{\log 1.6}{\log 5} \approx 0.2920$

94. $\log_3 4 = \dfrac{\log 4}{\log 3} \approx 1.2619$

95. $A = Pe^{rt}$

$\qquad A = 1450 e^{(0.03)(5)}$

$\qquad A \approx 1684.66$

The accrued amount is \$1684.66.

96. $A = Pe^{rt}$

$\qquad A = 940 e^{0.04(3)} = 940 e^{0.12} \approx 1059.85$

The investment grows to \$1059.85.

97. $\qquad 3^{2x} = 7$

$\qquad \log 3^{2x} = \log 7$

$\qquad 2x \log 3 = \log 7$

$\qquad\qquad x = \dfrac{\log 7}{2 \log 3} \approx 0.8856$

98. $\qquad 6^{3x} = 5$

$\qquad \log 6^{3x} = \log 5$

$\qquad 3x \log 6 = \log 5$

$\qquad\qquad x = \dfrac{\log 5}{3 \log 6} \approx 0.2994$

99.
$$3^{2x+1} = 6$$
$$\log 3^{2x+1} = \log 6$$
$$(2x+1)\log 3 = \log 6$$
$$2x\log 3 + \log 3 = \log 6$$
$$2x\log 3 = \log 6 - \log 3$$
$$x = \frac{\log 6 - \log 3}{2\log 3}$$
or
$$x = \frac{1}{2}\left(\frac{\log 6}{\log 3} - 1\right)$$
$$x \approx 0.3155$$

100.
$$4^{3x+2} = 9$$
$$\log 4^{3x+2} = \log 9$$
$$(3x+2)\log 4 = \log 9$$
$$3x\log 4 + 2\log 4 = \log 9$$
$$3x\log 4 = \log 9 - 2\log 4$$
$$x = \frac{\log 9 - 2\log 4}{3\log 4}$$
or
$$x = \frac{1}{3}\left(\frac{\log 9}{\log 4} - 2\right)$$
$$x \approx -0.1383$$

101.
$$5^{3x-5} = 4$$
$$\log 5^{3x-5} = \log 4$$
$$(3x-5)\log 5 = \log 4$$
$$3x\log 5 - 5\log 5 = \log 4$$
$$3x\log 5 = \log 4 + 5\log 5$$
$$x = \frac{\log 4 + 5\log 5}{3\log 5}$$
or
$$x = \frac{1}{3}\left(\frac{\log 4}{\log 5} + 5\right)$$
$$x \approx 1.9538$$

102.
$$8^{4x-2} = 3$$
$$\log 8^{4x-2} = \log 3$$
$$(4x-2)\log 8 = \log 3$$
$$4x\log 8 - 2\log 8 = \log 3$$
$$4x\log 8 = \log 3 + 2\log 8$$
$$x = \frac{\log 3 + 2\log 8}{4\log 8}$$
or
$$x = \frac{1}{4}\left(\frac{\log 3}{\log 8} + 2\right)$$
$$x \approx 0.6321$$

103.
$$5^{x-1} = \frac{1}{2}$$
$$\log 5^{x-1} = \log\frac{1}{2}$$
$$(x-1)\log 5 = \log\frac{1}{2}$$
$$x\log 5 - \log 5 = \log\frac{1}{2}$$
$$x\log 5 = \log\frac{1}{2} + \log 5$$
$$x = \frac{\log\frac{1}{2} + \log 5}{\log 5}$$
or
$$x = -\frac{\log 2}{\log 5} + 1$$
$$x \approx 0.5693$$

104.
$$4^{x+5} = \frac{2}{3}$$
$$\log 4^{x+5} = \log\frac{2}{3}$$
$$(x+5)\log 4 = \log\frac{2}{3}$$
$$x\log 4 + 5\log 4 = \log\frac{2}{3}$$
$$x\log 4 = \log\frac{2}{3} - 5\log 4$$
$$x = \frac{\log\frac{2}{3} - 5\log 4}{\log 4}$$
or
$$x = \frac{\log\frac{2}{3}}{\log 4} - 5$$
$$x \approx -5.2925$$

105.
$$\log_5 2 + \log_5 x = 2$$
$$\log_5 2x = 2$$
$$2x = 5^2$$
$$2x = 25$$
$$x = \frac{25}{2}$$

106.
$$\log_3 x + \log_3 10 = 2$$
$$\log_3(10x) = 2$$
$$10x = 3^2$$
$$10x = 9$$
$$x = \frac{9}{10}$$

107.
$$\log(5x) - \log(x+1) = 4$$
$$\log\frac{5x}{x+1} = 4$$
$$\frac{5x}{x+1} = 10^4$$
$$\frac{5x}{x+1} = 10,000$$
$$5x = 10,000x + 10,000$$
$$x = -1.0005$$

no solution, or $\varnothing$

108.
$$-\log_6(4x+7) + \log_6 x = 1$$
$$\log_6\frac{x}{4x+7} = 1$$
$$\frac{x}{4x+7} = 6$$
$$x = 6(4x+7)$$
$$x = 24x + 42$$
$$x = -\frac{42}{23}$$

$-\frac{42}{23}$ is rejected since $\log_6\left(-\frac{42}{23}\right)$ is undefined.

There is no solution, or $\varnothing$.

109.
$$\log_2 x + \log_2 2x - 3 = 1$$
$$\log_2(x \cdot 2x) = 4$$
$$2x^2 = 2^4$$
$$2x^2 = 16$$
$$x^2 = 8$$
$$x = \pm 2\sqrt{2}$$

$-2\sqrt{2}$ is rejected since $\log_2\left(-2\sqrt{2}\right)$ is undefined. The solution is $2\sqrt{2}$.

110.
$$\log_3(x^2 - 8x) = 2$$
$$3^2 = x^2 - 8x$$
$$0 = x^2 - 8x - 9$$
$$0 = (x-9)(x+1)$$
$$x = 9 \text{ or } x = -1$$

111. Let $y_0 = 27$, $y = 347$, $r = 11.4\% = 0.114$.
$$y = y_0 e^{kt}$$
$$347 = 27e^{(0.114)t}$$
$$\frac{347}{27} = e^{0.114t}$$
$$\ln\left(\frac{347}{27}\right) = 0.114t$$
$$\frac{1}{0.114}\ln\left(\frac{347}{27}\right) = t$$
$$22.4 \approx t$$

It will take approximately 22.4 years.

112.
$$y = y_0 e^{kt}$$
$$70,000,000 = 65,822,000e^{0.004t}$$
$$\frac{70,000,000}{65,822,000} = e^{0.004t}$$
$$\ln\frac{70,000,000}{65,822,000} = 0.004t$$
$$t = \frac{1}{0.004}\ln\frac{70,000,000}{65,822,000} \approx 15.4$$

It will take approximately 15.4 years.

113.
$$y = y_0 e^{kt}$$
$$2(22,600,000) = 22,600,000e^{0.007t}$$
$$2 = e^{0.007t}$$
$$\ln 2 = 0.007t$$
$$t = \frac{\ln 2}{0.007} \approx 99.0$$

It will take approximately 99.0 years.

114.
$$y = y_0 t^{kt}$$
$$2(7,746,400) = 7,746,400e^{0.016t}$$
$$2 = e^{0.016t}$$
$$\ln 2 = 0.016t$$
$$t = \frac{\ln 2}{0.016} \approx 43.3$$

It will take approximately 43.3 years.

115.
$$A = P\left(1 + \frac{r}{n}\right)^{nt}$$
$$10,000 = 5000\left(1 + \frac{0.08}{4}\right)^{4t}$$
$$2 = (1.02)^{4t}$$
$$\log 2 = \log 1.02^{4t}$$
$$\log 2 = 4t \log 1.02$$
$$t = \frac{\log 2}{4 \log 1.02} \approx 8.8$$

It will take 8.8 years.

116.
$$A = P\left(1 + \frac{r}{n}\right)^{nt}$$
$$10,000 = 6000\left(1 + \frac{0.06}{12}\right)^{12t}$$
$$\frac{5}{3} = (1.005)^{12t}$$
$$\log \frac{5}{3} = \log 1.005^{12t}$$
$$\log \frac{5}{3} = 12t \log 1.005$$
$$t = \frac{1}{12}\left(\frac{\log\left(\frac{5}{3}\right)}{\log(1.005)}\right) \approx 8.5$$

It was invested for approximately 8.5 years.

117. $Y_1 = e^x$, $Y_2 = 2$

$x \approx 0.69$

118. $Y_1 = 10^{0.3x}$, $Y_2 = 7$

$x \approx 2.82$

119. $3^x = \frac{1}{81}$
$$3^x = 3^{-4}$$
$$x = -4$$

120. $7^{4x} = 49$
$$7^{4x} = 7^2$$
$$4x = 2$$
$$x = \frac{1}{2}$$

121. $8^{3x-2} = 32$
$$(2^3)^{(3x-2)} = 2^5$$
$$2^{9x-6} = 2^5$$
$$9x - 6 = 5$$
$$9x = 11$$
$$x = \frac{11}{9}$$

122. $9^{x-2} = 27$
$$(3^2)^{(x-2)} = 3^3$$
$$3^{2x-4} = 3^3$$
$$2x - 4 = 3$$
$$2x = 7$$
$$x = \frac{7}{2}$$

123. $\log_4 4 = x$
$$4^x = 4^1$$
$$x = 1$$

124. $\log_3 x = 4$
$$3^4 = x$$
$$81 = x$$

125. $\log_5(x^2 - 4x) = 1$
$$5^1 = x^2 - 4x$$
$$0 = x^2 - 4x - 5$$
$$0 = (x-5)(x+1)$$
$$x - 5 = 0 \quad \text{or} \quad x + 1 = 0$$
$$x = 5 \qquad\qquad x = -1$$

Both check, so the solutions are 5 and -1.

126. $\log_4(3x - 1) = 2$
$$4^2 = 3x - 1$$
$$16 + 1 = 3x$$
$$\frac{17}{3} = x$$

127. $\ln x = -3.2$
$$e^{\ln x} = e^{-3.2}$$
$$x = e^{-3.2}$$

128. $\log_5 x + \log_5 10 = 2$

$\log_5 (10x) = 2$

$5^2 = 10x$

$\dfrac{25}{10} = x$

$\dfrac{5}{2} = x$

129. $\ln x - \ln 2 = 1$

$\ln \dfrac{x}{2} = 1$

$e^{\ln \frac{x}{2}} = e^1$

$\dfrac{x}{2} = e$

$x = 2e$

130. $\log_6 x - \log_6 (4x + 7) = 1$

$\log_6 \dfrac{x}{4x+7} = 1$

$6^1 = \dfrac{x}{4x+7}$

$24x + 42 = x$

$23x = -42$

$x = -\dfrac{42}{23}$

$-\dfrac{42}{23}$ is rejected since $\log_6 \left(-\dfrac{42}{23}\right)$ is undefined.

There is no solution, or $\varnothing$.

Chapter 9 Test

1. $f(x) = x$ and $g(x) = 2x - 3$

$(f \cdot g)(x) = f(x) \cdot g(x) = x(2x - 3) = 2x^2 - 3x$

2. $f(x) = x$ and $g(x) = 2x - 3$

$(f - g)(x) = f(x) - g(x)$

$= x - (2x - 3)$

$= -x + 3$

$= 3 - x$

3. $(f \circ h)(0) = f(h(0)) = f(5) = 5$

4. $(g \circ f)(x) = g(f(x)) = g(x) = x - 7$

5. $(g \circ h)(x) = g(h(x))$

$= g(x^2 - 6x + 5)$

$= x^2 - 6x + 5 - 7$

$= x^2 - 6x - 2$

6. $f(x) = 7x - 14, \ f^{-1}(x) = \dfrac{x + 14}{7}$

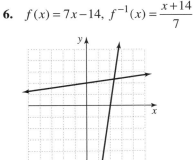

7. The graph represents a one-to-one function.

8. The graph does not represent a one-to-one function.

9. $y = 6 - 2x$ is one-to-one.

$x = 6 - 2y$

$2y = -x + 6$

$y = \dfrac{-x + 6}{2}$

$f^{-1}(x) = \dfrac{-x + 6}{2}$

10. $f = \{(0,0), (2,3), (-1,5)\}$ is one-to-one.

$f^{-1} = \{(0,0), (3,2), (5,-1)\}$

11. The function is not one-to-one.

12. $\log_3 6 + \log_3 4 = \log_3 (6 \cdot 4) = \log_3 24$

13. $\log_5 x + 3\log_5 x - \log_5 (x + 1)$

$= 4\log_5 x - \log_5 (x + 1)$

$= \log_5 x^4 - \log_5 (x + 1)$

$= \log_5 \dfrac{x^4}{x + 1}$

14. $\log_6 \dfrac{2x}{y^3} = \log_6 2x - \log_6 y^3$

$= \log_6 2 + \log_6 x - 3\log_6 y$

15. $\log_b \left(\dfrac{3}{25}\right) = \log_b 3 - \log_b 25$

$= \log_b 3 - \log_b 5^2$

$= \log_b 3 - 2\log_b 5$

$= 0.79 - 2(1.16)$

$= -1.53$

16. $\log_7 8 = \dfrac{\ln 8}{\ln 7} \approx 1.0686$

17. $8^{x-1} = \dfrac{1}{64}$

$8^{x-1} = 8^{-2}$

$x - 1 = -2$

$x = -1$

18. $3^{2x+5} = 4$

$\log 3^{2x+5} = \log 4$

$(2x+5)\log 3 = \log 4$

$2x = \dfrac{\log 4}{\log 3} - 5$

$x = \dfrac{1}{2}\left(\dfrac{\log 4}{\log 3} - 5\right)$

$x \approx -1.8691$

19. $\log_3 x = -2$

$x = 3^{-2}$

$x = \dfrac{1}{9}$

20. $\ln \sqrt{e} = x$

$\ln e^{1/2} = x$

$\dfrac{1}{2} = x$

21. $\log_8(3x-2) = 2$

$3x - 2 = 8^2$

$3x - 2 = 64$

$3x = 66$

$x = \dfrac{66}{3} = 22$

22. $\log_5 x + \log_5 3 = 2$

$\log_5(3x) = 2$

$3x = 5^2$

$3x = 25$

$x = \dfrac{25}{3}$

23. $\log_4(x+1) - \log_4(x-2) = 3$

$\log_4 \dfrac{x+1}{x-2} = 3$

$\dfrac{x+1}{x-2} = 4^3$

$\dfrac{x+1}{x-2} = 64$

$x + 1 = 64x - 128$

$129 = 63x$

$\dfrac{129}{63} = x$

$\dfrac{43}{21} = x$

24. $\ln(3x+7) = 1.31$

$3x + 7 = e^{1.31}$

$3x = e^{1.31} - 7$

$x = \dfrac{e^{1.31} - 7}{3} \approx -1.0979$

25. $y = \left(\dfrac{1}{2}\right)^x + 1$

26. $y = 3^x$ and $y = \log_3 x$

27. $A = \left(1 + \dfrac{r}{n}\right)^{nt}$, $P = 4000$, $t = 3$, $r = 0.09$,

and $n = 12$

$A = 4000\left(1 + \dfrac{0.09}{12}\right)^{12(3)}$

$= 4000(1.0075)^{36}$

≈ 5234.58

$\$5234.58$ will be in the account.

28. $A = \left(1 + \dfrac{r}{n}\right)^{nt}$, $P = 2000$, $A = 3000$

$r = 0.07$, $n = 2$

$3000 = 2000\left(1 + \dfrac{0.07}{2}\right)^{2t}$

$1.5 = (1.035)^{2t}$

$\log 1.5 = \log 1.035^{2t}$

$\log 1.5 = 2t \log 1.035$

$t = \dfrac{\log 1.5}{2 \log 1.035} \approx 5.9$

It would take 6 years.

29. Let $P = 3000$, $t = 10$.
Semiannually: Let $n = 2$, $r = 0.065$.

$A = P\left(1 + \dfrac{r}{n}\right)^{nt}$

$A = 3000\left(1 + \dfrac{0.065}{2}\right)^{2(10)}$

$A \approx 5688$

Monthly: Let $n = 12$, $r = 0.06$.

$A = 3000\left(1 + \dfrac{0.06}{12}\right)^{12(10)}$

$A \approx 5458$

The better investment is the 6.5% investment by
$\$5688 - \$5458 = \$230$.

30. $C = 150,000$, $r = 0.02$, $x = 20$

$y = C(1 - r)^x$

$y = 150,000(1 - 0.02)^{20}$

$y \approx 100,141$

The predicted population in 20 years is 100,141.

31. $C = 57,000$, $r = 0.026$, $x = 5$

$y = C(1 + r)^x$

$y = 57,000(1 + 0.026)^5$

$y \approx 64,805$

In 5 years, the predicted population is
64,805 animals.

32. $C = 400$, $r = 0.062$, $y = 1000$

$y = C(1 + r)^x$

$1000 = 400(1 + 0.062)^x$

$\dfrac{1000}{400} = 1.062^x$

$\ln\left(\dfrac{1000}{400}\right) = x \ln 1.062$

$x = \dfrac{\ln\left(\dfrac{1000}{400}\right)}{\ln 1.062}$

$x \approx 15$

It will take about 15 years to reach their goal.

33. Let $t = 0.5$.

$R(t) = 2.7^{-(1/3)t}$

$R(0.5) = 2.7^{-(1/3)(0.5)} \approx 0.85$

The probability is about 85%.

34. Let $t = 2$.

$R(t) = 2.7^{-(1/3)(2)} \approx 0.52$

The probability is about 52%.

35.

The solution is approximately 3.95.

Chapter 9 Cumulative Review

1. a. $(-8)(-1) = 8$

   b. $(-2)\dfrac{1}{6} = -\dfrac{1}{3}$

   c. $-1.2(0.3) = -0.36$

   d. $0(-11) = 0$

   e. $\left(\dfrac{1}{5}\right)\left(-\dfrac{10}{11}\right) = -\dfrac{2}{11}$

   f. $(7)(1)(-2)(-3) = 42$

   g. $8(-2)(0) = 0$

2. $\dfrac{1}{3}(x - 2) = \dfrac{1}{4}(x + 1)$

     $4(x - 2) = 3(x + 1)$

     $4x - 8 = 3x + 3$

         $x = 11$

3. $y = x^2$

4. $y = f(x) = -3x + 4, \ m = -3$

Perpendicular line: $m = \dfrac{1}{3}$, through $(-2, 6)$

$$y - y_1 = m(x - x_1)$$
$$y - 6 = \frac{1}{3}[x - (-2)]$$
$$y - 6 = \frac{1}{3}x + \frac{2}{3}$$
$$y = \frac{1}{3}x + \frac{20}{3}$$
$$f(x) = \frac{1}{3}x + \frac{20}{3}$$

5. Equation 2 is twice the opposite of equation 1 and equation 3 is one-half of equation 1. Therefore, the system is dependent. The solution is $\{(x, y, z) | x - 5y - 2z = 6\}$.

6. The angles labeled $y°$ and $(x - 40)°$ are alternate interior angles, so $y = x - 40$. The angles labeled $x°$ and $y°$ are supplementary, so $x + y = 180$.

$$\begin{cases} y = x - 40 \\ x + y = 180 \end{cases}$$

Replace y with $x - 40$ in the second equation.
$$x + (x - 40) = 180$$
$$2x = 220$$
$$x = 110$$
$$y = x - 40 = 110 - 40 = 70$$

7. a. $\dfrac{x^7}{x^4} = x^{7-4} = x^3$

b. $\dfrac{5^8}{5^2} = 5^{8-2} = 5^6$

c. $\dfrac{20x^6}{4x^5} = \dfrac{20}{4}x^{6-5} = 5x^1 = 5x$

d. $\dfrac{12y^{10}z^7}{14y^8z^7} = \dfrac{12}{14}y^{10-8}z^{7-7} = \dfrac{6}{7}y^2z^0 = \dfrac{6y^2}{7}$

8. a. $(4a^3)^2 = 4^2(a^3)^2 = 16a^6$

b. $\left(-\dfrac{2}{3}\right)^3 = \dfrac{(-2)^3}{3^3} = \dfrac{-8}{27} = -\dfrac{8}{27}$

c. $\left(\dfrac{4a^5}{b^3}\right)^3 = \dfrac{4^3(a^5)^3}{(b^3)^3} = \dfrac{64a^{15}}{b^9}$

d. $\left(\dfrac{3^{-2}}{x}\right)^{-3} = \dfrac{(3^{-2})^{-3}}{x^{-3}} = \dfrac{3^6}{x^{-3}} = 729x^3$

e. $(a^{-2}b^3c^{-4})^{-2} = (a^{-2})^{-2}(b^3)^{-2}(c^{-4})^{-2}$
$$= a^4b^{-6}c^8$$
$$= \dfrac{a^4c^8}{b^6}$$

9. a. $C(100) = \dfrac{2.6(100) + 10,000}{100}$
$$= 102.60$$
The cost is \$102.60 per disc for 100 discs.

b. $C(1000) = \dfrac{2.6(1000) + 10,000}{1000}$
$$= 12.60$$
The cost is \$12.60 per disc for 1000 discs.

10. a. $(3x - 1)^2 = (3x)^2 - 2(3x)(1) + 1^2$
$$= 9x^2 - 6x + 1$$

b. $\left(\dfrac{1}{2}x + 3\right)\left(\dfrac{1}{2}x - 3\right) = \left(\dfrac{1}{2}x\right)^2 - 3^2$
$$= \dfrac{1}{4}x^2 - 9$$

c. $(2x - 5)(6x + 7) = 12x^2 + 14x - 30x - 35$
$$= 12x^2 - 16x - 35$$

11. a. $\dfrac{x}{4} + \dfrac{5x}{4} = \dfrac{6x}{4} = \dfrac{3x}{2}$

b. $\dfrac{5}{7z^2} + \dfrac{x}{7z^2} = \dfrac{5 + x}{7z^2}$

c. $\dfrac{x^2}{x+7} - \dfrac{49}{x+7} = \dfrac{x^2-49}{x+7}$

$\qquad\qquad = \dfrac{(x+7)(x-7)}{x+7}$

$\qquad\qquad = x-7$

d. $\dfrac{x}{3y^2} - \dfrac{x+1}{3y^2} = \dfrac{x-x-1}{3y^2} = -\dfrac{1}{3y^2}$

12. $\dfrac{5}{x-2} + \dfrac{3}{x^2+4x+4} - \dfrac{6}{x+2}$

$= \dfrac{5}{x-2} + \dfrac{3}{(x+2)^2} - \dfrac{6}{x+2}$

$= \dfrac{5(x+2)^2 + 3(x-2) - 6(x-2)(x+2)}{(x-2)(x+2)(x+2)}$

$= \dfrac{-x^2+23x+38}{(x-2)(x+2)^2}$

13. $x^2-1\overline{\smash{\big)}\,3x^4+2x^3-8x+6}$

$\qquad\dfrac{3x^2+2x+3}{}$

$\qquad\quad\underline{3x^4-3x^2}$

$\qquad\qquad 2x^3+3x^2-8x$

$\qquad\qquad \underline{2x^3-2x}$

$\qquad\qquad\quad 3x^2-6x+6$

$\qquad\qquad\quad \underline{3x^2-3}$

$\qquad\qquad\qquad\quad -6x+9$

Solution: $3x^2+2x+3+\dfrac{-6x+9}{x^2-1}$

14. a. $\dfrac{\frac{a}{5}}{\frac{a-1}{10}} = \dfrac{a}{5}\cdot\dfrac{10}{a-1} = \dfrac{2a}{a-1}$

b. $\dfrac{\frac{3}{2+a}+\frac{6}{2-a}}{\frac{5}{a+2}-\frac{1}{a-2}} = \dfrac{\frac{3}{a+2}-\frac{6}{a-2}}{\frac{5}{a+2}-\frac{1}{a-2}}$

Multiply the numerator and the denominator
by $(a + 2)(a - 2)$.

$\dfrac{3(a-2)-6(a+2)}{5(a-2)-1(a+2)} = \dfrac{3a-6-6a-12}{5a-10-a-2}$

$\qquad\qquad\qquad\qquad = \dfrac{-3a-18}{4a-12}$

$\qquad\qquad\qquad\qquad = -\dfrac{3(a+6)}{4(a-3)}$

c. $\dfrac{x^{-1}+y^{-1}}{xy} = \dfrac{\frac{1}{x}+\frac{1}{y}}{xy} = \dfrac{\left(\frac{1}{x}+\frac{1}{y}\right)xy}{(xy)(xy)} = \dfrac{y+x}{x^2y^2}$

15. $\dfrac{2x}{2x-1}+\dfrac{1}{x} = \dfrac{1}{2x-1}$

$(2x-1)x\left(\dfrac{2x}{2x-1}+\dfrac{1}{x}\right) = (2x-1)x\left(\dfrac{1}{2x-1}\right)$

$\qquad\qquad 2x^2+2x-1 = x$

$\qquad\qquad 2x^2+x-1 = 0$

$\qquad\qquad (2x-1)(x+1) = 0$

$2x-1 = 0 \quad\text{or}\quad x+1 = 0$

$\quad 2x = 1 \qquad\qquad x = -1$

$\qquad x = \dfrac{1}{2}$

$x = \dfrac{1}{2}$ makes the denominator $2x - 1$ zero, so
the only solution is $x = -1$.

16. $\dfrac{x^3-8}{x-2} = \dfrac{(x-2)(x^2+2x+4)}{(x-2)}$

$\qquad\quad = x^2+2x+4$

17. Use distance = (rate)(time). Let c be the speed of
the current.

	d	r	t
Upstream	72	$30-c$	$1.5t$
Downstream	72	$30+c$	t

Upstream: $72 = 1.5t(30-c)$ or $t = \dfrac{72}{1.5(30-c)}$

Downstream: $72 = t(30+c)$ or $t = \dfrac{72}{30+c}$

$\dfrac{72}{1.5(30-c)} = \dfrac{72}{30+c}$

$\quad 72(30+c) = 72(1.5)(30-c)$

$\quad 2160+72c = 3240-108c$

$\qquad\quad 180c = 1080$

$\qquad\qquad c = 6$

The speed of the current is 6 miles per hour.

18. $2\,\lfloor 8 \quad -12 \quad -7$

$\qquad\qquad 16 \qquad 8$

$\qquad \overline{8 \qquad 4 \qquad 1}$

Solution: $8x + 4 + \dfrac{1}{x-2}$

19. a. $\sqrt[4]{81} = \sqrt[4]{3^4} = 3$

b. $\sqrt[5]{-243} = \sqrt[5]{(-3)^5} = -3$

c. $-\sqrt{25} = -\sqrt{5^2} = -5$

d. $\sqrt[4]{-81}$ is not a real number.

e. $\sqrt[3]{64x^3} = \sqrt[3]{4^3 x^3} = 4x$

20. $\dfrac{1}{a+5} = \dfrac{1}{3a+6} - \dfrac{a+2}{a^2+7a+10}$

$\dfrac{1}{a+5} = \dfrac{1}{3(a+2)} - \dfrac{a+2}{(a+2)(a+5)}$

$3(a+2) = a+5 - 3(a+2)$

$3a+6 = a+5 - 3a - 6$

$5a = -7$

$a = -\dfrac{7}{5}$

21. a. $\sqrt{x} \cdot \sqrt[4]{x} = x^{1/2} \cdot x^{1/4} = x^{3/4} = \sqrt[4]{x^3}$

b. $\dfrac{\sqrt{x}}{\sqrt[3]{x}} = \dfrac{x^{1/2}}{x^{1/3}} = x^{\frac{1}{2}-\frac{1}{3}} = x^{1/6} = \sqrt[6]{x}$

c. $\sqrt[3]{3} \cdot \sqrt{2} = 3^{1/3} \cdot 2^{1/2}$

$= 3^{2/6} \cdot 2^{3/6}$

$= 9^{1/6} \cdot 8^{1/6}$

$= 72^{1/6}$

$= \sqrt[6]{72}$

22. $y = kx$

$\dfrac{1}{2} = 12k$

$k = \dfrac{1}{24}, \; y = \dfrac{1}{24}x$

23. a. $\sqrt{3}\left(5 + \sqrt{30}\right) = 5\sqrt{3} + \sqrt{90} = 5\sqrt{3} + 3\sqrt{10}$

b. $\left(\sqrt{5} - \sqrt{6}\right)\left(\sqrt{7} + 1\right) = \sqrt{35} + \sqrt{5} - \sqrt{42} - \sqrt{6}$

c. $\left(7\sqrt{x} + 5\right)\left(3\sqrt{x} - \sqrt{5}\right)$

$= 21x - 7\sqrt{5x} + 15\sqrt{x} - 5\sqrt{5}$

d. $\left(4\sqrt{3} - 1\right)^2$

$= \left(4\sqrt{3}\right)^2 - 2\left(4\sqrt{3}\right)(1) + 1^2$

$= 16 \cdot 3 - 8\sqrt{3} + 1$

$= 49 - 8\sqrt{3}$

e. $\left(\sqrt{2x} - 5\right)\left(\sqrt{2x} + 5\right) = \left(\sqrt{2x}\right)^2 - 5^2$

$= 2x - 25$

f. $\left(\sqrt{x-3} + 5\right)^2 = \left(\sqrt{x-3}\right)^2 + 2\sqrt{x-3}(5) + 5^2$

$= x - 3 + 10\sqrt{x-3} + 25$

$= x + 22 + 10\sqrt{x-3}$

24. a. $\sqrt[3]{27} = \sqrt[3]{3^3} = 3$

b. $\sqrt[3]{-27} = \sqrt[3]{(-3)^3} = -3$

c. $\sqrt{\dfrac{9}{64}} = \sqrt{\left(\dfrac{3}{8}\right)^2} = \dfrac{3}{8}$

d. $\sqrt[4]{x^{12}} = x^3$

e. $\sqrt[3]{-125y^6} = -5y^2$

25. $\dfrac{\sqrt[4]{x}}{\sqrt[4]{81y^5}} = \dfrac{\sqrt[4]{x}}{\sqrt[4]{81y^5}} \cdot \dfrac{\sqrt[4]{y^3}}{\sqrt[4]{y^3}} = \dfrac{\sqrt[4]{xy^3}}{3y^2}$

26. a. $a^{1/4}\left(a^{3/4} - a^{7/4}\right) = a^{4/4} - a^{8/4} = a - a^2$

b. $(x^{1/2} - 3)(x^{1/2} + 5)$

$= x^{2/2} + 5x^{1/2} - 3x^{1/2} - 15$

$= x + 2x^{1/2} - 15$

27.
$$\sqrt{4-x} = x-2$$
$$\left(\sqrt{4-x}\right)^2 = (x-2)^2$$
$$4-x = x^2 - 4x + 4$$
$$0 = x^2 - 3x$$
$$0 = x(x-3)$$
$$x = 0 \quad \text{or} \quad x - 3 = 0$$
$$x = 3$$

$x = 0$ does not check, so the only solution is $x = 3$.

28. a.
$$\sqrt{\frac{54}{6}} = \sqrt{9} = 3$$

b.
$$\frac{\sqrt{108a^2}}{3\sqrt{3}} = \frac{1}{3}\sqrt{\frac{108a^2}{3}}$$
$$= \frac{1}{3}\sqrt{36a^2}$$
$$= \frac{1}{3}(6a)$$
$$= 2a$$

c.
$$\frac{3\sqrt[3]{81a^5b^{10}}}{\sqrt[3]{3b^4}} = 3\sqrt[3]{\frac{81a^5b^{10}}{3b^4}}$$
$$= 3\sqrt[3]{27a^5b^6}$$
$$= 9ab^2\sqrt[3]{a^2}$$

29.
$$3x^2 - 9x + 8 = 0$$
$$x^2 - 3x + \frac{8}{3} = 0$$
$$x^2 - 3x = -\frac{8}{3}$$
$$x^2 - 3x + \left(\frac{-3}{2}\right)^2 = -\frac{8}{3} + \left(\frac{-3}{2}\right)^2$$
$$x^2 - 3x + \frac{9}{4} = -\frac{8}{3} + \frac{9}{4}$$
$$\left(x - \frac{3}{2}\right)^2 = -\frac{5}{12}$$
$$x - \frac{3}{2} = \pm\sqrt{-\frac{5}{12}}$$
$$x - \frac{3}{2} = \pm\frac{i\sqrt{5}}{2\sqrt{3}}$$
$$x - \frac{3}{2} = \pm\frac{i\sqrt{15}}{6}$$
$$x = \frac{3}{2} \pm \frac{i\sqrt{15}}{6}$$
$$= \frac{9}{6} \pm \frac{i\sqrt{15}}{6}$$
$$= \frac{9 \pm i\sqrt{15}}{6}$$

The solutions are $\dfrac{9+i\sqrt{15}}{6}$ and $\dfrac{9-i\sqrt{15}}{6}$.

30. a.
$$\frac{\sqrt{20}}{3} + \frac{\sqrt{5}}{4} = \frac{2\sqrt{5}}{3} + \frac{\sqrt{5}}{4}$$
$$= \frac{8\sqrt{5} + 3\sqrt{5}}{12}$$
$$= \frac{11\sqrt{5}}{12}$$

b.
$$\sqrt[3]{\frac{24x}{27}} - \frac{\sqrt[3]{3x}}{2} = \frac{2\sqrt[3]{3x}}{3} - \frac{\sqrt[3]{3x}}{2}$$
$$= \frac{4\sqrt[3]{3x} - 3\sqrt[3]{3x}}{6}$$
$$= \frac{\sqrt[3]{3x}}{6}$$

31. $$\frac{3x}{x-2} - \frac{x+1}{x} = \frac{6}{x(x-2)}$$
$$3x(x) - (x+1)(x-2) = 6$$
$$3x^2 - x^2 + x + 2 = 6$$
$$2x^2 + x - 4 = 0$$
$$a = 2,\ b = 1,\ c = -4$$
$$x = \frac{-1 \pm \sqrt{1^2 - 4(2)(-4)}}{2(2)} = \frac{-1 \pm \sqrt{33}}{4}$$

32. $$\sqrt[3]{\frac{27}{m^4 n^8}} = \frac{\sqrt[3]{27}}{\sqrt[3]{m^4 n^8}}$$
$$= \frac{3}{mn^2 \sqrt[3]{mn^2}}$$
$$= \frac{3 \cdot \sqrt[3]{m^2 n}}{mn^2 \sqrt[3]{mn^2} \cdot \sqrt[3]{m^2 n}}$$
$$= \frac{3 \sqrt[3]{m^2 n}}{m^2 n^3}$$

33. $x^2 - 4x \le 0$
$$x(x - 4) = 0$$
$$x = 0,\ x = 4$$

Region	Test Point	$x(x-4) \le 0$	Result
$x < 0$	$x = -1$	$(-1)(-5) \le 0$	False
$0 < x < 4$	$x = 2$	$2(-2) \le 0$	True
$x > 4$	$x = 5$	$5(1) \le 0$	False

Solution: $[0, 4]$

34. $$c^2 = a^2 + b^2$$
$$8^2 = 4^2 + b^2$$
$$64 = 16 + b^2$$
$$48 = b^2$$
$$\pm 4\sqrt{3} = b$$
$b > 0$ so the length is $4\sqrt{3}$ inches.

35. $F(x) = (x-3)^2 + 1$

36. a. $i^8 = (i^2)^4 = (-1)^4 = 1$

b. $i^{21} = i(i^{20}) = i$

c. $i^{42} = i^2(i^{40}) = i^2 = -1$

d. $i^{-13} = \dfrac{1}{i^{13}} = \dfrac{1}{i(i^{12})} = \dfrac{1}{i} = \dfrac{i}{i^2} = -i$

37. $f(x) = x - 1$ and $g(x) = 2x - 3$

a. $(f+g)(x) = f(x) + g(x)$
$$= x - 1 + 2x - 3$$
$$= 3x - 4$$

b. $(f-g)(x) = f(x) - g(x)$
$$= x - 1 - (2x - 3)$$
$$= -x + 2$$

c. $(f \cdot g)(x) = f(x) \cdot g(x)$
$$= (x-1)(2x-3)$$
$$= 2x^2 - 5x + 3$$

d. $\left(\dfrac{f}{g}\right)(x) = \dfrac{f(x)}{g(x)} = \dfrac{x-1}{2x-3}$, where $x \ne \dfrac{3}{2}$.

38.
$$4x^2 + 8x - 1 = 0$$
$$x^2 + 2x - \frac{1}{4} = 0$$
$$x^2 + 2x = \frac{1}{4}$$
$$x^2 + 2x + \left(\frac{2}{2}\right)^2 = \frac{1}{4} + \left(\frac{2}{2}\right)^2$$
$$x^2 + 2x + 1 = \frac{1}{4} + 1$$
$$(x+1)^2 = \frac{5}{4}$$
$$x + 1 = \pm\sqrt{\frac{5}{4}}$$
$$x + 1 = \pm\frac{\sqrt{5}}{2}$$
$$x = -1 \pm \frac{\sqrt{5}}{2}$$
$$= \frac{-2 \pm \sqrt{5}}{2}$$

The solutions are $\dfrac{-2+\sqrt{5}}{2}$ and $\dfrac{-2-\sqrt{5}}{2}$.

39.
$$f(x) = x + 3$$
$$y = x + 3$$
$$x = y + 3$$
$$y = x - 3$$
$$f^{-1}(x) = x - 3$$

40.
$$\left(x - \frac{1}{2}\right)^2 = \frac{x}{2}$$
$$x^2 - x + \frac{1}{4} = \frac{1}{2}x$$
$$x^2 - \frac{3}{2}x + \frac{1}{4} = 0$$
$$4x^2 - 6x + 1 = 0$$
$$a = 4, b = -6, c = 1$$
$$x = \frac{-(-6) \pm \sqrt{(-6)^2 - 4(4)(1)}}{2(4)}$$
$$= \frac{6 \pm \sqrt{20}}{8}$$
$$= \frac{6 \pm 2\sqrt{5}}{8}$$
$$= \frac{3 \pm \sqrt{5}}{4}$$

The solutions are $\dfrac{3+\sqrt{5}}{4}$ and $\dfrac{3-\sqrt{5}}{4}$.

41. a. $\log_4 16 = \log_4 4^2 = 2$

b. $\log_{10} \dfrac{1}{10} = \log_{10} 10^{-1} = -1$

c. $\log_9 3 = \log_9 9^{1/2} = \dfrac{1}{2}$

42. $f(x) = -(x+1)^2 + 1$
Vertex: $(-1, 1)$
Axis of symmetry: $x = -1$

Chapter 10

Section 10.1 Practice Exercises

1. $x = \frac{1}{2}y^2$; $a = \frac{1}{2}$, $h = 0$, $k = 0$; vertex: (0, 0)

x	y
2	−2
$\frac{1}{2}$	−1
0	0
$\frac{1}{2}$	1
2	2

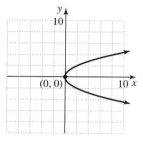

2. $x = -2(y+4)^2 - 1$; $a = -2$, $h = -1$, $k = -4$; vertex: (−1, −4)

x	y
−9	−6
3	−5
−1	−4
−3	−3
−9	−2

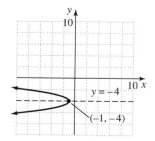

3.
$$y = -x^2 + 4x + 6$$
$$y - 6 = -x^2 + 4x$$
$$y - 6 = -(x^2 - 4x)$$
$$y - 6 - (+4) = -(x^2 - 4x + 4)$$
$$y - 10 = -(x - 2)^2$$
$$y = -(x - 2)^2 + 10$$

$a = -1$, $h = 2$, $k = 10$
vertex: (2, 10)

x	y
−1	1
0	6
1	9
2	10
3	9
4	6
5	1

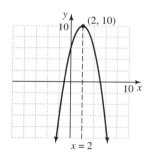

4. $x = 3y^2 + 6y + 4$
Find the vertex.
$$y = \frac{-b}{2a} = \frac{-6}{2(3)} = -1$$

$$x = 3(-1)^2 + 6(-1) + 4 = 3 - 6 + 4 = 1$$
vertex: (1, −1)
The axis of symmetry is the line $y = -1$.
Since $a > 0$, the parabola opens to the right.
$$x = 3(0)^2 + 6(0) + 4 = 4$$
The x-intercept is (4, 0).

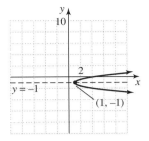

5. $x = y^2 + 4y - 1$

Solve the equation for y.

$y^2 + 4y + (-x - 1) = 0$

Here $a = 1$, $b = 4$, and $c = -x - 1$.

$y = \dfrac{-b \pm \sqrt{b^2 - 4ac}}{2a}$

$y = \dfrac{-4 \pm \sqrt{4^2 - 4(1)(-x-1)}}{2(1)}$

$= \dfrac{-4 \pm \sqrt{4^2 + 4(x+1)}}{2}$

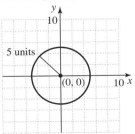

6. $\qquad x^2 + y^2 = 25$

$(x - 0)^2 + (y - 0)^2 = 5^2$

center: (0, 0); radius = 5

7. <u>Algebraic Solution</u>

$(x - 3)^2 + (y + 2)^2 = 4$

$h = 3$, $k = -2$, $r = \sqrt{4} = 2$

center: (3, −2)

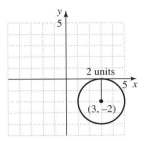

<u>Graphical Solution</u>

Solve the equation for y.

$(x - 3)^2 + (y + 2)^2 = 4$

$(y + 2)^2 = 4 - (x - 3)^2$

$y + 2 = \pm\sqrt{4 - (x - 3)^2}$

$y = -2 \pm \sqrt{4 - (x - 3)^2}$

8. $x^2 + y^2 = 20$

$y^2 = 20 - x^2$

$y = \pm\sqrt{20 - x^2}$

9. $\qquad x^2 + y^2 + 6x - 2y = 6$

$(x^2 + 6x) + (y^2 - 2y) = 6$

$(x^2 + 6x + 9) + (y^2 - 2y + 1) = 6 + 9 + 1$

$(x + 3)^2 + (y - 1)^2 = 16$

Center: (−3, 1); radius = $\sqrt{16} = 4$

10. Center: $(-2, -5)$; radius $= 9$

$(x-h)^2 + (y-k)^2 = r^2$

$h = -2$, $k = -5$, and $r = 9$.

The equation is $(x+2)^2 + (y+5)^2 = 81$.

Vocabulary, Readiness & Video Check 10.1

1. The circle, parabola, ellipse, and hyperbola are called the <u>conic sections</u>.

2. For a parabola that opens upward the lowest point is the <u>vertex</u>.

3. A <u>circle</u> is the set of all points in a plane that are the same distance from a fixed point. The fixed point is called the <u>center</u>.

4. The midpoint of a diameter of a circle is the <u>center</u>.

5. The distance from the center of a circle to any point of the circle is called the <u>radius</u>.

6. Twice a circle's radius is its <u>diameter</u>.

7. No, their graphs don't pass the vertical line test.

8. $x^2 + y^2 = r^2$

9. The formula for the standard form of a circle identifies the center and radius, so you just need to substitute these values into this formula and simplify.

10. Since the standard form of a circle involves a squared binomial for both x and y, we need to complete the square on both x and y.

Exercise Set 10.1

1. $y = x^2 - 7x + 5$; $a = 1$, upward

3. $x = -y^2 - y + 2$; $a = -1$, to the left

5. $y = -x^2 + 2x + 1$; $a = -1$, downward

7. $x = 3y^2$

   $x = 3(y-0)^2 + 0$

   $V(0, 0)$; opens to the right

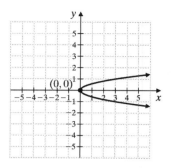

9. $x = -2y^2$

   $x = -2(y-0)^2 + 0$

   $V(0, 0)$; opens to the left

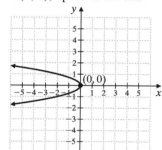

11. $y = -4x^2 = -4(x-0)^2 + 0$

    $V(0, 0)$; opens downward

13. $x = (y-2)^2 + 3$

    $V(3, 2)$; opens to the right

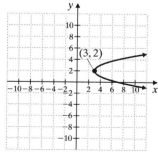

15. $y = -3(x-1)^2 + 5$

$V(1, 5)$; opens downward

17. $x = y^2 + 6y + 8$

$x = y^2 + 6y + 9 + 8 - 9$

$x = (y+3)^2 - 1$

$V(-1, -3)$; opens to the right

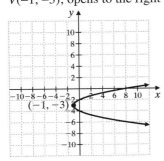

19. $y = x^2 + 10x + 20$

$y = x^2 + 10x + 25 + 20 - 25$

$y = (x+5)^2 - 5$

$V(-5, -5)$; opens upward

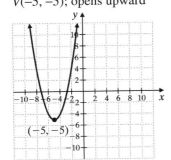

21. $x = -2y^2 + 4y + 6$

$x = -2(y^2 - 2y) + 6$

$x = -2(y^2 - 2y + 1) + 6 + 2$

$x = -2(y-1)^2 + 8$

$V(8, 1)$; opens to the left

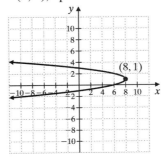

23. $x^2 + y^2 = 9$

   $(x-0)^2 + (y-0)^2 = 3^2$

  $C(0, 0)$ and $r = 3$

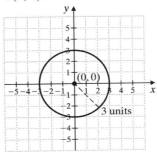

25. $x^2 + (y-2)^2 = 1$

   $(x-0)^2 + (y-2)^2 = 1^2$

  $C(0, 2)$ and $r = 1$

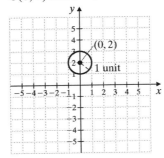

27. $(x-5)^2 + (y+2)^2 = 1$

$(x-5)^2 + (y+2)^2 = 1^2$

$C(5, -2)$ and $r = 1$

29. $x^2 + y^2 + 6y = 0$

$x^2 + y^2 + 6y + 9 = 0 + 9$

$(x-0)^2 + (y+3)^2 = 3^2$

$C(0, -3)$ and $r = 3$

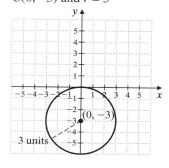

31. $x^2 + y^2 + 2x - 4y = 4$

$x^2 + 2x + 1 + y^2 - 4y + 4 = 4 + 1 + 4$

$(x+1)^2 + (y-2)^2 = 9$

$(x+1)^2 + (y-2)^2 = 3^2$

$C(-1, 2)$ and $r = 3$

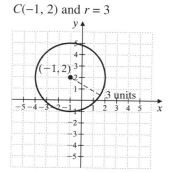

33. $(x+2)^2 + (y-3)^2 = 7$

$(x+2)^2 + (y-3)^2 = \left(\sqrt{7}\right)^2$

$C(-2, 3)$ and $r = \sqrt{7}$

35. $x^2 + y^2 - 4x - 8y - 2 = 0$

$x^2 - 4x + 4 + y^2 - 8y + 16 = 2 + 4 + 16$

$(x-2)^2 + (y-4)^2 = 22$

$(x-2)^2 + (y-4)^2 = \left(\sqrt{22}\right)^2$

$C(2, 4)$ and $r = \sqrt{22}$

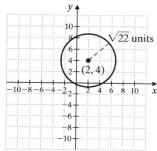

37. $3x^2 + 3y^2 = 75$

$x^2 + y^2 = 25$

$(x-0)^2 + (y-0)^2 = 5^2$

$C(0, 0)$ and $r = 5$

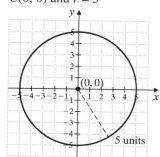

39. $6(x-4)^2 + 6(y-1)^2 = 24$

$(x-4)^2 + (y-1)^2 = 4$

$(x-4)^2 + (y-1)^2 = 2^2$

$C(4, 1)$ and $r = 2$

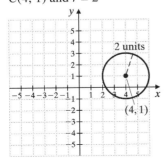

41. $4(x+1)^2 + 4(y-3)^2 = 12$

$(x+1)^2 + (y-3)^2 = 3$

$(x+1)^2 + (y-3)^2 = \left(\sqrt{3}\right)^2$

$C(-1, 3)$ and $r = \sqrt{3}$

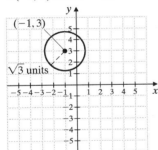

43. Center $(h, k) = (2, 3)$ and radius $r = 6$.

$(x-h)^2 + (y-k)^2 = r^2$

$(x-2)^2 + (y-3)^2 = 6^2$

$(x-2)^2 + (y-3)^2 = 36$

45. Center $(h, k) = (0, 0)$ and radius $r = \sqrt{3}$.

$(x-h)^2 + (y-k)^2 = r^2$

$(x-0)^2 + (y-0)^2 = \left(\sqrt{3}\right)^2$

$x^2 + y^2 = 3$

47. Center $(h, k) = (-5, 4)$ and radius $r = 3\sqrt{5}$.

$(x-h)^2 + (y-k)^2 = r^2$

$[x-(-5)]^2 + (y-4)^2 = \left(3\sqrt{5}\right)^2$

$(x+5)^2 + (y-4)^2 = 45$

49. $x = y^2 - 3$

$x = (y-0)^2 - 3$

Vertex: $(-3, 0)$

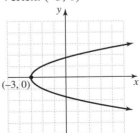

51. $y = (x-2)^2 - 2$

Vertex: $(2, -2)$

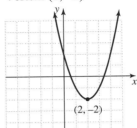

53. $x^2 + y^2 = 1$

Center: $(0, 0)$, radius $r = \sqrt{1} = 1$

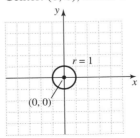

55. $x = (y+3)^2 - 1$

Vertex: $(-1, -3)$

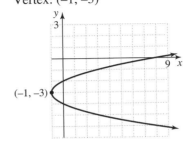

57. $(x-2)^2 + (y-2)^2 = 16$

Center: (2, 2), radius $r = \sqrt{16} = 4$

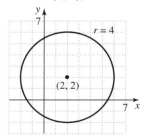

59. $x = -(y-1)^2$

Vertex: (0, 1)

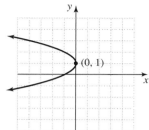

61. $(x-4)^2 + y^2 = 7$

Center: (4, 0), radius $r = \sqrt{7}$

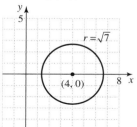

63. $y = 5(x+5)^2 + 3$

Vertex: (−5, 3)

65.
$$\frac{x^2}{8} + \frac{y^2}{8} = 2$$
$$8\left(\frac{x^2}{8} + \frac{y^2}{8}\right) = 8(2)$$
$$x^2 + y^2 = 16$$

Center: (0, 0), radius $r = \sqrt{16} = 4$

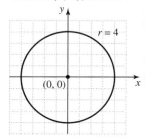

67.
$$y = x^2 + 7x + 6$$
$$y - 6 = x^2 + 7x$$
$$y - 6 + \frac{49}{4} = x^2 + 7x + \frac{49}{4}$$
$$y + \frac{25}{4} = \left(x + \frac{7}{2}\right)^2$$
$$y = \left(x + \frac{7}{2}\right)^2 - \frac{25}{4}$$

Vertex: $\left(-\frac{7}{2}, -\frac{25}{4}\right)$

69.
$$x^2 + y^2 + 2x + 12y - 12 = 0$$
$$(x^2 + 2x) + (y^2 + 12y) = 12$$
$$(x^2 + 2x + 1) + (y^2 + 12y + 36) = 12 + 1 + 36$$
$$(x+1)^2 + (y+6)^2 = 49$$

Center: (−1, −6), radius $r = \sqrt{49} = 7$

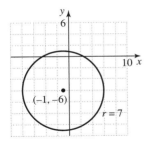

71.
$$x = y^2 + 8y - 4$$
$$x + 4 = y^2 + 8y$$
$$x + 4 + 16 = y^2 + 8y + 16$$
$$x + 20 = (y + 4)^2$$
$$x = (y + 4)^2 - 20$$
Vertex: $(-20, -4)$

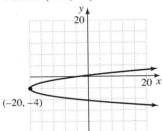

73.
$$x^2 - 10y + y^2 + 4 = 0$$
$$x^2 + (y^2 - 10y) = -4$$
$$x^2 + (y^2 - 10y + 25) = -4 + 25$$
$$x^2 + (y - 5)^2 = 21$$
Center: $(0, 5)$, radius $r = \sqrt{21}$

75.
$$x = -3y^2 + 30y$$
$$x = -3(y^2 - 10y)$$
$$x + [-3(25)] = -3(y^2 - 10y + 25)$$
$$x - 75 = -3(y - 5)^2$$
$$x = -3(y - 5)^2 + 75$$
Vertex: $(75, 5)$

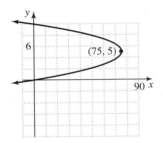

77.
$$5x^2 + 5y^2 = 25$$
$$x^2 + y^2 = 5$$
Center: $(0, 0)$, radius $r = \sqrt{5}$

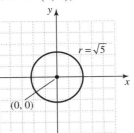

79.
$$y = 5x^2 - 20x + 16$$
$$y - 16 = 5(x^2 - 4x)$$
$$y - 16 + 5(4) = 5(x^2 - 4x + 4)$$
$$y + 4 = (x - 2)^2$$
$$y = (x - 2)^2 - 4$$
Vertex: $(2, -4)$

81. $y = 2x + 5$

83. $y = 3$

85. $\dfrac{1}{\sqrt{3}} = \dfrac{1 \cdot \sqrt{3}}{\sqrt{3} \cdot \sqrt{3}} = \dfrac{\sqrt{3}}{3}$

87. $\dfrac{4\sqrt{7}}{\sqrt{6}} = \dfrac{4\sqrt{7} \cdot \sqrt{6}}{\sqrt{6} \cdot \sqrt{6}} = \dfrac{4\sqrt{42}}{6} = \dfrac{2\sqrt{42}}{3}$

89. The vertex is $(1, -5)$.

91. a. $\text{radius} = \dfrac{1}{2}(\text{diameter})$

$\qquad = \dfrac{1}{2}(33 \text{ meters})$

$\qquad = 16.5 \text{ meters}$

b. $\text{circumference} = \pi(\text{diameter})$

$\qquad = \pi(33 \text{ meters})$

$\qquad \approx 103.67 \text{ meters}$

c. $\dfrac{103.67}{30} \approx 3.5$ meters apart

d. center: $(0, 16.5)$

e. $(x - 0)^2 + (y - 16.5)^2 = 16.5^2$

$\qquad x^2 + (y - 16.5)^2 = 16.5^2$

93. a. The radius was one-half of the diameter, or 125 feet.

b. $264 - 250 = 14$
The wheel was 14 feet above the ground.

c. $125 + 14 = 139$
The center of the wheel was 139 feet from the ground.

d. From the drawing, the center was at $(0, 139)$.

e. $(x - 0)^2 + (y - 139)^2 = 125^2$

$\qquad x^2 + (y - 139)^2 = 125^2$

95. answers may vary

97. Using A as $(0, 0)$, point B is at $(3, 1)$ and point C is at $(19, 13)$.

$d(B, C) = \sqrt{(x_2 - x_1)^2 + (y_2 - y_1)^2}$

$\qquad = \sqrt{(19 - 3)^2 + (13 - 1)^2}$

$\qquad = \sqrt{16^2 + 12^2}$

$\qquad = \sqrt{256 + 144}$

$\qquad = \sqrt{400}$

$\qquad = 20$

The distance across the lake is 20 meters.

99. $5x^2 + 5y^2 = 25$

$\qquad 5y^2 = 25 - 5x^2$

$\qquad y^2 = 5 - x^2$

$\qquad y = \pm\sqrt{5 - x^2}$

101. $y = 5x^2 - 20x + 16$

Section 10.2 Practice Exercises

1. $\dfrac{x^2}{25} + \dfrac{y^2}{4} = 1$

The equation is an ellipse with $a = 5$ and $b = 2$. The center is $(0, 0)$. The x-intercepts are $(5, 0)$ and $(-5, 0)$. The y-intercepts are $(2, 0)$ and $(-2, 0)$.

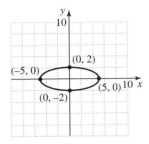

2. $9x^2 + 4y^2 = 36$

$$\frac{9x^2}{36} + \frac{4y^2}{36} = \frac{36}{36}$$

$$\frac{x^2}{4} + \frac{y^2}{9} = 1$$

This is an equation of an ellipse with $a = 2$ and $b = 3$. The ellipse has center $(0, 0)$, x-intercepts $(2, 0)$ and $(-2, 0)$, and y-intercepts $(3, 0)$ and $(-3, 0)$.

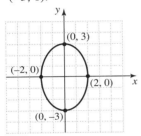

3. $\dfrac{(x-4)^2}{49} + \dfrac{(y+1)^2}{81} = 1$

This ellipse has center $(4, -1)$.
$a = 7$ and $b = 9$.
Find four points on the ellipse.
$(4 + 7, -1) = (11, -1)$
$(4 - 7, -1) = (-3, -1)$
$(4, -1 + 9) = (4, 8)$
$(4, -1 - 9) = (4, -10)$

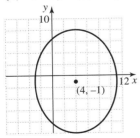

4. $\dfrac{x^2}{9} - \dfrac{y^2}{16} = 1$

This is a hyperbola with $a = 3$ and $b = 4$. It has center $(0, 0)$ and x-intercepts $(3, 0)$ and $(-3, 0)$. The asymptotes pass through $(3, 4)$, $(3, -4)$, $(-3, 4)$, and $(-3, -4)$.

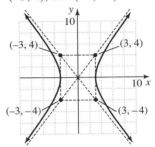

5. $9y^2 - 25x^2 = 225$

$$\frac{9y^2}{225} - \frac{25x^2}{225} = \frac{225}{225}$$

$$\frac{y^2}{25} - \frac{x^2}{9} = 1$$

This is a hyperbola with $a = 3$ and $b = 5$. The center is at $(0, 0)$ with y-intercepts $(0, 5)$ and $(0, -5)$. The asymptotes pass through $(3, 5)$, $(3, -5)$, $(-3, 5)$, and $(-3, -5)$.

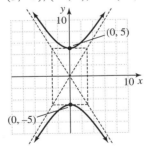

Vocabulary, Readiness & Video Check 10.2

1. A <u>hyperbola</u> is the set of points in a plane such that the absolute value of the differences of their distances from two fixed points is constant.

2. An <u>ellipse</u> is the set of points in a plane such that the sum of their distances from two fixed points is constant.

3. The two fixed points are each called a <u>focus</u>.

4. The point midway between the foci is called the <u>center</u>.

5. The graph of $\dfrac{x^2}{a^2} - \dfrac{y^2}{b^2} = 1$ is a <u>hyperbola</u> with center <u>(0, 0)</u> and <u>x</u>-intercepts of <u>(a, 0) and (−a, 0)</u>.

6. The graph of $\dfrac{x^2}{a^2} + \dfrac{y^2}{b^2} = 1$ is an <u>ellipse</u> with center <u>(0, 0)</u> and x-intercepts of <u>(a, 0) and (−a, 0)</u>.

7. a and b give us the location of 4 intercepts— $(a, 0)$, $(−a, 0)$, $(0, b)$ and $(0, −b)$ for $\dfrac{x^2}{a^2} + \dfrac{y^2}{b^2} = 1$ with center (0, 0). For Example 2, the values of a and b give us 4 points of the graph, just not intercepts. Here we move a distance of a units horizontally to the left and right of the center and b units above and below.

8. We use these points to draw asymptotes (also not part of the graph) which help us draw the correct shape of the hyperbola. The graph of a hyperbola gets closer and closer to the asymptotes without crossing them.

Exercise Set 10.2

1. $\dfrac{x^2}{16} + \dfrac{y^2}{4} = 1$ is an ellipse.

3. $x^2 − 5y^2 = 3$ is a hyperbola.

5. $-\dfrac{y^2}{25} + \dfrac{x^2}{36} = 1$ or

$\dfrac{x^2}{36} − \dfrac{y^2}{25} = 1$ is a hyperbola.

7. $\dfrac{x^2}{4} + \dfrac{y^2}{25} = 1$

$\dfrac{x^2}{2^2} + \dfrac{y^2}{5^2} = 1$

Center: (0, 0)
x-intercepts: (−2, 0), (2, 0)
y-intercepts: (0, −5), (0, 5)

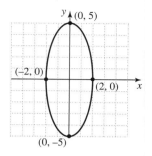

9. $\dfrac{x^2}{9} + y^2 = 1$

$\dfrac{x^2}{3^2} + \dfrac{y^2}{1^2} = 1$

Center: (0, 0)
x-intercepts: (−3, 0), (3, 0)
y-intercepts: (0, −1), (0, 1)

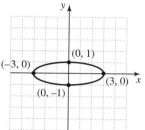

11. $9x^2 + y^2 = 36$

$\dfrac{x^2}{4} + \dfrac{y^2}{36} = 1$

$\dfrac{x^2}{2^2} + \dfrac{y^2}{6^2} = 1$

Center: (0, 0)
x-intercepts: (−2, 0), (2, 0)
y-intercepts: (0, −6), (0, 6)

13. $4x^2 + 25y^2 = 100$

$$\frac{x^2}{25} + \frac{y^2}{4} = 1$$

$$\frac{x^2}{5^2} + \frac{y^2}{2^2} = 1$$

Center: $(0, 0)$
x-intercepts: $(-5, 0)$, $(5, 0)$
y-intercepts: $(0, -2)$, $(0, 2)$

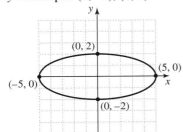

15. $\frac{(x+1)^2}{36} + \frac{(y-2)^2}{49} = 1$

$$\frac{(x+1)^2}{6^2} + \frac{(y-2)^2}{7^2} = 1$$

Center: $(-1, 2)$
Other points:
$(-1-6, 2) = (-7, 2)$
$(-1+6, 2) = (5, 2)$
$(-1, 2-7) = (-1, -5)$
$(-1, 2+7) = (-1, 9)$

17. $\frac{(x-1)^2}{4} + \frac{(y-1)^2}{25} = 1$

$$\frac{(x-1)^2}{2^2} + \frac{(y-1)^2}{5^2} = 1$$

Center: $(1, 1)$
Other points:
$(1-2, 1) = (-1, 1)$
$(1+2, 1) = (3, 1)$
$(1, 1-5) = (1, -4)$
$(1, 1+5) = (1, 6)$

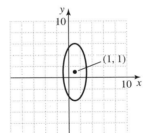

19. $\frac{x^2}{4} - \frac{y^2}{9} = 1$

$$\frac{x^2}{2^2} - \frac{y^2}{3^2} = 1$$

$a = 2$, $b = 3$

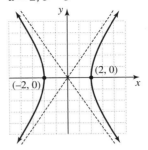

21. $\frac{y^2}{25} - \frac{x^2}{16} = 1$

$$\frac{y^2}{5^2} - \frac{x^2}{4^2} = 1$$

$a = 4$, $b = 5$

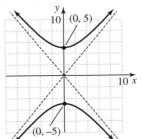

23. $x^2 - 4y^2 = 16$

$$\frac{x^2}{16} - \frac{y^2}{4} = 1$$

$$\frac{x^2}{4^2} - \frac{y^2}{2^2} = 1$$

$a = 4$, $b = 2$

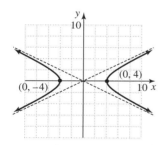

25. $16y^2 - x^2 = 16$

$$\frac{y^2}{1} - \frac{x^2}{16} = 1$$

$$\frac{y^2}{1^2} - \frac{x^2}{4^2} = 1$$

$a = 4,\ b = 1$

27. $\frac{y^2}{36} = 1 - x^2$

$$x^2 + \frac{y^2}{36} = 1$$

$$\frac{x^2}{1^2} + \frac{y^2}{6^2} = 1$$

$C(0, 0)$

x-intercepts: $(-1, 0), (1, 0)$

y-intercepts: $(0, -6), (0, 6)$

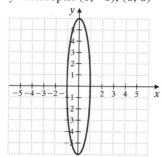

29. $4(x-1)^2 + 9(y+2)^2 = 36$

$$\frac{(x-1)^2}{9} + \frac{(y+2)^2}{4} = 1$$

$$\frac{(x-1)^2}{3^2} + \frac{(y+2)^2}{2^2} = 1$$

$C(1, -2)$

other points:

$(1-3, -2)$, or $(-2, -2)$

$(1+3, -2)$, or $(4, -2)$

$(1, -2-2)$, or $(1, -4)$

$(1, -2+2)$, or $(1, 0)$

31. $8x^2 + 2y^2 = 32$

$$\frac{8x^2}{32} + \frac{2y^2}{32} = 1$$

$$\frac{x^2}{4} + \frac{y^2}{16} = 1$$

$$\frac{x^2}{2^2} + \frac{y^2}{4^2} = 1$$

$C(0, 0)$

x-intercepts: $(-2, 0), (2, 0)$

y-intercepts: $(0, -4), (0, 4)$

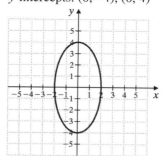

33. $25x^2 - y^2 = 25$

$$x^2 - \frac{y^2}{25} = 1$$

$$\frac{x^2}{1^2} - \frac{y^2}{5^2} = 1$$

$a = 1, b = 5$

35. $(x-7)^2 + (y-2)^2 = 4$

Circle; center $(7, 2)$, radius $r = \sqrt{4} = 2$

37. $y = x^2 + 12x + 36$

Parabola; $x = \dfrac{-b}{2a} = \dfrac{-12}{2(1)} = -6$

$y = (-6)^2 + 12(-6) + 36 = 0$

Vertex: $(-6, 0)$, opens upward

39. $\dfrac{y^2}{9} - \dfrac{x^2}{9} = 1$

$$\frac{y^2}{3^2} - \frac{x^2}{3^2} = 1$$

Hyperbola; center: $(0, 0)$
$a = 3, b = 3$
y-intercepts $(0, -3), (0, 3)$

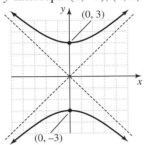

41. $\dfrac{x^2}{16} + \dfrac{y^2}{4} = 1$

$$\frac{x^2}{4^2} + \frac{y^2}{2^2} = 1$$

Ellipse; center: $(0, 0)$, $a = 4, b = 2$
x-intercepts $(-4, 0), (4, 0)$
y-intercepts $(0, -2), (0, 2)$

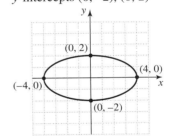

43. $x = y^2 + 4y - 1$

Parabola: $y = \dfrac{-b}{2a} = \dfrac{-4}{2(1)} = -2$

$x = (-2)^2 + 4(-2) - 1 = -5$

Vertex: $(-5, -2)$, opens to the right.

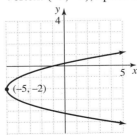

45. $9x^2 - 4y^2 = 36$

$$\frac{x^2}{4} - \frac{y^2}{9} = 1$$

$$\frac{x^2}{2^2} - \frac{y^2}{3^2} = 1$$

Hyperbola: center = (0, 0)
$a = 2$, $b = 3$
x-intercepts: (2, 0), (–2, 0)

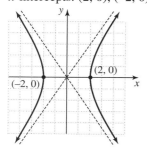

47. $\dfrac{(x-1)^2}{49} + \dfrac{(y+2)^2}{25} = 1$

$$\frac{(x-1)^2}{7^2} + \frac{(y+2)^2}{5^2} = 1$$

Ellipse; center: (1, –2)
$a = 7$, $b = 5$

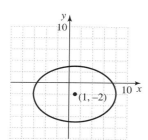

49. $\left(x + \dfrac{1}{2}\right)^2 + \left(y - \dfrac{1}{2}\right)^2 = 1$

Circle; center: $\left(-\dfrac{1}{2}, \dfrac{1}{2}\right)$, radius $r = 1$

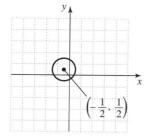

51. $(2x^3)(-4x^2) = -8x^5$

53. $-5x^2 + x^2 = -4x^2$

55. $\dfrac{x^2}{16} + \dfrac{y^2}{25} = 1$

$\sqrt{16} = 4$, so the distance between the
x-intercepts is $4 + 4 = 8$ units.
$\sqrt{25} = 5$, so the distance between the
y-intercepts is $5 + 5 = 10$ units.
The distance between the y-intercepts is longer
by $10 - 8 = 2$ units.

57. $4x^2 + y^2 = 16$

$$\frac{x^2}{4} + \frac{y^2}{16} = 1$$

$\sqrt{4} = 2$, so the distance between the x-intercepts
is $2 + 2 = 4$ units.
$\sqrt{16} = 4$, so the distance between the
y-intercepts is $4 + 4 = 8$ units.
The distance between the y-intercepts is longer
by $8 - 4 = 4$ units.

59. answers may vary

61. Circles: B, F
Ellipses: C, E, H
Hyperbolas: A, D, G

63. A: $c^2 = 36 + 13 = 49$; $c = \sqrt{49} = 7$

B: $c^2 = 4 - 4 = 0$; $c = \sqrt{0} = 0$

C: $c^2 = |25 - 16| = 9$; $c = \sqrt{9} = 3$

D: $c^2 = 39 + 25 = 64$; $c = \sqrt{64} = 8$

E: $c^2 = |81 - 17| = 64$; $c = \sqrt{64} = 8$

F: $c^2 = |36 - 36| = 0$; $c = \sqrt{0} = 0$

G: $c^2 = 65 + 16 = 81$; $c = \sqrt{81} = 9$

H: $c^2 = |144 - 140| = 4$; $c = \sqrt{4} = 2$

65. A: $e = \dfrac{7}{6}$

B: $e = \dfrac{0}{2} = 0$

C: $e = \dfrac{3}{5}$

D: $e = \dfrac{8}{5}$

Copyright © 2014 Pearson Education, Inc.

E: $e = \dfrac{8}{9}$

F: $e = \dfrac{0}{6} = 0$

G: $e = \dfrac{9}{4}$

H: $e = \dfrac{2}{12} = \dfrac{1}{6}$

67. They are equal to 0.

69. answers may vary

71. $a = 130,000,000 \Rightarrow a^2 = (130,000,000)^2$
$$= 1.69 \times 10^{16}$$

$b = 125,000,000 \Rightarrow b^2 = (125,000,000)^2$
$$= 1.5625 \times 10^{16}$$

Thus, the equation is
$$\frac{x^2}{1.69 \times 10^{16}} + \frac{y^2}{1.5625 \times 10^{16}} = 1.$$

73. $9x^2 + 4y^2 = 36$
$$4y^2 = 36 - 9x^2$$
$$y^2 = \frac{36 - 9x^2}{4}$$
$$y = \pm\sqrt{\frac{36 - 9x^2}{4}} = \pm\frac{\sqrt{36 - 9x^2}}{2}$$

75. $\dfrac{(x-1)^2}{4} - \dfrac{(y+1)^2}{25} = 1$
Center: $(1, -1)$
$a = 2, b = 5$

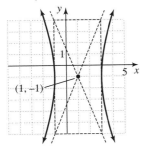

77. $\dfrac{y^2}{16} - \dfrac{(x+3)^2}{9} = 1$
Center: $(-3, 0)$
$a = 3, b = 4$

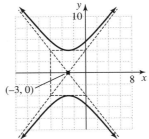

79. $\dfrac{(x+5)^2}{16} - \dfrac{(y+2)^2}{25} = 1$
Center: $(-5, -2)$
$a = 4, b = 5$

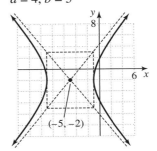

Integrated Review

1. $(x-7)^2 + (y-2)^2 = 4$
Circle; center: $(7, 2)$,
radius: $r = \sqrt{4} = 2$

2. $y = x^2 + 4$

Parabola; vertex: (0, 4)

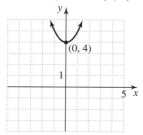

3. $y = x^2 + 12x + 36$

Parabola; $x = \dfrac{-b}{2a} = \dfrac{-12}{2(1)} = -6$

$y = (-6)^2 + 12(-6) + 36 = 0$

Vertex: (−6, 0)

4. $\dfrac{x^2}{4} + \dfrac{y^2}{9} = 1$

Ellipse; center: (0, 0)
$a = 2, b = 3$

5. $\dfrac{y^2}{9} - \dfrac{x^2}{9} = 1$

Hyperbola; center: (0, 0)
$a = 3, b = 3$

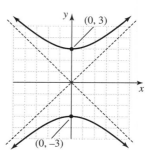

6. $\dfrac{x^2}{16} - \dfrac{y^2}{4} = 1$

Hyperbola; center: (0, 0)
$a = 4, b = 2$

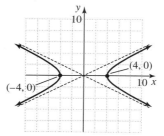

7. $\dfrac{x^2}{16} + \dfrac{y^2}{4} = 1$

Ellipse; center: (0, 0)
$a = 4, b = 2$

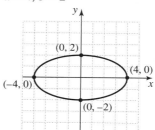

8. $x^2 + y^2 = 16$

Circle; center: (0, 0)
radius: $r = \sqrt{16} = 4$

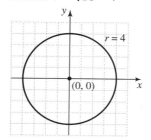

9. $x = y^2 + 4y - 1$

Parabola; $y = \dfrac{-b}{2a} = \dfrac{-4}{2(1)} = -2$

$x = (-2)^2 + 4(-2) - 1 = -5$
Vertex: $(-5, -2)$

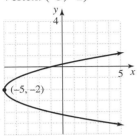

10. $x = -y^2 + 6y$

Parabola; $y = \dfrac{-b}{2a} = \dfrac{-6}{2(-1)} = 3$

$x = -(3)^2 + 6(3) = 9$
Vertex: $(9, 3)$

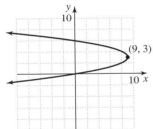

11. $9x^2 - 4y^2 = 36$

$\dfrac{x^2}{4} - \dfrac{y^2}{9} = 1$

Hyperbola; center: $(0, 0)$
$a = 2, b = 3$

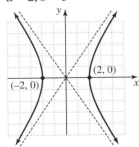

12. $9x^2 + 4y^2 = 36$

$\dfrac{x^2}{4} + \dfrac{y^2}{9} = 1$

Ellipse; center: $(0, 0)$
$a = 2, b = 3$

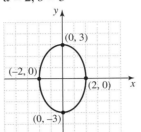

13. $\dfrac{(x-1)^2}{49} + \dfrac{(y+2)^2}{25} = 1$

Ellipse; center: $(1, -2)$,
$a = 7, b = 5$

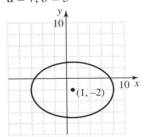

14. $y^2 = x^2 + 16$

$y^2 - x^2 = 16$

$\dfrac{y^2}{16} - \dfrac{x^2}{16} = 1$

Hyperbola; center: $(0, 0)$
$a = 4, b = 4$

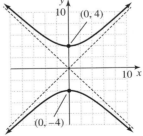

15. $\left(x+\dfrac{1}{2}\right)^2+\left(y-\dfrac{1}{2}\right)^2=1$

Circle; center: $\left(-\dfrac{1}{2},\dfrac{1}{2}\right)$, radius: $r=\sqrt{1}=1$

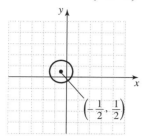

$\left(-\dfrac{1}{2},\dfrac{1}{2}\right)$

Section 10.3 Practice Exercises

1. $\begin{cases} y=-\sqrt{x} \\ x^2+y^2=20 \end{cases}$

Substitute $-\sqrt{x}$ for y in the second equation.
$$x^2+\left(-\sqrt{x}\right)^2=20$$
$$x^2+x=20$$
$$x^2+x-20=0$$
$$(x+5)(x-4)=0$$
$$x=-5 \ \text{ or } \ x=4$$
Let $x=-5$.
$$y=-\sqrt{-5} \qquad \text{Not a real number}$$
Let $x=4$.
$$y=-\sqrt{4}=-2$$
The solution is $(4,-2)$.

2. $\begin{cases} x^2-4y=4 \\ x+y=-1 \end{cases}$

Solve $x+y=-1$ for y.
$$y=-x-1$$
Replace y with $-x-1$ in the first equation and solve for x.
$$x^2-4(-x-1)=4$$
$$x^2+4x+4=4$$
$$x^2+4x=0$$
$$x(x+4)=0$$
$$x=0 \ \text{ or } \ x=-4$$
Let $x=0$, $\qquad$ Let $x=-4$,
$y=-0-1=-1$ $\qquad$ $y=-(-4)-1=3$
The solutions are $(0,-1)$ and $(-4,3)$.

3. $\begin{cases} x^2+y^2=9 \\ x-y=5 \end{cases}$

Solve the second equation for x.
$$x=y+5$$
Let $x=y+5$ in the first equation.
$$(y+5)^2+y^2=9$$
$$y^2+10y+25+y^2=9$$
$$2y^2+10y+16=0$$
$$y^2+5y+8=0$$
By the quadratic formula,
$$y=\frac{-5\pm\sqrt{5^2-4(1)(8)}}{2(1)}=\frac{-5\pm\sqrt{-7}}{2}$$
$\sqrt{-7}$ is not a real number. There is no real solution, or $\varnothing$.

4. $\begin{cases} x^2+4y^2=16 \\ x^2-y^2=1 \end{cases}$

Add the opposite of the second equation to the first.
$$\begin{array}{r} x^2+4y^2=16 \\ \underline{-x^2\ +y^2=-1} \\ 0\ +5y^2=15 \end{array}$$
$$y^2=3$$
$$y=\pm\sqrt{3}$$
Let $y=\sqrt{3}$. $\qquad\qquad$ Let $y=-\sqrt{3}$.
$x^2-\left(\sqrt{3}\right)^2=1$ $\qquad$ $x^2-\left(-\sqrt{3}\right)^2=1$
$\quad x^2-3=1$ $\qquad\qquad$ $\quad x^2-3=1$
$\qquad x^2=4$ $\qquad\qquad\quad$ $x^2=4$
$\qquad x=\pm2$ $\qquad\qquad\quad$ $x=\pm2$
The solutions are $\left(2,\sqrt{3}\right)$, $\left(2,-\sqrt{3}\right)$, $\left(-2,\sqrt{3}\right)$, and $\left(-2,-\sqrt{3}\right)$.

Vocabulary, Readiness & Video Check 10.3

1. Solving for y would either introduce tedious fractions (2nd equation) or a square root (1st equation) into the calculations.

2. When you multiply the left side of the equation by this number, do not forget to also multiply the right side.

Exercise Set 10.3

1. $\begin{cases} x^2 + y^2 = 25 & (1) \\ 4x + 3y = 0 & (2) \end{cases}$

Solve E2 for y.

$3y = -4x$

$y = -\dfrac{4x}{3}$

Substitute into E1.

$x^2 + \left(-\dfrac{4x}{3}\right)^2 = 25$

$x^2 + \dfrac{16x^2}{9} = 25$

$9\left(x^2 + \dfrac{16x^2}{9}\right) = 9(25)$

$9x^2 + 16x^2 = 225$

$25x^2 = 225$

$x^2 = 9$

$x = \pm\sqrt{9} = \pm 3$

$x = 3: y = -\dfrac{4(3)}{3} = -4$

$x = -3: y = -\dfrac{4(-3)}{3} = 4$

The solutions are $(3, -4)$ and $(-3, 4)$.

3. $\begin{cases} x^2 + 4y^2 = 10 & (1) \\ y = x & (2) \end{cases}$

Substitute x for y in E1.

$x^2 + 4x^2 = 10$

$5x^2 = 10$

$x^2 = 2$

$x = \pm\sqrt{2}$

Substitute these values into E2.

$x = \sqrt{2}: y = x = \sqrt{2}$

$x = -\sqrt{2}: y = x = -\sqrt{2}$

The solutions are $\left(\sqrt{2}, \sqrt{2}\right)$ and $\left(-\sqrt{2}, -\sqrt{2}\right)$.

5. $\begin{cases} y^2 = 4 - x & (1) \\ x - 2y = 4 & (2) \end{cases}$

Solve E2 for x.

$x = 2y + 4$

Substitute into E1.

$y^2 = 4 - (2y + 4)$

$y^2 = -2y$

$y^2 + 2y = 0$

$y(y + 2) = 0$

$y = 0$ or $y + 2 = 0$

$\qquad\qquad\qquad y = -2$

Substitute these values into the equation $x = 2y + 4$.

$y = 0: x = 2(0) + 4 = 4$

$y = -2: x = 2(-2) + 4 = 0$

The solutions are $(4, 0)$ and $(0, -2)$.

7. $\begin{cases} x^2 + y^2 = 9 & (1) \\ 16x^2 - 4y^2 = 64 & (2) \end{cases}$

Multiply E1 by 4 and add to E2.

$\quad 4x^2 + 4y^2 = 36$

$\underline{\quad 16x^2 - 4y^2 = 64 \quad}$

$20x^2 \qquad\quad = 100$

$\qquad\quad x^2 = 5$

$\qquad\quad x = \pm\sqrt{5}$

Substitute 5 for x^2 into E1.

$5 + y^2 = 9$

$\quad y^2 = 4$

$\quad y = \pm 2$

The solutions are $\left(-\sqrt{5}, -2\right)$, $\left(-\sqrt{5}, 2\right)$, $\left(\sqrt{5}, -2\right)$, and $\left(\sqrt{5}, 2\right)$.

9. $\begin{cases} x^2 + 2y^2 = 2 & (1) \\ x - y = 2 & (2) \end{cases}$

Solve E2 for x: $x = y + 2$

Substitute into E1.

$(y + 2)^2 + 2y^2 = 2$

$y^2 + 4y + 4 + 2y^2 = 2$

$3y^2 + 4y + 2 = 0$

$y = \dfrac{-4 \pm \sqrt{(4)^2 - 4(3)(2)}}{2(3)} = \dfrac{-4 \pm \sqrt{-8}}{6}$

There are no real solutions. The solution is $\varnothing$.

11. $\begin{cases} y = x^2 - 3 & (1) \\ 4x - y = 6 & (2) \end{cases}$

Substitute $x^2 - 3$ for y in E2.

$4x - (x^2 - 3) = 6$

$\quad 4x - x^2 + 3 = 6$

$\qquad\qquad 0 = x^2 - 4x + 3$

$\qquad\qquad 0 = (x - 3)(x - 1)$

$x - 3 = 0 \ \text{ or } \ x - 1 = 0$

$\quad x = 3 \ \text{ or } \qquad x = 1$

Substitute these values into E1.

$x = 3: y = (3)^2 - 3 = 6$

$x = 1: y = (1)^2 - 3 = -2$

The solutions are (3, 6) and (1, –2).

13. $\begin{cases} y = x^2 & (1) \\ 3x + y = 10 & (2) \end{cases}$

Substitute x^2 for y in E2.

$\qquad 3x + x^2 = 10$

$x^2 + 3x - 10 = 0$

$(x + 5)(x - 2) = 0$

$x + 5 = 0 \ \text{ or } \ x - 2 = 0$

$\quad x = -5 \ \text{ or } \qquad x = 2$

Substitute these values into E1.

$x = -5: y = (-5)^2 = 25$

$x = 2: y = (2)^2 = 4$

The solutions are (–5, 25) and (2, 4).

15. $\begin{cases} y = 2x^2 + 1 & (1) \\ x + y = -1 & (2) \end{cases}$

Substitute $2x^2 + 1$ for y in E2.

$x + 2x^2 + 1 = -1$

$2x^2 + x + 2 = 0$

$x = \dfrac{-1 \pm \sqrt{(1)^2 - 4(2)(2)}}{2(2)} = \dfrac{-1 \pm \sqrt{-15}}{4}$

There are no real solutions. The solution is $\varnothing$.

17. $\begin{cases} y = x^2 - 4 & (1) \\ y = x^2 - 4x & (2) \end{cases}$

Substitute $x^2 - 4$ for y in E2.

$x^2 - 4 = x^2 - 4x$

$\quad -4 = -4x$

$\qquad 1 = x$

Substitute this value into E1.

$y = (1)^2 - 4 = -3$

The solution is (1, –3).

19. $\begin{cases} 2x^2 + 3y^2 = 14 & (1) \\ -x^2 + y^2 = 3 & (2) \end{cases}$

Multiply E2 by 2 and add to E1.

$\quad 2x^2 + 3y^2 = 14$

$\underline{-2x^2 + 2y^2 = 6}$

$\qquad\qquad 5y^2 = 20$

$\qquad\qquad\ y^2 = 4$

$\qquad\qquad\ \ y = \pm 2$

Substitute 4 for y^2 into E2.

$-x^2 + 4 = 3$

$\quad -x^2 = -1$

$\qquad x^2 = 1$

$\qquad x = \pm 1$

The solutions are (–1, –2), (–1, 2), (1, –2), and (1, 2).

21. $\begin{cases} x^2 + y^2 = 1 & (1) \\ x^2 + (y+3)^2 = 4 & (2) \end{cases}$

Multiply E1 by –1 and add to E2.

$\quad -x^2 - y^2 = -1$

$\underline{\quad x^2 + (y+3)^2 = 4}$

$\quad (y+3)^3 - y^2 = 3$

$\ y^2 + 6y + 9 - y^2 = 3$

$\qquad\qquad\qquad 6y = -6$

$\qquad\qquad\qquad\ y = -1$

Replace y with –1 in E1.

$x^2 + (-1)^2 = 1$

$\qquad x^2 = 0$

$\qquad\ x = 0$

The solution is (0, –1).

23. $\begin{cases} y = x^2 + 2 & (1) \\ y = -x^2 + 4 & (2) \end{cases}$

Add E1 and E2.

$\quad y = x^2 + 2$

$\underline{\quad y = -x^2 + 4}$

$2y = 6$

$\ y = 3$

Substitute this value into E1.

$$3 = x^2 + 2$$
$$1 = x^2$$
$$\pm 1 = x$$

The solutions are $(-1, 3)$ and $(1, 3)$.

25. $\begin{cases} 3x^2 + y^2 = 9 & (1) \\ 3x^2 - y^2 = 9 & (2) \end{cases}$

Add E1 and E2.
$$3x^2 + y^2 = 9$$
$$\underline{3x^2 - y^2 = 9}$$
$$6x^2 = 18$$
$$x^2 = 3$$
$$x = \pm\sqrt{3}$$

Substitute 3 for x^2 in E1.
$$3(3) + y^2 = 9$$
$$y^2 = 0$$
$$y = 0$$

The solutions are $\left(-\sqrt{3}, 0\right)$, $\left(\sqrt{3}, 0\right)$.

27. $\begin{cases} x^2 + 3y^2 = 6 & (1) \\ x^2 - 3y^2 = 10 & (2) \end{cases}$

Solve E2 for x^2: $x^2 = 3y^2 + 10$.

Substitute into E1.
$$(3y^2 + 10) + 3y^2 = 6$$
$$6y^2 = -4$$
$$y^2 = -\frac{2}{3}$$

There are no real solutions. The solution is $\varnothing$.

29. $\begin{cases} x^2 + y^2 = 36 & (1) \\ y = \frac{1}{6}x^2 - 6 & (2) \end{cases}$

Solve E1 for x^2: $x^2 = 36 - y^2$.

Substitute into E2.
$$y = \frac{1}{6}(36 - y^2) - 6$$
$$y = 6 - \frac{1}{6}y^2 - 6$$
$$6y = -y^2$$
$$y^2 + 6y = 0$$
$$y(y + 6) = 0$$
$$y = 0 \text{ or } y = -6$$

Substitute these values into the equation

$$x^2 = 36 - y^2.$$
$$y = 0 : x^2 = 36 - (0)^2$$
$$x^2 = 36$$
$$x = \pm 6$$
$$y = -6 : x^2 = 36 - (6)^2$$
$$x^2 = 0$$
$$x = 0$$

The solutions are $(-6, 0)$, $(6, 0)$ and $(0, -6)$.

31. $\begin{cases} y = \sqrt{x} \\ x^2 + y^2 = 12 \end{cases}$

Substitute.
$$x^2 + \left(\sqrt{x}\right)^2 = 12$$
$$x^2 + x = 12$$
$$x^2 + x - 12 = 0$$
$$(x + 4)(x - 3) = 0$$
$$x + 4 = 0 \quad \text{or} \quad x - 3 = 0$$
$$x = -4 \qquad\qquad x = 3$$
$$x = -4 : \ y = \sqrt{-4}$$
$$x = 3 : \ y = \sqrt{3}$$

Since $\sqrt{-4}$ is not a real number, the only solution is $\left(3, \sqrt{3}\right)$.

33. $x > -3$

35. $y < 2x - 1$

37. $P = x + (2x - 5) + (5x - 20) = (8x - 25)$ inches

39. $P = 2(x^2 + 3x + 1) + 2(x^2)$
$= 2x^2 + 6x + 2 + 2x^2$
$= (4x^2 + 6x + 2)$ meters

41. answers may vary

43. There are 0, 1, 2, 3, or 4 possible real solutions. answers may vary

45. Let x and y represent the numbers.
$$\begin{cases} x^2 + y^2 = 130 \\ x^2 - y^2 = 32 \end{cases}$$
Add the equations.
$$\begin{array}{r} x^2 + y^2 = 130 \\ \underline{x^2 - y^2 = 32} \\ 2x^2 \quad\quad = 162 \\ x^2 = 81 \\ x = \pm 9 \end{array}$$

Replace x^2 with 81 in the first equation.
$$81 + y^2 = 130$$
$$y^2 = 49$$
$$y = \pm 7$$
The numbers are –9 and –7, –9 and 7, 9 and –7, and 9 and 7.

47. Let x and y be the length and width.
$$\begin{cases} xy = 285 \\ 2x + 2y = 68 \end{cases}$$
Solve the first equation for y: $y = \dfrac{285}{x}$.

Substitute into the second equation.
$$2x + 2\left(\frac{285}{x}\right) = 68$$
$$x + \frac{285}{x} = 34$$
$$x^2 + 285 = 34x$$
$$x^2 - 34x + 285 = 0$$
$$(x - 19)(x - 15) = 0$$
$$x = 19 \text{ or } x = 15$$
Using $x = 19$, $y = \dfrac{285}{x} = \dfrac{285}{19} = 15$.
Using $x = 15$, $y = \dfrac{285}{x} = \dfrac{285}{15} = 19$.
The dimensions are 19 cm by 15 cm.

49. $\begin{cases} p = -0.01x^2 - 0.2x + 9 \\ p = 0.01x^2 - 0.1x + 3 \end{cases}$
Substitute.
$$-0.01x^2 - 0.2x + 9 = 0.01x^2 - 0.1x + 3$$
$$0 = 0.02x^2 + 0.1x - 6$$
$$0 = x^2 + 5x - 300$$
$$0 = (x + 20)(x - 15)$$
$x + 20 = 0 \quad$ or $\quad x - 15 = 0$
$\quad\quad x = -20 \quad$ or $\quad\quad x = 15$
Disregard the negative.
$$p = -0.01(15)^2 - 0.2(15) + 9$$
$$p = 3.75$$
The equilibrium quantity is 15,000 compact discs, and the corresponding price is $3.75.

51. $\begin{cases} x^2 + 4y^2 = 10 \\ y = x \end{cases}$

53. $\begin{cases} y = x^2 + 2 \\ y = -x^2 + 4 \end{cases}$

Section 10.4 Practice Exercises

1. $\dfrac{x^2}{36} + \dfrac{y^2}{16} \geq 1$

First graph the ellipse $\dfrac{x^2}{36} + \dfrac{y^2}{16} = 1$ as a solid curve. Choose (0, 0) as a test point.
$$\frac{x^2}{36} + \frac{y^2}{16} \geq 1$$
$$\frac{0^2}{36} + \frac{0^2}{16} \geq 1$$
$$0 \geq 1 \quad \text{False}$$
The solution set is the region that does not contain (0, 0).

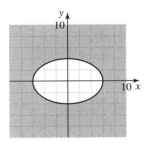

2. Graph the related equation $y = x^2 + 2$ and shade the region below the boundary curve since the inequality symbol is $\leq$.

3. $16y^2 > 9x^2 + 144$

    The related equation is $16y^2 = 9x^2 + 144$.

    $16y^2 - 9x^2 = 144$

    $$\frac{y^2}{9} - \frac{x^2}{16} = 1$$

    Graph the hyperbola as a dashed curve.
    Choose $(0, 0)$, $(0, 4)$, and $(0, -4)$ as test points.

    $(0, 0)$: $16(0)^2 > 9(0)^2 + 144$

    $\qquad\qquad 0 > 144$ False

    $(0, 4)$: $16(4)^2 > 9(0)^2 + 144$

    $\qquad\qquad 256 > 144$ True

    $(0, -4)$: $16(-4)^2 > 9(0)^2 + 144$

    $\qquad\qquad 256 > 144$ True

4. $\begin{cases} y \geq x^2 \\ y \leq -3x + 2 \end{cases}$

    Solve the related system $\begin{cases} y = x^2 \\ y = -3x + 2 \end{cases}$.

    Substitute $-3x + 2$ for y in the first equation.

$x^2 = -3x + 2$

$x^2 + 3x - 2 = 0$

$$x = \frac{-3 \pm \sqrt{3^2 - 4(1)(-2)}}{2(1)}$$

$$= \frac{-3 \pm \sqrt{17}}{2}$$

≈ 0.56 or -3.56

$y = -3x + 2 \approx -3(0.56) + 2 = 0.32$

$y \approx -3(-3.56) + 2 = 12.68$

The points of intersection are approximately $(0.56, 0.32)$ and $(-3.56, 12.68)$.

Graph $y = x^2$ and $y = -3x + 2$ as solid curves.
The region of the solution set is above the parabola but below the line.

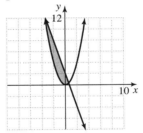

5. $\begin{cases} x^2 + y^2 < 16 \\ \dfrac{x^2}{4} - \dfrac{y^2}{9} < 1 \\ y < x + 3 \end{cases}$

    Graph $x^2 + y^2 = 16$, $\dfrac{x^2}{4} - \dfrac{y^2}{9} = 1$, and $y = x + 3$.

    The test point $(0, 0)$ gives true statements for all three inequalities; thus, the innermost region is the solution set.

    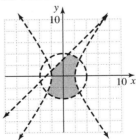

Vocabulary, Readiness & Video Check 10.4

1. For both, we graph the related equation to find the boundary and sketch it as a solid boundary for $\leq$ or $\geq$ and a dashed boundary for $<$ or $>$; also we choose a test point not on the boundary and shade that region if the test point is a solution of the original inequality or shade the other region if not.

2. A circle within a circle (either circle solid or dashed) where the inner circle is shaded inside and the outer circle is shaded outside; also, two non-intersecting circles (either circle solid or dashed), both shaded inside just to name a few examples.

Exercise Set 10.4

1. $y < x^2$

    First graph the parabola as a dashed curve.

Test Point	$y < x^2$; Result
$(0, 1)$	$1 < 0^2$; False

    Shade the region which does not contain $(0, 1)$.

    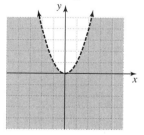

3. $x^2 + y^2 \geq 16$

    First graph the circle as a solid curve.

Test Point	$x^2 + y^2 \geq 16$; Result
$(0, 0)$	$0^2 + 0^2 \geq 16$; False

    Shade the region which does not contain $(0, 0)$.

    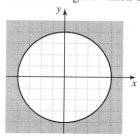

5. $\dfrac{x^2}{4} - y^2 < 1$

    First graph the hyperbola as a dashed curve.

Test Point	$\dfrac{x^2}{4} - y^2 < 1$; Result
$(-4, 0)$	$\dfrac{(-4)^2}{4} - 0^2 < 1$; False
$(0, 0)$	$\dfrac{(0)^2}{4} - 0^2 < 1$; True
$(4, 0)$	$\dfrac{(4)^2}{4} - 0^2 < 1$; False

    Shade the region containing $(0, 0)$.

7. $y > (x-1)^2 - 3$

    First graph the parabola as a dashed curve.

Test Point	$y > (x-1)^2 - 3$; Result
$(0, 0)$	$0 > (0-1)^2 - 3$; True

    Shade the region containing $(0, 0)$.

    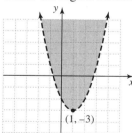

9. $x^2 + y^2 \leq 9$

    First graph the circle as a solid curve.

Test Point	$x^2 + y^2 \leq 9$; Result
$(0, 0)$	$0^2 + 0^2 \leq 9$; True

Shade the region containing (0, 0).

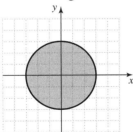

11. $y > -x^2 + 5$

First graph the parabola as a dashed curve.

Test Point	$y > -x^2 + 5$; Result
(0, 0)	$0 > -(0)^2 + 5$; False

Shade the region which does not contain (0, 0).

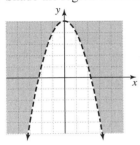

13. $\dfrac{x^2}{4} + \dfrac{y^2}{9} \leq 1$

First graph the ellipse as a solid curve.

Test Point	$\dfrac{x^2}{4} + \dfrac{y^2}{9} \leq 1$; Result
(0, 0)	$\dfrac{(0)^2}{4} + \dfrac{(0)^2}{9} \leq 1$; True

Shade the region containing (0, 0).

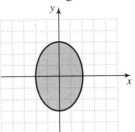

15. $\dfrac{y^2}{4} - x^2 \leq 1$

First graph the hyperbola as solid curves.

Test Point	$\dfrac{y^2}{4} - x^2 \leq 1$; Result
(0, −4)	$\dfrac{(-4)^2}{4} - 0^2 \leq 1$; False
(0, 0)	$\dfrac{(0)^2}{4} - 0^2 \leq 1$; True
(0, 4)	$\dfrac{(4)^2}{4} - 0^2 \leq 1$; False

Shade the region containing (0, 0).

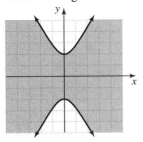

17. $y < (x-2)^2 + 1$

First graph the parabola as a dashed curve.

Test Point	$y < (x-2)^2 + 1$; Result
(0, 0)	$0 < (0-2)^2 + 1$; True

Shade the region containing (0, 0).

19. $y \leq x^2 + x - 2$

First graph the parabola as a solid curve.

Test Point	$y \leq x^2 + x - 2$; Result
(0, 0)	$0 \leq (0)^2 + (0) - 2$; False

Shade the region which does not contain (0, 0).

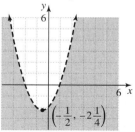

$$\left(-\frac{1}{2}, -2\frac{1}{4}\right)$$

21. $\begin{cases} 4x + 3y \geq 12 \\ x^2 + y^2 < 16 \end{cases}$

First graph $4x + 3y = 12$ as a solid line.

Test Point	$4x + 3y \geq 12$; Result
(0, 0)	$4(0) + 3(0) \geq 12$; False

Shade the region which does not contain (0, 0). Next, graph the circle $x^2 + y^2 = 16$ as a dashed curve.

Test Point	$x^2 + y^2 < 16$; Result
(0, 0)	$0^2 + 0^2 < 16$; True

Shade the region containing (0, 0). The solution to the system is the intersection.

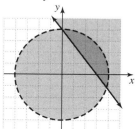

23. $\begin{cases} x^2 + y^2 \leq 9 \\ x^2 + y^2 \geq 1 \end{cases}$

First graph the circle with radius 3 as a solid curve.

Test Point	$x^2 + y^2 \leq 9$; Result
(0, 0)	$0^2 + 0^2 \leq 9$; True

Shade the region containing (0, 0). Next, graph the circle with 1 as a dashed curve.

Test Point	$x^2 + y^2 \geq 1$; Result
(0, 0)	$0^2 + 0^2 \geq 1$; False

Shade the region which does not contain (0, 0). The solution to the system is the intersection.

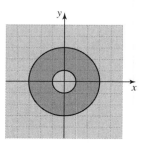

25. $\begin{cases} y > x^2 \\ y \geq 2x + 1 \end{cases}$

First graph the parabola as a dashed curve.

Test Point	$y > x^2$; Result
(0, 1)	$1 > 0^2$; True

Shade the region containing (0, 1). Next, graph $y = 2x + 1$ as a solid line.

Test Point	$y \geq 2x + 1$; Result
(0, 0)	$0 \geq 2(0) + 1$; False

Shade the region which does not contain (0, 0). The solution to the system is the intersection.

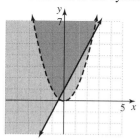

27. $\begin{cases} x^2 + y^2 > 9 \\ \quad\quad y > x^2 \end{cases}$

First graph the circle as a dashed curve.

Test Point	$x^2 + y^2 > 9$; Result
(0, 0)	$0^2 + 0^2 > 9$; False

Shade the region which does not contain (0, 0).
Next, graph the parabola as a dashed curve.

Test Point	$y > x^2$; Result
(0, 1)	$1 > 0^2$; True

Shade the region containing (0, 1). The solution
to the system is the intersection.

29. $\begin{cases} \dfrac{x^2}{4} + \dfrac{y^2}{9} \geq 1 \\ x^2 + y^2 \geq 4 \end{cases}$

First graph the ellipse as a solid curve.

Test Point	$\dfrac{x^2}{4} + \dfrac{y^2}{9} \geq 1$; Result
(0, 0)	$\dfrac{0^2}{4} + \dfrac{0^2}{9} \geq 1$; False

Shade the region which does not contain (0, 0).
Next, graph the circle as a solid curve.

Test Point	$x^2 + y^2 \geq 4$; Result
(0, 0)	$0^2 + 0^2 \geq 4$; False

Shade the region which does not contain (0, 0).
The solution to the system is the intersection.

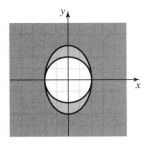

31. $\begin{cases} x^2 - y^2 \geq 1 \\ \quad\quad y \geq 0 \end{cases}$

First graph the hyperbola as solid curves.

Test Point	$x^2 - y^2 \geq 1$; Result
(−2, 0)	$(-2)^2 - 0^2 \geq 1$; True
(0, 0)	$0^2 - 0^2 \geq 1$; False
(2, 0)	$2^2 - 0^2 \geq 1$; True

Shade the region which does not contain (0, 0).
Next, graph $y = 0$ as a solid line.

Test Point	$y > 0$; Result
(0, 1)	$1 \geq 0$; True

Shade the region containing (0, 1). The solution
to the system is the intersection.

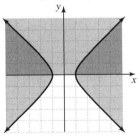

33. $\begin{cases} x + y \geq 1 \\ 2x + 3y < 1 \\ \quad\quad x > -3 \end{cases}$

First graph $x + y = 1$ as a solid line.

Test Point	$x + y \geq 1$; Result
(0, 0)	$0 + 0 \geq 1$; False

Shade the region which does not contain (0, 0).
Next, graph $2x + 3y = 1$ as a dashed line.

     581

Test Point	$2x + 3y < 1$; Result
$(0, 0)$	$2(0) 1+ 3(0) < 1$; True

Shade the region containing $(0, 0)$. Now graph the line $x = -3$ as a dashed line.

Test Point	$x > -3$; Result
$(0, 0)$	$0 > -3$; True

Shade the region containing $(0, 0)$. The solution to the system is the intersection.

35. $\begin{cases} x^2 - y^2 < 1 \\ \dfrac{x^2}{16} + y^2 \le 1 \\ x \ge -2 \end{cases}$

First graph the hyperbola as dashed curves.

Test Point	$x^2 - y^2 < 1$; Result
$(-2, 0)$	$(-2)^2 - 0^2 < 1$; False
$(0, 0)$	$0^2 - 0^2 < 1$; True
$(2, 0)$	$2^2 - 0^2 < 1$; False

Shade the region containing $(0, 0)$. Next, graph the ellipse as a solid curve.

Test Point	$\frac{x^2}{16} + y^2 \le 1$; Result
$(0, 0)$	$\frac{0^2}{16} + 0^2 \le 1$; True

Shade the region containing $(0, 0)$. Now graph the line $x = -2$ as a solid line.

Test Point	$x \ge -2$; Result
$(0, 0)$	$0 \ge -2$; True

Shade the region containing $(0, 0)$. The solution to the system is the intersection.

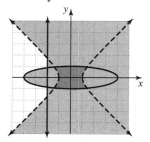

37. This is not a function because a vertical line can cross the graph in more than one place.

39. This is a function because a vertical line can cross the graph in no more than one place.

41. $f(x) = 3x^2 - 2$
$f(-1) = 3(-1)^2 - 2 = 3 - 2 = 1$

43. $f(x) = 3x^2 - 2$
$f(a) = 3(a)^2 - 2 = 3a^2 - 2$

45. answers may vary

47. $\begin{cases} y \le x^2 \\ y \ge x + 2 \\ x \ge 0 \\ y \ge 0 \end{cases}$

First graph $y = x^2$ as a solid curve.

Test Point	$y \le x^2$; Result
$(0, 1)$	$1 \le 0^2$; False

Shade the region which does not contain $(0, 1)$. Next, graph $y = x + 2$ as a solid line.

Test Point	$y \ge x + 2$; Result
$(0, 0)$	$0 \ge 0 + 2$; False

Shade the region which does not contain $(0, 0)$. Next graph the line $x = 0$ as a solid line, and shade to the right. Now graph the line $y = 0$ as a solid line, and shade above. The solution to the system is the intersection.

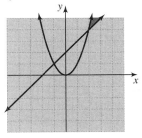

Chapter 10 Vocabulary Check

1. A <u>circle</u> is the set of all points in a plane that are the same distance from a fixed point, called the <u>center</u>.

2. A <u>nonlinear system of equations</u> is a system of equations at least one of which is not linear.

3. An <u>ellipse</u> is the set of points on a plane such that the sum of the distances of those points from two fixed points is a constant.

4. In a circle, the distance from the center to a point of the circle is called its <u>radius</u>.

5. A <u>hyperbola</u> is the set of points in a plane such that the absolute value of the difference of the distance from two fixed points is constant.

6. The circle, parabola, ellipse, and hyperbola are called the <u>conic sections</u>.

7. For a parabola that opens upward, the lowest point is the <u>vertex</u>.

8. Twice a circle's radius is its <u>diameter</u>.

Chapter 10 Review

1. center (–4, 4), radius 3
$$[x-(-4)]^2 + (y-4)^2 = 3^2$$
$$(x+4)^2 + (y-4)^2 = 9$$

2. center (5, 0), radius 5
$$(x-5)^2 + (y-0)^2 = 5^2$$
$$(x-5)^2 + y^2 = 25$$

3. center (–7, –9), radius $\sqrt{11}$
$$[x-(-7)]^2 + [y-(-9)]^2 = \left(\sqrt{11}\right)^2$$
$$(x+7)^2 + (y+9)^2 = 11$$

4. center (0, 0), radius $\dfrac{7}{2}$
$$(x-0)^2 + (y-0)^2 = \left(\dfrac{7}{2}\right)^2$$
$$x^2 + y^2 = \dfrac{49}{4}$$

5. $x^2 + y^2 = 7$
Circle; center (0, 0), radius $r = \sqrt{7}$

6. $x = 2(y-5)^2 + 4$
Parabola; vertex: (4, 5)

7. $x = -(y+2)^2 + 3$
Parabola; vertex: (3, –2)

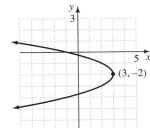

8. $(x-1)^2 + (y-2)^2 = 4$

Circle; center (1, 2), radius $r = \sqrt{4} = 2$

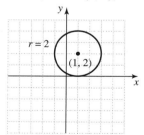

9. $y = -x^2 + 4x + 10$

Parabola; $x = \dfrac{-b}{2a} = \dfrac{-4}{2(-1)} = 2$

$y = -(2)^2 + 4(2) + 10 = 14$

Vertex: (2, 14)

10. $x = -y^2 - 4y + 6$

Parabola; $y = \dfrac{-b}{2a} = \dfrac{-(-4)}{2(-1)} = -2$

$x = -(-2)^2 - 4(-2) + 6 = 10$

Vertex: (10, −2)

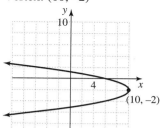

11. $x = \dfrac{1}{2}y^2 + 2y + 1$

Parabola; $y = \dfrac{-b}{2a} = \dfrac{-2}{2\left(\frac{1}{2}\right)} = -2$

$x = \dfrac{1}{2}(-2)^2 + 2(-2) + 1 = -1$

Vertex: (−1, −2)

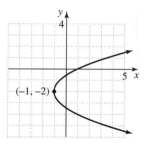

12. $y = -3x^2 + \dfrac{1}{2}x + 4$

Parabola; $x = \dfrac{-b}{2a} = \dfrac{-\frac{1}{2}}{2(-3)} = \dfrac{1}{12}$

$y = -3\left(\dfrac{1}{12}\right)^2 + \dfrac{1}{2}\left(\dfrac{1}{12}\right) + 4 = \dfrac{193}{48}$

Vertex: $\left(\dfrac{1}{12}, \dfrac{193}{48}\right)$

13.
$$x^2 + y^2 + 2x + y = \dfrac{3}{4}$$
$$(x^2 + 2x) + (y^2 + y) = \dfrac{3}{4}$$
$$(x^2 + 2x + 1) + \left(y^2 + y + \dfrac{1}{4}\right) = \dfrac{3}{4} + 1 + \dfrac{1}{4}$$
$$(x+1)^2 + \left(y + \dfrac{1}{2}\right)^2 = 2$$

Circle; center $\left(-1, -\dfrac{1}{2}\right)$, radius $r = \sqrt{2}$

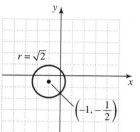

14.
$$x^2 + y^2 - 3y = \frac{7}{4}$$
$$x^2 + \left(y^2 - 3y + \frac{9}{4}\right) = \frac{7}{4} + \frac{9}{4}$$
$$x^2 + \left(y - \frac{3}{2}\right)^2 = 4$$

Circle; center $\left(0, \frac{3}{2}\right)$, radius $r = \sqrt{4} = 2$

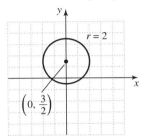

15.
$$4x^2 + 4y^2 + 16x + 8y = 1$$
$$(x^2 + 4x) + (y^2 + 2y) = \frac{1}{4}$$
$$(x^2 + 4x + 4) + (y^2 + 2y + 1) = \frac{1}{4} + 4 + 1$$
$$(x + 2)^2 + (y + 1)^2 = \frac{21}{4}$$

Circle; center $(-2, -1)$, radius $r = \sqrt{\frac{21}{4}} = \frac{\sqrt{21}}{2}$

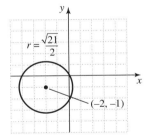

16.
$$3x^2 + 3y^2 + 18x - 12y = -12$$
$$x^2 + y^2 + 6x - 4y = -4$$
$$x^2 + 6x + 9 + y^2 - 4y + 4 = -4 + 9 + 4$$
$$(x + 3)^2 + (y - 2)^2 = 9$$
$$(x + 3)^2 + (y - 2)^2 = 3^2$$
$$C(-3, 2),\ r = 3$$

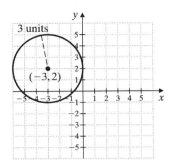

17.
$$x^2 - \frac{y^2}{4} = 1$$
$$\frac{x^2}{1^2} - \frac{y^2}{2^2} = 1$$
$$a = 1,\ b = 2$$

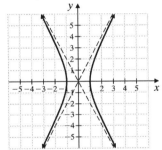

18.
$$x^2 + \frac{y^2}{4} = 1$$
$$\frac{x^2}{1^2} + \frac{y^2}{2^2} = 1$$
$$C(0, 0)$$
x-intercepts: $(-1, 0), (1, 0)$
y-intercepts: $(0, -2), (0, 2)$

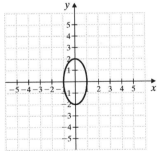

19.
$$4y^2 + 9x^2 = 36$$
$$\frac{y^2}{9} + \frac{x^2}{4} = 1$$
$$\frac{x^2}{2^2} + \frac{y^2}{3^2} = 1$$
$$C(0, 0)$$

x-intercepts: $(-2, 0)$, $(2, 0)$
y-intercepts: $(0, -3)$, $(0, 3)$

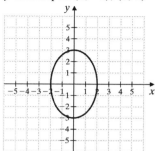

20. $-5x^2 + 25y^2 = 125$

$$-\frac{x^2}{25} + \frac{y^2}{5} = 1$$

$$\frac{y^2}{\left(\sqrt{5}\right)^2} - \frac{x^2}{5^2} = 1$$

$a = 5$, $b = \sqrt{5}$

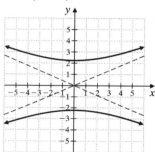

21. $x^2 - y^2 = 1$

$$\frac{x^2}{1^2} - \frac{y^2}{1^2} = 1$$

$a = 1$, $b = 1$

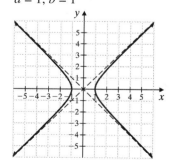

22. $\dfrac{(x+3)^2}{9} + \dfrac{(y-4)^2}{25} = 1$

$$\frac{(x+3)^2}{3^2} + \frac{(y-4)^2}{5^2} = 1$$

$C(-3, 4)$
other points:
$(-3 + 3, 4)$, or $(0, 4)$
$(-3 - 3, 4)$, or $(-6, 4)$
$(-3, 4 - 5)$, or $(-3, -1)$
$(-3, 4 + 5)$, or $(-3, 9)$

23. $y = x^2 + 9$

$V(0, 9)$
The parabola opens upward.

24. $36y^2 - 49x^2 = 1764$

$$\frac{y^2}{49} - \frac{x^2}{36} = 1$$

$$\frac{y^2}{7^2} - \frac{x^2}{6^2} = 1$$

$a = 6$, $b = 7$

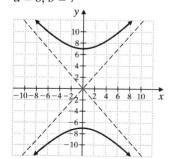

25. $x = 4y^2 - 16$

$V(-16, 0)$

The parabola opens to the right.

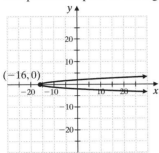

26. $y = x^2 + 4x + 6$

$y = (x^2 + 4x + 4) + 6 - 4$

$y = (x + 2)^2 + 2$

$V(-2, 2)$

The parabola opens upward.

27. $y^2 + 2(x - 1)^2 - 8 = 0$

$ y^2 + 2(x - 1)^2 = 8$

$\dfrac{y^2}{8} + \dfrac{(x - 1)^2}{4} = 1$

$\dfrac{y^2}{\left(\sqrt{8}\right)^2} + \dfrac{(x - 1)^2}{2^2} = 1$

$C(1, 0)$

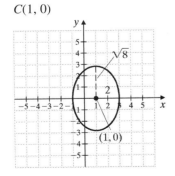

28. $x - 4y = y^2$

$ x = y^2 + 4y$

$ x = y^2 + 4y + 4 - 4$

$ x = (y + 2)^2 - 4$

$V(-4, -2)$

The parabola opens to the right.

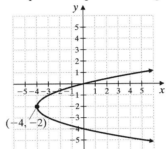

29. $x^2 - 4 = y^2$

$ x^2 - y^2 = 4$

$\dfrac{x^2}{4} - \dfrac{y^2}{4} = 1$

$\dfrac{x^2}{2^2} - \dfrac{y^2}{2^2} = 1$

$a = 2, \, b = 2$

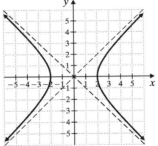

30. $x^2 = 4 - y^2$

$ x^2 + y^2 = 4$

$ x^2 + y^2 = 2^2$

$C(0, 0), \, r = 2$

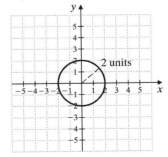

31.
$$36y^2 = 576 + 16x^2$$
$$36y^2 - 16x^2 = 576$$
$$\frac{y^2}{16} - \frac{x^2}{36} = 1$$
$$\frac{y^2}{4^2} - \frac{x^2}{6^2} = 1$$
$$a = 6, \ b = 4$$

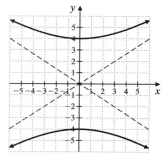

32.
$$3(x-7)^2 + 3(y+4)^2 = 1$$
$$(x-7)^2 + (y+4)^2 = \frac{1}{3}$$
$$(x-7)^2 + (y+4)^2 = \left(\sqrt{\frac{1}{3}}\right)^2$$
$$(x-7)^2 + (y+4)^2 = \left(\frac{\sqrt{3}}{3}\right)^2$$
$$C(7, -4)$$

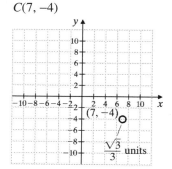

33.
$$\begin{cases} y = 2x - 4 \\ y^2 = 4x \end{cases}$$
Substitute.
$$y^2 = 4x$$
$$(2x-4)^2 = 4x$$
$$4x^2 - 16x + 16 = 4x$$
$$4x^2 - 16x + 16 = 4x$$
$$4x^2 - 20x + 16 = 0$$
$$4(x^2 - 5x + 4) = 0$$
$$4(x-4)(x-1) = 0$$
$$x - 4 = 0 \quad \text{or} \quad x - 1 = 0$$
$$x = 4 \qquad\qquad x = 1$$
$$x = 4: \ y = 2(4) - 4 = 4$$
$$x = 1: \ y = 2(1) - 4 = -2$$
The solutions are $(1, -2)$ and $(4, 4)$.

34.
$$\begin{cases} x^2 + y^2 = 4 \\ x - y = 4 \end{cases}$$
Solve equation 2 for x.
$$x = 4 + y$$
Substitute.
$$x^2 + y^2 = 4$$
$$(4 + y)^2 + y^2 = 4$$
$$16 + 8y + y^2 + y^2 = 4$$
$$2y^2 + 8y + 12 = 0$$
$$a = 2, \ b = 8, \ c = 12$$
$$y = \frac{-8 \pm \sqrt{8^2 - 4(2)(12)}}{2(2)} = \frac{-8 \pm \sqrt{-32}}{4}$$
Since $\sqrt{-32}$ is not a real number, there is no solution. The solution set is $\varnothing$.

35.
$$\begin{cases} y = x + 2 \\ y = x^2 \end{cases}$$
Substitute.
$$x + 2 = x^2$$
$$0 = x^2 - x - 2$$
$$0 = (x-2)(x+1)$$
$$x - 2 = 0 \quad \text{or} \quad x + 1 = 0$$
$$x = 2 \qquad\qquad x = -1$$
$$x = 2: \ y = 2^2 = 4$$
$$x = -1: \ y = (-1)^2 = 1$$
The solutions are $(-1, 1)$ and $(2, 4)$.

36. $\begin{cases} 4x - y^2 = 0 \\ 2x^2 + y^2 = 16 \end{cases}$

Add the equations.

$$2x^2 + 4x = 16$$
$$2x^2 + 4x - 16 = 0$$
$$2(x^2 + 2x - 8) = 0$$
$$2(x + 4)(x - 2) = 0$$
$$x + 4 = 0 \quad \text{or} \quad x - 2 = 0$$
$$x = -4 \qquad\qquad x = 2$$

$x = -4: \ 4(-4) - y^2 = 0$
$$-16 = y^2$$
$$\pm\sqrt{-16} = y$$

$x = 2: \ 4(2) - y^2 = 0$
$$8 = y^2$$
$$\pm\sqrt{8} = y$$
$$\pm 2\sqrt{2} = y$$

Since $\sqrt{-16}$ is not a real number, the solutions are $\left(2, 2\sqrt{2}\right)$ and $\left(2, -2\sqrt{2}\right)$.

37. $\begin{cases} x^2 + 4y^2 = 16 \\ x^2 + y^2 = 4 \end{cases}$

Multiply equation 2 by -1. Add the results.

$$x^2 + 4y^2 = 16$$
$$\underline{-x^2 - y^2 = -4}$$
$$3y^2 = 12$$
$$y^2 = 4$$
$$y = \pm\sqrt{4}$$
$$y = \pm 2$$

$y = -2: \ x^2 + (-2)^2 = 4$
$$x^2 = 0$$
$$x = 0$$

$y = 2: \ x^2 + (2)^2 = 4$
$$x^2 = 0$$
$$x = 0$$

The solutions are (0, 2) and (0, -2).

38. $\begin{cases} x^2 + 2y = 9 \\ 5x - 2y = 5 \end{cases}$

Add the equations.

$$x^2 + 5x = 14$$
$$x^2 + 5x - 14 = 0$$
$$(x + 7)(x - 2) = 0$$

$$x + 7 = 0 \quad \text{or} \quad x - 2 = 0$$
$$x = -7 \qquad\qquad x = 2$$
$$x = -7: \ 5(-7) - 2y = 5$$
$$-2y = 40$$
$$y = -20$$
$$x = 2: \ 5(2) - 2y = 5$$
$$-2y = -5$$
$$y = \frac{5}{2}$$

The solutions are $\left(2, \dfrac{5}{2}\right)$ and $(-7, -20)$.

39. $\begin{cases} y = 3x^2 + 5x - 4 \\ y = 3x^2 - x + 2 \end{cases}$

Substitute.

$$3x^2 + 5x - 4 = 3x^2 - x + 2$$
$$5x - 4 = -x + 2$$
$$6x = 6$$
$$x = 1$$

$x = 1: \ y = 3(1)^2 + 5(1) - 4 = 4$

The solution is (1, 4).

40. $\begin{cases} x^2 - 3y^2 = 1 \\ 4x^2 + 5y^2 = 21 \end{cases}$

Solve equation 1 for x^2.

$$x^2 = 1 + 3y^2$$

Substitute.

$$4(1 + 3y^2) + 5y^2 = 21$$
$$4 + 12y^2 + 5y^2 = 21$$
$$17y^2 = 17$$
$$y^2 = 1$$
$$y = \pm 1$$

$y = -1: \ x^2 = 1 + 3(-1)^2$
$$x^2 = 4$$
$$x = \pm 2$$

$y = 1: \ x^2 = 1 + 3(1)^2$
$$x^2 = 4$$
$$x = \pm 2$$

The solutions are $(-2, -1)$, $(-2, 1)$, $(2, -1)$, and $(2, 1)$.

41. Let x be the width and y be the length.
$$\begin{cases} xy = 150 \\ 2x + 2y = 50 \end{cases}$$
Solve equation 2 for x.
$$2x + 2y = 50$$
$$x + y = 25$$
$$x = 25 - y$$
Substitute.
$$xy = 150$$
$$(25 - y)y = 150$$
$$25y - y^2 = 150$$
$$0 = y^2 - 25y + 150$$
$$0 = (y - 15)(y - 10)$$
$$y - 15 = 0 \quad \text{or} \quad y - 10 = 0$$
$$y = 15 \qquad\qquad y = 10$$
$y = 15 : x = 25 - 15 = 10$
$y = 10 : x = 25 - 10 = 15$
The dimensions are length = 15 feet and width = 10 feet.

42. An ellipse and a hyperbola can intersect at a maximum of 4 points. Therefore, there are a maximum of 4 real number solutions.

43. $y \le -x^2 + 3$

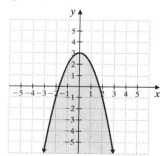

44. $x < y^2 - 1$

45. $x^2 + y^2 < 9$

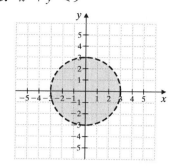

46. $\dfrac{x^2}{4} + \dfrac{y^2}{9} \ge 1$

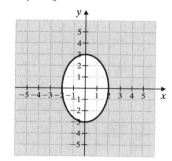

47. $\begin{cases} 3x + 4y \le 12 \\ x - 2y > 6 \end{cases}$

48. $\begin{cases} x^2 + y^2 \le 16 \\ x^2 + y^2 \ge 4 \end{cases}$

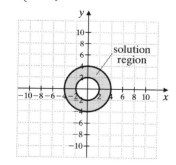

49. $\begin{cases} x^2 + y^2 < 4 \\ x^2 - y^2 \le 1 \end{cases}$

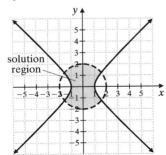

50. $\begin{cases} x^2 + y^2 < 4 \\ y \ge x^2 - 1 \\ x \ge 0 \end{cases}$

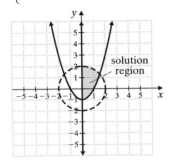

51. $C(-7, 8)$, $r = 5$

$$[x - (-7)]^2 + (y - 8)^2 = 5^2$$
$$(x + 7)^2 + (y - 8)^2 = 25$$

52. $y = x^2 + 6x + 9$

$y = (x + 3)^2$

$V(-3, 0)$
The parabola opens upward.

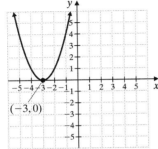

53. $x = y^2 + 6y + 9$

$x = (y + 3)^2$

$V(0, -3)$
The parabola opens to the right.

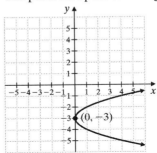

54. $\dfrac{y^2}{4} - \dfrac{x^2}{16} = 1$

$$\dfrac{y^2}{2^2} - \dfrac{x^2}{4^2} = 1$$

$a = 4$, $b = 2$

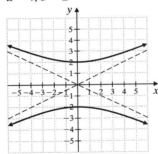

55. $\dfrac{y^2}{4} + \dfrac{x^2}{16} = 1$

$$\dfrac{y^2}{2^2} + \dfrac{x^2}{4^2} = 1$$

$C(0, 0)$
x-intercepts: $(-4, 0)$, $(4, 0)$
y-intercepts: $(0, -2)$, $(0, 2)$

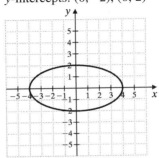

56. $\dfrac{(x-2)^2}{4}+(y-1)^2=1$

$\dfrac{(x-2)^2}{2^2}+\dfrac{(y-1)^2}{1^2}=1$

$C(2,1)$
other points:
$(2-2,1)$, or $(0,1)$
$(2+2,1)$, or $(4,1)$
$(2,1-1)$, or $(2,0)$
$(2,1+1)$, or $(2,2)$

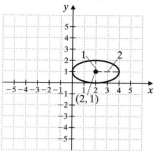

57. $y^2=x^2+6$

$y^2-x^2=6$

$\dfrac{y^2}{6}-\dfrac{x^2}{6}=1$

$\dfrac{y^2}{\left(\sqrt{6}\right)^2}-\dfrac{x^2}{\left(\sqrt{6}\right)^2}=1$

$a=\sqrt{6},\ b=\sqrt{6}$

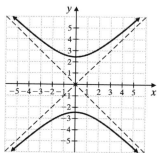

58. $y^2+(x-2)^2=10$

$(x-2)^2+y^2=\left(\sqrt{10}\right)^2$

$C(2,0),\ r=\sqrt{10}$

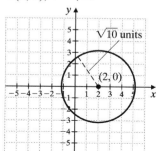

59. $3x^2+6x+3y^2=9$

$x^2+2x+y^2=3$

$(x^2+2x+1)+y^2=3+1$

$(x+1)^2+y^2=4$

$(x+1)^2+(y-0)^2=2^2$

$C(-1,0),\ r=2$

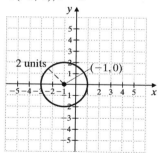

60. $x^2+y^2-8y=0$

$x^2+(y^2-8y+16)=0+16$

$x^2+(y-4)^2=16$

$(x-0)^2+(y-4)^2=4^2$

$C(0,4),\ r=4$

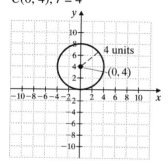

61. $6(x-2)^2 + 9(y+5)^2 = 36$

$$\frac{(x-2)^2}{6} + \frac{(y+5)^2}{4} = 1$$

$$\frac{(x-2)^2}{\left(\sqrt{6}\right)^2} + \frac{(y+5)^2}{2^2} = 1$$

$C(2, -5)$
other points:
$\left(2-\sqrt{6}, -5\right)$
$\left(2+\sqrt{6}, -5\right)$
$(2, -5-2)$, or $(2, -7)$
$(2, -5+2)$, or $(2, -3)$

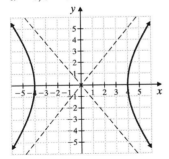

62. $\dfrac{x^2}{16} - \dfrac{y^2}{25} = 1$

$$\frac{x^2}{4^2} - \frac{y^2}{5^2} = 1$$

$a = 4, b = 5$

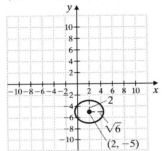

63. $\begin{cases} y = x^2 - 5x + 1 \\ y = -x + 6 \end{cases}$

Substitute.

$$x^2 - 5x + 1 = -x + 6$$
$$x^2 - 4x - 5 = 0$$
$$(x-5)(x+1) = 0$$
$$x - 5 = 0 \quad \text{or} \quad x + 1 = 0$$
$$x = 5 \qquad\qquad x = -1$$

$x = 5: y = -5 + 6 = 1$
$x = -1: y = -(-1) + 6 = 7$
The solutions are $(5, 1)$ and $(-1, 7)$.

64. $\begin{cases} x^2 + y^2 = 10 \\ 9x^2 + y^2 = 18 \end{cases}$

Multiply equation 1 by -1. Add the results.

$$-x^2 - y^2 = -10$$
$$\underline{9x^2 + y^2 = 18}$$
$$8x^2 \qquad = 8$$
$$x^2 = 1$$
$$x = \pm\sqrt{1}$$
$$x = \pm 1$$

$x = -1: (-1)^2 + y^2 = 10$
$$y^2 = 9$$
$$y = \pm 3$$

$x = 1: (1)^2 + y^2 = 10$
$$y^2 = 9$$
$$y = \pm 3$$

The solutions are $(-1, 3)$, $(-1, -3)$, $(1, 3)$, and $(1, -3)$.

65. $x^2 - y^2 < 1$

66. $\begin{cases} y > x^2 \\ x + y \geq 3 \end{cases}$

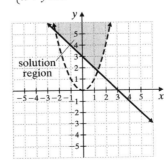

Chapter 10 Test

1. $x^2 + y^2 = 36$

    Circle; center: (0, 0), radius $r = \sqrt{36} = 6$

    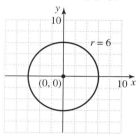

2. $x^2 - y^2 = 36$

    $\dfrac{x^2}{36} - \dfrac{y^2}{36} = 1$

    Hyperbola; center: (0, 0), $a = 6$, $b = 6$

    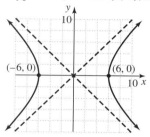

3. $16x^2 + 9y^2 = 144$

    $\dfrac{x^2}{9} + \dfrac{y^2}{16} = 1$

    Ellipse; center: (0, 0), $a = 3$, $b = 4$

    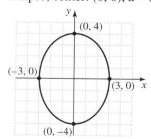

4. $y = x^2 - 8x + 16$

    $y = (x - 4)^2$

    Parabola; vertex: (4, 0)

5. $x^2 + y^2 + 6x = 16$

    $(x^2 + 6x) + y^2 = 16$

    $(x^2 + 6x + 9) + y^2 = 16 + 9$

    $(x + 3)^2 + y^2 = 25$

    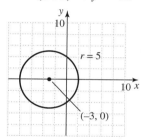

6. $x = y^2 + 8y - 3$

    $x + 16 = (y^2 + 8y + 16) - 3$

    $x = (y + 4)^2 - 19$

    Parabola; vertex: (−4, −19)

    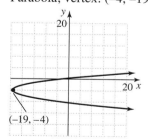

7. $\dfrac{(x-4)^2}{16}+\dfrac{(y-3)^2}{9}=1$

Ellipse: center: (4, 3), $a = 4$, $b = 3$

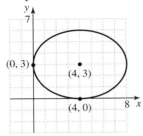

8. $y^2 - x^2 = 1$

Hyperbola: center: (0, 0), $a = 1$, $b = 1$

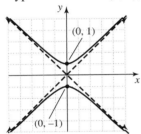

9. $\begin{cases} x^2 + y^2 = 169 \\ 5x + 12y = 0 \end{cases}$

$12y = -5x$

$y = -\dfrac{5x}{12}$

Substitute.

$x^2 + \left(-\dfrac{5x}{12}\right)^2 = 169$

$x^2 + \dfrac{25x^2}{144} = 169$

$\dfrac{169x^2}{144} = 169$

$\dfrac{x^2}{144} = 1$

$x^2 = 144$ so $x = \pm 12$.

Substitute back.

$x = 12$: $y = -\dfrac{5}{12}(12) = -5$

$x = -12$: $y = -\dfrac{5}{12}(-12) = 5$

The solutions are (12, −5) and (−12, 5).

10. $\begin{cases} x^2 + y^2 = 26 \\ x^2 - 2y^2 = 23 \end{cases}$

Multiply the second equation by −1. Add the results.

$x^2 + y^2 = 26$

$\underline{-x^2 + 2y^2 = -23}$

$3y^2 = 3$

$y^2 = 1$

$y = \pm 1$

$y = -1$: $x^2 + (-1)^2 = 26$

$x^2 = 25$

$x = \pm 5$

$y = 1$: $x^2 + 1^2 = 26$

$x^2 = 25$

$x = \pm 5$

The solutions are (−5, −1), (−5, 1), (5, −1), and (5, 1).

11. $\begin{cases} y = x^2 - 5x + 6 \\ y = 2x \end{cases}$

Substitute.

$x^2 - 5x + 6 = 2x$

$x^2 - 7x + 6 = 0$

$(x - 6)(x - 1) = 0$

$x - 6 = 0$ or $x - 1 = 0$

$x = 6$ $x = 1$

$x = 6$: $y = 2(6) = 12$

$x = 1$: $y = 2(1) = 2$

The solutions are (6, 12) and (1, 2).

12. $\begin{cases} x^2 + 4y^2 = 5 \\ y = x \end{cases}$

Substitute.

$x^2 + 4x^2 = 5$

$5x^2 = 5$

$x^2 = 1$

$x = \pm 1$

$x = -1$: $y = -1$

$x = 1$: $y = 1$

The solutions are (−1, −1) and (1, 1).

13. $\begin{cases} 2x + 5y \geq 10 \\ y \geq x^2 + 1 \end{cases}$

First graph $\begin{cases} 2x + 5y = 10 \\ y = x^2 + 1 \end{cases}$ or

$\begin{cases} y = -\dfrac{2}{5}x + 2 \\ y = 1 \cdot (x - 0)^2 + 1 \end{cases}$

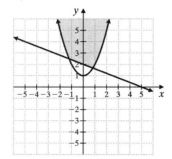

14. $\begin{cases} \dfrac{x^2}{4} + y^2 \leq 1 \\ x + y > 1 \end{cases}$

First graph $\begin{cases} \dfrac{x^2}{2^2} + \dfrac{y^2}{1^2} = 1 \\ x + y = 1 \end{cases}$

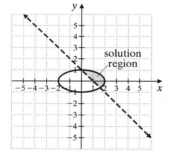

15. $\begin{cases} x^2 + y^2 > 1 \\ \dfrac{x^2}{4} - y^2 \geq 1 \end{cases}$

First graph $\begin{cases} x^2 + y^2 = 1 \\ \dfrac{x^2}{2^2} - \dfrac{y^2}{1^2} = 1 \end{cases}$

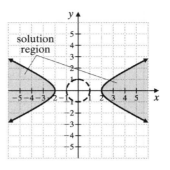

16. $\begin{cases} x^2 + y^2 \geq 4 \\ x^2 + y^2 < 16 \\ y \geq 0 \end{cases}$

First graph $\begin{cases} x^2 + y^2 = 2^2 \\ x^2 + y^2 = 4^2 \\ y = 0 \end{cases}$

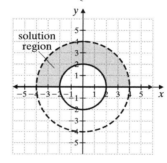

17. Graph B; vertex in second quadrant, opens to the right.

18. $100x^2 + 225y^2 = 22{,}500$

$\dfrac{x^2}{225} + \dfrac{y^2}{100} = 1$

$\dfrac{x^2}{15^2} + \dfrac{y^2}{10^2} = 1$

Height = 10 feet
Width = 2(15) = 30 feet

Chapter 10 Cumulative Review

1. $4 \cdot (9y) = (4 \cdot 9)y = 36y$

2. $3x + 4 > 1 \quad$ and $\quad 2x - 5 \leq 9$
$ 3x > -3 \quad$ and $\quad 2x \leq 14$
$ x > -1 \quad$ and $\quad x \leq 7$
$-1 < x \leq 7$
$(-1, 7]$

3. $x = -2y$

$y = -\dfrac{1}{2}x$

x	$y = -\frac{1}{2}x$
-2	1
0	0
4	-2

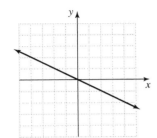

4. $(3, 2), (1, -4)$

$m = \dfrac{-4-2}{1-3} = \dfrac{-6}{-2} = 3$

5. $\begin{cases} 3x + \dfrac{y}{2} = 2 \quad (1) \\ 6x + y = 5 \quad (2) \end{cases}$

Multiply E1 by -2 and add to E2.

$-6x - y = -4$

$\underline{6x + y = 5}$

$0 = 1$

This is a false statement. Therefore, the solution is $\varnothing$.

6. Let x = speed of one plane. Then
$x + 25$ = speed of the other plane.

$d_{\text{plane }1} + d_{\text{plane }2} = 650$ miles

$2x + 2(x + 25) = 650$

$2x + 2x + 50 = 650$

$4x = 600$

$x = 150$

$x + 25 = 150 + 25 = 175$

The planes are traveling at 150 mph and 175 mph.

7. a. $(5x^2)^3 = 5^3(x^2)^3 = 125x^6$

b. $\left(\dfrac{2}{3}\right)^3 = \dfrac{2^3}{3^3} = \dfrac{8}{27}$

c. $\left(\dfrac{3p^4}{q^5}\right)^2 = \dfrac{3^2(p^4)^2}{(q^5)^2} = \dfrac{9p^8}{q^{10}}$

d. $\left(\dfrac{2^{-3}}{y}\right)^{-2} = \dfrac{(2^{-3})^{-2}}{y^{-2}}$

$= 2^6 y^2$

$= 64y^2$

e. $(x^{-5}y^2z^{-1})^7 = (x^{-5})^7(y^2)^7(z^{-1})^7$

$= x^{-35}y^{14}z^{-7}$

$= \dfrac{y^{14}}{x^{35}z^7}$

8. a. $\dfrac{4^8}{4^3} = 4^{8-3} = 4^5$

b. $\dfrac{y^{11}}{y^5} = y^{11-5} = y^6$

c. $\dfrac{32x^7}{4x^6} = \dfrac{32}{4}x^{7-6} = 8x$

d. $\dfrac{18a^{12}b^6}{12a^8b^6} = \dfrac{18}{12}a^{12-8}b^{6-6} = \dfrac{3}{2}a^4b^0 = \dfrac{3a^4}{2}$

9. $2x^2 = \dfrac{17}{3}x + 1$

$3(2x^2) = 3\left(\dfrac{17}{3}x + 1\right)$

$6x^2 = 17x + 3$

$6x^2 - 17x - 3 = 0$

$(6x + 1)(x - 3) = 0$

$6x + 1 = 0 \quad$ or $\quad x - 3 = 0$

$6x = -1 \quad$ or $\qquad x = 3$

$x = -\dfrac{1}{6}$

The solutions are $-\dfrac{1}{6}$ and 3.

10. a. $3y^2 + 14y + 15 = (3y + 5)(y + 3)$

b. $20a^5 + 54a^4 + 10a^3$

$= 2a^3(10a^2 + 27a + 5)$

$= 2a^3(2a + 5)(5a + 1)$

c. $(y-3)^2 - 2(y-3) - 8$

Let $u = y - 3$. Then $u^2 = (y-3)^2$ and

$$u^2 - 2u - 8 = (u-4)(u+2)$$
$$= [(y-3)-4][(y-3)+2]$$
$$= (y-7)(y-1)$$

11. $\dfrac{7}{x-1} + \dfrac{10x}{x^2-1} - \dfrac{5}{x+1}$

$$= \dfrac{7}{x-1} + \dfrac{10x}{(x+1)(x-1)} - \dfrac{5}{x+1}$$

$$= \dfrac{7(x+1)+10x-5(x-1)}{(x+1)(x-1)}$$

$$= \dfrac{7x+7+10x-5x+5}{(x+1)(x-1)}$$

$$= \dfrac{12x+12}{(x+1)(x-1)}$$

$$= \dfrac{12(x+1)}{(x+1)(x-1)}$$

$$= \dfrac{12}{x-1}$$

12. $\dfrac{2}{3a-15} - \dfrac{a}{25-a^2}$

$$= \dfrac{2}{3(a-5)} + \dfrac{a}{a^2-25}$$

$$= \dfrac{2}{3(a-5)} + \dfrac{a}{(a+5)(a-5)}$$

$$= \dfrac{2(a+5)+3a}{3(a+5)(a-5)}$$

$$= \dfrac{2a+10+3a}{3(a+5)(a-5)}$$

$$= \dfrac{5a+10}{3(a+5)(a-5)}$$

$$= \dfrac{5(a+2)}{3(a+5)(a-5)}$$

13. a. $\dfrac{\frac{2x}{27y^2}}{\frac{6x^2}{9}} = \dfrac{2x}{27y^2} \cdot \dfrac{9}{6x^2} = \dfrac{1}{3y^2} \cdot \dfrac{1}{3x} = \dfrac{1}{9xy^2}$

b. $\dfrac{\frac{5x}{x+2}}{\frac{10}{x-2}} = \dfrac{5x}{x+2} \cdot \dfrac{x-2}{10} = \dfrac{x(x-2)}{2(x+2)}$

c. $\dfrac{\frac{x}{y^2}+\frac{1}{y}}{\frac{y}{x^2}+\frac{1}{x}} = \dfrac{\left(\frac{x}{y^2}+\frac{1}{y}\right)x^2y^2}{\left(\frac{y}{x^2}+\frac{1}{x}\right)x^2y^2}$

$$= \dfrac{x^3+x^2y}{y^3+xy^2}$$

$$= \dfrac{x^2(x+y)}{y^2(y+x)}$$

$$= \dfrac{x^2}{y^2}$$

14. a. $(a^{-1}-b^{-1})^{-1} = \left(\dfrac{1}{a}-\dfrac{1}{b}\right)^{-1}$

$$= \left(\dfrac{b-a}{ab}\right)^{-1}$$

$$= \dfrac{ab}{b-a}$$

b. $\dfrac{2-\frac{1}{x}}{4x-\frac{1}{x}} = \dfrac{\left(2-\frac{1}{x}\right)x}{\left(4x-\frac{1}{x}\right)x}$

$$= \dfrac{2x-1}{4x^2-1}$$

$$= \dfrac{2x-1}{(2x+1)(2x-1)}$$

$$= \dfrac{1}{2x+1}$$

15. $\begin{array}{r} 2x-5 \\ x+2 \overline{)\,2x^2-x-10} \end{array}$

$$\begin{array}{r} \underline{2x^2+4x} \\ -5x-10 \\ \underline{-5x-10} \\ 0 \end{array}$$

Answer: $2x-5$

16. $\dfrac{2}{x+3} = \dfrac{1}{x^2-9} - \dfrac{1}{x-3}$

$$\dfrac{2}{x+3} = \dfrac{1}{(x+3)(x-3)} - \dfrac{1}{x-3}$$

$$2(x-3) = 1 - 1(x+3)$$

$$2x-6 = 1-x-3$$

$$2x-6 = -x-2$$

$$3x = 4$$

$$x = \dfrac{4}{3}$$

17.

$$4\,\lfloor\, 4 \quad -25 \quad 35 \quad 0 \quad 17 \quad 0 \quad 0$$
$$\underline{ 16 \quad -36 \quad -4 \quad -16 \quad 4 \quad 16}$$
$$ 4 \quad -9 \quad -1 \quad -4 \quad 1 \quad 4 \quad 16$$

Thus, $P(4) = 16$.

18. $y = \dfrac{k}{x}$

$$3 = \dfrac{k}{\dfrac{2}{3}}$$

$$k = 3\left(\dfrac{2}{3}\right) = 2$$

Thus, the equation is $y = \dfrac{2}{x}$.

19.

$$\dfrac{2x}{x-3} + \dfrac{6-2x}{x^2-9} = \dfrac{x}{x+3}$$

$$\dfrac{2x}{x-3} + \dfrac{-2(x-3)}{(x+3)(x-3)} = \dfrac{x}{x+3}$$

$$\dfrac{2x}{x-3} - \dfrac{2}{x+3} = \dfrac{x}{x+3}$$

$$\dfrac{2x}{x-3} = \dfrac{x}{x+3} + \dfrac{2}{x+3}$$

$$\dfrac{2x}{x-3} = \dfrac{x+2}{x+3}$$

$$2x(x+3) = (x+2)(x-3)$$

$$2x^2 + 6x = x^2 - x - 6$$

$$x^2 + 7x + 6 = 0$$

$$(x+6)(x+1) = 0$$

$$x+6 = 0 \quad \text{or} \quad x+1 = 0$$

$$x = -6 \quad \text{or} \quad x = -1$$

The solutions are -6 and -1.

20. **a.** $\sqrt[5]{-32} = -2$ because $(-2)^5 = -32$.

    b. $\sqrt[4]{625} = 5$ because $5^4 = 625$.

    c. $-\sqrt{36} = -6$ because $6^2 = 36$.

    d. $-\sqrt[3]{-27x^3} = -(-3x) = 3x$

    e. $\sqrt{144y^2} = 12y$

21. Let $t = $ time it will take together.

$$\dfrac{1}{4} + \dfrac{1}{5} = \dfrac{1}{t}$$

$$20t\left(\dfrac{1}{4} + \dfrac{1}{5}\right) = 20t\left(\dfrac{1}{t}\right)$$

$$5t + 4t = 20$$

$$9t = 20$$

$$t = \dfrac{20}{9} = 2\dfrac{2}{9}$$

It will take them $2\dfrac{2}{9}$ hours. No, they can not finish before the movie starts.

22. **a.** $\dfrac{\sqrt{32}}{\sqrt{4}} = \sqrt{\dfrac{32}{4}} = \sqrt{8} = \sqrt{4 \cdot 2} = 2\sqrt{2}$

    b. $\dfrac{\sqrt[3]{240y^2}}{5\sqrt[3]{3y^{-4}}} = \dfrac{1}{5}\sqrt[3]{\dfrac{240y^2}{3y^{-4}}}$

$$= \dfrac{1}{5}\sqrt[3]{80y^6}$$

$$= \dfrac{1}{5}\sqrt[3]{8y^6 \cdot 10}$$

$$= \dfrac{2y^3\sqrt[3]{10}}{5}$$

    c. $\dfrac{\sqrt[5]{64x^9y^2}}{\sqrt[5]{2x^2y^{-8}}} = \sqrt[5]{\dfrac{64x^9y^2}{2x^2y^{-8}}}$

$$= \sqrt[5]{32x^7y^{10}}$$

$$= \sqrt[5]{32x^5y^{10} \cdot x^2}$$

$$= 2xy^2\sqrt[5]{x^2}$$

23. **a.** $\sqrt[3]{1} = 1$

    b. $\sqrt[3]{-64} = -4$

    c. $\sqrt[3]{\dfrac{8}{125}} = \dfrac{\sqrt[3]{8}}{\sqrt[3]{125}} = \dfrac{2}{5}$

    d. $\sqrt[3]{x^6} = x^2$

    e. $\sqrt[3]{-27x^9} = -3x^3$

24. a.
$$\sqrt{5}\left(2+\sqrt{15}\right)=2\sqrt{5}+\sqrt{5}\cdot\sqrt{15}$$
$$=2\sqrt{5}+\sqrt{75}$$
$$=2\sqrt{5}+5\sqrt{3}$$

b.
$$\left(\sqrt{3}-\sqrt{5}\right)\left(\sqrt{7}-1\right)$$
$$=\sqrt{3}\cdot\sqrt{7}-\sqrt{3}\cdot1-\sqrt{5}\cdot\sqrt{7}+\sqrt{5}\cdot1$$
$$=\sqrt{21}-\sqrt{3}-\sqrt{35}+\sqrt{5}$$

c.
$$\left(2\sqrt{5}-1\right)^2=\left(2\sqrt{5}\right)^2-2\cdot2\sqrt{5}\cdot1+1^2$$
$$=4(5)-4\sqrt{5}+1$$
$$=21-4\sqrt{5}$$

d.
$$\left(3\sqrt{2}+5\right)\left(3\sqrt{2}-5\right)=\left(3\sqrt{2}\right)^2-5^2$$
$$=9(2)-25$$
$$=18-25$$
$$=-7$$

25. a.
$$z^{2/3}\left(z^{1/3}-z^5\right)=z^{2/3+1/3}-z^{2/3+5}$$
$$=z^{3/3}-z^{2/3+15/3}$$
$$=z-z^{17/3}$$

b.
$$(x^{1/3}-5)(x^{1/3}+2)$$
$$=x^{1/3}\cdot x^{1/3}+2x^{1/3}-5x^{1/3}-5(2)$$
$$=x^{2/3}-3x^{1/3}-10$$

26.
$$\frac{-2}{\sqrt{3}+3}=\frac{-2\left(\sqrt{3}-3\right)}{\left(\sqrt{3}+3\right)\left(\sqrt{3}-3\right)}$$
$$=\frac{-2\left(\sqrt{3}-3\right)}{\left(\sqrt{3}\right)^2-3^2}$$
$$=\frac{-2\left(\sqrt{3}-3\right)}{3-9}$$
$$=\frac{-2\left(\sqrt{3}-3\right)}{-6}$$
$$=\frac{\sqrt{3}-3}{3}$$

27. a.
$$\frac{\sqrt{20}}{\sqrt{5}}=\sqrt{\frac{20}{5}}=\sqrt{4}=2$$

b.
$$\frac{\sqrt{50x}}{2\sqrt{2}}=\frac{1}{2}\sqrt{\frac{50x}{2}}=\frac{1}{2}\sqrt{25x}=\frac{5\sqrt{x}}{2}$$

c.
$$\frac{7\sqrt[3]{48x^4y^8}}{\sqrt[3]{6y^2}}=7\sqrt[3]{\frac{48x^4y^8}{6y^2}}$$
$$=7\sqrt[3]{8x^4y^6}$$
$$=7\sqrt[3]{8x^3y^6\cdot x}$$
$$=7\cdot2xy^2\sqrt[3]{x}$$
$$=14xy^2\sqrt[3]{x}$$

d.
$$\frac{2\sqrt[4]{32a^8b^6}}{\sqrt[4]{a^{-1}b^2}}=2\sqrt[4]{\frac{32a^8b^6}{a^{-1}b^2}}$$
$$=2\sqrt[4]{32a^9b^4}$$
$$=2\sqrt[4]{16a^8b^4\cdot2a}$$
$$=2\cdot2a^2b\sqrt[4]{2a}$$
$$=4a^2b\sqrt[4]{2a}$$

28.
$$\sqrt{2x-3}=x-3$$
$$\left(\sqrt{2x-3}\right)^2=(x-3)^2$$
$$2x-3=x^2-6x+9$$
$$0=x^2-8x+12$$
$$0=(x-6)(x-2)$$
$$x-6=0 \text{ or } x-2=0$$
$$x=6 \text{ or } \quad x=2$$
Discard 2 as an extraneous solution. The solution is 6.

29. a.
$$\frac{\sqrt{45}}{4}-\frac{\sqrt{5}}{3}=\frac{3\sqrt{5}}{4}-\frac{\sqrt{5}}{3}$$
$$=\frac{9\sqrt{5}-4\sqrt{5}}{12}$$
$$=\frac{5\sqrt{5}}{12}$$

b.
$$\sqrt[3]{\frac{7x}{8}}+2\sqrt[3]{7x}=\frac{\sqrt[3]{7x}}{2}+2\sqrt[3]{7x}$$
$$=\frac{\sqrt[3]{7x}}{2}+\frac{4\sqrt[3]{7x}}{2}$$
$$=\frac{5\sqrt[3]{7x}}{2}$$

30.
$$9x^2-6x=-4$$
$$9x^2-6x+4=0$$
$$a=9, b=-6, c=4$$
$$b^2-4ac=(-6)^2-4(9)(4)$$
$$=36-144$$
$$=-108$$
Two complex but not real solutions

31. $\sqrt{\dfrac{7x}{3y}} = \dfrac{\sqrt{7x}}{\sqrt{3y}} = \dfrac{\sqrt{7x} \cdot \sqrt{3y}}{\sqrt{3y} \cdot \sqrt{3y}} = \dfrac{\sqrt{21xy}}{3y}$

32. $\dfrac{4}{x-2} - \dfrac{x}{x+2} = \dfrac{16}{x^2 - 4}$

$\dfrac{4}{x-2} - \dfrac{x}{x+2} = \dfrac{16}{(x+2)(x-2)}$

$4(x+2) - x(x-2) = 16$

$4x + 8 - x^2 + 2x = 16$

$0 = x^2 - 6x + 8$

$0 = (x-4)(x-2)$

$x - 4 = 0 \ \text{ or } \ x - 2 = 0$

$x = 4 \ \text{ or } \quad x = 2$

Discard the solution 2 as extraneous. The solution is 4.

33. $\sqrt{2x-3} = 9$

$\left(\sqrt{2x-3}\right)^2 = 9^2$

$2x - 3 = 81$

$2x = 84$

$x = 42$

The solution is 42.

34. $x^3 + 2x^2 - 4x \ge 8$

$x^3 + 2x^2 - 4x - 8 \ge 0$

$x^2(x+2) - 4(x+2) \ge 0$

$(x+2)(x^2 - 4) \ge 0$

$(x+2)(x+2)(x-2) \ge 0$

$(x+2)^2(x-2) \ge 0$

$(x+2)^2 = 0 \ \text{ or } \ x - 2 = 0$

$x + 2 = 0 \ \text{ or } \quad x = 2$

$x = -2$

Region	Test Point	$(x+2)^2(x-2) \ge 0$ Result
A: $(-\infty, -2)$	-3	$(-1)^2(-5) \ge 0$ False
B: $(-2, 2)$	0	$(2)^2(-2) \ge 0$ False
C: $(2, \infty)$	3	$(5)^2(1) \ge 0$ True

Solution: $[2, \infty)$

35. a. $i^7 = i^4 \cdot i^3 = 1 \cdot (-i) = -i$

b. $i^{20} = (i^4)^5 = 1^5 = 1$

c. $i^{46} = i^{44} \cdot i^2 = (i^4)^{11} \cdot (-1) = 1^{11}(-1) = -1$

d. $i^{-12} = \dfrac{1}{i^{12}} = \dfrac{1}{(i^4)^3} = \dfrac{1}{1^3} = 1$

36. $f(x) = (x+2)^2 - 1$

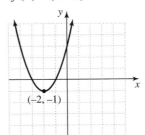

37. $p^2 + 2p = 4$

$p^2 + 2p + \left(\dfrac{2}{2}\right)^2 = 4 + 1$

$p^2 + 2p + 1 = 5$

$(p+1)^2 = 5$

$p + 1 = \pm\sqrt{5}$

$p = -1 \pm \sqrt{5}$

The solutions are $-1 + \sqrt{5}$ and $-1 - \sqrt{5}$.

38. $f(x) = -x^2 - 6x + 4$

The maximum will occur at the vertex.

$x = \dfrac{-b}{2a} = \dfrac{-(-6)}{2(-1)} = -3$

$f(-3) = -(-3)^2 - 6(-3) + 4 = 13$

The maximum value is 13.

39.
$$\frac{1}{4}m^2 - m + \frac{1}{2} = 0$$
$$4\left(\frac{1}{4}m^2 - m + \frac{1}{2}\right) = 4(0)$$
$$m^2 - 4m + 2 = 0$$
$$a = 1, b = -4, c = 2$$
$$m = \frac{-(-4) \pm \sqrt{(-4)^2 - 4(1)(2)}}{2(1)}$$
$$= \frac{4 \pm \sqrt{16-8}}{2}$$
$$= \frac{4 \pm \sqrt{8}}{2}$$
$$= \frac{4 \pm 2\sqrt{2}}{2}$$
$$= 2 \pm \sqrt{2}$$

The solutions are $2+\sqrt{2}$ and $2-\sqrt{2}$.

40.
$$f(x) = \frac{x+1}{2}$$
$$y = \frac{x+1}{2}$$
$$x = \frac{y+1}{2}$$
$$2x = y+1$$
$$2x-1 = y$$
$$f^{-1}(x) = 2x-1$$

41.
$$p^4 - 3p^2 - 4 = 0$$
$$(p^2 - 4)(p^2 + 1) = 0$$
$$(p+2)(p-2)(p^2+1) = 0$$

$$p+2 = 0 \quad \text{or} \quad p-2 = 0 \quad \text{or} \quad p^2 + 1 = 0$$
$$p = -2 \quad \text{or} \quad p = 2 \quad \text{or} \quad p^2 = -1$$
$$p = \pm i$$

The solutions are -2, 2, $-i$, and i.

42. $f(x) = x^2 - 3x + 2$
$$g(x) = -3x + 5$$

a.
$$(f \circ g)(x) = f[g(x)]$$
$$= f(-3x+5)$$
$$= (-3x+5)^2 - 3(-3x+5) + 2$$
$$= 9x^2 - 30x + 25 + 9x - 15 + 2$$
$$= 9x^2 - 21x + 12$$

b.
$$(f \circ g)(-2) = f[g(-2)]$$
$$= f[-3(-2)+5]$$
$$= f(11)$$
$$= (11)^2 - 3(11) + 2$$
$$= 121 - 33 + 2$$
$$= 90$$

c.
$$(g \circ f)(x) = g[f(x)]$$
$$= g(x^2 - 3x + 2)$$
$$= -3(x^2 - 3x + 2) + 5$$
$$= -3x^2 + 9x - 6 + 5$$
$$= -3x^2 + 9x - 1$$

d.
$$(g \circ f)(5) = g[f(5)]$$
$$= g[(5)^2 - 3(5) + 2]$$
$$= g(12)$$
$$= -3(12) + 5$$
$$= -36 + 5$$
$$= -31$$

43.
$$\frac{x+2}{x-3} \le 0$$
$$x + 2 = 0 \quad \text{or} \quad x - 3 = 0$$
$$x = -2 \quad \text{or} \quad x = 3$$

Region	Test Point	$\frac{x+2}{x-3} \le 0$ Result
A: $(-\infty, -2)$	-3	$\frac{-1}{-6} \le 0$; False
B: $(-2, 3)$	0	$\frac{2}{-3} \le 0$; True
C: $(3, \infty)$	4	$\frac{6}{1} \le 0$; False

Solution: $[-2, 3)$

44. $4x^2 + 9y^2 = 36$
$$\frac{x^2}{9} + \frac{y^2}{4} = 1$$
Ellipse: center $(0, 0)$, $a = 3$, $b = 2$

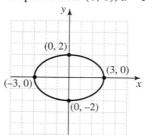

45. $g(x) = \dfrac{1}{2}(x+2)^2 + 5$

Vertex: $(-2, 5)$, axis: $x = -2$

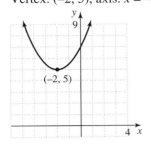

46. a. $64^x = 4$

$(4^2)^x = 4$

$4^{2x} = 4$

$2x = 1$

$x = \dfrac{1}{2}$

b. $125^{x-3} = 25$

$(5^3)^{x-3} = 5^2$

$5^{3x-9} = 5^2$

$3x - 9 = 2$

$3x = 11$

$x = \dfrac{11}{3}$

c. $\dfrac{1}{81} = 3^{2x}$

$3^{-4} = 3^{2x}$

$-4 = 2x$

$-\dfrac{4}{2} = x$

$-2 = x$

47. $f(x) = x^2 - 4x - 12$

$x = \dfrac{-b}{2a} = \dfrac{-(-4)}{2(1)} = 2$

$f(2) = (2)^2 - 4(2) - 12 = -16$

Vertex: $(2, -16)$

48. $\begin{cases} x + 2y < 8 \\ y \geq x^2 \end{cases}$

First, graph $x + 2y = 8$ as a dashed line.

Test Point	$x + 2y < 8$; Result
$(0, 0)$	$0 + 2(0) < 8$; True

Shade the region containing $(0, 0)$. Next, graph the parabola $y = x^2$ as a solid curve.

Test Point	$y \geq x^2$; Result
$(0, 1)$	$1 \geq 0^2$; True

Shade the region containing $(0, 1)$. The solution to the system is the intersection.

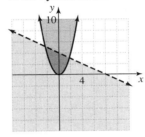

49. $(2, -5)$, $(1, -4)$

$d = \sqrt{[-4 - (-5)]^2 + (1 - 2)^2}$

$= \sqrt{1^2 + (-1)^2}$

$= \sqrt{2} \approx 1.414$

50. $\begin{cases} x^2 + y^2 = 36 & (1) \\ \quad\quad\; y = x + 6 & (2) \end{cases}$

Substitute $x + 6$ for y in E1.

$x^2 + (x+6)^2 = 36$

$x^2 + (x^2 + 12x + 36) = 36$

$2x^2 + 12x = 0$

$2x(x + 6) = 0$

$2x = 0$ or $x + 6 = 0$

$x = 0$ or $\quad\quad x = -6$

Use these values in E2 to find y.

$x = 0: y = 0 + 6 = 6$

$x = -6: y = -6 + 6 = 0$

The solutions are $(0, 6)$ and $(-6, 0)$.

Chapter 11

1. $a_n = 5 + n^2$

$a_1 = 5 + 1^2 = 5 + 1 = 6$

$a_2 = 5 + 2^2 = 5 + 4 = 9$

$a_3 = 5 + 3^2 = 5 + 9 = 14$

$a_4 = 5 + 4^2 = 5 + 16 = 21$

$a_5 = 5 + 5^2 = 5 + 25 = 30$

Thus, the first five terms of the sequence are 6, 9, 14, 21, and 30.

2. $a_n = \dfrac{(-1)^n}{5n}$

a. $a_1 = \dfrac{(-1)^1}{5(1)} = -\dfrac{1}{5}$

b. $a_4 = \dfrac{(-1)^4}{5(4)} = \dfrac{1}{20}$

c. $a_{30} = \dfrac{(-1)^{30}}{5(30)} = \dfrac{1}{150}$

d. $a_{19} = \dfrac{(-1)^{19}}{5(19)} = -\dfrac{1}{95}$

3. a. 1, 3, 5, 7, …

These numbers are the first four odd natural numbers, so a general term might be $a_n = 2n - 1$.

b. 3, 9, 27, 81, …

These numbers are all powers of 3 ($3 = 3^1$, $9 = 3^2$, $27 = 3^3$, and $81 = 3^4$), so a general term might be $a_n = 3^n$.

c. $\dfrac{1}{2}, \dfrac{2}{3}, \dfrac{3}{4}, \dfrac{4}{5}, \ldots$

The numerators are the first four natural numbers and each denominator is one greater than the numerator, so a general term might be $a_n = \dfrac{n}{n+1}$.

d. $-\dfrac{1}{2}, -\dfrac{1}{3}, -\dfrac{1}{4}, -\dfrac{1}{5}, \ldots$

The denominators are consecutive natural numbers beginning with 2 and each term is negative, so a general term might be

$a_n = -\dfrac{1}{n+1}$.

4. $v_n = 3950(0.8)^n$

$v_3 = 3950(0.8)^3$

$\quad = 3950(0.512)$

$\quad = 2022.4$

The value of the copier after three years is $2022.40.

Vocabulary, Readiness & Video Check 11.1

1. The nth term of the sequence a_n is called the <u>general</u> term.

2. A <u>finite</u> sequence is a function whose domain is $\{1, 2, 3, 4, \ldots, n\}$ where n is some natural number.

3. An <u>infinite</u> sequence is a function whose domain is $\{1, 2, 3, 4, \ldots\}$.

4. $a_n = 7^n$

$a_1 = 7^1 = 7$

5. $a_n = \dfrac{(-1)^n}{n}$

$a_1 = \dfrac{(-1)^1}{1} = -1$

6. $a_n = (-1)^n \cdot n^4$

$a_1 = (-1)^1 \cdot 1^4 = -1$

7. A sequence is a <u>function</u> whose <u>domain</u> is the set of natural numbers. We use $\underline{a_n}$ to mean the general term of a sequence.

8. If the negative is inside the parentheses, such as $(-2)^n$, it is also raised to the power. Since a_2 and a_4 would then have even powers, they would then be positive terms—but all original terms are negative.

9. $a_9 = 0.10(2)^{9-1} = \$25.60$

Exercise Set 11.1

1. $a_n = n + 4$
$a_1 = 1 + 4 = 5$
$a_2 = 2 + 4 = 6$
$a_3 = 3 + 4 = 7$
$a_4 = 4 + 4 = 8$
$a_5 = 5 + 4 = 9$
Thus, the first five terms of the sequence
$a_n = n + 4$ are 5, 6, 7, 8, 9.

3. $a_n = (-1)^n$

$a_1 = (-1)^1 = -1$
$a_2 = (-1)^2 = 1$
$a_3 = (-1)^3 = -1$
$a_4 = (-1)^4 = 1$
$a_5 = (-1)^5 = -1$
Thus, the first five terms of the sequence
$a_n = (-1)^n$ are $-1, 1, -1, 1, -1$.

5. $a_n = \dfrac{1}{n+3}$

$a_1 = \dfrac{1}{1+3} = \dfrac{1}{4}$
$a_2 = \dfrac{1}{2+3} = \dfrac{1}{5}$
$a_3 = \dfrac{1}{3+3} = \dfrac{1}{6}$
$a_4 = \dfrac{1}{4+3} = \dfrac{1}{7}$
$a_5 = \dfrac{1}{5+3} = \dfrac{1}{8}$
Thus, the first five terms of the sequence
$a_n = \dfrac{1}{n+3}$ are $\dfrac{1}{4}, \dfrac{1}{5}, \dfrac{1}{6}, \dfrac{1}{7}, \dfrac{1}{8}$.

7. $a_n = 2n$
$a_1 = 2(1) = 2$
$a_2 = 2(2) = 4$
$a_3 = 2(3) = 6$
$a_4 = 2(4) = 8$
$a_5 = 2(5) = 10$
Thus, the first five terms of the sequence
$a_n = 2n$ are 2, 4, 6, 8, 10.

9. $a_n = -n^2$
$a_1 = -1^2 = -1$
$a_2 = -2^2 = -4$
$a_3 = -3^2 = -9$
$a_4 = -4^2 = -16$
$a_5 = -5^2 = -25$
Thus, the first five terms of the sequence
$a_n = n^2$ are $-1, -4, -8, -16, -25$.

11. $a_n = 2^n$
$a_1 = 2^1 = 2$
$a_2 = 2^2 = 4$
$a_3 = 2^3 = 8$
$a_4 = 2^4 = 16$
$a_5 = 2^5 = 32$
Thus, the first five terms of the sequence
$a_n = 2^n$ are 2, 4, 8, 16, 32.

13. $a_n = 2n + 5$
$a_1 = 2(1) + 5 = 2 + 5 = 7$
$a_2 = 2(2) + 5 = 4 + 5 = 9$
$a_3 = 2(3) + 5 = 6 + 5 = 11$
$a_4 = 2(4) + 5 = 8 + 5 = 13$
$a_5 = 2(5) + 5 = 10 + 5 = 15$
Thus, the first five terms of the sequence
$a_n = 2n + 5$ are 7, 9, 11, 13, 15.

15. $a_n = (-1)^n n^2$
$a_1 = (-1)^1 (1)^2 = -1(1) = -1$
$a_2 = (-1)^2 (2)^2 = 1(4) = 4$
$a_3 = (-1)^3 (3)^2 = -1(9) = -9$
$a_4 = (-1)^4 (4)^2 = 1(16) = 16$
$a_5 = (-1)^5 (5)^2 = -1(25) = -25$
Thus, the first five terms of the sequence
$a_n = (-1)^n n^2$ are $-1, 4, -9, 16, -25$.

17. $a_n = 3n^2$
$a_5 = 3(5)^2 = 3(25) = 75$

19. $a_n = 6n - 2$
$a_{20} = 6(20) - 2 = 120 - 2 = 118$

21. $a_n = \dfrac{n+3}{n}$

$a_{15} = \dfrac{15+3}{15} = \dfrac{18}{15} = \dfrac{6}{5}$

23. $a_n = (-3)^n$

$a_6 = (-3)^6 = 729$

25. $a_n = \dfrac{n-2}{n+1}$

$a_6 = \dfrac{6-2}{6+1} = \dfrac{4}{7}$

27. $a_n = \dfrac{(-1)^n}{n}$

$a_8 = \dfrac{(-1)^8}{8} = \dfrac{1}{8}$

29. $a_n = -n^2 + 5$

$a_{10} = -10^2 + 5 = -100 + 5 = -95$

31. $a_n = \dfrac{(-1)^n}{n+6}$

$a_{19} = \dfrac{(-1)^{19}}{19+6} = -\dfrac{1}{25}$

33. $3, 7, 11, 15$, or $4(1) - 1, 4(2) - 1, 4(3) - 1,$
$4(4) - 1$. In general, $a_n = 4n - 1$.

35. $-2, -4, -8, -16$, or $-2, -2^2, -2^3, -2^4$

In general, $a_n = -2^n$.

37. $\dfrac{1}{3}, \dfrac{1}{9}, \dfrac{1}{27}, \dfrac{1}{81}$, or $\dfrac{1}{3}, \dfrac{1}{3^2}, \dfrac{1}{3^3}, \dfrac{1}{3^4}$

In general, $a_n = \dfrac{1}{3^n}$.

39. $a_n = 32n - 16$

$a_2 = 32(2) - 16 = 64 - 16 = 48$ ft
$a_3 = 32(3) - 16 = 96 - 16 = 80$ ft
$a_4 = 32(4) - 16 = 128 - 16 = 112$ ft

41. $0.10, 0.20, 0.40$, or $0.10, 0.10(2), 0.10(2)^2$

In general, $a_n = 0.10(2)^{n-1}$

$a_{14} = 0.10(2)^{13} = \$819.20$

43. $a_n = 75(2)^{n-1}$

$a_6 = 75(2)^5 = 75(32) = 2400$ cases
$a_1 = 75(2)^0 = 75(1) = 75$ cases

45. $a_n = \dfrac{1}{2} a_{n-1}$ for $n > 1, a_1 = 800$

In 2000, $n = 1$ and $a_1 = 800$.

In 2001, $n = 2$ and $a_2 = \dfrac{1}{2}(800) = 400$.

In 2002, $n = 3$ and $a_3 = \dfrac{1}{2}(400) = 200$.

In 2003, $n = 4$ and $a_4 = \dfrac{1}{2}(200) = 100$.

In 2004, $n = 5$ and $a_5 = \dfrac{1}{2}(100) = 50$.

The population estimate for 2004 is 50 sparrows.
Continuing the sequence:

In 2005, $n = 6$ and $a_6 = \dfrac{1}{2}(50) = 25$.

In 2006, $n = 7$ and $a_7 = \dfrac{1}{2}(25) \approx 12$.

In 2007, $n = 8$ and $a_8 = \dfrac{1}{2}(12) = 6$.

In 2008, $n = 9$ and $a_9 = \dfrac{1}{2}(6) = 3$.

In 2009, $n = 10$ and $a_{10} = \dfrac{1}{2}(3) \approx 1$.

In 2010, $n = 11$ and $a_{11} = \dfrac{1}{2}(1) \approx 0$.

The population is estimated to become extinct in 2010.

47. $f(x) = (x-1)^2 + 3$

49. $f(x) = 2(x+4)^2 + 2$

51. $(-4, -1)$ and $(-7, -3)$

$$d = \sqrt{[-7-(-4)]^2 + [-3-(-1)]^2}$$
$$= \sqrt{(-7+4)^2 + (-3+1)^2}$$
$$= \sqrt{(-3)^2 + (-2)^2}$$
$$= \sqrt{9+4}$$
$$= \sqrt{13} \text{ units}$$

53. $(2, -7)$ and $(-3, -3)$

$$d = \sqrt{(-3-2)^2 + [-3-(-7)]^2}$$
$$= \sqrt{(-5)^2 + (-3+7)^2}$$
$$= \sqrt{(-5)^2 + (4)^2}$$
$$= \sqrt{25+16}$$
$$= \sqrt{41} \text{ units}$$

55. $a_n = \dfrac{1}{\sqrt{n}}$

n	$u(n)$
1	1
2	.70711
3	.57735
4	.5
5	.44721
6	.40825
7	.37796

$u(n) = 1/\sqrt{(n)}$

Thus, the first five terms of the sequence

$a_n = \dfrac{1}{\sqrt{n}}$ are 1, 0.7071, 0.5774, 0.5, 0.4472.

57. $a_n = \left(1+\dfrac{1}{n}\right)^n$

n	$u(n)$
1	2
2	2.25
3	2.3704
4	2.4414
5	2.4883
6	2.5216
7	2.5465

$u(n) = (1+1/n)^n$

Thus, the first five terms of the sequence

$a_n = \left(1+\dfrac{1}{n}\right)^n$ are 2, 2.25, 2.3704, 2.4414, 2.4883.

Section 11.2 Practice Exercises

1. $a_1 = 4$
$a_2 = 4+5 = 9$
$a_3 = 9+5 = 14$
$a_4 = 14+5 = 19$
$a_5 = 19+5 = 24$
The first five terms are 4, 9, 14, 19, 24.

2. a. $a_n = a_1 + (n-1)d$
Here, $a_1 = 2$ and $d = -3$.
$a_n = 2 + (n-1)(-3) = 2 - 3n + 3 = 5 - 3n$

b. $a_n = 5 - 3n$
$a_{12} = 5 - 3 \cdot 12 = 5 - 36 = -31$

3. Since the sequence is arithmetic, the ninth term is $a_9 = a_1 + (9-1)d = a_1 + 8d$.
a_1 is the first term of the sequence, so $a_1 = 3$. d is the constant difference, so
$d = a_2 - a_1 = 9 - 3 = 6$. Thus,
$a_9 = a_1 + 8d = 3 + 8 \cdot 6 = 51$.

4. We need to find a_1 and d. The given facts, $a_3 = 23$ and $a_8 = 63$, lead to a system of linear equations.
$$\begin{cases} a_3 = a_1 + (3-1)d \\ a_8 = a_1 + (8-1)d \end{cases} \text{ or } \begin{cases} 23 = a_1 + 2d \\ 63 = a_1 + 7d \end{cases}$$
We solve the system by elimination. Multiply both sides of the second equation by -1.
$$\begin{cases} 23 = a_1 + 2d \\ -1(63) = -1(a_1 + 7d) \end{cases} \text{ or } \begin{cases} 23 = a_1 + 2d \\ -63 = -a_1 - 7d \end{cases}$$
$$\overline{ -40 = -5d}$$
$$8 = d$$
To find a_1, let $d = 8$ in $23 = a_1 + 2d$.
$23 = a_1 + 2(8)$
$23 = a_1 + 16$
$7 = a_1$
Thus, $a_1 = 7$ and $d = 8$, so
$a_n = 7 + (n-1)(8) = 7 + 8n - 8 = -1 + 8n$ and
$a_6 = -1 + 8 \cdot 6 = 47$.

5. The first term, a_1, is 57,000, and d is 2200.
$a_n = 57,000 + (n-1)(2200)$
$ = 54,800 + 2200n$
$a_3 = 54,800 + 2200 \cdot 3 = 61,400$
The salary for the third year is \$61,400.

6. $a_1 = 8$

$a_2 = 8(-3) = -24$
$a_3 = -24(-3) = 72$
$a_4 = 72(-3) = -216$
The first four terms are 8, -24, 72, and -216.

7. $a_n = a_1 r^{n-1}$

Here, $a_1 = 64$ and $r = \dfrac{1}{4}$.

Evaluate a_n for $n = 7$.

$a_7 = 64\left(\dfrac{1}{4}\right)^{7-1}$

$= 64\left(\dfrac{1}{4}\right)^6$

$= 64\left(\dfrac{1}{4096}\right)$

$= \dfrac{1}{64}$

8. Since the sequence is geometric and $a_1 = -3$, the seventh term must be $a_1 r^{7-1}$, or $-3r^6$. r is the common ratio of terms, so r must be $\dfrac{6}{-3}$, or -2.

$a_7 = -3r^6$
$a_7 = -3(-2)^6 = -192$

9. Notice that $\dfrac{27}{4} \div \dfrac{9}{2} = \dfrac{3}{2}$, so $r = \dfrac{3}{2}$.

$a_2 = a_1\left(\dfrac{3}{2}\right)^{2-1}$

$\dfrac{9}{2} = a_1\left(\dfrac{3}{2}\right)^1$, or $a_1 = 3$

The first term is 3, and the common ration is $\dfrac{3}{2}$.

10. Since the culture is reduced by one-half each day, the population sizes are modeled by a geometric sequence. Here, $a_1 = 4800$ and $r = \dfrac{1}{2}$.

$a_n = a_1 r^{n-1} = 4800\left(\dfrac{1}{2}\right)^{n-1}$

$a_7 = 4800\left(\dfrac{1}{2}\right)^{7-1} = 75$

The bacterial culture should measure 75 units at the beginning of day 7.

Vocabulary, Readiness & Video Check 11.2

1. A geometric sequence is one in which each term (after the first) is obtained by multiplying the preceding term by a constant r. The constant r is called the common ratio.

2. An arithmetic sequence is one in which each term (after the first) differs from the preceding term by a constant amount d. The constant d is called the common difference.

3. The general term of an arithmetic sequence is $a_n = a_1 + (n-1)d$ where a_1 is the first term and d is the common difference.

4. The general term of a geometric sequence is $a_n = a_1 r^{n-1}$ where a_1 is the first term and r is the common ratio.

5. If there is a common difference between each term and its preceding term in a sequence, it's an arithmetic sequence.

6. An arithmetic sequence has a common difference between terms—you add the same number to go from term to term. A geometric sequence has a common ratio between terms—you multiply by the same number to go from term to term.

Exercise Set 11.2

1. $a_n = a_1 + (n-1)d$

$a_1 = 4;\ d = 2$
$a_1 = 4$
$a_2 = 4 + (2-1)2 = 6$
$a_3 = 4 + (3-1)2 = 8$
$a_4 = 4 + (4-1)2 = 10$
$a_5 = 4 + (5-1)2 = 12$
The first five terms are 4, 6, 8, 10, 12.

3. $a_n = a_1 + (n-1)d$

$a_1 = 6,\ d = -2$
$a_1 = 6$
$a_2 = 6 + (2-1)(-2) = 4$
$a_3 = 6 + (3-1)(-2) = 2$
$a_4 = 6 + (4-1)(-2) = 0$
$a_5 = 6 + (5-1)(-2) = -2$
The first five terms are 6, 4, 2, 0, -2.

5. $a_n = a_1 r^{n-1}$

$a_1 = 1,\ r = 3$

$a_1 = 1(3)^{1-1} = 1$

$a_2 = 1(3)^{2-1} = 3$

$a_3 = 1(3)^{3-1} = 9$

$a_4 = 1(3)^{4-1} = 27$

$a_5 = 1(3)^{5-1} = 81$

The first five terms are 1, 3, 9, 27, 81.

7. $a_n = a_1 r^{n-1}$

$a_1 = 48,\ r = \dfrac{1}{2}$

$a_1 = 48\left(\dfrac{1}{2}\right)^{1-1} = 48$

$a_2 = 48\left(\dfrac{1}{2}\right)^{2-1} = 24$

$a_3 = 48\left(\dfrac{1}{2}\right)^{3-1} = 12$

$a_4 = 48\left(\dfrac{1}{2}\right)^{4-1} = 6$

$a_5 = 48\left(\dfrac{1}{2}\right)^{5-1} = 3$

The first five terms are 48, 24, 12, 6, 3.

9. $a_n = a_1 + (n-1)d$

$a_1 = 12,\ d = 3$

$a_n = 12 + (n-1)3$

$a_8 = 12 + 7(3) = 12 + 21 = 33$

11. $a_n = a_1 r^{n-1}$

$a_1 = 7,\ d = -5$

$a_n = a_1 r^{n-1}$

$a_4 = 7(-5)^3 = 7(-125) = -875$

13. $a_n = a_1 + (n-1)d$

$a_1 = -4,\ d = -4$

$a_n = -4 + (n-1)(-4)$

$a_{15} = -4 + 14(-4) = -4 - 56 = -60$

15. 0, 12, 24

$a_1 = 0$ and $d = 12$

$a_n = 0 + (n-1)12$

$a_9 = 8(12) = 96$

17. 20, 18, 16

$a_1 = 20$ and $d = -2$

$a_n = 20 + (n-1)(-2)$

$a_{25} = 20 + 24(-2) = 20 - 48 = -28$

19. 2, −10, 50

$a_1 = 2$ and $r = -5$

$a_n = 2(-5)^{n-1}$

$a_5 = 2(-5)^4 = 2(625) = 1250$

21. $a_4 = 19,\ a_{15} = 52$

$\begin{cases} a_4 = a_1 + (4-1)d \\ a_{15} = a_1 + (15-1)d \end{cases}$ or

$\begin{cases} 19 = a_1 + 3d \\ 52 = a_1 + 14d \end{cases}$

$\begin{cases} -19 = -a_1 - 3d \\ \ \ 52 = a_1 + 14d \end{cases}$

Adding yields $33 = 11d$ or $d = 3$. Then
$a_1 = 19 - 3(3) = 10$.

$\begin{aligned} a_n &= 10 + (n-1)3 \\ &= 10 + 3n - 3 \\ &= 7 + 3n \end{aligned}$

and $\begin{aligned} a_8 &= 7 + 3(8) \\ &= 7 + 24 \\ &= 31 \end{aligned}$

23. $a_2 = -1,\ a_4 = 5$

$\begin{cases} a_2 = a_1 + (2-1)d \\ a_4 = a_1 + (4-1)d \end{cases}$ or

$\begin{cases} -1 = a_1 + d \\ \ \ 5 = a_1 + 3d \end{cases}$

$\begin{cases} 1 = -a_1 - d \\ 5 = a_1 + 3d \end{cases}$

Adding yields $6 = 2d$ or $d = 3$. Then
$a_1 = -1 - 3 = -4$.

$\begin{aligned} a_n &= -4 + (n-1)3 \\ &= -4 + 3n - 3 \\ &= -7 + 3n \end{aligned}$

and $\begin{aligned} a_9 &= -7 + 3(9) \\ &= -7 + 27 \\ &= 20 \end{aligned}$

25. $a_2 = -\dfrac{4}{3}$ and $a_3 = \dfrac{8}{3}$

Notice that $\dfrac{8}{3} \div \dfrac{-4}{3} = \dfrac{8}{3} \cdot -\dfrac{3}{4} = -2$, so $r = -2$.

Then

$a_2 = a_1(-2)^{2-1}$

$-\dfrac{4}{3} = a_1(-2)$

$\dfrac{2}{3} = a_1.$

The first term is $\dfrac{2}{3}$ and the common ratio is -2.

27. answers may vary

29. $2, 4, 6$ is an arithmetic sequence.

$a_1 = 2$ and $d = 2$

31. $5, 10, 20$ is a geometric sequence.

$a_1 = 5$ and $r = 2$

33. $\dfrac{1}{2}, \dfrac{1}{10}, \dfrac{1}{50}$ is a geometric sequence.

$a_1 = \dfrac{1}{2}$ and $r = \dfrac{1}{5}$

35. $x, 5x, 25x$ is a geometric sequence.

$a_1 = x$ and $r = 5$

37. $p, p+4, p+8$ is an arithmetic sequence.

$a_1 = p$ and $d = 4$

39. $a_1 = 14$ and $d = \dfrac{1}{4}$

$a_n = 14 + (n-1)\dfrac{1}{4}$

$a_{21} = 14 + 20\left(\dfrac{1}{4}\right) = 14 + 5 = 19$

41. $a_1 = 3$ and $r = -\dfrac{2}{3}$

$a_n = 3\left(-\dfrac{2}{3}\right)^{n-1}$

$a_4 = 3\left(-\dfrac{2}{3}\right)^3 = 3\left(-\dfrac{8}{27}\right) = -\dfrac{8}{9}$

43. $\dfrac{3}{2}, 2, \dfrac{5}{2}, \ldots$

$a_1 = \dfrac{3}{2}$ and $d = \dfrac{1}{2}$

$a_n = \dfrac{3}{2} + (n-1)\dfrac{1}{2}$

$a_{15} = \dfrac{3}{2} + 14\left(\dfrac{1}{2}\right) = \dfrac{17}{2}$

45. $24, 8, \dfrac{8}{3}, \ldots$

$a_1 = 24$ and $r = \dfrac{1}{3}$

$a_n = 24\left(\dfrac{1}{3}\right)^{n-1}$

$a_6 = 24\left(\dfrac{1}{3}\right)^5 = 24\left(\dfrac{1}{243}\right) = \dfrac{8}{81}$

47. $a_3 = 2$, $a_{17} = -40$

$\begin{cases} a_3 = a_1 + (3-1)d \\ a_{17} = a_1 + (17-1)d \end{cases}$ or

$\begin{cases} 2 = a_1 + 2d \\ -40 = a_1 + 16d \end{cases}$

$\begin{cases} -2 = -a_1 - 2d \\ -40 = a_1 + 16d \end{cases}$

Adding yields $-42 = 14d$ or $d = -3$. Then $a_1 = 2 - 2(-3) = 8.$

$a_n = 8 + (n-1)(-3) = 8 - 3n + 3 = 11 - 3n$

and

$a_{10} = 11 - 3(10) = 11 - 30 = -19$

49. $54, 58, 62$

$a_1 = 54$ and $d = 4$

$a_n = 54 + (n-1)4$

$a_{20} = 54 + 19(4) = 54 + 76 = 130$

The general term of the sequence is $a_n = 4n + 50$. There are 130 seats in the twentieth row.

51. $a_1 = 6$ and $r = 3$

$a_n = 6(3)^{n-1} = 2 \cdot 3 \cdot (3)^{n-1} = 2(3)^n$

The general term of the sequence is $a_n = 6(3)^{n-1}$ or $a_n = 2(3)^n$.

53. $a_1 = 486$ and $r = \dfrac{1}{3}$

Initial Height $= a_1 = 486\left(\dfrac{1}{3}\right)^{1-1} = 486$

Rebound $1 = a_2 = 486\left(\dfrac{1}{3}\right)^{2-1} = 162$

Rebound $2 = a_3 = 486\left(\dfrac{1}{3}\right)^{3-1} = 54$

Rebound $3 = a_4 = 486\left(\dfrac{1}{3}\right)^{4-1} = 18$

Rebound $4 = a_5 = 486\left(\dfrac{1}{3}\right)^{5-1} = 6$

The first five terms of the sequence are 486, 162, 54, 18, 6.

The general term is $a_n = 486\left(\dfrac{1}{3}\right)^{n-1}$ or

$a_n = \dfrac{486}{3^{n-1}}$. Since $a_6 = 2$ and $a_7 = \dfrac{2}{3}$, a_7 is the

first term less than 1. Since a_7 corresponds to the 6th bounce, it takes 6 bounces for the ball to rebound less than 1 foot.

55. $a_1 = 4000$ and $d = 125$
$a_n = 4000 + (n-1)125$ or
$a_n = 3875 + 125n$
$a_{12} = 4000 + 11(125) = 5375$
His salary for his last month of training is \$5375.

57. $a_1 = 400$ and $r = \dfrac{1}{2}$

12 hours = 4(3 hours), so we seek the fourth term after a_1, namely a_5.

$a_n = a_1 r^{n-1}$

$a_5 = 400\left(\dfrac{1}{2}\right)^4 = \dfrac{400}{16} = 25$

25 grams of the radioactive material remain after 12 hours.

59. $\dfrac{1}{3(1)} + \dfrac{1}{3(2)} + \dfrac{1}{3(3)} = \dfrac{1}{3} + \dfrac{1}{6} + \dfrac{1}{9}$

$= \dfrac{6}{18} + \dfrac{3}{18} + \dfrac{2}{18}$

$= \dfrac{11}{18}$

61. $3^0 + 3^1 + 3^2 + 3^3 = 1 + 3 + 9 + 27 = 40$

63. $\dfrac{8-1}{8+1} + \dfrac{8-2}{8+2} + \dfrac{8-3}{8+3} = \dfrac{7}{9} + \dfrac{6}{10} + \dfrac{5}{11}$

$= \dfrac{770}{990} + \dfrac{594}{990} + \dfrac{450}{990}$

$= \dfrac{1814}{990}$

$= \dfrac{907}{495}$

65. $a_1 = \$11,782.40$, $r = 0.5$

The first four terms of the sequence are \$11,782.40, \$5891.20, \$2945.60, \$1472.80.

67. $a_1 = 19.652$, $d = -0.034$

The first four terms of the sequence are 19.652, 19.618, 19.584, 19.550.

69. answers may vary

Section 11.3 Practice Exercises

1. a. $\displaystyle\sum_{i=0}^{4} \dfrac{i-3}{4} = \dfrac{0-3}{4} + \dfrac{1-3}{4} + \dfrac{2-3}{4} + \dfrac{3-3}{4} + \dfrac{4-3}{4}$

$= \left(-\dfrac{3}{4}\right) + \left(-\dfrac{2}{4}\right) + \left(-\dfrac{1}{4}\right) + 0 + \dfrac{1}{4}$

$= -\dfrac{5}{4}$, or $-1\dfrac{1}{4}$

b. $\displaystyle\sum_{i=2}^{5} 3^i = 3^2 + 3^3 + 3^4 + 3^5$

$= 9 + 27 + 81 + 243$

$= 360$

2. a. Since the difference of each term and the preceding term is 5, the terms correspond to the first six terms of the arithmetic sequence $a_n = 5 + (n-1)5 = 5n$. Thus, in summation notation,

$5 + 10 + 15 + 20 + 25 + 30 = \displaystyle\sum_{i=1}^{6} 5i$.

b. Since each term is the product of the preceding term and $\dfrac{1}{5}$, these terms correspond to the first four terms of the geometric sequence $a_n = \dfrac{1}{5}\left(\dfrac{1}{5}\right)^{n-1} = \left(\dfrac{1}{5}\right)^n$.

In summation notation,

$$\frac{1}{5} + \frac{1}{25} + \frac{1}{125} + \frac{1}{625} = \sum_{i=1}^{4}\left(\frac{1}{5}\right)^i.$$

3. $S_4 = \displaystyle\sum_{i=1}^{4} \frac{2+3i}{i^2}$

$\qquad = \dfrac{2+3\cdot1}{1^2} + \dfrac{2+3\cdot2}{2^2} + \dfrac{2+3\cdot3}{3^2} + \dfrac{2+3\cdot4}{4^2}$

$\qquad = \dfrac{5}{1} + \dfrac{8}{4} + \dfrac{11}{9} + \dfrac{14}{16}$

$\qquad = 5 + 2 + \dfrac{11}{9} + \dfrac{7}{8}$

$\qquad = \dfrac{655}{72}$ or $9\dfrac{7}{72}$

4. $S_5 = \displaystyle\sum_{i=1}^{5} i(2i-1)$

$\qquad = 1(2\cdot1-1) + 2(2\cdot2-1) + 3(2\cdot3-1)$
$\qquad\quad + 4(2\cdot4-1) + 5(2\cdot5-1)$

$\qquad = 1 + 6 + 15 + 28 + 45$

$\qquad = 95$

There are 95 plants after 5 years.

Vocabulary, Readiness & Video Check 11.3

1. A series is an <u>infinite</u> series if it is the sum of all the terms of an infinite sequence.

2. A series is a <u>finite</u> series if it is the sum of a finite number of terms.

3. A shorthand notation for denoting a series when the general term of the sequence is known is called <u>summation</u> notation.

4. In the notation $\displaystyle\sum_{i=1}^{7}(5i-2)$, the Σ is the Greek uppercase letter <u>sigma</u> and the i is called the <u>index of summation</u>.

5. The sum of the first n terms of a sequence is a finite series known as a <u>partial sum</u>.

6. For the notation in Exercise 4 above, the beginning value of i is <u>1</u> and the ending value of i is <u>7</u>.

7. sigma/sum, index of summation, beginning value of i, ending value of i, and general term of the sequence

8. the sum of the first 7 terms of the sequence

Exercise Set 11.3

1. $\displaystyle\sum_{i=1}^{4}(i-3) = (1-3) + (2-3) + (3-3) + (4-3)$

$\qquad\qquad\qquad = -2 + (-1) + 0 + 1$

$\qquad\qquad\qquad = -2$

3. $\displaystyle\sum_{i=4}^{7}(2i+4) = [2(4)+4] + [2(5)+4] + [2(6)+4]$

$\qquad\qquad\qquad\qquad\qquad + [2(7)+4]$

$\qquad\qquad\qquad = 12 + 14 + 16 + 18$

$\qquad\qquad\qquad = 60$

5. $\displaystyle\sum_{i=2}^{4}(i^2-3) = (2^2-3) + (3^2-3) + (4^2-3)$

$\qquad\qquad\qquad = 1 + 6 + 13$

$\qquad\qquad\qquad = 20$

7. $\displaystyle\sum_{i=1}^{3}\left(\frac{1}{i+5}\right) = \dfrac{1}{1+5} + \dfrac{1}{2+5} + \dfrac{1}{3+5}$

$\qquad\qquad\qquad = \dfrac{1}{6} + \dfrac{1}{7} + \dfrac{1}{8}$

$\qquad\qquad\qquad = \dfrac{28}{168} + \dfrac{24}{168} + \dfrac{21}{168}$

$\qquad\qquad\qquad = \dfrac{73}{168}$

9. $\displaystyle\sum_{i=1}^{3}\frac{1}{6i} = \dfrac{1}{6(1)} + \dfrac{1}{6(2)} + \dfrac{1}{6(3)}$

$\qquad\qquad = \dfrac{1}{6} + \dfrac{1}{12} + \dfrac{1}{18}$

$\qquad\qquad = \dfrac{6+3+2}{36}$

$\qquad\qquad = \dfrac{11}{36}$

11. $\displaystyle\sum_{i=2}^{6} 3i = 3(2) + 3(3) + 3(4) + 3(5) + 3(6)$

$\qquad\qquad = 6 + 9 + 12 + 15 + 18$
$\qquad\qquad = 60$

13. $\displaystyle\sum_{i=3}^{5} i(i+2) = 3(3+2) + 4(4+2) + 5(5+2)$

$\qquad\qquad\qquad = 15 + 24 + 35$
$\qquad\qquad\qquad = 74$

15. $\displaystyle\sum_{i=1}^{5} 2^i = 2^1 + 2^2 + 2^3 + 2^4 + 2^5$

$\qquad\qquad = 2 + 4 + 8 + 16 + 32$
$\qquad\qquad = 62$

17. $\displaystyle\sum_{i=1}^{4} \frac{4i}{i+3} = \frac{4(1)}{1+3} + \frac{4(2)}{2+3} + \frac{4(3)}{3+3} + \frac{4(4)}{4+3}$

$\qquad\qquad = 1 + \dfrac{8}{5} + 2 + \dfrac{16}{7}$

$\qquad\qquad = \dfrac{105}{35} + \dfrac{56}{35} + \dfrac{80}{35}$

$\qquad\qquad = \dfrac{241}{35}$

19. $1 + 3 + 5 + 7 + 9$
$a_1 = 1,\ d = 2$
$a_n = 1 + (n-1)2 = 2n - 1$
$\displaystyle\sum_{i=1}^{5}(2i - 1)$

21. $4 + 12 + 36 + 108 = 4 + 4(3) + 4(3)^2 + 4(3)^3$

$\qquad\qquad\qquad\quad = \displaystyle\sum_{i=1}^{4} 4(3)^{i-1}$

23. $12 + 9 + 6 + 3 + 0 + (-3)$
$a_1 = 12,\ d = -3$
$a_n = 12 + (n-1)(-3) = -3n + 15$
$\displaystyle\sum_{i=1}^{6}(-3i + 15)$

25. $12 + 4 + \dfrac{4}{3} + \dfrac{4}{9} = \dfrac{4}{3^{-1}} + \dfrac{4}{3^0} + \dfrac{4}{3} + \dfrac{4}{3^2}$

$\qquad\qquad\qquad = \displaystyle\sum_{i=1}^{4} \dfrac{4}{3^{i-2}}$

27. $1 + 4 + 9 + 16 + 25 + 36 + 49$

$\quad = 1^2 + 2^2 + 3^2 + 4^2 + 5^2 + 6^2 + 7^2$

$\quad = \displaystyle\sum_{i=1}^{7} i^2$

29. $a_n = (n+2)(n-5)$

$S_2 = \displaystyle\sum_{i=1}^{2}(i+2)(i-5)$

$\qquad = (1+2)(1-5) + (2+2)(2-5)$
$\qquad = 3(-4) + 4(-3)$
$\qquad = -12 - 12$
$\qquad = -24$

31. $a_n = (-1)^n$

$S_6 = \displaystyle\sum_{i=1}^{6}(-1)^i$

$\qquad = (-1)^1 + (-1)^2 + (-1)^3 + (-1)^4 + (-1)^5$
$\qquad\qquad\qquad\qquad + (-1)^6$
$\qquad = -1 + 1 + (-1) + 1 + (-1) + 1$
$\qquad = 0$

33. $a_n = (n+3)(n+1)$

$S_4 = \displaystyle\sum_{i=1}^{4}(i+3)(i+1)$

$\qquad = (1+3)(1+1) + (2+3)(2+1) + (3+3)(3+1)$
$\qquad\qquad\qquad\qquad + (4+3)(4+1)$
$\qquad = 4(2) + 5(3) + 6(4) + 7(5)$
$\qquad = 8 + 15 + 24 + 35$
$\qquad = 82$

35. $a_n = -2n$

$S_4 = \displaystyle\sum_{i=1}^{4}(-2i)$

$\qquad = -2(1) + (-2)(2) + (-2)(3) + (-2)(4)$
$\qquad = -2 - 4 - 6 - 8$
$\qquad = -20$

37. $a_n = -\dfrac{n}{3}$

$S_3 = \displaystyle\sum_{i=1}^{3} -\dfrac{i}{3} = -\dfrac{1}{3} - \dfrac{2}{3} - \dfrac{3}{3} = -2$

39. $1, 2, 3, \ldots, 10$

$a_n = n$

$S_{10} = \sum_{i=1}^{10} i = 1 + 2 + 3 + \ldots + 10 = 55$

A total of 55 trees were planted.

41. $a_n = 6 \cdot 2^{n-1}$

$S_5 = \sum_{i=1}^{5} 6 \cdot 2^{i-1}$

$= 6(2)^0 + 6(2)^1 + 6(2)^2 + 6(2)^3 + 6(2)^4$

$= 6 + 12 + 24 + 48 + 96$

$= 186$

Adding in the original 6 units, there will be 192 fungus units at the end of the fifth day.

43. $a_4 = (4+1)(4+3) = 5(7) = 35$

In the fourth year, 35 species were born.

$S_4 = \sum_{i=1}^{4} (i+1)(i+3)$

$= 2(4) + 3(5) + 4(6) + 5(7)$

$= 8 + 15 + 24 + 12$

$= 82$

There were 82 species born in the first four years.

45. $a_n = (n+1)(n+2)$

$a_4 = (4+1)(4+2) = 5(6) = 30$

30 opossums were killed in the fourth month.

$S_4 = \sum_{i=1}^{4} (i+1)(i+2)$

$= 2(3) + (3)(4) + (4)(5) + (5)(6)$

$= 6 + 12 + 20 + 30$

$= 68$

68 opossums were killed in the four months.

47. $a_n = 100(0.5)^n$

$a_4 = 100(0.5)^4 = 6.25$

The decay in the fourth year is 6.25 pounds.

$S_4 = \sum_{i=1}^{4} 100(0.5)^i$

$= 100(0.5)^1 + 100(0.5)^2 + 100(0.5)^3$
$ + 100(0.5)^4$

$= 100(0.5) + 100(0.25) + 100(0.125)$
$ + 100(0.0625)$

$= 50 + 25 + 12.5 + 6.25$

$= 93.75$

The decay over the four years is 93.75 pounds.

49. $a_1 = 40$ and $r = \dfrac{4}{5}$

$a_5 = 40 \left(\dfrac{4}{5} \right)^4 = 16.384$

The length of the fifth swing is approximately 16.4 inches.

$S_5 = \sum_{i=1}^{5} 40 \left(\dfrac{4}{5} \right)^{i-1}$

$= 40 \left(\dfrac{4}{5} \right)^0 + 40 \left(\dfrac{4}{5} \right)^1 + 40 \left(\dfrac{4}{5} \right)^2 + 40 \left(\dfrac{4}{5} \right)^3$
$ + 40 \left(\dfrac{4}{5} \right)^4$

$= 40 + 32 + 25.6 + 20.48 + 16.384$

$= 134.464$

The pendulum swings about 134.5 inches in five swings.

51. $\dfrac{5}{1 - \frac{1}{2}} = \dfrac{5}{\frac{1}{2}} = 5 \cdot \dfrac{2}{1} = 10$

53. $\dfrac{\frac{1}{3}}{1 - \frac{1}{10}} = \dfrac{\frac{1}{3}}{\frac{9}{10}} = \dfrac{1}{3} \cdot \dfrac{10}{9} = \dfrac{10}{27}$

55. $\dfrac{3(1 - 2^4)}{1 - 2} = \dfrac{3(1 - 16)}{-1} = \dfrac{3(-15)}{-1} = \dfrac{-45}{-1} = 45$

57. $\dfrac{10}{2} (3 + 15) = \dfrac{10}{2} (18) = \dfrac{180}{2} = 90$

59. a. $\sum_{i=1}^{7} (i + i^2)$

$= (1 + 1^2) + (2 + 2^2) + (3 + 3^2) + (4 + 4^2)$
$ + (5 + 5^2) + (6 + 6^2) + (7 + 7^2)$

$= 2 + 6 + 12 + 20 + 30 + 42 + 56$

b. $\displaystyle\sum_{i=1}^{7} i + \sum_{i=1}^{7} i^2$

$= (1 + 2 + 3 + 4 + 5 + 6 + 7)$
$\quad + (1 + 4 + 9 + 16 + 25 + 36 + 49)$

c. answers may vary

d. True; answers may vary.

Integrated Review

1. $a_n = n - 3$
$a_1 = 1 - 3 = -2$
$a_2 = 2 - 3 = -1$
$a_3 = 3 - 3 = 0$
$a_4 = 4 - 3 = 1$
$a_5 = 5 - 3 = 2$
Therefore, the first five terms are $-2, -1, 0, 1, 2$.

2. $a_n = \dfrac{7}{1 + n}$
$a_1 = \dfrac{7}{1 + 1} = \dfrac{7}{2}$
$a_2 = \dfrac{7}{1 + 2} = \dfrac{7}{3}$
$a_3 = \dfrac{7}{1 + 3} = \dfrac{7}{4}$
$a_4 = \dfrac{7}{1 + 4} = \dfrac{7}{5}$
$a_5 = \dfrac{7}{1 + 5} = \dfrac{7}{6}$
The first five terms are $\dfrac{7}{2}, \dfrac{7}{3}, \dfrac{7}{4}, \dfrac{7}{5}$, and $\dfrac{7}{6}$.

3. $a_n = 3^{n-1}$
$a_1 = 3^{1-1} = 3^0 = 1$
$a_2 = 3^{2-1} = 3^1 = 3$
$a_3 = 3^{3-1} = 3^2 = 9$
$a_4 = 3^{4-1} = 3^3 = 27$
$a_5 = 3^{5-1} = 3^4 = 81$
The first five terms are 1, 3, 9, 27, and 81.

4. $a_n = n^2 - 5$
$a_1 = 1^2 - 5 = 1 - 5 = -4$
$a_2 = 2^2 - 5 = 4 - 5 = -1$
$a_3 = 3^2 - 5 = 9 - 5 = 4$
$a_4 = 4^2 - 5 = 16 - 5 = 11$
$a_5 = 5^2 - 5 = 25 - 5 = 20$
The first five terms are $-4, -1, 4, 11$, and 20.

5. $(-2)^n; a_6$
$a_6 = (-2)^6 = 64$

6. $-n^2 + 2; a_4$
$a_4 = -(4)^2 + 2 = -16 + 2 = -14$

7. $\dfrac{(-1)^n}{n}; a_{40}$
$a_{40} = \dfrac{(-1)^{40}}{40} = \dfrac{1}{40}$

8. $\dfrac{(-1)^n}{2n}; a_{41}$
$a_{41} = \dfrac{(-1)^{41}}{2(41)} = \dfrac{-1}{82} = -\dfrac{1}{82}$

9. $a_1 = 7; d = -3$
$a_1 = 7$
$a_2 = 7 - 3 = 4$
$a_3 = 4 - 3 = 1$
$a_4 = 1 - 3 = -2$
$a_5 = -2 - 3 = -5$
The first five terms are 7, 4, 1, -2, -5.

10. $a_1 = -3; r = 5$
$a_1 = -3$
$a_2 = -3(5) = -15$
$a_3 = -15(5) = -75$
$a_4 = -75(5) = -375$
$a_5 = -375(5) = -1875$
The first five terms are $-3, -15, -75, -375,$ -1875.

11. $a_1 = 45;\ r = \dfrac{1}{3}$

$a_1 = 45$

$a_2 = 45\left(\dfrac{1}{3}\right) = 15$

$a_3 = 15\left(\dfrac{1}{3}\right) = 5$

$a_4 = 5\left(\dfrac{1}{3}\right) = \dfrac{5}{3}$

$a_5 = \dfrac{5}{3}\left(\dfrac{1}{3}\right) = \dfrac{5}{9}$

The first five terms are 45, 15, 5, $\dfrac{5}{3}, \dfrac{5}{9}$.

12. $a_1 = -12;\ d = 10$

$a_1 = -12$
$a_2 = -12 + 10 = -2$
$a_3 = -2 + 10 = 8$
$a_4 = 8 + 10 = 18$
$a_5 = 18 + 10 = 28$

The first five terms are $-12, -2, 8, 18, 28$.

13. $a_1 = 20;\ d = 9$

$a_n = a_1 + (n-1)d$

$a_{10} = 20 + (10-1)9$
$\phantom{a_{10}} = 20 + 81$
$\phantom{a_{10}} = 101$

14. $a_1 = 64;\ r = \dfrac{3}{4}$

$a_n = a_1 r^{n-1}$

$a_6 = 64\left(\dfrac{3}{4}\right)^{6-1}$

$ = 64\left(\dfrac{3}{4}\right)^{5}$

$ = 64\left(\dfrac{243}{1024}\right)$

$ = \dfrac{243}{16}$

15. $a_1 = 6;\ r = \dfrac{-12}{6} = -2$

$a_n = a_1 r^{n-1}$

$a_7 = 6(-2)^{7-1} = 6(-2)^6 = 6(64) = 384$

16. $a_1 = -100;\ d = -85 - (-100) = 15$

$a_n = a_1 + (n-1)d$

$a_{20} = -100 + (20-1)(15)$
$\phantom{a_{20}} = -100 + (19)(15)$
$\phantom{a_{20}} = -100 + 285$
$\phantom{a_{20}} = 185$

17. $a_4 = -5,\ a_{10} = -35$

$a_n = a_1 + (n-1)d$

$\begin{cases} a_4 = a_1 + (4-1)d \\ a_{10} = a_1 + (10-1)d \end{cases}$

$\begin{cases} -5 = a_1 + 3d \\ -35 = a_1 + 9d \end{cases}$

Multiply eq. 2 by -1, then add the equations.

$\begin{cases} -5 = a_1 + 3d \\ (-1)(-35) = -1(a_1 + 9d) \end{cases}$

$\begin{cases} -5 = a_1 + 3d \\ 35 = -a_1 - 9d \end{cases}$

$30 = -6d$
$-5 = d$

To find a_1, let $d = -5$ in

$-5 = a_1 + 3d$
$-5 = a_1 + 3(-5)$
$10 = a_1$

Thus, $a_1 = 10$ and $d = -5$, so

$a_n = 10 + (n-1)(-5) = -5n + 15$

$a_5 = -5(5) + 15 = -10$

18. $a_4 = 1;\ a_7 = \dfrac{1}{8}$

$a_n = a_1 r^{n-1}$

$a_4 = a_1 r^{4-1}$ so $1 = a_1 r^3$

$a_7 = a_1 r^{7-1}$ so $\dfrac{1}{8} = a_1 r^6$

Since $a_1 r^6 = (a_1 r^3) r^3$, $\dfrac{1}{8} = 1 \cdot r^3$ and $r = \dfrac{1}{2}$.

$a_5 = a_4 \cdot r$ so $a_5 = 1 \cdot \dfrac{1}{2} = \dfrac{1}{2}$

19. $\displaystyle\sum_{i=1}^{4} 5i = 5(1) + 5(2) + 5(3) + 5(4)$

$ = 5 + 10 + 15 + 20$
$ = 50$

20. $\displaystyle\sum_{i=1}^{7}(3i+2)$

$\begin{aligned}
&= (3(1)+2)+(3(2)+2)+(3(3)+2)\\
&\quad+(3(4)+2)+(3(5)+2)+(3(6)+2)\\
&\quad+(3(7)+2)\\
&= 5+8+11+14+17+20+23\\
&= 98
\end{aligned}$

21. $\displaystyle\sum_{i=3}^{7}2^{i-4}$

$\begin{aligned}
&= 2^{3-4}+2^{4-4}+2^{5-4}+2^{6-4}+2^{7-4}\\
&= 2^{-1}+2^{0}+2^{1}+2^{2}+2^{3}\\
&= \frac{1}{2}+1+2+4+8\\
&= 15\frac{1}{2}\\
&= \frac{31}{2}
\end{aligned}$

22. $\displaystyle\sum_{i=2}^{5}\frac{i}{i+1}=\frac{2}{2+1}+\frac{3}{3+1}+\frac{4}{4+1}+\frac{5}{5+1}$

$\begin{aligned}
&= \frac{2}{3}+\frac{3}{4}+\frac{4}{5}+\frac{5}{6}\\
&= \frac{61}{20}
\end{aligned}$

23. $S_3 = \displaystyle\sum_{i=1}^{3}i(i-4)$

$\begin{aligned}
&= 1(1-4)+2(2-4)+3(3-4)\\
&= -3-4-3\\
&= -10
\end{aligned}$

24. $S_{10} = \displaystyle\sum_{i=1}^{10}(-1)^{i}(i+1)$

$\begin{aligned}
&= (-1)^{1}(1+1)+(-1)^{2}(2+1)\\
&\quad+(-1)^{3}(3+1)+(-1)^{4}(4+1)\\
&\quad+(-1)^{5}(5+1)+(-1)^{6}(6+1)\\
&\quad+(-1)^{7}(7+1)+(-1)^{8}(8+1)\\
&\quad+(-1)^{9}(9+1)+(-1)^{10}(10+1)\\
&= -2+3-4+5-6+7-8+9-10+11\\
&= 5
\end{aligned}$

Section 11.4 Practice Exercises

1. 2, 9, 16, 23, 30, ...
Use the formula for S_n of an arithmetic sequence, replacing n with 5, a_1 with 2, and a_n with 30.

$S_n = \dfrac{n}{2}(a_1+a_n)$

$S_5 = \dfrac{5}{2}(2+30)=\dfrac{5}{2}(32)=80$

2. Because 1, 2, 3, ..., 50 is an arithmetic sequence, use the formula for S_n with $n = 50$, $a_1 = 1$, and $a_n = 50$.

$S_n = \dfrac{n}{2}(a_1+a_n)$

$S_5 = \dfrac{50}{2}(1+50)=25(51)=1275$

3. The list 6, 7, ..., 15 is the first 10 terms of an arithmetic sequence. Use the formula for S_n with $n = 10$, $a_1 = 6$, and $a_n = 15$.

$S_{10} = \dfrac{10}{2}(6+15)=5(21)=105$

There are a total of 105 blocks of ice.

4. 32, 8, 2, $\dfrac{1}{2}$, $\dfrac{1}{8}$

Use the formula for the partial sum S_n of the terms of a geometric sequence. Here, $n = 5$, the first term $a_1 = 32$, and the common ratio $r = \dfrac{1}{4}$.

$\begin{aligned}
S_n &= \frac{a_1(1-r^n)}{1-r}\\[2mm]
S_5 &= \frac{32\left[1-\left(\frac{1}{4}\right)^5\right]}{1-\frac{1}{4}}\\[2mm]
&= \frac{32\left(1-\frac{1}{1024}\right)}{\frac{3}{4}}\\[2mm]
&= \frac{32-\frac{1}{32}}{\frac{3}{4}}\\[2mm]
&= \frac{\frac{1023}{32}}{\frac{3}{4}}\\[2mm]
&= \frac{1023}{32}\cdot\frac{4}{3}\\[2mm]
&= \frac{341}{8} \text{ or } 42\frac{5}{8}
\end{aligned}$

5. The donations are modeled by the first seven terms of a geometric sequence. Evaluate S_n when $n = 7$, $a_1 = 250,000$, and $r = 0.8$.

$$S_7 = \frac{250,000[1-(0.8)^7]}{1-0.8} = 987,856$$

The total amount donated during the seven years is $987,856.

6. $7, \dfrac{7}{4}, \dfrac{7}{16}, \dfrac{7}{64}, \ldots$

For this geometric sequence $r = \dfrac{1}{4}$. Since $|r| < 1$, use the formula for S_∞ of a geometric sequence with $a_1 = 7$ and $r = \dfrac{1}{4}$.

$$S_\infty = \frac{a_1}{1-r} = \frac{7}{1-\frac{1}{4}} = \frac{7}{\frac{3}{4}} = \frac{28}{3} = 9\frac{1}{3}$$

7. We must find the sum of the terms of an infinite geometric sequence whose first term, a_1, is 36 and whose common ratio, r, is 0.96. Since $|r| < 1$, we may use the formula for S_∞.

$$S_\infty = \frac{a_1}{1-r} = \frac{36}{1-0.96} = \frac{36}{0.04} = 900$$

The ball travels a total distance of 900 inches before it comes to a rest.

Vocabulary, Readiness & Video Check 11.4

1. Each term after the first is 5 more than the preceding term; the sequence is <u>arithmetic</u>.

2. Each term after the first is 2 times the preceding term; the sequence is <u>geometric</u>.

3. Each term after the first is -3 times the preceding term; the sequence is <u>geometric</u>.

4. Each term after the first is 2 more than the preceding term; the sequence is <u>arithmetic</u>.

5. Each term after the first is 7 more than the preceding term; the sequence is <u>arithmetic</u>.

6. Each term after the first is -1 times the preceding term; the sequence is <u>geometric</u>.

7. Use the general term formula from Section 11.2 for the general term of an arithmetic sequence: $a_n = a_1 + (n-1)d$.

8. It would be a sequence in which every number is the same since you would multiply each term by 1 to get the next term; to find a partial sum of n terms, just multiply the first term by n since that term, which just repeats, will be added n times.

9. The common ratio r is 3 for this sequence so that $|r| \geq 1$, or $|3| \geq 1$; S_∞ doesn't exist if $|r| \geq 1$.

Exercise Set 11.4

1. $1, 3, 5, 7, \ldots$

$d = 2$; $a_6 = 1 + (6-1)(2) = 11$

$$S_6 = \frac{6}{2}(1+11) = 3(12) = 36$$

3. $4, 12, 36, \ldots$

$a_1 = 4$, $r = 3$, $n = 5$

$$S_5 = \frac{4(1-3^5)}{1-3} = 484$$

5. $3, 6, 9, \ldots$

$d = 3$; $a_6 = 3 + (6-1)(3) = 18$

$$S_6 = \frac{6}{2}(3+18) = 3(21) = 63$$

7. $2, \dfrac{2}{5}, \dfrac{2}{25}, \ldots$

$a_1 = 2$, $r = \dfrac{1}{5}$, $n = 4$

$$S_4 = \frac{2\left[1-\left(\frac{1}{5}\right)^4\right]}{1-\frac{1}{5}} = \frac{\frac{1248}{625}}{\frac{4}{5}} = \frac{312}{125}$$

9. $1, 2, 3, \ldots, 10$

The first term is 1 and the tenth term is 10.

$$S_{10} = \frac{10}{2}(1+10) = 5(11) = 55$$

11. $1, 2, 3, 7$

The first term is 1 and the fourth term is 7.

$$S_4 = \frac{4}{2}(1+7) = 2(8) = 16$$

13. 12, 6, 3, ...

$$a_1 = 12, \quad r = \frac{1}{2}$$

$$S_\infty = \frac{12}{1 - \frac{1}{2}} = \frac{12}{\frac{1}{2}} = 12 \cdot \frac{2}{1} = 24$$

15. $\frac{1}{10}, \frac{1}{100}, \frac{1}{1000}, ...$

$$a_1 = \frac{1}{10}, \quad r = \frac{1}{10}$$

$$S_\infty = \frac{\frac{1}{10}}{1 - \frac{1}{10}} = \frac{\frac{1}{10}}{\frac{9}{10}} = \frac{1}{10} \cdot \frac{10}{9} = \frac{1}{9}$$

17. $-10, -5, -\frac{5}{2}, ...$

$$a_1 = -10, \quad r = \frac{1}{2}$$

$$S_\infty = \frac{-10}{1 - \frac{1}{2}} = \frac{-10}{\frac{1}{2}} = -10 \cdot \frac{2}{1} = -20$$

19. $2, -\frac{1}{4}, \frac{1}{32}, ...$

$$a_1 = 2, \quad r = -\frac{1}{8}$$

$$S_\infty = \frac{2}{1 - \left(-\frac{1}{8}\right)} = \frac{2}{\frac{9}{8}} = 2 \cdot \frac{8}{9} = \frac{16}{9}$$

21. $\frac{2}{3}, -\frac{1}{3}, \frac{1}{6}, ...$

$$a_1 = \frac{2}{3}, \quad r = -\frac{1}{2}$$

$$S_\infty = \frac{\frac{2}{3}}{1 - \left(-\frac{1}{2}\right)} = \frac{\frac{2}{3}}{\frac{3}{2}} = \frac{2}{3} \cdot \frac{2}{3} = \frac{4}{9}$$

23. $-4, 1, 6, ..., 41$

The first term is -4 and the tenth term is 41.

$$S_{10} = \frac{10}{2}(-4 + 41) = 5(37) = 185$$

25. $3, \frac{3}{2}, \frac{3}{4}, ...$

$$a_1 = 3, r = \frac{1}{2}, n = 7$$

$$S_7 = \frac{3\left[1 - \left(\frac{1}{2}\right)^7\right]}{1 - \frac{1}{2}} = \frac{381}{64}$$

27. $-12, 6, -3, ...$

$$a_1 = -12, r = -\frac{1}{2}, n = 5$$

$$S_5 = \frac{-12\left[1 - \left(-\frac{1}{2}\right)^5\right]}{1 - \left(-\frac{1}{2}\right)} = -\frac{33}{4} = -8.25$$

29. $\frac{1}{2}, \frac{1}{4}, 0, ..., -\frac{17}{4}$

The first term is $\frac{1}{2}$ and the twentieth term is $-\frac{17}{4}$.

$$S_{20} = \frac{20}{2}\left(\frac{1}{2} - \frac{17}{4}\right) = 10\left(\frac{-15}{4}\right) = -\frac{75}{2}$$

31. $a_1 = 8, r = -\frac{2}{3}, n = 3$

$$S_3 = \frac{8\left[1 - \left(-\frac{2}{3}\right)^3\right]}{1 - \left(-\frac{2}{3}\right)} = \frac{56}{9}$$

33. The first five terms are 4000, 3950, 3900, 3850, 3800.

$$a_1 = 4000, d = -50, n = 12$$

$$a_{12} = 4000 + 11(-50) = 3450$$

3450 cars will be sold in month 12.

$$S_{12} = \frac{12}{2}(4000 + 3450) = 44,700$$

44,700 cars will be sold in the first year.

    619

35. Firm *A*:
The first term is 22,000 and the tenth term is 31,000.

$$S_{10} = \frac{10}{2}(22,000 + 31,000)$$
$$= \$265,000$$

Firm *B*:
The first term is 20,000 and the tenth term is 30,800.

$$S_{10} = \frac{10}{2}(20,000 + 30,800)$$
$$= \$254,000$$

Thus, Firm *A* is making the more profitable offer.

37. $a_1 = 30,000,\ r = 1.10,\ n = 4$

$a_4 = 30,000(1.10)^{4-1} = 39,930$
She made $39,930 during her fourth year of business.

$$S_4 = \frac{30,000(1 - 1.10^4)}{1 - 1.10} = 139,230$$

She made $139,230 during the first four years of business.

39. $a_1 = 30,\ r = 0.9,\ n = 5$

$a_5 = 30(0.9)^{5-1} = 19.683$
Approximately 20 minutes to assemble the first computer.

$$S_5 = \frac{30(1 - 0.9^5)}{1 - 0.9} = 122.853$$

Approximately 123 minutes to assemble the first 5 computers.

41. $a_1 = 20,\ r = \dfrac{4}{5}$

$$S_\infty = \frac{20}{1 - \frac{4}{5}} = 100$$

We double the number (to account for the flight up as well as down) and subtract 20 (since the first bounce was preceded by only a downward flight). Thus, the ball travels $2(100) - 20 = 180$ feet.

43. Player *A*:
The first term is 1 and the ninth term is 9.

$$S_9 = \frac{9}{2}(1 + 9) = 45 \text{ points}$$

Player *B*:
The first term is 10 and the sixth term is 15.

$$S_6 = \frac{6}{2}(10 + 15) = 75 \text{ points}$$

45. The first term is 200 and the twentieth is $200 - 19(5) = 105$.

$$S_{20} = \frac{20}{2}(200 + 105) = 3050$$

Thus, $3050 rent is paid for 20 days during the holiday rush.

47. $a_1 = 0.01,\ r = 2,\ n = 30$

$$S_3 = \frac{0.01\left[1 - 2^{30}\right]}{1 - 2} = 10,737,418.23$$

He would pay $10,737,418.23 in room and board for the 30 days.

49. $6 \cdot 5 \cdot 4 \cdot 3 \cdot 2 \cdot 1 = 720$

51. $\dfrac{3 \cdot 2 \cdot 1}{2 \cdot 1} = \dfrac{3 \cdot \cancel{2} \cdot \cancel{1}}{\cancel{2} \cdot \cancel{1}} = 3$

53. $(x + 5)^2 = x^2 + 2 \cdot x \cdot 5 + 5^2 = x^2 + 10x + 25$

55. $(2x - 1)^3 = (2x - 1)^2(2x - 1)$
$= (4x^2 - 4x + 1)(2x - 1)$
$= 8x^3 - 4x^2 + 2x - 8x^2 + 4x - 1$
$= 8x^3 - 12x^2 + 6x - 1$

57. $0.8\overline{88} = 0.8 + 0.08 + 0.008 + \cdots$
$= \dfrac{8}{10} + \dfrac{8}{100} + \dfrac{8}{1000} + \cdots$

This is a geometric series with $a_1 = \dfrac{8}{10},\ r = \dfrac{1}{10}$.

$$S_\infty = \frac{\frac{8}{10}}{1 - \frac{1}{10}} = \frac{\frac{8}{10}}{\frac{9}{10}} = \frac{8}{10} \cdot \frac{10}{9} = \frac{8}{9}$$

59. answers may vary

Section 11.5 Practice Exercises

1. $(p+r)^7$

   The $n = 7$ row of Pascal's triangle is

   1   7   21   35   35   21   7   1

   Using the $n = 7$ row of Pascal's triangle as the coefficients, $(p + r)^7$ can be expanded as

   $$p^7 + 7p^6r + 21p^5r^2 + 35p^4r^3 + 35p^3r^4 + 21p^2r^5 + 7pr^6 + r^7$$

2. **a.** $\dfrac{6!}{7!} = \dfrac{6 \cdot 5 \cdot 4 \cdot 3 \cdot 2 \cdot 1}{7 \cdot 6 \cdot 5 \cdot 4 \cdot 3 \cdot 2 \cdot 1} = \dfrac{1}{7}$

   b. $\dfrac{8!}{4!2!} = \dfrac{8 \cdot 7 \cdot 6 \cdot 5 \cdot 4!}{4! \cdot 2 \cdot 1}$

   $\qquad\quad = \dfrac{8 \cdot 7 \cdot 6 \cdot 5}{2 \cdot 1}$

   $\qquad\quad = 4 \cdot 7 \cdot 6 \cdot 5$

   $\qquad\quad = 840$

   c. $\dfrac{5!}{4!1!} = \dfrac{5 \cdot 4 \cdot 3 \cdot 2 \cdot 1}{4 \cdot 3 \cdot 2 \cdot 1 \cdot 1} = 5$

   d. $\dfrac{9!}{9!0!} = \dfrac{9!}{9! \cdot 1} = 1$

3. $(a+b)^9$

   Let $n = 9$ in the binomial formula.

   $$(a+b)^9 = a^9 + \frac{9}{1!}a^8b + \frac{9 \cdot 8}{2!}a^7b^2 + \frac{9 \cdot 8 \cdot 7}{3!}a^6b^3 + \frac{9 \cdot 8 \cdot 7 \cdot 6}{4!}a^5b^4 + \frac{9 \cdot 8 \cdot 7 \cdot 6 \cdot 5}{5!}a^4b^5 + \frac{9 \cdot 8 \cdot 7 \cdot 6 \cdot 5 \cdot 4}{6!}a^3b^6$$
   $$+ \frac{9 \cdot 8 \cdot 7 \cdot 6 \cdot 5 \cdot 4 \cdot 3}{7!}a^2b^7 + \frac{9 \cdot 8 \cdot 7 \cdot 6 \cdot 5 \cdot 4 \cdot 3 \cdot 2}{8!}ab^8 + b^9$$
   $$= a^9 + 9a^8b + 36a^7b^2 + 84a^6b^3 + 126a^5b^4 + 126a^4b^5 + 84a^3b^6 + 36a^2b^7 + 9ab^8 + b^9$$

4. $(a+5)^3$

   Replace b with 5 in the binomial formula.

   $$(a+5)^3 = a^3 + \frac{3}{1!}a^2(5) + \frac{3 \cdot 2}{2!}a(5)^2 + (5)^3$$
   $$= a^3 + 3a^2(5) + 3a(25) + 125$$
   $$= a^3 + 15a^2 + 75a + 125$$

5. $(3x-2y)^3$

   Let $a = 3x$ and $b = -2y$ in the binomial formula.

   $$(3x-2y)^3 = (3x)^3 + \frac{3}{1!}(3x)^2(-2y) + \frac{3 \cdot 2}{2!}(3x)(-2y)^2 + (-2y)^3$$
   $$= 27x^3 + 3(9x^2)(-2y) + 3(3x)(4y^2) - 8y^3$$
   $$= 27x^3 - 54x^2y + 36xy^2 - 8y^3$$

6. $(x-4y)^{11}$

Use the formula with $n = 11$, $a = x$, $b = -4y$, and $r + 1 = 7$. Notice that, since $r + 1 = 7$, $r = 6$.

$$\frac{n!}{r!(n-r)!}a^{n-r}b^r = \frac{11!}{6!5!}x^5(-4y)^6 = 462x^5(4096y^6) = 1,892,352x^5y^6$$

Vocabulary, Readiness & Video Check 11.5

1. $0! = \underline{1}$

2. $1! = \underline{1}$

3. $4! = 4 \cdot 3 \cdot 2 \cdot 1 = \underline{24}$

4. $2! = 2 \cdot 1 = \underline{2}$

5. $3!0! = 3 \cdot 2 \cdot 1 \cdot 1 = \underline{6}$

6. $0!2! = 1 \cdot 2 \cdot 1 = \underline{2}$

7. Pascal's triangle gives you the coefficients of the terms of the expanded binomial; also, the power tells you how many terms the expansion has (1 more than the power on the binomial).

8. $4 \cdot 3 \cdot 2 \cdot 1 = 24$; $0! = 1$

9. The theorem is in terms of $(a+b)^n$, so if your binomial is in the form $(a-b)^n$, then remember to think of it as $(a+(-b))^n$, so your second term is $-b$.

10. We are using the formula for the $(r + 1)$st term in a binomial—so if $r + 1 = 4$ then $r = 3$.

Exercise Set 11.5

1. $(m+n)^3 = m^3 + 3m^2n + 3mn^2 + n^3$

3. $(c+d)^5 = c^5 + 5c^4d + 10c^3d^2 + 10c^2d^3 + 5cd^4 + d^5$

5. $(y-x)^5 = \left[y+(-x)\right]^5$
$$= y^5 - 5y^4x + 10y^3x^2 - 10y^2x^3 + 5yx^4 - x^5$$

7. answers may vary

9. $\dfrac{8!}{7!} = \dfrac{8 \cdot 7!}{7!} = 8$

11. $\dfrac{7!}{5!} = \dfrac{7 \cdot 6 \cdot 5!}{5!} = 7 \cdot 6 = 42$

13. $\dfrac{10!}{7!2!} = \dfrac{10 \cdot 9 \cdot 8 \cdot 7!}{7!2!} = \dfrac{10 \cdot 9 \cdot 8}{2 \cdot 1} = 360$

15. $\dfrac{8!}{6!0!} = \dfrac{8 \cdot 7 \cdot 6!}{6!1} = 8 \cdot 7 = 56$

17. Let $n = 7$ in the binomial theorem.

$$(a+b)^7 = a^7 + \frac{7}{1!}a^6b + \frac{7\cdot6}{2!}a^5b^2 + \frac{7\cdot6\cdot5}{3!}a^4b^3 + \frac{7\cdot6\cdot5\cdot4}{4!}a^3b^4 + \frac{7\cdot6\cdot5\cdot4\cdot3}{5!}a^2b^5 + \frac{7\cdot6\cdot5\cdot4\cdot3\cdot2}{6!}ab^6 + b^7$$

$$= a^7 + 7a^6b + 21a^5b^2 + 35a^4b^3 + 35a^3b^4 + 21a^2b^5 + 7ab^6 + b^7$$

19. Let $b = 2b$ and $n = 5$ in the binomial theorem.

$$(a+2b)^5 = a^5 + \frac{5}{1!}a^4(2b) + \frac{5\cdot4}{2!}a^3(2b)^2 + \frac{5\cdot4\cdot3}{3!}a^2(2b)^3 + \frac{5\cdot4\cdot3\cdot2}{4!}a(2b)^4 + (2b)^5$$

$$= a^5 + 10a^4b + 40a^3b^2 + 80a^2b^3 + 80ab^4 + 32b^5$$

21. Let $a = q$, $b = r$, and $n = 9$ in the binomial theorem.

$$(q+r)^2 = q^9 + \frac{9}{1!}q^8r + \frac{9\cdot8}{2!}q^7r^2 + \frac{9\cdot8\cdot7}{3!}q^6r^3 + \frac{9\cdot8\cdot7\cdot6}{4!}q^5r^4 + \frac{9\cdot8\cdot7\cdot6\cdot5}{5!}q^4r^5 + \frac{9\cdot8\cdot7\cdot6\cdot5\cdot4}{6!}q^3r^6$$

$$+ \frac{9\cdot8\cdot7\cdot6\cdot5\cdot4\cdot3}{7!}q^2r^7 + \frac{9\cdot8\cdot7\cdot6\cdot5\cdot4\cdot3\cdot2}{8!}qr^8 + r^9$$

$$= q^9 + 9q^8r + 36q^7r^2 + 84q^6r^3 + 126q^5r^4 + 126q^4r^5 + 84q^3r^6 + 36q^2r^7 + 9qr^8 + r^9$$

23. Let $a = 4a$ and $n = 5$ in the binomial theorem.

$$(4a+b)^5 = (4a)^5 + \frac{5}{1!}(4a)^4b + \frac{5\cdot4}{2!}(4a)^3b^2 + \frac{5\cdot4\cdot3}{3!}(4a)^2b^3 + \frac{5\cdot4\cdot3\cdot2}{4!}(4a)b^4 + b^5$$

$$= 1024a^5 + 1280a^4b + 640a^3b^2 + 160a^2b^3 + 20ab^4 + b^5$$

25. Let $a = 5a$, $b = -2b$, and $n = 4$ in the binomial theorem.

$$(5a-2b)^4 = (5a)^4 + \frac{4}{1!}(5a)^3(-2b) + \frac{4\cdot3}{2!}(5a)^2(-2b)^2 + \frac{4\cdot3\cdot2}{3!}(5a)(-2b)^3 + (-2b)^4$$

$$= 625a^4 - 1000a^3b + 600a^2b^2 - 160ab^3 + 16b^4$$

27. Let $a = 2a$, $b = 3b$, and $n = 3$ in the binomial theorem.

$$(2a+3b)^3 = (2a)^3 + \frac{3}{1!}(2a)^2(3b) + \frac{3\cdot2}{2!}(2a)(3b)^2 + (3b)^3$$

$$= 8a^3 + 36a^2b + 54ab^2 + 27b^3$$

29. Let $a = x$, $b = 2$, and $n = 5$ in the binomial theorem.

$$(x+2)^5 = x^5 + \frac{5}{1!}x^4(2) + \frac{5\cdot4}{2!}x^3(2)^2 + \frac{5\cdot4\cdot3}{3!}x^2(2)^3 + \frac{5\cdot4\cdot3\cdot2}{4!}x(2)^4 + (2)^5$$

$$= x^5 + 10x^4 + 40x^3 + 80x^2 + 80x + 32$$

31. 5th term of $(c-d)^5$ corresponds to $r = 4$:

$$\frac{5!}{4!(5-4)!}c^{5-4}(-d)^4 = 5cd^4$$

33. 8th term of $(2c+d)^7$ corresponds to $r = 7$:

$$\frac{7!}{7!(7-7)!}(2c)^{7-7}(d)^7 = d^7$$

35. 4th term of $(2r-s)^5$ corresponds to $r = 3$:

$$\frac{5!}{3!(5-3)!}(2r)^{5-3}(-s)^3 = -40r^2s^3$$

37. 3rd term of $(x+y)^4$ corresponds to $r = 2$: $\dfrac{4!}{2!(4-2)!}(x)^{4-2}(y)^2 = 6x^2y^2$

39. 2nd term of $(a+3b)^{10}$ corresponds to $r = 1$: $\dfrac{10!}{1!(10-1)!}(a)^{10-1}(3b)^1 = 30a^9b$

41. $f(x) = |x|$

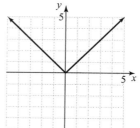

Not one-to-one

43. $H(x) = 2x + 3$

One-to-one

45. $f(x) = x^2 + 3$

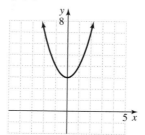

Not one-to-one

47. $(\sqrt{x} + \sqrt{3})^5$

Use the binomial theorem with $n = 5$, $a = \sqrt{x}$, , and $b = \sqrt{3}$.

$$\left(\sqrt{x} + \sqrt{3}\right)^5 = \left(\sqrt{x}\right)^5 + 5\left(\sqrt{x}\right)^4\left(\sqrt{3}\right) + 10\left(\sqrt{x}\right)^3\left(\sqrt{3}\right)^2 + 10\left(\sqrt{x}\right)^2\left(\sqrt{3}\right)^3 + 5\left(\sqrt{x}\right)\left(\sqrt{3}\right)^4 + \left(\sqrt{3}\right)^5$$
$$= x^2\sqrt{x} + 5\sqrt{3}x^2 + 30x\sqrt{x} + 30\sqrt{3}x + 45\sqrt{x} + 9\sqrt{3}$$

49. $\dbinom{9}{5} = \dfrac{9!}{5!(9-5)!}$

$= \dfrac{9!}{5!4!}$

$= \dfrac{9 \cdot 8 \cdot 7 \cdot 6 \cdot 5 \cdot 4 \cdot 3 \cdot 2 \cdot 1}{(5 \cdot 4 \cdot 3 \cdot 2 \cdot 1) \cdot (4 \cdot 3 \cdot 2 \cdot 1)}$

$= 126$

51. $\dbinom{8}{2} = \dfrac{8!}{2!(8-2)!}$

$= \dfrac{8!}{2!6!}$

$= \dfrac{8 \cdot 7 \cdot 6 \cdot 5 \cdot 4 \cdot 3 \cdot 2 \cdot 1}{(2 \cdot 1) \cdot (6 \cdot 5 \cdot 4 \cdot 3 \cdot 2 \cdot 1)}$

$= 28$

53. answers may vary

Chapter 11 Vocabulary Check

1. A <u>finite sequence</u> is a function whose domain is the set of natural numbers $\{1, 2, 3, ..., n\}$, where n is some natural number.

2. The <u>factorial of n</u>, written $n!$, is the product of the first n consecutive natural numbers.

3. An <u>infinite sequence</u> is a function whose domain is the set of natural numbers.

4. A <u>geometric sequence</u> is a sequence in which each term (after the first) is obtained by multiplying the preceding term by a constant amount r. The constant r is called the <u>common ratio</u> of the sequence.

5. A sum of the terms of a sequence is called a <u>series</u>.

6. The nth term of the sequence a_n is called the <u>general term</u>.

7. An <u>arithmetic sequence</u> is a sequence in which each term (after the first) differs from the preceding term by a constant amount d. The constant d is called the <u>common difference</u> of the sequence.

8. A triangular array of the coefficients of the terms of the expansions of $(a+b)^n$ is called <u>Pascal's triangle</u>.

Chapter 11 Review

1. $a_n = -3n^2$

$a_1 = -3(1)^2 = -3$
$a_2 = -3(2)^2 = -12$
$a_3 = -3(3)^2 = -27$
$a_4 = -3(4)^2 = -48$
$a_5 = -3(5)^2 = -75$

2. $a_n = n^2 + 2n$

$a_1 = 1^2 + 2(1) = 3$
$a_2 = 2^2 + 2(2) = 8$
$a_3 = 3^2 + 2(3) = 15$
$a_4 = 4^2 + 2(4) = 24$
$a_5 = 5^2 + 2(5) = 35$

3. $a_n = \dfrac{(-1)^n}{100}$

$a_{100} = \dfrac{(-1)^{100}}{100} = \dfrac{1}{100}$

4. $a_n = \dfrac{2n}{(-1)^n}$

$a_{50} = \dfrac{2(50)}{(-1)^{50}} = 100$

5. $\dfrac{1}{6 \cdot 1}, \dfrac{1}{6 \cdot 2}, \dfrac{1}{6 \cdot 3}, \ldots$

In general, $a_n = \dfrac{1}{6n}$.

6. $-1, 4, -9, 16, \ldots$

$a_n = (-1)^n n^2$

7. $a_n = 32n - 16$

$a_5 = 32(5) - 16 = 144$ feet
$a_6 = 32(6) - 16 = 176$ feet
$a_7 = 32(7) - 16 = 208$ feet

8. Since the measure is 80 at the end of the first day is 80, the measures at the ends of the first 5 days are:

80

$2 \cdot 80 = 160$

$2 \cdot 160 = 320$

$2 \cdot 320 = 640$

$2 \cdot 640 = 1280$

The general terms for the measure at the end of the nth day is $80 \cdot 2^{n-1}$.

$80 \cdot 2^{n-1} \geq 10,000$

$2^{n-1} \geq 125$

Since $2^6 = 64$ and $2^7 = 128$, $n - 1 = 7$, so $n = 8$. It will take 8 days before the yeast culture measures at least 10,000.

9. 2006: $a_1 = 660,000$

2007: $a_2 = 660,000(2) = 1,320,000$

2008: $a_3 = 1,320,000(2) = 2,640,000$

2009: $a_4 = 2,640,000(2) = 5,280,000$

2010: $a_5 = 5,280,000(2) = 10,560,000$

There will be 10,560,000 acres of infested trees in 2010.

10. $a_n = 50 + (n-1)8$

$a_1 = 50$

$a_2 = 50 + 8 = 58$

$a_3 = 50 + 2(8) = 66$

$a_4 = 50 + 3(8) = 74$

$a_5 = 50 + 4(8) = 82$

$a_6 = 50 + 5(8) = 90$

$a_7 = 50 + 6(8) = 98$

$a_8 = 50 + 7(8) = 106$

$a_9 = 50 + 8(8) = 114$

$a_{10} = 50 + 9(8) = 122$

There are 122 seats in the tenth row.

11. $a_1 = -2, r = \frac{2}{3}$

$a_1 = -2$

$a_2 = -2\left(\frac{2}{3}\right) = -\frac{4}{3}$

$a_3 = \left(-\frac{4}{3}\right)\left(\frac{2}{3}\right) = -\frac{8}{9}$

$a_4 = \left(-\frac{8}{9}\right)\left(\frac{2}{3}\right) = -\frac{16}{27}$

$a_5 = \left(-\frac{16}{27}\right)\left(\frac{2}{3}\right) = -\frac{32}{81}$

12. $a_n = 12 + (n-1)(-1.5)$

$a_1 = 12$

$a_2 = 12 + (1)(-1.5) = 10.5$

$a_3 = 12 + 2(-1.5) = 9$

$a_4 = 12 + 3(-1.5) = 7.5$

$a_5 = 12 + 4(-1.5) = 6$

13. $a_n = -5 + (n-1)4$

$a_{30} = 5 + (30-1)4 = 111$

14. $a_n = 2 + (n-1)\frac{3}{4}$

$a_{11} = 2 + 10\left(\frac{3}{4}\right) = \frac{19}{2}$

15. 12, 7, 2,...

$a_1 = 12, d = -5, n = 20$

$a_{20} = 12 + (20-1)(-5) = -83$

16. $a_n = a_1 r^{n-1}, a_1 = 4, r = \frac{3}{2}$

$a_6 = 4\left(\frac{3}{2}\right)^{6-1} = \frac{243}{8}$

17. $a_4 = 18, a_{20} = 98$

$\begin{cases} a_4 = a_1 + (4-1)d \\ a_{20} = a_1 + (20-1)d \end{cases}$

$\begin{cases} 18 = a_1 + 3d \\ 98 = a_1 + 19d \end{cases}$

$\begin{cases} -18 = -a_1 - 3d \\ 98 = a_1 + 19d \end{cases}$

Adding yields $80 = 16d$ or $d = 5$. Then $a_1 = 18 - 3(5) = 3$.

18. $a_3 = -48, a_4 = 192$

$r = \frac{a_4}{a_3} = \frac{192}{-48} = -4$

$a_3 = a_1 r^{3-1}$

$-48 = a_1(-4)^2$

$-48 = 16a_1$

$-3 = a_1$

$r = -4, a_1 = -3$

19. $\dfrac{3}{10}, \dfrac{3}{10^2}, \dfrac{3}{10^3}, \dots$

In general, $a_n = \dfrac{3}{10^n}$

20. 50, 58, 66, ...

$a_n = 50 + (n-1)8$ or $a_n = 42 + 8n$

21. $\dfrac{8}{3}$, 4, 6, ...

Geometric; $a_1 = \dfrac{8}{3}$,

$r = \dfrac{4}{\frac{8}{3}} = 4 \cdot \dfrac{3}{8} = \dfrac{12}{8} = \dfrac{3}{2}$

22. −10.5, −6.1, −1.7

Arithmetic; $a_1 = -10.5$,

$d = -6.1 - (-10.5) = 4.4$

23. $7x, -14x, 28x$

Geometric; $a_1 = 7x$, $r = -2$

24. neither

25. $a_1 = 8, \ r = 0.75$

$a_1 = 8$

$a_2 = 8(0.75) = 6$

$a_3 = 8(0.75)^2 = 4.5$

$a_4 = 8(0.75)^3 \approx 3.4$

$a_5 = 8(0.75)^4 \approx 2.5$

$a_6 = 8(0.75)^5 \approx 1.9$

Yes, a ball that rebounds to a height of 2.5 feet after the fifth bounce is good, since $2.5 \geq 1.9$.

26. $a_1 = 25, \ d = -4$

$a_n = a_1 + (n-1)d$

$a_n = 25 + (n-1)(-4) = 29 - 4n$

$a_7 = 25 + 6(-4) = 1$

Continuing the progression as far as possible leaves 1 can in the top row.

27. $a_1 = 1, \ r = 2$

$a_n = 2^{n-1}$

$a_{10} = 2^9 = 512$

$a_{30} = 2^{29} = 536{,}870{,}912$

You save \$512 on the tenth day and \$536,870,912 on the thirtieth day.

28. $a_n = a_1 r^{n-1}, \ a_1 = 30, \ r = 0.7$

$a_5 = 30(0.7)^4 = 7.203$

The length is 7.203 inches on the fifth swing.

29. $a_1 = 900, \ d = 150$

$a_n = 900 + (n-1)150 = 150_n + 750$

$a_6 = 900 + (6-1)150 = 1650$

Her salary is \$1650 per month at the end of training.

30. $\dfrac{1}{512}, \dfrac{1}{256}, \dfrac{1}{128}, \dots$

first fold: $a_1 = \dfrac{1}{256}, \ r = 2$

$a_{15} = \dfrac{1}{256}(2)^{15-1} = 64$

After 15 folds, the thickness is 64 inches.

31. $\displaystyle\sum_{i=1}^{5}(2i-1) = \big[2(1)-1\big] + \big[2(2)-1\big] + \big[2(3)-1\big]$

$\qquad\qquad\qquad + \big[2(4)-1\big] + \big[2(5)-1\big]$

$\qquad = 1 + 3 + 5 + 7 + 9$

$\qquad = 25$

32. $\displaystyle\sum_{i=1}^{5} i(i+2) = 1(1+2) + 2(2+2) + 3(3+2)$

$\qquad\qquad\qquad + 4(4+2) + 5(5+2)$

$\qquad = 3 + 8 + 15 + 24 + 35$

$\qquad = 85$

33. $\displaystyle\sum_{i=2}^{4} \dfrac{(-1)^i}{2i} = \dfrac{(-1)^2}{2(2)} + \dfrac{(-1)^3}{2(3)} + \dfrac{(-1)^4}{2(4)}$

$\qquad = \dfrac{1}{4} - \dfrac{1}{6} + \dfrac{1}{8}$

$\qquad = \dfrac{5}{24}$

34. $\displaystyle\sum_{i=3}^{5} 5(-1)^{i-1} = 5(-1)^{3-1} + 5(-1)^{4-1} + 5(-1)^{5-1}$

$\qquad = 5(1) + 5(-1) + 5(1)$

$\qquad = 5 - 5 + 5$

$\qquad = 5$

35. $1 + 3 + 9 + 27 + 81 + 243$

$\qquad = 3^0 + 3^1 + 3^2 + 3^3 + 3^4 + 3^5$

$\qquad = \displaystyle\sum_{i=1}^{6} 3^{i-1}$

36. $6 + 2 + (-2) + (-6) + (-10) + (-14) + (-18)$

$a_1 = 6, \ d = -4$

$a_n = 6 + (n-1)(-4)$

$\displaystyle\sum_{i=1}^{7}[6 + (i-1)(-4)]$

37. $\dfrac{1}{4} + \dfrac{1}{16} + \dfrac{1}{64} + \dfrac{1}{256} = \dfrac{1}{4^1} + \dfrac{1}{4^2} + \dfrac{1}{4^3} + \dfrac{1}{4^4}$

$\qquad = \displaystyle\sum_{i=1}^{4} \dfrac{1}{4^i}$

38. $1 + \left(-\dfrac{3}{2}\right) + \dfrac{9}{4} = \left(-\dfrac{3}{2}\right)^0 + \left(-\dfrac{3}{2}\right)^1 + \left(-\dfrac{3}{2}\right)^2$

$\qquad = \displaystyle\sum_{i=1}^{3}\left(-\dfrac{3}{2}\right)^{i-1}$

39. $a_1 = 20, \ r = 2$

$a_n = 20(2)^n$ represents the number of yeast, where n represents the number of 8-hour periods. Since $48 = 6(8)$ here, $n = 6$.

$a_6 = 20(2)^6 = 1280$

There are 1280 yeast after 48 hours.

40. $a_n = n^2 + 2n - 1$

$a_4 = (4)^2 + 2(4) - 1 = 23$

$S_4 = \displaystyle\sum_{i=1}^{4}(i^2 + 2i - 1)$

$\qquad = (1 + 2 - 1) + (4 + 4 - 1) + (9 + 6 - 1)$
$\qquad\qquad + (16 + 8 - 1)$
$\qquad = 46$

23 cranes are born in the fourth year and 46 cranes are born in the first four years.

41. For Job A: $a_1 = 39,500, \ d = 2200$;

$a_5 = 39,500 + (5-1)2200 = \$48,330$

For Job B: $a_1 = 41,000, \ d = 1400$

$a_5 = 41,000 + (5-1)1400 = \$46,600$

For the fifth year, Job A has a higher salary.

42. $a_n = 200(0.5)^n$

$a_3 = 200(0.5)^3 = 25$

$S_3 = \displaystyle\sum_{i=1}^{3} 200(0.5)^i$

$\qquad = 200(0.5) + 200(0.5)^2 + 200(0.5)^3$
$\qquad = 175$

25 kilograms decay in the third year and 175 kilograms decay in the first three years.

43. $a_n = (n-3)(n+2)$

$S_4 = \displaystyle\sum_{i=1}^{4}(i-3)(i+2)$

$\qquad = (1-3)(1+2) + (2-3)(2+2)$
$\qquad\qquad + (3-3)(3+2) + (4-3)(4+2)$
$\qquad = -6 - 4 + 0 + 6$
$\qquad = -4$

44. $a_n = n^2$

$S_6 = \displaystyle\sum_{i=1}^{6} i^2$

$\qquad = (1)^2 + (2)^2 + (3)^2 + (4)^2 + (5)^2 + (6)^2$
$\qquad = 91$

45. $a_n = -8 + (n-1)3 = 3n - 11$

$S_5 = \displaystyle\sum_{i=1}^{5}(3i - 11)$

$\qquad = [3(1) - 11] + [3(2) - 11] + [3(3) - 11]$
$\qquad\qquad + [3(4) - 11] + [3(5) - 11]$
$\qquad = -8 - 5 - 2 + 1 + 4$
$\qquad = -10$

46. $a_n = 5(4)^{n-1}$

$S_3 = \displaystyle\sum_{i=1}^{3} 5(4)^{i-1} = 5(4)^0 + 5(4)^1 + 5(4)^2 = 105$

47. 15, 19, 23, ...

$a_1 = 15, \ d = 4, \ a_6 = 15 + (6-1)4 = 35$

$S_6 = \dfrac{6}{2}[15 + 35] = 150$

48. $5, -10, 20, \ldots$

$a_1 = 5, \ r = -2$

$$S_n = \frac{a_1(1 - r^n)}{1 - r}$$

$$S_9 = \frac{5(1 - (-2)^9)}{1 - (-2)} = 855$$

49. $a_1 = 1, \ d = 2, \ n = 30, \ a_{30} = 1 + (30 - 1)2 = 59$

$$S_{30} = \frac{30}{2}[1 + 59] = 900$$

50. $7, 14, 21, 28, \ldots$

$a_n = 7 + (n - 1)7$

$a_{20} = 7 + (20 - 1)7 = 140$

$$S_{20} = \frac{20}{2}(7 + 140) = 1470$$

51. $8, 5, 2, \ldots$

$a_1 = 8, \ d = -3, \ n = 20$

$a_{20} = 8 + (20 - 1)(-3) = -49$

$$S_{20} = \frac{20}{2}[8 + (-49)]$$
$$= -410$$

52. $\dfrac{3}{4}, \dfrac{9}{4}, \dfrac{27}{4}, \ldots$

$a_1 = \dfrac{3}{4}, \ r = 3$

$$S_8 = \frac{\frac{3}{4}(1 - 3^8)}{1 - 3} = 2460$$

53. $a_1 = 6, \ r = 5$

$$S_4 = \frac{6(1 - 5^4)}{1 - 5} = 936$$

54. $a_1 = -3, \ d = -6$

$a_n = -3 + (n - 1)(-6)$

$a_{100} = -3 + (100 - 1)(-6) = -597$

$$S_{100} = \frac{100}{2}(-3 + (-597)) = -30,000$$

55. $5, \dfrac{5}{2}, \dfrac{5}{4}, \ldots$

$a_1 = 5, \ r = \dfrac{1}{2}$

$$S_\infty = \frac{5}{1 - \frac{1}{2}} = 10$$

56. $18, -2, \dfrac{2}{9}, \ldots$

$a_1 = 18, \ r = -\dfrac{1}{9}$

$$S_\infty = \frac{18}{1 + \frac{1}{9}} = \frac{81}{5}$$

57. $-20, -4, -\dfrac{4}{5}, \ldots$

$a_1 = -20, \ r = \dfrac{1}{5}$

$$S_\infty = \frac{-20}{1 - \frac{1}{5}} = -25$$

58. $0.2, 0.02, 0.002, \ldots$

$a_1 = 0.2 = \dfrac{1}{5}, \ r = \dfrac{1}{10}$

$$S_\infty = \frac{\frac{1}{5}}{1 - \frac{1}{10}} = \frac{2}{9}$$

59. $a_1 = 20,000, \ r = 1.15, \ n = 4$

$a_4 = 20,000(1.15)^{4-1} = 30,418$

$$S_4 = \frac{20,000(1 - 1.15^4)}{1 - 1.15} = 99,868$$

He earned \$30,418 during the fourth year and \$99,868 over the four years.

60. $a_n = 40(0.8)^{n-1}$

$a_4 = 40(0.8)^{4-1} = 20.48$

$$S_4 = \frac{40(1 - 0.8^4)}{1 - 0.8} = 118.08$$

He takes 20 minutes to assemble the fourth television and 118 minutes to assemble the first four televisions.

61. $a_1 = 100, d = -7, n = 7$

$a_7 = 100 + (7 - 1)(-7) = 58$

$$S_7 = \frac{7}{2}(100 + 58) = 553$$

The rent for the seventh day is \$58 and the rent for 7 days is \$553.

62. $a_1 = 15$, $r = 0.8$

$$S_\infty = \frac{15}{1-0.8} = 75 \text{ feet downward}$$

$$a_1 = 15(0.8) = 12, \; r = 0.8$$

$$S_\infty = \frac{12}{1-0.8} = 60 \text{ feet upward}$$

The total distance is 135 feet.

63. $1800, \; 600, \; 200, \ldots$

$$a_1 = 1800, \; r = \frac{1}{3}, \; n = 6$$

$$S_6 = 1800 \frac{\left(1-\left(\frac{1}{3}\right)^6\right)}{1-\frac{1}{3}} \approx 2696$$

Approximately 2696 mosquitoes were killed during the first six days after the spraying.

64. $1800, \; 600, \; 200, \; \ldots$

For which n is $a_n < 1$?

$$a_n = 1800\left(\frac{1}{3}\right)^{n-1} < 1$$

$$\left(\frac{1}{3}\right)^{n-1} < \frac{1}{1800}$$

$$(n-1)\log\frac{1}{3} < \log\frac{1}{1800}$$

$$(n-1)\log 3^{-1} < \log 1800^{-1}$$

$$(n-1)(-\log 3) < -\log 1800$$

$$n-1 > \frac{-\log 1800}{-\log 3}$$

$$n > 1 + \frac{\log 1800}{\log 3}$$

$$n > 7.8$$

No longer effective on the 8th day

$$S_8 = \frac{1800\left(1-\left(\frac{1}{3}\right)^8\right)}{1-\frac{1}{3}} \approx 2700$$

About 2700 mosquitoes were killed.

65. $0.5\overline{55} = 0.5 + 0.05 + 0.005 + \cdots$

$$a_1 = 0.5, \; r = 0.1$$

$$S_\infty = \frac{0.5}{1-0.1} = \frac{0.5}{0.9} = \frac{5}{9}$$

66. 27, 30, 33, …

$$a_n = 27 + (n-1)(3)$$
$$a_{20} = 27 + (20-1)(3) = 84$$
$$S_{20} = \frac{20}{2}(27 + 84) = 1110$$

There are 1110 seats in the theater.

67. $(x+z)^5 = x^5 + 5x^4 z + 10x^3 z^2 + 10x^2 z^3 + 5xz^4 + z^5$

68. $(y-r)^6 = y^6 + 6y^5(-r) + 15y^4(-r)^2 + 20y^3(-r)^3 + 15y^2(-r)^4 + 6y(-r)^5 + (-r)^6$
$$= y^6 - 6y^5 r + 15y^4 r^2 - 20y^3 r^3 + 15y^2 r^4 - 6yr^5 + r^6$$

69. $(2x+y)^4 = (2x)^4 + 4(2x)^3 y + 6(2x)^2 y^2 + 4(2x)y^3 + y^4$
$$= 16x^4 + 32x^3 y + 24x^2 y^2 + 8xy^3 + y^4$$

70. $(3y-z)^4 = (3y)^4 + 4(3y)^3(-z) + 6(3y)^2(-z)^2 + 4(3y)(-z)^3 + (-z)^4$
$$= 81y^4 - 108y^3 z + 54y^2 z^2 - 12yz^3 + z^4$$

71. $(b+c)^8 = b^8 + \dfrac{8}{1!}b^7 c + \dfrac{8 \cdot 7}{2!}b^6 c^2 + \dfrac{8 \cdot 7 \cdot 6}{3!}b^5 c^3 + \dfrac{8 \cdot 7 \cdot 6 \cdot 5}{4!}b^4 c^4 + \dfrac{8 \cdot 7 \cdot 6 \cdot 5 \cdot 4}{5!}b^3 c^5$
$$+ \frac{8 \cdot 7 \cdot 6 \cdot 5 \cdot 4 \cdot 3}{6!}b^2 c^6 + \frac{8 \cdot 7 \cdot 6 \cdot 5 \cdot 4 \cdot 3 \cdot 2}{7!}bc^7 + c^8$$
$$= b^8 + 8b^7 c + 28b^6 c^2 + 56b^5 c^3 + 70b^4 c^4 + 56b^3 c^5 + 28b^2 c^6 + 8bc^7 + c^8$$

72. $(x-w)^7 = x^7 + \dfrac{7}{1!}x^6(-w) + \dfrac{7 \cdot 6}{2!}x^5(-w)^2 + \dfrac{7 \cdot 6 \cdot 5}{3!}x^4(-w)^3 + \dfrac{7 \cdot 6 \cdot 5 \cdot 4}{4!}x^3(-w)^4 + \dfrac{7 \cdot 6 \cdot 5 \cdot 4 \cdot 3}{5!}x^2(-w)^5$
$$+ \frac{7 \cdot 6 \cdot 5 \cdot 4 \cdot 3 \cdot 2}{6!}x(-w)^6 + (-w)^7$$
$$= x^7 - 7x^6 w + 21x^5 w^2 - 35x^4 w^3 + 35x^3 w^4 - 21x^2 w^5 + 7xw^6 - w^7$$

73. $(4m-n)^4 = (4m)^4 + \dfrac{4}{1!}(4m)^3(-n) + \dfrac{4 \cdot 3}{2!}(4m)^2(-n)^2 + \dfrac{4 \cdot 3 \cdot 2}{3!}(4m)(-n)^3 + (-n)^4$
$$= 256m^4 - 256m^3 n + 96m^2 n^2 - 16mn^3 + n^4$$

74. $(p-2r)^5 = p^5 + \dfrac{5}{1!}p^4(-2r) + \dfrac{5 \cdot 4}{2!}p^3(-2r)^2 + \dfrac{5 \cdot 4 \cdot 3}{3!}p^2(-2r)^3 + \dfrac{5 \cdot 4 \cdot 3 \cdot 2}{4!}p(-2r)^4 + (-2r)^5$
$$= p^5 - 10p^4 r + 40p^3 r^2 - 80p^2 r^3 + 80pr^4 - 32r^5$$

75. The 4th term corresponds to $r = 3$.

$$\frac{7!}{3!(7-3)!}a^{7-3}b^3 = 35a^4 b^3$$

76. The 11th term corresponds to $r = 10$.

$$\frac{10!}{10!0!}y^{10-10}(2z)^{10} = 1024z^{10}$$

77. $\displaystyle\sum_{i=1}^{4} i^2(i+1) = 1^2(1+1) + 2^2(2+1) + 3^2(3+1) + 4^2(4+1)$

$$= 2 + 12 + 36 + 80$$
$$= 130$$

78. 14, 8, 2, ...

$a_1 = 14, \ d = -6, \ n = 15$

$a_{15} = 14 + (15-1)(-6) = -70$

79. 27, 9, 3, 1, ...

$a_1 = 27, \ r = \dfrac{1}{3}$

$S_\infty = \dfrac{27}{1 - \dfrac{1}{3}} = 40.5$

80. $(2x-3)^4 = (2x)^4 + 4(2x)^3(-3) + 6(2x)^2(-3)^2 + 4(2x)(-3)^3 + (-3)^4$

$$= 16x^4 - 96x^3 + 216x^2 - 216x + 81$$

Chapter 11 Test

1. $a_n = \dfrac{(-1)^n}{n+4}$

$a_1 = \dfrac{(-1)^1}{1+4} = -\dfrac{1}{5}$

$a_2 = \dfrac{(-1)^2}{2+4} = \dfrac{1}{6}$

$a_3 = \dfrac{(-1)^3}{3+4} = -\dfrac{1}{7}$

$a_4 = \dfrac{(-1)^4}{4+4} = \dfrac{1}{8}$

$a_5 = \dfrac{(-1)^5}{5+4} = -\dfrac{1}{9}$

2. $a_n = 10 + 3(n-1)$

$a_{80} = 10 + 3(80-1) = 247$

3. $\dfrac{2}{5}, \dfrac{2}{25}, \dfrac{2}{125}, ...$

In general, $a_n = \dfrac{2}{5}\left(\dfrac{1}{5}\right)^{n-1}$ or $a_n = \dfrac{2}{5^n}$.

4. $(-1)^1 9 \cdot 1, (-1)^2 9 \cdot 2, ..., a_n = (-1)^n 9n$

5. $a_n = 5(2)^{n-1}, S_5 = \dfrac{5(1-2^5)}{1-2} = 155$

6. $a_n = 18 + (n-1)(-2)$

   $a_1 = 18$, $a_{30} = 18 + (30-1)(-2) = -40$

   $S_{30} = \dfrac{30}{2}[18 - 40] = -330$

7. $a_1 = 24$, $r = \dfrac{1}{6}$

   $S_\infty = \dfrac{24}{1 - \frac{1}{6}} = \dfrac{144}{5}$

8. $\dfrac{3}{2}, -\dfrac{3}{4}, \dfrac{3}{8}, \ldots$

   $a_1 = \dfrac{3}{2}$, $r = -\dfrac{1}{2}$

   $S_\infty = \dfrac{\frac{3}{2}}{1 - \left(-\frac{1}{2}\right)} = 1$

9. $\displaystyle\sum_{i=1}^{4} i(i-2) = 1(1-2) + 2(2-2) + 3(3-2) + 4(4-2)$

   $\qquad\qquad = -1 + 0 + 3 + 8 - 20 + 40 - 80$

   $\qquad\qquad = 10$

10. $\displaystyle\sum_{i=2}^{4} 5(2)^i (-1)^{i-1} = 5(2)^2(-1)^{2-1} + 5(2)^3(-1)^{3-1} + 5(2)^4(-1)^{4-1} = -20 + 40 - 80 = -60$

11. $(a-b)^6 = a^6 - 6a^5b + 15a^4b^2 - 20a^3b^3 + 15a^2b^4 - 6ab^5 + b^6$

12. $(2x+y)^5 = (2x)^5 + \dfrac{5}{1!}(2x)^4 y + \dfrac{5\cdot4}{2!}(2x)^3 y^2 + \dfrac{5\cdot4\cdot3}{3!}(2x)^2 y^3 + \dfrac{5\cdot4\cdot3\cdot2}{4!}(2x)y^4 + y^5$

   $\qquad\quad = 32x^5 + 80x^4 y + 80x^3 y^2 + 40x^2 y^3 + 10xy^4 + y^5$

13. $a_n = 250 + 75(n-1)$

   $a_{10} = 250 + 75(10-1) = 925$

   There were 925 people in the town at the beginning of the tenth year.

   $a_1 = 250 + 75(1-1) = 250$

   There were 250 people in the town at the beginning of the first year.

14. 1, 3, 5, ...

   $a_1 = 1$, $d = 2$, $n = 8$

   $a_8 = 1 + (8-1)2 = 15$

   $1 + 3 + 5 + 7 + 9 + 11 + 13 + 15$

   $S_8 = \dfrac{8}{2}[1 + 15] = 64$

   There were 64 shrubs planted in the 8 rows.

15. $a_1 = 80$, $r = \dfrac{3}{4}$, $n = 4$

$a_4 = 80\left(\dfrac{3}{4}\right)^{4-1} = 33.75$

The arc length is 33.75 cm on the 4th swing.

$S_4 = \dfrac{80\left(1 - \left(\frac{3}{4}\right)^4\right)}{1 - \frac{3}{4}} = 218.75$

The total of the arc lengths is 218.75 cm for the first 4 swings.

16. $a_1 = 80$, $r = \dfrac{3}{4}$

$S_\infty = \dfrac{80}{1 - \frac{3}{4}} = 320$

The total of the arc lengths is 320 cm before the pendulum comes to rest.

17. 16, 48, 80,...

$a_{10} = 16 + (10 - 1)32 = 304$

He falls 304 feet during the 10th second.

$S_{10} = \dfrac{10}{2}[16 + 304] = 1600$

He falls 1600 feet during the first 10 seconds.

18. $0.42\overline{42} = 0.42 + 0.0042 + 0.000042$

$a_1 = 0.42 = \dfrac{42}{100}$, $r = 0.01 = \dfrac{1}{100}$

$S_\infty = \dfrac{\frac{42}{100}}{1 - \frac{1}{100}} = \dfrac{42}{100} \cdot \dfrac{100}{99} = \dfrac{14}{33}$

Thus, $0.42\overline{42} = \dfrac{14}{33}$.

Chapter 11 Cumulative Review

1. a. $\dfrac{20}{-4} = -5$

b. $\dfrac{-9}{-3} = 3$

c. $-\dfrac{3}{8} \div 3 = -\dfrac{3}{8} \cdot \dfrac{1}{3} = -\dfrac{1}{8}$

d. $\dfrac{-40}{10} = -4$

e. $\dfrac{-1}{10} \div \dfrac{-2}{5} = \dfrac{1}{10} \cdot \dfrac{5}{2} = \dfrac{1}{4}$

f. $\dfrac{8}{0}$ is undefined.

2. a. $3a - (4a + 3) = 3a - 4a - 3 = -a - 3$

b. $(5x - 3) + (2x + 6) = 7x + 3$

c. $4(2x - 5) - 3(5x + 1) = 8x - 20 - 15x - 3$
$= -7x - 23$

3. Let x = the original price, then
$x - 0.08x = 2162$
$0.92x = 2162$
$x = 2350$
The original price is \$2350.

4. Let x = the price before taxes, then
$x + 0.06x = 344.50$
$1.06x = 344.50$
$x = 325$
The price before taxes was \$325.

5. $3y - 2x = 7$
$3y = 2x + 7$
$y = \dfrac{1}{3}(2x + 7)$
$y = \dfrac{2x}{3} + \dfrac{7}{3}$

6. If the line is to be parallel, then the slope has to be the same as the slope of the given line.

Therefore, $m = \dfrac{3}{2}$.

$(y - (-2)) = \dfrac{3}{2}(x - 3)$

$y + 2 = \dfrac{3}{2}(x - 3)$

$y = \dfrac{3}{2}x - \dfrac{13}{2}$

$f(x) = \dfrac{3}{2}x - \dfrac{13}{2}$

7. a. $(3x^6)(5x) = 3 \cdot 5x^{6+1} = 15x^7$

b. $(-2.4x^3 p^2)(4xp^{10}) = -2.4 \cdot 4x^{3+1}p^{2+10}$
$= -9.6x^4 p^{12}$

8.
$$y^3 + 5y^2 - y = 5$$
$$y^3 + 5y^2 - y - 5 = 0$$
$$(y^3 + 5y^2) + (-y - 5) = 0$$
$$y^2(y + 5) - 1(y + 5) = 0$$
$$(y^2 - 1)(y + 5) = 0$$
$$(y + 1)(y - 1)(y + 5) = 0$$
$$y = -5, -1, 1$$

9.
$$\underline{-2}\,|\,1 \quad -2 \quad -11 \quad 0 \quad 34$$
$$\quad -2 \quad 8 \quad 6 \quad -12$$
$$\overline{1 \quad -4 \quad -3 \quad 6 \quad 22}$$

Answer: $x^3 - 4x^2 - 3x + 6 + \dfrac{22}{x+2}$

10.
$$\frac{5}{3a-6} - \frac{a}{a-2} + \frac{3+2a}{5a-10}$$
$$= \frac{5}{3(a-2)} - \frac{a}{a-2} + \frac{3+2a}{5(a-2)}$$
$$= \frac{5 \cdot 5 - 3 \cdot 5a + 3(3+2a)}{3 \cdot 5(a-2)}$$
$$= \frac{25 - 15a + 9 + 6a}{15(a-2)}$$
$$= \frac{34 - 9a}{15(a-2)}$$

11. a. $\sqrt{50} = \sqrt{2}\sqrt{25} = 5\sqrt{2}$

    b. $\sqrt[3]{24} = \sqrt[3]{8}\sqrt[3]{3} = 2\sqrt[3]{3}$

    c. $\sqrt{26} = \sqrt{26}$

    d. $\sqrt[4]{32} = \sqrt[4]{16}\sqrt[4]{2} = 2\sqrt[4]{2}$

12.
$$\sqrt{3x+6} - \sqrt{7x-6} = 0$$
$$\sqrt{3x+6} = \sqrt{7x-6}$$
$$\left(\sqrt{3x+6}\right)^2 = \left(\sqrt{7x-6}\right)^2$$
$$3x + 6 = 7x - 6$$
$$-4x = -12$$
$$x = 3$$

13.
$$2420 = 2000(1+r)^2$$
$$\frac{2420}{2000} = (1+r)^2$$
$$\frac{121}{100} = (1+r)^2$$
$$\pm\sqrt{\frac{121}{100}} = 1+r$$
$$\pm\frac{11}{10} = 1+r$$
$$-1 \pm \frac{11}{10} = r$$

Discard the negative value.
$$r = -1 + \frac{11}{10} = \frac{1}{10} = 0.10$$
The interest rate is 10%.

14. a. $\sqrt[3]{\dfrac{4}{3x}} = \dfrac{\sqrt[3]{4}}{\sqrt[3]{3x}} = \left(\dfrac{\sqrt[3]{9x^2}}{\sqrt[3]{9x^2}}\right) = \dfrac{\sqrt[3]{36x^2}}{3x}$

    b. $\dfrac{\sqrt{2}+1}{\sqrt{2}-1} = \dfrac{\sqrt{2}+1}{\sqrt{2}-1} \cdot \left(\dfrac{\sqrt{2}+1}{\sqrt{2}+1}\right)$
$$= \frac{2 + 2\sqrt{2} + 1}{2 - 1}$$
$$= 3 + 2\sqrt{2}$$

15.
$$(x-3)^2 - 3(x-3) - 4 = 0$$
$$x^2 - 6x + 9 - 3x + 9 - 4 = 0$$
$$x^2 - 9x + 14 = 0$$
$$(x-2)(x-7) = 0$$
$$x = 2, 7$$

16.
$$\frac{10}{(2x+4)^2} - \frac{1}{2x+4} = 3$$
$$10 - (2x+4) = 3(2x+4)^2$$
$$10 - 2x - 4 = 3(4x^2 + 16x + 16)$$
$$-2x + 6 = 12x^2 + 48x + 48$$
$$12x^2 + 50x + 42 = 0$$
$$6x^2 + 25x + 21 = 0$$
$$(6x+7)(x+3) = 0$$
$$x = -\frac{7}{6}, -3$$

17. $\dfrac{5}{x+1} < -2$

$x + 1 = 0$

$x = -1$

Solve $\dfrac{5}{x+1} = -2$.

$(x+1)\dfrac{5}{x+1} = (x+1)(-2)$

$5 = -2x - 2$

$7 = -2x$

$-\dfrac{7}{2} = x$

Region	Test Point	$\dfrac{5}{x+1} < -2$; Result
$\left(-\infty, -\dfrac{7}{2}\right)$	$x = -6$	$\dfrac{5}{-5} < -2$; False
$\left(-\dfrac{7}{2}, -1\right)$	$x = -2$	$\dfrac{5}{-1} < -2$; True
$(-1, \infty)$	$x = 4$	$\dfrac{5}{5} < -2$; False

The solution set is $\left(-\dfrac{7}{2}, -1\right)$.

18. $f(x) = (x+2)^2 - 6$

Axis of symmetry: $x = -2$

vertex: $(-2, -6)$

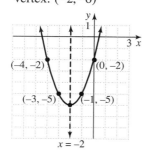

19. $f(t) = -16t^2 + 20t$

The maximum height occurs at the vertex.

$t = \dfrac{-20}{2(-16)} = \dfrac{5}{8}$

$f\left(\dfrac{5}{8}\right) = -16\left(\dfrac{5}{8}\right)^2 + 20\left(\dfrac{5}{8}\right) = \dfrac{25}{4}$

The maximum height of $\dfrac{25}{4}$ feet occurs at

$\dfrac{5}{8}$ second.

20. $f(x) = x^2 + 3x - 18$

$a = 1, b = 3, c = -18$

$x = \dfrac{-3}{2(1)} = -\dfrac{3}{2}$

$f\left(-\dfrac{3}{2}\right) = \left(-\dfrac{3}{2}\right)^2 + 3\left(\dfrac{3}{2}\right) - 18 = -\dfrac{81}{4}$

The vertex is $\left(-\dfrac{3}{2}, -\dfrac{81}{4}\right)$.

21. a. $(f \circ g)(2) = f(g(2)) = f(5) = 5^2 = 25$

$(g \circ f)(2) = g(f(2)) = g(4) = 4 + 3 = 7$

b. $(f \circ g)(x) = f(x+3)$

$= (x+3)^2$

$= x^2 + 6x + 9$

$(g \circ f)(x) = g(x^2) = x^2 + 3$

22. $f(x) = -2x + 3$

$y = -2x + 3$

$x = -2y + 3$

$x - 3 = -2y$

$\dfrac{x-3}{-2} = y$

$f^{-1}(x) = -\dfrac{x-3}{2}$ or $f^{-1}(x) = \dfrac{3-x}{2}$

23. $f^{-1} = \{(1,0), (7,-2), (-6,3), (4,4)\}$

24. a. $(f \circ g)(2) = f(g(2)) = f(3) = 3^2 - 2 = 7$

$(g \circ f)(2) = g(f(2)) = g(2) = 2 + 1 = 3$

b. $(f \circ g)(x) = f(x+1)$
$$= (x+1)^2 - 2$$
$$= x^2 + 2x - 1$$
$$(g \circ f)(x) = g(x^2 - 2) = x^2 - 2 + 1 = x^2 - 1$$

25. a. $2^x = 16$
$$2^x = 2^4$$
$$x = 4$$

    b. $9^x = 27$
$$(3^2)^x = 3^3$$
$$2x = 3$$
$$x = \frac{3}{2}$$

    c. $4^{x+3} = 8^x$
$$(2^2)^{x+3} = (2^3)^x$$
$$2^{2x+6} = 2^{3x}$$
$$2x + 6 = 3x$$
$$x = 6$$

    d.

        The solution is approximately 1.431.

26. a. $\log_2 32 = x$
$$2^x = 32$$
$$2^x = 2^5$$
$$x = 5$$

    b. $\log_4 \frac{1}{64} = x$
$$4^x = \frac{1}{64}$$
$$4^x = 4^{-3}$$
$$x = -3$$

    c. $\log_{1/2} x = 5$
$$\left(\frac{1}{2}\right)^5 = x$$
$$x = \frac{1}{32}$$

27. a. $\log_3 3^2 = 2$

b. $\log_7 7^{-1} = -1$

c. $5^{\log_5 3} = 3$

d. $2^{\log_2 6} = 6$

28. a. $4^x = 64$
$$(2^2)^x = 2^6$$
$$2x = 6$$
$$x = 3$$

    b. $8^x = 32$
$$(2^3)^x = 2^5$$
$$3x = 5$$
$$x = \frac{5}{3}$$

    c. $9^{x+4} = 243^x$
$$(3^2)^{x+4} = (3^5)^x$$
$$3^{2x+8} = 3^{5x}$$
$$2x + 8 = 5x$$
$$8 = 3x$$
$$x = \frac{8}{3}$$

29. a. $\log_{11} 10 + \log_{11} 3 = \log_{11}(10 \cdot 3) = \log_{11} 30$

    b. $\log_3 \frac{1}{2} + \log_3 12 = \log_3 \left(\frac{1}{2} \cdot 12\right) = \log_3 6$

    c. $\log_2(x+2) + \log_2 x = \log_2[(x+2)x]$
$$= \log_2(x^2 + 2x)$$

30. a. $\log 100{,}000 = \log_{10} 10^5 = 5$

    b. $\log 10^{-3} = \log_{10} 10^{-3} = -3$

    c. $\ln \sqrt[5]{e} = \ln e^{1/5} = \frac{1}{5}$

    d. $\ln e^4 = 4$

31. $A = Pe^{rt}$
$$A = 1600e^{0.09(5)} \approx 2509.30$$
$2509.30 is owed after 5 years.

32. a. $\log_6 5 + \log_6 4 = \log_6 (5 \cdot 4) = \log_6 20$

b. $\log_8 12 - \log_8 4 = \log_8 \dfrac{12}{4} = \log_8 3$

c. $2\log_2 x + 3\log_2 x - 2\log_2 (x-1)$

$= 5\log_2 x - \log_2 (x-1)^2$

$= \log_2 x^5 - \log_2 (x-1)^2$

$= \log_2 \dfrac{x^5}{(x-1)^2}$

33. $3^x = 7$

$\log 3^x = \log 7$

$x \log 3 = \log 7$

$x = \dfrac{\log 7}{\log 3} \approx 1.7712$

34. $10{,}000 = 5000\left(1 + \dfrac{0.02}{4}\right)^{4t}$

$2 = (1.005)^{4t}$

$\ln 2 = \ln 1.005^{4t}$

$\ln 2 = 4t \ln(1.005)$

$t = \dfrac{\ln 2}{4 \ln 1.005} \approx 34.7$

It takes 34.7 years.

35. $\log_4 (x-2) = 2$

$4^2 = x - 2$

$x - 2 = 16$

$x = 18$

36. $\log_4 10 - \log_4 x = 2$

$\log_4 \dfrac{10}{x} = 2$

$4^2 = \dfrac{10}{x}$

$16 = \dfrac{10}{x}$

$16x = 10$

$x = \dfrac{5}{8}$

37. $\dfrac{x^2}{16} - \dfrac{y^2}{25} = 1$

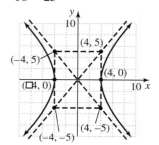

38. $(8, 5), (-2, 4)$

$d = \sqrt{(-2-8)^2 + (4-5)^2} = \sqrt{101}$ units

39. $\begin{cases} y = \sqrt{x} \\ x^2 + y^2 = 6 \end{cases}$

Replace y with $\sqrt{x}$ in the first equation.

$(x)^2 + \left(\sqrt{x}\right)^2 = 6$

$x^2 + x - 6 = 0$

$(x+3)(x-2) = 0$

$x = -3 (\text{discard})$ or $x = 2$

$x = 2: \ y = \sqrt{x} = \sqrt{2}$

$\left(2, \sqrt{2}\right)$

40. $\begin{cases} x^2 + y^2 = 36 \\ x - y = 6 \Rightarrow x = y + 6 \end{cases}$

Replace x with $y + 6$ in the first equation.

$(y+6)^2 + y^2 = 36$

$2y^2 + 12y = 0$

$2y(y+6) = 0$

$y = 0$ or $y = -6$

$x = 0 + 6 = 6$ $x = -6 + 6 = 0$

$(0, -6); (6, 0)$

41. $\dfrac{x^2}{9} + \dfrac{y^2}{16} \le 1$

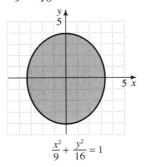

$$\dfrac{x^2}{9} + \dfrac{y^2}{16} = 1$$

42.

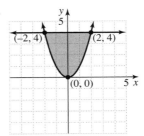

43. $a_n = n^2 - 1$

$a_1 = 1^2 - 1 = 0$

$a_2 = 2^2 - 1 = 3$

$a_3 = 3^2 - 1 = 8$

$a_4 = 4^2 - 1 = 15$

$a_5 = 5^2 - 1 = 24$

44. $a_n = \dfrac{n}{n+4}$

$a_8 = \dfrac{8}{8+4} = \dfrac{8}{12} = \dfrac{2}{3}$

45. $a_1 = 2,\ d = 9 - 2 = 7$

$a_{11} = 2 + (11-1)(7) = 72$

46. $a_1 = 2,\ r = \dfrac{10}{2} = 5$

$a_6 = 2(5)^{6-1} = 2(5)^5 = 6250$

47. a. $\displaystyle\sum_{i=0}^{6} \dfrac{i-2}{2} = \dfrac{0-2}{2} + \dfrac{1-2}{2} + \dfrac{2-2}{2} + \dfrac{3-2}{2} + \dfrac{4-2}{2} + \dfrac{5-2}{2} + \dfrac{6-2}{2}$

$$= -1 - \dfrac{1}{2} + 0 + \dfrac{1}{2} + 1 + \dfrac{3}{2} + 2$$

$$= \dfrac{7}{2}$$

b. $\displaystyle\sum_{i=3}^{5} 2^i = 2^3 + 2^4 + 2^5 = 8 + 16 + 32 = 56$

48. a. $\displaystyle\sum_{i=0}^{4} i(i+1) = 0(0+1) + 1(1+1) + 2(2+1) + 3(3+1) + 4(4+1)$

$$= 0 + 2 + 6 + 12 + 20$$
$$= 40$$

b. $\displaystyle\sum_{i=0}^{3} 2^i = 2^0 + 2^1 + 2^2 + 2^3 = 1 + 2 + 4 + 8 = 15$

49. $a_1 = 1,\ a_{30} = 30$

$$S_n = \frac{n}{2}(a_1 + a_n) = \frac{30}{2}(1 + 30) = 465$$

50. $(x - y)^6$ where $a = x$, $b = -y$, $n = 6$, and $r = 2$.

$$\frac{6!}{2!(6-2)!}x^{6-2}(-y)^2 = 15x^4 y^2$$

The third term in the expansion of $(x - y)^6$ is $15x^4 y^2$.

Appendices

Appendix A.3 Exercise Set

1. $V = lwh$
    $ = (6 \text{ in.})(4 \text{ in.})(3 \text{ in.})$
    $ = 72 \text{ cu in.}$
    $SA = 2lh + 2wh + 2lw$
    $ = 2(6 \text{ in.})(3 \text{ in.}) + 2(4 \text{ in.})(3 \text{ in.}) + 2(6 \text{ in.})(4 \text{ in.})$
    $ = 36 \text{ sq in.} + 24 \text{ sq in.} + 48 \text{ sq in.}$
    $ = 108 \text{ sq in.}$

3. $V = s^3 = (8 \text{ cm})^3 = 512 \text{ cu cm}$

    $SA = 6s^2 = 6(8 \text{ cm})^2 = 384 \text{ sq cm}$

5. $V = \dfrac{1}{3}\pi r^2 h$

    $ = \dfrac{1}{3}\pi(2 \text{ yd})^2(3 \text{ yd})$

    $ = 4\pi \text{ cu yd}$

    $ \approx 4\left(\dfrac{22}{7}\right) \text{ cu yd}$

    $ \approx \dfrac{88}{7} \text{ cu yd}$

    $ \approx 12\dfrac{4}{7} \text{ cu yd}$

    $SA = \pi r\sqrt{r^2 + h^2} + \pi r^2$
    $ = \pi(2 \text{ yd})\sqrt{(2 \text{ yd})^2 + (3 \text{ yd})^2} + \pi(2 \text{ yd})^2$
    $ = \pi(2 \text{ yd})\left(\sqrt{13} \text{ yd}\right) + \pi(4 \text{ sq yd})$
    $ = \left(2\sqrt{13} + 4\right)\pi \text{ sq yd}$
    $ \approx 3.14\left(2\sqrt{13} + 4\right) \text{ sq yd}$
    $ \approx 35.20 \text{ sq yd}$

7. $V = \dfrac{4}{3}\pi r^3$

    $ = \dfrac{4}{3}\pi(5 \text{ in.})^3$

    $ = \dfrac{500}{3}\pi \text{ cu in.}$

    $ \approx \dfrac{500}{3}\left(\dfrac{22}{7}\right) \text{ cu in.}$

    $ \approx 523\dfrac{17}{21} \text{ cu in.}$

$SA = 4\pi r^2$
$= 4\pi(5 \text{ in.})^2$
$= 100\pi \text{ sq in.}$
$\approx 100\left(\dfrac{22}{7}\right) \text{ sq in.}$
$\approx 314\dfrac{2}{7} \text{ sq in.}$

9. $V = \dfrac{1}{3}s^2h = \dfrac{1}{3}(6 \text{ cm})^2(4 \text{ cm}) = 48 \text{ cu cm}$

$SA = B + \dfrac{1}{2}pl$
$= (6 \text{ cm})^2 + \dfrac{1}{2}(24 \text{ cm})(5 \text{ cm})$
$= 36 \text{ sq cm} + 60 \text{ sq cm}$
$= 96 \text{ sq cm}$

11. $V = s^3$
$= \left(1\dfrac{1}{3} \text{ in.}\right)^3$
$= \left(\dfrac{4}{3} \text{ in.}\right)^3$
$= \dfrac{64}{27} \text{ cu in.}$
$= 2\dfrac{10}{27} \text{ cu in.}$

13. $SA = 2lh + 2wh + 2lw$
$= 2(2 \text{ ft})(1.4 \text{ ft}) + 2(2 \text{ ft})(3 \text{ ft}) + 2(1.4 \text{ ft})(3 \text{ ft})$
$= 5.6 \text{ sq ft} + 12 \text{ sq ft} + 8.4 \text{ sq ft}$
$= 26 \text{ sq ft}$

15. $V = \dfrac{1}{3}s^2h$
$= \dfrac{1}{3}(5 \text{ in.})^2(1.3 \text{ in.})$
$= \dfrac{1}{3}(25 \text{ sq in.})\left(\dfrac{13}{10} \text{ in.}\right)$
$= \dfrac{65}{6} \text{ cu in.}$
$= 10\dfrac{5}{6} \text{ cu in.}$

17. $V = \dfrac{1}{3}s^2h = \dfrac{1}{3}(12 \text{ cm})^2(20 \text{ cm}) = 960 \text{ cu cm}$

19. $SA = 4\pi r^2 = 4\pi(7 \text{ in.})^2 = 196\pi \text{ sq in.}$

21. $V = (2 \text{ ft})\left(2\dfrac{1}{2} \text{ ft}\right)\left(1\dfrac{1}{2} \text{ ft}\right) = 7\dfrac{1}{2} \text{ cu ft}$

23. $V = \dfrac{1}{3}\pi r^2h$
$\approx \dfrac{1}{3}\left(\dfrac{22}{7}\right)(2 \text{ cm})^2(3 \text{ cm})$
$= \dfrac{88}{7} \text{ cu cm}$
$\approx 12\dfrac{4}{7} \text{ cu cm}$

Appendix B Exercise Set

1. $f(x) = 3|x|$
Find and plot ordered-pair solutions.

x	$f(x) = 3|x|$
-1	$3|-1| = 3$
0	$3|0| = 0$
1	$3|1| = 3$

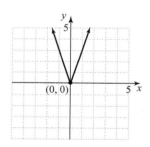

3. $f(x) = \dfrac{1}{4}|x|$
Find and plot ordered-pair solutions.

x	$f(x) = \tfrac{1}{4}|x|$
-4	1
0	0
4	1

5. $g(x) = 2|x| + 3$
Write in the form $g(x) = a|x - h| + k$.
$g(x) = 2|x - 0| + 3$

- vertex is $(h, k) = (0, 3)$

- since $a > 0$, V-shape opens up

- since $|a| = |2| = 2 > 1$, the graph is narrower than $y = |x|$

x	$g(x) = 2|x| + 3$
-1	$2|-1| + 3 = 5$
0	$2|0| + 3 = 3$
1	$2|1| + 3 = 5$

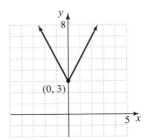

7. $h(x) = -\dfrac{1}{2}|x|$

- vertex is $(h, k) = (0, 0)$

- since $a < 0$, V-shape opens down

- since $|a| = \left|-\dfrac{1}{2}\right| = \dfrac{1}{2} < 1$, the graph is wider than $y = |x|$

x	$h(x) = -\frac{1}{2}|x|$
-2	$-\frac{1}{2}|-2| = -1$
0	$-\frac{1}{2}|0| = 0$
2	$-\frac{1}{2}|2| = -1$

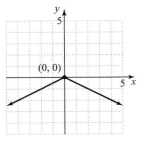

9. $f(x) = 4|x - 1|$
Write in the form $f(x) = a|x - h| + k$.
$f(x) = 4|x - 1| + 0$

- vertex is $(h, k) = (1, 0)$

- since $a > 0$, V-shape opens up

- since $|a| = |4| = 4 > 1$, the graph is narrower than $y = |x|$

x	$f(x) = 4|x - 1|$
-2	$4|-2 - 1| = 12$
0	$4|0 - 1| = 4$
2	$4|2 - 1| = 4$

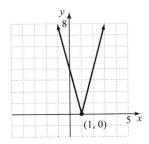

11. $g(x) = -\dfrac{1}{3}|x| - 2$
Write in the form $g(x) = a|x - h| + k$.
$g(x) = -\dfrac{1}{3}|x - 0| + (-2)$

- vertex is $(h, k) = (0, -2)$

- since $a < 0$, V-shape opens down

- since $|a| = \left|-\dfrac{1}{3}\right| = \dfrac{1}{3} < 1$, the graph is wider than $y = |x|$

x	$g(x) = -\frac{1}{3}\lvert x\rvert - 2$
-3	$-\frac{1}{3}\lvert-3\rvert - 2 = -3$
0	$-\frac{1}{3}\lvert0\rvert - 2 = -2$
3	$-\frac{1}{3}\lvert3\rvert - 2 = -3$

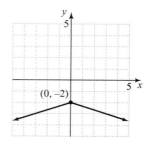

13. $f(x) = -2\lvert x - 3\rvert + 4$
This function is already written in the form
$f(x) = a\lvert x - h\rvert + k$.

- vertex is $(h, k) = (3, 4)$

- since $a < 0$, V-shape opens down

- since $\lvert a\rvert = \lvert-2\rvert = 2 > 1$, the graph is narrower than $y = \lvert x\rvert$

x	$f(x) = -2\lvert x - 3\rvert + 4$
-1	$-2\lvert-1 - 3\rvert + 4 = -4$
1	$-2\lvert1 - 3\rvert + 4 = 0$
4	$-2\lvert4 - 3\rvert + 4 = 2$

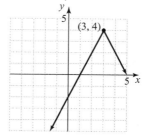

15. $f(x) = \frac{2}{3}\lvert x + 2\rvert - 5$

Write in the form $f(x) = a\lvert x - h\rvert + k$.

$f(x) = \frac{2}{3}\lvert x - (-2)\rvert + (-5)$

- vertex is $(h, k) = (-2, -5)$

- since $a > 0$, V-shape opens up

- since $\lvert a\rvert = \left\lvert\frac{2}{3}\right\rvert = \frac{2}{3} < 1$, the graph is wider than $y = \lvert x\rvert$

x	$f(x) = \frac{2}{3}\lvert x + 2\rvert - 5$
-5	$\frac{2}{3}\lvert-5 + 2\rvert - 5 = -3$
-2	$\frac{2}{3}\lvert-2 + 2\rvert - 5 = -5$
1	$\frac{2}{3}\lvert1 + 2\rvert - 5 = -3$

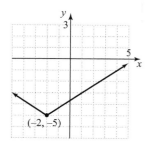

Appendix C

Vocabulary, Readiness & Video Check Appendix C

1. $\begin{vmatrix} 7 & 2 \\ 0 & 8 \end{vmatrix} = 56$

2. $\begin{vmatrix} 6 & 0 \\ 1 & 2 \end{vmatrix} = 12$

3. $\begin{vmatrix} -4 & 2 \\ 0 & 8 \end{vmatrix} = -32$

4. $\begin{vmatrix} 5 & 0 \\ 3 & -5 \end{vmatrix} = -25$

5. $\begin{vmatrix} -2 & 0 \\ 3 & -10 \end{vmatrix} = 20$

6. $\begin{vmatrix} -1 & 4 \\ 0 & -18 \end{vmatrix} = 18$

Appendix C Exercise Set

1. $\begin{vmatrix} 3 & 5 \\ -1 & 7 \end{vmatrix} = ad - bc = 3(7) - 5(-1) = 21 + 5 = 26$

3. $\begin{vmatrix} 9 & -2 \\ 4 & -3 \end{vmatrix} = ad - bc$

$$= 9(-3) - (-2)(4)$$
$$= -27 + 8$$
$$= -19$$

5. $\begin{vmatrix} -2 & 9 \\ 4 & -18 \end{vmatrix} = ad - bc$

$$= -2(-18) - 9(4)$$
$$= 36 - 36$$
$$= 0$$

7. $\begin{vmatrix} \frac{3}{4} & \frac{5}{2} \\ -\frac{1}{6} & \frac{7}{3} \end{vmatrix} = ad - bc$

$$= \left(\frac{3}{4}\right)\left(\frac{7}{3}\right) - \left(\frac{5}{2}\right)\left(-\frac{1}{6}\right)$$
$$= \frac{7}{4} + \frac{5}{12}$$
$$= \frac{21}{12} + \frac{5}{12}$$
$$= \frac{26}{12}$$
$$= \frac{13}{6}$$

9. $\begin{cases} 2y - 4 = 0 \\ x + 2y = 5 \end{cases}$ or $\begin{cases} 0x + 2y = 4 \\ 1x + 2y = 5 \end{cases}$

$D = \begin{vmatrix} 0 & 2 \\ 1 & 2 \end{vmatrix} = 0 - 2 = -2$

$D_x = \begin{vmatrix} 4 & 2 \\ 5 & 2 \end{vmatrix} = 8 - 10 = -2$

$D_y = \begin{vmatrix} 0 & 4 \\ 1 & 5 \end{vmatrix} = 0 - 4 = -4$

$x = \dfrac{D_x}{D} = \dfrac{-2}{-2} = 1$

$y = \dfrac{D_y}{D} = \dfrac{-4}{-2} = 2$

The solution is (1, 2).

11. $\begin{cases} 3x + y = 1 \\ 2y = 2 - 6x \end{cases}$ or $\begin{cases} 3x + 1y = 1 \\ 6x + 2y = 2 \end{cases}$

$D = \begin{vmatrix} 3 & 1 \\ 6 & 2 \end{vmatrix} = 6 - 6 = 0$

Since $D = 0$, Cramer's Rule cannot be used.
Notice that equation (2) is equation (1)
multiplied by 2.
The solution is $\{(x, y) | 3x + y = 1\}$.

13. $\begin{cases} 5x - 2y = 27 \\ -3x + 5y = 18 \end{cases}$

$D = \begin{vmatrix} 5 & -2 \\ -3 & 5 \end{vmatrix} = 25 - 6 = 19$

$D_x = \begin{vmatrix} 27 & -2 \\ 18 & 5 \end{vmatrix} = 135 + 36 = 171$

$D_y = \begin{vmatrix} 5 & 27 \\ -3 & 18 \end{vmatrix} = 90 + 81 = 171$

$x = \dfrac{D_x}{D} = \dfrac{171}{19} = 9$ and $y = \dfrac{D_y}{D} = \dfrac{171}{19} = 9$

The solution is (9, 9).

15. $\begin{cases} 2x - 5y = 4 \\ x + 2y = -7 \end{cases}$

$D = \begin{vmatrix} 2 & -5 \\ 1 & 2 \end{vmatrix} = 4 + 5 = 9$

$D_x = \begin{vmatrix} 4 & -5 \\ -7 & 2 \end{vmatrix} = 8 - 35 = -27$

$D_y = \begin{vmatrix} 2 & 4 \\ 1 & -7 \end{vmatrix} = -14 - 4 = -18$

$x = \dfrac{D_x}{D} = \dfrac{-27}{9} = -3$ and $y = \dfrac{D_y}{D} = \dfrac{-18}{9} = -2$

The solution is (−3, −2).

17. $\begin{cases} \dfrac{2}{3}x - \dfrac{3}{4}y = -1 \\ -\dfrac{1}{6}x + \dfrac{3}{4}y = \dfrac{5}{2} \end{cases}$

$D = \begin{vmatrix} \frac{2}{3} & -\frac{3}{4} \\ -\frac{1}{6} & \frac{3}{4} \end{vmatrix} = \dfrac{1}{2} - \dfrac{1}{8} = \dfrac{3}{8}$

$D_x = \begin{vmatrix} -1 & -\frac{3}{4} \\ \frac{5}{2} & \frac{3}{4} \end{vmatrix} = -\dfrac{3}{4} + \dfrac{15}{8} = \dfrac{9}{8}$

$D_y = \begin{vmatrix} \frac{2}{3} & -1 \\ -\frac{1}{6} & \frac{5}{2} \end{vmatrix} = \dfrac{10}{6} - \dfrac{1}{6} = \dfrac{9}{6}$

$x = \dfrac{D_x}{D} = \dfrac{\frac{9}{8}}{\frac{3}{8}} = 3$ and $y = \dfrac{D_y}{D} = \dfrac{\frac{9}{6}}{\frac{3}{8}} = 4$

The solution is (3, 4).

19. Expand by first row.

$$\begin{vmatrix} 2 & 1 & 0 \\ 0 & 5 & -3 \\ 4 & 0 & 2 \end{vmatrix} = 2\begin{vmatrix} 5 & -3 \\ 0 & 2 \end{vmatrix} - 1\begin{vmatrix} 0 & -3 \\ 4 & 2 \end{vmatrix} + 0\begin{vmatrix} 0 & 5 \\ 4 & 0 \end{vmatrix}$$
$$= 2(10-0) - 1(0+12) + 0$$
$$= 20 - 12$$
$$= 8$$

21. Expand by third column.

$$\begin{vmatrix} 4 & -6 & 0 \\ -2 & 3 & 0 \\ 4 & -6 & 1 \end{vmatrix} = 0\begin{vmatrix} -2 & 3 \\ 4 & -6 \end{vmatrix} - 0\begin{vmatrix} 4 & -6 \\ 4 & -6 \end{vmatrix} + 1\begin{vmatrix} 4 & -6 \\ -2 & 3 \end{vmatrix}$$
$$= 0 - 0 + 1(12 - 12)$$
$$= 0$$

23. Expand by first row.

$$\begin{vmatrix} 1 & 0 & 4 \\ 1 & -1 & 2 \\ 3 & 2 & 1 \end{vmatrix} = 1\begin{vmatrix} -1 & 2 \\ 2 & 1 \end{vmatrix} - 0\begin{vmatrix} 1 & 2 \\ 3 & 1 \end{vmatrix} + 4\begin{vmatrix} 1 & -1 \\ 3 & 2 \end{vmatrix}$$
$$= 1(-1-4) - 0 + 4(2+3)$$
$$= -5 + 20$$
$$= 15$$

25. Expand by first row.

$$\begin{vmatrix} 3 & 6 & -3 \\ -1 & -2 & 3 \\ 4 & -1 & 6 \end{vmatrix} = 3\begin{vmatrix} -2 & 3 \\ -1 & 6 \end{vmatrix} - 6\begin{vmatrix} -1 & 3 \\ 4 & 6 \end{vmatrix} - 3\begin{vmatrix} -1 & -2 \\ 4 & -1 \end{vmatrix}$$
$$= 3(-12+3) - 6(-6-12) - 3(1+8)$$
$$= -27 + 108 - 27$$
$$= 54$$

27. $\begin{cases} 3x + z = -1 \\ -x - 3y + z = 7 \\ 3y + z = 5 \end{cases}$

$$D = \begin{vmatrix} 3 & 0 & 1 \\ -1 & -3 & 1 \\ 0 & 3 & 1 \end{vmatrix} = 3\begin{vmatrix} -3 & 1 \\ 3 & 1 \end{vmatrix} - 0\begin{vmatrix} -1 & 1 \\ 0 & 1 \end{vmatrix} + 1\begin{vmatrix} -1 & -3 \\ 0 & 3 \end{vmatrix}$$
$$= 3(-3-3) - 0 + 1(-3-0)$$
$$= -18 - 3$$
$$= -21$$

$$D_x = \begin{vmatrix} -1 & 0 & 1 \\ 7 & -3 & 1 \\ 5 & 3 & 1 \end{vmatrix}$$
$$= -1\begin{vmatrix} -3 & 1 \\ 3 & 1 \end{vmatrix} - 0\begin{vmatrix} 7 & 1 \\ 5 & 1 \end{vmatrix} + 1\begin{vmatrix} 7 & -3 \\ 5 & 3 \end{vmatrix}$$
$$= -1(-3-3) - 0 + 1|21+15|$$
$$= 6 + 36$$
$$= 42$$

$$D_y = \begin{vmatrix} 3 & -1 & 1 \\ -1 & 7 & 1 \\ 0 & 5 & 1 \end{vmatrix} = 3\begin{vmatrix} 7 & 1 \\ 5 & 1 \end{vmatrix} + 1\begin{vmatrix} -1 & 1 \\ 5 & 1 \end{vmatrix} + 0\begin{vmatrix} -1 & 1 \\ 7 & 1 \end{vmatrix}$$
$$= 3(7-5) + 1(-1-5) + 0$$
$$= 6 - 6$$
$$= 0$$

$$D_z = \begin{vmatrix} 3 & 0 & -1 \\ -1 & -3 & 7 \\ 0 & 3 & 5 \end{vmatrix}$$
$$= 3\begin{vmatrix} -3 & 7 \\ 3 & 5 \end{vmatrix} - 0\begin{vmatrix} -1 & 7 \\ 0 & 5 \end{vmatrix} - 1\begin{vmatrix} -1 & -3 \\ 0 & 3 \end{vmatrix}$$
$$= 3(-15-21) - 0 - 1(-3-0)$$
$$= -108 + 3$$
$$= -105$$

$$x = \frac{D_x}{D} = \frac{42}{-21} = -2, \quad y = \frac{D_y}{D} = \frac{0}{-21} = 0,$$
$$z = \frac{D_z}{D} = \frac{-105}{-21} = 5$$

The solution is $(-2, 0, 5)$.

29. $\begin{cases} x + y + z = 8 \\ 2x - y - z = 10 \\ x - 2y + 3z = 22 \end{cases}$

$$D = \begin{vmatrix} 1 & 1 & 1 \\ 2 & -1 & -1 \\ 1 & -2 & 3 \end{vmatrix}$$
$$= 1\begin{vmatrix} -1 & -1 \\ -2 & 3 \end{vmatrix} - 1\begin{vmatrix} 2 & -1 \\ 1 & 3 \end{vmatrix} + 1\begin{vmatrix} 2 & -1 \\ 1 & -2 \end{vmatrix}$$
$$= 1(-3-2) - 1(6+1) + 1(-4+1)$$
$$= -5 - 7 - 3$$
$$= -15$$

$$D_x = \begin{vmatrix} 8 & 1 & 1 \\ 10 & -1 & -1 \\ 22 & -2 & 3 \end{vmatrix}$$
$$= 8\begin{vmatrix} -1 & -1 \\ -2 & 3 \end{vmatrix} - 10\begin{vmatrix} 1 & 1 \\ -2 & 3 \end{vmatrix} + 22\begin{vmatrix} 1 & 1 \\ -1 & -1 \end{vmatrix}$$
$$= 8(-3-2) - 10(3+2) + 22(-1+1)$$
$$= -40 - 50 + 0$$
$$= -90$$

$$D_y = \begin{vmatrix} 1 & 8 & 1 \\ 2 & 10 & -1 \\ 1 & 22 & 3 \end{vmatrix}$$
$$= 1\begin{vmatrix} 10 & -1 \\ 22 & 3 \end{vmatrix} - 8\begin{vmatrix} 2 & -1 \\ 1 & 3 \end{vmatrix} + 1\begin{vmatrix} 2 & 10 \\ 1 & 22 \end{vmatrix}$$
$$= 1(30+22) - 8(6+1) + 1(44-10)$$
$$= 52 - 56 + 34$$
$$= 30$$

$$D_z = \begin{vmatrix} 1 & 1 & 8 \\ 2 & -1 & 10 \\ 1 & -2 & 22 \end{vmatrix}$$

$$= 1\begin{vmatrix} -1 & 10 \\ -2 & 22 \end{vmatrix} - 1\begin{vmatrix} 2 & 10 \\ 1 & 22 \end{vmatrix} + 8\begin{vmatrix} 2 & -1 \\ 1 & -2 \end{vmatrix}$$

$$= 1(-22+20) - 1(44-10) + 8(-4+1)$$

$$= -2 - 34 - 24$$

$$= -60$$

$$x = \frac{D_x}{D} = \frac{-90}{-15} = 6, \quad y = \frac{D_y}{D} = \frac{30}{-15} = -2,$$

$$z = \frac{D_z}{D} = \frac{-60}{-15} = 4$$

The solution is $(6, -2, 4)$.

31. $\begin{cases} 2x + 2y + z = 1 \\ -x + y + 2z = 3 \\ x + 2y + 4z = 0 \end{cases}$

$$D = \begin{vmatrix} 2 & 2 & 1 \\ -1 & 1 & 2 \\ 1 & 2 & 4 \end{vmatrix} = 2\begin{vmatrix} 1 & 2 \\ 2 & 4 \end{vmatrix} - 2\begin{vmatrix} -1 & 2 \\ 1 & 4 \end{vmatrix} + 1\begin{vmatrix} -1 & 1 \\ 1 & 2 \end{vmatrix}$$

$$= 2(4-4) - 2(-4-2) + 1(-2-1)$$

$$= 0 + 12 - 3$$

$$= 9$$

$$D_x = \begin{vmatrix} 1 & 2 & 1 \\ 3 & 1 & 2 \\ 0 & 2 & 4 \end{vmatrix} = 1\begin{vmatrix} 1 & 2 \\ 2 & 4 \end{vmatrix} - 3\begin{vmatrix} 2 & 1 \\ 2 & 4 \end{vmatrix} + 0$$

$$= 1(4-4) - 3(8-2)$$

$$= -18$$

$$D_y = \begin{vmatrix} 2 & 1 & 1 \\ -1 & 3 & 2 \\ 1 & 0 & 4 \end{vmatrix} = 2\begin{vmatrix} 3 & 2 \\ 0 & 4 \end{vmatrix} - 1\begin{vmatrix} -1 & 2 \\ 1 & 4 \end{vmatrix} + 1\begin{vmatrix} -1 & 3 \\ 1 & 0 \end{vmatrix}$$

$$= 2(12-0) - 1(-4-2) + 1(0-3)$$

$$= 24 + 6 - 3$$

$$= 27$$

$$D_z = \begin{vmatrix} 2 & 2 & 1 \\ -1 & 1 & 3 \\ 1 & 2 & 0 \end{vmatrix} = 2\begin{vmatrix} 1 & 3 \\ 2 & 0 \end{vmatrix} - 2\begin{vmatrix} -1 & 3 \\ 1 & 0 \end{vmatrix} + 1\begin{vmatrix} -1 & 1 \\ 1 & 2 \end{vmatrix}$$

$$= 2(0-6) - 2(0-3) + 1(-2-1)$$

$$= -12 + 6 - 3$$

$$= -9$$

$$x = \frac{D_x}{D} = \frac{-18}{9} = -2, \quad y = \frac{D_y}{D} = \frac{27}{9} = 3,$$

$$z = \frac{D_z}{D} = \frac{-9}{9} = -1$$

The solution is $(-2, 3, -1)$.

33. $\begin{cases} x - 2y + z = -5 \\ 3y + 2z = 4 \\ 3x - y = -2 \end{cases}$

$$D = \begin{vmatrix} 1 & -2 & 1 \\ 0 & 3 & 2 \\ 3 & -1 & 0 \end{vmatrix} = 1\begin{vmatrix} 3 & 2 \\ -1 & 0 \end{vmatrix} + 2\begin{vmatrix} 0 & 2 \\ 3 & 0 \end{vmatrix} + 1\begin{vmatrix} 0 & 3 \\ 3 & -1 \end{vmatrix}$$

$$= 1(0+2) + 2(0-6) + 1(0-9)$$

$$= 2 - 12 - 9$$

$$= -19$$

$$D_x = \begin{vmatrix} -5 & -2 & 1 \\ 4 & 3 & 2 \\ -2 & -1 & 0 \end{vmatrix}$$

$$= -5\begin{vmatrix} 3 & 2 \\ -1 & 0 \end{vmatrix} + 2\begin{vmatrix} 4 & 2 \\ -2 & 0 \end{vmatrix} + 1\begin{vmatrix} 4 & 3 \\ -2 & -1 \end{vmatrix}$$

$$= -5(0+2) + 2(0+4) + 1(-4+6)$$

$$= -10 + 8 + 2$$

$$= 0$$

$$D_y = \begin{vmatrix} 1 & -5 & 1 \\ 0 & 4 & 2 \\ 3 & -2 & 0 \end{vmatrix} = 1\begin{vmatrix} 4 & 2 \\ -2 & 0 \end{vmatrix} + 5\begin{vmatrix} 0 & 2 \\ 3 & 0 \end{vmatrix} + 1\begin{vmatrix} 0 & 4 \\ 3 & -2 \end{vmatrix}$$

$$= 1(0+4) + 5(0-6) + 1(0-12)$$

$$= 4 - 30 - 12$$

$$= -38$$

$$D_z = \begin{vmatrix} 1 & -2 & -5 \\ 0 & 3 & 4 \\ 3 & -1 & -2 \end{vmatrix}$$

$$= 1\begin{vmatrix} 3 & 4 \\ -1 & -2 \end{vmatrix} + 2\begin{vmatrix} 0 & 4 \\ 3 & -2 \end{vmatrix} - 5\begin{vmatrix} 0 & 3 \\ 3 & -1 \end{vmatrix}$$

$$= 1(-6+4) + 2(0-12) - 5(0-9)$$

$$= -2 - 24 + 45$$

$$= 19$$

$$x = \frac{D_x}{D} = \frac{0}{-19} = 0, \quad y = \frac{D_y}{D} = \frac{-38}{-19} = 2,$$

$$z = \frac{D_z}{D} = \frac{19}{-19} = -1$$

The solution is $(0, 2, -1)$.

35. $\begin{vmatrix} 1 & x \\ 2 & 7 \end{vmatrix} = -3$

$$(1)(7) - x \cdot 2 = -3$$

$$7 - 2x = -3$$

$$-2x = -10$$

$$x = 5$$

37. 0; answers may vary

       647

Appendix D

Viewing Window and Interpreting Window Settings Exercise Set

1. Yes, since every coordinate is between −10 and 10.

3. No, since −11 is less than −10.

5. answers may vary. Any values such that Xmin < −90, Ymin < −80, Xmax > 55, and Ymax > 80.

7. answers may vary. Any values such that Xmin < −11, Ymin < −5, Xmax > 7, and Ymax > 2.

9. answers may vary. Any values such that Xmin < 50, Ymin < −50, Xmax > 200, and Ymax > 200.

11. Xmin = −12   Ymin = −12
     Xmax = 12   Ymax = 12
      Xscl = 3   Yscl = 3

13. Xmin = −9   Ymin = −12
     Xmax = 9   Ymax = 12
      Xscl = 1   Yscl = 2

15. Xmin = −10   Ymin = −25
     Xmax = 10   Ymax = 25
      Xscl = 2   Yscl = 5

17. Xmin = −10   Ymin = −30
     Xmax = 10   Ymax = 30
      Xscl = 1   Yscl = 3

19. Xmin = −20   Ymin = −30
     Xmax = 30   Ymax = 50
      Xscl = 5   Yscl = 10

Graphing Equations and Square Viewing Window Exercise Set

1. Setting A:

   Setting B:

   Setting B shows all intercepts.

3. Setting A:

   Setting B:

   Setting B shows all intercepts.

5. Setting A:

   Setting B:

   Setting B shows all intercepts.

7. $3x = 5y$

   $y = \dfrac{3}{5}x$

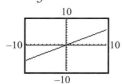

9. $9x - 5y = 30$
$-5y = -9x + 30$
$y = \dfrac{9}{5}x - 6$

11. $y = -7$

13. $x + 10y = -5$
$10y = -x - 5$
$y = -\dfrac{1}{10}x - \dfrac{1}{2}$

15. $y = \sqrt{x}$

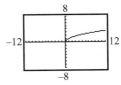

17. $y = x^2 + 2x + 1$

19. $y = |x|$

21. $x + 2y = 30$
$2y = -x + 30$
$y = -\dfrac{1}{2}x + 15$

Standard window:

Adjusted window:

Practice Final Exam

1. $\sqrt{216} = \sqrt{36 \cdot 6} = \sqrt{36} \cdot \sqrt{6} = 6\sqrt{6}$

2. $\dfrac{(4-\sqrt{16}) - (-7-20)}{-2(1-4)^2} = \dfrac{(4-4) - (-27)}{-2(-3)^2}$

$\qquad\qquad = \dfrac{0+27}{-2(9)}$

$\qquad\qquad = \dfrac{27}{-18}$

$\qquad\qquad = -\dfrac{3}{2}$

3. $\left(\dfrac{1}{125}\right)^{-1/3} = 125^{1/3} = \sqrt[3]{125} = \sqrt[3]{5^3} = 5$

4. $(-9x)^{-2} = \dfrac{1}{(-9x)^2} = \dfrac{1}{(-9)^2 x^2} = \dfrac{1}{81x^2}$

5. $\dfrac{\frac{5}{x} - \frac{7}{3x}}{\frac{9}{8x} - \frac{1}{x}} = \dfrac{24x\left(\frac{5}{x} - \frac{7}{3x}\right)}{24x\left(\frac{9}{8x} - \frac{1}{x}\right)}$

$\qquad\qquad = \dfrac{24x\left(\frac{5}{x}\right) - 24x\left(\frac{7}{3x}\right)}{24x\left(\frac{9}{8x}\right) - 24x\left(\frac{1}{x}\right)}$

$\qquad\qquad = \dfrac{24 \cdot 5 - 8 \cdot 7}{3 \cdot 9 - 24}$

$\qquad\qquad = \dfrac{120 - 56}{27 - 24}$

$\qquad\qquad = \dfrac{64}{3}$

6. $\left(\dfrac{64c^{4/3}}{a^{-2/3}b^{5/6}}\right)^{1/2} = \dfrac{64^{1/2}\, c^{\frac{4}{3} \cdot \frac{1}{2}}}{a^{-\frac{2}{3} \cdot \frac{1}{2}}b^{\frac{5}{6} \cdot \frac{1}{2}}}$

$\qquad\qquad = \dfrac{8c^{2/3}}{a^{-1/3}b^{5/12}}$

$\qquad\qquad = \dfrac{8a^{1/3}c^{2/3}}{b^{5/12}}$

7. a.

b.

The regression equation is
$y = 295.07x + 6800.4$.

c.

The population of Florida is predicted to be 21,553.9 thousand in 2020.

8. $3x^2y - 27y^3 = 3y(x^2 - 9y^2)$

$\qquad\qquad = 3y[x^2 - (3y)^2]$

$\qquad\qquad = 3y(x + 3y)(x - 3y)$

9. $16y^3 - 2 = 2(8y^3 - 1)$

$\qquad\qquad = 2[(2y)^3 - 1^3]$

$\qquad\qquad = 2(2y - 1)[(2y)^2 + 2y \cdot 1 + 1^2]$

$\qquad\qquad = 2(2y - 1)(4y^2 + 2y + 1)$

10. $x^2y - 9y - 3x^2 + 27 = y(x^2 - 9) - 3(x^2 - 9)$

$\qquad\qquad = (x^2 - 9)(y - 3)$

$\qquad\qquad = (x + 3)(x - 3)(y - 3)$

11. $(4x^3y - 3x - 4) - (9x^3y + 8x + 5)$

$\qquad = 4x^3y - 3x - 4 - 9x^3y - 8x - 5$

$\qquad = 4x^3y - 9x^3y - 3x - 8x - 4 - 5$

$\qquad = -5x^3y - 11x - 9$

12. $(6m + n)^2 = (6m)^2 + 2(6m)(n) + n^2$

$\qquad\qquad = 36m^2 + 12mn + n^2$

13. $(2x - 1)(x^2 - 6x + 4)$

$\qquad = 2x(x^2 - 6x + 4) - 1(x^2 - 6x + 4)$

$\qquad = 2x^3 - 12x^2 + 8x - x^2 + 6x - 4$

$\qquad = 2x^3 - 13x^2 + 14x - 4$

14. $\dfrac{3x^2-12}{x^2+2x-8} \div \dfrac{6x+18}{x+4} = \dfrac{3x^2-12}{x^2+2x-8} \cdot \dfrac{x+4}{6x+18}$

$\qquad = \dfrac{3(x^2-4)}{(x+4)(x-2)} \cdot \dfrac{x+4}{6(x+3)}$

$\qquad = \dfrac{3(x+2)(x-2)(x+4)}{(x+4)(x-2)\cdot 6(x+3)}$

$\qquad = \dfrac{x+2}{2(x+3)}$

15. $\dfrac{2x^2+7}{2x^4-18x^2} - \dfrac{6x+7}{2x^4-18x^2} = \dfrac{(2x^2+7)-(6x+7)}{2x^4-18x^2}$

$\qquad = \dfrac{2x^2+7-6x-7}{2x^4-18x^2}$

$\qquad = \dfrac{2x^2-6x}{2x^2(x^2-9)}$

$\qquad = \dfrac{2x(x-3)}{2x^2(x+3)(x-3)}$

$\qquad = \dfrac{1}{x(x+3)}$

16. $\dfrac{3}{x^2-x-6} + \dfrac{2}{x^2-5x+6}$

$\qquad = \dfrac{3}{(x-3)(x+2)} + \dfrac{2}{(x-2)(x-3)}$

$\qquad = \dfrac{3(x-2)}{(x-3)(x+2)(x-2)} + \dfrac{2(x+2)}{(x-2)(x-3)(x+2)}$

$\qquad = \dfrac{3(x-2)+2(x+2)}{(x-3)(x+2)(x-2)}$

$\qquad = \dfrac{3x-6+2x+4}{(x-3)(x+2)(x-2)}$

$\qquad = \dfrac{5x-2}{(x-3)(x+2)(x-2)}$

17. $\sqrt{125x^3} - 3\sqrt{20x^3} = \sqrt{25x^2\cdot 5x} - 3\sqrt{4x^2\cdot 5x}$

$\qquad = 5x\sqrt{5x} - 3\cdot 2x\sqrt{5x}$

$\qquad = 5x\sqrt{5x} - 6x\sqrt{5x}$

$\qquad = (5x-6x)\sqrt{5x}$

$\qquad = -x\sqrt{5x}$

18. $\left(\sqrt{5}+5\right)\left(\sqrt{5}-5\right) = \left(\sqrt{5}\right)^2 - 5^2 = 5-25 = -20$

19. $2x+1\overline{\smash{\big)}\,4x^3+0x^2-5x+0}$ with quotient $2x^2-x-2$

$\qquad \underline{4x^3+2x^2}$

$\qquad\quad -2x^2-5x$

$\qquad\quad \underline{-2x^2\ -x}$

$\qquad\qquad\quad -4x+0$

$\qquad\qquad\quad \underline{-4x-2}$

$\qquad\qquad\qquad\quad 2$

$(4x^3-5x)\div(2x+1) = 2x^2-x-2+\dfrac{2}{2x+1}$

20. Algebraic Solution:

$15x+26 = -2(x+1)-1$

$15x+26 = -2x-2-1$

$15x+26 = -2x-3$

$17x+26 = -3$

$17x = -29$

$x = -\dfrac{29}{17} \approx -1.70588$

Graphical Solution:

Graph $y_1 = 15x+26$ and $y_2 = -2(x+1)-1$.

The solution is $-\dfrac{29}{17}$.

21. $|6x-5|-3 = -2$

$\quad |6x-5| = 1$

$6x-5 = -1 \quad$ or $\quad 6x-5 = 1$

$6x = 4 \qquad$ or $\qquad 6x = 6$

$x = \dfrac{4}{6} \qquad$ or $\qquad x = 1$

$x = \dfrac{2}{3}$

Both solutions check.

22.
$$3n(7n - 20) = 96$$
$$21n^2 - 60n = 96$$
$$21n^2 - 60n - 96 = 0$$
$$3(7n^2 - 20n - 32) = 0$$
$$3(7n + 8)(n - 4) = 0$$
$$7n + 8 = 0 \quad \text{or} \quad n - 4 = 0$$
$$n = -\frac{8}{7} \quad \text{or} \quad n = 4$$
Both solutions check.

23.
$$-3 < 2(x - 3) \le 4$$
$$-3 < 2x - 6 \le 4$$
$$-3 + 6 < 2x - 6 + 6 \le 4 + 6$$
$$3 < 2x \le 10$$
$$\frac{3}{2} < \frac{2x}{2} \le \frac{10}{2}$$
$$\frac{3}{2} < x \le 5$$
$$\left(\frac{3}{2}, 5\right]$$

24. $|3x + 1| > 5$
$$3x + 1 < -5 \quad \text{or} \quad 3x + 1 > 5$$
$$3x < -6 \qquad\qquad 3x > 4$$
$$x < -2 \qquad\qquad x > \frac{4}{3}$$
$$(-\infty, -2) \cup \left(\frac{4}{3}, \infty\right)$$

25.
$$\frac{x^2 + 8}{x} - 1 = \frac{2(x + 4)}{x}$$
$$x\left(\frac{x^2 + 8}{x} - 1\right) = x\left(\frac{2(x + 4)}{x}\right)$$
$$x^2 + 8 - x = 2(x + 4)$$
$$x^2 - x + 8 = 2x + 8$$
$$x^2 - 3x = 0$$
$$x(x - 3) = 0$$
$$x = 0 \quad \text{or} \quad x - 3 = 0$$
$$x = 3$$
The only solution is 3.

26.
$$y^2 - 3y = 5$$
$$y^2 - 3y - 5 = 0$$
$$y = \frac{-(-3) \pm \sqrt{(-3)^2 - 4(1)(-5)}}{2(1)}$$
$$y = \frac{3 \pm \sqrt{9 + 20}}{2}$$
$$y = \frac{3 \pm \sqrt{29}}{2}$$

27.
$$x = \sqrt{x - 2} + 2$$
$$x - 2 = \sqrt{x - 2}$$
$$(x - 2)^2 = \left(\sqrt{x - 2}\right)^2$$
$$x^2 - 4x + 4 = x - 2$$
$$x^2 - 5x + 6 = 0$$
$$(x - 2)(x - 3) = 0$$
$$x - 2 = 0 \quad \text{or} \quad x - 3 = 0$$
$$x = 2 \quad \text{or} \qquad x = 3$$

28.
$$2x^2 - 7x > 15$$
$$2x^2 - 7x - 15 > 0$$
$$(2x + 3)(x - 5) > 0$$
$$2x + 3 = 0 \quad \text{or} \quad x - 5 = 0$$
$$x = -\frac{3}{2} \quad \text{or} \qquad x = 5$$

Region	Test Point	$2x^2 - 7x > 15$ Result
$\left(-\infty, -\frac{3}{2}\right)$	-3	$39 > 15$ True
$\left(-\frac{3}{2}, 5\right)$	0	$0 > 15$ False
$(5, \infty)$	6	$30 > 15$ True

$$\left(-\infty, -\frac{3}{2}\right) \cup (5, \infty)$$

29. $\begin{cases} \dfrac{x}{2}+\dfrac{y}{4}=-\dfrac{3}{4} \\ x+\dfrac{3}{4}y=-4 \end{cases}$

$\begin{cases} -8\left(\dfrac{x}{2}+\dfrac{y}{4}\right)=-8\left(-\dfrac{3}{4}\right) \\ 4\left(x+\dfrac{3}{4}y\right)=4(-4) \end{cases}$

$\begin{array}{r} -4x-2y=6 \\ 4x+3y=-16 \\ \hline y=-10 \end{array}$

Let $y=-10$ in the second equation.

$x+\dfrac{3}{4}y=-4$

$x+\dfrac{3}{4}(-10)=-4$

$x-\dfrac{15}{2}=-\dfrac{8}{2}$

$x=\dfrac{7}{2}$

The solution is $\left(\dfrac{7}{2},-10\right)$.

30. $4x+6y=7$

$6y=-4x+7$

$y=-\dfrac{2}{3}x+\dfrac{7}{6}$

slope $=-\dfrac{2}{3}$, y-intercept $\left(0,\dfrac{7}{6}\right)$

Plot points: $\left(0,\dfrac{7}{6}\right),\left(\dfrac{7}{4},0\right),\left(-3,\dfrac{19}{6}\right)$

31. $2x-y>5$

Graph $2x-y=5$ with a dashed line because the inequality is $>$.

x	$2x-y=5$	y
0	$2(0)-y=5$	-5
1	$2(1)-y=5$	-3
3	$2(3)-y=5$	1

Test point $(0,0)$: $2(0)-0 \overset{?}{>} 5$

$0>5$ False

Shade the region that does not include $(0,0)$.

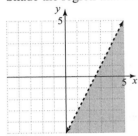

32. $y=-3$

The graph of $y=-3$ is a horizontal line with a y-intercept of $(0,-3)$.

33. $g(x)=-|x+2|-1$

x	$g(x)=-|x+2|-1$	$g(x)$
-5	$-|-5+2|-1=-4$	-4
-4	$-|-4+2|-1=-3$	-3
-3	$-|-3+2|-1=-2$	-2
-2	$-|-2+2|-1=-1$	-1
-1	$-|-1+2|-1=-2$	-2
0	$-|0+2|-1=-3$	-3
1	$-|1+2|-1=-4$	-4

Domain: All real numbers, $(-\infty,\infty)$

Range: $(-\infty, -1]$

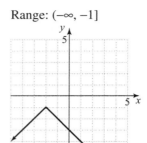

34. $h(x) = x^2 - 4x + 4$

x-intercept: Let $h(x) = 0$ and solve for x.

$$0 = x^2 - 4x + 4$$
$$0 = (x-2)^2$$
$$x - 2 = 0$$
$$x = 2$$

x-intercept: $(2, 0)$

y-intercept: Let $x = 0$.

$$h(0) = 0^2 - 4(0) + 4 = 4$$

y-intercept: $(0, 4)$

x-coordinate of vertex:

$$-\frac{b}{2a} = -\frac{-4}{2(1)} = 2$$

vertex: $(2, 0)$

35. $f(x) = \begin{cases} -\dfrac{1}{2}x & \text{if} \quad x \le 0 \\ 2x - 3 & \text{if} \quad x > 0 \end{cases}$

If $x \le 0$

x	$-\frac{1}{2}x$	$f(x)$
0	$-\frac{1}{2}(0)$	0
-2	$-\frac{1}{2}(-2)$	1
-4	$-\frac{1}{2}(-4)$	2

If $x > 0$

x	$2x - 3$	$f(x)$
1	$2(1) - 3$	-1
2	$2(2) - 3$	1
3	$2(3) - 3$	3

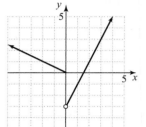

Domain: $(-\infty, \infty)$
Range: $(-3, \infty)$

36. through $(4, -2)$ and $(6, -3)$

$$\text{slope} = m = \frac{y_2 - y_1}{x_2 - x_1} = \frac{-3 - (-2)}{6 - 4} = \frac{-1}{2}$$

$$y - y_1 = m(x - x_1)$$
$$y - (-2) = -\frac{1}{2}(x - 4)$$
$$y + 2 = -\frac{1}{2}x + 2$$
$$y = -\frac{1}{2}x$$
$$f(x) = -\frac{1}{2}x$$

37. through $(-1, 2)$ and perpendicular to $3x - y = 4$
Find the slope of $3x - y = 4$ by writing the equation in slope-intercept form.

$$3x - y = 4$$
$$-y = -3x + 4$$
$$y = 3x - 4$$

The slope is 3. The slope of a line perpendicular to this line is $-\frac{1}{3}$.

Substitute $m = -\frac{1}{3}$ and $(x_1, y_1) = (-1, 2)$ in the equation:

Copyright © 2014 Pearson Education, Inc.

$$y - y_1 = m(x - x_1)$$
$$y - 2 = -\frac{1}{3}[x - (-1)]$$
$$y - 2 = -\frac{1}{3}(x + 1)$$
$$y - 2 = -\frac{1}{3}x - \frac{1}{3}$$
$$y = -\frac{1}{3}x + \frac{5}{3}$$
$$f(x) = -\frac{1}{3}x + \frac{5}{3}$$

38. $(x_1, y_1) = (-6, 3); \ (x_2, y_2) = (-8, -7)$

$$d = \sqrt{(x_2 - x_1)^2 + (y_2 - y_1)^2}$$
$$= \sqrt{[-8 - (-6)]^2 + (-7 - 3)^2}$$
$$= \sqrt{(-2)^2 + (-10)^2}$$
$$= \sqrt{4 + 100}$$
$$= \sqrt{104}$$
$$= 2\sqrt{26} \text{ units}$$

39. $(x_1, y_1) = (-2, -5); (x_2, y_2) = (-6, 12)$

$$\text{midpoint} = \left(\frac{x_1 + x_2}{2}, \frac{y_1 + y_2}{2}\right)$$
$$= \left(\frac{-2 + (-6)}{2}, \frac{-5 + 12}{2}\right)$$
$$= \left(\frac{-8}{2}, \frac{7}{2}\right)$$
$$= \left(-4, \frac{7}{2}\right)$$

40. $\sqrt{\dfrac{9}{y}} = \dfrac{\sqrt{9}}{\sqrt{y}} = \dfrac{\sqrt{9}}{\sqrt{y}} \cdot \dfrac{\sqrt{y}}{\sqrt{y}} = \dfrac{\sqrt{9} \cdot \sqrt{y}}{\sqrt{y} \cdot \sqrt{y}} = \dfrac{3\sqrt{y}}{y}$

41.
$$\frac{4 - \sqrt{x}}{4 + 2\sqrt{x}} = \frac{4 - \sqrt{x}}{4 + 2\sqrt{x}} \cdot \frac{4 - 2\sqrt{x}}{4 - 2\sqrt{x}}$$
$$= \frac{\left(4 - \sqrt{x}\right)\left(4 - 2\sqrt{x}\right)}{\left(4 + 2\sqrt{x}\right)\left(4 - 2\sqrt{x}\right)}$$
$$= \frac{16 - 12\sqrt{x} + 2x}{16 - 4x}$$
$$= \frac{2\left(8 - 6\sqrt{x} + x\right)}{2(8 - 2x)}$$
$$= \frac{8 - 6\sqrt{x} + x}{8 - 2x}$$

42. The graph of y_1 is below the graph of y_2 for x-values less than -3. The solution set for $y_1 < y_2$ is $(-\infty, -3)$.

43. Let x be the amount of money international travelers spend in New York. Then $x + 4$ is the amount of money international travelers spend in California and $2x - 1$ is the amount of money international travelers spend in Florida.
$$x + (x + 4) + (2x - 1) = 39$$
$$4x + 3 = 39$$
$$4x = 36$$
$$x = 9$$
$x + 4 = 9 + 4 = 13$
$2x - 1 = 2(9) - 1 = 18 - 1 = 17$
International travelers spend \$9 billion in New York, \$13 billion in California, and \$17 billion in Florida.

44. Let $x =$ the number.
$$(x + 1) \cdot \left(2 \cdot \frac{1}{x}\right) = \frac{12}{5}$$
$$\frac{2}{x}(x + 1) = \frac{12}{5}$$
$$5x\left[\frac{2}{x}(x + 1)\right] = 5x\left(\frac{12}{5}\right)$$
$$10(x + 1) = x \cdot 12$$
$$10x + 10 = 12x$$
$$10 = 2x$$
$$5 = x$$
The number is 5.

45. a. Let $x =$ the number of years after 1990.

The data appear to be quadratic.

b.

The quadratic regression equation is
$$y = 0.016x^2 - 0.417x + 16.581.$$

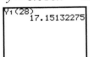

The birth rate in 2018 is predicted to be 17.15 births per 1000 people.

46. Use the Pythagorean Theorem.

$$c^2 = a^2 + b^2$$
$$20^2 = x^2 + (x+8)^2$$
$$400 = x^2 + x^2 + 16x + 64$$
$$0 = 2x^2 + 16x - 336$$
$$0 = 2(x^2 + 8x - 168)$$

$$x = \frac{-8 \pm \sqrt{8^2 - 4(1)(-168)}}{2(1)}$$

$$x = \frac{-8 \pm \sqrt{736}}{2}$$

$x \approx -17.6$ or $x \approx 9.6$
Discard a negative distance.
$x + 8 + x = 9.6 + 8 + 9.6 = 27.2$
$27.2 - 20 = 7.2$ or about 7
A person saves about 7 feet.

47. a. Find the vertex.

$$s(t) = -16t^2 + 32t + 256$$

t-value: $\dfrac{-b}{2a} = \dfrac{-32}{2(-16)} = 1$

$s(t)$-value:

$$s(1) = -16(1)^2 + 32(1) + 256 = 272$$

The maximum height is 272 feet.

b. Let $s(t) = 0$ and solve for t.

$$0 = -16t^2 + 32t + 256$$
$$0 = -16(t^2 - 2t - 16)$$

$$t = \frac{-(-2) \pm \sqrt{(-2)^2 - 4(1)(-16)}}{2(1)}$$

$$t = \frac{2 \pm \sqrt{68}}{2}$$

$$t = \frac{2 \pm 2\sqrt{17}}{2}$$

$$t = 1 \pm \sqrt{17}$$

$t \approx -3.12$ or $t \approx 5.12$
Discard a negative time.
The stone will hit the water in
approximately 5.12 seconds.

48. Let x = amount of 10% solution to add to
mixture.

solution	amount of solution	amount of fructose
10%	x	$0.10x$
20%	$20 - x$	$0.20(20 - x)$
17.5%	20	$0.175(20)$

$$0.10x + 0.20(20 - x) = 0.175(20)$$
$$0.10x + 4 - 0.20x = 3.5$$
$$-0.10x = -0.5$$
$$x = 5$$

$20 - x = 15$
Therefore, mix 5 gallons of 10% solution with
15 gallons of 20% solution.

49. $-\sqrt{-8} = -\sqrt{4 \cdot (-1) \cdot 2} = -\sqrt{4} \cdot \sqrt{-1} \cdot \sqrt{2} = -2i\sqrt{2}$

50. $(12 - 6i) - (12 - 3i) = 12 - 6i - 12 + 3i$
$ = 12 - 12 - 6i + 3i$
$ = 0 - 3i$
$ = -3i$

51. $(4 + 3i)^2 = (4 + 3i)(4 + 3i)$
$ = 16 + 12i + 12i + 9i^2$
$ = 16 + 24i - 9$
$ = 7 + 24i$

52. $\dfrac{1+4i}{1-i} = \dfrac{1+4i}{1-i} \cdot \dfrac{1+i}{1+i}$

$\phantom{\dfrac{1+4i}{1-i}} = \dfrac{(1+4i)(1+i)}{(1-i)(1+i)}$

$\phantom{\dfrac{1+4i}{1-i}} = \dfrac{1 + 5i + 4i^2}{1 - i^2}$

$\phantom{\dfrac{1+4i}{1-i}} = \dfrac{1 + 5i - 4}{1 - (-1)}$

$\phantom{\dfrac{1+4i}{1-i}} = \dfrac{-3 + 5i}{2}$

$\phantom{\dfrac{1+4i}{1-i}} = -\dfrac{3}{2} + \dfrac{5}{2}i$

53. $g(x) = x - 7$ and $h(x) = x^2 - 6x + 5$

$(g \circ h)(x) = (x^2 - 6x + 5) - 7 = x^2 - 6x - 2$

54. $f(x) = 6 - 2x$ is a one-to-one function since there
is only one $f(x)$ value for each x-value.

Inverse:

$y = 6 - 2x \qquad \Rightarrow \qquad x = 6 - 2y$
$$x + 2y = 6$$
$$2y = -x + 6$$
$$y = \frac{-x + 6}{2}$$
$$f^{-1}(x) = \frac{-x + 6}{2}$$

55. $\log_5 x + 3\log_5 x - \log_5 (x+1)$

$= \log_5 x + \log_5 x^3 - \log_5 (x+1)$

$= \log_5 x \cdot x^3 - \log_5 (x+1)$

$= \log_5 x^4 - \log_5 (x+1)$

$= \log_5 \dfrac{x^4}{x+1}$

56. $8^{x-1} = \dfrac{1}{64}$

$(2^3)^{x-1} = \dfrac{1}{2^6}$

$2^{3(x-1)} = 2^{-6}$

$3(x-1) = -6$

$3x - 3 = -6$

$3x = -3$

$x = -1$

57. $3^{2x+5} = 4$

$\log 3^{2x+5} = \log 4$

$(2x+5)\log 3 = \log 4$

$2x+5 = \dfrac{\log 4}{\log 3}$

$2x = \dfrac{\log 4}{\log 3} - 5$

$x = \dfrac{1}{2}\left(\dfrac{\log 4}{\log 3} - 5\right)$

$x \approx -1.8691$

58. $\log_8 (3x-2) = 2$

$8^2 = 3x - 2$

$64 = 3x - 2$

$66 = 3x$

$22 = x$

59. $\log_4 (x+1) - \log_4 (x-2) = 3$

$\log_4 \dfrac{x+1}{x-2} = 3$

$4^3 = \dfrac{x+1}{x-2}$

$64 = \dfrac{x+1}{x-2}$

$64(x-2) = x+1$

$64x - 128 = x + 1$

$63x = 129$

$x = \dfrac{129}{63} = \dfrac{43}{21}$

60. Graph $y_1 = e^{0.2x}$ and $y_2 = e^{-0.4x} + 2$.

The solution is approximately 3.95.

61. $y = \left(\dfrac{1}{2}\right)^x + 1$

x	$\left(\dfrac{1}{2}\right)^x + 1$	y
-3	$\left(\dfrac{1}{2}\right)^{-3} + 1 = 9$	9
-2	$\left(\dfrac{1}{2}\right)^{-2} + 1 = 5$	5
-1	$\left(\dfrac{1}{2}\right)^{-1} + 1 = 3$	3
0	$\left(\dfrac{1}{2}\right)^{0} + 1 = 2$	2
1	$\left(\dfrac{1}{2}\right)^{1} + 1 = 1\frac{1}{2}$	$1\frac{1}{2}$
2	$\left(\dfrac{1}{2}\right)^{2} + 1 = 1\frac{1}{4}$	$1\frac{1}{4}$
3	$\left(\dfrac{1}{2}\right)^{3} + 1 = 1\frac{1}{8}$	$1\frac{1}{8}$

62. $C = 57{,}000$, $r = 0.026$, $x = 5$

$y = C(1+r)^x$

$y = 57{,}000(1+0.026)^5$

$y \approx 64{,}805$

In 5 years, the population is predicted to be 64,805 prairie dogs.

63. $x^2 - y^2 = 36$
$x^2 - y^2 = 6^2$

hyperbola, with x-intercepts $(-6, 0)$, $(6, 0)$

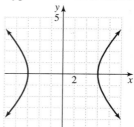

64. $16x^2 + 9y^2 = 144$
$$\frac{16x^2}{144} + \frac{9y^2}{144} = \frac{144}{144}$$
$$\frac{x^2}{9} + \frac{y^2}{16} = 1$$

Ellipse, x-intercepts $(-3, 0)$, $(3, 0)$
        y-intercepts $(0, -4)$, $(0, 4)$

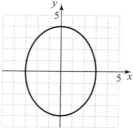

65. $\qquad x^2 + y^2 + 6x = 16$
$(x^2 + 6x + 9) + y^2 = 16 + 9$
$\qquad (x + 3)^2 + y^2 = 25$
$[x - (-3)]^2 + (y - 0)^2 = 5^2$

circle with center $(-3, 0)$ and radius 5

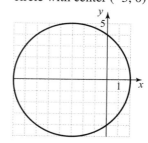

66. $\begin{cases} x^2 + y^2 = 26 \\ x^2 - 2y^2 = 23 \end{cases}$

Multiply equation (2) by -1 and add the equations.

$\qquad x^2 + y^2 = 26$
$\underline{-x^2 + 2y^2 = -23}$
$\qquad\qquad 3y^2 = 3$
$\qquad\qquad\; y^2 = 1$
$\qquad\qquad\;\; y = \pm 1$

Substitute $y = -1$ and $y = 1$ into equation (1).

$x^2 + (-1)^2 = 26$
$\qquad\quad x^2 = 25$
$\qquad\quad\; x = \pm 5$

$x^2 + 1^2 = 26$
$\qquad\; x^2 = 25$
$\qquad\;\; x = \pm 5$

The solutions are $(-5, -1)$, $(-5, 1)$, $(5, -1)$, $(5, 1)$.

67. $a_n = \dfrac{(-1)^n}{n+4}$

$a_1 = \dfrac{(-1)^1}{1+4} = -\dfrac{1}{5}$

$a_2 = \dfrac{(-1)^2}{2+4} = \dfrac{1}{6}$

$a_3 = \dfrac{(-1)^3}{3+4} = -\dfrac{1}{7}$

$a_4 = \dfrac{(-1)^4}{4+4} = \dfrac{1}{8}$

$a_5 = \dfrac{(-1)^5}{5+4} = -\dfrac{1}{9}$

The first five terms are $-\dfrac{1}{5}, \dfrac{1}{6}, -\dfrac{1}{7}, \dfrac{1}{8}, -\dfrac{1}{9}$.

68. $a_n = 5(2)^{n-1}$

$a_1 = 5(2)^{1-1} = 5(2)^0 = 5$
$r = 2$
$n = 5$

$S_n = \dfrac{a_1(1 - r^n)}{1 - r}$

$S_5 = \dfrac{5(1 - 2^5)}{1 - 2} = \dfrac{5(1 - 32)}{-1} = 155$

69. Sequence $\dfrac{3}{2}, -\dfrac{3}{4}, \dfrac{3}{8}, \ldots$

$a_1 = \dfrac{3}{2}, \ r = -\dfrac{1}{2}$

$S_\infty = \dfrac{a_1}{1-r} = \dfrac{\frac{3}{2}}{1-\left(-\frac{1}{2}\right)} = \dfrac{\frac{3}{2}}{\frac{3}{2}} = 1$

70. $\displaystyle\sum_{i=1}^{4} i(i-2) = 1(1-2) + 2(2-2) + 3(3-2) + 4(4-2)$

$\qquad\qquad = 1(-1) + 2(0) + 3(1) + 4(2)$

$\qquad\qquad = -1 + 0 + 3 + 8$

$\qquad\qquad = 10$

71. $(2x+y)^5 = \dbinom{5}{0}(2x)^5 + \dbinom{5}{1}(2x)^4(y) + \dbinom{5}{2}(2x)^3(y)^2 + \dbinom{5}{3}(2x)^2(y)^3 + \dbinom{5}{4}(2x)^1(y)^4 + \dbinom{5}{5}y^5$

$\qquad\qquad = 2^5 x^5 + 5 \cdot 2^4 x^4 y + 10 \cdot 2^3 x^3 y^2 + 10 \cdot 2^2 x^2 y^3 + 5 \cdot 2xy^4 + y^5$

$\qquad\qquad = 32x^5 + 80x^4 y + 80x^3 y^2 + 40x^2 y^3 + 10xy^4 + y^5$